A book that analyzes and explains the mysteries of the universe from a new perspective

New Mechanics and New Electromagnetism of Axiomatized

Verification and Application

Wang Jianhua (China)

AMERICAN ACADEMIC PRESS

AMERICAN ACADEMIC PRESS

By AMERICAN ACADEMIC PRESS

201 Main Street

Salt Lake City

UT 84111 USA

Email manu@AcademicPress.us

Visit us at http://www.AcademicPress.us

ISBN: 979-8-3370-8915-7

Distributed to the trade by National Book Network Suite 200, 4501 Forbes Boulevard, Lanham, MD 20706

10 9 8 7 6 5 4 3 2 1

Brief introduction

In the last hundred years, there has not been any substantial breakthrough in fundamental physics, and all advances in physics have been refinements and additions to previous fundamental physics. There are many theoretical difficulties and unsolved mysteries in the field of physics to this day.

At the beginning of the 20th century, the world-famous mathematician Hilbert suggested deducing all the theorems of physics by means of mathematical axiomatization. However for more than 100 years mathematicians and physicists have not accomplished this great scientific dream. Many people are skeptical about the suggestion of axiomatizing physics.

The author has established a new axiomatic system of physics consisting of six axioms based on physical experiments. Many famous theorems of physics can be deduced from the new axiomatic system. For example:

(1) Based on the laws of motion of the system of mass points, it can be deduced that the trajectories of the planets around the sun are eggshell curves, not elliptical curves.

(2) The new universal gravitation formula, viz:

$$F = K\frac{m_1 m_2}{r^3}$$

The K in the equation is the new gravitational constant.

(3) In the inertial system S', the force F' on the object is:

$$F' = \left(1 + \frac{v' + u}{\sqrt{c^2 - (v' + u)^2}}\right) ma$$

In the formula, v' is the velocity of the object in the inertial frame S', and u is the velocity of the inertial frame S', in the cosmic space reference frame. The velocity v' is in the same direction as the velocity u.

(4) The new formula for the electric field strength, viz:

$$E = -\frac{1}{4\pi\varepsilon_0}\frac{q}{r^2}\sqrt{1 - \frac{v^2}{c^2}\sin\theta}$$

(5) A new formula for the magnetic field force, viz:

$$F = \frac{\mu_0}{4\pi}\frac{q_1 v_1 \cdot q_2 v_2}{r^2}\sin\theta\cos\beta$$

In addition, the authors have designed three very simple optical experiments based on the principle of lasers and the theory of interference of light, which will directly verify whether the principle of special relativity of the invariance of the speed of light is correct or incorrect.

This book is the initial realization of mathematician Hilbert's scientific dream. It will inspire and broaden the reader's horizons in observing and studying the mysteries of the universe, and promote the progress of mechanics and electrodynamics.

Due to the limitations of the author's level, there are inevitably deficiencies and errors in the book. Experts and scholars are invited to criticize and correct them.

This book is aimed at readers who have studied Newtonian mechanics and classical electromagnetism.

About the author: Jianhua Wang, male, born on January 11, 1960. Bachelor of Engineering. Currently living in Jinan City, Shandong Province, China.

Publication of the scholarly monograph NEW THEORY OF PLANETARY MOTION AND NEW FORMULA OF UNIVERSAL GRAVITATION in 2022.Publisher American Academic Press. ISBN 978-1631815836.

Author email: wangjianhua600@aliyun.com

Contents

Preamble

In the last hundred years, there has not been any substantial breakthrough in fundamental physics, and all advances in physics have been refinements and additions to previous fundamental physics. There are many theoretical difficulties and unsolved mysteries in the field of physics to this day.

The contradictions and dilemmas currently facing physics imply that the existing fundamental physics is not yet perfect. There must be new and deeper theoretical systems in the field of fundamental physics that have not yet been discovered.

The German mathematician HilbertD (1862.1.23~1943.2.14) was a great mathematician of the twentieth century.In 1900, Hilbert proposed 23 most important mathematical problems at the Congress of Mathematicians in Paris, hoping that the mathematicians of the twentieth century would study them. These are the famous Hilbert's 23 problems.

The 6th mathematical problem posed by Hilbert was about the axiomatization of physics. Hilbert suggested that the entire theorem of physics be deduced by mathematical axiomatization.

Axiomatization originally originated in the field of mathematics, and the Euclidean geometric system is an example of successful axiomatization, with its five axioms serving as the basis and starting point of argument for the entire geometric system.

The core of axiomatization in physics lies in the search for a set of fundamental assumptions or axioms that serve as the basis and starting point for the entire theoretical system of physics. Then from these most basic axioms, other theorems of physics are deduced through logical reasoning and deduction.

The authors summarize six axioms of physics based on physical experiments and establish an axiomatized new physics on the basis of these six axioms.

The main issues analyzed and discussed in this book are as follows.

(1) The orbits of planets moving around the sun are eggshell curves, not elliptical curves.

Since both the planets and the sun move around the center of mass O of the solar system, the motion of the solar system belongs to the motion of system of mass points. According to the laws of motion of the system of mass points and the formula for the centripetal force of a curve, it can be determined by mathematical derivation that the trajectory of the planet's motion around the sun is an eggshell curve, not an elliptical curve. The new orbit of planetary motion is shown in the Figure 0-1.

(2) The new universal gravitation formula

In the classical theory, the magnitude of the universal gravitation $\boldsymbol{F'}$ is:

$$\boldsymbol{F' = G\frac{m_1 m_2}{r^2}}$$

The new universal gravitation formula derived by the authors, based on the laws of motion of the system of mass points, is:

$$\boldsymbol{F = K\frac{m_1 m_2}{r^3}}$$

The $\boldsymbol{K}$ is the new gravitational constant. The relationship between the new gravitational constant $\boldsymbol{K}$ and the old gravitational constant $\boldsymbol{G}$ is:

$$\boldsymbol{G = K\frac{1}{r}}$$

The following two conclusions can be drawn from the above equation.

Conclusion 1. If the distance $\boldsymbol{r}$ between the Sun and the planets is greater, then the ratio of the Newtonian gravitational force $\boldsymbol{F'}$ to the actual gravitational force $\boldsymbol{F}$ will be smaller. From this it can be determined that if the cosmic probe is farther away from the sun, then astronomers will assume that the resistance the cosmic probe will experience at the edge of the solar system will be greater. This is the reason why spacecraft deviate from their theoretical orbits.

Conclusion 2, if the distance $\boldsymbol{r}$ between two microscopic particles is smaller, then the ratio of the Newtonian gravitational force $\boldsymbol{F'}$ to the actual gravitational force $\boldsymbol{F}$ will be larger. From this physicists believe that there is a strong nuclear force acting between microscopic particles inside the nucleus of an atom.

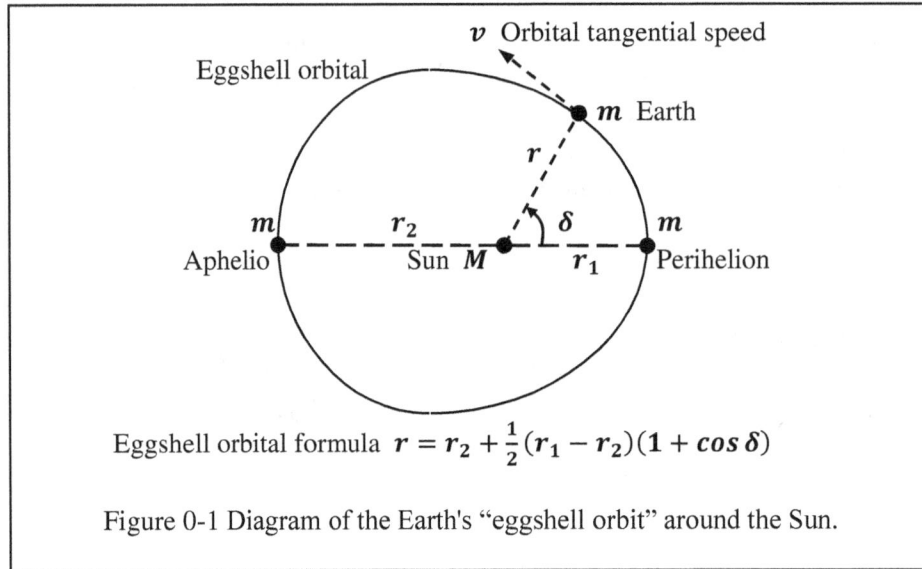

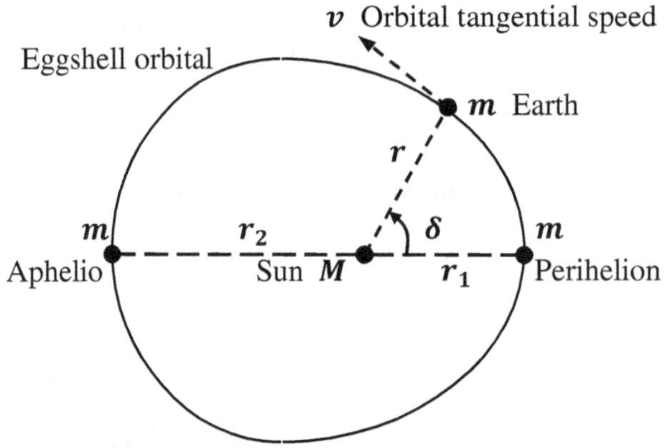

Eggshell orbital formula $\boldsymbol{r = r_2 + \frac{1}{2}(r_1 - r_2)(1 + \cos\delta)}$

Figure 0-1 Diagram of the Earth's "eggshell orbit" around the Sun.

(3) New tangential angular momentum conservation theorem.

The tangential velocity $\boldsymbol{v}$ of curvilinear motion can be decomposed into transverse velocity $\boldsymbol{v_\varphi}$ and radial velocity $\boldsymbol{v_r}$, i.e:

$$\boldsymbol{v} = \sqrt{\boldsymbol{v_\varphi^2 + v_r^2}} = \sqrt{(\boldsymbol{v\sin\theta})^2 + (\boldsymbol{v\cos\theta})^2}$$

Multiplying the above equation by the distance $\boldsymbol{r}$ and the intrinsic mass $\boldsymbol{m}$ yields the new tangential angular momentum conservation theorem, viz:

$$\boldsymbol{H_v = rmv} = \sqrt{(\boldsymbol{rmv\sin\theta})^2 + (\boldsymbol{rm\cos\theta})^2} = \boldsymbol{C}$$

The $\boldsymbol{C}$ in the formula is a constant. The above equation shows that the old theorem of conservation of angular momentum $\boldsymbol{L_v = rmv\sin\theta = C}$ is not compatible with the change in tangential velocity $\boldsymbol{v}$ is contradictory.

(4) New force formula.

The authors derive a new force formula based on the new axiom of the invariance of the speed of light and the theorem of conservation of momentum, viz:

$$\boldsymbol{F} = \left(\boldsymbol{1} + \frac{\boldsymbol{v}}{\sqrt{\boldsymbol{c^2 - v^2}}}\right)\boldsymbol{ma}$$

If the object's displacement velocity $\boldsymbol{v}$ is much less than the speed of light $\boldsymbol{c}$, then the above equation changes to Newton's second law $\boldsymbol{F = ma}$.

(5) The intrinsic mass $\boldsymbol{m}$ of an object is the set of energy element masses $\boldsymbol{m_h}$.

In modern physics, the mass $\boldsymbol{m}$ of an object has two sources.

The first source is the energy $\boldsymbol{\varepsilon}$, which is the source of the photon mass $\boldsymbol{m_c}$.

The second source is the "Higgs field", which is the source of the mass $\boldsymbol{m}$ of an object.

The authors define the energy $\boldsymbol{\varepsilon_h}$ contained in Planck's constant $\boldsymbol{h}$ as the "energy element", and the mass $\boldsymbol{m_h}$ of the energy element $\boldsymbol{\varepsilon_h}$ as the energy element mass. The authors believe that both the mass $\boldsymbol{m_c}$ of a photon and the mass $\boldsymbol{m}$ of an object have their origin in the set of energy-element masses $\boldsymbol{m_h}$. In this book the "Higgs field" is not the source of the mass of an object.

(6) Formula for the conservation of the intrinsic mass m of an object.

The intrinsic mass m of an object is the set of energy element masses m_h, i.e.:

$$m = \sum\sum m_h$$

Multiplying the above equation by the spiral speed of light $c = \sqrt{v_\omega^2 + v^2}$ for the energy element ε_h gives the relational equation:

$$mc = m\sqrt{v_\omega^2 + v^2}$$

The v in the formula is the speed of motion of the object in the cosmic vacuum frame of reference. The above equation can be equivalently varied as:

$$m = \sqrt{\left(m\frac{v_\omega}{c}\right)^2 + \left(m\frac{v}{c}\right)^2} = \sqrt{m_\omega^2 + m_v^2}$$

The above equation is the formula for the conservation of the intrinsic mass m of an object. The above equation shows that the intrinsic mass m of an object can be decomposed into an internal rotational mass m_ω and a displacement mass m_v, i.e.:

$$m_\omega = m\sqrt{1 - \frac{v^2}{c^2}}, \qquad m_v = m\frac{v}{c}$$

The following two conclusions can be deduced from the above equations.

Conclusion 1 If the velocity $v = 0$, then the internal rotational mass m_ω is equal to the intrinsic mass m (i.e., $m_\omega = m$), and the displaced mass $m_v = 0$.

Conclusion 2 If the velocity $v = c$, then the internal rotational mass $m_\omega = 0$ and the displaced mass m_v is equal to the intrinsic mass m (i.e. $m_\omega = m$) . Since the photon's velocity is always equal to the speed of light c, the photon's internal rotational mass $m_\omega = 0$ and the photon's displaced mass m_v equals the intrinsic mass m (i.e., $m_v = m$).

(7) Physicists' interpretation of the mass velocity formula is wrong.

The special relativistic mass velocity formula is:

$$m = \frac{m_0}{\sqrt{1 - \frac{v^2}{c^2}}}$$

Comparing the above formula with the internal rotating mass formula $m_\omega = m\sqrt{1 - \frac{v^2}{c^2}}$, it can be determined that the rest mass m_0 of special relativity is essentially the internal rotating mass m_ω (i.e., $m_0 = m_\omega$), while the moving mass m of special relativity is essentially the intrinsic mass of the object.

Since there is no formula for the conservation of intrinsic mass m in contemporary physics and there are no physical quantities for internal rotational mass m_ω and displacement mass m_v, physicists have interpreted the mass-velocity formula incorrectly.

(8) Based on the new axiom of the invariance of the speed of light and the physical experiment of the change of positive and negative electrons into photons by collision, the authors deduced that the electron has the structure of a hollow circular tube, and preliminarily explained the mechanism of the formation of the electron circular tube.

(9) Conservation formula for electric charge q.

Since the intrinsic mass m_e of an electron is the set of the masses m_h of the energy elements, i.e:

$$m_e = \sum m_h$$

And the energetic element ε_h in the electron mass m_e always moves at the spiral speed of light c, viz:

$$c = \sqrt{v_\omega^2 + v^2}$$

Since v is the displacement velocity of the electron, the internal rotation velocity v_ω is the spin velocity of the electron. Multiplying both sides of the equal sign by the charge q gives the relational equation:

$$qc = q\sqrt{v_\omega^2 + v^2}$$

The v in the formula is the speed of motion of electric charge q in the cosmic vacuum frame of reference. The above equation can be equivalently varied as:

$$q = \sqrt{\left(q\frac{v_\omega}{c}\right)^2 + \left(q\frac{v}{c}\right)^2} = \sqrt{q_\omega^2 + q_v^2}$$

The above equation is the electric charge conservation equation. The above equation shows that electric charge q can be decomposed into internal rotational electric charge q_ω and magnetic electric charge q_v, i.e.:

$$q_\omega = q\sqrt{1 - \frac{v^2}{c^2}}, \qquad q_v = q\frac{v}{c}$$

The above equations are the internal rotational electric charge equation and the magnetic electric charge equation. The internal rotational electric charge q_ω generates a kinematic electric field and the magnetic electric charge q_v generates a magnetic electric field. The following two conclusions can be deduced from the above equations.

Conclusion 1 If the electric charge velocity $v = 0$, then the internal rotational electric charge q_ω is equal to the stationary electric charge q_0 (i.e., $q_\omega = q_0$) and the magnetic electric charge $q_v = 0$.

Conclusion 2 If the velocity $v = c$, then the internal rotational electric charge $q_\omega = 0$ and the magnetic electric charge q_v is equal to the stationary electric charge q_0 (i.e. $q_v = q_0$). Since the speed of the photon is always equal to the speed of light c, the magnetic electric charge q_v of the photon is equal to the stationary electric charge q_0.

Since the photon has a magnetic electric charge q_v, the direction of propagation of light is deflected in a magnetic field.

(10) Based on the axiom of directional force of the energy element ε_h, the authors succeeded in theoretically explaining why there are repulsive forces between electric charges of the same kind and suction forces between positive and negative electric charges.

(11) Based on the axiom of the invariance of the speed of light and the vortex field formula $e^y\left(\frac{e}{r}\right)^{D_k kr} = 1$, the authors derived a formula for the electric field force between the kinematic electric charge q_1 and the kinematic electric charge q_2, viz:

$$F_\omega = -\frac{1}{4\pi\varepsilon_0}\sqrt{1 - \frac{v_1^2}{c^2}}\sqrt{1 - \frac{v_2^2}{c^2}} \cdot \frac{q_1 q_2}{r^2}\sin\theta\cos\beta, \quad r > 1$$

The v_1 and v_2 in the equation are the velocities of the motion electric charge q_1 and q_2, respectively, θ is the angle between the distance r and the velocity v_1, and β is the angle between the velocity v_1 and the velocity v_2.

An equation for the magnetic field force between the motion electric charge q_1 and the motion electric charge q_2 is derived, viz:

(12) The authors derived the Lorentz Force formula from the electric charge conservation formula, viz:

$$F_B = qvB\sin\theta$$

A new Faraday's law of electromagnetic induction was derived, viz:

$$\oint E_v dl = -\iint \frac{\partial B}{\partial t} dS$$

The E_v in the equation is the time-varying magnetic electric field strength, not the time-varying electromagnetic field E.

(13) The author derived the Biot-Savart Law, Ampere's Law, and many other classical laws of electrodynamics from the equation of conservation of electric charge.

(14) The authors proved theoretically that the displacement electric current $I_D = \varepsilon_0 \iint \frac{\partial E}{\partial t} dS$ does not exist according to the electric charge conservation formula.

(15) The authors have designed 2 very simple new optical experiments based on the laser principle and the

interference theory of light. From this, it can be directly verified that the principle of constant speed of light of special relativity is wrong.

(16) The authors have discovered 2 new formulas that are not currently found in mathematical theory.

New mathematical formula 1: the mathematical formula for the eggshell curve, viz:

$$r = r_2 + \frac{1}{2}(r_1 - r_2)(1 + \cos\delta)$$

New Math Equation 2: The Vortex Field Equation, i.e:

$$e^y \left(\frac{e}{r}\right)^{D_k k r} = 1 \qquad 0 < r < +\infty$$

The above equation is the equirectangular vortex field equation. In the formula r is the polar radius, y is the vortex potential, k is the source of the vortex field and D_k is the constant associated with the vortex source k.

Chapter 1 The Axiomatic System of the New Physics

Axiom 1: The Axiom of the Absolute Frame of Reference

The vastness of the cosmic space contains many celestial bodies and matter. The author defines the three-dimensional vacuum $\boldsymbol{V = xyz}$ of the universe, which does not contain any celestial bodies and matter, as absolute space, and the cosmic vacuum frame of reference $\boldsymbol{S(x, y, z)}$ as the absolute frame of reference. This definition is the axiom of absolute reference system.

The new physics holds that the absolute frame of reference $\boldsymbol{S}$ is the absolutely stationary frame of reference, which is free from motion and bending variations, and that all frames of reference $\boldsymbol{S'}$ moving at uniform speed in the cosmic vacuum are inertial frames. Alternatively, the cosmic vacuum frame of reference $\boldsymbol{S}$ is the unique absolutely stationary frame of reference, while there are innumerable inertial frames of reference $\boldsymbol{S'}$.

Axiom 2: Axiom of Absoluteness of Time Difference Δt

Any event takes a certain amount of time from start to finish. Assume that all inertial frames of reference have equal unit time. If observer A observes a time difference of $\boldsymbol{\Delta t}$ in the absolute frame of reference $\boldsymbol{S}$ and observer B observes a time difference of $\boldsymbol{\Delta t'}$ in the inertial frame of reference $\boldsymbol{S'}$, then the two time differences are absolutely equal, viz:

$$\boldsymbol{\Delta t = \Delta t'}$$

The above formula is the axiom of absoluteness of time difference. The above equation shows that time $\boldsymbol{t}$ and space V are two physical quantities independent of each other.

Axiom 3: Axiom of the smallest energy element

According to quantum mechanics, the energy $\boldsymbol{\varepsilon}$ of a photon is:

$$\boldsymbol{\varepsilon = hv}$$

When the light wave frequency $\boldsymbol{v = 1}$, the energy $\boldsymbol{\varepsilon_h}$ contained in Planck's constant $\boldsymbol{h}$ is:

$$\boldsymbol{\varepsilon_h = h \cdot 1s = 6.626 \times 10^{-34} J = 4.135 \times 10^{-15} eV} \tag{1.3.1}$$

This book defines the energy $\boldsymbol{\varepsilon_h}$ contained in Planck's constant $\boldsymbol{h}$ as the energy element. The above formula is the energy element formula.

The new physics considers the energy element $\boldsymbol{\varepsilon_h}$ to be the smallest indivisible particle of energy. According to the law of conservation of energy, the energy element $\boldsymbol{\varepsilon_h}$ can neither be eliminated nor created from nothing. This book calls the above view the axiom of the smallest energy element.

The following three theorems can be deduced from the axiom of the smallest energy element.

Theorem 3-1

The ratio of the energy element $\boldsymbol{\varepsilon_h}$ to the square of the speed of light is equal to the mass $\boldsymbol{m_h}$ of the energy element, i.e:

$$\boldsymbol{m_h = \frac{\varepsilon_h}{c^2}} \tag{1.3.2}$$

Theorem 3-2

The mass m_h of the energy element ε_h is the smallest mass particle.

Since the energy element ε_h is the smallest energy particle and the energy element ε_h contains the mass m_h, the mass m_h of the energy element ε_h is the smallest mass particle.

An electron is one of the elementary particles. The mass $m_e = 9.11 \times 10^{-31} kg$ of an electron and the energy $\varepsilon_e = 5.11 \times 10^5 eV$ of an electron (electron volts). From this it can be determined that the mass m_{eV} possessed per unit electron volt eV is:

$$m_{eV} = \frac{m_e}{\varepsilon_e} = \frac{9.11 \times 10^{-31}}{5.11 \times 10^5} = 1.78 \times 10^{-36} kg$$

According to the above equation, the mass m_h possessed by the energy element ε_h is:

$$m_h = m_{eV}\varepsilon_h = 1.78 \times 10^{-36} \times 4.135 \times 10^{-15} = 7.35 \times 10^{-51} kg$$

The mass m_h of the energy element ε_h is the physical quantity that produces gravitational.

Theorem 3-3

The intrinsic mass m of an object is composed of the energy element ε_h.

Since the photon is composed of the energy element ε_h, and the collision of positive and negative electrons changes into two γ-photons, the mass m_e of the electron is composed of the mass m_h of the energy element. From this it can be determined that the intrinsic mass m_i of the elementary particle i is composed of the energy element ε_h, i.e:

$$m_i = \sum m_h \quad i = 1, 2, 3, \cdots, n \tag{1.3.3}$$

The above formula is the formula for the intrinsic mass of elementary particle i. Since the object intrinsic mass m is composed of the intrinsic mass m_i of the elementary particles, the object intrinsic mass m is composed of the mass m_h of the energy element, i.e.:

$$m = \sum m_i = \sum \sum m_h \tag{1.3.4}$$

The above formula is the formula for the intrinsic mass of the object. Since the above formula contains only the energy element ε_h does not contain the object's speed v of motion, the intrinsic mass m of the object in the new physics is a constant independent of the speed v. The mass m of an object in motion in current physics is a function of the velocity v, i.e.:

$$m = \frac{m_0}{\sqrt{1 - \frac{v^2}{c^2}}} \tag{1.3.5}$$

The m_0 in the formula is the rest mass of the object.

Axiom 4: Axiom of invariance of the speed of light for the energy element ε_h

The meaning of the axiom of invariance of the speed of light for the energy element ε_h is that in the cosmic vacuum frame of reference S, the speed of motion of the energy element ε_h is always equal to the speed of light c in all directions.

The following eight theorems can be deduced from the axiom of invariance of the speed of light for the energy element ε_h.

Theorem 4-1

The energy element ε_h does not require a medium for its propagation in the cosmic vacuum frame of reference S. Just as light does not require a medium to propagate in the cosmic vacuum.

Theorem 4-2

The speed of light $\boldsymbol{c}$ of the energy element $\boldsymbol{\varepsilon_h}$ is independent of the state of motion of the energy emitting source.

Theorem 4-3

In the inertial frame of reference $\boldsymbol{S'}$, the velocity of motion $\boldsymbol{c'}$ of the energy element $\boldsymbol{\varepsilon_h}$ satisfies the Galilean velocity transformation equation.

Assume that the inertial frame of reference $\boldsymbol{S'}$ is moving with velocity $\boldsymbol{u}$ in the cosmic vacuum frame of reference $\boldsymbol{S}$. When the direction of motion of the energy element $\boldsymbol{\varepsilon_h}$ is the same as, or opposite to, the direction of motion of the inertial frame of reference $\boldsymbol{S'}$, the velocity $\boldsymbol{c'}$ of the energy element $\boldsymbol{\varepsilon_h}$ in the inertial frame of reference $\boldsymbol{S'}$ is:

$$\boldsymbol{c'} = \boldsymbol{c} \mp \boldsymbol{u} \tag{1.4.1}$$

Theorem 4-4

The energy element $\boldsymbol{\varepsilon_h}$ has an internal rotational velocity $\boldsymbol{v_\omega}$ (i.e., a circular velocity) inside the elementary particle intrinsic mass $\boldsymbol{m_i}$.

Suppose $\boldsymbol{v}$ is the displacement velocity of the object in the cosmic space reference system. Since the intrinsic mass $\boldsymbol{m}$ of the object consists of the elementary particle intrinsic mass $\boldsymbol{m_i}$, and the intrinsic mass $\boldsymbol{m_i}$ consists of the energy element mass $\boldsymbol{m_h}$, the energy element $\boldsymbol{\varepsilon_h}$ in the elementary particle intrinsic mass $\boldsymbol{m_i}$ is also displaced forward synchronously with the velocity $\boldsymbol{v}$ in the cosmic space reference system.

In the cosmic space reference system, since the energy element $\boldsymbol{\varepsilon_h}$ can only move within the elementary particle intrinsic mass $\boldsymbol{m_i}$, and the velocity of the energy element $\boldsymbol{\varepsilon_h}$ is always equal to the speed of light $\boldsymbol{c}$, the energy element $\boldsymbol{\varepsilon_h}$ has a circular velocity $\boldsymbol{v_\omega}$ within the elementary particle $\boldsymbol{m_i}$ that rotates around the center point O of the intrinsic mass $\boldsymbol{m_i}$. This circular velocity $\boldsymbol{v_\omega}$ is perpendicular to the displacement velocity $\boldsymbol{v}$.

In order to distinguish the spin velocity of an object, this book defines the circumferential velocity $\boldsymbol{v_\omega}$ of the energy element $\boldsymbol{\varepsilon_h}$ as the internal rotational velocity of the energy element $\boldsymbol{\varepsilon_h}$.

Theorem 4-5

The energy element $\boldsymbol{\varepsilon_h}$ rotates at the speed of light $\sqrt{\boldsymbol{v_\omega^2} + \boldsymbol{v^2}}$ of the spiral within the intrinsic mass $\boldsymbol{m_i}$ of the elementary particle, i.e:

$$\boldsymbol{c} = \sqrt{\boldsymbol{v_\omega^2} + \boldsymbol{v^2}} \tag{1.4.2}$$

Since the speed of light $\boldsymbol{c}$ in the formula is not a linear motion, the above formula is the spiral speed of light formula for the energy element $\boldsymbol{\varepsilon_h}$.

Since the energy element $\boldsymbol{\varepsilon_h}$ is always moving at the speed of light $\boldsymbol{c}$ in the cosmic vacuum frame of reference $\boldsymbol{S}$, and the electron is composed of the energy element $\boldsymbol{\varepsilon_h}$, and the displacement velocity $\boldsymbol{v}$ of the electron is less than the speed of light $\boldsymbol{c}$, the rotational velocity of the energy element $\boldsymbol{\varepsilon_h}$ within the electron is equal to the helical speed of light $\boldsymbol{c} = \sqrt{\boldsymbol{v_\omega^2} + \boldsymbol{v^2}}$.

The spiral speed of light $\boldsymbol{c}$ of the energy element $\boldsymbol{\varepsilon_h}$ is responsible for the formation of the intrinsic mass $\boldsymbol{m_i}$ of the microscopic elementary particles. When the internal rotational velocity $\boldsymbol{v_\omega} = \boldsymbol{0}$ of the energy element $\boldsymbol{\varepsilon_h}$, then the matter consisting of the energy element $\boldsymbol{\varepsilon_h}$ is a photon.

The motion of the energy element $\boldsymbol{\varepsilon_h}$ rotating around the center O of the elementary particle is analogous to the motion of the Moon around the Earth. Assume that $\boldsymbol{\delta}$ is the angle between the displacement speed $\boldsymbol{v}$ and the Spiral speed of light $\boldsymbol{c} = \sqrt{\boldsymbol{v_\omega^2} + \boldsymbol{v^2}}$. The relationship between the Spiral speed of light $\boldsymbol{c}$, the displacement speed $\boldsymbol{v}$, and the internal rotation speed $\boldsymbol{v_\omega}$ is shown in Figure 1-1.

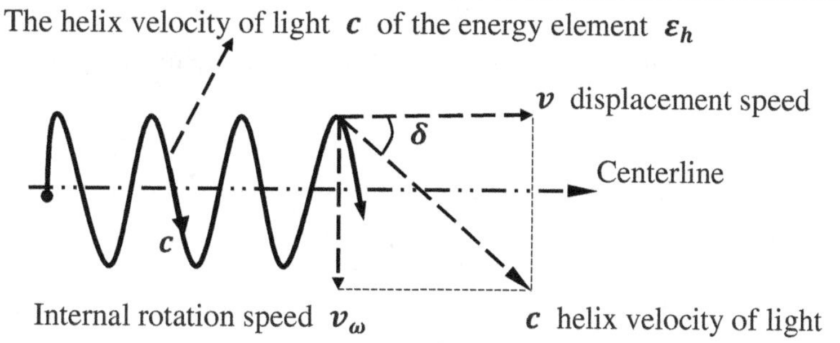

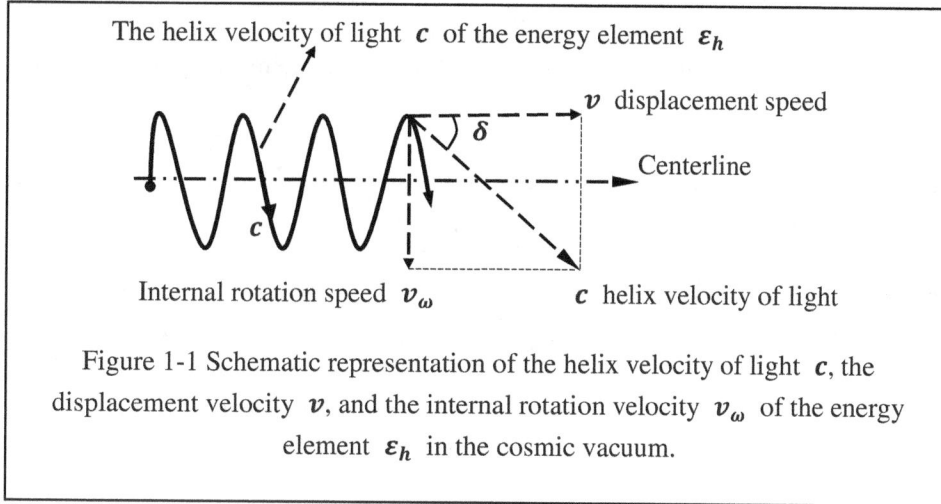

Figure 1-1 Schematic representation of the helix velocity of light c, the displacement velocity v, and the internal rotation velocity v_ω of the energy element ε_h in the cosmic vacuum.

Many important laws in physics can be derived from the spiral speed of light formula $c = \sqrt{v_\omega^2 + v^2}$. For example, the law of conservation of momentum in mechanics, the magnetic flux density formula in electromagnetism, the Biot-Saval law, and Faraday's law of electromagnetic induction, to name a few.

Theorem 4-6

Mass-energy equivalence formula.

Squaring the spiral speed of light $c = \sqrt{v_\omega^2 + v^2}$, and multiplying it by equation (1.3.4) yields the relational equation:

$$mc^2 = \sum\sum m_h v_\omega^2 + \sum\sum m_h v^2 \tag{1.4.3}$$

The mc^2 in equation is the mass-energy equivalence formula from classical theory, i.e:

$$\varepsilon = mc^2 \tag{1.4.4}$$

Since the velocity v_ω is the internal rotation velocity of the energy element ε_h, this book defines mv_ω^2 as the internal rotation mass-energy equivalence ε_ω, i.e:

$$\varepsilon_\omega = mv_\omega^2 = \sum\sum m_h v_\omega^2 \tag{1.4.5}$$

Since the velocity v is the displacement velocity of the energy element ε_h, this book defines mv^2 as the displacement mass-energy equivalence ε_v, i.e:

$$\varepsilon_v = mv^2 = \sum\sum m_h v^2 \tag{1.4.6}$$

Substituting the above equation and equation (1.4.5) into equation (1.4.3) gives the relationship equation:

$$\varepsilon = mc^2 = mv_\omega^2 + mv^2 \tag{1.4.7}$$

The above equation shows that the mass-energy equivalence $\varepsilon = mc^2$ of an object contains the internal rotation mass-energy equivalence $\varepsilon_\omega = mv_\omega^2$ and the displacement mass-energy equivalence $\varepsilon_v = mv^2$.

Theorem 4-7

Conservation formula for helical light speed momentum of an object

In physics, momentum P is equal to the product of the mass m of an object and the displacement velocity v, i.e:

$$P = mv$$

Multiplying the spiral speed of light $c = \sqrt{v_\omega^2 + v^2}$, by the intrinsic mass m of the object, yields the relational equation:

$$mc = \sqrt{(mv_\omega)^2 + (mv)^2} \tag{1.4.8}$$

This book defines mc as the object's helical light speed momentum $\boldsymbol{P_c}$, i.e:

$$\boldsymbol{P_c = mc} \tag{1.4.9}$$

Since the velocity $\boldsymbol{v_\omega}$ is the internal rotational velocity of the energy element $\boldsymbol{\varepsilon_h}$, this book defines $\boldsymbol{mv_\omega}$ as the object's internal rotational momentum $\boldsymbol{P_\omega}$, i.e:

$$\boldsymbol{P_\omega = mv_\omega} \tag{1.4.10}$$

Since the velocity $\boldsymbol{v}$ is the displacement velocity of the energy element $\boldsymbol{\varepsilon_h}$, this book defines $\boldsymbol{mv}$ as the displacement momentum $\boldsymbol{P_v}$ of the object, i.e:

$$\boldsymbol{P_v = mv} \tag{1.4.11}$$

Substituting the above equation and (1.4.10) into equation (1.4.8) gives the relationship equation:

$$\boldsymbol{P_c = mc = \sqrt{(mv_\omega)^2 + (mv)^2} = \sqrt{(P_\omega)^2 + (P_v)^2}} \tag{1.4.12}$$

The above equation is the conservation equation for the helical light speed momentum of an object. The above equation shows that the helical light speed momentum of the object is independent of the external force $\boldsymbol{F}$. The helical light speed momentum $\boldsymbol{P_c = mc}$ contains the internal rotational momentum $\boldsymbol{P_\omega = mv_\omega}$ and the displacement momentum $\boldsymbol{P_v = mv}$.

Theorem 4-8

Equation for conservation of displacement momentum $\boldsymbol{P_v}$ of an object

Assume that the system of mass points consists of multiple mass points. At time $\boldsymbol{t}$, the helical light speed momentum $\boldsymbol{P_c}$ of the system of mass points is:

$$\boldsymbol{P_c = \sqrt{\left(\sum P_{i\omega}\right)^2 + \left(\sum P_{iv}\right)^2}} \tag{1.4.13}$$

At time $\boldsymbol{t'}$, the helical light speed momentum $\boldsymbol{P_c}$ of the system of mass points is:

$$\boldsymbol{P_c = \sqrt{\left(\sum P'_{i\omega}\right)^2 + \left(\sum P'_{iv}\right)^2}} \tag{1.4.14}$$

When the time changes from $\boldsymbol{t}$ to $\boldsymbol{t'}$, the change $\boldsymbol{dP_v}$ in displacement momentum $\boldsymbol{P_v}$ is:

$$\boldsymbol{dP_v = \sum P'_{iv} - \sum P_{iv}} \tag{1.4.15}$$

In Newtonian mechanics, the differential expression for the Newtonian force $\boldsymbol{F}$ is:

$$\boldsymbol{F = \frac{dP}{dt} = m\frac{dv}{dt} + v\frac{dm}{dt}} \tag{1.4.16}$$

When the mass $\boldsymbol{m}$ of the system of mass points is kept constant, the above equation changes to:

$$\boldsymbol{F = ma} \tag{1.4.17}$$

Note that the intrinsic mass $\boldsymbol{m}$ of the object is not related to the change in velocity $\boldsymbol{v}$. Assuming that the total mass of system of mass points $\boldsymbol{m = \sum_i^n m_i}$ remains constant, take the derivative of equation (1.4.15) with respect to time $\boldsymbol{t}$ to obtain the relationship equation:

$$\boldsymbol{F = \frac{dP_v}{dt} = \frac{d}{dt}\left(\sum P'_{iv} - \sum P_{iv}\right)} \tag{1.4.18}$$

If the external force $\boldsymbol{F \neq 0}$ on the system of mass points, i.e:

$$\boldsymbol{F = \frac{d}{dt}\left(\sum P'_{iv} - \sum P_{iv}\right) \neq 0} \tag{1.4.19}$$

The relationship equation is obtained from the above equation:

$$\boldsymbol{\sum P_{iv} - \sum P'_{iv} \neq 0} \tag{0.4.20}$$

If the system of mass points is subjected to an external force $\boldsymbol{F = 0}$, i.e:

$$\boldsymbol{F = \frac{d}{dt}\left(\sum P'_{iv} - \sum P_{iv}\right) = 0} \tag{1.4.21}$$

The relationship equation is obtained from the above equation:

$$\sum P_{iv} - \sum P'_{iv} = 0 \quad (1.4.22)$$

The above equation is the equation for the conservation of displacement momentum P_v of system of mass points. The above equation shows that if the system of mass points is not subjected to external force F or the sum of external forces applied $\sum_i^n F_i = 0$, then the total displacement momentum P_v of the system of mass points remains constant, i.e.:

$$P_v = \sum P_{iv} = \sum P'_{iv} \quad (1.4.23)$$

According to the mass-velocity formula of special relativity, the relationship between the moving mass m and the velocity v of an object is:

$$m = \frac{m_0}{\sqrt{1 - \frac{v^2}{c^2}}}$$

The m_0 in the formula is the rest mass. Subject to the constraints of the above equation, physicists simply cannot derive the conservation of momentum theorem.

Axiom 5: Directional force axiom for the energy element ε_h

The implication of the directional force axiom for energy elements is that there is a suction force between energy elements ε_h moving in the same direction and a repulsion force between energy elements ε_h moving in the opposite direction.

The following three theorems can be deduced from the directional force axiom for the energy element ε_h.

Theorem 5-1

There are three different directional forces between the energy elements ε_h.

Assume that the energy elements ε_{hA} and ε_{hB} do not move in the same straight line.

If the energy elements ε_{hA} and ε_{hB} move in the same direction, then the directional force $f_{\varepsilon h}$ between the energy elements ε_{hA} and ε_{hB} is the suction force $f_{\varepsilon h(+)}$, i.e.:

$$f_{\varepsilon h(+)} > 0 \quad (1.5.1)$$

If the energy elements ε_{hA} and ε_{hB} move in opposite directions, then the directional force $f_{\varepsilon h}$ between the energy elements ε_{hA} and ε_{hB} is the repulsive force $f_{\varepsilon h(-)}$, i.e:

$$f_{\varepsilon h(-)} < 0 \quad (1.5.2)$$

If the directions of motion of the energy elements ε_{hA} and ε_{hB} are perpendicular to each other, then the directional force $f_{\varepsilon h}$ between the energy elements ε_{hA} and ε_{hB} is equal to 0, i.e:

$$f_{\varepsilon h} = 0 \quad (1.5.3)$$

Note that the gravitational force F and the directional force $f_{\varepsilon h}$ are two forces of different nature. The gravitational force F is the force between the masses of the energy elements m_h, and the directional force $f_{\varepsilon h}$ is the force between the directions of motion of the energy elements ε_h. Although there is no directional force $f_{\varepsilon h}$ between energy elements ε_h whose directions of motion are perpendicular to each other, there is a gravitational force F.

Theorem 5-2

The directional force between energy elements ε_h is the source of the electric and magnetic field forces.

Since the electron is composed of energy element ε_h and energy element ε_h is moving within the electron, the directional force $f_{\varepsilon h}$ between the energy elements ε_h is the source of the electric and magnetic field forces.

Theorem 5-3

Electromagnetic Fields are not matter; Electromagnetic Fields are a physical property of electric charge.

Theorem 5-4

The directional forces between the energy elements ε_h interact directly with each other and do not depend on any medium interaction.

Axiom 6: Axiom of conservation of the vortex field

The meaning of the axiom of conservation of the vortex field is that the vortex fields all vary according to the equation of conservation of the vortex field, viz:

$$e^{y}\left(\frac{e}{r}\right)^{D_k k r} = 1 \qquad 0 < r < +\infty \tag{1.6.1}$$

The above equation is the conservation equation for the vortex field. In the formula, y is the vortex potential of the vortex field, r is the polar radius of the vortex field, k is the source of the vortex field, and D_k is the constant associated with the source k of the vortex field.

Many important experimental laws of physics (e.g., the law of gravity, the electromagnetic field force equation, the wave equation for gravitational waves and the wave equation for electromagnetic waves, etc.) can be derived from the above equation.

The new physics of axiomatization is based on the six axioms mentioned above.

Chapter 2 The Earth's orbit around the Sun is an "eggshell curve"

Introduction: If we consider the solar system as a system of mass points in motion, then the planets and the sun are mass points moving around the center O of mass of the solar system. According to the laws of motion of the system of mass points and the formula for the centripetal force, it can be determined by mathematical derivation that the orbit of a planet orbiting the Sun is not an elliptical curve, but rather an "eggshell curve" with a greater curvature at perihelion and a smaller curvature at aphelion.

2.1 The orbit in which the Sun and Earth revolve around the barycenter O of the solar system

2.1.1 Model of the Sun and Earth orbiting simultaneously around the barycenter O of the solar system

In classical theory, the orbits of the planets revolving around the Sun are elliptical curves, as shown in Figure 2-1.

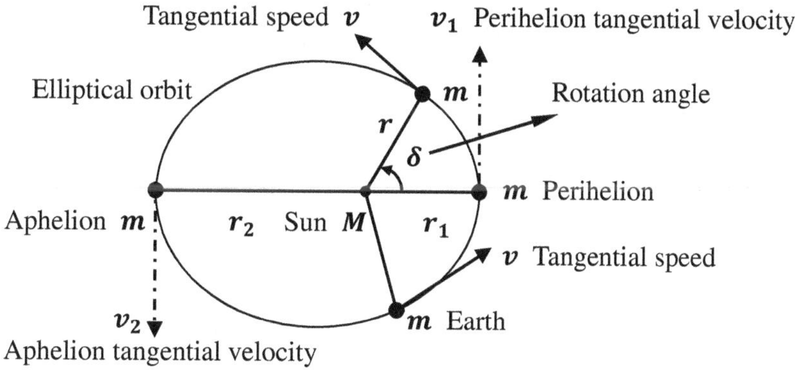

Figure 2-1 The Earth's elliptical orbit around the Sun.

The polar equation of an ellipse is:

$$r = \frac{p}{1 + e\cos\delta}$$

Theoretically, the solar system is a system of mass points consisting of the Sun, Earth, Jupiter, and eight other planets. Although the Sun has a large mass and volume, it is considered a point of mass within the system of points of mass of the solar system. Therefore, the Sun revolves around the center O of mass of the solar system.

Assume that the point O is the center of mass of the solar system. Since the Sun is not the center O of mass of the solar system, the solar frame of reference and the center O of mass frame of reference of the solar system, are two different frame of references.

Current theory assumes that the Sun is at rest at one of the focal points of the ellipse. The Sun is the center point of gravitational and the planets all revolve around the stationary Sun. This book refers to the motion of the planets revolving around the focal point of the ellipse as “elliptical motion”

In theory, since both the sun and planets revolve around the center O of mass of the solar system, and the focal point of the ellipse is not the center O of mass of the solar system, the center of the universal gravitational force $\boldsymbol{F}$ should be the center O of mass, not the sun (focal point of the ellipse).

In theory, the orbit of the Earth revolving around the barycenter O of the solar system is different from the orbit of the Earth revolving around the Sun. In addition, the orbit of the Earth revolving around the moving Sun is different from the elliptical orbit of the Earth revolving around the stationary Sun. Classical physics typically uses elliptical orbits to derive Newton's law of gravitational. So far, no physicist has derived the universal law of gravitational from the perspective of the barycenter O of the solar system.

To simplify the theoretical analysis, assume that the system of mass points in the solar system only includes two mass points, the Sun and the Earth. This book refers to the system of mass points composed of the Sun and the Earth as the "Sun-Earth system." Let $\boldsymbol{M}$ represent the mass of the Sun and $\boldsymbol{m}$ represent the mass of the Earth. The total mass of the Sun-Earth system, denoted as $\sum \boldsymbol{m_i}$, is:

$$\sum \boldsymbol{m_i} = \boldsymbol{M} + \boldsymbol{m} \qquad (2.1.1)$$

Assuming $\boldsymbol{R_O}$ is the distance from the Earth to the center O of mass; $\boldsymbol{\delta}$ is the angle of Earth's revolution around the center O of mass. Assuming $\boldsymbol{v_O}$ is the tangential velocity of Earth's revolution around the center O of mass; $\boldsymbol{\theta}$ is the angle between the tangential velocity $\boldsymbol{v_O}$ and the distance $\boldsymbol{R_O}$. Assuming $\boldsymbol{R_S}$ is the distance from the Sun to the center O of mass of the solar system; $\boldsymbol{v_S}$ is the tangential velocity of the Sun's revolution around the center O of mass of the solar system. As shown in Figure 2-2.

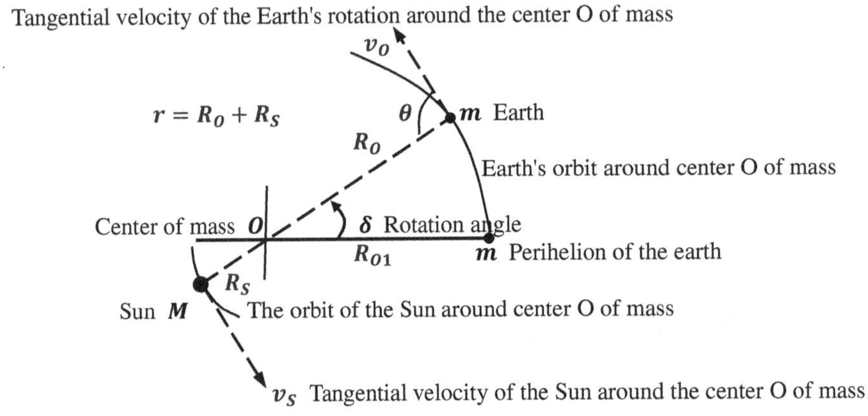

Figure 2-2 Schematic of the Sun and Earth orbiting the center O of mass.

According to the kinematics of the system of mass points, both the sun and the earth revolve around the center O of mass. The distance $\boldsymbol{r}$ from the earth to the sun is:

$$\boldsymbol{r} = \boldsymbol{R_O} + \boldsymbol{R_S} \qquad (2.1.2)$$

The universal gravitational force $\boldsymbol{F}$ between the Earth and the Sun not only exerts an attractive force $\boldsymbol{F_m}$ on the Earth, but also exerts an attractive force $\boldsymbol{F_S}$ on the Sun. Therefore, it can be determined that the gravitational force $\boldsymbol{F}$ between the Earth and the Sun will cause the distances $\boldsymbol{R_O}$ from the Earth to the center O of mass and $\boldsymbol{R_S}$ from the Sun to the center O of mass to change in a certain proportion simultaneously. In addition, the distance $\boldsymbol{r}$ from the Earth to the Sun also changes in the same proportion simultaneously.

At the moment of perihelion, the distance between the center O of mass and the Earth and the Sun is at its minimum. At the moment of aphelion, the distance between the center O of mass and the Earth and the Sun is at its maximum. As the Earth moves from perihelion towards aphelion, the distance from the center O of mass to the Earth and the Sun gradually increases from the minimum value. Conversely, as the Earth moves from aphelion towards perihelion, the distance from the center O of mass to the Earth and the Sun gradually decreases from the maximum value.

Although the value of the variation ΔR_S of the distance R_S from the Sun to the center O of mass is very small, it is not possible to analyze and derive a new gravitational formula if the value of the variation ΔR_S is ignored during the theoretical analysis.

In theory, since the sun always revolves around the center O of mass, and the sun is stationary at the focus of the ellipse, analyzing the motion of planets using the center O of mass as a reference frame is more accurate and scientific than using the focus of the ellipse as a reference frame.

2.1.2 The distance from the center O of mass to the Earth and the Sun are respectively R_O and R_S

Assuming point O is the center of mass of the system of mass points; assuming the center O of mass is moving at a constant speed. Let r_C be the distance from the origin C of the coordinate system to the center O of mass. Let m_i be the mass of mass point i; r_i be the distance from mass point i to the origin C. Let $\sum m_i$ be the total mass of the system of mass points. According to the definition of the center of mass, the relationship formula can be obtained:

$$r_C \sum m_i = \sum m_i r_i \qquad (2.1.3)$$

Assuming that the origin C of the coordinate system coincides with the center O of mass, meaning the distance from the center O of mass to the origin C is $r_C = 0$. The formula above then becomes:

$$\sum m_i r_i = 0 \qquad (2.1.4)$$

When the origin C of the coordinate system coincides with the center O of mass of the solar-terrestrial system, the distance $r_C = 0$ from the center O of mass to the origin C. Based on the above equation, the relationship equation can be obtained:

$$mR_O = MR_S \qquad (2.1.5)$$

Since the motion of the solar system around the galactic center of mass can be regarded as uniform, the combined force F acting on the center O of mass of the solar system is always equal to zero, i.e. $F = 0$. According to the kinematics of a point mass, the sun and the earth always revolve around the center O of mass of the solar system.

The multiplication of equation (2.1.2) with the solar mass M results in the relation equ:

$$Mr = MR_O + MR_S \qquad (2.1.6)$$

Substituting equation (2.1.5) into the above equation and removing MR_S gives the relational equation:

$$R_O = \frac{M}{M+m} r \qquad (2.1.7)$$

The R_O in the formula is the distance between the center O of mass and the Earth, and r is the distance between the Earth and the Sun. Note that the elliptical orbit of the Earth does not have the above relational formula. Multiplying equation (2.1.2) by the Earth's mass m gives the relational formula:

$$mr = mR_O + mR_S \qquad (2.1.8)$$

Substituting equation (2.1.5) into the above equation and removing mR_O, we can obtain the relational equation:

$$R_S = \frac{m}{M+m} r \qquad (2.1.9)$$

The R_S in the formula is the distance between the center O of mass and the Sun, and r is the distance between the Earth and the Sun. Note that the Earth's elliptical orbit does not have the above relationship formula.

2.1.3 Distance between the center O of mass and the Earth and the Sun at the moments of perihelion and aphelion

At the moment of perihelion. Assume that R_{O1} is the distance from the Earth's perihelion to the center O of mass; R_{S1} is the distance from the Sun's perigee to the center O of mass. Assume that v_{O1} is the tangential velocity of the Earth's revolution around the center O of mass at perihelion; assume that v_{S1} is the tangential velocity of the Sun's revolution around the center O of mass at perigee. This is shown in Figure 2-3.

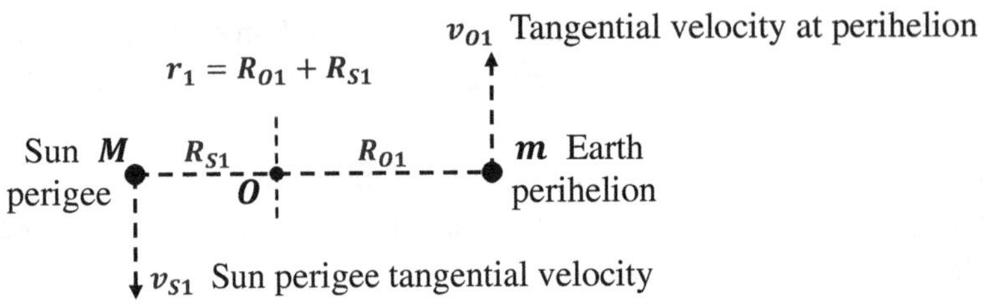

Figure 2-3 Distance between the Sun M and the Earth m to the center O of mass at the Moment of Perihelion.

The distance r_1 from the Earth's perihelion to the Sun is:

$$r_1 = R_{01} + R_{S1} \tag{2.1.10}$$

At the moment of perihelion, according to equation (2.1.5), the relation equation can be obtained:

$$mR_{01} = MR_{S1} \tag{2.1.11}$$

According to equation (2.1.7), the distance R_{01} from the Earth's perihelion to the center O of mass is:

$$R_{01} = \frac{M}{M+m} r_1 \tag{2.1.12}$$

According to equation (2.1.9), the distance R_{S1} from the solar perigee to the center O of mass is:

$$R_{S1} = \frac{m}{M+m} r_1 \tag{2.1.13}$$

Note that the perihelion of the Earth's elliptical orbit does not have the above formula and equation (2.1.12). Theoretically, the distance r_1 from the Earth's perihelion to the Sun can be described by two different frames of reference.

One is described by the frame of reference of the center O of mass of the solar system, i.e. at the moment of perihelion the sum of the distances from the center O of mass to the Earth and to the Sun is $r_1 = R_{01} + R_{S1}$.

The other is described in terms of the solar (elliptical focus) frame of reference, i.e., the distance r'_1 from the elliptical focus to perihelion at the moment of perihelion. At this point, the center O of mass of the solar system moves around the sun since the sun (elliptical focus) is stationary. Note that the elliptical orbits of the planets do not conform to the laws of motion of the system of mass points.

Although the solar frame of reference and the frame of reference of the center O of mass of the solar system are two different frames of reference, the perihelion distance r_1 of the frame of reference of the center O of mass is equal to the perihelion distance r'_1 of the solar frame of reference, i.e:

$$r = r'_1$$

Note that the perihelion-to-sun distance r'_1 of a planet's elliptical orbit does not have the relationship of the formula $r'_1 = R_{01} + R_{S1}$.

At the moment of apogee. Assume that R_{02} is the distance from the Earth's apogee to the center O of mass; R_{S2} is the distance from the Sun's apogee to the center O of mass. Assume that v_{02} is the tangential velocity of the Earth's apogee around the center O of mass; and v_{S2} is the tangential velocity of the Sun's apogee around the center O of mass. This is shown in Figure 2-4.

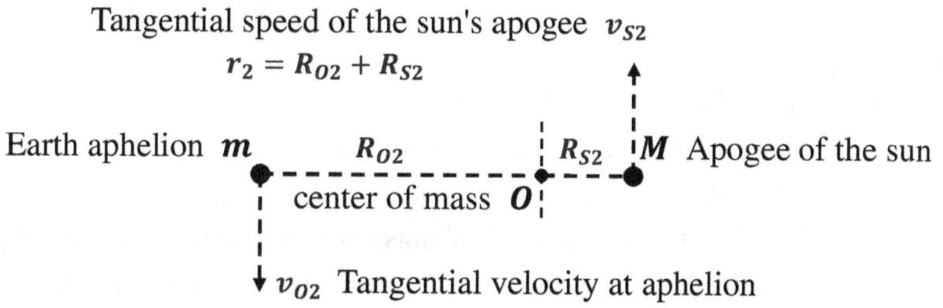

Figure 2-4 At the moment of aphelion, the distance between the Sun M and the Earth m to the center O of mass.

In the frame of reference of the center O of mass of the solar system, the distance r_2 from the Earth's aphelion to the Sun is:

$$r_2 = R_{O2} + R_{S2} \tag{2.1.14}$$

At the moment of aphelion, according to equation (2.1.5), the relation equation can be obtained:

$$mR_{O2} = MR_{S2} \tag{2.1.15}$$

According to equation (2.1.7), the distance R_{O2} from the Earth's aphelion to the center O of mass is:

$$R_{O2} = \frac{M}{M+m} r_2 \tag{2.1.16}$$

According to equation (2.1.9), the distance R_{S2} from the solar apogee to the center O of mass is:

$$R_{S2} = \frac{m}{M+m} r_2 \tag{2.1.17}$$

Note that the aphelion of the Earth's elliptical orbit does not have the above formula and equation (2.1.16). Although the solar frame of reference and the center O of mass frame of reference of the solar system are two different frames of reference, the aphelion distance r_2 of the center O of mass frame of reference is equal to the aphelion distance r_2' of the solar frame of reference, i.e:

$$r_2 = r_2'$$

Note that the distance r_2' from the aphelion of an elliptical orbit to the Sun does not have the relationship of the formula $r_2' = R_{O2} + R_{S2}$.

2.1.4 Orbits and perimeters of the Earth's orbits around the center O of mass

The difference value ΔR_O between the distance from the Earth's perihelion and aphelion to the center O of mass is:

$$\Delta R_O = R_{O2} - R_{O1} \tag{2.1.18}$$

According to Figure 2-2. assume that perihelion is the starting point of the rotation angle δ. If the Earth rotates counterclockwise around the center O of mass, then the distance R_O from the Earth to the center O of mass can be expressed as:

$$R_O = R_{O2} + \frac{1}{2}(R_{O1} - R_{O2})(1 + \cos\delta) \tag{2.1.19}$$

The above equation is the formula for the distance of the Earth's revolution around the center O of mass. If the angle of rotation $\delta = 0$, then the Earth is at perihelion and the distance from perihelion to the center O of mass is R_{O1}. When the rotation angle $\delta = 180^0$, the Earth is at aphelion. The distance from the aphelion to the center O of mass is R_{O2}..

It is important to note that scientists often use conic curves to analyze and describe the motion of the planets. Conic sections include elliptic curves, parabolas, hyperbolas, and circles. In a plane coordinate system, each corresponds to a quadratic curve. Conics are also known as quadratic curves.

Physicists believe that the orbits of the planets around the sun are elliptical curves, and that the sun is one of the focal points of the elliptical orbits of the planets. If the velocity of the planet increases to a certain value, then the orbit of the planet's motion will follow a parabola or hyperbola. The motion of artificial satellites or spacecraft follows this law. Conics are the pattern of motion of cosmic objects. Planets attracted by gravitational can only move along conic curves; they cannot move along any other curves.

However, the orbit of the Earth around the center O of mass, equation (2.1.19), is a new kind of conic curve. Since the orbit of equation (2.1.19) resembles the profile of the long axis of an eggshell, the curve of equation (2.1.19) is referred to in this book as the "eggshell curve"..

If the distance R_{O1} from the perihelion to the center O of mass is equal to the distance R_{O2} from the aphelion to the center O of mass, i.e., $R_{O1} = R_{O2} = R$, then equation (2.1.19) becomes the equation for a circular orbit, i.e., $R_O = R$. At this point, the circular radius R_O remains the same regardless of any change in the rotation angle δ. From this it can be seen that the circular orbit is a special case of the motion orbit of formula (2.1.19).

As the Earth moves from perihelion to aphelion, the rotation angle δ varies in the interval 0 to π. Thus, the circumference L_O of the Earth's orbit around the center O of mass is:

$$L_O = 2\int_0^{\pi} R_O d\delta \qquad (2.1.20)$$

Substituting equation (2.1.19) into the above equation gives the relational equation:

$$L_O = 2\int_0^{\pi} \left(R_{O2} + \frac{1}{2}(R_{O1} - R_{O2})(1 + \cos\delta)\right) d\delta$$

Solve the integral equation above to obtain the relational equation:

$$L_O = \pi(R_{O1} + R_{O2}) \qquad (2.1.21)$$

The above formula is the formula for the circumference of the Earth's orbit around the center O of mass. The above formula shows that half of the sum of the distances from the center O of mass to perihelion and aphelion is equal to the average distance from the center O of mass to the Earth $\widetilde{R_O}$, i.e:

$$\widetilde{R_O} = \frac{1}{2}(R_{O1} + R_{O2})$$

If the distances from the center O of mass to perihelion and aphelion are equal, i.e., $R_{O1} = R_{O2} = R$, then equation (2.1.21) becomes a formula for the circumference of a circular orbit, i.e., $L_O = 2\pi R$. From this it can be seen that a circular orbit is a special case of the orbit of motion of equation (2.1.19). Since equation (2.1.19) is not an elliptic curve, the Earth's orbit around the center O of mass is neither circular nor elliptic.

2.1.5 Orbits of the Earth around the Sun and Perimeters of Orbits

Multiply equation (2.1.19) by the relational equation $\frac{M+m}{M}$ to obtain the relational equation:

$$\frac{M+m}{M} R_O = \frac{M+m}{M}\left(R_{O2} + \frac{1}{2}(R_{O1} - R_{O2})(1 + \cos\delta)\right)$$

Substituting equation (2.1.7) $R_O = \frac{M}{M+m} r$, equation (2.1.12) $R_{O1} = \frac{M}{M+m} r_1$, and equation $R_{O2} = \frac{M}{M+m} r_2$ into the above equations, we can obtain the relationship equation:

$$r = r_2 + \frac{1}{2}(r_1 - r_2)(1 + \cos\delta) \qquad (2.1.22)$$

The formula above is the formula for the distance the Earth travels around the Sun. The δ in the formula is the angle at which the Earth rotates counterclockwise around the Sun. When the angle of rotation $\delta = 0$, the Earth is at perihelion and the distance from perihelion to the Sun is r_1. If the rotation angle is $\delta = 180^0$, then the Earth is at aphelion. The distance from aphelion to the sun is r_2.

When the distance from the Sun to the Earth's perihelion and aphelion is the same, i.e. $r_1 = r_2 = r$, the above formula becomes the formula for a circular orbit. At this point, no matter how the rotation angle δ changes, the circular radius r always remains the same. From this point of view, the circular orbit is a special case of the above formula.

Equation (2.1.22) can be expressed as a quadratic curve, i.e:

$$\begin{cases} x = \left(r_2 + \frac{1}{2}(r_1 - r_2)(1 + \cos\delta)\right)\cos\delta \\ y = \frac{1}{2}(r_1 + r_2)\sin\delta \end{cases} \quad (2.1.23)$$

As the Earth moves from perihelion to aphelion. Since the interval of variation of the rotation angle δ is from 0 to π, the circumference L of the Earth's orbit around the Sun is:

$$L = 2\int_0^{\pi} r d\delta = 2\int_0^{\pi}\left(r_2 + \frac{1}{2}(r_1 - r_2)(1 + \cos\delta)\right)d\delta \quad (2.1.24)$$

Solving the above equation gives the relational equation:

$$L = \pi(r_1 + r_2) \quad (2.1.25)$$

The above formula is the formula for the circumference of the Earth's orbit around the Sun. Since the interval of variation of the angle of revolution is from 0 to 2π, the average distance $\tilde{r}$ of the Earth's revolution around the Sun is:

$$\tilde{r} = \frac{L}{2\pi} = \frac{1}{2}(r_1 + r_2) \quad (2.1.26)$$

The above equation shows that half of the sum of the distances from the Sun to perihelion and aphelion is equal to the mean distance $\tilde{r}$ from the Sun to the Earth. If the distances from the Sun to perihelion and aphelion are equal, i.e. $r_1 = r_2 = r$, equation (2.1.25) becomes a formula for the circumference of a circular orbit, i.e. $L = 2\pi r$. From this it can be seen that a circular orbit is a special case of the orbit of motion of equation (2.1.22). Since equation (2.1.22) is not an elliptical curve, the Earth's orbit around the Sun is neither circular nor elliptical.

Multiply equation (2.1.25) by $\frac{M}{M+m}$ to obtain the relational equation:

$$\frac{M}{M+m}L = \pi\left(\frac{M}{M+m}r_1 + \frac{M}{M+m}r_2\right)$$

Substituting equation (2.1.12) $R_{O1} = \frac{M}{M+m}r_1$ and equation (2.1.16) $R_{O2} = \frac{M}{M+m}r_2$ into the above equation gives the relationship equation:

$$\frac{M}{M+m}L = \pi(R_{O1} + R_{O2}) = L_O \quad (2.1.27)$$

The above equation is equation (2.1.21). The above equation shows that the circumference L of the Earth's orbit around the Sun magnifies the circumference L_O of the Earth's orbit around the center O of mass by a factor of $\frac{M+m}{M}$. Note that the Earth's elliptical orbit does not have the above relational formula.

2.1.6 Orbital perimeter of the Moon around the center of mass O' of the Earth-Moon system

According to equation (2.1.12) and equation (2.1.16), the relation equation can be obtained:

$$R_{O1} = \frac{M}{M+m}r_1, \qquad R_{O2} = \frac{M}{M+m}r_2$$

If we know the perihelion distance r_1 and the aphelion distance r_2 from the Earth to the Sun, we can calculate the orbital circumference L_O of the Earth's orbit around the center O of mass using equation (2.1.21) $L_O = \pi(R_{O1} + R_{O2})$.

Since the mass m of the Earth , is much smaller than the mass M of the Sun , it can be assumed that the distance R_O from the Earth to the center O of mass , is equal to the distance r from the Earth to the Sun,, i.e. $R_{O1} \approx r_1$, and $R_{O2} \approx r_2$. Consequently, the circumference L_O of the Earth's orbit around the center O of mass, is equal to the circumference of the Earth's orbit around the Sun , i.e., :

$$L_O \approx L = \pi(r_1 + r_2) \quad (2.1.28)$$

Based on astronomical observations of the Earth and Moon:

Mass of the Earth $m = 5.965 \times 10^{24} \cdot kg$;

Mass of the Moon $m' = 0.07349 \times 10^{24} \cdot kg$;

Perigee distance of the moon $r_1 = 363300 \cdot km$;

The apogee distance of the moon $r_2 = 405493 \cdot km$;
The average period of the moon's rotation $T = 27.32$ days;
The average speed of the moon's rotation $\tilde{v} = 1.023 \cdot km/s$;
According to the above data, the circumference L_ρ of the moon's elliptical orbit is:

$$L_\rho = vT = 2414738 \cdot km$$

According to equation (2.1.25), the circumference L of the moon's orbit around the Earth is

$$L = \pi(r_1 + r_2) = 2415234 \cdot km$$

According to equation (2.1.21), the orbital circumference L_O of the Moon around the center O' of mass of the Earth-Moon system is:

$$L_O = \pi(R_{O1} + R_{O2})$$

According to the data for the Earth, the ratio of the mass m of the Earth to the mass $(m + m')$ of the Earth-Moon system is:

$$\frac{m}{m + m'} = \frac{5.965}{5.965 + 0.7369} = 0.9878 \qquad (2.1.29)$$

According to equation (2.1.27), the circumference L of the Moon's orbit around the Earth enlarges the circumference L_O of the Moon's orbit around the Earth-Moon center O' of mass by the factor $\frac{m}{m+m'}$, i.e:

$$L_O = \frac{m}{m + m'} L \qquad (2.1.30)$$

Substituting equation (2.1.29) and the circumference of the Moon's orbit around the Earth, $L = 2415234 \cdot km$, into the above equation gives the relational equation:

$$L_O = \frac{m}{m + m'} L = 2385768 \cdot km \qquad (2.1.31)$$

The difference value ΔL between the circumference L of the Moon's eggshell orbit around the Earth and the circumference L_O of the Moon's eggshell orbit around the center of mass O' of the Earth-Moon system is:

$$\Delta L = L - L_O = 2415234 - 2385768 = 29466 \cdot km \qquad (2.1.32)$$

The difference value $\Delta L'$ between the circumference L_ρ of the Moon's elliptical orbit around the Earth, and the circumference L_O of the Moon's eggshell orbit around the center of mass O' of the Earth-Moon system, is:

$$\Delta L' = L_\rho - L_O = 2414738 - 2385768 = 28970 \cdot km \qquad (2.1.33)$$

From the above equation and equation (2.1.32), it can be determined that the difference value between the circumference L_ρ of the moon's elliptical orbit and the circumference L_O of the moon's eggshell orbit is about $30000 \cdot km$.

In summary, since the mass m of the Earth is much smaller than the mass M of the Sun, the circumference L_O of the Earth's orbit around the center O of mass, of the solar system, is approximately equal to the circumference L of the Earth's orbit around the Sun.

However, since the ratio of the Moon's mass m' to the Earth's mass m is equal to 1%, the difference value ΔL between the circumference L_O of the Moon's eggshell orbit and the circumference L_ρ of the Moon's elliptical orbit is greater than 1%. Obviously, this large difference value in distance is not theoretically negligible.

2.2 The orbit of the Earth around the center O of mass of the solar system is an "eggshell curve"

2.2.1 "Tangential velocity v_O", "Radial velocity v_{OR}" and "Transverse velocity $v_{O\varphi}$" of the Earth's revolution around the center O of mass of the Solar System

According to mass-point dynamics. The Earth's motion is curvilinear if the combined force F on the Earth is in a direction other than the direction of motion of the Earth. The trajectory of the Earth is sandwiched between the combined force F and the velocity v (direction of motion). The direction of the velocity v is tangent to the orbit.

The direction of the combined force $\boldsymbol{F}$ points to the concave side of the trajectory.

According to mass-point kinematics, the Earth orbits around the center O of mass of the solar system. Assume that $\boldsymbol{R_O}$ is the distance from the center O of mass to orbital point A. Assume that $\boldsymbol{dt}$ is the time it takes the Earth to move from orbital point A to orbital point B. Assume that $\boldsymbol{\varphi}$ is the angular velocity of point A. Assume that $\boldsymbol{dR_O}$ is the change in radial displacement; $\boldsymbol{dR_{O\varphi}}$ is the change in lateral displacement. This is shown in Figure 2-5.

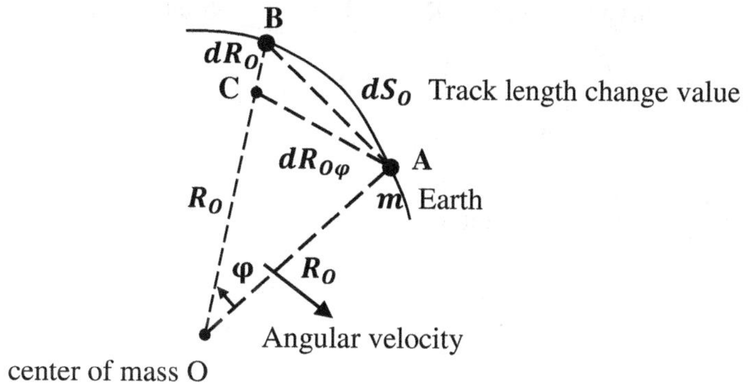

Figure 2-5. Schematic diagram of the change of the orbit length of the earth moving around the center O of mass.

When the change $\boldsymbol{dS_O}$ in the length of the orbit is very small, the ABC triangle in Figure 2-5 can be considered a right triangle. According to the law of curvilinear motion of a point of mass, as the Earth moves from point A to point B, the change $\boldsymbol{dS_O}$ in the length of the orbit, is:

$$\boldsymbol{dS_O} = \sqrt{(\boldsymbol{dR_O})^2 + (\boldsymbol{dR_{O\varphi}})^2}$$

Taking the derivative of the above equation with respect to time $\boldsymbol{t}$ gives the relational equation:

$$\frac{\boldsymbol{ds_O}}{\boldsymbol{dt}} = \sqrt{\left(\frac{\boldsymbol{dR_O}}{\boldsymbol{dt}}\right)^2 + \left(\frac{\boldsymbol{dR_{O\varphi}}}{\boldsymbol{dt}}\right)^2} \qquad (2.2.1)$$

In this book, the velocity $\boldsymbol{v_O} = \frac{\boldsymbol{ds_O}}{\boldsymbol{dt}}$ is defined as the tangential velocity of the Earth's rotation around the center O of mass. Define the velocity $\boldsymbol{v_{OR}} = \frac{\boldsymbol{dR_O}}{\boldsymbol{dt}}$ as the radial velocity of the Earth's rotation around the center O of mass. Define the velocity $\boldsymbol{v_{O\varphi}} = \frac{\boldsymbol{dR_{O\varphi}}}{\boldsymbol{dt}}$ as the transverse velocity of the Earth's rotation around the center O of mass. Note that the direction of the radial velocity $\boldsymbol{v_{OR}}$ is toward the center O of mass, or in the opposite direction. The direction of the transverse velocity $\boldsymbol{v_{O\varphi}}$ is perpendicular to the distance $\boldsymbol{R_O}$. Thus, the above equation can be expressed as:

$$\boldsymbol{v_O} = \sqrt{(\boldsymbol{v_{O\varphi}})^2 + (\boldsymbol{v_{OR}})^2} \qquad (2.2.2)$$

The above equation shows that the tangential velocity $\boldsymbol{v_O}$ of the Earth's revolution around the center O of mass can be decomposed into two velocity components perpendicular to each other. Note that the elliptical orbit of the Earth does not have the above relational equation.

By the nature of curvilinear motion, every point on a curve corresponds to a circle of curvature, and every circle of curvature has a certain curvature $\boldsymbol{K}$ and radius of curvature $\boldsymbol{\rho}$.

Furthermore, the tangential velocity $\boldsymbol{v}$ at each point of the curve corresponds to a circle of curvature. Alternatively, the tangential velocity $\boldsymbol{v}$ corresponds to the radius $\boldsymbol{\rho}$ and curvature $\boldsymbol{K}$ of the circle of curvature at that point. Different tangential velocities $\boldsymbol{v}$ usually correspond to different circles of curvature.

Assume that $\boldsymbol{\rho_O}$ is the radius of curvature corresponding to the tangential velocity $\boldsymbol{v_O}$; $\boldsymbol{O_{vO}}$ is the circular center of the circle of curvature corresponding to the radius of curvature $\boldsymbol{\rho_O}$, as shown in Figure 2-6.

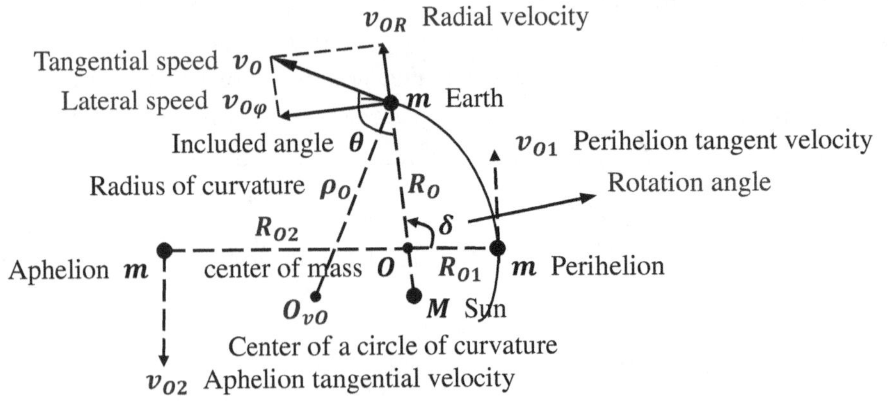

Figure 2-6. the center O_{vO} and radius ρ_O of the Earth's circle of curvature as it rotates around its center O of mass.

Although the Earth always orbits around the center O of mass, the centripetal acceleration a_O at each orbital point usually points not to the center O of mass, but to the center O_{vO} of the point's circle of curvature.

Alternatively, the tangential velocity v_O of the Earth's rotation around the center O of mass is always perpendicular to the radius of curvature ρ_O at that point. The tangential velocity v_O is usually not perpendicular to the distance R_O from the Earth to the center O of mass.

Only the tangential velocities v_O, of both perihelion and aphelion are perpendicular to the distance R_O, from the Earth to the center O of mass. This is shown in Figure 2-6.

Let θ be the angle between the tangential velocity v_O and the distance R_O. The tangential velocity v_O can be decomposed into two orthogonal velocity components. The first velocity component is the radial velocity v_{OR} along the direction of the distance R_O; the second velocity component is the transverse velocity $v_{O\varphi}$ perpendicular to the distance R_O. This is shown in Figure 2-6.

According to Figure 2-6. As the Earth moves gradually from perihelion to aphelion, the radial velocity v_{OR} is:

$$v_{OR} = v_O \cos\theta \tag{2.2.3}$$

When the rotation angle $\delta = 0$, (or $\delta = 180^0$). Since the pinch angle $\theta = 90^0$, the radial velocity $v_{OR} = 0$, so the tangential velocity v_O is perpendicular to the distance R_O. as shown in Figure 2-6.

Assume that the angular velocity possessed by the distance R_O is $\varphi = d\delta/dt$. according to Figure 2-6. As the Earth moves gradually from perihelion to aphelion, the transverse velocity $v_{O\varphi}$ is:

$$v_{O\varphi} = v_O \sin\theta = R_O\,\varphi \tag{2.2.4}$$

The transverse velocity $v_{O\varphi}$ is the circular velocity with the radius of the distance R_O. When the rotation angle $\delta = 0$, (or $\delta = 180^0$). Since the angle $\theta = 90^0$, the transverse velocity $v_{O\varphi}$ is equal to the tangential velocity v_O, i.e., $v_{O\varphi} = v_O$, so the tangential velocity v_O is perpendicular to the distance R_O. as shown in Figure 2-6.

The tangential velocity v_O is equal to the vectorial sum of the transverse velocity $v_{O\varphi}$ and the radial velocity v_{OR}, i.e:

$$v_O = \sqrt{v_{O\varphi}^2 + v_{OR}^2}$$

The formula above is equation (2.2.2). Note that the elliptical orbit of the Earth does not have the above equation.

2.2.2 The tangential velocities v_O of perihelion and aphelion are perpendicular to the distance R_O of the two points from the center of mass O

Suppose R_{O1} is the distance from the center O of mass to the Earth's perihelion. Since the distance r_1 from the Earth's perihelion to the Sun is the minimum distance of the Earth's revolution around the Sun, the distance R_{O1} from the perihelion to the center O of mass is also the minimum distance of the Earth's revolution around the center O of mass. According to equation (2.2.2), the tangential velocity v_{O1} of the Earth's perihelion in revolution around the center O of mass is:

$$v_{O1} = \sqrt{v_{O\varphi1}^2 + v_{OR1}^2} \qquad (2.2.5)$$

The $v_{O\varphi1}$ in equation is the transverse velocity of the perihelion's revolution around the center O of mass. v_{OR1} is the radial velocity of the perihelion's revolution around the center O of mass.

As the Earth gradually approaches perihelion from aphelion. Since the distance R_O from the Earth to the center O of mass is gradually decreasing, the radial velocity of the Earth before it reaches perihelion is $v_{OR} = \frac{dR_O}{dt} < 0$. When the Earth leaves perihelion. Since the distance R_O from the Earth to the center O of mass begins to gradually increase, the radial velocity $v_{OR} > 0$ after the Earth leaves perihelion.

Since the radial velocity v_{OR} before and after perihelion have opposite positive and negative signs, the radial velocity v_{OR1} at perihelion is:

$$v_{OR1} = \frac{dR_O}{dt} = 0 \qquad (2.2.6)$$

Suppose φ_1 is the angular velocity of the perihelion's revolution around the center O of mass. According to equation (2.2.4), the transverse velocity $v_{O\varphi1}$ of the perihelion's revolution around the center O of mass is:

$$v_{O\varphi1} = R_{O1}\varphi_1 \qquad (2.2.7)$$

Substituting the above equation and equation (2.2.6) into equation (2.2.5) gives the relationship equation:

$$v_{O1} = v_{O\varphi1} = R_{O1}\varphi_1 \qquad (2.2.8)$$

Since the transverse velocity of the perihelion $v_{O\varphi1}$ is equal to the tangential velocity of the perihelion v_{O1}, the tangential velocity of the perihelion v_{O1} is perpendicular to the distance R_{O1} from the perihelion to the center O of mass. This is shown in Figure 2-7.

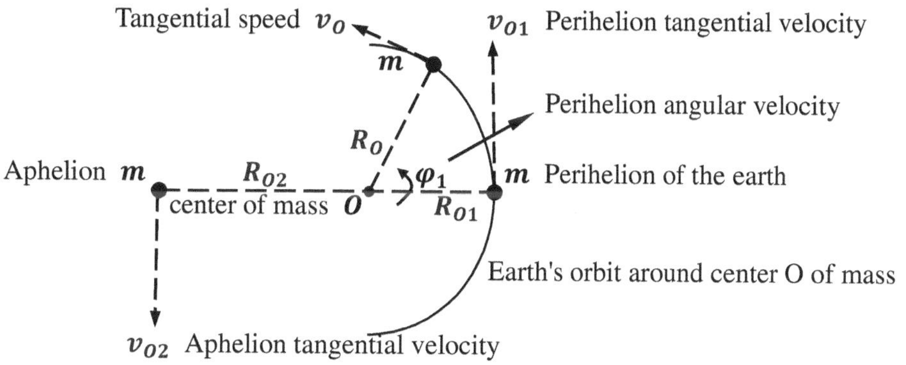

Figure 2-7. Perihelion tangential velocity v_{O1} and aphelion tangential velocity v_{O2} of the Earth's orbit around center O of mass.

Since the Earth rotates around the center O of mass at perihelion, and the tangential velocity v_{O1} at perihelion is perpendicular to the distance R_{O1} from perihelion to the center O of mass, the center O of mass is the center of the perihelion curvature circle, and the distance R_{O1} is the radius of the perihelion curvature circle. This is shown in Figure 2-7.

Let R_{O2} be the distance from the center O of mass to the Earth's aphelion. Since the distance r_2 from the Earth's aphelion to the Sun is the maximum distance of the Earth's orbit around the Sun, the distance R_{O2} from the aphelion to the center O of mass is also the maximum distance of the Earth's orbit around the center O of mass. According to equation (2.2.2), the tangential velocity v_{O2} of the Earth's aphelion around the center O of mass is:

$$v_{O2} = \sqrt{v_{O\varphi 2}^2 + v_{OR2}^2} \qquad (2.2.9)$$

The $v_{O\varphi 2}$ in equation is the transverse velocity of the aphelion's revolution around the center O of mass, and v_{OR2} is the radial velocity of the aphelion's revolution around the center O of mass.

As the Earth moves from perihelion to aphelion. As the distance R_O from the Earth to the center O of mass gradually increases, the radial velocity of the Earth before it reaches the aphelion is $v_{OR} = \frac{dR_O}{dt} > 0$.

When the Earth moves away from the aphelion. Since the distance R_O from the Earth to the center O of mass begins to decrease gradually, the radial velocity $v_{OR} < 0$ after the Earth leaves the aphelion is:

Since the radial velocity v_{OR} before and after the aphelion has opposite positive and negative signs, the radial velocity v_{OR2} at the aphelion is:

$$v_{OR2} = \frac{dR_O}{dt} = 0 \qquad (2.2.10)$$

Assume that φ_2 is the angular velocity of the aphelion as it revolves around the center O of mass. According to equation (2.2.4), the transverse velocity $v_{O\varphi 2}$ of the aphelion's revolution around the center O of mass is:

$$v_{O\varphi 2} = R_{O2}\varphi_2 \qquad (2.2.11)$$

Substituting the above equation and equation (2.2.10) into equation (2.2.9) gives the relational equation:

$$v_{O2} = v_{O\varphi 2} = R_{O2}\varphi_2 \qquad (2.2.12)$$

Since the transverse velocity $v_{O\varphi 2}$ of the aphelion is equal to the tangential velocity v_2 of the aphelion, the tangential velocity v_{O2} of the aphelion is perpendicular to the distance R_{O2} from the aphelion to the center O of mass. This is shown in Figure 2-7.

Since the Earth rotates around the center O of mass at aphelion, and the tangential velocity of the aphelion $v_{O2}$is perpendicular to the distance R_{O2} from the aphelion to the center O of mass, the center O of mass is the center of the aphelion's circle of curvature, and the distance R_{O2} is the radius of the aphelion's circle of curvature. This is shown in Figure 2-7.

In summary, the center O of mass is the center of the perihelion and aphelion curvature circles. The distances R_{O1} and R_{O2} from the center O of mass to the perihelion and aphelion points are the radii of the perihelion and aphelion curvature circles, respectively.

2.2.3 The orbit of the Earth around its center O of mass is an "eggshell curve"

According to equation (2.1.12) and equation (2.1.16), the relation equation can be obtained:

$$R_{O1} = \frac{M}{M+m} r_1, \quad R_{O2} = \frac{M}{M+m} r_2$$

Note that the elliptical orbit of the Earth does not have the above relationship. Since the rotation of the Earth around the center O of mass at perihelion and aphelion is circular, and the distance r_1 from the Sun to perihelion, is less than the distance r_2 from the Sun to aphelion, the radius R_{O1} of the circle of curvature at perihelion, is less than that at aphelion, i.e. $R_{O1} < R_{O2}$.

Since the center O of mass is the center of the perihelion curvature circle and the aphelion curvature circle, the curvature K_1 at the perihelion of the Earth's revolution around the center O of mass is, according to the definition of curvature, the following:

$$K_1 = \frac{1}{R_{O1}} \qquad (2.2.13)$$

The curvature K_2 at the aphelion of the Earth's revolution around the center O of mass is:

$$K_2 = \frac{1}{R_{O2}} \qquad (2.2.14)$$

Since the radius of curvature R_{O1} at perihelion is smaller than the radius of curvature R_{O2} at aphelion, the curvature K_1 at perihelion is larger than the curvature K_2 at aphelion, i.e. $K_1 > K_2$. From this it can be seen that the Earth's orbit around the center O of mass, is analogous to an "eggshell curve". This is shown in Figure 2-8.

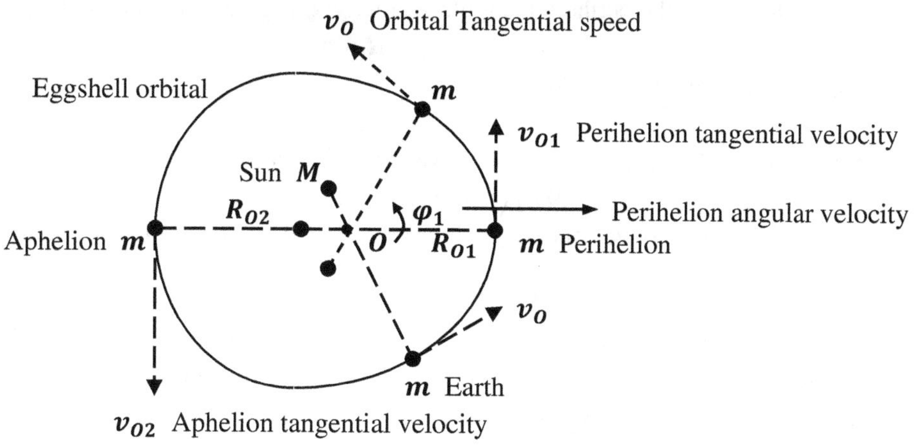

Figure 2-8 Schematic of the Earth's orbit around the center O of mass.

It is important to note. In the eggshell orbit of the Earth around the center O of mass, only the tangential velocities v_{O1} and v_{O2} at the perihelion and aphelion points are perpendicular to the distance from the center O of mass to the Earth. The tangential velocities v_O at other points in the eggshell orbit are not perpendicular to the distance R_O from the center O of mass to the Earth.

2.3 Velocity and centripetal force of the Earth's rotation around the center O of mass

2.3.1 Relationship between tangential velocity v_{O1} and v_1 at Earth's perihelion

At the moment of perihelion, according to equation (2.1.5) $mR_O = MR_S$, the relation equation can be obtained:

$$mR_{O1} = MR_{S1}$$

The R_{O1} in the formula is the distance from perihelion to center O of mass, and R_{S1} is the distance from the sun's perigee to the center of mass O. In the frame of reference of center O of mass, the time of the Earth's orbit around center O of mass should be equal to the time of the Sun's orbit around center O of mass. Otherwise, the center O of mass would not always be on the line from the Earth to the Sun. From this it can be seen that both the Sun and the Earth rotate around the center O of mass with the same angular velocity.

Assume that φ_1 is the angular velocity of the Earth's perihelion as it revolves around the center O of mass. Multiplying the above equation by the angular velocity φ_1 gives the relational equation:

$$mR_{O1}\varphi_1 = MR_{S1}\varphi_1 \qquad (2.3.1)$$

According to the equation for linear velocity $v = r\varphi$, the tangential velocity v_{O1} of the Earth's perihelion rotating around the center O of mass is:

$$v_{O1} = R_{O1}\varphi_1 \qquad (2.3.2)$$

The tangential velocity v_{S1} of the Sun's perigee around the center O of mass is:

$$v_{S1} = R_{S1}\varphi_1 \qquad (2.3.3)$$

Substituting the above two formulas into equation (2.3.1) gives the relational equation:

$$mv_{O1} = Mv_{S1} \qquad (2.3.4)$$

According to the parallel frame of reference, the angular velocity φ_1 of the Earth's perihelion revolution around the center O of mass is also the angular velocity of the Earth's perihelion revolution around the Sun.

According to equation (2.1.10), the distance r_1 from perihelion to the Sun is:

$$r_1 = R_{O1} + R_{S1}$$

Multiplying the above equation by the perihelion angular velocity φ_1. gives the relational equation:

$$r_1\varphi_1 = R_{O1}\varphi_1 + R_{S1}\varphi_1$$

The above formula can be changed to the following form:

$$v_1 = v_{O1} + v_{S1} \tag{2.3.5}$$

Multiplying the above equation by the solar mass M gives the relationship equation:

$$Mv_1 = Mv_{O1} + Mv_{S1} \tag{2.3.6}$$

Substitute equation (2.3.4) into the above equation and remove Mv_{S1} to obtain the relational equation:

$$v_{O1} = \frac{M}{M+m} v_1 \tag{2.3.7}$$

Note that the perihelion of the Earth's elliptical orbit does not have the above relationship. The above equation is the relationship between the tangential velocity v_{O1} and v_1 at Earth's perihelion. Obviously, the tangential velocity v_{O1} amplifies the tangential velocity v_1 by a factor of $\frac{M}{M+m}$.

Multiplying equation (2.1.12) $R_{O1} = \frac{M}{M+m} r_1$ by the perihelion angular velocity φ_1 gives the relation equation:

$$R_{O1}\varphi_1 = \frac{M}{M+m} r_1\varphi_1$$

Since the tangential velocity $v_{O1} = R_{O1}\varphi_1$ of the Earth's perihelion revolving around the center O of mass, as well as the tangential velocity $v_1 = r_1\varphi_1$ of the Earth's perihelion revolving around the Sun, the above equation is equation (2.3.7).

According to equation (2.3.7), the relation between the tangential velocity v_{O2} and v_2 at the Earth's aphelion is:

$$v_{O2} = \frac{M}{M+m} v_2 \tag{2.3.8}$$

Note: The aphelion of the Earth's elliptical orbit does not have the above relation. From the above equation, it is clear that the tangential velocity v_{O2} at aphelion magnifies the tangential velocity v_2 by a factor of $\frac{M}{M+m}$.

2.3.2 Tangential velocity v_O and mean velocity $\tilde{v_O}$ of the Earth's revolution around the center O of mass

Since the velocity v_{O1} at Earth's perihelion is the maximum velocity and the velocity v_{O2} at Earth's aphelion is the minimum velocity, the difference value Δv_O between the perihelion velocity v_{O1} and the aphelion velocity v_{O2} is:

$$\Delta v_O = v_{O1} - v_{O2} \tag{2.3.9}$$

The tangential velocity v_O of the Earth's rotation around the center O of mass can be expressed as:

$$v_O = v_{O2} + \frac{1}{2}(v_{O1} - v_{O2})(1 + \cos\delta) \tag{2.3.10}$$

In this book, the above formula is referred to as the formula for the tangential velocity of the Earth's rotation around the center O of mass. The δ in the formula is the counterclockwise rotation angle of the Earth's rotation around the center O of mass.

When the rotation angle $\delta = 0$, the Earth is at perihelion, and the tangential velocity at perihelion is v_{O1}. When the rotation angle $\delta = 180^0$, the Earth is at aphelion, and the tangential velocity at aphelion around the center O of mass is v_{O2}.

If the tangential velocity v_{O1} at the perihelion is equal to the tangential velocity v_{O2} at the aphelion, i.e. $v_{O1} = v_{O2} = v_O$, the above formula becomes the formula for the circular velocity. At this point, no matter how the rotation angle δ changes, the tangential velocity v_O always remains the same.

Note: The current conic contains only elliptic lines, parabolas, and hyperbolas, not eggshell curves.

As the Earth moves from perihelion to aphelion, the rotation angle δ varies over an interval from 0 to π. Within this interval of variation, the integral $\sum v_O$ of the tangential velocity v_O is:

$$\sum v_O = \int_0^{\pi} v_O d\delta \tag{2.3.11}$$

Substituting the tangential velocity v_O from equation (2.3.10) into the above equation gives the relational equation:

$$\sum v_O = \int_0^{\pi} \left(v_{O2} + \frac{1}{2}(v_{O1} - v_{O2})(1 + \cos\delta) \right) d\delta$$

After solving the above equation, the relation equation is obtained:

$$\sum v_O = \frac{1}{2}\pi(v_{O1} + v_{O2}) \tag{2.3.12}$$

The mean velocity $\widetilde{v_O}$ of the Earth's revolution around the center O of mass in the interval 0 to π of the variation of the angle of rotation is:

$$\widetilde{v_O} = \frac{1}{\pi}\sum v_O = \frac{1}{2}(v_{O1} + v_{O2}) \tag{2.3.13}$$

The above equation is the formula for the mean velocity of the Earth's revolution around the center O of mass. The above equation shows that half of the sum of the perihelion and aphelion velocities of the Earth's revolution around the center O of mass is equal to the average velocity of the Earth's revolution around the center O of mass $\widetilde{v_O}$. Note that the Earth's elliptical orbit does not have the above relationship formula.

2.3.3 The centripetal force F_{vO} of the Earth's rotation around the center O of mass consists of two components

According to the previous Figure 2-6, the tangential velocity v_O is perpendicular to the radius of curvature ρ_O of the circle of curvature. According to the centripetal force formula, the centripetal force F_{vO} of the Earth rotating around the center O of mass is:

$$F_{vO} = m\frac{v_O^2}{\rho_O} \tag{2.3.14}$$

Note that the Earth's elliptical orbit does not have the centripetal force F_{vO} of the Earth's revolution around the center O of mass.

The centripetal force F_{vO} of the Earth's rotation around the center O of mass usually points to the center of the circle of curvature O_{vO}, not to the center O of mass. Only at perihelion and aphelion do the centripetal forces F_{vO} point to the center O of mass. In this case, the center O of mass and the center of the circle O_{vO} coincide.

In addition, the radius ρ_O of curvature of the circle of curvature is usually not equal to the distance R_O from the Earth to the center O of mass, i.e. $\rho_O \neq R_O$. Only the radius ρ_O of curvature of the perihelion and aphelion points is equal to the distance R_O from the Earth to the center O of mass.

According to equation (2.2.2), the tangential velocity v_O of the Earth's rotation around the center O of mass can be decomposed into the transverse velocity $v_{O\varphi}$ and the radial velocity v_{OR}, i.e:

$$v_O = \sqrt{v_{O\varphi}^2 + v_{OR}^2} \tag{2.3.15}$$

Substituting the above equation into equation (2.3.14) gives the relational equation:

$$F_{vO} = m\frac{v_{O\varphi}^2 + v_{OR}^2}{\rho_O} \tag{2.3.16}$$

The above equation shows that the centripetal force F_{vO} of the Earth's revolution around the center O of mass consists of two component forces.

The first component force is the centripetal force $m\frac{v_{O\varphi}^2}{\rho_O}$ generated by the transverse velocity $v_{O\varphi}$. This centripetal force is directed toward the center O of mass.

The second component force is the centripetal force $m\frac{v_{OR}^2}{\rho_O}$ generated by the radial velocity v_{OR}. This centripetal force is perpendicular to the distance R_O from the Earth to the center O of mass.

Note that since the perihelion has the largest angular velocity φ_1 and the aphelion has the smallest angular velocity φ_2, the angular velocity difference value $\Delta\varphi$ between the perihelion and the aphelion is:

$$\Delta\varphi = \varphi_1 - \varphi_2 \tag{2.3.17}$$

The angular velocity φ of the Earth's rotation around the center O of mass can be expressed as

$$\varphi = \varphi_2 + \frac{1}{2}(\varphi_1 - \varphi_2)(1 + \cos\delta) \tag{2.3.18}$$

The above formula is the "Formula for the angular velocity of the Earth's revolution around the center of mass O". When the angle of revolution is $\delta = 0$, the Earth is in the position of perihelion, and the angular velocity of perihelion around the center O of mass is φ_1. When the angle of revolution is $\delta = 180^0$, the Earth is in the position of aphelion, and the angular velocity of aphelion around the center O of mass is φ_2.

If the perihelion angular velocity φ_1 of the Earth's revolution around the center O of mass is equal to the aphelion angular velocity φ_2, i.e., $\varphi_1 = \varphi_2 = \varphi$, then the above formula becomes the formula for the angular velocity of uniform circular motion. At this point, no matter how the angle of rotation δ changes, the angular velocity φ remains constant.

As the Earth moves gradually from perihelion to aphelion, the tangential velocity v_O and the angular velocity φ decrease. On the contrary, as the Earth moves from aphelion to perihelion, the tangential velocity v_O and the angular velocity φ gradually increase.

2.4 The orbit of the sun around the center O of mass of the solar system is an "eggshell curve"

2.4.1 The tangential velocity v_S of the sun's rotation around the center O of mass can be decomposed into a radial velocity v_{SR} and a transverse velocity $v_{S\varphi}$

According to the kinematics of a point mass, both the Earth and the Sun revolve around the center O of mass of the solar system. The tangential velocity v_O of the Earth's revolution around the center O of mass can be decomposed into two orthogonal velocity components.

Similarly, the tangential velocity v_S of the Sun around the center O of mass can be decomposed into two orthogonal velocity components. This is shown in Figure 2-9.

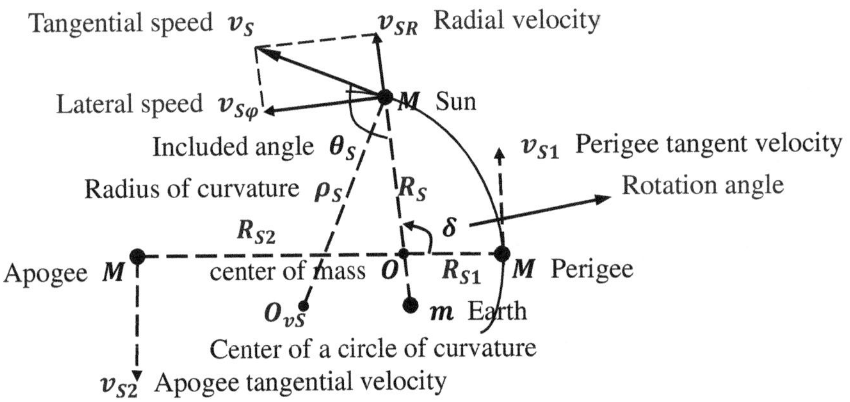

Figure 2-9. The center O_{vS} and radius ρ_S of the Sun's circle of curvature as it rotates about its center O of mass.

Assume that R_S is the distance from the sun to the center O of mass. Assume that θ_S is the angle between distance R_S and tangential velocity v_S. If the sun moves from perigee to apogee, then the radial velocity v_{SR} of the tangential velocity v_S projected onto the distance R_S is:

$$v_{SR} = v_S \cos \theta_S \tag{2.4.1}$$

When the angle $\theta_S = 90^0$, the radial velocity $v_{SR} = 0$ of the Sun's revolution around the center O of mass. At this time, the tangential velocity v_S of the Sun's revolution around the center O of mass, is perpendicular to the distance R_S, as shown in Figure 2-9.

Assume that φ is the angular velocity of the Earth and Sun as they rotate about the center O of mass. The tangential velocity v_S of the Sun produces a transverse velocity $v_{S\varphi}$ of :

$$v_{S\varphi} = v_S \sin \theta_S = R_S \varphi \tag{2.4.2}$$

The transverse velocity $v_{S\varphi}$ is perpendicular to the distance R_S.

Note that the transverse velocity $v_{S\varphi}$ is the circular velocity that takes the distance R_S as the radius and the center O of mass as the center of the circle. When the angle $\theta_S = 90^0$, the transverse velocity $v_{S\varphi}$ is equal to the tangential velocity v_S, i.e., $v_{S\varphi} = v_S$. In this case, the tangential velocity v_S is perpendicular to the distance R_S. This is shown in Figure 2-9.

The vectorial sum of the transverse velocity $v_{S\varphi}$ and the radial velocity v_{SR} is equal to the tangential velocity v_S of the Sun around the center O of mass, i.e:

$$v_S = \sqrt{v_{S\varphi}^2 + v_{SR}^2} \tag{2.4.3}$$

Suppose O_{vS} is the center of the circle of curvature corresponding to the tangential velocity v_S; ρ_S is the radius of curvature corresponding to the tangential velocity v_S. In this case, the tangential velocity v_S is always perpendicular to the radius of curvature ρ_S. The tangential velocity v_S is usually not perpendicular to the distance R_S from the Sun to the center O of mass. Only the perigee and apogee tangential velocities v_S are perpendicular to the distance R_S from the Sun to the center O of mass. In this case, the center O of mass coincides with the center of the circle O_{vS}. There is a variable angle θ_S between the tangential velocity v_S and the distance R_S. This is shown in Figure 2-9.

Although the Sun always orbits around the center O of mass, the tangential velocity v_S of the Sun's orbit around the center O of mass is usually perpendicular to the radius of curvature ρ_S and not perpendicular to the Sun's distance from the center O of mass, R_S. This is shown in Figure 2-9.

2.4.2 Tangential velocity at perigee and apogee of the Sun perpendicular to the distance from the center O of mass to the Sun

Since the distance r_1 from perihelion to the Sun is the minimum distance, the distance R_{S1} from the Sun's perigee to the center O of mass is also the minimum distance. According to equation (2.4.3), the tangential velocity v_{S1} of the Sun's perigee around the center O of mass is given by:

$$v_{S1} = \sqrt{v_{S\varphi1}^2 + v_{SR1}^2} \tag{2.4.4}$$

The v_{SR1} in the equation is the radial velocity at solar perigee and $v_{S\varphi1}$ is the transverse velocity at solar perigee.

According to the definition of velocity, the radial velocity $v_{SR} = \frac{dR_S}{dt}$ of the Sun's revolution around the center O of mass. The distance R_S from the Sun to the center O of mass decreases as the Sun approaches perigee. At this point, the radial velocity $v_{SR} < 0$. As the Sun moves away from perigee, the distance R_S from the Sun to the center O of mass gradually increases. At this point, the radial velocity $v_{SR} > 0$. Since the solar radial velocity v_{SR} has the opposite sign before and after perigee, the radial velocity v_{SR1} at solar perigee is :

$$v_{SR1} = \frac{dR_S}{dt} = 0 \tag{2.4.5}$$

Let φ_1 be the angular velocity at solar perigee. According to equation (2.4.2), the transverse velocity $v_{S\varphi1}$ at solar perigee is:

$$v_{S\varphi1} = R_{S1}\varphi_1 \tag{2.4.6}$$

Substituting the above equation and equation (2.4.5) into equation (2.4.4) gives the relationship equation:

$$v_{S1} = v_{S\varphi 1} = R_{S1}\varphi_1 \tag{2.4.7}$$

Since the transverse velocity $v_{S\varphi 1}$ of the Sun's perigee is equal to the tangential velocity v_{S1} of perigee, the tangential velocity v_{S1} of the Sun's perigee is perpendicular to the distance R_{S1} from the Sun's perigee to the center O of mass. This is shown in Figure 2-10.

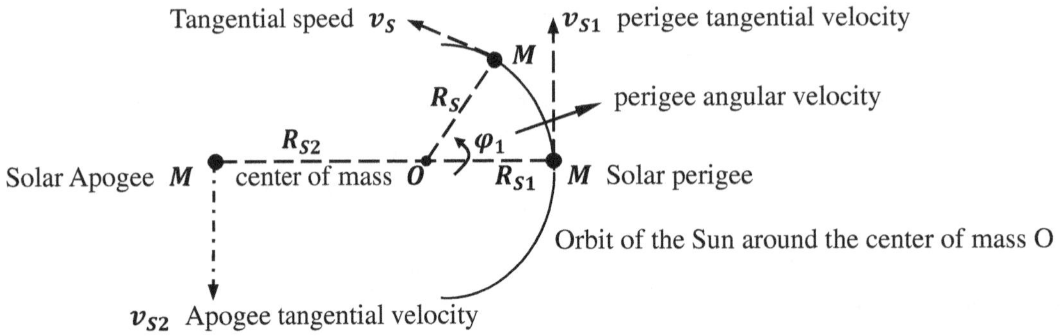

Figure 2-10. Tangential velocities v_{S1} and v_{S2} of perigee and apogee of the Sun rotating about the center O of mass.

Since the tangential velocity v_{S1} of solar perigee is perpendicular to the distance R_{S1}, the center O of mass is the center of curvature (center of the circle) of solar perigee and the distance R_{S1} is the radius of the perigee circle. This is shown in Figure 2-10.

Suppose R_{S2} is the distance from the Sun's apogee to the center O of mass; φ_2 is the angular velocity of the apogee; and v_{S2} is the tangential velocity of the apogee.

Simulating the analytical process above, it can be determined that the radial velocity of the solar apogee $v_{SR2} = 0$. The transverse velocity of the solar apogee $v_{S\varphi 2}$ is equal to the tangential velocity v_{S2}, viz:

$$v_{S2} = v_{S\varphi 2} = R_{S2}\varphi_2 \tag{2.4.8}$$

The above equation shows that the tangential velocity v_{S2} at the apogee of the Sun is perpendicular to the distance R_{S2} from the apogee to the center O of mass.

2.4.3 The orbit of the Sun around the center O of mass of the solar system is an "eggshell orbit"

According to equation (2.1.13) and equation (2.1.17), the relation equation can be obtained:

$$R_{S1} = \frac{m}{M+m}r_1, \quad R_{S2} = \frac{m}{M+m}r_2$$

Since the distance r_1 from perihelion to the Sun is smaller than the distance r_1 from aphelion to the Sun, the radius of curvature R_{S1} of the Sun's perigee is smaller than the radius of curvature R_{S2} of its apogee. That is, $R_{S1} < R_{S2}$.

Since the tangential velocity v_{S2} of the solar apogee is perpendicular to the distance R_{S2} from the apogee to the center O of mass, the center O of mass is the center of the circle of curvature of the solar apogee, and the distance R_{S2} is the radius of the circle of curvature of the solar apogee. This is shown in Figure 2-11.

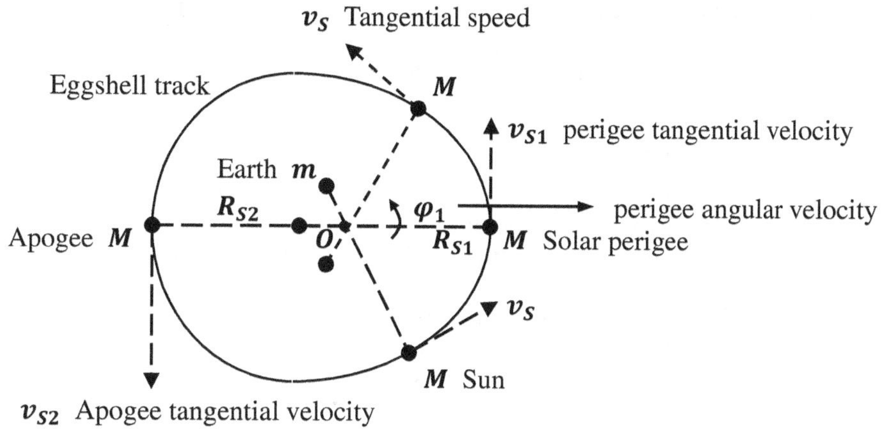

Figure 2-11. Schematic diagram of the orbit of the sun moving around the center O of mass.

Since the center of the circle of curvature of both solar perigee and solar apogee is the center O of mass, according to the definition of curvature, the curvature K_{S1} of solar perigee is:

$$K_{S1} = \frac{1}{R_{S1}} \tag{2.4.9}$$

The curvature K_{S2} of the solar apogee is:

$$K_{S2} = \frac{1}{R_{S2}} \tag{2.4.10}$$

Since the radius of curvature R_{S1} of perigee is smaller than the radius of curvature R_{S2} of apogee, i.e., $R_{S1} < R_{S2}$, the curvature K_{S1} of perigee is larger than the curvature K_{S2} of apogee, i.e., $K_{S1} > K_{S2}$, and it can be stated that the Sun's orbit around the center O of mass of the Solar System is similar to an eggshell curve. In this book, the Sun's orbit around the center O of mass of the solar system is defined as the Sun's eggshell orbit. This is shown in Figure 2-11.

Note that in the eggshell orbit of the Sun, only the tangential velocity v_{S1} at perigee and v_{S2} at apogee are perpendicular to the distance of the Sun from the center O of mass.

2.4.4 Relationship between the tangential velocity v_{S1} of the Sun's perigee and the tangential velocity v_1 of the Earth's perihelion

According to equation (2.3.4), the relation equation can be obtained:

$$mv_{01} = Mv_{S1} \tag{2.4.11}$$

According to equation (2.3.5), the relation equation can be obtained:

$$v_1 = v_{01} + v_{S1}$$

Multiplying the above equation by the mass m of the Earth gives the relationship equation:

$$mv_1 = mv_{01} + mv_{S1} \tag{2.4.12}$$

Substitute equation (2.4.11) into the above equation and remove mv_{01} to obtain the relational equation:

$$v_{S1} = \frac{m}{M+m}v_1 \tag{2.4.13}$$

In the formula, v_{S1} is the tangential velocity at perigee of the Sun; v_1 is the tangential velocity at perihelion of the Earth's orbit around the Sun.

According to equation (2.4.13), the tangential velocity v_{S2} at solar apogee is:

$$v_{S2} = \frac{m}{M+m}v_2 \tag{2.4.14}$$

The v_{S2} in the formula is the tangential velocity at the Sun's apogee, and v_2 is the tangential velocity at the apogee of the Earth's orbit around the Sun.

2.4.5 The ratio of the tangential velocity v_O and the tangential velocity v_S of the Earth and the Sun revolving around the center O of mass of the solar system is a constant

According to equation (2.1.5), the relation equation can be obtained:

$$mR_O = MR_S \tag{2.4.15}$$

Take the derivative of the above equation with respect to time t to obtain the relational equation:

$$m\frac{dR_O}{dt} = M\frac{dR_S}{dt} \tag{2.4.16}$$

The velocity $v_{OR} = \frac{dR_O}{dt}$ is the radial velocity of the Earth's orbit around the center O of mass of the solar system; the velocity $v_{SR} = \frac{dR_S}{dt}$ is the radial velocity of the Sun's orbit around the center O of mass of the solar system, and the above formula can be changed to:

$$mv_{OR} = Mv_{SR} \tag{2.4.17}$$

Squaring the above equation gives the relationship equation:

$$(mv_{OR})^2 = (Mv_{SR})^2 \tag{2.4.18}$$

Assume that φ is the angular velocity of the Earth and Sun as they revolve around the center O of mass of the solar system. Multiply equation (2.4.15) by the angular velocity φ to obtain the relational equation:

$$mR_O\varphi = MR_S\varphi \tag{2.4.19}$$

According to equation (2.2.4), the transverse velocity $v_{O\varphi}$ of the Earth's rotation around the center O of mass is:

$$v_{O\varphi} = R_O\varphi$$

According to equation (2.4.2), the transverse velocity $v_{S\varphi}$ of the Sun's revolution around the center O of mass of the solar system is:

$$v_{S\varphi} = R_S\varphi$$

According to equation (2.4.15), the ratio of the above two equations is:

$$\frac{v_{O\varphi}}{v_{S\varphi}} = \frac{R_O}{R_S} = \frac{M}{m} \tag{2.4.20}$$

The above equation shows that the ratio of the transverse velocity $v_{O\varphi}$ and the transverse velocity $v_{S\varphi}$ of the Earth and the Sun revolving around the center O of mass of the solar system is a constant. Squaring the above equation gives the relational equation:

$$(mv_{O\varphi})^2 = (Mv_{S\varphi})^2 \tag{2.4.21}$$

Add the above equation to equation (2.4.18) to obtain the relational equation:

$$m^2(v_{O\varphi}^2 + v_{OR}^2)^2 = M^2(v_{S\varphi}^2 + v_{SR}^2)^2 \tag{2.4.22}$$

According to equation (2.2.2), the relation equation can be obtained:

$$v_O = \sqrt{v_{O\varphi}^2 + v_{OR}^2}$$

According to equation (2.4.3), the tangential velocity v_S of the Sun's revolution around the center O of mass of the solar system is:

$$v_S = \sqrt{v_{S\varphi}^2 + v_{SR}^2}$$

Substituting the above two formulas into equation (2.4.22) gives the relational equation:

$$\frac{v_O}{v_S} = \frac{M}{m} \tag{2.4.23}$$

The above equation shows that the ratio of the tangential velocity v_O to the tangential velocity v_S is constant for both the Earth and the Sun revolving around the center O of mass of the solar system. We can also prove the above formula based on the center of mass momentum of the system of mass points.

In the frame of reference of the center O of mass of the solar system, since the velocity $v_C = 0$ of the center O of mass of the solar system, the momentum $P_C = 0$ of the center O of mass of the solar system, i.e:

$$P_C = (M + m)v_C = 0 \tag{2.4.24}$$

Equation (2.4.23) can be varied as follows:

$$mv_O = Mv_S \qquad (2.4.25)$$

Since the momentum mv_O of the Earth's revolution around the center O of mass of the solar system is equal to the momentum Mv_S of the Sun's revolution around the center O of mass of the solar system, the momentum P_C of the center O of mass of the solar system is:

$$P_C = mv_O - Mv_S = 0$$

The equation above is equation (2.4.24).

2.4.6 The tangential velocity v_O of the Earth's revolution around the center O of mass is parallel to the tangential velocity v_S of the Sun's revolution around the center O of mass of the solar system

According to equation (2.2.4), the transverse velocity $v_{O\varphi}$ of the Earth's rotation around the center O of mass is:

$$v_{O\varphi} = v_O \sin\theta = R_O\varphi$$

The θ in the equation is the angle between the tangential velocity v_O and the distance R_O from the Earth to the center O of mass, as shown in Figure 2-12.

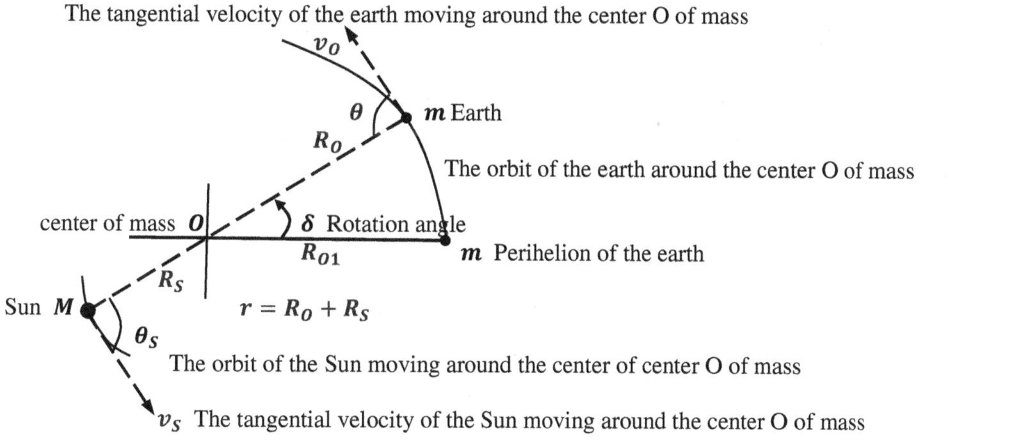

Figure 2-12 Schematic diagram of the movement of the earth and the sun around the center O of mass.

According to equation (2.4.2), the transverse velocity $v_{S\varphi}$ of the Sun's revolution around the center O of mass is:

$$v_{S\varphi} = v_S \sin\theta_S = R_S\varphi$$

The θ_S in the equation is the angle between the tangential velocity v_S and the distance R_S from the Sun to the center O of mass, as shown in Figure 2-12. Divide the above two formulas to obtain the relational formula:

$$\frac{v_O \sin\theta}{v_S \sin\theta_S} = \frac{R_O}{R_S} \qquad (2.4.26)$$

According to equation (2.4.20) and equation (2.4.23), the relation equation can be obtained:

$$\frac{v_{O\varphi}}{v_{S\varphi}} = \frac{R_O}{R_S} = \frac{v_O}{v_S} = \frac{M}{m}$$

Based on the above equation, the relationship equation can be obtained:

$$\frac{v_S R_O}{v_O R_S} = 1$$

Substituting the above equation into equation (2.4.26) gives the relational equation:

$$\frac{\sin\theta}{\sin\theta_S} = 1 \qquad (2.4.27)$$

Since the angle θ is equal to the angle θ_S (i.e. $\theta = \theta_S$), the tangential velocity v_O of the Earth's orbit around the center O of mass is parallel to the tangential velocity v_S of the Sun's orbit around the center O of mass. The tangential velocity v of the Earth's orbit around the Sun can thus be determined as:

$$v = v_O + v_S \tag{2.4.28}$$

According to the above relational equation, the tangential velocities v_1 and v_2 at the perihelion and aphelion of the Earth's orbit around the Sun are respectively:

$$v_1 = v_{O1} + v_{S1}, \qquad v_2 = v_{O2} + v_{S2} \tag{2.4.29}$$

v_{O1} and v_{S1} in the formula are the tangential velocities of the Earth and Sun rotating about the center O of mass at the time of perihelion. The v_{O2} and v_{S2} in the formula are the tangential velocities of the Earth and Sun rotating about the center O of mass at the time of aphelion.

Since the angle $\theta = \theta_S$, the transverse velocity $v_{O\varphi}$ of the Earth's revolution around the center O of mass is:

$$v_{O\varphi} = v_O \sin\theta = R_O\varphi \tag{2.4.30}$$

The lateral velocity $v_{S\varphi}$ of the sun moving around the center of mass O is:

$$v_{S\varphi} \sin\theta = v_S = R_S\varphi \tag{2.4.31}$$

2.5 Earth's Orbit around the Sun and Centripetal Force

2.5.1 Radial velocity v_r and transverse velocity v_φ of the Earth's orbit around the Sun

Suppose dt is the time it takes for the Earth to move from point A to point B. Suppose dS is the length of the Earth's displacement; r is the distance from the Earth to the Sun. Assume dr is the change in radial displacement; dr_φ is the change in lateral displacement. As shown in Figure 2-13.

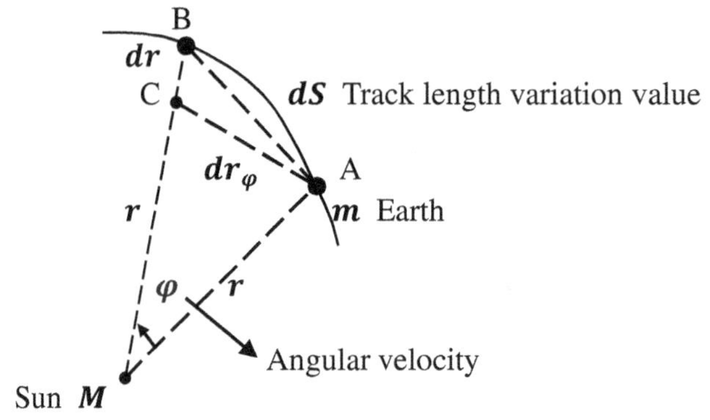

Figure 2-13 Schematic representation of the change in the length of the Earth's orbit around the Sun.

When the length dS of the earth's displacement is small, the ABC triangle can be considered a right triangle. When the earth moves from point A to point B, the length dS of the earth's displacement is:

$$dS = \sqrt{(dr)^2 + (dr_\varphi)^2}$$

Take the derivative of the above equation with respect to time t to obtain the relational equation:

$$\frac{dS}{dt} = \sqrt{\left(\frac{dr}{dt}\right)^2 + \left(\frac{dr_\varphi}{dt}\right)^2} \tag{2.5.1}$$

In this book, the velocity $v = dS/dt$ is defined as the tangential velocity of the Earth's orbit around the Sun. Define the velocity $v_r = dr/dt$ as the radial velocity of the Earth's orbit around the Sun; the direction of the radial velocity v_r is toward the Sun or in the opposite direction. Define the velocity $v_\varphi = dr_\varphi/dt$ as the transverse velocity of the Earth's revolution around the Sun; the direction of the transverse velocity v_φ is perpendicular to the distance r from the Earth to the Sun. Thus, the above equation can be expressed as:

$$v = \sqrt{v_\varphi^2 + v_r^2} \qquad (2.5.2)$$

The above equation shows that the tangential velocity v of the Earth's orbit around the Sun can be decomposed into two velocity components that are perpendicular to each other.

2.5.2 Relationship between tangential velocity v, radial velocity v_r, and transverse velocity v_φ of the Earth's orbit around the Sun

As the Earth moves from perihelion to aphelion. Let θ be the angle between the tangential velocity v and the distance r, as shown in Figure 2-14.

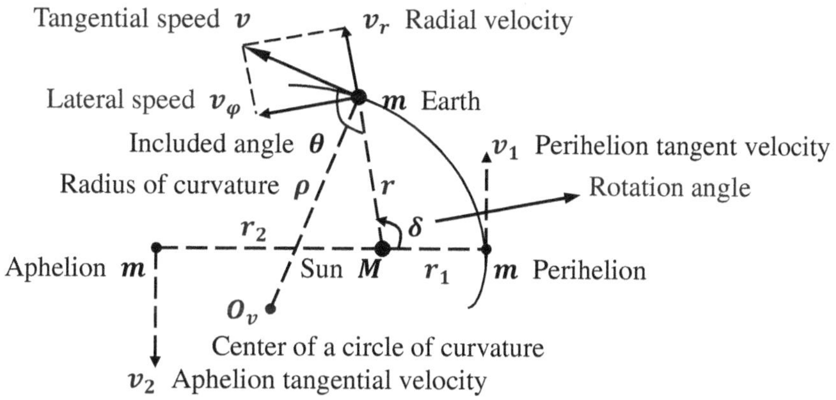

Figure 2-14 Center O_v and Radius of Curvature ρ of the Earth's Circle of Curvature Moving Toward Aphelion.

The tangential velocity v of the earth's orbit around the sun produces a radial velocity v_r at a distance r of:

$$v_r = v\cos\theta \qquad (2.5.3)$$

Suppose φ is the angular velocity of the Earth's revolution around the Sun. The tangential velocity v of the Earth produces a transverse velocity v_φ of:

$$v_\varphi = v\sin\theta = r\varphi \qquad (2.5.4)$$

The transverse velocity v_φ in the formula is perpendicular to the distance from the Earth to the Sun r. When the angle $\theta = 90^0$, the transverse velocity v_φ is equal to the tangential velocity v of the Earth's orbit around the Sun, i.e. $v_\varphi = v$. At this point, the tangential velocity v is perpendicular to the distance r.

The tangential velocity v of the Earth's orbit around the Sun is equal to the vectorial sum of the transverse velocity v_φ and the radial velocity v_r, i.e:

$$v = \sqrt{v_\varphi^2 + v_r^2}$$

The above equation is equation (2.5.2). The tangential velocity v at each point in the Earth's orbit around the Sun corresponds to a circle of curvature and a radius of curvature ρ. The tangential velocity v is usually perpendicular to the radius of curvature ρ rather than to the distance r from the Earth to the Sun. Only the tangential velocities v at perihelion and aphelion are perpendicular to the distance r. In this case, the angle $\theta = 90^0$. This is shown in Figure 2-14.

Although the Earth always revolves around the Sun, the centripetal acceleration a at any point in the Earth's orbit is not toward the Sun, but toward the center O_v of the circle of curvature at that point. This is shown in Figure 2-14.

It should be noted that since the Sun revolves around the center O of mass of the solar system, the centripetal acceleration a of the Earth's perihelion and aphelion does not point toward the Sun.

2.5.3 Tangential velocity v_1 perpendicular to perihelion at distance r_1 from Earth to Sun

Assume that r_1 is the distance from Earth's perihelion to the Sun; v_1 is the tangential velocity at perihelion; v_{r1} is the radial velocity at perihelion; and $v_{\varphi 1}$ is the transverse velocity at perihelion. According to equation (2.5.2), the relation equation can be obtained:

$$v_1 = \sqrt{v_{\varphi 1}^2 + v_{r1}^2} \tag{2.5.5}$$

According to the definition of velocity, the radial velocity $v_r = dr/dt$ of the Earth's revolution around the Sun. As the Earth moves closer to perihelion, the distance r from the Earth to the Sun decreases. At this point, the Earth's radial velocity $v_r < 0$. As the Earth moves away from perihelion, the Earth's distance r from the Sun gradually increases. At this point, the Earth's radial velocity $v_r > 0$. Since the radial velocity v_r has opposite positive and negative signs before and after perihelion, the Earth's radial velocity v_{r1} at perihelion is:

$$v_{r1} = 0 \tag{2.5.6}$$

Let φ_1 be the angular velocity of the perihelion. According to equation (2.5.4), the relation equation can be obtained:

$$v_{\varphi 1} = r_1 \varphi_1 \tag{2.5.7}$$

Substituting the above equation and equation (2.5.6) into equation (2.5.5) gives the relationship equation:

$$v_1 = v_{\varphi 1} = r_1 \varphi_1 \tag{2.5.8}$$

Since the transverse velocity $v_{\varphi 1}$ of perihelion is equal to the tangential velocity v_1, the tangential velocity v_1 of perihelion is perpendicular to the distance r_1 from perihelion to the Sun. This is shown in Figure 2-15.

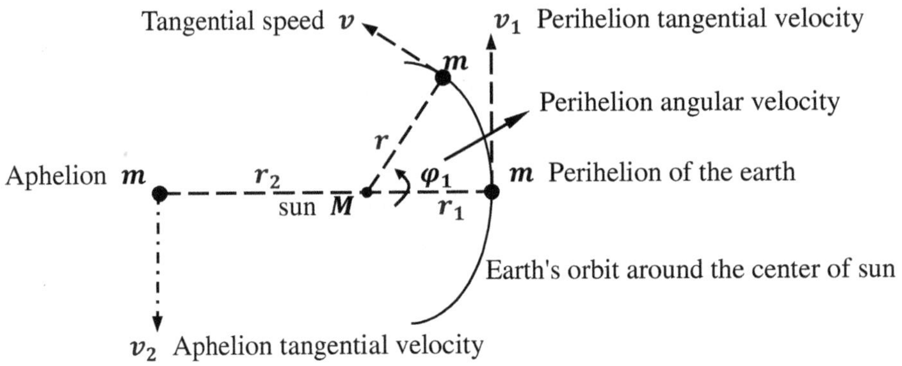

Figure 2-15 Tangential velocity v_1 at perihelion and v_2 at aphelion for the Earth's motion around the Sun.

Suppose r_2 is the distance from the Earth's aphelion to the Sun; v_2 is the tangential velocity of the aphelion; v_{r2} is the radial velocity of the aphelion; and $v_{\varphi 2}$ is the transverse velocity of the aphelion.

Simulating the above analytical process, it can be determined that the radial velocity $v_{r2} = 0$ of the aphelion. The transverse velocity $v_{\varphi 2}$ of the aphelion , is equal to the tangential velocity v_2 , i.e:

$$v_2 = v_{\varphi 2} = r_2 \varphi_2 \tag{2.5.9}$$

The φ_2 in the formula is the angular velocity of the aphelion. Since the transverse velocity $v_{\varphi 2}$ of the aphelion is equal to the tangential velocity v_2, the tangential velocity v_2 of the aphelion is perpendicular to the distance r_2 from the aphelion to the sun. This is shown in Figure 2-15.

2.5.4 The Earth's orbit around the Sun is an "eggshell" curve

The Earth's orbit can be divided into the Earth's orbit around the center O of mass of the solar system and the Earth's orbit around the Sun. Theoretically, the Earth's orbit around the center O of mass can be transformed into the

Earth's orbit around the Sun. To determine the centripetal force of the Earth's orbit around the Sun, the radii of curvature of the Earth's perihelion and aphelion must be known.

Theoretically, the centripetal force F_{vO} of the Earth's revolution around the center O of mass should be equal to the centripetal force F_v of the Earth's revolution around the Sun, i.e. $F_{vO} = F_v$. According to the formula for centripetal force, $F = mv^2/r$, the centripetal force F_{vO} of the Earth's revolution around the center O of mass can be equivalently converted to the centripetal force F_v of the Earth's revolution around the Sun.

If the tangential velocity v_{O1} of the perihelion, and the radius R_{O1} of curvature , are known, then according to the centripetal force formula $F = mv^2/r$, the centripetal force F_{O1} of the Earth's perihelion as it revolves around the center O of mass, can be equivalently converted to the centripetal force F_1 of the Earth around the Sun. From this, the radius ρ_1 of curvature of the Earth's perihelion as it orbits the Sun, can be determined.

Assume that φ_1 is the angular velocity of the Earth's perihelion as it revolves around the center O of mass. According to equation (2.2.8), the tangential velocity v_{O1} of the Earth's perihelion as it revolves around the center O of mass is:

$$v_{O1} = R_{O1}\varphi_1 \tag{2.5.10}$$

Assume that F_{vO1} is the perihelion centripetal force of the Earth's revolution around the center O of mass. Since the perihelion tangential velocity v_{O1} of the Earth around the center O of mass is perpendicular to the distance R_{O1} from the perihelion to the center O of mass, the perihelion centripetal force F_{vO1} of the Earth's revolution around the center O of mass, is given by the centripetal force formula, i.e.:

$$F_{vO1} = m\frac{v_{O1}^2}{R_{O1}} \tag{2.5.11}$$

The centripetal force F_{vO1} points to the center O of mass. The distance R_{O1} from the perihelion to the center O of mass is the radius of curvature. According to equation (2.1.12), the distance R_{O1} is:

$$R_{O1} = \frac{M}{M+m}r_1 \tag{2.5.12}$$

According to equation (2.3.7), the tangential velocity v_{O1} at the perihelion of the Earth's revolution around the center O of mass is:

$$v_{O1} = \frac{M}{M+m}v_1$$

The velocity v_1 in the formula is the perihelion tangential velocity of the Earth's orbit around the Sun. Substituting the above equation and equation (2.5.12) into equation (2.5.11) gives the relational equation:

$$F_{vO1} = m\frac{v_{O1}^2}{R_{O1}} = m\frac{M}{M+m}\frac{v_1^2}{r_1} \tag{2.5.13}$$

The above equation is the formula for the perihelion centripetal force. The perihelion centripetal force F_{vO1} points to the center O of mass.

Suppose O_{v1} is the center of the circle of curvature at the perihelion of the Earth's orbit around the Sun; ρ_1 is the radius of curvature at the perihelion of the Earth's orbit around the Sun. Note that ρ_1 is not the radius $\rho' = \frac{b^2}{a}$ of curvature of the perihelion of the Earth's elliptical orbit, nor is it the radius R_{O1} of curvature of the perihelion of the Earth's revolution around the center O of mass. Suppose that F_{v1} is the centripetal force at the perihelion of the Earth's revolution around the Sun. This is shown in Figure 2-16.

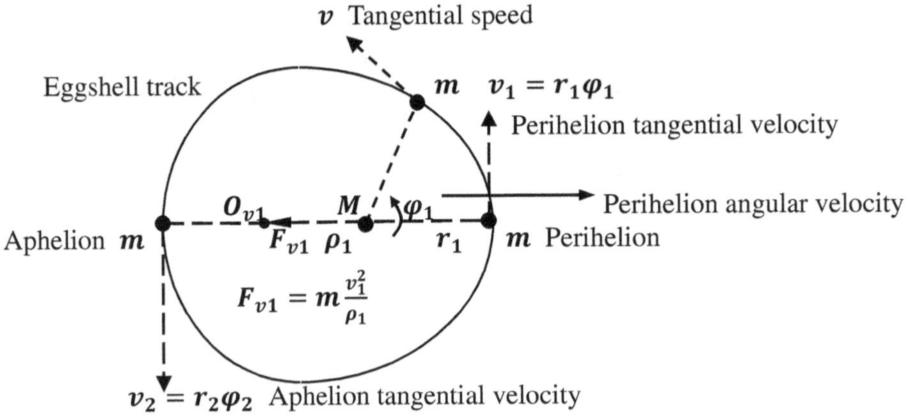

Figure 2-16 The center O_{v1} and the radius of curvature ρ_1 of the circle of curvature of the Earth's perihelion around the Sun.

Since the tangential velocity v_1 at the perihelion of the Earth's orbit around the Sun is perpendicular to the distance r_1 from the Earth to the Sun, the centripetal force F_{v1} at the perihelion of the Earth's orbit around the Sun is given by the centripetal force formula:

$$F_{v1} = m\frac{v_1^2}{\rho_1} \qquad (2.5.14)$$

Note that the perihelion centripetal force F_{v1} of the Earth's orbit around the Sun, points to the center O_{v1} of the perihelion curvature circle. Since there is uniqueness in the gravitational force F between the Earth's perihelion and the Sun, and the gravitational force F is the source of the perihelion centripetal force, the perihelion centripetal force F_{vO1} of the Earth's rotation about the center O of mass, is equal to the perihelion centripetal force F_{v1} of the Earth's rotation about the Sun, i.e.:

$$F_{vO1} = m\frac{v_{O1}^2}{R_{O1}} = m\frac{M}{M+m}\frac{v_1^2}{r_1} = m\frac{v_1^2}{\rho_1} = F_{v1} \qquad (2.5.15)$$

The above equation shows that in the center-of-mass frame of reference O, the centripetal force F_{vO1} at the perihelion points to the center-of-mass O, and the radius of curvature of the centripetal force F_{vO1} is R_{O1}. In the solar frame of reference, the perihelion centripetal force F_{v1} points to the center O_{v1} of curvature of the perihelion circle of curvature, and the radius of curvature of the centripetal force F_{v1} is ρ_1.

According to the above equation, the radius ρ_1 of curvature at the perihelion of the Earth's orbit around the Sun is:

$$\rho_1 = \frac{M+m}{M}r_1 \qquad (2.5.16)$$

Since the radius of curvature ρ_1 of perihelion is greater than the distance r_1 from perihelion to the Sun (i.e. $\rho_1 > r_1$), the center O_{v1} of the perihelion curvature circle, is not the Sun. In other words, the center O_{v1} of the perihelion curvature circle lies on the extension line beyond the distance r_1 from perihelion to the Sun. At this point, the perihelion centripetal force F_{v1}, is pointing toward the center O_{v1} of the perihelion curvature circle; it is not pointing toward the Sun (the focus of the ellipse). This is shown in Figure 2-16.

Assume that F_{vO2} is the aphelion centripetal force of the Earth's revolution around the center O of mass. Simulating the above analytical process and based on equation (2.5.13), the aphelion centripetal force F_{vO2} for the Earth's revolution around the center O of mass can be determined as:

$$F_{vO2} = m\frac{v_{O2}^2}{R_{O2}} = m\frac{M}{M+m}\frac{v_2^2}{r_2} \qquad (2.5.17)$$

The above equation is the formula for the centripetal force at the aphelion.The aphelion centripetal force F_{vO2}

points toward the center O of mass.

Assume that O_{v2} is the center of the aphelion circle of curvature of the Earth's orbit around the Sun; ρ_2 is the radius of the aphelion circle of curvature. Note that ρ_2 is not the radius $\rho' = \frac{b^2}{a}$ of curvature of the aphelion of the Earth's elliptical orbit. Assume that F_{v2} is the centripetal force at the aphelion of the Earth's orbit around the Sun. Since the aphelion tangential velocity v_2 of the Earth's orbit around the Sun is perpendicular to the distance r_2 from the aphelion to the Sun, the aphelion centripetal force F_{v2} of the Earth's orbit around the Sun is given by the centripetal force formula:

$$F_{v2} = m\frac{v_2^2}{\rho_2} \quad (2.5.18)$$

Note that the aphelion centripetal force F_{v2} points to the center O_{v2} of the aphelion curvature circle, not points to the Sun (the focus of the ellipse). This is shown in Figure 2-17.

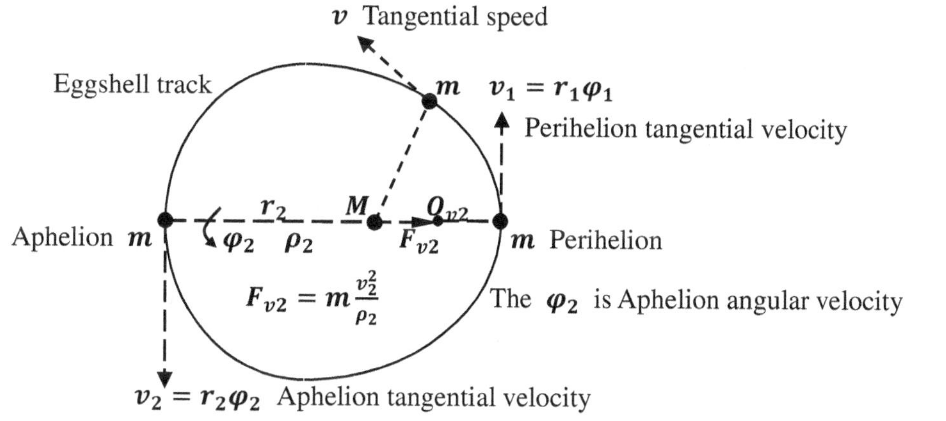

Figure 2-17 The center O_{v2} and the radius of curvature ρ_2 of the circle of curvature at the aphelion of the Earth's revolution around the Sun.

Due to the uniqueness of the gravitational force F between the Earth's aphelion and the Sun, which is the source of the centripetal force F_{vO2} of the aphelion, the centripetal force of the aphelion, for the Earth's revolution around the center O of mass, is equal to the centripetal force F_{v2} of the aphelion, for the Earth's revolution around the Sun, i.e:

$$F_{vO2} = m\frac{v_{O2}^2}{R_{O2}} = m\frac{M}{M+m}\frac{v_2^2}{r_2} = m\frac{v_2^2}{\rho_2} = F_{v2} \quad (2.5.19)$$

The above equations show that in the center-of-mass frame of reference, the centripetal force F_{vO2} at the aphelion points to the center-of-mass O, and the radius of curvature of the centripetal force F_{vO2} is R_{O2}. In the solar frame of reference, the centripetal force F_{v2} at the aphelion points to the center O_{v2} of curvature of the aphelion circle of curvature, and the radius of curvature of the centripetal force F_{v2} is ρ_2. This is shown in Figure 2-17.

According to the above equation, the radius of curvature ρ_2 at the aphelion of the Earth's orbit around the Sun is:

$$\rho_2 = \frac{M+m}{M} r_2 \quad (2.5.20)$$

Since the radius of curvature ρ_2 of the aphelion is greater than the distance r_2 from the aphelion to the Sun (i.e. $\rho_2 > r_2$), the center O_{v2} of the circle of curvature of the aphelion is not the Sun. In other words, the center O_{v2} of the circle of curvature of the aphelion is on the extension line beyond the distance r_2 from the aphelion to the Sun. At this point, the centripetal force F_{v2} at the aphelion points toward the center O_{v2} of the aphelion's circle of curvature, not toward the Sun (the focus of the ellipse). This is shown in Figure 2-17.

Compare equation (2.5.20) with equation (2.5.16). Since the distance $r_1 < r_2$, the radius of curvature ρ_1 at the perihelion of the Earth's orbit around the Sun's yanggong is smaller than the radius of curvature ρ_2 at the aphelion, i.e. $\rho_1 < \rho_2$. According to the definition of curvature, the curvature K_1 of the Earth at the perihelion of the Earth's orbit around the Sun is:

$$K_1 = \frac{1}{\rho_1} \tag{2.5.21}$$

The curvature K_2 at the aphelion of the Earth's orbit around the Sun is:

$$K_2 = \frac{1}{\rho_2} \tag{2.5.22}$$

Since the radius of curvature $\rho_1 < \rho_2$, the curvature K_1 at perihelion is greater than the curvature K_2 at aphelion, i.e. $K_1 > K_2$. From this, we can determine that the orbit of the Earth around the Sun is an "eggshell orbit" rather than an elliptical orbit. The elliptical orbit of the Earth does not conform to the laws of motion of the system of mass points.

It should be noted that in the Earth's orbit around the Sun, only the perihelion tangential velocity v_1 and the aphelion tangential velocity v_2 are perpendicular to the Earth-Sun distance r. The tangential velocities v at other points in the orbit are not perpendicular to the Earth-Sun distance r.

According to Kepler's first law, the Earth's orbit around the Sun is an elliptical orbit, and the Sun is located at one of the foci of the ellipse. According to the elliptical nature, the curvatures K_1 and K_2 at the two ends of the long axis of the Earth's elliptical orbit are equal, i.e. $K_1 = K_2$. From the above analysis and discussion, it can be seen that the elliptical orbit of the planet does not conform to the laws of motion of the system of mass points.

2.5.5 The centripetal force F_v of the Earth's orbit around the Sun can be decomposed into two components

Assume that v is the tangential velocity of the Earth's revolution around the Sun; ρ is the radius of the circle of curvature corresponding to the tangential velocity v. Since the tangential velocity v is perpendicular to the radius of curvature ρ, according to the centripetal force formula, the centripetal force F_v generated by the tangential velocity v is:

$$F_v = m\frac{v^2}{\rho} \tag{2.5.23}$$

The direction of the centripetal force F_v points to the center of the circle of curvature O_v, not points to the Sun, nor points to the center O of mass. The radius of curvature ρ is not equal to the distance r from the Earth to the Sun , i.e. $\rho \neq r$. Nor is it equal to the distance R_O from the Earth to the center O of mass, i.e. $\rho \neq R_O$.

According to equation (2.5.16) and equation (2.5.20), the radius ρ_1 of curvature at the perihelion and ρ_2 at the aphelion of the Earth's orbit around the Sun are:

$$\rho_1 = \frac{M+m}{M}r_1, \qquad \rho_2 = \frac{M+m}{M}r_2$$

According to the above two formulas, the radius of curvature ρ of the Earth's orbit around the Sun can theoretically be expressed as two mathematical formulas.

The first mathematical expression of the radius of curvature ρ is:

$$\rho = \frac{M+m}{M}r\sin\theta \tag{2.5.24}$$

Since the tangential velocity v has a radial component velocity in the direction of the distance r; and a transverse component velocity in the direction perpendicular to the distance r, the entrainment angle θ is used to calculate the magnitude of the radial and transverse velocities, not to calculate the magnitude of the distance r component. It can be seen that the above formula is incorrect.

Substituting the above equation into equation (2.5.23) gives the relationship equation:

$$F_v = \frac{M}{M+m}m\frac{v^2}{r\sin\theta} \tag{2.5.25}$$

If angle $\theta = 0$ between tangential velocity v and distance r, or angle $\theta = 180^0$, then centripetal force $F_v = \infty$. Obviously, the above formula is not a centripetal force formula.

The second mathematical expression for the radius of curvature ρ is:

$$\rho = \frac{M+m}{M}r \tag{2.5.26}$$

Substituting the above equation into equation (2.5.23) gives the relational equation:

$$F_v = \frac{M}{M+m}m\frac{v^2}{r} \tag{2.5.27}$$

The above formula is the formula for the centripetal force of the Earth's orbit around the Sun. According to equation (2.5.2), the tangential velocity v of the Earth's orbit around the Sun can be decomposed into the transverse velocity v_φ and the radial velocity v_r, i.e:

$$v = \sqrt{v_\varphi^2 + v_r^2}$$

Substituting the above equation into equation (2.5.23) gives the relational equation:

$$F_v = m\frac{v_\varphi^2 + v_r^2}{\rho} \tag{2.5.28}$$

The above equation shows that the centripetal force F_v of the Earth's revolution around the Sun can be decomposed into two components.

The first component is the radial centripetal force $F_{vr} = mv_\varphi^2/\rho$ caused by the transverse velocity v_φ. The direction of the radial centripetal force F_{vr} points to the center of the circle of curvature.

The second component is the transverse centripetal force $F_{v\varphi} = mv_r^2/\rho$ caused by the radial velocity v_r. The direction of the transverse centripetal force $F_{v\varphi}$ is perpendicular to the distance r from the Earth to the Sun.

2.5.6 The center of the circle of curvature at perihelion and aphelion of an elliptical orbit is neither the Sun nor the center O of mass

The current theory is that the Earth's orbit around the Sun is an elliptical curve, with the Sun at one of the foci of the ellipse. The radii of curvature ρ' of the Earth at perihelion and aphelion are equal.

Suppose a is half the long axis of the ellipse, b is half the short axis of the ellipse, and c is half the focal length of the ellipse. The radius of curvature ρ' of the elliptical orbit is:

$$\rho' = \frac{(2ar - r^2)^{\frac{3}{2}}}{ab} \tag{2.5.29}$$

The distance from the perihelion of the elliptical orbit to the Sun is $r_1 = a - c$, and the distance from the aphelion to the Sun is $r_2 = a + c$. Substituting these two formulas into the above equation gives the radii of curvature ρ' of the perihelion and aphelion of the elliptical orbit:

$$\rho' = \frac{b^2}{a} \tag{2.5.30}$$

Since half of the short axis of the ellipse $b = \sqrt{a^2 - c^2} = \sqrt{r_1 r_2}$ and half of the long axis $a = (r_1 + r_2)/2$, the above equation becomes:

$$\rho' = \frac{b^2}{a} = \frac{2r_1 r_2}{r_1 + r_2} \tag{2.5.31}$$

The ρ' in the equation is the radius of curvature of the elliptical orbit at perihelion and aphelion.

However, since both the Sun and the Earth revolve around the center O of mass of the solar system, which is not the focal point of the ellipse, the radii of curvature at perihelion and aphelion of the Earth's eggshell orbit around the Sun are, according to equation (2.5.16) and equation:

$$\rho_1 = \frac{M+m}{M}r_1, \qquad \rho_2 = \frac{M+m}{M}r_2$$

The above formula shows that the radii of curvature of perihelion and aphelion of the Earth's orbit around the Sun are not equal under the condition that both the Sun and the Earth orbit around the center O of mass of the solar system.

Let O_1' be the center of the circle of curvature at the perihelion of the elliptical orbit. According to equation (2.5.31), we obtain the following relational equation:

$$\rho' = \frac{2r_1r_2}{r_1 + r_2} > r_1$$

Since the radius of curvature ρ' is greater than the distance r_1, the center of curvature O_1' of the perihelion circle of an elliptical orbit, is not the Sun. The center of curvature O_1' of the perihelion of the elliptical orbit lies on an extension line beyond the distance r_1 from the perihelion to the Sun. This is shown in Figure 2-18.

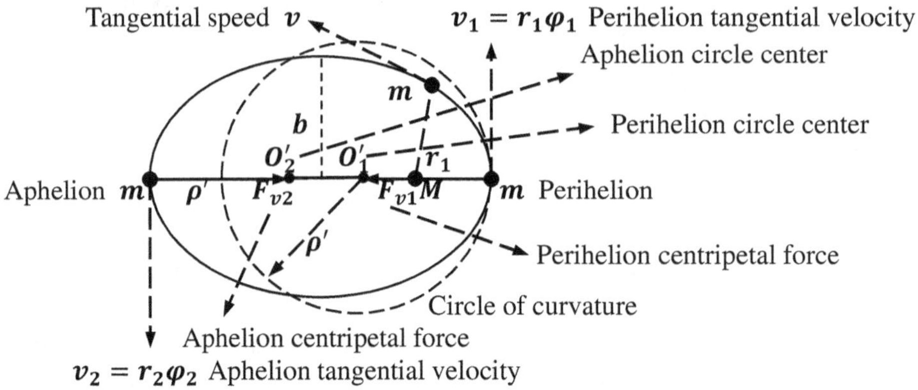

Figure 2-18 Radii of curvature ρ' of elliptical perihelion and aphelion and centers of curvature circles O_1' and O_2'.

According to equation (2.1.12), the distance R_{O1} from the perihelion to the center O of mass is:

$$R_{O1} = \frac{M}{M + m} r_1$$

Since the distance $r_2 > r_1$, the following relation equation can be obtained according to equation (2.5.31):

$$\rho' = \frac{2r_1r_2}{r_1 + r_2} > \frac{M}{M + m} r_1$$

Since the radius of curvature ρ' is greater than the distance R_{O1}, the center O_1' of the circle of curvature at perihelion of the Earth's elliptical orbit is neither the Sun nor the center O of mass.

Let O_2' be the center of the circle of curvature at the aphelion of an elliptical orbit. The following relational equation can be obtained from equation (2.5.31):

$$\rho' = \frac{2r_1r_2}{r_1 + r_2} < r_2$$

Since the radius of curvature ρ' is less than the distance r_2, the center O_2' of the circle of curvature at the aphelion of an elliptical orbit is not the Sun. The center of curvature O_2' of the aphelion lies within the distance between the aphelion and the Sun. This is shown in Figure 2-18.

Equation (2.1.16) gives the distance R_{O2} from the aphelion to the center of mass O:

$$R_{O2} = \frac{M}{M + m} r_2$$

Since the distance $r_2 > r_1$, the following relation equation can be obtained according to equation (2.5.31):

$$\rho' = \frac{2r_1r_2}{r_1 + r_2} \neq \frac{M}{M + m} r_2$$

Since the radius of curvature ρ' is not equal to the distance R_{O2}, the center O_1' of the circle of curvature at the aphelion of an elliptical orbit is neither the Sun nor the center O of mass.

In summary, since the Sun and the Earth always revolve around the center O of mass of the solar system, and since the centers of curvature of the perihelion and aphelion of the Earth's elliptical orbit are not the center O of mass, the Earth's elliptical orbit does not conform to the law of motion of the system of mass points.

Chapter 3 The causes of gravitational changes

Introduction: This chapter analyzes and discusses the causes of the variation of gravitational from the perspective of the Milky Way Galaxy frame of reference. The authors found that the distance $\boldsymbol{r}$ between the planets and the Sun varies periodically with the motion of the center O of mass of the solar system. It is for this reason that the force $\boldsymbol{F}$ of gravitational, and the velocity $\boldsymbol{v}$, of the planet's orbit change periodically. The author thus explains, theoretically and accurately, why the orbits of the planets around the sun are all in the same plane.

3.1 The velocity of the Earth around the center O of mass and the velocity of the Earth around the Sun

3.1.1 Relationship between the tangential velocity v_O of the Earth's orbit around the center O of mass and the tangential velocity v of the Earth's orbit around the Sun

According to equation (2.4.25), the momentum $\boldsymbol{Mv_S}$ of the Sun around the center O of mass of the solar system is equal to the momentum $\boldsymbol{mv_O}$ of the Earth around the center O of mass, i.e:

$$\boldsymbol{mv_O = Mv_S} \tag{3.1.1}$$

According to equation (2.4.28), the tangential velocity $\boldsymbol{v}$ of the Earth's orbit around the Sun is:

$$v = v_O + v_S$$

Multiplying the above equation by the solar mass $\boldsymbol{M}$ gives the relationship equation:

$$\boldsymbol{Mv = Mv_O + Mv_S} \tag{3.1.2}$$

Substituting equation (3.1.1) into the above equation and eliminating $\boldsymbol{Mv_S}$ gives the relationship equation:

$$v_O = \frac{M}{M+m} v \tag{3.1.3}$$

The above equation shows that the tangential velocity $\boldsymbol{v}$ of the Earth's orbit around the Sun amplifies the tangential velocity $\boldsymbol{v_O}$ of the sphere's orbit around the center O of mass by a factor of $(\boldsymbol{M+m})/\boldsymbol{M}$.

Based on the above equation, the perihelion tangential velocity $\boldsymbol{v_{O1}}$ and the aphelion tangential velocity $\boldsymbol{v_{O2}}$ of the Earth's revolution around the center O of mass can be determined:

$$v_{O1} = \frac{M}{M+m} v_1, \qquad v_{O2} = \frac{M}{M+m} v_2 \tag{3.1.4}$$

Note that the perihelion and aphelion of the Earth's elliptical orbit do not have the above relationship.

In addition, multiplying the velocity equation $\boldsymbol{v = v_O + v_S}$ by the Earth's mass $\boldsymbol{m}$ gives the relation equation:

$$\boldsymbol{mv = mv_O + mv_S} \tag{3.1.5}$$

Substituting equation (3.1.1) into the above equation and eliminating $\boldsymbol{mv_O}$ gives the relational equation:

$$v_S = \frac{m}{M+m} v \tag{3.1.6}$$

Based on the above equation, the tangential velocity $\boldsymbol{v_{S1}}$ at perigee and the tangential velocity $\boldsymbol{v_{S2}}$ at apogee of the Sun's revolution around the center O of mass of the solar system can be determined:

$$v_{S1} = \frac{m}{M+m} v_1, \qquad v_{S2} = \frac{m}{M+m} v_2 \tag{3.1.7}$$

The $\boldsymbol{v_1}$ in the formula is the tangential velocity of the Earth's orbit around the Sun at perihelion, and $\boldsymbol{v_2}$ is the tangential velocity of the Earth's orbit around the Sun at aphelion.

3.1.2 Equation for the velocity of the earth around the sun

According to equation (2.3.10), the tangential velocity v_O of the Earth's rotation around the center O of mass is:

$$v_O = v_{O2} + \frac{1}{2}(v_{O1} - v_{O2})(1 + \cos\delta) \quad (3.1.8)$$

Multiply the above equation by $(M + m)/M$ to get the relationship equation:

$$\frac{M+m}{M} v_O = \frac{M+m}{M}\left(v_{O2} + \frac{1}{2}(v_{O1} - v_{O2})(1 + \cos\delta)\right) \quad (3.1.9)$$

According to equation (3.1.3) and equation (3.1.4), the tangential velocity v of the Earth's orbit around the Sun, the perihelion tangential velocity v_1 and the aphelion tangential velocity v_2 are given:

$$v = \frac{M+m}{M} v_O, \qquad v_1 = \frac{M+m}{M} v_{O1}, \qquad v_2 = \frac{M+m}{M} v_{O2}$$

Substituting the above equation into equation (3.1.9) gives the relationship equation:

$$v = v_2 + \frac{1}{2}(v_1 - v_2)(1 + \cos\delta) \quad (3.1.10)$$

The above formula is the formula for the speed of the Earth's rotation around the Sun. The δ in the formula is the angle of the Earth's counterclockwise rotation around the Sun. When the angle of rotation is $\delta = 0$, the Earth is at perihelion, and the tangential velocity of perihelion is v_1. When the angle of revolution is $\delta = 180^0$, the Earth is in the position of aphelion, and the tangential velocity of aphelion is v_2.

When the perihelion velocity v_1 of the Earth's orbit around the Sun is equal to the aphelion velocity v_2, i.e., $v_1 = v_2 = v_O$, the above formula becomes the circumferential velocity formula, i.e., $v = v_O$. At this point, the circumferential velocity v_O remains the same regardless of the change in the Earth's rotation angle δ around the Sun.

Taking the derivative of equation (3.1.10) with respect to time t gives the following relational equation:

$$a = \frac{dv}{dt} = \frac{dv}{d\delta}\frac{d\delta}{dt} = -\frac{1}{2}(v_1 - v_2)\varphi \sin\delta \quad (3.1.11)$$

The formula above is the formula for the tangential acceleration of the Earth as it revolves around the Sun. a is the tangential acceleration, not the centripetal acceleration $a_r = m\frac{v^2}{r}$. $\varphi = d\delta/dt$ is the angular velocity of the Earth as it rotates counterclockwise around its center O of mass.

When the rotation angle $\delta = 0$, the Earth is at perihelion. The tangential acceleration of the Earth at perihelion $a_1 = 0$. When the rotation angle is $0 < \delta < 180^0$, the tangential velocity v of the Earth's revolution around the Sun becomes smaller and smaller because the tangential acceleration $a < 0$.

When the rotation angle $\delta = 180^0$, the Earth is in the position of aphelion. The tangential acceleration $a_2 = 0$. When the rotation angle is $180^0 < \delta < 360^0$, the tangential velocity v of the Earth's revolution around the Sun gradually becomes larger because the tangential acceleration $a > 0$.

The tangential acceleration a of the Earth's revolution around the Sun is maximum when the rotation angles are $\delta = 90^0$ and $\delta = 270^0$.

3.1.3 Mean velocity of the Earth's orbit around the center O of mass $\tilde{v_O}$ and mean velocity of the Earth's orbit around the Sun $\tilde{v}$

As the Earth moves from perihelion to aphelion, the angle of rotation δ varies over the interval 0 to π. The tangential velocity v_O of the Earth's revolution around the center O of mass is integrated over the interval 0 to π as:

$$\int_0^{\pi} v_O d\delta = \int_0^{\pi}\left(v_{O2} + \frac{1}{2}(v_{O1} - v_{O2})(1 + \cos\delta)\right) d\delta \quad (3.1.12)$$

Solving the above equation gives the relational equation:

$$\int_0^{\pi} v_O d\delta = \frac{1}{2}\pi(v_{O1} + v_{O2}) \quad (3.1.13)$$

The mean velocity $\widetilde{v_O}$ of the Earth's revolution around the center O of mass in the interval 0 to π of the variation of the angle of rotation δ is:

$$\widetilde{v_O} = \frac{1}{\pi}\int_0^{\pi} v_O d\delta = \frac{1}{2}(v_{O1} + v_{O2}) \tag{3.1.14}$$

The above equation is the formula for the mean velocity of the Earth's rotation around the center O of mass. From the above formula, the mean velocity of the Earth's rotation around the center of mass $\widetilde{v_O}$ is equal to half the sum of the perihelion velocity v_{O1} and the aphelion velocity v_{O2}.

Multiply the above equation by $(M+m)/M$ to obtain the relationship equation:

$$\frac{M+m}{M}\widetilde{v_O} = \frac{1}{2}\left(\frac{M+m}{M}v_{O1} + \frac{M+m}{M}v_{O2}\right) \tag{3.1.15}$$

Substituting equation (3.1.4) into the above equation gives the mean velocity $\tilde{v}$ of the Earth's orbit around the Sun:

$$\tilde{v} = \frac{1}{2}(v_1 + v_2) \tag{3.1.16}$$

The above formula is the formula for the average speed of the Earth's rotation around the Sun. From the above formula, it can be seen that the average speed $\tilde{v}$ of the Earth's rotation around the Sun is equal to half of the sum of the perihelion speed v_1 and the aphelion speed v_2.

Note that the mean velocity $\tilde{v}$ of the Earth's orbit around the Sun, is equal to the mean velocity $\widetilde{v_O}$ of the Earth's orbit around the center O of mass multiplied by the coefficient $(M+m)/M$, i.e:

$$\tilde{v} = \frac{M+m}{M}\widetilde{v_O} \tag{3.1.17}$$

3.1.4 Relationship between the perihelion and aphelion velocities v_1 and v_2 of the Earth's orbit around the Sun and the Earth's rotation period T

Although we do not know the exact position of the center O of mass of the solar system, the distances r_1 and r_2 from the Sun to the Earth's perihelion and aphelion can be obtained by calculations from astronomical observations.According to equation (2.1.25), the circumference L of the Earth's orbit around the Sun is:

$$L = \pi(r_1 + r_2)$$

Assuming that T is the period of the Earth's orbit around the Sun, the average velocity $\tilde{v}$ of the Earth's motion around the Sun is:

$$\tilde{v} = \frac{L}{T} = \frac{\pi(r_1 + r_2)}{T} \tag{3.1.18}$$

The above equation is the formula for the average velocity of the Earth's orbit around the Sun. According to the law of conservation of angular momentum, the angular momentum L_v of the Earth's orbit around the Sun is

$$L_v = rmv\sin\theta = C$$

The C in the formula is a constant. According to the above equation, the angular momentum L_{v1} at the perihelion of the Earth's orbit around the Sun and L_{v2} at the aphelion are equal, i.e:

$$L_v = rmv\sin\theta = r_1 m v_1 \sin\theta_1 = r_2 m v_2 \sin\theta_2 = C \tag{3.1.19}$$

Since the perihelion velocity v_1 and the aphelion velocity v_2 of the Earth's rotation around the Sun are both perpendicular to the distance r from the Earth to the Sun, the angle $\theta_1 = 90^0$ between the perihelion velocity v_1 and the distance r_1, and the angle $\theta_2 = 90^0$ between the aphelion velocity v_2 and the distance r_2. Thus the above equation becomes the following equation:

$$r_1 v_1 = r_2 v_2 \tag{3.1.20}$$

Substituting the above equation into equation (3.1.16) gives the relationship equation:

$$\tilde{v} = \frac{1}{2}\left(v_1 + \frac{r_1 v_1}{r_2}\right) \tag{3.1.21}$$

The $\tilde{v}$ in the formula is the average speed of the Earth's revolution around the Sun. Simplifying the above formula gives the relational formula:

$$v_1 = \frac{2r_2}{r_1 + r_2}\tilde{v} \tag{3.1.22}$$

Substituting the average velocity $\tilde{v}$ from equation (3.1.18) into the above equation gives the relationship

equation:

$$v_1 = \frac{2\pi r_2}{T} \tag{3.1.23}$$

The above formula is the formula for the velocity of the Earth at perihelion. By simulating the analytical process above, the relational formula can be obtained:

$$v_2 = \frac{2\pi r_1}{T} \tag{3.1.24}$$

The above formula is the formula for the speed at the Earth's aphelion.

3.1.5 Mean velocity of the Earth's orbit around the Sun $\tilde{v}$ in agreement with actual observations

The perimeter of the ellipse L' is given by:

$$L' = 2\pi a(1 - 0.25e^2)$$

The a in the formula is half the long axis of the ellipse; e is the eccentricity of the ellipse. The mean linear velocity $\tilde{v}'$ of the Earth's elliptical orbit is:

$$\tilde{v}' = \frac{L'}{T} = \frac{2\pi a(1 - 0.25e^2)}{T} \tag{3.1.25}$$

According to the data of the Earth's motion, the total length of the Earth's elliptical orbit $L' = 0.94 \times 10^8 m$. The period of the Earth's revolution around the Sun $T = 365.2564$ days. Substituting the two data into the above equation, we obtain the average linear velocity $\tilde{v}'$ of the Earth's elliptical orbit:

$$\tilde{v}' = \frac{L'}{T} = \frac{94 \times 10^8}{365.2564 \times 24 \times 3600} = 29.786km/s \tag{3.1.26}$$

According to equation (3.1.18), the average velocity $\tilde{v}$ of the eggshell orbit is:

$$\tilde{v} = \frac{L}{T} = \frac{\pi(r_1 + r_2)}{T}$$

Comparing the above equation with equation (3.1.25), we can see that the average velocity $\tilde{v}'$ of the elliptical orbit, is not equal to the average velocity $\tilde{v}$ of the eggshell orbit , i.e. $\tilde{v}' \neq \tilde{v}$.

According to astronomical observations. The distance $r_1 = 147098 \cdot km$ from the Earth's perihelion to the Sun, the distance $r_2 = 152098 \cdot km$ from the Earth's aphelion to the Sun, and the period of the Earth's revolution $T = 365.24$.

The velocity $v_1 = \frac{2\pi r_2}{T} = 30.284 \cdot km/s$ of the Earth's perihelion, according to equation (3.1.23), and the velocity $v_2 = \frac{2\pi r_1}{T} = 29.2883 \cdot km/s$ of the Earth's aphelion, according to equation (3.1.24).

According to equation (3.1.16), the average speed $\tilde{v}$ of the Earth's orbit around the Sun is:

$$\tilde{v} = \frac{1}{2}(v_1 + v_2) = \frac{1}{2}(30.284 + 29.2883) = 29.786km/s$$

Comparing the mean velocity $\tilde{v}$ in the above equation with the mean velocity $\tilde{v}'$ for elliptical orbits in equation (3.1.26), it can be seen that the mean velocity $\tilde{v}$ for eggshell orbits is consistent with astronomical observations.

If the distances r_1 and r_2 from the Sun to the planet's perihelion and aphelion are known, and the period T of the planet's orbit around the Sun is known, then the velocity v_1 of the planet at perihelion, is given by equation (3.1.23):

$$v_1 = \frac{2\pi r_2}{T}$$

According to equation (3.1.24), the velocity v_2 at the planet's aphelion is:

$$v_2 = \frac{2\pi r_1}{T}$$

The mean velocity $\tilde{v}$ of the planet's eggshell orbit and $\tilde{v}'$ of the planet's elliptical orbit. as shown in Table 1.

Table 1	r_1 $k \cdot km$	r_2 $k \cdot km$	T day	v_1 km/s	v_2 km/s	$\widetilde{v}$ km/s	$\widetilde{v}'$ km/s	$\widetilde{v}' - \widetilde{v}$ km/s
Mercury	46001	69817	87.969	57.716	38.0280	47.872	47.890	**0.018**
Venus	107476	108942	224.70	35.258	34.7836	35.021	35.03	**0.009**
Earth	147098	152098	365.24	30.284	29.2883	29.786	29.786	**0**
Mars	206620	249230	686.98	26.383	21.8723	24.128	24.131	**0.003**
Jupiter	740574	816521	4329.6	13.715	12.4390	13.077	13.070	**-0.007**
Saturn	1353573	1513326	10832	10.160	9.0874	9.624	9.690	**0.066**
Uranus	2748938	3004420	30667	7.125	6.5187	6.822	6.835	**0.013**
Neptune	4452941	4553946	60327	5.490	5.3679	5.429	5.478	**0.049**
Pluto	4436825	7375928	90581	5.922	3.5621	4.742	4.749	**0.007**
Moon	363.300 Perigee distance	405.493 Apogee distance	27.32 orbital period	1.0794 Perigee velocity	0.9670 Apogee velocity	1.232 Average eggshell orbital speed	1.023 Average speed of elliptical orbit	**0.002**

Table 1: Difference between the mean velocity $\widetilde{v}'$ of an elliptical orbit and the mean velocity $\widetilde{v}$ of an eggshell orbit $\Delta v = \widetilde{v}' - \widetilde{v}$

According to Table 1, the difference value $\Delta v = \widetilde{v}' - \widetilde{v}$ between the mean velocity $\widetilde{v}'$ of a planetary elliptical orbit and the mean velocity $\widetilde{v}$ of a planetary eggshell orbit is small.

3.2 Derivation of the tangential angular momentum conservation theorem

3.2.1 Nature of Kepler's Third Law (Periodic Law)

The German astronomer Kepler analyzed and calculated the data from the astronomical observations of the Danish astronomer Tigu and the tables of the constellations, and he summarized Kepler's three laws of planetary motion. In 1609 he published the first two laws of planetary motion in the journal Nouvelle Astronomie, and in 1618 he proposed the third law (the law of periodicity) in his book The Harmony of the Universe, viz:

$$\frac{a^3}{T^2} = \frac{GM}{4\pi^2} = k \tag{3.2.1}$$

In the formula, a is half the long axis of the ellipse; T is the period of the planet's rotation; G is the universal gravitational constant; M is the mass of the central object, and k is the constant corresponding to the mass M. Using the centripetal force formula and Newton's universal gravitational force formula, the above formula can be derived.

According to equation (2.5.16), the radius ρ_1 of curvature at perihelion of the Earth's eggshell orbit is:

$$\rho_1 = \frac{M+m}{M} r_1$$

According to the centripetal force formula, the centripetal force F_{v1} at the perihelion of the Earth's orbit around the Sun is:

$$F_{v1} = m\frac{v_1^2}{\rho_1} = \frac{Mm}{M+m}\frac{v_1^2}{r_1} \tag{3.2.2}$$

Substituting the perihelion velocity v_1 from equation (3.1.23) into the above equation gives the relationship equation:

$$F_{v1} = m\frac{v_1^2}{\rho_1} = \frac{Mm}{M+m}\frac{4\pi^2 r_2^2}{T^2 r_1} \tag{3.2.3}$$

The above formula is the formula for the centripetal force at perihelion of the Earth's orbit around the Sun.

According to equation (2.5.20), the radius ρ_2 of curvature at the aphelion of the Earth's orbit around the Sun is:

$$\rho_2 = \frac{M+m}{M} r_2$$

According to the centripetal force formula, the centripetal force F_{v2} at the aphelion of the Earth's orbit around the Sun is:

$$F_{v2} = m\frac{v_2^2}{\rho_2} = \frac{Mm}{M+m}\frac{v_2^2}{r_2} \tag{3.2.4}$$

Substituting the aphelion velocity v_2 from equation (3.1.24) into the above equation gives the relation equation:

$$F_{v2} = m\frac{v_2^2}{\rho_2} = \frac{Mm}{M+m}\frac{4\pi^2 r_1^2}{T^2 r_2} \tag{3.2.5}$$

The above equation is the formula for the centripetal force at the aphelion of the Earth's orbit around the Sun.

Since both the perihelion centripetal force F_{v1} and the aphelion centripetal force F_{v2} are equal to the gravitational force, the relation equation can be obtained:

$$\begin{cases} F_{v1} = \dfrac{Mm}{M+m}\dfrac{4\pi^2 r_2^2}{T^2 r_1} = G\dfrac{Mm}{r_1^2} \\ F_{v2} = \dfrac{Mm}{M+m}\dfrac{4\pi^2 r_1^2}{T^2 r_2} = G\dfrac{Mm}{r_2^2} \end{cases} \tag{3.2.6}$$

Simplify the above equation to obtain the relational equation:

$$\frac{r_2^2 r_1}{T^2} = \frac{G(M+m)}{4\pi^2}, \qquad \frac{r_1^2 r_2}{T^2} = \frac{G(M+m)}{4\pi^2} \tag{3.2.7}$$

Since the planetary mass m is much smaller than the solar mass M, the above equation can be simplified to the following equation:

$$\frac{G(M+m)}{4\pi^2} \approx \frac{GM}{4\pi^2} = k \tag{3.2.8}$$

Substituting the above equation into equation (3.2.7) gives the relationship equation:

$$r_1 = r_2$$

The above result is clearly inconsistent with the objective facts. It can be determined that Kepler's third law (periodic law) belongs to the formula for the period of circular motion. Kepler's third law, i.e., equation (3.2.7), should be expressed as follows:

$$\frac{r^3}{T^2} = \frac{GM}{4\pi^2} = k \tag{3.2.9}$$

The r in the formula is the radius of the circular orbit, T is the period of circular motion, G is the gravitational constant, and M is the solar mass.

Since physicists consider planetary orbits to be elliptical curves, they use half the long axis of the ellipse $a = (r_1 + r_2)/2$ instead of the radius of the circle r, i.e:

$$a = \frac{1}{2}(r_1 + r_2) = r$$

Substituting the above equation into equation (3.2.9) gives the relationship equation:

$$\frac{a^3}{T^2} = \frac{GM}{4\pi^2} = k \tag{3.2.10}$$

The equation above is Kepler's third law.

In summary, Kepler's third law is a periodic law that replaces the circular radius r with half the long axis of the ellipse $a = (r_1 + r_2)/2$.

It should be noted that Kepler's Third Law does not correspond to the objective fact that the Earth and the Sun move around the center O of mass of the solar system. Since Kepler's third law is derived from Newton's gravitational formula, Newton's gravitational formula does not correspond to the objective fact that the Earth and the Sun rotate around the center O of mass of the solar system.

3.2.2 Newton's Formula for Universal Gravitational Contradicts the Law of Conservation of Angular Momentum

According to the classical theory, the Sun is stationary at the elliptical focal position and the Sun is the center of gravitational. Since the perihelion and aphelion velocities are both perpendicular to the distance from the Sun to the Earth, the Earth's perihelion centripetal force $\boldsymbol{F_{v1}}$ and aphelion centripetal force $\boldsymbol{F_{v2}}$ are:

$$\boldsymbol{F_{v1}=m\frac{v_1^2}{r_1}},\qquad \boldsymbol{F_{v2}=m\frac{v_2^2}{r_2}}\qquad (3.2.11)$$

Since both the perihelion centripetal force $\boldsymbol{F_{v1}}$ and the aphelion centripetal force $\boldsymbol{F_{v2}}$ are equal to the Newtonian gravitational force $\boldsymbol{F=G\frac{Mm}{r^2}}$, the relation equation can be obtained:

$$\begin{cases} \boldsymbol{F_{v1}=m\frac{v_1^2}{r_1}=G\frac{Mm}{r_1^2}} \\ \boldsymbol{F_{v2}=m\frac{v_2^2}{r_2}=G\frac{Mm}{r_2^2}} \end{cases}\qquad (3.2.12)$$

Simplify the above equation to obtain the relational equation:

$$\boldsymbol{v_1=\sqrt{\frac{GM}{r_1}}},\qquad \boldsymbol{v_2=\sqrt{\frac{GM}{r_2}}}\qquad (3.2.13)$$

According to the law of conservation of angular momentum, the angular momentum $\boldsymbol{L_v}$ of the Earth's orbit around the Sun is:

$$\boldsymbol{L_v=rmv\sin\theta=C}$$

The C in the equation is a constant. According to the above equation, the perihelion angular momentum $\boldsymbol{L_{v1}}$ and the aphelion angular momentum $\boldsymbol{L_{v2}}$ are equal, i.e:

$$\boldsymbol{L_v=rmv\sin\theta=r_1mv_1\sin\theta_1=r_2mv_2\sin\theta_2=C}\qquad (3.2.14)$$

Since the angle $\boldsymbol{\theta_1=\theta_2=90^0}$ between the velocity $\boldsymbol{v_1}$ and $\boldsymbol{v_2}$ and the distance $\boldsymbol{r}$, the relation equation can be obtained according to the above equation:

$$\boldsymbol{r_1v_1-r_2v_2=0}\qquad (3.2.15)$$

We can use the above equation to verify that Newton's formula for gravitational is correct or incorrect.

According to the planetary motion data provided by astronomy, the mass of the Sun $\boldsymbol{M=1.989\times10^{30}\cdot kg}$; Newton's gravitational constant $\boldsymbol{G=6.673\times10^{-11}\cdot Nm^2/kg^2}$. According to formula (3.2.13), the perihelion velocity $\boldsymbol{v_1}$ of the planet, and the aphelion velocity $\boldsymbol{v_2}$, as shown in Table 2..

Table 2	r_1 $k\cdot km$	r_2 $k\cdot km$	$v_1=\sqrt{\frac{GM}{r_1}}$ km/s	$v_2=\sqrt{\frac{GM}{r_2}}$ km/s	$r_1v_1-r_2v_2=0$ The difference should be equal to **0**
Mercury	46001	69817	53.7157	43.6018	**-573170.0**
Venus	107476	108942	35.1422	34.9049	**-25666.5**
Earth	147098	152098	30.0387	29.5408	**-74463.9**
Mars	206620	249230	25.3454	23.0773	**-514688.9**
Jupiter	740574	816521	13.3875	12.7497	**-495963.3**
Saturn	1353573	1513326	9.9025	9.3652	**-768844.0**
Uranus	2748938	3004420	6.9487	6.6467	**-867932.9**
Neptune	4452941	4553946	5.4596	5.3987	**-274111.6**
Pluto	4436825	7375928	5.4695	4.2421	**-7022209.8**
Moon	363.300 Perigee distance	405.493 Apogee distance	1.0467 Perigee velocity	0.9908 Apogee velocity	**-21.5**

Table 2: The angular momentum of the perihelion and aphelion of the planet should satisfy $\boldsymbol{r_1v_1-r_2v_2=0}$.

Note that the orbits of the planets around the Sun in Table 2 are elliptical.

From Table 2, the difference value between the product r_1v_1 and the product r_2v_2 is very large. If the difference value $r_1v_1 - r_2v_2$ is multiplied by the planetary mass m, the difference value $r_1mv_1 - r_2mv_2$ is even larger. It can be seen that the perihelion velocity v_1 and aphelion velocity v_2 calculated according to Newton's gravitational formula $F = GMm/r^2$ are wrong.

In addition, according to equation (3.1.23) and equation (3.1.24), the perihelion velocity $v_1 = \frac{2\pi r_2}{T}$ of the planet and the aphelion velocity $v_2 = \frac{2\pi r_1}{T}$ are given in Table 3.

Table 3	r_1 $k \cdot km$	r_2 $k \cdot km$	T day	$v_1 = \frac{2\pi r_2}{T}$ Km/s	$v_2 = \frac{2\pi r_1}{T}$ km/s	$r_1v_1 - r_2v_2 = 0$ The difference should be equal to **0**
Mercury	46001	69817	87.969	57.7162	38.0280	**2.0**
Venus	107476	108942	224.70	35.2581	34.7836	**4.6**
Earth	147098	152098	365.24	30.2839	29.2883	**9.3**
Mars	206620	249230	686.98	26.3829	21.8723	**1.5**
Jupiter	740574	816521	4329.6	13.7147	12.4390	**45.5**
Saturn	1353573	1513326	10832	10.1600	9.0874	**103.0**
Uranus	2748938	3004420	30667	7.1245	6.5187	**-103.9**
Neptune	4452941	4553946	60327	5.4896	5.3679	**-261.8**
Pluto	4436825	7375928	90581	5.9217	3.5621	**-243.5**
Moon	363.300 Perigee distance	405.493 Apogee distance	27.32 orbital period	1.0794 Perigee velocity	0.9670 Apogee velocity	**0.03**

Table 3: New angular momentum of the planet at perihelion and aphelion should satisfy $r_1v_1 - r_2v_2 = 0$.

Note that the orbits of the planets around the Sun in Table 3 are egg shell orbits.

As can be seen in Table 3, the difference value between the product r_1v_1 and the product r_2v_2 is very small. This difference value is caused by the rounding of the decimal point in the calculation. Obviously, by multiplying equation (3.1.23) and equation (3.1.24) by distance r_1 and distance r_2, respectively, the relation equation can be obtained:

$$\begin{cases} r_1v_1 = \dfrac{2\pi r_2 r_1}{T} \\ r_2v_2 = \dfrac{2\pi r_1 r_2}{T} \end{cases} \tag{3.2.16}$$

The above equation shows that the difference value $r_1v_1 - r_2v_2 = 0$.

In summary, comparing Table 3 with Table 2, we can see that the difference value $r_1v_1 - r_2v_2$ of elliptical orbits in Table 2, is very large, and the difference value $r_1v_1 - r_2v_2$ of eggshell orbits in Table 3, is very small.

3.2.3 Derivation of the tangential angular momentum conservation theorem

According to equation (3.2.11), the centripetal force F_{v1} at perihelion and F_{v2} at aphelion are:

$$F_{v1} = m\frac{v_1^2}{r_1}, \qquad F_{v2} = m\frac{v_2^2}{r_2}$$

The above equation can be varied as:

$$\begin{cases} F_{v1} = m\dfrac{v_1^2 r_1^2}{r_1^3} \\ F_{v2} = m\dfrac{v_2^2 r_2^2}{r_2^3} \end{cases} \tag{3.2.17}$$

Multiplying the numerator and denominator in the equation by the solar mass M gives the relationship equation:

$$\begin{cases} F_{v1} = \dfrac{v_1^2 r_1^2}{M} \cdot \dfrac{Mm}{r_1^3} \\ F_{v2} = \dfrac{v_2^2 r_2^2}{M} \cdot \dfrac{Mm}{r_2^3} \end{cases} \tag{3.2.18}$$

Since the centripetal force at perihelion and aphelion is equal to the gravitational force, the gravitational force F_1 at perihelion and F_2 at aphelion are:

$$\begin{cases} F_1 = F_{v1} = \dfrac{v_1^2 r_1^2}{M} \cdot \dfrac{Mm}{r_1^3} \\ F_2 = F_{v2} = \dfrac{v_2^2 r_2^2}{M} \cdot \dfrac{Mm}{r_2^3} \end{cases} \tag{3.2.19}$$

According to the above equation, the gravitational force formula is:

$$F = \frac{v^2 r^2}{M} \cdot \frac{Mm}{r^3} = K \frac{Mm}{r^3} \tag{3.2.20}$$

The K in the formula is the new universal gravitational constant. According to the above equation, the universal gravitational constant K for the Earth's revolution around the Sun is:

$$K = \frac{v^2 r^2}{M} \tag{3.2.21}$$

The above equation can be varied as:

$$\sqrt{KM} = vr \tag{3.2.22}$$

Multiplying the above equation by the Earth's mass m gives the relationship equation:

$$m\sqrt{KM} = mvr \tag{3.2.23}$$

Since v is the tangential velocity of the Earth's orbit around the Sun, the above equation is the equation for the conservation of the tangential angular momentum of the Earth around the Sun. Since the universal gravitational constant K, the Earth's mass m and the Sun's mass M are constants, the tangential angular momentum H_v of an object m rotating around the center O can be expressed as:

$$H_v = mvr = mr_1v_1 = mr_2v_2 = \mathrm{C} \tag{3.2.24}$$

This book calls the above formula the tangential angular momentum conservation formula. C in the formula is a constant. Note that the above formula is not the angular momentum conservation formula (3.2.14).

3.3 As the center O of mass moves in the Earth's orbital plane OXY, the distance R_0 from the center O of mass to the Earth will vary periodically

3.3.1 When the center O of mass is at rest, the orbit of the Earth around the center O of mass is circular

It is assumed that the center O of mass lies in the Earth's orbital plane, the X and Y axes of the frame of reference of the center O of mass lie in the Earth's orbital plane OXY, and the Z axis is perpendicular to the Earth's orbital plane OXY.

Assume that v'_C is the velocity of the center O of mass of the solar system as it orbits the galactic center; v'_{CX} is the velocity component of the velocity v'_Cprojected on the X-axis; v'_{CY} is the velocity component of the velocity v'_C projected on the Y-axis; and v'_{CZ} is the velocity component of the velocity v'_C' projected on the Z-axis.

According to the rules of velocity arithmetic, the galactic velocity v'_C can be expressed as:

$$v'_C = \sqrt{v'^2_{CX} + v'^2_{CY} + v'^2_{CZ}} \tag{3.3.1}$$

Note that the projected velocities $(v'_{CX}, v'_{CY}, v'_{CZ})$ are the velocities of the galactic frame of reference, not the solar system center-of-mass O frame of reference.

Suppose the center O of mass is at rest in the galactic frame of reference. Since the galactic velocity $\boldsymbol{v}'_C = \mathbf{0}$ of the center O of mass, the projected velocity is $\boldsymbol{v}'_{CX} = \boldsymbol{v}'_{CY} = \boldsymbol{v}'_{CZ} = \mathbf{0}$.

Assume that $\boldsymbol{v}'_{mo}$ is the circumferential speed of the Earth's rotation around the stationary center O of mass; and $\boldsymbol{R}_0$ is the distance from the Earth to the stationary center O of mass. According to the centripetal force formula, the centripetal force $\boldsymbol{F}_{vm0}$ of the Earth's rotation around the stationary center O of mass is:

$$\boldsymbol{F}_{vm0} = \boldsymbol{m}\frac{\boldsymbol{v}'^2_{m0}}{\boldsymbol{R}_0} \tag{3.3.2}$$

The centripetal force $\boldsymbol{F}_{vm0}$ points toward the resting center O of mass, as shown in Figure 3-1.

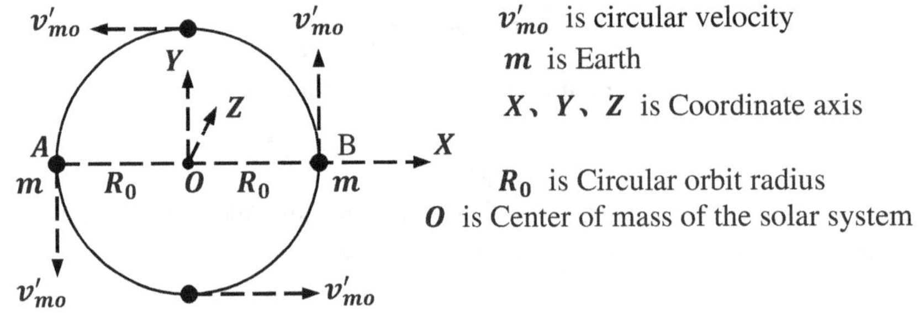

Figure 3-1 Schematic representation of the circular orbit and the circular velocity $\boldsymbol{v}'_{mo}$ of the Earth when the center O of mass of the solar system is at rest.

Assume that $\boldsymbol{r}_0$ is the distance from the Earth to the Sun. According to equation (2.1.7), the distance $\boldsymbol{r}_0$ is:

$$\boldsymbol{r}_0 = \frac{\boldsymbol{M}+\boldsymbol{m}}{\boldsymbol{M}}\boldsymbol{R}_0 \tag{3.3.3}$$

According to Newton's formula for gravitational, the gravitational force $\boldsymbol{F}$ between the Earth and the Sun is:

$$\boldsymbol{F} = \boldsymbol{G}\frac{\boldsymbol{Mm}}{\boldsymbol{r}_0^2} \tag{3.3.4}$$

Since the Earth and the Sun always revolve around the center O of mass, and the gravitational force $\boldsymbol{F}$ is the source of the centripetal force, the centripetal force $\boldsymbol{F}_{vm0}$ of the Earth revolving around the stationary center O of mass should be equal to the gravitational force $\boldsymbol{F}$, i.e:

$$\boldsymbol{F} = \boldsymbol{F}_{vm0} \tag{3.3.5}$$

If the state of the Sun and the Earth revolving around the stationary center O of mass remains constant, then we cannot find the reason why the distance $\boldsymbol{R}_0$ of the Earth to the center O of mass would vary periodically. At this time, the Earth's orbit around the stationary center O of mass is a circular orbit. This is shown in Figure 3-1.

3.3.2 If the motion of the center O of mass is perpendicular to the plane OXY of the Earth's orbit, then the Earth's orbit around the center O of mass is a spiral orbit of constant radius R_0

Assume that the velocity component $\boldsymbol{v}'_{CX} = \boldsymbol{v}'_{CY} = \mathbf{0}$ and the velocity component $\boldsymbol{v}'_{CZ} = \boldsymbol{v}'_C$, i.e. the center O of mass of the solar system moves along the Z axis.

Since the center O of mass moves in a direction perpendicular to the Earth's orbital plane OXY, the centripetal force $\boldsymbol{F}_{vO}$ is the same at every point on the Earth's orbit. In the galactic frame of reference, the Earth's orbit around the center O of mass is a spiral orbit of equal radius $\boldsymbol{R}_0$. The projection of the spiral orbit onto the OXY plane is still a circular orbit. This is shown in Figure 3-2.

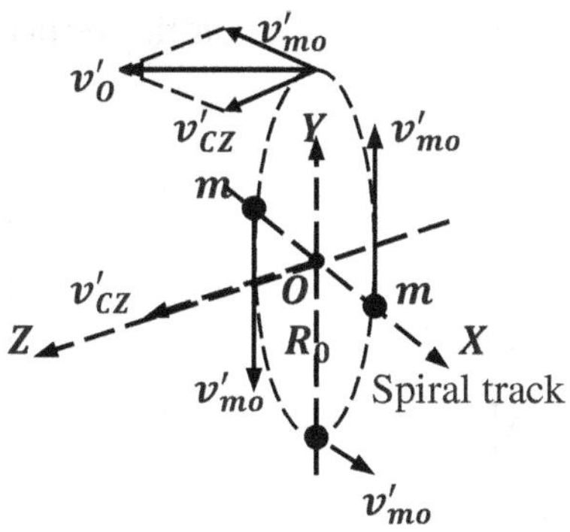
v'_{mo} is circular velocity
v'_O is orbit tangent speed
v'_{CZ} is center O of mass velocity
m is Earth
O is center of mass of the solar system
R_0 is radius of spiral track
Spiral track

Figure 3-2 Schematic of the tangential velocity v'_O of the Earth's spiral orbit and the velocity v'_{CZ} of the center of mass O of the solar system moving along the Z-axis.

The tangential velocity v'_O at any point on the Earth's spiral orbit is:

$$v'_O = \sqrt{v'^2_{CZ} + v'^2_{m0}} \tag{3.3.6}$$

In the formula, v'_{CZ} is the velocity of the center O of mass along the z-axis; v'_{mo} is the velocity perpendicular to the z-axis. Since the tangential velocity v'_O is perpendicular to the distance R_0 from the Earth to the center O of mass, the centripetal force F_{vO} of the Earth's revolution around the center O of mass is:

$$F_{vO} = m\frac{v'^2_O}{R_O} = m\frac{v'^2_{CZ} + v'^2_{m0}}{R_O} \tag{3.3.7}$$

The centripetal force F_{vO} is directed toward the center O of mass. The direction of motion of the center O of mass is perpendicular to the Earth's orbital plane OXY. as shown in Figure 3-2.

If the state of the Sun and the Earth revolving around the center O of mass remains the same, then we cannot find out the reason why the distance R_0 of the Earth to the center O of mass varies periodically. At this point, the Earth's orbit around the moving center O of mass is a spiral orbit. The center O of mass is the center of the spiral orbit and R_0 is the radius of the spiral orbit. This is shown in Figure 3-2 (the spiral orbit is not shown in the figure).

3.3.3 Velocities of the Solar System center O of mass and the Earth in the Galactic Frame of reference

Why does the gravitational force F between a planet and the Sun vary periodically? Physics has yet to provide a convincing explanation.

General relativity uses the "curvature of space-time" to explain changes in gravitational. General relativity says that space is curved around massive stars. General relativity describes the force of gravitational F, as the geometric curvature of "spacetime", which relates the energy and momentum emitted by matter through Einstein's field equations. At low speeds, general relativity reverts to Newtonian gravitational.

The authors use the galactic velocity v'_C of the center O of mass of the solar system to analyze and study the problem of gravitational variation from the point of view of the galactic frame of reference. The authors find that the distance r between the planet and the Sun varies periodically with the motion of the center O of mass. It is for this reason that the gravitational force F and the orbital velocity v of the planet remain periodically varying.

Assume that the velocity component $v'_{CX} \neq 0$, the velocity component $v'_{CY} \neq 0$ and the velocity component $v'_C = 0$, i.e., the center O of mass of the solar system is moving in the OXY plane of the Earth's orbit. Let v'_{CXY} be the velocity of the center O of mass moving in the plane OXY. The velocity v'_{CXY} at this point is:

$$\mathrm{v}'_{\mathrm{CXY}} = \sqrt{\mathrm{v}'^2_{\mathrm{CX}} + \mathrm{v}'^2_{\mathrm{CY}}} = v'_C \tag{3.3.8}$$

Suppose the horizontal line AB is a straight line connecting the perihelion point A and the aphelion point B. Suppose a line aOd in the plane OXY is perpendicular to the horizontal line AB. Suppose that the center O of mass moves down the vertical line $\boldsymbol{aOd}$ at the velocity $\boldsymbol{v'_{CXY}}$, as shown in Figure 3-3.

Suppose point $\boldsymbol{a}$ and point $\boldsymbol{b}$ are located on the Earth's orbit, where point $\boldsymbol{a}$ is on the vertical line $\boldsymbol{aOd}$ and point $\boldsymbol{b}$ is very close to point $\boldsymbol{a}$. This is shown in Figure 3-3.

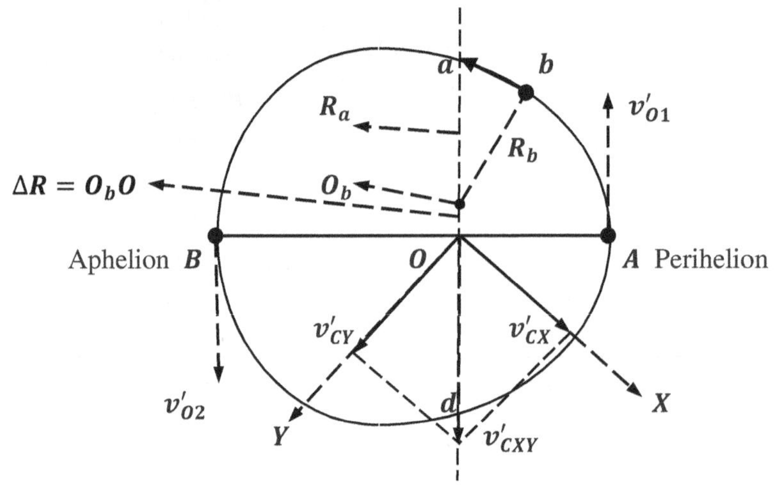

Figure 3-3 Schematic diagram of the center O of mass of the solar system moving downward perpendicular to the horizontal line AB with velocity $\boldsymbol{v'_{CXY}}$.

Note that all velocities in Figure 3-3 are in the Galactic frame of reference.

A note about the symbols in Figure 3-3;

The $\boldsymbol{v'_C}$ is the speed of rotation of the center O of mass of the solar system around the galactic center.

The $\boldsymbol{v'_{CX}}$ is the velocity component of $\boldsymbol{v'_C}$ projected on the x-axis;

The $\boldsymbol{v'_{CY}}$ is the velocity component of the velocity $\boldsymbol{v'_C}$ projected on the Y axis;

The $\boldsymbol{v'_{CXY}}$ is the velocity component of the velocity $\boldsymbol{v'_C}$ projected onto the orbital plane OXY;

The $\boldsymbol{v'_{01}}$ is the galactic frame of reference velocity possessed by the Earth at the perihelion position;

The $\boldsymbol{v'_{02}}$ is the galactic frame of reference velocity possessed by the Earth at the aphelion position;

Point O is the position of the center of mass of the solar system corresponding to point $\boldsymbol{a}$;

Point $\boldsymbol{O_b}$ is the position of the center of mass of the solar system corresponding to point $\boldsymbol{b}$;

The $\boldsymbol{R_a}$ is the distance from point $\boldsymbol{a}$ to the center O of mass;

The $\boldsymbol{R_b}$ is the distance from point $\boldsymbol{b}$ to the center O of mass;

The $\Delta \boldsymbol{R} = \boldsymbol{O_bO}$ is the distance displaced vertically downward from the center O of mass;

3.3.4 Reasons why the distance R_O from the Earth to the center O of mass increases when the Earth moves in an orbit above the horizontal line AB

Assume that point $\boldsymbol{O_b}$ is the position of the center of mass of the solar system corresponding to point $\boldsymbol{b}$; assume that point O is the position of the center of mass of the solar system corresponding to point $\boldsymbol{a}$. As shown in Figure 3-3.

The motion of the Earth around the center O of mass can be divided into motion above the horizontal line AB and motion below the horizontal line AB.

When the Earth moves in an orbit above the horizontal line AB, suppose the Earth moves from point $\boldsymbol{b}$ to point $\boldsymbol{a}$. If the projected velocity $\boldsymbol{v'_{CXY}} = \boldsymbol{0}$, then the Earth's orbit around the center O of mass is a circular orbit. At this point, point $\boldsymbol{O_b}$ is the center of the circular orbit; the distance $\boldsymbol{R_b}$ is the radius of the circular orbit, i.e., the distances from the center of mass $\boldsymbol{O_b}$ to both point $\boldsymbol{b}$ and point $\boldsymbol{a}$ are equal to the radius $\boldsymbol{R_b}$. This is shown in Figure 3-3.

If the projected velocity $\boldsymbol{v'_{CXY}} \neq \boldsymbol{0}$, then the center O of mass moves vertically downward in the orbital plane OXY along the line $\boldsymbol{aOd}$. Since it takes time for the Earth to move from point $\boldsymbol{b}$ to point $\boldsymbol{a}$, the center of mass $\boldsymbol{O_b}$

corresponding to point $\boldsymbol{b}$ and the center O of mass corresponding to point $\boldsymbol{a}$ are not the same point.

As the Earth begins to move from orbital point $\boldsymbol{b}$ to orbital point $\boldsymbol{a}$, the center of mass of the solar system begins to move downward along the straight line $\boldsymbol{aOd}$ from the position of point $\boldsymbol{O_b}$ because the projected velocity $\boldsymbol{v'_{CXY}} \neq \boldsymbol{0}$. When the Earth reaches point $\boldsymbol{a}$ from point $\boldsymbol{b}$, the center of mass of the solar system correspondingly descends from the position of point $\boldsymbol{O_b}$ to the position of point O. At this point, the value of the distance $\boldsymbol{R_b}$ increases slightly. The value by which the distance $\boldsymbol{R_b}$ increases is $\boldsymbol{\Delta R}$. This is shown in Figure 3-3.

At this point, the distance $\boldsymbol{R_a}$ from point $\boldsymbol{a}$ to the center O of mass is:

$$\boldsymbol{R_a = R_b + \Delta R} \quad (3.3.9)$$

Note that the distance change value $\boldsymbol{\Delta R}$ is caused by the downward motion of the center of mass of the solar system along the line $\boldsymbol{aOd}$. In other words, the centers of mass of the solar system corresponding to distances $\boldsymbol{R_b}$ and $\boldsymbol{R_a}$ are not the same point. The center of mass of the solar system corresponding to the distance $\boldsymbol{R_b}$ is the point $\boldsymbol{O_b}$, and the center of mass of the solar system corresponding to the distance $\boldsymbol{R_a}$ is the point O. Obviously, the position of point O is the new position of the center of mass of the solar system after it has moved downward from the position of point $\boldsymbol{O_b}$.

3.3.5 Reasons why the distance R_O from the Earth to the center O of mass decreases when the Earth moves in an orbit below the horizontal line AB

Suppose point $\boldsymbol{d}$ and point $\boldsymbol{c}$ are located on the Earth's orbit, where point $\boldsymbol{d}$ is on the vertical line $\boldsymbol{aOd}$ and point $\boldsymbol{c}$ is very close to point $\boldsymbol{d}$. Assume that point $\boldsymbol{O_c}$ is the position of the center of mass of the solar system corresponding to point $\boldsymbol{c}$; assume that point O is the position of the center of mass of the solar system corresponding to point $\boldsymbol{d}$. As shown in Figure 3-4.

When the Earth moves in an orbit below the horizontal line AB, suppose the Earth moves from point $\boldsymbol{c}$ to point $\boldsymbol{d}$. If the projected velocity $\boldsymbol{v'_{CXY} = 0}$, then the Earth's orbit around the center O of mass is a circular orbit. At this point, point $\boldsymbol{O_c}$ is the center of the circular orbit; the distance $\boldsymbol{R_c}$ is the radius of the circular orbit, i.e., the distances from the center of mass $\boldsymbol{O_c}$ to both point $\boldsymbol{c}$ and point $\boldsymbol{d}$ are equal to the radius $\boldsymbol{R_c}$. This is shown in Figure 3-4.

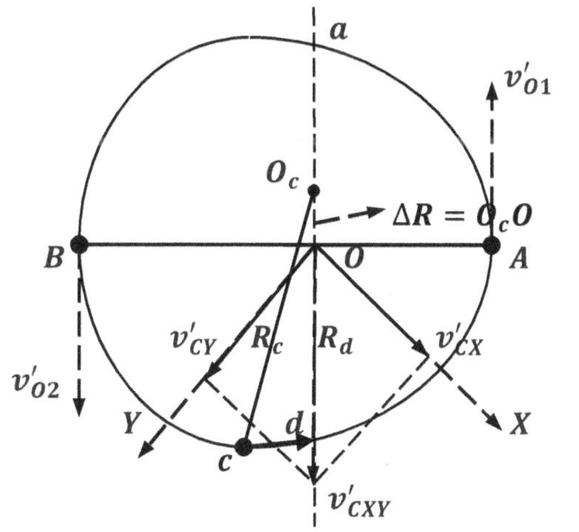

Figure 3-4 Schematic diagram of the center O of mass of the solar system moving downward perpendicular to the horizontal line AB with velocity $\boldsymbol{v'_{CXY}}$.

Note that all velocities in Figure 3-4 are in the Galactic frame of reference.

A note about the symbols in Figure 3-4;

The $\boldsymbol{v'_C}$ is the speed of rotation of the center O of mass of the solar system around the galactic center.

The $\boldsymbol{v'_{CX}}$ is the velocity component of $\boldsymbol{v'_C}$ projected on the x-axis;

The $\boldsymbol{v'_{CY}}$ is the velocity component of the velocity $\boldsymbol{v'_C}$ projected on the Y axis;

The v'_{CXY} is the velocity component of the velocity v'_C projected onto the orbital plane OXY;
The v'_{01} is the galactic frame of reference velocity possessed by the Earth at the perihelion position;
The v'_{02} is the galactic frame of reference velocity possessed by the Earth at the aphelion position;
Point O is the position of the center of mass of the solar system corresponding to point d;
Point O_c is the position of the center of mass of the solar system corresponding to point c;
The R_d is the distance from point d to the center O of mass;
The R_c is the distance from point c to the center O of mass;
The $\Delta R = O_cO$ is the distance displaced vertically downward from the center O of mass;

If the projected velocity $v'_{CXY} \neq 0$, then the center O of mass moves vertically downward in the orbital plane OXY along the line aOd. Since it takes time for the Earth to move from point c to point d, the center of mass O_c corresponding to point c and the center O of mass corresponding to point d are not the same point.

As the Earth begins to move from orbital point c to orbital point d, the center of mass of the solar system begins to move downward along the straight line aOd from the position of point O_c because the projected velocity $v'_{CXY} \neq 0$. When the Earth reaches point d from point c, the center of mass of the solar system correspondingly descends from the position of point O_c to the position of point O. At this point, the value of distance R_c shrinks slightly. The reduced value of distance R_c is ΔR, as shown in Figure 3-4.

At this point, the distance R_d from point d to the center O of mass is:

$$R_d = R_c - \Delta R \tag{3.3.10}$$

Note that the distance change value ΔR is caused by the downward motion of the center of mass of the solar system along the line aOd. In other words, the centers of mass of the solar system corresponding to distances R_c and R_d are not the same point. The center of mass of the solar system corresponding to the distance R_c is the point O_c, and the center of mass of the solar system corresponding to the distance R_d is the point O. Obviously, the position of point O is the new position of the center of mass of the solar system after it has moved downward from the position of point O_c.

3.4 Causes of periodic variations in the gravitational force *F* between the planets and the Sun

3.4.1 Angle β between the velocity v_{CZ}' (Z-axis) and the velocity v_{CXY}' in the galactic reference frame

Since the center O of mass of the solar system moves around the galactic center, the orbits of the planets around the center O of mass are three-dimensional spatial spiral curves. A video URL of a planetary spiral orbit is shown below:

https://www.iqiyi.com/w_19s10865zp.html

Assume that the center O of mass of the solar system is located in the orbital plane OXY of the Earth. Assume that the projected velocity of the velocity v'_C in the X, Y, and Z coordinate axes of the frame of reference of the center O of mass is $v'_{CX} \neq 0$, $v'_{CY} \neq 0$, and $v'_{CZ} \neq 0$, respectively.

Assume that v'_{CXY} is the projected velocity of the velocity v'_C in the Earth's orbital plane OXY; assume that the projected velocity v'_{CZ} on the Z coordinate axis is perpendicular to the orbital plane OXY.

Assume that the horizontal line AB is a straight line connecting the perihelion point A and the aphelion point B. Assume that the straight line aOd in the plane OXY is perpendicular to the horizontal line AB, as shown in Figure 3-5.

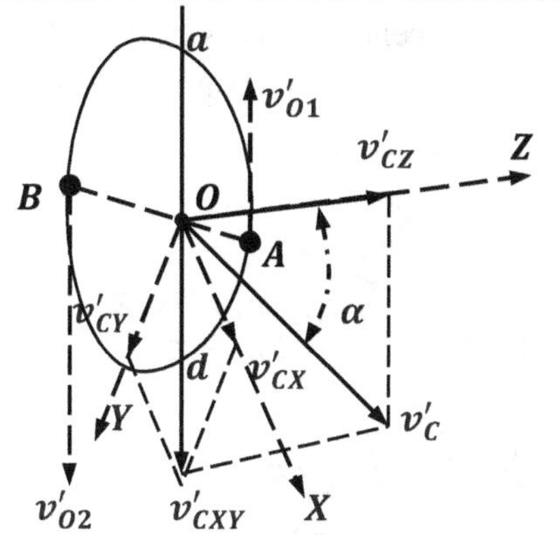

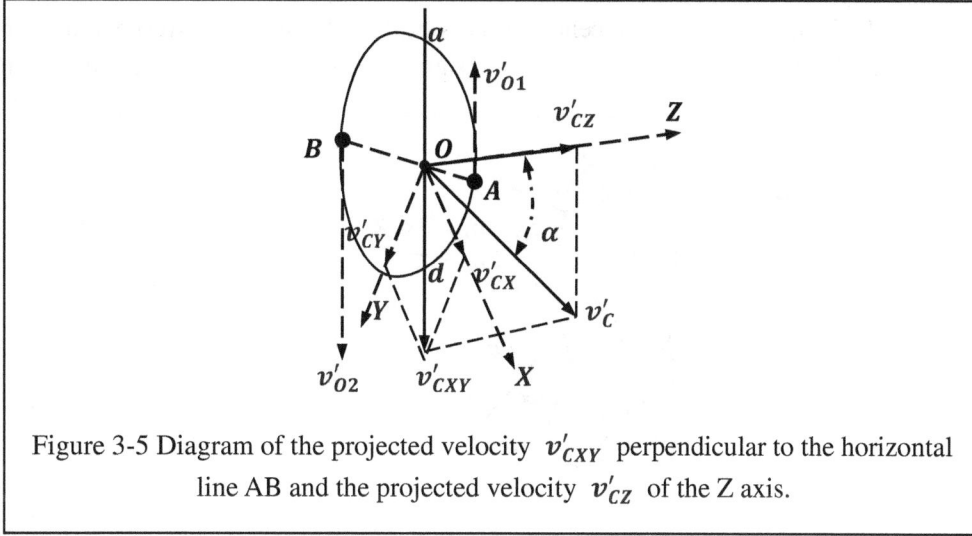

Figure 3-5 Diagram of the projected velocity v'_{CXY} perpendicular to the horizontal line AB and the projected velocity v'_{CZ} of the Z axis.

Note that all velocities in Figure 3-5 are in the Galactic frame of reference.

A note about the symbols in Figure 3-5;

A is perihelion; **B** is aphelion.

The v'_C is the speed of rotation of the center O of mass of the solar system around the galactic center.

The v'_{CX} is the velocity component of v'_C projected on the x-axis;

The v'_{CY} is the velocity component of the velocity v'_C projected on the Y axis;

The v'_{CXY} is the velocity component of the velocity v'_C projected onto the orbital plane OXY;

The v'_{01} is the galactic frame of reference velocity possessed by the Earth at the perihelion position;

The v'_{02} is the galactic frame of reference velocity possessed by the Earth at the aphelion position;

The α is the angle between the velocity v'_C of the center O of mass and the Z coordinate axis.

The aOd line is the orbit of the center O of mass moving vertically downward, and the aOd line is perpendicular to the horizontal line AB;

In the galactic frame of reference, the projected velocity v'_{CXY} in the orbital plane OXY of the Earth is:

$$v'_{CXY} = \sqrt{v'^2_{CX} + v'^2_{CY}} \quad (3.4.1)$$

In the galactic frame of reference, the velocity v'_C of the center O of mass of the solar system is:

$$v'_C = \sqrt{v'^2_{CXY} + v'^2_{CZ}} \quad (3.4.2)$$

Note that the velocity v'_C lies in the plane defined by the two velocities v'_{CXY} and v'_{CZ}.

The plane defined by the velocities v'_{CZ} and v'_{CXY} is perpendicular to the orbital plane OXY. Let α be the angle between the galactic velocity v'_C and the Z-axis. The projected velocity v'_{CXY} of velocity v'_C projected onto the Earth's orbital plane OXY is:

$$v'_{CXY} = v'_C \sin\alpha \quad (3.4.3)$$

The projected velocity v'_{CZ} of velocity v'_C projected on the Z-axis is:

$$v'_{CZ} = v'_C \cos\alpha \quad (3.4.4)$$

Since the orbit of the Earth around the center O of mass is a spiral orbit in three dimensions, the velocity $v'_{CZ} \neq 0$. The angle $\alpha \neq 90^0$ between the velocity v'_C and the Z-axis.

3.4.2 The reason why the gravitational force *F* between the Earth and the Sun decreases as the Earth moves in an orbit above the horizontal line AB

Suppose point a and point b are located on the Earth's orbit, where point a is on the vertical line aOd and point $\boldsymbol{b}$ is very close to point $\boldsymbol{a}$. This is shown in Figure 3-6.

Assume that point $\boldsymbol{O_b}$ is the position of the center of mass of the solar system corresponding to point $\boldsymbol{b}$; assume that point O is the position of the center of mass of the solar system corresponding to point $\boldsymbol{a}$. As shown in Figure 3-6.

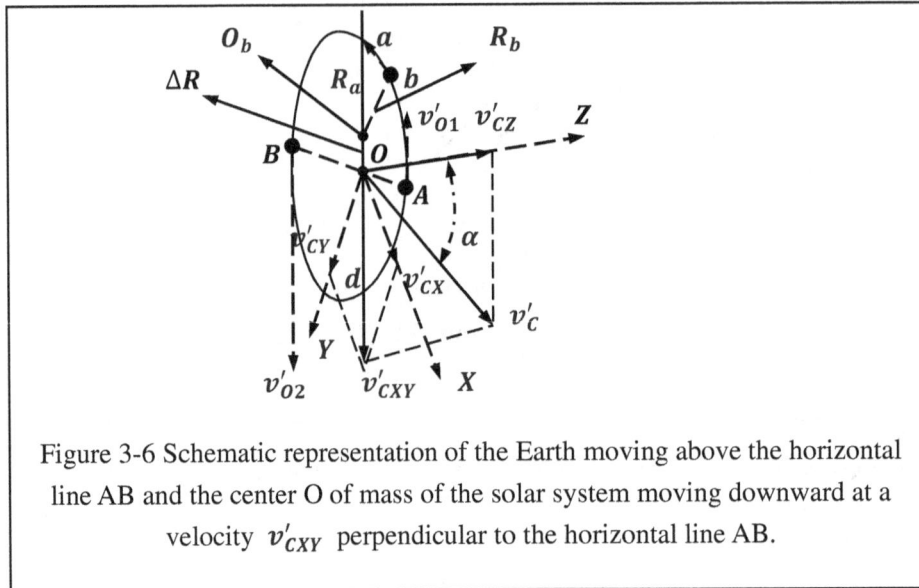

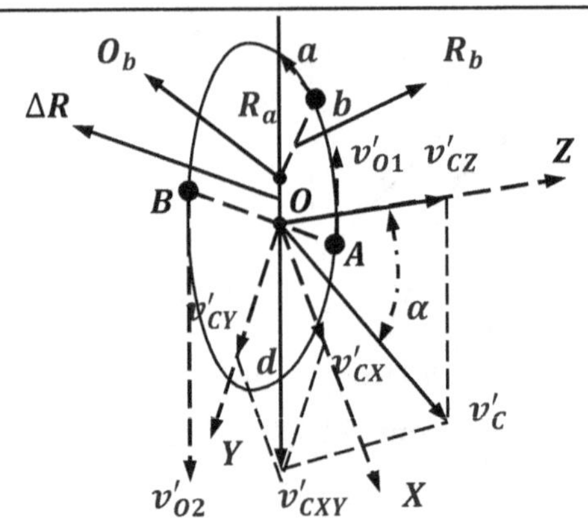

Figure 3-6 Schematic representation of the Earth moving above the horizontal line AB and the center O of mass of the solar system moving downward at a velocity $\boldsymbol{v'_{CXY}}$ perpendicular to the horizontal line AB.

A note about the symbols in Figure 3-6;

Point O is the position of the center of mass of the solar system corresponding to point $\boldsymbol{a}$;

Point $\boldsymbol{O_b}$ is the position of the center of mass of the solar system corresponding to point $\boldsymbol{b}$;

The $\boldsymbol{R_a}$ is the distance from point $\boldsymbol{a}$ to the center O of mass;

The $\boldsymbol{R_b}$ is the distance from point $\boldsymbol{b}$ to the center O of mass;

The $\Delta \boldsymbol{R} = \boldsymbol{O_b O}$ is the distance displaced vertically downward from the center O of mass;

As the Earth moves from point $\boldsymbol{b}$ to point $\boldsymbol{a}$. Since the velocity $\boldsymbol{v'_{CXY}} \neq \boldsymbol{0}$, the center O of mass moves vertically downward along the line $\boldsymbol{aOd}$ in the orbital plane OXY. Since it takes time for the Earth to move from point $\boldsymbol{b}$ to point $\boldsymbol{a}$, the center of mass $\boldsymbol{O_b}$ corresponding to point $\boldsymbol{b}$ and the center O of mass corresponding to point $\boldsymbol{a}$ are not the same point.

As the Earth moves from point $\boldsymbol{b}$ to point $\boldsymbol{a}$, the center of mass of the solar system moves from point $\boldsymbol{O_b}$ to the position of point O. At this point, the value of the distance $\boldsymbol{R_b}$ increases slightly. The value by which the distance $\boldsymbol{R_b}$ increases is $\Delta\boldsymbol{R}$. This is shown in Figure 3-6.

At this point, the distance $\boldsymbol{R_a}$ from point $\boldsymbol{a}$ to the center O of mass is:

$$\boldsymbol{R_a} = \boldsymbol{R_b} + \Delta\boldsymbol{R} \tag{3.4.5}$$

Note that the distance change value $\Delta\boldsymbol{R}$ is caused by the downward motion of the center of mass of the solar system along the line $\boldsymbol{aOd}$. In other words, the centers of mass of the solar system corresponding to distances $\boldsymbol{R_b}$ and $\boldsymbol{R_a}$ are not the same point. The center of mass of the solar system corresponding to the distance $\boldsymbol{R_b}$ is the point $\boldsymbol{O_b}$, and the center of mass of the solar system corresponding to the distance $\boldsymbol{R_a}$ is the point O. Obviously, the position of point O is the new position of the center of mass of the solar system after it has moved downward from the position of point $\boldsymbol{O_b}$.

According to equation (2.1.7), the relationship between the distance $\boldsymbol{R_O}$ from the Earth to the center O of mass and the distance $\boldsymbol{r}$ from the Earth to the Sun is:

$$\boldsymbol{R_O} = \frac{\boldsymbol{M}}{\boldsymbol{M}+\boldsymbol{m}}\boldsymbol{r}$$

According to the above equation, the distance $\boldsymbol{r_b}$ from the Earth's orbital point $\boldsymbol{b}$ to the Sun is:

$$r_b = \frac{M+m}{M} R_b \tag{3.4.6}$$

At the orbital point b, the gravitational force F_b between the Earth and the Sun is:

$$F_b = G\frac{Mm}{r_b^2} \tag{3.4.7}$$

According to equation (3.4.6), the distance r_a from the orbital point a to the Sun is:

$$r_a = \frac{M+m}{M} R_a \tag{3.4.8}$$

At the orbital point a, the gravitational force F_a between the Earth and the Sun is:

$$F_a = G\frac{Mm}{r_a^2} \tag{3.4.9}$$

Compare the distance r_a in equation (3.4.8) with the distance r_b in equation (3.4.6). Since the distance $r_a > r_b$, the gravitational force F_a at point a is less than the gravitational force F_b at point b, i.e., $F_a < F_b$.

As the Earth moves from perihelion A to aphelion B, since the centripetal force F_{vO} is greater than the gravitational force F, the Earth gradually moves away from the Sun.

The Earth moves from orbital point b to point a by overcoming the resistance of Newtonian gravitational. Since the kinetic energy $E_a = mv_{Oa}'^2/2$ of point a, is less than the kinetic energy $E_b = mv_{Ob}'^2/2$ of point b, the galactic velocity v_{Oa}' of point a, is less than the galactic velocity v_{Ob}' of point b.

As the Earth moves from orbital point b toward aphelion B, since the tangential velocity v_O of the Earth gradually decreases, the centripetal force F_{vO} of the Earth's revolution around the center O of mass also gradually decreases. When the Earth reaches the aphelion B, the centripetal force F_{vO} at the aphelion is equal to the gravitational force F at the aphelion, so the distance R_O from the Earth to the center O of mass no longer increases.

Let v_{O2}' be the galactic velocity of the Earth's aphelion. Since the Earth also has a velocity in the galactic frame of reference, the Earth's aphelion should also have two different velocities in cosmic space.

Velocity 1 is the velocity v_{O2} of the Earth's aphelion as it revolves around the center O of mass in the solar system's center O of mass frame of reference.

Velocity 2 is the velocity v_{O2}' of the Earth's aphelion as it orbits around the center O of mass in the galactic frame of reference.

Since the galactic velocity v_{O2}' at the aphelion is vertically downwards, the velocity v_{O2}' is in the same direction as the galactic velocity v_{CXY}' in the Earth's orbital plane OXY. And the fact that the velocity v_{O2}' is in the same direction as v_{CXY}' is exactly why the velocity v_{O2} is smaller at the aphelion. Since the velocity $v_{O2} > 0$, the aphelion velocity v_{O2}' is greater than the velocity v_{CXY}', i.e. $v_{O2}' > v_{CXY}'$. In the frame of reference of the center O of mass, the velocity v_{O2} of the Earth at the aphelion rotating around the center O of mass is:

$$v_{O2} = v_{O2}' - v_{CXY}' \tag{3.4.10}$$

Note that the velocity v_{O2} and the velocity v_{O2}' at the aphelion are two velocities with different meanings. The former, v_{O2}, belongs to the velocity of the center O of mass frame of reference of the solar system , and the latter, v_{O2}', belongs to the velocity of the galactic center frame of reference.

If the velocity v_{O2}' and the velocity v_{CXY}' are considered as the velocities of two trains moving parallel to each other, then the velocity v_{O2} of the Earth's aphelion as it revolves around the center O of mass is the relative velocity of the two trains.

According to the centripetal force formula, the centripetal force F_{vO} of the Earth's rotation around the center O of mass is:

$$F_{vO} = m\frac{(v_O \sin\theta)^2}{R_O} \tag{3.4.11}$$

The θ in the equation is the angle between the tangential velocity v_O and the distance R_O, and $v_O \sin\theta$ is the transverse velocity of the Earth's rotation around the center O of mass. The direction of the centripetal force F_{vO} points toward the center O of mass.

Since the angle $\theta_2 = 90^0$ between the tangential velocity v_{O2} of the aphelion and the distance R_{O2}, the centripetal force F_{v02} of the aphelion is:

$$F_{v02} = m\frac{v_{02}^2}{R_{02}} = m\frac{(v'_{02} - v'_{CXY})^2}{R_{02}} \qquad (3.4.12)$$

Note that the projected velocity v'_{CXY} is smaller than the aphelion galactic velocity v'_{02}, i.e., the galactic velocity $v'_{CXY} < v'_{02}$. Only in this case the aphelion velocity v_{02} is minimized for the Earth rotation around the center O of mass.

Since the aphelion velocity v_{02} is minimum, the aphelion distance R_{02} is maximum, and the aphelion centripetal force F_{v02} is equal to the aphelion universal gravitational force F_2, both the aphelion centripetal force F_{v02} and the aphelion gravitational force F_2 are minimum.

3.4.3 The reason why the gravitational force F between the Earth and the Sun increases as the Earth moves in an orbit below the horizontal line AB

Suppose point d and point c are located on the Earth's orbit, where point d is on the vertical line aOd and point c is very close to point d. This is shown in Figure 3-7.

Assume that point O_c is the position of the center of mass of the solar system corresponding to point c; assume that point O is the position of the center of mass of the solar system corresponding to point d. As shown in Figure 3-7.

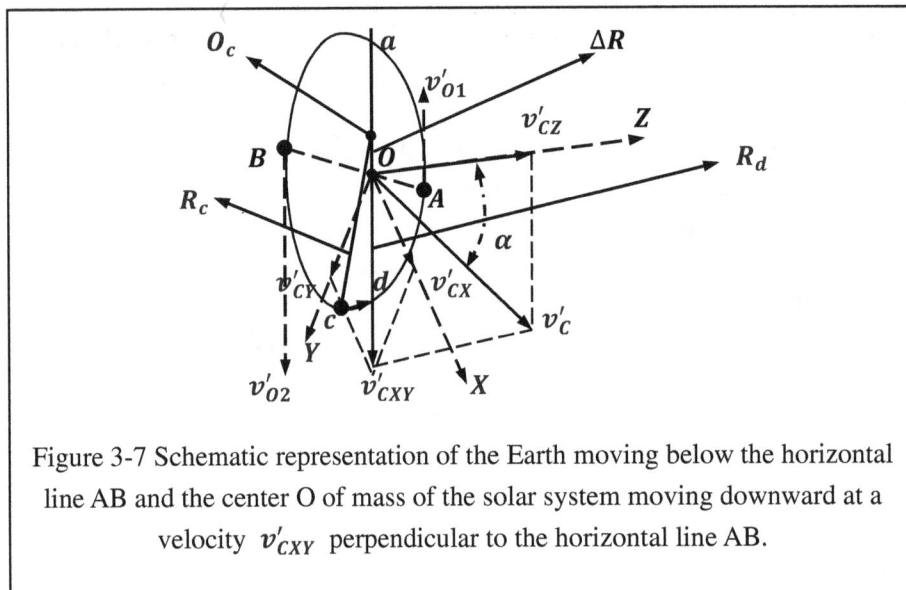

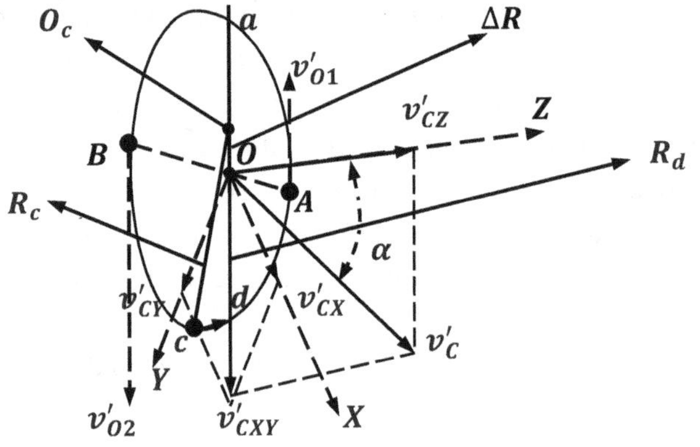

Figure 3-7 Schematic representation of the Earth moving below the horizontal line AB and the center O of mass of the solar system moving downward at a velocity v'_{CXY} perpendicular to the horizontal line AB.

A note about the symbols in Figure 3-7;

Point O is the position of the center of mass of the solar system corresponding to point d;

Point O_c is the position of the center of mass of the solar system corresponding to point c;

The R_d is the distance from point d to the center O of mass;

The R_c is the distance from point c to the center O of mass;

The $\Delta R = O_cO$ is the distance displaced vertically downward from the center O of mass;

As the Earth moves from point c to point d. Since the velocity $v'_{CXY} \neq 0$, the center O of mass moves vertically downward along the line aOd in the orbital plane OXY. Since it takes time for the Earth to move from point c to point d, the center of mass O_c corresponding to point c and the center O of mass corresponding to point d are not the same point.

As the Earth moves from point c to point d, the center of mass of the solar system moves from point O_c to the position of point O. At this point, the value of the distance R_c decreases slightly. The value by which the distance R_c decreases is ΔR. This is shown in Figure 3-7.

At this point, the distance R_d from point d to the center O of mass is:

$$R_d = R_c - \Delta R \tag{3.4.13}$$

Note that the distance change value ΔR is caused by the downward motion of the center of mass of the solar system along the line aOd. In other words, the centers of mass of the solar system corresponding to distances R_c and R_d are not the same point. The center of mass of the solar system corresponding to the distance R_c is the point O_c, and the center of mass of the solar system corresponding to the distance R_d is the point O. Obviously, the position of point O is the new position of the center of mass of the solar system after it has moved downward from the position of point O_c.

According to equation (2.1.7), the relationship between the distance R_O from the Earth to the center O of mass and the distance r from the Earth to the Sun is:

$$R_O = \frac{M}{M+m} r$$

According to the above equation, the distance r_c from the Earth's orbital point c to the Sun is:

$$r_c = \frac{M+m}{M} R_c \tag{3.4.14}$$

At the orbital point c, the gravitational force F_c between the Earth and the Sun is:

$$F_c = G\frac{Mm}{r_c^2} \tag{3.4.15}$$

According to equation (3.4.14), the distance r_d from the orbital point d to the Sun is:

$$r_d = \frac{M+m}{M} R_d \tag{3.4.16}$$

At the orbital point d, the gravitational force F_d between the Earth and the Sun is:

$$F_d = G\frac{Mm}{r_d^2} \tag{3.4.17}$$

Compare the distance r_d in equation (3.4.16) with the distance r_c in equation (3.4.14). Since the distance $r_d < r_c$, the gravitational force F_d at point d is greater than the gravitational force F_c at point c, i.e., $F_d > F_c$.

As the Earth moves from aphelion B to perihelion A, since the centripetal force F_{vO} is less than the gravitational force F, the Earth gradually moves closer to the Sun.

As the Earth moves from orbital point c to point d, the kinetic energy $E_d = mv_{Od}'^2/2$ of point d is greater than the kinetic energy $E_c = mv_{Oc}'^2/2$ of point c. Therefore, the galactic velocity v_{Od}' of point d is greater than the galactic velocity v_{Oc}' of point c.

As the Earth moves from orbital point c toward perihelion A, since the tangential velocity v_O of the Earth gradually increases, the centripetal force F_{vO} of the Earth's revolution around the center O of mass also gradually increases. When the Earth reaches perihelion A, the distance R_O from the Earth to the center O of mass no longer decreases because the centripetal force F_{vO} is equal to the gravitational force F of the perihelion.

Let v_{O1}' be the galactic velocity of the Earth's perihelion. Since the Earth also has a velocity in the galactic frame of reference, the Earth's perihelion should also have two different velocities in cosmic space.

Velocity 1 is the velocity v_{O1} of the Earth's perihelion as it revolves around the center O of mass in the solar system's center O of mass frame of reference.

Velocity 2 is the velocity v_{O1}' of the Earth's perihelion as it orbits around the center O of mass in the Galactic frame of reference.

Since the galactic velocity v_{O1}' at perihelion is vertically upward, the velocity v_{O1}' is in the opposite direction of the galactic velocity v_{CXY}' in the plane of Earth's orbit, OXY. And the fact that the velocity v_{O1}' is in the opposite direction to v_{CXY}' is exactly why the perihelion velocity v_{O1} is larger. In the center-of-mass O frame of reference, the velocity v_{O1} of the Earth at perihelion revolving around center-of-mass O is:

$$v_{O1} = v_{O1}' + v_{CXY}' \tag{3.4.18}$$

Note that the perihelion velocity v_{O1} and the velocity v_{O1}' are two velocities with different meanings. The former, v_{O1}, belongs to the velocity of the center O of mass frame of reference of the solar system, and the latter, v_{O1}', belongs to the velocity of the galactic center frame of reference.

If the velocities v'_{01} and v'_{CXY} are considered as the velocities of two parallel moving trains, then the velocity v_{01} of the Earth's perihelion as it revolves around the center O of mass is the relative velocity of the two trains.

According to the centripetal force formula, the centripetal force F_{vO} of the Earth's rotation around the center O of mass is:

$$F_{vO} = m\frac{(v_O \sin\theta)^2}{R_O} \tag{3.4.19}$$

The θ in the equation is the angle between the tangential velocity v_O and the distance R_O, and $v_O \sin\theta$ is the transverse velocity of the Earth's rotation around the center O of mass. The direction of the centripetal force F_{vO} points toward the center O of mass.

Since the angle $\theta_1 = 90^0$ between the tangential velocity v_{01} of the perihelion and the distance R_{01}, the centripetal force F_{v01} at the perihelion is:

$$F_{v01} = m\frac{v_{01}^2}{R_{01}} = m\frac{(v'_{01} + v'_{CXY})^2}{R_{01}} \tag{3.4.20}$$

Note that the velocity v'_{CXY} is in the opposite direction to the velocity v'_{01}, and only in this case is the perihelion velocity v_{01} of the Earth's revolution around the center O of mass maximized.

Since the perihelion velocity v_{01} is maximum, the perihelion distance R_{01} is minimum, and the perihelion centripetal force F_{v01} is equal to the perihelion gravitational force F_1, the perihelion centripetal force F_{v01} and the perihelion gravitational force F_1 are both maximum.

Classical theory suggests that the gravitational force F not only changes the direction of the tangential velocity v in the Earth's orbit, but also causes the tangential velocity v to increase (or decrease) continuously. However, physicists have never answered the question of why the gravitational force F changes periodically.

This book argues that the motion of the center O of mass around the galactic center causes the center O of mass to move perpendicular to the horizontal line AB in the Earth's orbital plane OXY, thus causing the distance R_O from the Earth to the center O of mass to constantly change. It is this change in the distance R_O that causes the gravitational force F between the Earth and the Sun to change periodically.

3.5 Reasons why the eight planets of the solar system orbit the sun on the same thin plane

3.5.1 Planetary Orbit Stabilization Formulas

Since the galactic velocity v'_{01} at perihelion and the galactic velocity v'_{02} at aphelion are both perpendicular to the horizontal line AB, and since the galactic velocity v'_{CXY} is also perpendicular to the horizontal line AB, both the velocity v'_{01} and the velocity v'_{02} are parallel to the velocity v'_{CXY}.

According to equation (3.4.18), the perihelion velocity v_{01} of the Earth's revolution around the center O of mass is:

$$v_{01} = v'_{01} + v'_{CXY} \tag{3.5.1}$$

According to equation (3.4.10), the aphelion tangential velocity v_{02} for the Earth's revolution around the center O of mass is:

$$v_{02} = v'_{02} - v'_{CXY} \tag{3.5.2}$$

Equation (3.5.1) minus the above equation gives the relational equation:

$$(v_{01} - v_{02}) + (v'_{02} - v'_{01}) = 2v'_{CXY} \tag{3.5.3}$$

Since the aphelion velocity v'_{02} is in the same direction as the velocity v'_{CXY} and the perihelion velocity v'_{01} is in the opposite direction to the velocity v'_{CXY}, the aphelion velocity v'_{02} is always greater than the perihelion velocity v'_{01}, i.e. $v'_{02} > v'_{01}$. If the velocity $v'_{02} < v'_{01}$, then the above formula does not hold.

Add equation (3.5.1) and equation (3.5.2) to obtain the relational equation:

$$v_{O1}+v_{O2}=v'_{O1}+v'_{O2} \tag{3.5.4}$$

The above formula can be modified to:

$$v_{O1}-v'_{O2}=v'_{O1}-v_{O2}=k$$

The k in the equation is a constant. Based on the above equation, the relational equation can be obtained:

$$v'_{O2}=v_{O1}-k, \qquad v'_{O1}=v_{O2}+k$$

Substituting the above equation into equation (3.5.3) gives the relationship equation:

$$v'_{CXY}=v_{O1}-v_{O2}-k \tag{3.5.5}$$

If constant $k=v_{O1}-v_{O2}$, then velocity $v'_{CXY}=0$. At this point, the planetary orbit is a circular orbit. If constant $k<v_{O1}-v_{O2}$, then the planetary orbit is an eggshell orbit due to the velocity $v'_{CXY}\neq 0$.

As the constant k becomes gradually smaller from its maximum value $k=v_{O1}-v_{O2}$. Since the velocity $v'_{CXY}=v_{O1}-v_{O2}-k$ becomes gradually larger, the deviation of the planetary orbit from the circular orbit becomes gradually larger.

When the constant $k=0$, equation (3.5.5) becomes the following equation:

$$v'_{CXY}=v_{O1}-v_{O2} \tag{3.5.6}$$

In this book, the above formula is referred to as the planetary orbit stabilization formula. The above formula shows that when the difference value in velocity between perihelion and aphelion $(v_{O1}-v_{O2})$ is equal to the projected velocity v'_{CXY}. Since the velocity v'_{CXY} is maximum, the orbit of the planet around the center O of mass has the greatest deviation from a circular orbit. At this point, the orbit of the planet around the center O of mass is a stable orbit.

In the center-of-mass O frame of reference. The velocity difference value between the planet's perihelion and aphelion is $\Delta v_O=v_{O1}-v_{O2}>0$. In the Galactic Center frame of reference, the galactic velocity difference value between the planet's perihelion and aphelion is $\Delta v'_O=v'_{O2}-v'_{O1}>0$. From the point of view of the coordinate transformations of the inertial system, the two velocity differences value should be equal, i.e, $\Delta v_O-\Delta v'_O=0$. From this we can obtain the relational equation:

$$v_{O1}-v_{O2}=v'_{O2}-v'_{O1} \tag{3.5.7}$$

Substituting the above equation into equation (3.5.3) gives the relationship equation:

$$v'_{CXY}=v_{O1}-v_{O2}=v'_{O2}-v'_{O1} \tag{3.5.8}$$

The above equation is equation (3.5.6). The above equation shows that the smaller the projected velocity v'_{CXY} in the planetary orbital plane OXY, the smaller the difference value in velocity between perihelion and aphelion of the planet's orbit $\Delta v_O=v_{O1}-v_{O2}$.

If the projected velocity $v'_{CXY}=0$, the planetary orbit is circular, since the perihelion and aphelion of the planet's revolution around the center O of mass have equal velocities, i.e. $v_{O1}=v_{O2}$.

Add equation (3.5.7) to equation (3.5.4), or subtract, to obtain the relationship equation:

$$v_{O1}=v'_{O2}, \qquad v_{O2}=v'_{O1} \tag{3.5.9}$$

The above equation shows that if the perihelion velocity v_{O1} is equal to the aphelion galactic velocity v'_{O2}, and the aphelion rate v_{O2} is equal to the perihelion galactic velocity v'_{O1}, then the planetary orbit at that time is a stable orbit.

Note that the perihelion velocity v_{O1} and the perihelion velocity v'_{O1} are two different velocities. The former, v_{O1}, belongs to the velocity of the frame of reference of the center O of mass of the solar system, and the latter, v'_{O1}, belongs to the velocity of the frame of reference of the galactic center.

If the galactic velocity v'_{O1} and the galactic velocity v'_{CXY} are regarded as the velocities of two cars in opposite directions, then the perihelion velocity v_{O1} is equal to the relative velocity between the two cars, i.e. $v_{O1}=v'_{O1}+v'_{CXY}$.

If the galactic velocity v'_{O2} and the galactic velocity v'_{CXY} are considered as the velocities of two cars in the same direction, then the velocity v_{O2} is equal to the relative velocity between the two cars, i.e. $v_{O2}=v'_{O2}-v'_{CXY}$

3.5.2 Velocity v_{CXY}' in the planetary orbital plane OXY

According to equations (2.3.7) and (2.3.8), the perihelion velocity $\boldsymbol{v_{01}}$ and the aphelion velocity $\boldsymbol{v_{02}}$ of the Earth are as follows:

$$v_{01} = \frac{M}{M+m} v_1, \qquad v_{02} = \frac{M}{M+m} v_2$$

Substituting the above two formulas into equation (3.5.6) gives the relational equation:

$$v_{CXY}' = \frac{M}{M+m}(v_1 - v_2) \tag{3.5.10}$$

Since the mass $\boldsymbol{m}$ of the Earth, is very small compared to the mass $\boldsymbol{M}$ of the Sun, the above equation can be simplified to:

$$v_{Cxy}' = v_1 - v_2 \tag{3.5.11}$$

According to equations (3.1.23) and (3.1.24), the perihelion velocity $\boldsymbol{v_1}$ and the aphelion velocity $\boldsymbol{v_2}$ of the Earth's orbit around the Sun are given:

$$v_1 = \frac{2\pi r_2}{T}, \qquad v_2 = \frac{2\pi r_1}{T}$$

According to the law of conservation of angular momentum, the angular momentum $\boldsymbol{L_{v1} = r_1 m v_1}$ at the perihelion of the Earth's orbit around the Sun, , is equal to the angular momentum $\boldsymbol{L_{v2} = r_2 m v_2}$ at the aphelion, i.e:

$$r_1 m v_1 = r_2 m v_2 \tag{3.5.12}$$

Since the perihelion-to-sun distance $\boldsymbol{r_1 = 1.471 \times 10^8 \cdot km}$, the Earth's perihelion velocity $\boldsymbol{v_1 = 30.287 \cdot km/s}$, and the aphelion-to-sun distance $\boldsymbol{r_2 = 1.521 \times 10^8 \cdot km}$. Substituting the data into the above equation, we obtain the Earth's aphelion velocity $\boldsymbol{v_2}$:

$$v_2 = \frac{v_1 r_1}{r_2} = \frac{30.287 \times 1.471 \times 10^8}{1.521 \times 10^8} = 29.291 km/s$$

Substituting the perihelion velocity $\boldsymbol{v_1}$ and the aphelion velocity $\boldsymbol{v_2}$ into equation (3.5.11) gives the projected velocity $\boldsymbol{v_{CXY}'}$, i.e:

$$v_{CXY}' = v_1 - v_2 = 0.996 km/s \tag{3.5.13}$$

The above equation shows that the projected velocity $\boldsymbol{v_{CXY}'}$ in the Earth's orbital plane OXY is approximately equal to $\boldsymbol{0.996 \cdot km/s}$.

The galactic velocity $\boldsymbol{v_c'}$ of the center of mass O of the solar system is projected onto the planetary orbital plane OXY with a velocity $\boldsymbol{v_{CXY}' = v_1 - v_2}$, as well as the perihelion and aphelion velocities $\boldsymbol{v_1}$ and $\boldsymbol{v_2}$ of the planet's revolution around the Sun. As shown in Table 4.

Table 4	r_1 $k \cdot km$	r_2 $k \cdot km$	T day	v_1 km/s	v_2 km/s	v_{CXY}' km/s
Mercury	46001	69817	87.969	57.716	38.028	**19.688**
Venus	107476	108942	224.70	35.258	34.784	**0.474**
Earth	147098	152098	365.24	30.287	29.291	**0.996**
Mars	206620	249230	686.98	26.383	21.872	**4.511**
Jupiter	740574	816521	4329.6	13.715	12.439	**1.276**
Saturn	1353573	1513326	10832	10.160	9.087	**1.073**
Uranus	2748938	3004420	30667	7.125	6.519	**0.606**
Neptune	4452941	4553946	60327	5.490	5.368	**0.122**
Pluto	4436825	7375928	90581	5.923	3.562	**2.361**
Moon	363.300	405.493	27.32	1.0794	0.9670	**0.111**
	Perigee distance	Apogee distance	orbital period	Perigee velocity	Apogee velocity	Projection speed

Table 4: The projected velocity $v_{CXY}' = v_1 - v_2$ of the galactic velocity v_c' on the planetary orbital plane OXY

According to astronomical data:

Earth mass $\boldsymbol{M = 5.965 \times 10^{24} \cdot kg}$;

Moon mass $\boldsymbol{m = 0.0349 \times 10^{24} \cdot kg}$.

The ratio of the mass of the Earth, $\boldsymbol{M}$, to the mass $\boldsymbol{(M+m)}$ is:

$$\boldsymbol{\frac{M}{M+m} = \frac{5.965 \times 10^{24}}{(5.965 + 0.07349) \times 10^{24}} = 0.9878} \qquad (3.5.14)$$

Suppose $\boldsymbol{O'}$ is the center of mass of the Earth-Moon system. In the solar frame of reference. Suppose $\boldsymbol{v'_c}$ is the velocity of the center of mass $\boldsymbol{O'}$ as it orbits the Sun (velocity $\boldsymbol{v'_c \approx 30 \cdot km/s}$). The lunar perigee velocity $\boldsymbol{v_1 = 1.079 \cdot km/s}$ and the lunar apogee velocity $\boldsymbol{v_2 = 0.967 \cdot km/s}$.

According to equation (3.5.10) . The center-of-mass velocity of the Earth-Moon system, $\boldsymbol{v'_C}$, the projected velocity $\boldsymbol{v'_{CXY}}$ in the plane OXY of the lunar orbit is:

$$\boldsymbol{v'_{CXY} = 0.9878 \times (1.079 - 0.967) = 0.111 \cdot km/s}$$

Note that the projected velocity $\boldsymbol{v'_{CXY}}$ belongs to the velocity of the solar system's center-of-mass O frame of reference; it does not belong to the velocity of the Galactic Center frame of reference, nor does it belong to the velocity of the Earth's frame of reference.

3.5.3 Reasons why the planets of the solar system all orbit the sun in the same plane

According to news reports on the Internet, a team of researchers at the National Astronomical Observatory of Japan (NAOJ) has accurately measured some of the fundamental distances in astronomy using the **VLBI** Radio Object Detection Network **(VERA)** and other advanced radio telescopes. The latest astronomical observations and calculations show that the distance from the Sun to the center of the Milky Way Galaxy is $\boldsymbol{26,100}$ light-years, and that the solar system revolves around the center of the Milky Way at a speed of $\boldsymbol{v'_C = 250 \cdot km/s}$.

In addition, scientists at the Australian Astronomical Observatory **(AAO)** have conducted a detailed analysis of observations from the Gaia telescope. Their results suggest that the Sun rotates around the center of the Milky Way at a speed of $\boldsymbol{v'_C = 250 \cdot km/s}$. The results were published in the Astrophysical Journal in November 2016.

According to equation (3.4.3), the velocity $\boldsymbol{v'_C}$ of the center O of mass of the solar system, the projected velocity $\boldsymbol{v'_{CXY}}$ in the planetary orbital plane OXY is:

$$\boldsymbol{v'_{CXY} = v'_C \sin \alpha}$$

The angle $\boldsymbol{\alpha}$ in the equation is the angle between the velocity $\boldsymbol{v'_C}$ and the projected velocity $\boldsymbol{v'_{CZ}}$ (Z axis). Alternatively, the angle $\boldsymbol{\alpha}$ is the angle between the vertical line (Z axis) of the planetary orbital plane OXY and the direction of motion of the center of mass O of the solar system. Note that the projected velocity $\boldsymbol{v'_{CZ}}$ (Z axis) is perpendicular to the projected velocity $\boldsymbol{v'_{CXY}}$.

According to Table 4, the projected velocity $\boldsymbol{v'_{CXY} = 0.996 \cdot km/s}$ in the Earth's orbital plane OXY. Since the center O of mass of the Solar System is very close to the center of mass of the Sun, the velocity $\boldsymbol{v'_C}$ of the center O of mass can be regarded as the velocity of the Sun as it orbits the center of the Galaxy. The angle $\boldsymbol{\alpha}$ between the velocity $\boldsymbol{v'_C}$ and the Z coordinate axis is:

$$\boldsymbol{\sin \alpha = \frac{v'_{CXY}}{v'_C} = \frac{0.996}{250} = 0.004} \qquad (3.5.15)$$

According to the trigonometric table, the angle $\boldsymbol{\alpha \approx 0.25^0}$. This result shows that the angle $\boldsymbol{\alpha}$ between the velocity $\boldsymbol{v'_C}$ and the Z-axis is approximately equal to $\boldsymbol{0.25^0}$ degrees. In other words, the Earth's orbital plane OXY appears to be perpendicular to the direction of the Sun's motion.

It is assumed that the projected velocities $\boldsymbol{v'_{CXY}}$ are all within the planetary orbital plane OXY. The galactic velocity $\boldsymbol{v'_C}$ at the center O of mass of the solar system, and the angle $\boldsymbol{\alpha}$ between it and the z-axis of the eight planetary frames of reference, are shown in Table 5.

Table 5	r_1 $k \cdot km$	r_2 $k \cdot km$	T day	v_1 km/s	v_2 km/s	v'_{CXY} km/s	$\sin\alpha$	α angle
Mercury	46001	69817	87.969	57.716	38.028	19.688	**0.0820**	4.7^0
Venus	107476	108942	224.70	35.258	34.784	0.474	**0.0020**	0.1^0
Earth	147098	152098	365.24	30.284	29.288	0.996	**0.0042**	0.25^0
Mars	206620	249230	686.98	26.383	21.872	4.511	**0.0188**	1.1^0
Jupiter	740574	816521	4329.6	13.715	12.439	1.276	**0.0053**	0.3^0
Saturn	1353573	1513326	10832	10.160	9.087	1.073	**0.0045**	0.3^0
Uranus	2748938	3004420	30667	7.125	6.519	0.606	**0.0025**	0.2^0
Neptune	4452941	4553946	60327	5.490	5.368	0.122	**0.0005**	0.0^0
Pluto	4436825	7375928	90581	5.923	3.562	2.361	**0.0098**	0.6^0
Halley Comet	87664	5250890	27485	13.893	0.232	13.661	**0.0569**	3.3^0
Moon	363.300 Perigee distance	405.493 Apogee distance	27.32 orbital period	1.0794 Perigee velocity	0.9670 Apogee velocity	0.111 Projection speed	**0.0037**	0.2^0

Table 5 The angle α between the orbital plane of the planet and the direction of motion of the center O of mass of the solar system,

According to the data on the Earth's motion, the center of mass O' of the Earth-Moon system revolves around the Sun at an average speed $v'_C = 29.786 \cdot km/s$. According to Table 5, the solar system velocity v'_C of the center of mass O' has a projected velocity $v'_{CXY} = 0.111 \cdot km/s$ in the lunar orbital plane OXY. According to equation (3.5.15), the relational equation is obtained:

$$\sin\alpha = \frac{v'_{CXY}}{v'_C} = \frac{0.111}{29.786} = 0.0037 \qquad (3.5.16)$$

The angle α in the formula is the angle between the vertical line (Z-axis) of the lunar orbital plane OXY and the direction of motion of the Earth. According to the trigonometric table, the angle $\alpha \approx 0.2^0$. This result shows that the angle α between the velocity v'_C and the Z-axis is approximately equal to 0.2^0 degrees.

In other words, the orbital plane OXY of the Moon appears to be perpendicular to the direction of the Earth's motion. Since the Moon's orbital plane OXY is perpendicular to the direction of the Earth's rotation around the Sun, the Moon is similar to the Sun in that it rises in the east and sets in the west.

According to Table 5, the velocity v'_C of the Sun's orbit around the Galactic Center has the largest angle α with the Z-axis of Mercury, i.e., $\alpha \approx 4.7^0$; and the angle α with the Z-axis of Halley's Comet is the second largest, i.e., $\alpha \approx 3.3^0$.

Visually, the planets of the Solar System orbit the Sun almost in one plane. In other words, the orbital planes OXY of the planets are almost always perpendicular to the speed v'_C of the Sun's rotation around the galactic center.

According to photographs of the Milky Way published by U.S. scientists, the stars orbiting around the center of the Milky Way also move on a plane that is thicker in the center and thinner at the edges. In other words, near the center of the Milky Way, the angle α between the plane of the orbits of its stars and the direction of galactic motion is larger. This is similar to the orbit of Mercury. Farther from the galactic center, the angle α between the plane of its stellar orbit and the direction of galactic motion is smaller. This is similar to the orbit of Neptune.

Chapter 4 Newton's Formula for Gravitational Contradicts the Law of Conservation of Mechanical Energy

Introduction: According to the law of conservation of angular momentum, the angular momentum of the Earth's perihelion and aphelion are equal. However, according to Newton's formula for universal gravitation and the law of conservation of angular momentum, the authors found by mathematical derivation that the distances from the Sun to perihelion and aphelion are equal. This result clearly contradicts the objective facts. Thus, it can be stated that Newton's formula for universal gravitation contradicts the law of conservation of angular momentum.

4.1 The centripetal force F_v' at the perihelion and aphelion of an elliptical orbit is not equal to the gravitational force F

4.1.1 The gravitational force F between the Earth and the center O of mass can be decomposed into the tangential gravitational force f_{Ov} and the normal gravitational force f_{On}

According to classical mechanics, the sun (the focal point of the ellipse) is the center toward which gravitational points. This view is not correct. In our solar system, the Sun and planets move around the center O of mass of the solar system.

The gravitational force F on the planets and the Sun is directed toward the center O of mass of the solar system, and the gravitational forces between the Sun and the planets are equal in magnitude and opposite in direction. From this, it can be determined that the center O of mass of the solar system is the center of gravitational pointing, and the Sun (the focus of the ellipse) is not the center of gravitational pointing.

According to mass-point kinematics, if the direction of the combined force F on the Earth does not coincide with the direction of the Earth's motion, then the Earth's motion is curvilinear. The trajectory of the Earth's motion lies between the combined force F and the velocity v. The velocity v is tangent to the Earth's orbit; the direction of the combined force F points to the concave side of the orbit.

Suppose F is the gravitational force between the Earth and the center O of mass. The universal gravitational force F can be decomposed into a tangential gravitational force f_{Ov} and a normal gravitational force f_{On}. The tangential gravitational force f_{Ov} is in the same or opposite direction to the tangential velocity v_O, and the normal gravitational force f_{On} is perpendicular to the tangential velocity v_O. In other words, the normal gravitational force f_{On} points to the center O_{vO} of the circle of curvature at the orbital point, and the tangential gravitational force f_{Ov} is perpendicular to the radius ρ_O of the circle of curvature, as shown in Figure 4-1.

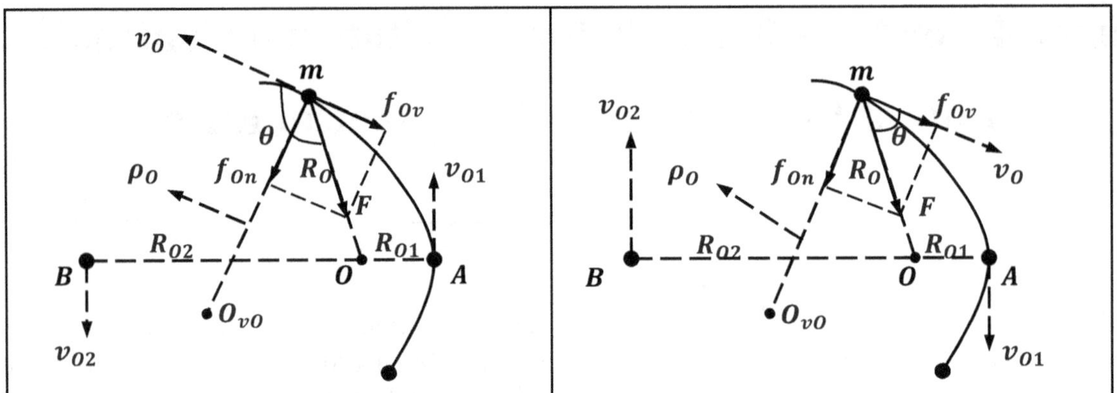

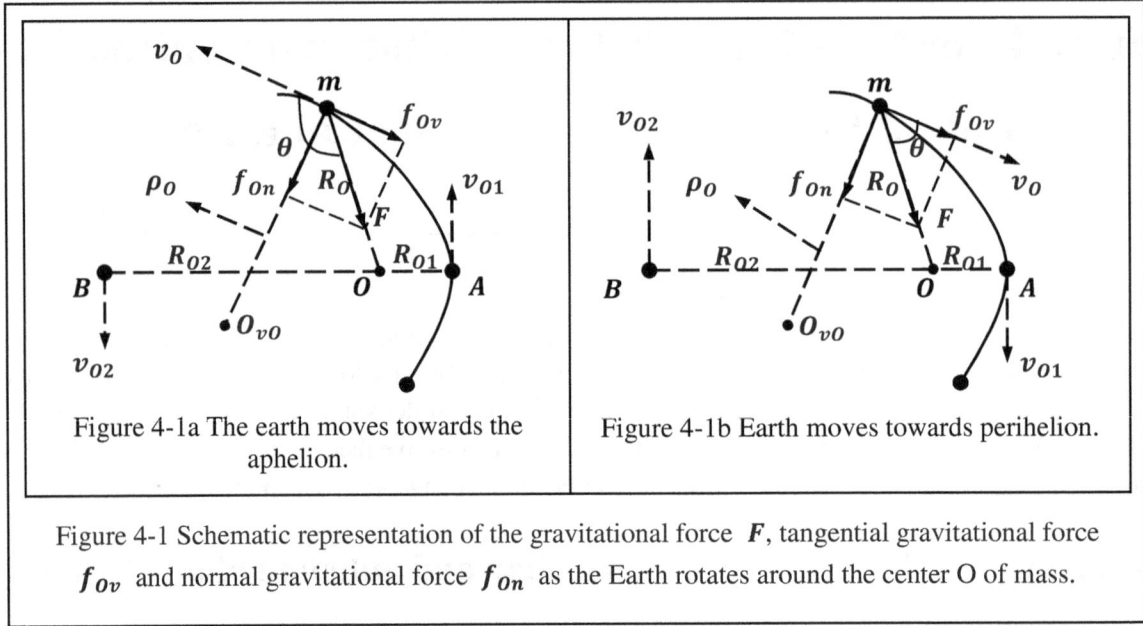

Figure 4-1a The earth moves towards the aphelion.

Figure 4-1b Earth moves towards perihelion.

Figure 4-1 Schematic representation of the gravitational force $\boldsymbol{F}$, tangential gravitational force $\boldsymbol{f_{Ov}}$ and normal gravitational force $\boldsymbol{f_{On}}$ as the Earth rotates around the center O of mass.

A note about the symbols in Figure 4-1;

O is the center of mass of the solar-terrestrial system, A is the perihelion and B is the aphelion;

$\boldsymbol{v_O}$ is the tangential velocity of the Earth at a point on its rotational orbit around the center O of mass of the Solar System;

$\boldsymbol{O_{vO}}$ is the center of the circle of curvature corresponding to the tangential velocity $\boldsymbol{v_O}$;

$\boldsymbol{\rho_O}$ is the radius of the circle of curvature;

$\boldsymbol{F}$ is the gravitational force of the Earth's revolution around the center O of mass of the solar system;

$\boldsymbol{f_{Ov}}$ is the tangential universal gravitational force of the Earth's revolution about the center O of mass;

$\boldsymbol{f_{On}}$ is the normal universal gravitational force of the Earth's revolution about the center O of mass;

$\boldsymbol{v_{O1}}$ is the perihelion tangential velocity of the Earth's revolution around the center O of mass;

$\boldsymbol{v_{O2}}$ is the tangential velocity at the far point of the Earth's revolution around the center O of mass;

$\boldsymbol{R_O}$ is the distance from the Earth to the center O of mass;

$\boldsymbol{\theta}$ is the angle between the tangential velocity $\boldsymbol{v_O}$ and the distance $\boldsymbol{R_O}$;

$\boldsymbol{R_{O1}}$ is the distance from the center O of mass to perihelion;

$\boldsymbol{R_{O2}}$ is the distance from the center O of mass to the aphelion;

The tangential gravitational force $\boldsymbol{f_{Ov}}$ is related to the universal gravitational force $\boldsymbol{F}$ by the equation:

$$\boldsymbol{f_{Ov} = F\cos\theta} \qquad (4.1.1)$$

The formula for the relationship between the normal gravitational force $\boldsymbol{f_{On}}$ and the universal gravitational force $\boldsymbol{F}$ is:

$$\boldsymbol{f_{On} = F\sin\theta} \qquad (4.1.2)$$

Although the Earth always revolves around the center O of mass, the normal gravitational force $\boldsymbol{f_{On}}$ at each point in the Earth's orbit is usually directed not toward the center O of mass but toward the center $\boldsymbol{O_{vO}}$ of the curvature circle corresponding to that point. From this, it can be determined that the tangential velocity $\boldsymbol{v_O}$ is usually not perpendicular to the Earth's distance $\boldsymbol{R_O}$, from the center O of mass. This is shown in Figure 4-1.

The vectorial sum of the tangential gravitational force $\boldsymbol{f_{Ov}}$ and the normal gravitational force $\boldsymbol{f_{On}}$ is equal to the universal gravitational force $\boldsymbol{F}$, i.e:

$$\boldsymbol{F = \sqrt{f_{Ov}^2 + f_{On}^2}} \qquad (4.1.3)$$

When the Earth moves from perihelion to aphelion. Since the angle $\boldsymbol{\theta} > \mathbf{90^0}$ between the gravitational force $\boldsymbol{F}$ and the tangential velocity $\boldsymbol{v_O}$, the tangential gravitational force $\boldsymbol{f_{Ov}} < \mathbf{0}$. Since the direction of the tangential gravitational force $\boldsymbol{f_{Ov}}$ is opposite to the direction of the tangential velocity $\boldsymbol{v_O}$, the tangential velocity $\boldsymbol{v_O}$ gradually decreases. As shown in Figure 4-1a.

When the Earth moves from aphelion to perihelion. Since the angle $\boldsymbol{\theta} < \mathbf{90^0}$ between the gravitational force $\boldsymbol{F}$ and the tangential velocity $\boldsymbol{v_O}$, the tangential gravitational force $\boldsymbol{f_{Ov}} > \mathbf{0}$. Since the direction of the tangential gravitational force $\boldsymbol{f_{Ov}}$ is the same as the direction of the tangential velocity $\boldsymbol{v_O}$, the tangential velocity $\boldsymbol{v_O}$ gradually increases. As shown in Figure 4-1b.

Since the angle $\boldsymbol{\theta_1} = \mathbf{90^0}$ between the tangential velocity $\boldsymbol{v_{O1}}$ of the perihelion and the distance $\boldsymbol{R_{O1}}$, the tangential gravitational force of the perihelion $\boldsymbol{f_{Ov1}} = \mathbf{0}$. According to equation (4.1.3), the normal gravitational force $\boldsymbol{f_{On1}}$ of the perihelion of the Earth's revolution around the center O of mass, is equal to the gravitational force $\boldsymbol{F_1}$ of the perihelion, namely:

$$\boldsymbol{f_{On1}} = \boldsymbol{F_1} \qquad (\mathbf{4.1.4})$$

The above equation shows that the perihelion normal gravitational force $\boldsymbol{f_{On1}}$ of the Earth's revolution around the center O of mass is equal to the perihelion universal gravitational force $\boldsymbol{F_1}$.

Since the angle $\boldsymbol{\theta_2} = \mathbf{90^0}$ between the tangential velocity $\boldsymbol{v_2}$ and the distance $\boldsymbol{r_2}$ of the aphelion, the tangential gravitational force of the aphelion $\boldsymbol{f_{Ov2}} = \mathbf{0}$. According to equation (4.1.3), the normal gravitational force $\boldsymbol{f_{On2}}$ of the aphelion of the Earth's revolution around the center O of mass , is equal to the gravitational force $\boldsymbol{F_2}$ of the aphelion, i.e:

$$\boldsymbol{f_{On2}} = \boldsymbol{F_2}$$

The above equation shows that the normal gravitational force $\boldsymbol{f_{On2}}$ at the aphelion of the Earth's revolution around the center O of mass is equal to the universal gravitational force $\boldsymbol{F_2}$ at the aphelion.

4.1.2 The normal gravitational force f_n of the Earth's orbit around the Sun, is equal to the normal gravitational force f_{On} of the Earth's orbit around the center of mass

There are two types of orbits. One is the Earth's orbit around the center O of mass of the Solar System, and the other is the Earth's orbit around the Sun. Theoretically, the Earth's orbit around the Sun is another manifestation of the Earth's orbit around the center O of mass of the solar system. In the Sun-Earth system, the combined force $\boldsymbol{F}$ acting on the Earth is the gravitational force between the Earth and the Sun.

The gravitational force $\boldsymbol{F}$ of the Earth's rotation around the Sun can be decomposed into a tangential gravitational force $\boldsymbol{f_v}$ and a normal gravitational force $\boldsymbol{f_n}$. The tangential gravitational force $\boldsymbol{f_v}$ is in the same or opposite direction as the tangential velocity $\boldsymbol{v}$, and the normal gravitational force $\boldsymbol{f_n}$ is perpendicular to the tangential velocity $\boldsymbol{v}$. In other words, the normal gravitational force $\boldsymbol{f_n}$ points toward the center $\boldsymbol{O_v}$ of the circle of curvature at the orbital point, and the tangential gravitational force $\boldsymbol{f_v}$ is perpendicular to the radius $\boldsymbol{\rho_v}$ of the circle of curvature. This is shown in Figure 4-2.

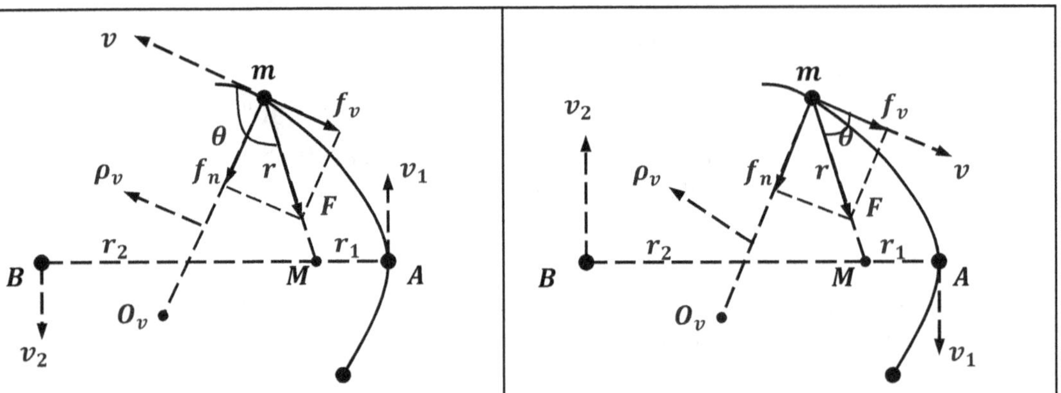

Figure 4-2a The earth moves towards the aphelion.

Figure 4-2b Earth moves towards perihelion.

Figure 4-2 Schematic of the gravitational force $\boldsymbol{F}$, tangential gravitational $\boldsymbol{f_v}$, and normal gravitational $\boldsymbol{f_n}$ as the Earth rotates around the Sun.

A note about the symbols in Figure 4-2;
The $\boldsymbol{M}$ is the sun, A is the perihelion and B is the aphelion;
The $\boldsymbol{v}$ is the tangential velocity of the Earth at a point on its orbit around the Sun;
The $\boldsymbol{O_v}$ is the center of the circle of curvature corresponding to the tangential velocity $\boldsymbol{v}$;
The $\boldsymbol{\rho_v}$ is the radius of the circle of curvature;
The $\boldsymbol{F}$ is the gravitational force of the Earth's rotation around the Sun;
The $\boldsymbol{f_v}$ is the tangential universal gravitational force of the Earth's revolution around the Sun;
The $\boldsymbol{f_n}$ is the normal universal gravitational force of the Earth's rotation around the Sun;
The $\boldsymbol{v_1}$ is the perihelion tangential velocity of the Earth's orbit around the Sun;
The $\boldsymbol{v_2}$ is the tangential velocity at the far point of the Earth's orbit around the Sun;
The $\boldsymbol{r}$ is the distance from the Earth to the Sun;
The $\boldsymbol{\theta}$ is the angle between the tangential velocity $\boldsymbol{v}$ and the distance $\boldsymbol{r}$;
The $\boldsymbol{r_1}$ is the distance from the Sun to perihelion;
The $\boldsymbol{r_2}$ is the distance from the Sun to aphelion;

The tangential gravitational force $\boldsymbol{f_v}$ is related to the universal gravitational force $\boldsymbol{F}$ by the equation:

$$\boldsymbol{f_v = F\cos\theta = f_{Ov}} \tag{4.1.5}$$

The above equation is equation (4.1.1). The above equation shows that the tangential gravitational force $\boldsymbol{f_v}$ of the Earth's orbit around the Sun is equal to the tangential gravitational force $\boldsymbol{f_{Ov}}$ of the Earth's orbit around the center O of mass.

The formula for the relationship between the normal gravitational force $\boldsymbol{f_n}$ and the universal gravitational force $\boldsymbol{F}$ is:

$$\boldsymbol{f_n = F\sin\theta = f_{On}} \tag{4.1.6}$$

The above equation is equation (4.1.2). The above equation shows that the normal gravitational force $\boldsymbol{f_n}$ of the Earth's orbit around the Sun is equal to the normal gravitational force $\boldsymbol{f_{On}}$ of the Earth's orbit around the center O of mass.

Although the Earth revolves around the Sun, the normal gravitational force $\boldsymbol{f_n}$ at any point in the Earth's orbit is usually not directed toward the Sun, but toward the center $\boldsymbol{O_v}$ of the circle of curvature corresponding to that point.

From this, the tangential velocity $\boldsymbol{v}$ is usually not perpendicular to the distance $\boldsymbol{r}$ between the Earth and the Sun, as shown in Figure 4-2.

The vectorial sum of the tangential gravitational force $\boldsymbol{f_v}$ and the normal gravitational force $\boldsymbol{f_n}$ of the Earth's orbit around the Sun is equal to the universal gravitational force $\boldsymbol{F}$, i.e:

$$\boldsymbol{F} = \sqrt{\boldsymbol{f_v^2} + \boldsymbol{f_n^2}} = \sqrt{\boldsymbol{f_{Ov}^2} + \boldsymbol{f_{On}^2}} \quad (4.1.7)$$

When the Earth moves from perihelion to aphelion. Since the angle $\boldsymbol{\theta} > 90^0$ between the gravitational force $\boldsymbol{F}$ and the tangential velocity $\boldsymbol{v}$, the tangential gravitational force $\boldsymbol{f_v} < 0$. Since the direction of the tangential gravitational force $\boldsymbol{f_v}$ is opposite to the direction of the tangential velocity $\boldsymbol{v}$, the tangential velocity $\boldsymbol{v}$ gradually decreases. As shown in Figure 4-2a.

When the Earth moves from aphelion to perihelion. Since the angle $\boldsymbol{\theta} < 90^0$ between the gravitational force $\boldsymbol{F}$ and the tangential velocity $\boldsymbol{v}$, the tangential gravitational force $\boldsymbol{f_v} > 0$. Since the direction of the tangential gravitational force $\boldsymbol{f_v}$ is the same as the direction of the tangential velocity $\boldsymbol{v}$, the tangential velocity $\boldsymbol{v}$ gradually increases. As shown in Figure 4-2b.

Since the angle $\boldsymbol{\theta_1} = 90^0$ between the tangential velocity $\boldsymbol{v_1}$ of the perihelion and the distance $\boldsymbol{r_1}$, the tangential gravitational force of the perihelion $\boldsymbol{f_{v1}} = 0$. According to equation (4.1.7), the normal gravitational force $\boldsymbol{f_{n1}}$ of the perihelion of the Earth's revolution around the Sun, is equal to the gravitational force $\boldsymbol{F_1}$ of the perihelion, viz:

$$\boldsymbol{f_{n1}} = \boldsymbol{F_1}$$

Comparing the above equation with equation (4.1.4), the relationship equation can be obtained:

$$\boldsymbol{f_{n1}} = \boldsymbol{f_{On1}} = \boldsymbol{F_1} \quad (4.1.8)$$

The above equation shows that the normal gravitational force $\boldsymbol{f_{n1}}$ at the perihelion of the Earth's revolution around the Sun is equal to the normal gravitational force $\boldsymbol{f_{On1}}$ at the perihelion of the Earth's revolution around the center O of mass, and both are equal to the gravitational force at the perihelion $\boldsymbol{F_1}$.

Since the angle $\boldsymbol{\theta_2} = 90^0$ between the tangential velocity $\boldsymbol{v_2}$ and the distance $\boldsymbol{r_2}$ of the aphelion, the tangential gravitational force of the aphelion $\boldsymbol{f_{v2}} = 0$. According to equation (4.1.7), the normal gravitational force $\boldsymbol{f_{n2}}$ of the aphelion of the Earth's orbit around the Sun, is equal to the aphelion gravitational force $\boldsymbol{F_2}$, i.e:

$$\boldsymbol{f_{n2}} = \boldsymbol{f_{On2}} = \boldsymbol{F_2} \quad (4.1.9)$$

The above equation shows that the normal gravitational force $\boldsymbol{f_{n2}}$ at the aphelion of the Earth's orbit around the Sun is equal to the normal gravitational force $\boldsymbol{f_{On2}}$ at the aphelion of the Earth's orbit around the center O of mass. Both are equal to the universal gravitational force $\boldsymbol{F_2}$ of the aphelion.

4.1.3 Perihelion centripetal force F_{v1} and aphelion centripetal force F_{v2} of the Earth's orbit around the Sun

According to equation (2.5.13), the centripetal force $\boldsymbol{F_{vO1}}$ at the perihelion of the Earth's revolution around the center O of mass of the solar system is:

$$\boldsymbol{F_{vO1}} = \boldsymbol{m}\frac{\boldsymbol{v_{O1}^2}}{\boldsymbol{R_{O1}}} = \frac{\boldsymbol{M}}{\boldsymbol{M} + \boldsymbol{m}}\boldsymbol{m}\frac{\boldsymbol{v_1^2}}{\boldsymbol{r_1}} \quad (4.1.10)$$

$\boldsymbol{v_{O1}}$ in the formula is the tangential velocity at perih This is shown in Figure elion of the Earth's revolution around the center O of mass of the solar system, and $\boldsymbol{v_1}$ is the tangential velocity at perihelion of the Earth's revolution around the Sun. $\boldsymbol{R_{O1}}$ in the formula is the distance from perihelion to the center O of mass, and $\boldsymbol{r_1}$ is the distance from perihelion to the Sun. The centripetal force $\boldsymbol{F_{vO1}}$ at perihelion, points toward the center O of mass. This is shown in Figure 4-3.

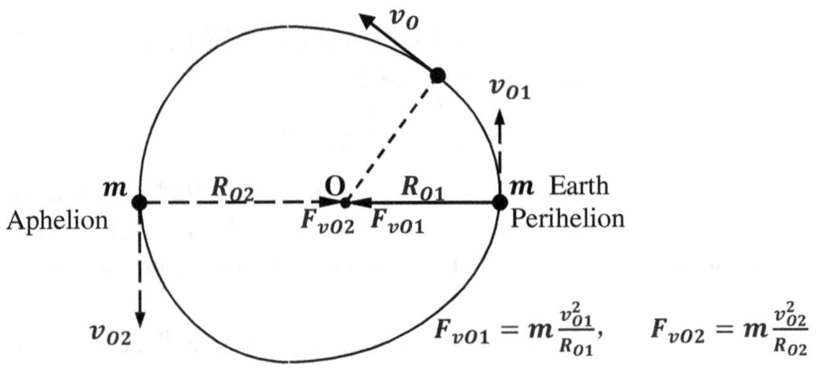

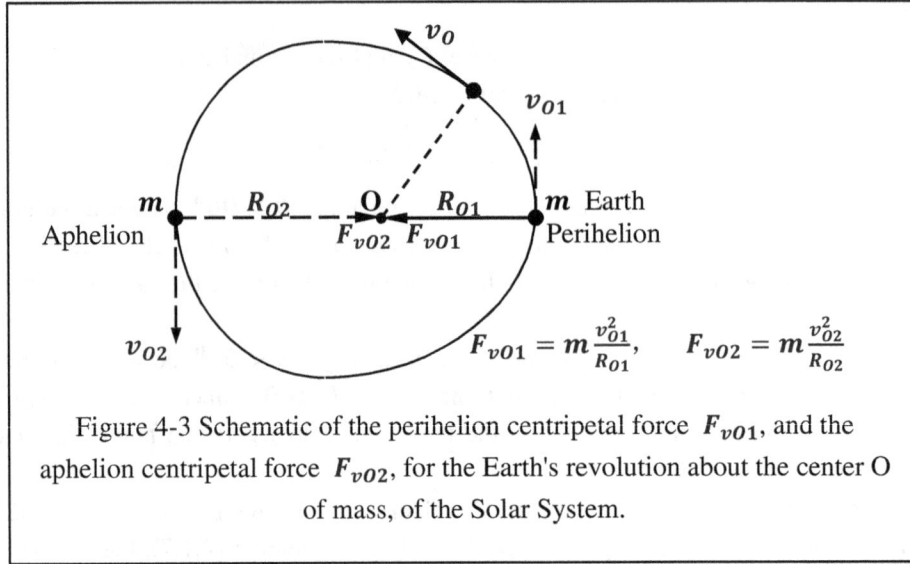

Figure 4-3 Schematic of the perihelion centripetal force F_{vO1}, and the aphelion centripetal force F_{vO2}, for the Earth's revolution about the center O of mass, of the Solar System.

According to equation (2.5.14), the centripetal force F_{v1} at the perihelion of the Earth's revolution around the Sun is:

$$F_{v1} = m\frac{v_1^2}{\rho_1} \quad (4.1.11)$$

The ρ_1 in the formula is the radius of the perihelion curvature circle of the Earth's revolution around the Sun. The perihelion centripetal force F_{v1} points to the center O_{v1} of the perihelion curvature circle.

Since the universal gravitational force F_1 between the Earth's perihelion and the Sun is unique, and the universal gravitational force F_1 is the source of the perihelion centripetal forces F_{vO1} and F_{v1}, the perihelion centripetal forces F_{vO1} and F_{v1} are both equal to the universal gravitational force F_1, namely:

$$F_{v1} = F_{vO1} = F_1 \quad (4.1.12)$$

Based on the above equation, the relationship equation can be obtained:

$$F_{v1} = m\frac{v_1^2}{\rho_1} = m\frac{v_{O1}^2}{R_{O1}} = \frac{M}{M+m}m\frac{v_1^2}{r_1} = F_1 \quad (4.1.13)$$

The above formula is the formula for the centripetal force at perihelion. Based on the above formula, the relation formula can be obtained:

$$\rho_1 = \frac{M+m}{M}r_1$$

From the above equation, the radius ρ_1 of the circle of curvature at perihelion of the Earth's orbit around the Sun is greater than the distance r_1 from perihelion to the Sun.

According to equation (2.5.17), the centripetal force F_{vO2} at the aphelion of the Earth's revolution around the center O of mass of the solar system is:

$$F_{vO2} = m\frac{v_{O2}^2}{R_{O2}} = \frac{M}{M+m}m\frac{v_2^2}{r_2} \quad (4.1.14)$$

The v_{O2} in the formula is the tangential velocity at the aphelion of the Earth's revolution around the center O of mass of the solar system, and v_2 is the tangential velocity at the aphelion of the Earth's revolution around the Sun. R_{O2} in the formula is the distance from the aphelion to the center O of mass, and r_2 is the distance from the aphelion to the Sun. The centripetal force F_{vO1} at the aphelion points toward the center O of mass. This is shown in Figure 4-3.

According to equation (2.5.18), the centripetal force F_{v2} at the aphelion of the Earth's orbit around the Sun is:

$$F_{v2} = m\frac{v_2^2}{\rho_2}$$

The ρ_2 in the formula is the radius of the circle of curvature at the aphelion of the Earth's orbit around the Sun.

The centripetal force $\boldsymbol{F_{v2}}$ at aphelion points to the center $\boldsymbol{O_{v2}}$ of the circle of curvature at aphelion.

Since the gravitational force $\boldsymbol{F_2}$ between the Earth's aphelion and the Sun is unique, and the gravitational force $\boldsymbol{F_2}$ is the source of the centripetal forces $\boldsymbol{F_{vO2}}$ and $\boldsymbol{F_{v2}}$ at the aphelion, the centripetal forces $\boldsymbol{F_{vO2}}$ and $\boldsymbol{F_{v2}}$ at the aphelion are both equal to the gravitational force $\boldsymbol{F_2}$, i.e.:

$$\boldsymbol{F_{v2} = F_{vO2} = F_2} \tag{4.1.15}$$

Based on the above equation, the relationship equation can be obtained:

$$\boldsymbol{F_{v2} = m\frac{v_2^2}{\rho_2} = m\frac{v_{O2}^2}{R_{O2}} = \frac{M}{M+m}m\frac{v_2^2}{r_2} = F_2} \tag{4.1.16}$$

The above formula is the formula for the centripetal force at Aphelion. Based on the above formula, the relation formula can be obtained:

$$\boldsymbol{\rho_2 = \frac{M+m}{M}r_2}$$

From the above equation, the radius $\boldsymbol{\rho_2}$ of the circle of curvature at the aphelion of the Earth's orbit around the Sun is greater than the distance $\boldsymbol{r_2}$ from the aphelion to the Sun.

4.1.4 The centripetal forces at perihelion and aphelion of the Earth's elliptical orbit are not part of the centripetal force of the Earth's revolution about the center O of mass of the solar system, nor are they part of the centripetal force of the Earth's revolution about the Sun

There is no formula given by physicists to calculate the centripetal force at the perihelion and aphelion of an elliptical orbit. According to the theory of curves, to calculate the magnitude of the centripetal force $\boldsymbol{F'_{v1}}$ at the perihelion of an elliptical orbit (or $\boldsymbol{F'_{v2}}$ at the aphelion), one should first determine the circle of curvature and the radius of curvature of the perihelion and aphelion.

Suppose$\boldsymbol{v_1}$ is the tangential velocity at perihelion of the Earth's elliptical orbit and $\boldsymbol{r_1}$ is the distance from perihelion to the Sun. Suppose $\boldsymbol{v_2}$ is the tangential velocity at aphelion of the Earth's elliptical orbit and $\boldsymbol{r_2}$ is the distance from aphelion to the Sun. According to equation (2.5.31), the radii of curvature $\boldsymbol{\rho'}$ at the perihelion and aphelion of the elliptical orbit are:

$$\boldsymbol{\rho' = \frac{b^2}{a} = \frac{2r_1r_2}{r_1+r_2}} \tag{4.1.17}$$

According to the centripetal force formula, the centripetal force $\boldsymbol{F'_{v1}}$ at the perihelion and $\boldsymbol{F'_{v2}}$ at the aphelion of an elliptical orbit are:

$$\begin{cases} \boldsymbol{F'_{v1} = m\frac{v_1^2}{\rho'} = \frac{r_1+r_2}{2r_2}m\frac{v_1^2}{r_1}} \\ \boldsymbol{F'_{v2} = m\frac{v_2^2}{\rho'} = \frac{r_1+r_2}{2r_1}m\frac{v_2^2}{r_2}} \end{cases} \tag{4.1.18}$$

Compare the above equation with equation (4.1.13) and equation (4.1.16). It is clear that the centripetal force at perihelion and aphelion of an elliptical orbit cannot be converted to the centripetal force at perihelion and aphelion of the Earth's revolution around the center O of mass of the solar system because of the constant $\boldsymbol{\frac{r_1+r_2}{2r_2} \neq \frac{M}{M+m}}$ in the above equation.

It should be noted that since the centers of the circles of curvature at perihelion and aphelion of elliptical orbits are neither the Sun nor the center O of mass, the centripetal forces $\boldsymbol{F'_{v1}}$ and $\boldsymbol{F'_{v2}}$ in equation (4.1.18) are not part of the centripetal force of the Earth's revolution around the Sun, nor of the Earth's revolution around the center O of mass of the solar system.

According to equation (4.1.18), the centripetal force $\boldsymbol{F'_{v1}}$ at perihelion contains the constant $\boldsymbol{\frac{r_1+r_2}{2r_2}}$. This means that the centripetal force $\boldsymbol{F'_{v1}}$ at the perihelion of an elliptical orbit is related to the distance $\boldsymbol{r_2}$ from the aphelion to the Sun. Obviously, physicists cannot explain why the centripetal force $\boldsymbol{F'_{v1}}$ at perihelion, is related to the distance $\boldsymbol{r_2}$ from aphelion to the Sun.

4.1.5 The centripetal force F_{v1}' at the perihelion and F_{v2}' at the aphelion of an elliptical orbit does not satisfy the centripetal force equation

According to equation (2.5.16) and equation (2.5.20), the radius ρ_1 of the circle of curvature at perihelion and ρ_2 of the circle of curvature at aphelion of the Earth's orbit around the Sun are:

$$\rho_1 = \frac{M+m}{M} r_1, \qquad \rho_2 = \frac{M+m}{M} r_2 \tag{4.1.19}$$

According to the centripetal force formula, the centripetal force F_{v1} at perihelion and F_{v2} at aphelion of the Earth's orbit around the Sun are:

$$\begin{cases} F_{v1} = m\dfrac{v_1^2}{\rho_1} = m\dfrac{M}{M+m}\dfrac{v_1^2}{r_1} \\ F_{v2} = m\dfrac{v_2^2}{\rho_2} = m\dfrac{M}{M+m}\dfrac{v_2^2}{r_2} \end{cases} \tag{4.1.20}$$

Note that the centripetal forces F_{v1} and F_{v2} satisfy the centripetal force equation. According to equation (4.1.17), the radii ρ' of the circles of curvature at the perihelion and aphelion of the elliptical orbit are:

$$\rho' = \frac{2r_1r_2}{r_1+r_2}$$

According to equation (4.1.18), the centripetal force F'_{v1} at perihelion and F'_{v2} at aphelion of an elliptical orbit are:

$$\begin{cases} F'_{v1} = m\dfrac{v_1^2}{\rho'} = \dfrac{r_1+r_2}{2r_2} m\dfrac{v_1^2}{r_1} \\ F'_{v2} = m\dfrac{v_2^2}{\rho'} = \dfrac{r_1+r_2}{2r_1} m\dfrac{v_2^2}{r_2} \end{cases}$$

Since the radius of the circle of curvature ρ' at the perihelion and aphelion of the elliptical orbit is not equal to the radius of the circle of curvature at the perihelion and aphelion of the Earth's revolution around the Sun, i.e, $\rho' \neq \rho_1$ and $\rho' \neq \rho_2$, the centripetal forces F'_{v1} and F'_{v2} at the perihelion and aphelion of the elliptical orbit do not satisfy the centripetal force formula. We can prove this conclusion theoretically.

The centripetal force F'_{v1} at the perihelion in equation (4.1.18) can be changed to the following form:

$$F'_{v1} = \frac{1}{2}m\frac{v_1^2}{r_1} + \frac{1}{2}m\frac{v_1^2}{r_2} \tag{4.1.21}$$

The above equation shows that the centripetal force F'_{v1} at the perihelion of an elliptical orbit consists of two centripetal force components.

If the circle of curvature is not considered, then the first centripetal force component $\frac{1}{2}m\frac{v_1^2}{r_1}$ to the right of the equals sign satisfies the centripetal force equation.

However. The second centripetal force component $\frac{1}{2}m\frac{v_1^2}{r_2}$, does not satisfy the centripetal force equation because the velocity v_1 and the radius r_2 do not correspond. Furthermore, physicists cannot explain to which type of centripetal force $\frac{1}{2}m\frac{v_1^2}{r_2}$ belongs. From this it can be seen that the centripetal force F'_{v1} at the perihelion of an elliptical orbit does not satisfy the centripetal force formula.

Similarly, the centripetal force F'_{v2} at the aphelion of an elliptical orbit can be changed to the following form:

$$F'_{v2} = \frac{1}{2}m\frac{v_2^2}{r_2} + \frac{1}{2}m\frac{v_2^2}{r_1} \tag{4.1.22}$$

Since the velocity v_2 and the radius r_1 do not correspond, the second centripetal force component $\frac{1}{2}m\frac{v_2^2}{r_1}$ to the right of the equals sign does not satisfy the centripetal force formula. From this it can be seen that the centripetal force F'_{v2} at the aphelion of the elliptical orbit also does not satisfy the centripetal force formula.

4.2 Kinetic energy of the Sun and Earth in revolution about the center O of mass

4.2.1 The kinetic energy E_v of the Earth's orbit around the Sun is greater than the kinetic energy E_{vO} of the Earth's orbit around the center O of mass of the solar system

According to the kinetic energy formula, the kinetic energy $\boldsymbol{E_v}$ of the Earth's revolution around the Sun is:

$$E_v = \frac{1}{2}mv^2 \tag{4.2.1}$$

The $\boldsymbol{v}$ in the equation is the tangential velocity of the Earth's orbit around the Sun. According to equation (3.1.3), the tangential velocity $\boldsymbol{v_O}$ of the Earth's revolution around the center O of mass of the solar system is:

$$v_O = \frac{M}{M+m}v$$

The kinetic energy $\boldsymbol{E_{vO}}$ of the Earth's revolution around the center O of mass of the solar system is:

$$E_{vO} = \frac{1}{2}mv_O^2 = \frac{1}{2}m\left(\frac{M}{M+m}v\right)^2 \tag{4.2.2}$$

The above equation is the kinetic energy equation for the Earth's revolution around the center O of mass of the solar system. Substituting the kinetic energy $\boldsymbol{E_v}$ of the Earth's revolution around the Sun into the above equation, we obtain the relational equation:

$$E_{vO} = \left(\frac{M}{M+m}\right)^2 E_v \tag{4.2.3}$$

Since the constant $\frac{M}{M+m} < 1$, the kinetic energy $\boldsymbol{E_v}$ of the Earth's revolution around the Sun, is greater than the kinetic energy $\boldsymbol{E_{vO}}$ of the Earth's revolution around the solar system's center O of mass.

4.2.2 Sum E_O of the kinetic energies of the Sun and the Earth revolving around the center of mass O

According to equation (3.1.6), the tangential velocity $\boldsymbol{v_S}$ of the Sun's revolution around the center O of mass of the solar system is:

$$v_S = \frac{m}{M+m}v$$

According to the kinetic energy formula, the kinetic energy $\boldsymbol{E_{vS}}$ of the Sun in revolution around the center O of mass of the solar system is:

$$E_{vS} = \frac{1}{2}Mv_S^2 = \frac{1}{2}M\left(\frac{m}{M+m}v\right)^2 \tag{4.2.4}$$

The above equation is the kinetic energy equation for the revolution of the Sun around the center O of mass of the Solar System. The addition of the kinetic energies $\boldsymbol{E_{vO}}$ and $\boldsymbol{E_{vS}}$ of the Earth and the Sun around the center O of mass of the solar system gives the relational equation:

$$E_O = E_{vO} + E_{vS} = \frac{1}{2}m\left(\frac{M}{M+m}\right)^2 v^2 + \frac{1}{2}M\left(\frac{m}{M+m}\right)^2 v^2$$

Simplifying the above equation gives the relational equation:

$$E_O = \frac{1}{2}\frac{M}{M+m}mv^2 \tag{4.2.5}$$

The above equation is the formula for the sum of the kinetic energies of the Sun and the Earth revolving around the center O of mass of the solar system.

Note that the sum of the kinetic energy $\boldsymbol{E_O}$ of the Sun and the Earth around the center O of mass of the solar system is not the kinetic energy $\boldsymbol{E_{vO}}$ of the Earth's revolution around the center O of mass of the solar system, nor is it the kinetic energy of the Sun-Earth system around the Milky Way. Substituting the kinetic energy $\boldsymbol{E_v}$ of the Earth's revolution around the Sun from equation (4.2.1) into the above equation gives the relational equation:

$$E_O = \frac{M}{M+m} E_v$$

Since the constant $\frac{M}{M+m} < 1$, the kinetic energy $E_O < E_v$. This means that the kinetic energy E_v of the Earth's revolution around the Sun, is greater than the sum E_O of the kinetic energies of the Sun's and Earth's revolutions around the center O of mass.

4.2.3 The difference value ΔE_v in kinetic energy between the perihelion and aphelion of the Earth's orbit around the Sun, is greater than the difference value ΔE_{vO} in kinetic energy between the perihelion and aphelion of the Earth's orbit around the center O of mass of the solar system

According to the kinetic energy formula, the kinetic energy E_{v1} at perihelion and E_{v2} at aphelion of the Earth's orbit around the Sun are:

$$E_{v1} = \frac{1}{2} m v_1^2, \qquad E_{v2} = \frac{1}{2} m v_2^2$$

The difference value between the kinetic energy E_{v1} at perihelion and E_{v2} at aphelion of the Earth's revolution around the Sun is

$$\Delta E_v = E_{v1} - E_{v2} = \frac{1}{2} m (v_1^2 - v_2^2) \qquad (4.2.6)$$

According to equation (4.2.3), the kinetic energy E_{vO1} at the perihelion and E_{vO2} at the aphelion of the Earth's revolution around the center O of mass are:

$$E_{vO1} = \left(\frac{M}{M+m}\right)^2 E_{v1}, \qquad E_{vO2} = \left(\frac{M}{M+m}\right)^2 E_{v2}$$

The difference value ΔE_{vO} between the kinetic energy E_{vO1} at perihelion and E_{vO2} at aphelion of the Earth's revolution around the center O of mass is:

$$\Delta E_{vO} = E_{vO1} - E_{vO2} = \left(\frac{M}{M+m}\right)^2 (E_{v1} - E_{v2}) \qquad (4.2.7)$$

Substituting equation (4.2.6) into the above equation gives the relational equation:

$$\Delta E_{vO} = \left(\frac{M}{M+m}\right)^2 \frac{1}{2} m (v_1^2 - v_2^2) = \left(\frac{M}{M+m}\right)^2 \Delta E_v \qquad (4.2.8)$$

The above equation shows that the maximum value ΔE_v of the change in kinetic energy of the Earth's revolution around the Sun, is greater than the maximum value ΔE_{vO} of the change in kinetic energy of the Earth's revolution around the center O of mass.

4.2.4 The kinetic energy of the Earth around the Sun is greater than the sum of the kinetic energies of the Sun and the Earth around the center O of mass

According to equation (4.2.5), the sum E_{O1} of the kinetic energies at the perihelion of the Sun and Earth revolving around the center O of mass is given by:

$$E_{O1} = \frac{1}{2} \frac{M}{M+m} m v_1^2 \qquad (4.2.9)$$

In addition, the sum E_{O2} of the kinetic energies of the Sun and the Earth at the aphelion of the revolution around the center O of mass is:

$$E_{O2} = \frac{1}{2} \frac{M}{M+m} m v_2^2 \qquad (4.2.10)$$

The difference value ΔE_O in kinetic energy between the perihelion and aphelion of the Sun and Earth revolving around the center O of mass is:

$$\Delta E_O = E_{O1} - E_{O2} = \frac{M}{M+m} \cdot \frac{1}{2} m (v_1^2 - v_2^2) \qquad (4.2.11)$$

Substituting equation (4.2.6) into the above equation gives the relational equation:

$$\Delta E_O = \frac{M}{M+m}\Delta E_v \tag{4.2.12}$$

From the above equation, the maximum value ΔE_v of the change in kinetic energy of the Earth's revolution around the Sun, is greater than the maximum value ΔE_O of the change in kinetic energy of the Sun and the Earth's revolution around the center O of mass.

To summarize: The maximum value ΔE_v of the kinetic energy change of the Earth's revolution around the Sun, is greater than the maximum value ΔE_{vO} of the kinetic energy change of the Earth's revolution around the center O of mass, and greater than the maximum value ΔE_O of the kinetic energy change of the Sun's and Earth's revolutions around the center O of mass.

4.3 Newton's Formula for Universal Gravitation Contradicts the Law of Conservation of Mechanical Energy

4.3.1 The mechanical energy W_1 at perihelion of the Earth's orbit around the Sun, is equal to the mechanical energy W_2 at aphelion

As the Earth moves from aphelion to perihelion, the work done by the Sun's gravitational force is positive, because the work done by the Sun's gravitational force makes the Earth's perihelion velocity v_1 greater. The work done A' by Newton's gravitational force F' is:

$$A' = \int F'dl = -\int_{r_2}^{r_1} G\frac{Mm}{r^2}dr = GMm\left(\frac{1}{r_1} - \frac{1}{r_2}\right) \tag{4.3.1}$$

The work A' done by gravitational is the same whether the Earth rotates around the Sun or around the center O of mass. In other words, the work A' done by gravitational is unique.

Suppose v_1 is the tangential velocity of perihelion of the Earth's orbit around the Sun; v_2 is the tangential velocity of aphelion. According to the kinetic energy formula, the kinetic energy E_{v1} at perihelion and E_{v2} at aphelion of the Earth's orbit around the Sun are:

$$E_{v1} = \frac{1}{2}mv_1^2\ , \qquad E_{v2} = \frac{1}{2}mv_2^2$$

As the Earth moves from aphelion to perihelion, the difference value between the kinetic energy E_{v1} at perihelion and E_{v2} at aphelion is:

$$\Delta E_v = E_{v1} = E_{v2} = \frac{1}{2}m(v_1^2 - v_2^2) \tag{4.3.2}$$

According to Newton's formula for gravitational potential energy, the gravitational potential energy E'_{P1} for perihelion and E'_{P2} for aphelion are as follows:

$$\begin{cases} E'_{P1} = \int_{\infty}^{r_1} G\frac{Mm}{r_1^2}dr = -\frac{GMm}{r_1} \\ E'_{P2} = \int_{\infty}^{r_2} G\frac{Mm}{r_2^2}dr = -\frac{GMm}{r_2} \end{cases} \tag{4.3.3}$$

According to the mechanical energy formula, the mechanical energy W_1 at the perihelion of the Earth's orbit around the Sun is:

$$W_1 = E_{v1} + E'_{P1} = \frac{1}{2}mv_1^2 - \frac{GMm}{r_1} \tag{4.3.4}$$

The mechanical energy W_2 at the aphelion of the Earth's orbit around the Sun is:

$$W_2 = E_{v2} + E'_{P2} = \frac{1}{2}mv_2^2 - \frac{GMm}{r_2} \tag{4.3.5}$$

According to the law of conservation of mechanical energy, the mechanical energy W_1 at perihelion is equal to the mechanical energy W_2 at aphelion, i.e:

$$W_1 - W_2 = \frac{1}{2}m(v_1^2 - v_2^2) - GMm\left(\frac{1}{r_1} - \frac{1}{r_2}\right) = 0 \qquad (4.3.6)$$

The above equation shows that the maximum value $\frac{1}{2}m(v_1^2 - v_2^2)$ of the change in kinetic energy of the Earth's revolution around the Sun, is equal to the work done by the gravitational force, i.e:

$$\frac{1}{2}m(v_1^2 - v_2^2) = GMm\left(\frac{1}{r_1} - \frac{1}{r_2}\right) = A' \qquad (4.3.7)$$

Since the velocities v_1 and v_2 belong to the tangential velocities at perihelion and aphelion of the Earth's orbit around the Sun, we can use the above equation to prove that elliptical orbits do not satisfy the law of conservation of mechanical energy.

4.3.2 Elliptical orbits do not obey the law of conservation of mechanical energy

Since the centripetal force at both perihelion and aphelion is equal to the gravitational force, the centripetal force F'_{v1} at perihelion of an elliptical orbit, is equal to the gravitational force at perihelion, and the centripetal force F'_{v2} at aphelion, is equal to the gravitational force at aphelion, viz:

$$\begin{cases} F'_{v1} = m\dfrac{v_1^2}{\rho'} = \dfrac{GMm}{r_1^2} \\ F'_{v2} = m\dfrac{v_2^2}{\rho'} = \dfrac{GMm}{r_2^2} \end{cases} \qquad (4.3.8)$$

Based on the above equation, the relationship equation can be obtained:

$$\begin{cases} \dfrac{1}{2}mv_1^2 = \dfrac{1}{2}\dfrac{GMm}{r_1^2}\rho' \\ \dfrac{1}{2}mv_2^2 = \dfrac{1}{2}\dfrac{GMm}{r_2^2}\rho' \end{cases} \qquad (4.3.9)$$

According to the above equation, the difference value between the perihelion kinetic energy and the aphelion kinetic energy is:

$$\frac{1}{2}m(v_1^2 - v_2^2) = \frac{1}{2}GMm\rho\left(\frac{1}{r_1^2} - \frac{1}{r_2^2}\right) \qquad (4.3.10)$$

According to equation (4.1.17), the radii ρ' of the circles of curvature at the perihelion and aphelion of the elliptical orbit are:

$$\rho' = \frac{2r_1r_2}{r_1 + r_2}$$

Substituting the above equation into equation (4.3.10) gives the relational equation:

$$\frac{1}{2}m(v_1^2 - v_2^2) = GMm\frac{r_1r_2}{r_1 + r_2}\left(\frac{1}{r_1^2} - \frac{1}{r_2^2}\right) \qquad (4.3.11)$$

Comparing the above equation with equation (4.3.7), we can obtain the relational equation which is:

$$\frac{1}{2}m(v_1^2 - v_2^2) = GMm\left(\frac{1}{r_1} - \frac{1}{r_2}\right) \neq GMm\frac{r_1r_2}{r_1 + r_2}\left(\frac{1}{r_1^2} - \frac{1}{r_2^2}\right) \qquad (4.3.12)$$

The above equation shows that the centripetal force in an elliptical orbit does not satisfy the law of conservation of mechanical energy.

4.3.3 Mechanical energy of the Sun and Earth in revolution around the center of mass O

As the Earth moves from perihelion to aphelion, the difference value in kinetic energy ΔE_v between perihelion and aphelion of the Earth's revolution around the Sun is given by equation (4.3.2):

$$\Delta E_v = E_{v1} - E_{v2} = \frac{1}{2}m(v_1^2 - v_2^2)$$

According to equation (4.3.6), the difference value $\Delta W = W_1 - W_2$ between the mechanical energy at perihelion and aphelion of the Earth's orbit around the Sun is given by:

$$\Delta W = W_1 - W_2 = \frac{1}{2}m(v_1^2 - v_2^2) - Gm\left(\frac{1}{r_1} - \frac{1}{r_2}\right) = 0$$

According to equation (4.2.11), the difference value ΔE_O between the kinetic energies at perihelion and aphelion of the Sun and Earth revolving around the center O of mass is given by:

$$\Delta E_O = E_{O1} - E_{O2} = \frac{M}{M+m} \cdot \frac{1}{2}m(v_1^2 - v_2^2) = \frac{M}{M+m}\Delta E_v$$

The above equation shows that the difference value ΔE_v in kinetic energy is greater than the difference value ΔE_O in kinetic energy.

According to the law of conservation of mechanical energy, the difference value $\Delta W_O = W_{O1} - W_{O2}$ between the mechanical energy at perihelion and aphelion of the Sun and Earth revolving around the center O of mass of the solar system is given by:

$$\Delta W_O = W_{O1} - W_{O2} = \frac{M}{M+m} \cdot \frac{1}{2}m(v_1^2 - v_2^2) - GMm\left(\frac{1}{r_1} - \frac{1}{r_2}\right) = 0$$

According to the law of conservation of energy, the difference value ΔW_O in mechanical energy between points A and B of the planets' orbits is unique. Since both the Sun and the planets revolve around the center O of mass of the solar system, the difference value in mechanical energy should be the difference value $\Delta W_O = W_{O1} - W_{O2}$ in mechanical energy between the Sun and the Earth revolving around the center O of mass, and not the difference value $\Delta W = W_1 - W_2$ in mechanical energy between the Earth's revolutions around the Sun. We can prove this conclusion by the following analysis.

Assume that the Earth and the Sun have equal masses and that the center O of mass is located at the midpoint of the distance from the Sun to the Earth. Since the Sun and the Earth revolve around the center O of mass with equal and opposite velocities v, the kinetic energy E_O of the Sun and the Earth revolving around the center O of mass is:

$$E_O = \frac{1}{2}mv^2 + \frac{1}{2}mv^2 = mv^2 \tag{4.3.13}$$

Since the sun and the earth are moving in opposite directions, the relative speed of the earth's orbit around the sun is $2v$. The kinetic energy E_v of the earth's orbit around the sun is:

$$E_v = \frac{1}{2}m(2v)^2 = 2mv^2 \tag{4.3.14}$$

Obviously, the kinetic energy E_v is greater than the kinetic energy E_O. From this it can be determined that the mechanical energy in the law of conservation of mechanical energy should be the mechanical energy of the Sun and Earth system revolving around the center O of mass, not the mechanical energy of the Earth revolving around the Sun.

4.3.4 Newton's Formula for Universal Gravitation Contradicts the Law of Conservation of Mechanical Energy

At the moment of perihelion, according to equation (4.2.9), the kinetic energy E_{O1} of the Sun and Earth revolving around the center O of mass is:

$$E_{O1} = \frac{1}{2}\frac{M}{M+m}mv_1^2$$

At the moment of perihelion, the mechanical energy W_{O1} of the Earth and Sun revolving around the center O of mass is:

$$\Delta W_O = W_{O1} = E_{O1} + +E'_{P1} = \frac{1}{2}\frac{M}{M+m}mv_1^2 - \frac{GMm}{r_1} \tag{4.3.15}$$

At the moment of aphelion, according to equation (4.2.10), the kinetic energy E_{O2} of the Sun and Earth revolving around the center O of mass is:

$$E_{O2} = \frac{1}{2}\frac{M}{M+m}mv_2^2$$

At the moment of aphelion, the mechanical energy W_{O2} of the Sun and Earth revolving around the center O of mass is:

$$W_{O2}=\frac{1}{2}\frac{M}{M+m}mv_2^2-\frac{GMm}{r_2} \tag{4.3.16}$$

According to the law of conservation of mechanical energy, the mechanical energy W_{O1} of the Sun and Earth at the moment of perihelion is equal to the mechanical energy W_{O2} of the Sun and Earth at the moment of aphelion, i.e:

$$\Delta W_O=W_{O1}-W_{O2}=\frac{1}{2}\frac{Mm}{M+m}(v_1^2-v_2^2)-GMm\left(\frac{1}{r_1}-\frac{1}{r_2}\right)=0 \tag{4.3.17}$$

The difference value ΔE_O between the kinetic energy of the perihelion and aphelion of the Sun and Earth revolving around the center O of mass is:

$$\Delta E_O=\frac{1}{2}\frac{Mm}{M+m}(v_1^2-v_2^2)$$

According to equation (4.3.17), the difference value ΔE_O in kinetic energy, is equal to the work $GMm\left(\frac{1}{r_1}-\frac{1}{r_2}\right)$, done by Newton's gravitational force F', i.e. $\Delta E_O=GMm\left(\frac{1}{r_1}-\frac{1}{r_2}\right)$. Since the tangential velocities v_1 and v_2, of perihelion and aphelion are perpendicular to the distance from the Earth to the Sun, we can verify that the formula for Newton's gravitational force is correct or incorrect in terms of the centripetal forces of the Sun and Earth around the center O of mass.

According to equation (4.1.10), the centripetal force F_{vO1} at the perihelion of the Earth's revolution around the center O of mass of the solar system is:

$$F_{vO1}=m\frac{v_{O1}^2}{R_{O1}}=m\frac{M}{M+m}\cdot\frac{v_1^2}{r_1}$$

Since the centripetal force F_{vO1} at perihelion is equal to the gravitational force F_1' at perihelion, the relation equation can be obtained:

$$F_{vO1}=m\frac{M}{M+m}\cdot\frac{v_1^2}{r_1}=G\frac{Mm}{r_1^2} \tag{4.3.18}$$

According to the above equation, the square of the velocity v_1 at the perihelion is:

$$v_1^2=\frac{G(M+m)}{r_1} \tag{4.3.19}$$

According to equation (4.1.14), the centripetal force F_{vO2} at the aphelion of the Earth's revolution around the center O of mass of the solar system is:

$$F_{vO2}=m\frac{v_{O2}^2}{R_{O2}}=m\frac{M}{M+m}\cdot\frac{v_2^2}{r_2}$$

Since the centripetal force F_{vO2} at the aphelion is equal to the gravitational force F_2' at the aphelion, the relation equation can be obtained:

$$F_{vO2}=m\frac{M}{M+m}\cdot\frac{v_2^2}{r_2}=G\frac{Mm}{r_2^2} \tag{4.3.20}$$

According to the above equation, the square of the velocity v_2 at the aphelion is:

$$v_2^2=\frac{G(M+m)}{r_2} \tag{4.3.21}$$

Substituting the above equation and equation (4.3.19) into equation (4.3.17) gives the relationship equation:

$$\Delta W_O=W_{O1}-W_{O2}=\frac{1}{2}\frac{Mm}{M+m}G(M+m)\left(\frac{1}{r_1}-\frac{1}{r_2}\right)-GMm\left(\frac{1}{r_1}-\frac{1}{r_2}\right)$$

Simplifying the above equation gives the relational equation:

$$\Delta W_O=W_{O1}-W_{O2}=-\frac{1}{2}GMm\left(\frac{1}{r_1}-\frac{1}{r_2}\right)\neq 0 \tag{4.3.22}$$

The above equation shows that the mechanical energy W_{O1} at perihelion is not equal to the mechanical energy W_{O2} at aphelion, i.e. $W_{O1}\neq W_{O2}$. From this it can be seen that Newton's formula for gravitational contradicts the law of conservation of mechanical energy.

Chapter 5 Conservation theorem for tangential angular momentum

Introduction: According to Newtonian mechanics, the law of conservation of angular momentum is:

$$L_v = rmv\sin\theta = C$$

According to the curve motion law of the mass point, the tangential velocity v of the According to the law of curvilinear motion of the mass point, the tangential velocity of the mass point can be decomposed into transverse and radial velocities, i.e:

$$v = \sqrt{v_\varphi^2 + v_r^2} = \sqrt{(v\sin\theta)^2 + (v\cos\theta)^2}$$

Multiplying the above equation by the distance r and the mass m gives the relationship equation:

$$rmv = \sqrt{(rmv\sin\theta)^2 + (rmv\cos\theta)^2}$$

Since the above equation includes angular momentum $L_v = rmv\sin\theta$, the conservation of angular momentum is false.

5.1 Tangential angular momentum H_v of the Earth around the Sun

5.1.1 Definition of tangential angular momentum H_v

Suppose that v is the tangential velocity of the mass m rotating about the origin O, and r is the distance from the mass m to the origin O. Suppose that θ is the angle between the tangential velocity v and the distance r. Both the tangential velocity v and the distance r are in the plane OXY. This is shown in Figure 5-1.

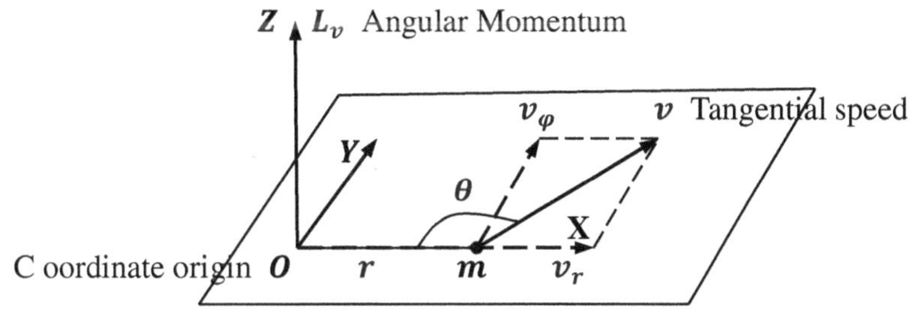

Figure 5-1 Diagram of angular momentum L_v, tangential velocity v, transverse velocity v_φ, and radial velocity v_r of a mass m rotating about the origin O.

When the combined force on the mass m is zero. According to the conservation of angular momentum, the angular momentum L_v of the mass m rotating about the origin O is:

$$L_v = rmv\sin\theta = C \quad (5.1.1)$$

The C in the formula is a constant. The angular momentum L_v is at the origin O. The direction of the angular momentum L_v is perpendicular to the plane defined by both the velocity v and the distance r. This is shown in Figure 5-1.

The law of conservation of angular momentum is a constraint that must be satisfied for the motion of celestial bodies. Through theoretical analysis, the authors found that the law of conservation of angular momentum contradicts the formula for calculating velocity.

In order to analyze and study the laws of motion of the planets, the authors define the product $\boldsymbol{rmv}$ as the tangential angular momentum $\boldsymbol{H_v}$, viz:

$$\boldsymbol{H_v = rmv} \tag{5.1.2}$$

The above formula is the tangential angular momentum formula. $\boldsymbol{H_v}$ in the formula is the tangential angular momentum of the mass $\boldsymbol{m}$ rotating around the origin O.

According to equation (2.5.2), the tangential velocity $\boldsymbol{v}$ of the Earth $\boldsymbol{m}$ revolving around the Sun can be decomposed into the transverse velocity $\boldsymbol{v_\varphi}$ and radial velocity $\boldsymbol{v_r}$, viz:

$$\boldsymbol{v = \sqrt{v_\varphi^2 + v_r^2} = \sqrt{(v\sin\theta)^2 + (v\cos\theta)^2}} \tag{5.1.3}$$

Substituting the above equation into equation (5.1.2) gives the relationship equation:

$$\boldsymbol{H_v = rmv = \sqrt{(rmv_\varphi)^2 + (rmv_r)^2} = \sqrt{(rmv\sin\theta)^2 + (rmv\cos\theta)^2}} \tag{5.1.4}$$

The above equation shows that the tangential angular momentum $\boldsymbol{H_v = rmv}$ can be decomposed into two perpendicular components.

Because of the transverse velocity $\boldsymbol{v_\varphi = v\sin\theta}$, the authors define $\boldsymbol{rmv\sin\theta}$ as the transverse angular momentum $\boldsymbol{H_{v\varphi}}$, viz:

$$\boldsymbol{H_{v\varphi} = rmv\sin\theta} \tag{5.1.5}$$

Because of the radial velocity $\boldsymbol{v_r = v\cos\theta}$, the authors define $\boldsymbol{rmv\cos\theta}$ as the radial angular momentum $\boldsymbol{H_{vr}}$, viz:

$$\boldsymbol{H_{vr} = rmv\cos\theta} \tag{5.1.6}$$

5.1.2 Position and direction of tangential angular momentum H_v

According to classical mechanics, the formula for angular momentum is:

$$\boldsymbol{L_v = rmv\sin\theta}$$

In classical mechanics, the angular momentum $\boldsymbol{L_v}$ is located at the origin O (center of rotation). The angular momentum $\boldsymbol{L_v}$ is perpendicular to the plane defined by the velocity $\boldsymbol{v}$ and the radius $\boldsymbol{r}$, as shown in Figure 5-1.

It is clear that the mathematical expressions for both the transverse angular momentum $\boldsymbol{H_{v\varphi}}$ and the angular momentum $\boldsymbol{L_v}$ are identical, i.e:

$$\boldsymbol{L_v = H_{v\varphi} = rmv\sin\theta} \tag{5.1.7}$$

Since tangential angular momentum $\boldsymbol{H_v}$ contains two variables, velocity $\boldsymbol{v}$ and distance $\boldsymbol{r}$, according to the classical theory, tangential angular momentum $\boldsymbol{H_v}$ should be located at the origin O. And the direction of tangential angular momentum $\boldsymbol{H_v}$ is perpendicular to the plane defined by both velocity $\boldsymbol{v}$ and distance $\boldsymbol{r}$.

It is important to note that because of the radial velocity $\boldsymbol{v_r = v\cos\theta}$, the radial angular momentum $\boldsymbol{H_{vr} \neq 0}$. And the angle between the radial angular momentum $\boldsymbol{H_{vr}}$ and the tangential angular momentum $\boldsymbol{H_v}$ is $(180^0 - \boldsymbol{\theta})$. This is shown in Figure 5-2.

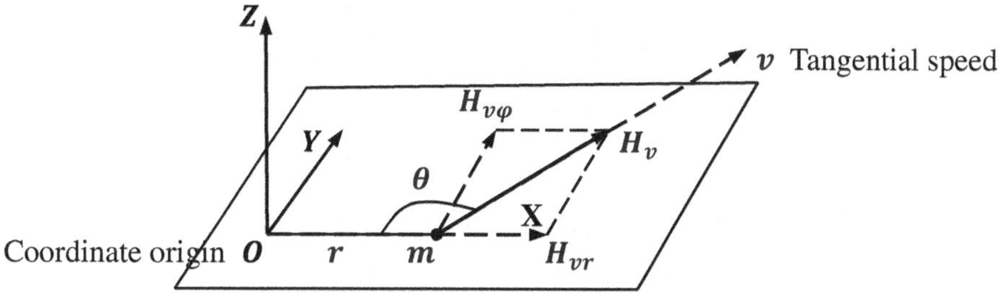

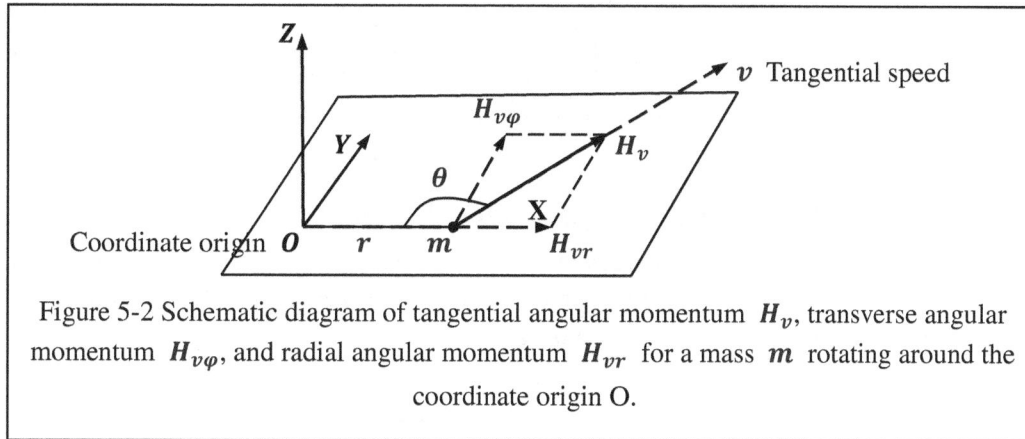

Figure 5-2 Schematic diagram of tangential angular momentum H_v, transverse angular momentum $H_{v\varphi}$, and radial angular momentum H_{vr} for a mass m rotating around the coordinate origin O.

However, if the tangential angular momentum H_v is located at the origin O and is perpendicular to the plane defined by both the velocity v and the distance r, then the radial angular momentum $H_{vr} = 0$, since the angle between the radial angular momentum H_{vr} and the tangential angular momentum H_v is 90^0. However, the radial angular momentum $H_{vr} \neq 0$.

From this, it can be determined that the tangential angular momentum H_v should not be at the origin O, but should be located at the mass m. The direction of the tangential angular momentum H_v is the same as the direction of the tangential velocity v. as shown in Figure 5-2.

According to the rules of vector arithmetic, the product of vector a and vector b is of two types.

One type is the dot product $a \cdot b = ab\cos\theta$ of vector a and vector b. The dot product $a \cdot b$ of a vector is a scalar.

The other type is the forked product $a \times b = ab\sin\theta$ of vector a and vector b. The forked product $a \times b$ of a vector is a vector. The fork product $a \times b$ of a vector is perpendicular to the plane defined by the vector a and the vector b.

In vectorial operations, the tangential angular momentum $H_v = rmv$ is not part of the dot product $a \cdot b$ of the vectors, i.e:

$$rmv \neq mr \cdot v = mrv\cos\theta \tag{5.1.8}$$

The tangential angular momentum $H_v = rmv$ is also not part of the fork product $a \times b$ of the vectors, i.e:

$$rmv \neq mr \times v = mrv\sin\theta \tag{5.1.9}$$

Since the direction of tangential angular momentum H_v is the same as the direction of tangential velocity v, the distance r in tangential angular momentum H_v can only be a scalar.

It should be noted that if the distance r is considered as a vector, two unexplained problems arise.

The first open question is to which type of vector product does the tangential angular momentum $H_v = rmv$ belong?

Tangential angular momentum $H_v = rmv$ is the product of distance r and velocity v. If distance r is considered a vector, then tangential angular momentum $H_v = rmv$ is neither the dot product $a \cdot b = ab\cos\theta$ of vectors nor the fork product $a \times b = ab\sin\theta$ of vectors.

The second unexplained problem is that the radial angular momentum $H_{vr} = rmv\cos\theta$ is a vector, not a scalar.

According to the rules of vector arithmetic, the radial angular momentum $H_{vr} = rmv\cos\theta$ should be a scalar. However, since the radial angular momentum H_{vr} always points along the distance r to the origin O (or in the opposite direction), the radial angular momentum H_{vr} is a vector, not a scalar.

The two difficult questions above can be avoided if the distance r is considered as a scalar.

5.1.3 Tangential angular momentum H_{vO} of the Earth around the center O of mass of the solar system

According to equation (2.2.2), the tangential velocity v_O of the Earth's rotation around the center O of mass can be decomposed into the transverse velocity $v_{O\varphi}$ and the radial velocity v_{OR}, i.e:

$$v_O = \sqrt{(v_{O\varphi})^2 + (v_{OR})^2} \tag{5.1.10}$$

Multiplying the above equation by the Earth's mass m and the distance R_O from the Earth to the center O of mass gives the relational equation:

$$H_{vO} = R_O m v_O = \sqrt{(R_O m v_{O\varphi})^2 + (R_O m v_{OR})^2} \tag{5.1.11}$$

The above equation is the formula for the tangential angular momentum of the Earth's revolution about the center O of mass. H_{vO} in the equation is the tangential angular momentum of the Earth's revolution about the center O of mass. The above equation shows that the tangential angular momentum H_{vO} can be decomposed into two components. This is shown in Figure 5-3.

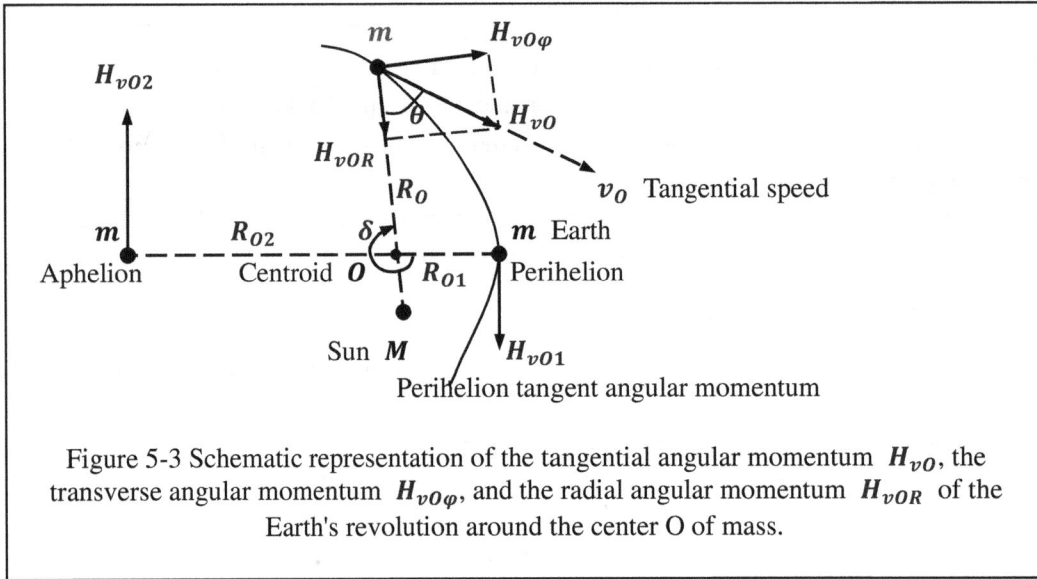

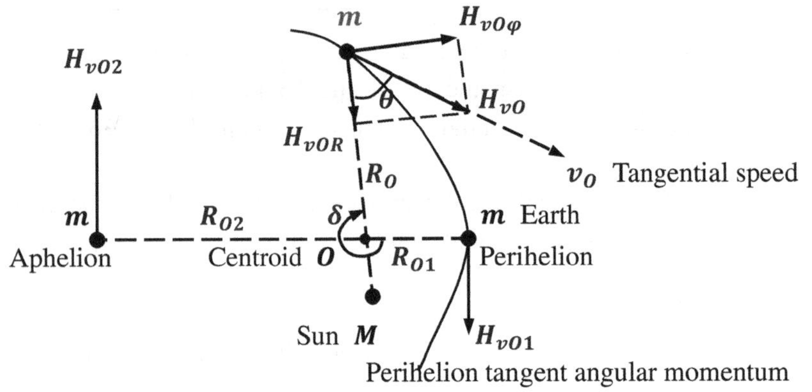

Figure 5-3 Schematic representation of the tangential angular momentum H_{vO}, the transverse angular momentum $H_{vO\varphi}$, and the radial angular momentum H_{vOR} of the Earth's revolution around the center O of mass.

According to equation (2.2.4), the transverse velocity $v_{O\varphi} = v_O \sin\theta$ of the Earth's rotation around the center O of mass. This book defines $R_O m v_{O\varphi}$ in the root symbol of equation (5.1.11) as the transverse angular momentum $H_{vO\varphi}$ of the Earth's rotation around the center O of mass, i.e:

$$H_{vO\varphi} = R_O m v_{O\varphi} = R_O m v_O \sin\theta \tag{5.1.12}$$

The above equation is the equation for the transverse angular momentum of the Earth as it revolves around the center O of mass. The transverse angular momentum $H_{vO\varphi}$ is located on the Earth. The direction of the transverse angular momentum $H_{vO\varphi}$ is perpendicular to the distance R_O from the Earth to the center O of mass. This is shown in Figure 5-3.

According to equation (2.2.3), the radial velocity $v_{OR} = v_O \cos\theta$ of the Earth's rotation around the center O of mass. This book defines $R_O m v_{OR}$ in the root symbol of equation (5.1.11) as the radial angular momentum H_{vOR} of the Earth's rotation around the center O of mass, i.e:

$$H_{vOR} = R_O m v_{OR} = R_O m v_O \cos\theta \tag{5.1.13}$$

The above equation is the formula for the radial angular momentum of the Earth as it revolves around the center O of mass. The radial angular momentum H_{vOR} is located on the Earth. The direction of the radial angular momentum H_{vOR} is parallel to the distance R_O from the Earth to the center O of mass. This is shown in Figure 5-3.

The vectorial sum of the transverse angular momentum $H_{vO\varphi}$ and the radial angular momentum H_{vOR} is equal to the tangential angular momentum H_{vO} of the Earth's rotation around the center O of mass, i.e:

$$H_{vO} = \sqrt{H_{vO\varphi}^2 + H_{vOR}^2} \tag{5.1.14}$$

When the angle between the velocity v_O and the distance R_O is $\theta = 90^0$, Due to radial angular momentum

$H_{vOR} = R_O m v_O \cos\theta = 0$, therefore transverse angular momentum $H_{vO\varphi} = H_{vO}$. At this point, the tangential angular momentum H_{vO} is perpendicular to the distance R_O.

5.1.4 Tangential angular momentum H_S of the Sun in revolution about the center O of mass of the solar system

According to equation (2.4.3), the tangential velocity v_S of the Sun's revolution around the center O of mass of the solar system can be decomposed into a transverse velocity $v_{S\varphi}$ and a radial velocity v_{SR}, namely:

$$v_S = \sqrt{v_{S\varphi}^2 + v_{SR}^2} \qquad (5.1.15)$$

Multiply the above equation by the distance R_S and the solar mass M to obtain the relation:

$$H_{vS} = R_S M v_S = \sqrt{\left(R_S M v_{S\varphi}\right)^2 + (R_S M v_{SR})^2} \qquad (5.1.16)$$

The above formula is the formula for the tangential angular momentum of the sun as it revolves around the center O of mass of the solar system. H_{vS} in the formula is the tangential angular momentum of the sun as it revolves around the center O of mass of the solar system, and R_S is the distance from the sun to the center O of mass. The above equation shows that the tangential angular momentum H_{vS} can be decomposed into two mutually perpendicular components.

The transverse velocity $v_{S\varphi} = v_S \sin\theta$ of the Sun in revolution around the center O of mass of the solar system according to equation (2.4.2). This book defines $R_S m v_{S\varphi}$ in the root symbol of equation (5.1.16) as the transverse angular momentum $H_{vS\varphi}$ of the Sun, i.e:

$$H_{vS\varphi} = R_S m v_{S\varphi} = R_S m v_S \sin\theta \qquad (5.1.17)$$

The above equation is the formula for the angular momentum of the Sun as it revolves around the center O of mass of the solar system. The transverse angular momentum $H_{vS\varphi}$ is located on the Sun. The direction of the angular momentum $H_{vS\varphi}$ is perpendicular to the distance R_S from the Sun to the center O of mass of the solar system.

According to equation (2.4.1), the radial velocity of the sun's revolution around the center O of mass of the solar system is $v_{SR} = v_S \cos\theta$. This book defines $R_S m v_{SR}$ in the root symbol of equation (5.1.16) as the radial angular momentum H_{vSR} of the sun's revolution around the center O of mass of the solar system, i.e:

$$H_{vSR} = R_S m v_{SR} = R_S m v_S \cos\theta \qquad (5.1.18)$$

The above equation is the formula for the radial angular momentum of the Sun. The radial angular momentum H_{vSR} is located on the sun. The direction of the radial angular momentum H_{vSR} is parallel to the distance R_S from the Sun to the center O of mass of the solar system.

The vectorial sum of the sun's transverse angular momentum $H_{vS\varphi}$ and radial angular momentum H_{vSR} is equal to the sun's tangential angular momentum H_{vS}, i.e:

$$H_{vS} = \sqrt{H_{vS\varphi}^2 + H_{vSR}^2} \qquad (5.1.19)$$

When the angle $\theta = 90^0$ between the velocity v_S and the distance R_S. Due to the radial angular momentum $H_{vSR} = R_S m v_S \cos\theta = 0$ of the sun, the transverse angular momentum $H_{vS\varphi} = H_{vS}$ of the sun. At this time the tangential angular momentum H_{vS} is perpendicular to the distance R_S.

5.1.5 The sum H_O of the tangential angular momentum of the Sun and Earth revolving around the center O of mass of the solar system

According to equation (3.1.3), the tangential velocity v_O of the Earth's rotation around the center O of mass is:

$$v_O = \frac{M}{M+m} v$$

According to equation (2.1.7), the distance R_O from the Earth to the center O of mass is:

$$R_O = \frac{M}{M+m} r$$

Multiplying the distance R_O by the speed v_O and then by the mass m of the Earth gives the relationship equation:

$$H_{vO} = R_O m v_O = \left(\frac{M}{M+m}\right)^2 rmv \qquad (5.1.20)$$

The above equation is the equation for the tangential angular momentum of the Earth's revolution around the center O of mass. Substituting equation (5.1.2) into the above equation gives the relational equation:

$$H_{vO} = \left(\frac{M}{M+m}\right)^2 H_v \qquad (5.1.21)$$

The above equation shows that the tangential angular momentum H_v of the Earth's orbit around the Sun is greater than the tangential angular momentum H_{vO} of the Earth's orbit around the center O of mass of the solar system.

According to equation (3.1.6), the tangential velocity v_S of the Sun's revolution around the center O of mass of the solar system is:

$$v_S = \frac{m}{M+m} v$$

According to equation (2.1.9), the distance R_S from the Sun to the center O of mass is:

$$R_S = \frac{m}{M+m} r$$

Multiplying the distance R_S, by the velocity v_S, and then by the solar mass M, gives the relational equation:

$$H_{vS} = R_S M v_S = \left(\frac{m}{M+m}\right)^2 rMv \qquad (5.1.22)$$

The above equation is the formula for the tangential angular momentum of the Sun as it revolves around the center O of mass of the solar system. Since the tangential velocity v_S of the Sun is in the opposite direction to the tangential velocity v_O of the Earth, the tangential angular momentums H_{vO} and H_{vS} are in the opposite direction. Substituting the above equation into equation (5.1.20) gives the following equation:

$$H_O = H_{vO} + H_{vS} = \frac{Mm}{M+m} rv \qquad (5.1.23)$$

The above formula is the formula for the sum of the tangential angular momentum of the Sun and the Earth revolving around the center O of mass of the solar system. Substituting the tangential angular momentum $H_v = rmv$ of the Earth's revolution around the Sun into the above formula gives the relational formula:

$$H_O = \frac{M}{M+m} H_v \qquad (5.1.24)$$

Because of the coefficient $\frac{M}{M+m} < 1$, the sum of the tangential angular momentum H_O is smaller than the tangential angular momentum H_v of the earth revolving around the sun, i.e. $H_O < H_v$.

Divide equation (5.1.20) by equation (5.1.22) to obtain the relational equation:

$$\frac{H_{vO}}{H_{vS}} = \frac{M}{m} \qquad (5.1.25)$$

From the above equation, the ratio of the tangential angular momentum H_{vO} of the Earth's revolution around the center O of mass of the solar system to the tangential angular momentum H_{vS} of the Sun's revolution around the center O of mass of the solar system is a constant.

5.1.6 Tangential angular momentum H_v of the Earth around the Sun

According to equation (5.1.4), the tangential angular momentum H_v of the Earth's revolution around the Sun is:

$$H_v = rmv = \sqrt{(rmv\,\text{si}\,\theta)^2 + (rmv\cos\theta)^2} \qquad (5.1.26)$$

The above equation is the formula for the tangential angular momentum of the Earth's orbit around the Sun. The above equation shows that the tangential angular momentum H_v can be decomposed into the transverse angular momentum $H_{v\varphi} = rmv\sin\theta$ and the radial angular momentum $H_{vr} = rmv\cos\theta$. This is shown in Figure 5-4.

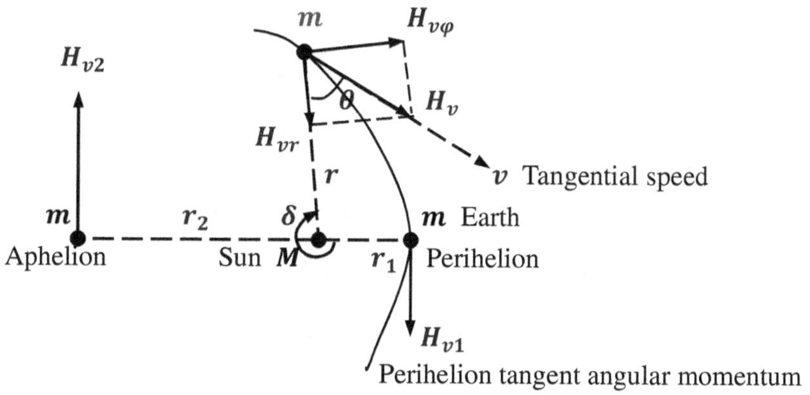

Figure 5-4 Schematic representation of the tangential angular momentum H_v, the transverse angular momentum $H_{v\varphi}$, and the radial angular momentum H_{vr} of the Earth's orbit around the Sun.

The transverse angular momentum $H_{v\varphi}$ of the Earth's orbit around the Sun is:

$$H_{v\varphi} = rmv_{\varphi} = rmv\sin\theta \tag{5.1.27}$$

The above equation is the formula for the angular momentum of the Earth's revolution around the Sun. The transverse angular momentum $H_{v\varphi}$ is on the Earth. The direction of the angular momentum $H_{v\varphi}$ is perpendicular to the distance r between the Earth and the Sun. It is shown in Figure 5-4.

The radial angular momentum H_{vr} of the Earth's orbit around the Sun is:

$$H_{vr} = rmv_r = rmv\cos\theta \tag{5.1.28}$$

The above equation is the formula for the radial angular momentum of the Earth's orbit around the Sun. The radial angular momentum H_{vr} is located on the Earth. The direction of the radial momentum H_{vr}, is parallel to the distance r from the Earth to the Sun. This is shown in Figure 5-4.

The vector sum of the transverse angular momentum $H_{v\varphi}$ and the radial angular momentum H_{vr} of the Earth's orbit around the Sun is:

$$H_v = \sqrt{H_{v\varphi}^2 + H_{vr}^2} \tag{5.1.29}$$

When the angle $\theta = 90^0$ between the velocity v and the distance r. Due to the radial angular momentum $H_{vr} = 0$, therefore, the transverse angular momentum $H_{v\varphi} = H_v$. At this time, the tangential angular momentum H_v is perpendicular to the distance r.

5.2 Conservation equation for tangential angular momentum

5.2.1 Tangential angular momentum H_{v1} at perihelion of the Earth's orbit around the Sun

Suppose v_1 is the tangential velocity at perihelion of the Earth's orbit around the Sun, and r_1 is the distance from perihelion to the Sun; suppose θ_1 is the angle between velocity v_1 and distance r_1. Since the tangential velocity v_1 at perihelion is perpendicular to the distance r_1, the angle $\theta_1 = 90^0$.

According to equation (5.1.26), the tangential angular momentum H_{v1} at perihelion of the Earth's orbit around the Sun is:

$$H_{v1} = r_1 m v_1 \tag{5.2.1}$$

Note that the tangential angular momentum H_{v1} is on the Earth. The direction of tangential momentum H_{v1} is the same as the direction of velocity v_1 at perihelion, as shown in Figure 5-5.

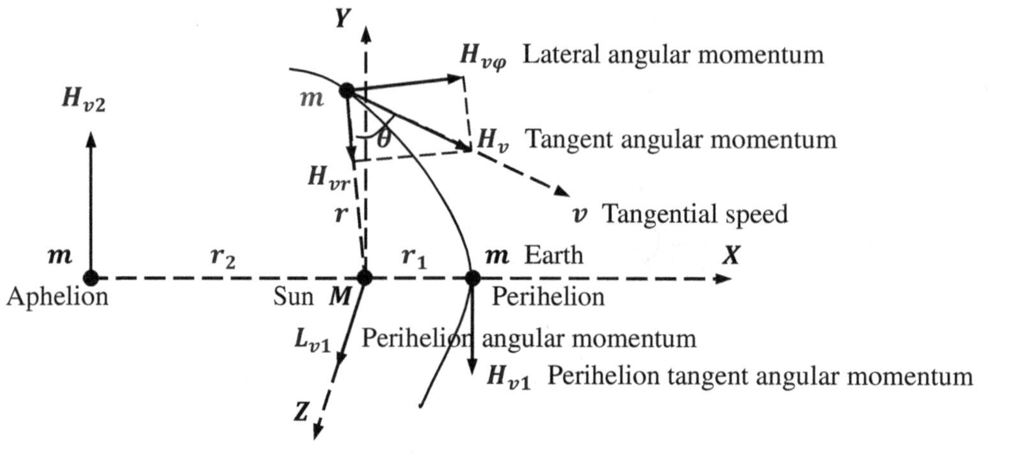

Figure 5-5 Diagram of the tangential angular momentum H_{v1} at perihelion of the Earth's orbit around the Sun equal to the angular momentum L_{v1} at perihelion.

According to the angular momentum formula $L_v = rmv\sin\theta$, the angular momentum L_{v1} at perihelion of the Earth's orbit around the Sun is:

$$L_{v1} = r_1 m v_1 \tag{5.2.2}$$

Note that the angular momentum L_{v1} of perihelion is on the Sun. The direction of angular momentum L_{v1} at perihelion is perpendicular to the plane of the Earth's orbit. Although the perihelion angular momentum L_{v1} and the perihelion tangential angular momentum H_{v1} are at different locations, they are equal in value, i.e:

$$H_{v1} = L_{v1} = r_1 m v_1 \tag{5.2.3}$$

Suppose v_2 is the tangential velocity of the aphelion of the Earth's orbit around the Sun; r_2 is the distance from the aphelion to the Sun. The tangential angular momentum H_{v2} of the aphelion of the Earth's orbit around the Sun is:

$$H_{v2} = r_2 m v_2 \tag{5.2.4}$$

The tangential angular momentum H_{v2} of the aphelion is located on the Earth. The direction of the tangential angular momentum H_{v2} is the same as the direction of the tangential velocity v_2. as shown in Figure 5-5.

The angular momentum L_{v2} at the aphelion of the Earth's orbit around the Sun is:

$$L_{v2} = r_2 m v_2 \tag{5.2.5}$$

The angular momentum L_{v2} of the aphelion is on the Sun. The direction of the angular momentum L_{v2} is perpendicular to the plane of the Earth's orbit. Although the angular momentum L_{v2} of the aphelion and the tangential angular momentum H_{v2} of the aphelion are at different locations, they are equal in value, i.e:

$$H_{v2} = L_{v2} = r_2 m v_2 \tag{5.2.6}$$

5.2.2 Conservation equation for tangential angular momentum

Since the angle $\theta_1 = 90^0$ of the perihelion,the angle $\theta_2 = 90^0$ of the aphelion, the angular momentum L_{v1} of the perihelion is equal to the angular momentum L_{v2} of the aphelion according to the law of conservation of angular momentum, viz:

$$L_v = rmv\sin\theta = r_1 m v_1 = r_2 m v_2 = C \tag{5.2.7}$$

The C in the equation is a constant. Note that the angular momentum $L_v = rmv\sin\theta = C$ is on the sun and that the angular momentum L_v is perpendicular to the plane defined by both the velocity v and the distance r.

According to equation (5.2.3) and equation (5.2.6), the relation equation can be obtained:

$$H_{v1} = H_{v2} = r_1 m v_1 = r_2 m v_2 = C \tag{5.2.8}$$

Based on the above equation, the relationship equation can be obtained:

$$H_v = rmv = r_1 m v_1 = r_2 m v_2 = C \tag{5.2.9}$$

The above formula is a conservation formula for tangential angular momentum. Note that the distance r in the formula is a scalar, not a vector.

According to Newtonian mechanics, the angular momentum $\boldsymbol{L_v} = \boldsymbol{rmv}\sin\boldsymbol{\theta} = \boldsymbol{C}$ of the Earth's orbit around the Sun is on the Sun, and the angular momentum $\boldsymbol{L_v}$ is perpendicular to the plane defined by the velocity $\boldsymbol{v}$ and the distance $\boldsymbol{r}$.

However, the tangential angular momentum $\boldsymbol{H_v}$ of the Earth's orbit around the Sun is neither at the Sun nor at the center O of mass, but on the Earth. The direction of the tangential angular momentum $\boldsymbol{H_v}$ is the same as the direction of the tangential velocity $\boldsymbol{v}$.

Comparing equation (5.2.7) with equation (5.2.9), it can be seen that equation $\boldsymbol{L_v} = \boldsymbol{rmv}\sin\boldsymbol{\theta} = \boldsymbol{C}$ for the conservation of angular momentum contradicts equation $\boldsymbol{H_v} = \boldsymbol{rmv} = \boldsymbol{C}$ for the conservation of tangential angular momentum.

5.2.3 Equation of conservation of tangential angular momentum for a system of mass points

Assume that O is the center of mass of System of mass points, and assume that System of mass points contains $\boldsymbol{n}$ mass points rotating around the center O of mass. Assume that $\boldsymbol{m_i}$ is the mass of mass $\boldsymbol{m_i}$. The total mass $\boldsymbol{M}$ of System of mass points is:

$$\boldsymbol{M} = \sum_{i=1}^{n} \boldsymbol{m_i}$$

Theoretically, a system of mass points can be transformed into a system of mass points containing only two mass points. Suppose $(\boldsymbol{M} - \boldsymbol{m_i})$ is the mass of the synthetic mass $\boldsymbol{M_i}$; suppose $\boldsymbol{v_i}$ is the tangential velocity of the mass $\boldsymbol{m_i}$ rotating around the synthetic mass $\boldsymbol{M_i}$; and suppose $\boldsymbol{r_i}$ is the distance from the mass $\boldsymbol{m_i}$ to the synthetic mass $\boldsymbol{M_i}$.

When the combined force $\boldsymbol{F}$ on the system of mass points is always zero. The conservation equation for the tangential angular momentum of the mass $\boldsymbol{m_i}$ rotating around the synthesized mass $\boldsymbol{M_i}$ is:

$$\boldsymbol{H_{vi}} = \boldsymbol{r_i m_i v_i} = \boldsymbol{C} \qquad (5.2.10)$$

The above equation is a conservation equation for the tangential angular momentum of a system of mass points. C in the equation is a constant. $\boldsymbol{H_{vi}}$ in the formula is the tangential angular momentum of the mass $\boldsymbol{m_i}$ rotating about the synthesized mass $\boldsymbol{M_i}$, not the tangential angular momentum of the mass$\boldsymbol{m_i}$ rotating about the center O of mass

The tangential angular momentum $\boldsymbol{H_{vi}}$ is neither on the synthesized mass $\boldsymbol{M_i}$ nor on the center O of mass, but on the mass $\boldsymbol{m_i}$. The direction of tangential angular momentum $\boldsymbol{H_{vi}}$ is the same as the direction of tangential velocity $\boldsymbol{v_i}$.

5.2.4 Equation for the conservation of tangential angular momentum of a mass m_i rotating about a center of mass O

Assume that $\boldsymbol{R_{Oi}}$ is the distance from mass $\boldsymbol{m_i}$ to the center O of mass, and $\boldsymbol{v_{Oi}}$ is the tangential velocity of mass $\boldsymbol{m_i}$ as it rotates around the center O of mass. According to the conservation of tangential angular momentum formula (5.2.9), the tangential angular momentum $\boldsymbol{H_{Oi}}$ of mass $\boldsymbol{m_i}$ as it revolves around center O of mass is:

$$\boldsymbol{H_{Oi}} = \boldsymbol{R_{Oi} m_i v_{Oi}} = \boldsymbol{C} \qquad (5.2.11)$$

The above equation is the equation for the conservation of tangential angular momentum of a mass $\boldsymbol{m_i}$ revolving around a center O of mass. C in the formula is a constant. Notice that the tangential angular momentum $\boldsymbol{H_{Oi}}$ is not at the center O of mass. The position and direction of the tangential angular momentum $\boldsymbol{H_{Oi}}$ are the same as the position and direction of the tangential velocity $\boldsymbol{v_{Oi}}$.

When the combined force of the system of mass points is zero, the sum $\boldsymbol{H_O}$ of the tangential angular momentum of $\boldsymbol{n}$ mass points rotating around the center O of mass is given by the tangential angular momentum conservation formula:

$$\boldsymbol{H_O} = \sum_{i=1}^{n} \boldsymbol{R_{Oi} m_i v_{Oi}} = \boldsymbol{C} \qquad (5.2.12)$$

The above equation is the formula for the conservation of tangential angular momentum for $\boldsymbol{n}$ points of mass revolving around the center O of mass. Note that the sum $\boldsymbol{H_O}$ of the tangential angular momentum is not at the center O of mass.

5.3 Equation for the conservation of tangential angular momentum for the Sun and Earth orbiting the center O of mass of the solar system

5.3.1 Equation of conservation of tangential angular momentum for the rotation of the Earth around the center O of mass of the Solar System

Since the gravitational force $\boldsymbol{F}$ between the Earth and the Sun is an internal force of the solar system, the gravitational force $\boldsymbol{F}$ cannot change the velocity and kinetic energy of the center O of mass of the solar system as it revolves around the galactic center.

Since the gravitational force $\boldsymbol{F}$ between the Earth and the Sun can change the speed and momentum of the Earth around the Sun, the gravitational force $\boldsymbol{F}$ can change the tangential angular momentum of the Earth and the Sun around the center O of mass.

The Earth revolves not only around the Sun, but also around the center O of mass. Since $\boldsymbol{v}$ is the velocity of the Earth's revolution around the Sun and $\boldsymbol{r}$ is the distance from the Earth to the Sun, the conservation of tangential angular momentum of the Earth's revolution around the Sun is given by equation (5.2.9):

$$\boldsymbol{H_v = rmv = C}$$

The C in the equation is a constant. According to equation (3.1.3), the tangential velocity $\boldsymbol{v_O}$ of the Earth's rotation around the center O of mass is:

$$\boldsymbol{v_O = \frac{M}{M+m}v}$$

According to equation (2.1.7), the distance $\boldsymbol{R_O}$ from the Earth to the center O of mass is:

$$\boldsymbol{R_O = \frac{M}{M+m}r}$$

Multiplying the distance $\boldsymbol{R_O}$ by the speed $\boldsymbol{v_O}$ and then by the mass $\boldsymbol{m}$ of the Earth gives the relationship equation:

$$\boldsymbol{H_{vO} = R_O m v_O = \left(\frac{M}{M+m}\right)^2 rmv} \qquad \textbf{(5.3.1)}$$

The $\boldsymbol{H_{vO}}$ in the formula is the tangential angular momentum of the Earth's revolution around the center O of mass. The above equation is the equation for the tangential angular momentum of the Earth's revolution around the center O of mass. Substituting equation (5.2.9) $\boldsymbol{H_v = rmv}$ into the above formula yields the relational formula:

$$\boldsymbol{H_{vO} = \left(\frac{M}{M+m}\right)^2 H_v = D} \qquad \textbf{(5.3.2)}$$

The D in the formula is a constant. Since the tangential angular momentum $\boldsymbol{H_v = rmv}$ of the Earth's revolution around the Sun is equal to the constant C, the tangential angular momentum $\boldsymbol{H_{vO}}$ of the Earth's revolution around the center O of mass is equal to the constant D. The above equation shows that the tangential angular momentum $\boldsymbol{H_v}$ of the Earth's revolution around the Sun is greater than the tangential angular momentum $\boldsymbol{H_{vO}}$ of the Earth's revolution around the center O of mass.

5.3.2 Conservation formulas for the tangential angular momentum of the Sun and Earth revolving around the center O of mass of the solar system

According to equation (3.1.6), the tangential velocity $\boldsymbol{v_S}$ of the Sun's revolution around the center O of mass of the solar system is:

$$v_S = \frac{m}{M+m} v$$

According to equation (2.1.9), the distance R_S from the Sun to the center O of mass of the solar system is:

$$R_S = \frac{m}{M+m} r$$

Multiplying the distance R_S by the velocity v_S and then by the solar mass M gives the relational equation:

$$H_S = R_S M v_S = \left(\frac{m}{M+m}\right)^2 rMv = \frac{mM}{(M+m)^2} rmv \qquad (5.3.3)$$

The H_S in the formula is the tangential angular momentum of the Sun's revolution around the center O of mass of the solar system. The above formula is the formula for the tangential angular momentum of the Sun's revolution around the center O of mass of the solar system. Since the tangential angular momentum $H_v = rmv$ is equal to the constant C, the tangential angular momentum H_S of the Sun's revolution around the center O of mass is also constant.

Since the direction of the Earth's tangential velocity v_O is opposite to the direction of the Sun's tangential velocity v_S, the direction of the tangential angular momentum H_{vO} is opposite to the direction of the tangential angular momentum H_S. Adding the tangential angular momentums H_{vO} and H_S gives the relational equation:

$$H_O = H_{vO} + H_S = \frac{M}{M+m} rmv = J \qquad (5.3.4)$$

The J in the formula is a constant. The above equation is the formula for the conservation of tangential momentum of the Sun and Earth as they revolve around the center O of mass of the Solar System. Note that the tangential angular momentum H_O of the Sun and Earth revolving around the center O of mass of the solar system is not equal to the tangential angular momentum $H_v = rmv$ of the Earth revolving around the Sun, nor is it equal to the tangential angular momentum $H_{vO} = R_O m v_O$ of the Earth's motion around the center O of mass.

5.3.3 The tangential angular momentum at perihelion and aphelion satisfies the conservation equation for tangential angular momentum

According to equation (2.5.26), the radius ρ of the circle of curvature of the Earth's orbit around the Sun is:

$$\rho = \frac{M+m}{M} r$$

According to the tangential angular momentum formula $H_v = rmv$, the tangential angular momentum $H_{\rho 1}$ of the Earth's perihelion rotating about the center O_{v1} of the circle of curvature is:

$$H_{\rho 1} = \rho_1 m v_1 = \frac{M+m}{M} r_1 m v_1 \qquad (5.3.5)$$

The radius ρ_1 of the circle of curvature in the equation is the distance from perihelion to the center O_{v1} of the circle of curvature. This is shown in Figure 5-6a.

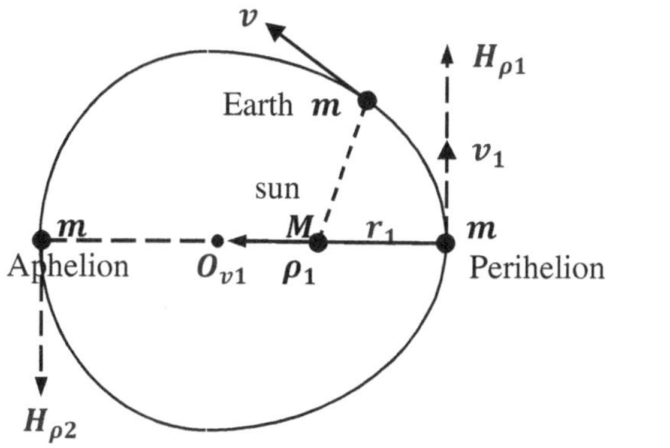

Figure 5-6a The center of curvature O_{v1} and the radius of curvature ρ_1 of the perihelion.

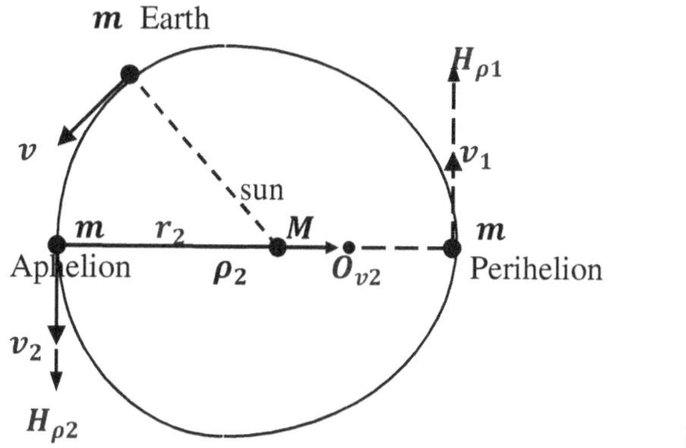

Figure 5-6b The center of curvature O_{v2} and the radius of curvature ρ_2 of the aphelion point.

According to equation (5.3.5), the tangential angular momentum $H_{\rho 2}$ of the Earth's aphelion rotating about the center O_{v2} of the circle of curvature is:

$$H_{\rho 2} = \rho_2 m v_2 = \frac{M+m}{M} r_2 m v_2 \tag{5.3.6}$$

The radius ρ_2 of the circle of curvature in the equation is the distance from the aphelion to the center O_{v2} of the circle of curvature. This is shown in Figure 5-6b.

According to the conservation formula for tangential angular momentum (5.2.9), tangential angular momentum $r_1 m v_1 = r_2 m v_2 = C$. Assuming that H_ρ is the tangential angular momentum of the Earth's revolution around the center O_v of the circle of curvature, then one obtains the relation equation:

$$H_\rho = H_{\rho 1} = H_{\rho 2} = \frac{M+m}{M} rmv = U \tag{5.3.7}$$

The **U** in the formula is a constant. The above formula shows that the tangential angular momentum of the perihelion and aphelion rotating about the center of the circle of curvature satisfies the conservation of tangential angular momentum.

5.3.4 Angular momentum at perihelion and aphelion of elliptical orbits Failure to satisfy the conservation of angular momentum formula

According to equation (2.5.31), the radii ρ' of the circles of curvature at the perihelion and aphelion of an elliptical orbit are:

$$\rho' = \frac{2r_1 r_2}{r_1 + r_2}$$

Suppose O'_{v1} is the center of the circle of curvature of the perihelion of an elliptical orbit. Since $\frac{2r_1 r_2}{r_1+r_2} > r_1$ in the above equation, the radius of curvature $\rho'(AO'_{v1})$ of the perihelion of an elliptical orbit is greater than the distance r_1 from the perihelion to the Sun, i.e. $\rho' > r_1$.

From this it can be seen that the center O'_{v1} of the circle of curvature of the perihelion of the elliptical orbit, lies on the extension line beyond the distance r_1 from the perihelion to the Sun. This is shown in Figure 5-7.

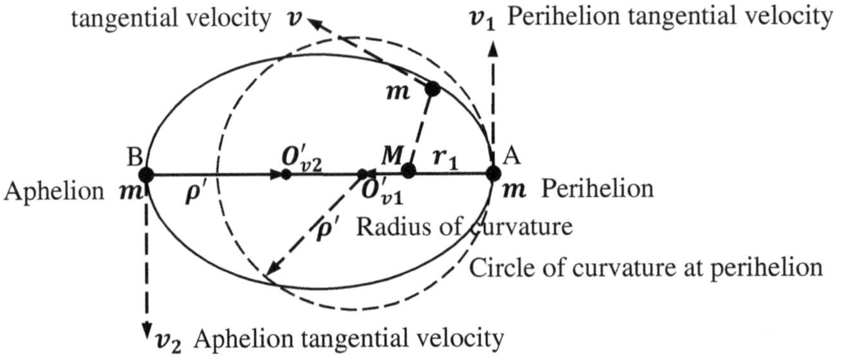

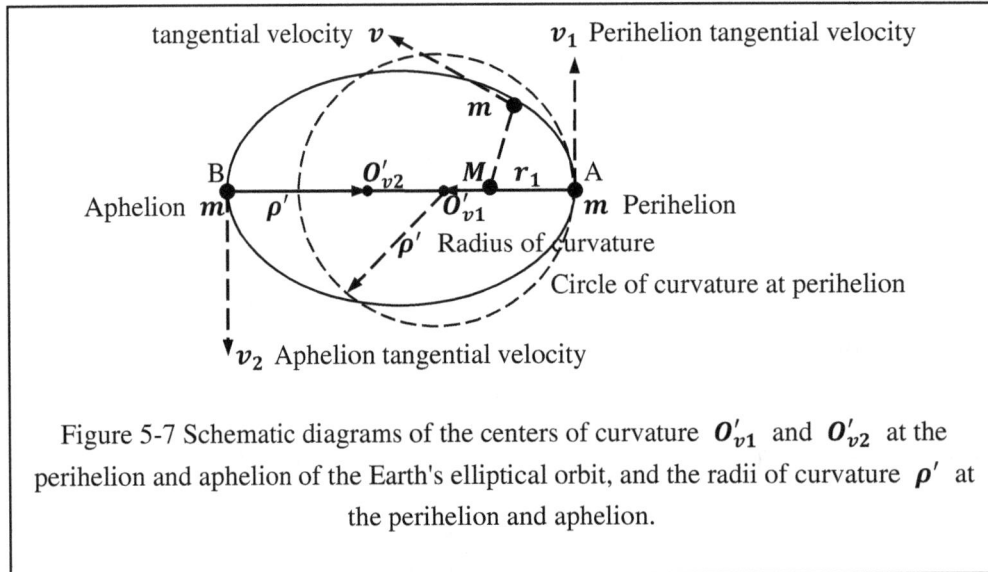

Figure 5-7 Schematic diagrams of the centers of curvature O'_{v1} and O'_{v2} at the perihelion and aphelion of the Earth's elliptical orbit, and the radii of curvature ρ' at the perihelion and aphelion.

Suppose O'_{v2} is the center of the circle of curvature of the aphelion of an elliptical orbit. Since $\frac{2r_1 r_2}{r_1+r_2} < r_2$ in the above formula, the radius of curvature ρ' (BO'_{v2}) of the aphelion of an elliptical orbit is smaller than the distance r_2 from the aphelion to the Sun, i.e. $\rho' < r_2$.

From this it can be determined that the center O'_{v2} of the circle of curvature of the aphelion of the elliptical orbit is located at a point within the distance r_2 from the aphelion to the Sun. This is shown in Figure 5-7.

According to the angular momentum formula, the perihelion of the Earth's elliptical orbit orbits around the center of curvature O'_{v1} with angular momentum $L'_{\rho 1} = \rho' m v_1 \sin\theta$. Since the angle $\theta = 90^0$ between the perihelion velocity v_1 and the distance r_1, the angular momentum $L'_{\rho 1}$ of the perihelion is:

$$L'_{\rho 1} = \rho' m v_1$$

Note that the radius ρ' of curvature of the perihelion of an elliptical orbit, is not equal to the distance r_1 from the perihelion to the Sun. This is shown in Figure 5-7. According to the above equation, the angular momentum $L'_{\rho 2}$ at the aphelion of the Earth's elliptical orbit is:

$$L'_{\rho 2} = \rho' m v_2$$

Note that the radius ρ' of curvature of the aphelion of an elliptical orbit, is not equal to the distance r_2 from the aphelion to the Sun. This is shown in Figure 5-7.

According to the law of conservation of angular momentum, the angular momentum $L'_{\rho 1}$ at the perihelion of an elliptical orbit is equal to the angular momentum $L'_{\rho 2}$ at the aphelion, i.e:

$$\rho' m v_1 = \rho' m v_2 \tag{5.3.8}$$

Based on the above equation, the relationship equation can be obtained:

$$v_1 = v_2$$

The above results clearly do not correspond to the objective facts. It can be seen that the angular momentum $L'_{\rho 1}$ at perihelion and $L'_{\rho 2}$ at aphelion of an elliptical orbit do not satisfy the law of conservation of angular momentum, i.e. $L'_{\rho 1} \neq L'_{\rho 2}$.

5.4 Equation for the conservation of tangential angular momentum in the solar system

5.4.1 Tangential velocity v_{Si} of the synthetic sun M_{Si} around the center O of mass of the solar system

Suppose there are n planets in the solar system. Suppose m_S is the mass of the Sun and m_i is the mass of planet m_i. The mass M of the solar system is:

$$M = m_S + \sum_{i=1}^{n} m_i \tag{5.4.1}$$

Let M_{Si} be the mass of the synthesized Sun M_{Si}, i.e:

$$M_{Si} = M - m_i = m_S + \sum_{j=1}^{n-1} m_j \tag{5.4.2}$$

Suppose R_{Oi} is the distance from planet m_i to the center O of mass of the solar system, and R_{Si} is the distance from the synthetic sun M_{Si} to the center O of mass of the solar system, as shown in Figure 5-8.

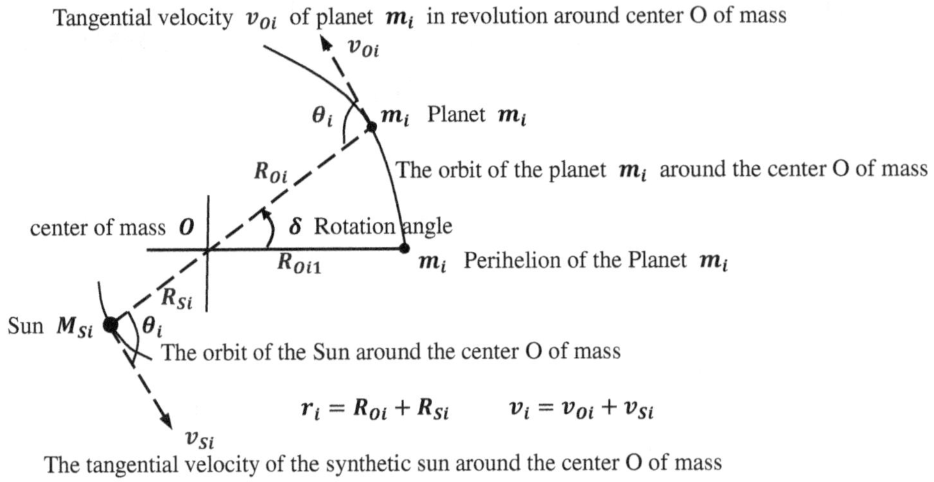

Figure 5-8 Diagram of the tangential velocity v_{Oi} and distance R_{Oi} for the motion of planet m_i around the center O of mass of the solar system.

The distance r_i (similar to the distance r from the Earth to the Sun) from planet m_i to the synthetic Sun M_{Si} is:

$$r_i = R_{Oi} + R_{Si} \tag{5.4.3}$$

In the solar system center O of mass frame of reference, since the velocity $v_C = 0$ of the center O of mass, the momentum $P_C = 0$ of the center O of mass, ie:

$$P_C = Mv_C = 0 \tag{5.4.4}$$

According to equation (3.1.1), the momentum mv_O of the Earth's revolution around the center O of mass of the solar system is equal to the momentum Mv_S of the Sun's revolution around the center O of mass of the solar system, i.e:

$$mv_O = Mv_S \tag{5.4.5}$$

The above equation is consistent with the momentum $P_C = 0$ equation.

Assume that v_{Si} is the tangential velocity of the synthetic sun M_{Si} as it orbits around the center O of mass of the solar system, and v_{Oi} is the tangential velocity of the planet m_i as it orbits around the center O of mass of the solar system, as shown in Figure 5-8.

According to equation (5.4.5), the momentum $m_i v_{Oi}$ of the planet m_i revolving around the center O of mass of the solar system is equal to the momentum $M_{Si} v_{Si}$ of the synthetic sun M_{Si} revolving around the center O of mass of the solar system, i.e:

$$m_i v_{Oi} = M_{Si} v_{Si} \tag{5.4.6}$$

The above equation is consistent with the formula for the momentum $P_C = 0$ of the center O of mass. Note that the tangential velocity v_{Si} of the synthetic sun M_{Si} in revolution about the center O of mass of the solar system is similar to the tangential velocity of the sun in revolution about the center O of mass of the solar system. The tangential velocity v_{Si} of the synthetic Sun M_{Si} is synthesized from the tangential velocities of the other planets and the Sun.

Since the direction of the tangential velocity v_{Oi} is opposite to the direction of the tangential velocity v_{Si}, the tangential velocity v_i of the planet m_i moving around the synthetic sun M_{Si} is:

$$v_i = v_{Oi} + v_{Si} \tag{5.4.7}$$

Multiplying the above equation by the mass M_{Si} of the synthetic sun M_{Si} gives the relationship equation:

$$M_{Si} v_i = M_{Si} v_{Oi} + M_{Si} v_{Si} \tag{5.4.8}$$

Substituting equation (5.4.6) into the above equation and eliminating $M_{Si} v_{Si}$ gives the relational equation:

$$v_{Oi} = \frac{M_{Si}}{M} v_i = \frac{M - m_i}{M} v_i \tag{5.4.9}$$

The v_{oi} in the equation is the tangential velocity of the planet m_i rotating around the center O of mass of the solar system, and v_i is the tangential velocity of the planet m_i rotating around the synthetic sun M_{si}.

Multiplying equation (5.4.7) by the planetary mass m_i gives the relationship equation:

$$m_i v_i = m_i v_{oi} + m_i v_{si} \tag{5.4.10}$$

Substituting equation (5.4.6) into the above equation and eliminating $m_i v_{oi}$ gives the relational equation:

$$v_{si} = \frac{m_i}{M} v_i \tag{5.4.11}$$

The v_{si} in the equation is the tangential velocity of the synthetic sun M_{si} rotating around the center O of mass of the solar system.

5.4.2 Sum H_{Oi} of the tangential angular momentum of the synthetic sun M_{Si} and planet m_i around the center O of mass of the solar system

According to equation (5.1.20), the tangential angular momentum H_{vO} of the Earth's revolution around the center O of mass of the solar system is:

$$H_{vO} = R_O m v_O = \left(\frac{M}{M+m}\right)^2 rmv \tag{5.4.12}$$

The M in the formula is the mass of the Sun, r is the distance from the Earth to the Sun, and v is the tangential velocity of the Earth's revolution around the Sun.

If the mass M_{Si} of the synthetic sun M_{Si} is considered to be the solar mass M and the planetary mass m_i is considered to be the Earth mass m, then the tangential angular momentum H_{vOi} of the planet m_i rotating around the center O of mass of the solar system is given by the above equation:

$$H_{vOi} = \left(\frac{M_{Si}}{M}\right)^2 r_i m_i v_i \tag{5.4.13}$$

The above equation is the formula for the tangential angular momentum of planet m_i as it orbits the center O of mass of the solar system. Note that M in the formula is the mass of the solar system, r_i is the distance from planet m_i to the center of mass of the synthetic sun M_{Si}, and v_i is the tangential velocity of planet m_i as it orbits the synthetic sun M_{Si}. This is shown in Figure 5-8.

According to equation (5.1.20), the tangential angular momentum H_{vS} of the Sun revolving around the center O of mass of the solar system is:

$$H_{vS} = \left(\frac{m}{M+m}\right)^2 rMv \tag{5.4.14}$$

The M in the formula is the mass of the Sun, r is the distance from the Earth to the Sun, and v is the tangential velocity of the Earth's revolution around the Sun.

According to the above equation, the tangential angular momentum H_{vSi} of the synthetic sun M_{Si} revolving around the center O of mass of the solar system is:

$$H_{vSi} = \left(\frac{m_i}{M}\right)^2 r_i M_{Si} v_i \tag{5.4.15}$$

The above equation is the formula for the tangential angular momentum of the synthetic sun M_{Si} revolving around the center O of mass. Dividing the above equation by equation (5.4.13) gives the relational equation:

$$\frac{H_{vOi}}{H_{vSi}} = \frac{M_{Si}}{m_i} \tag{5.4.16}$$

The above equation shows that the ratio of the tangential angular momentum H_{vOi} of the planet m_i to the tangential angular momentum H_{vSi} of the synthetic sun M_{Si} is a constant.

Add equation (5.4.13) and equation (5.4.15) to obtain the relational equation:

$$H_{Oi} = H_{vOi} + H_{vSi} = \frac{M_{Si}}{M} r_i m_i v_i \tag{5.4.17}$$

The sum H_{Oi} of the tangential angular momentum in the equation is analogous to the sum H_O of the tangential angular momentum of the Sun and the Earth rotating around the center O of mass of the solar system.

5.4.3 Equation for the conservation of tangential angular momentum in the solar system revolving around the center of mass O

Suppose there are $\boldsymbol{n}$ planets in the solar system. Suppose $\boldsymbol{m_i}$ is the mass of planet $\boldsymbol{m_i}$; $\boldsymbol{R_{Oi}}$ is the distance from planet $\boldsymbol{m_i}$ to the center O of mass of the solar system; and $\boldsymbol{v_{Oi}}$ is the tangential velocity of planet $\boldsymbol{m_i}$ as it orbits the center O of mass of the solar system. According to the tangential angular momentum formula $\boldsymbol{H_v = rmv}$, the tangential angular momentum $\boldsymbol{H_{vOi}}$ of planet $\boldsymbol{m_i}$ around the center O of mass is:

$$\boldsymbol{H_{vOi} = R_{Oi} m_i v_{Oi}} \qquad (5.4.18)$$

Note that the tangential angular momentum $\boldsymbol{H_{vOi}}$ is on the planet $\boldsymbol{m_i}$, not on the center O of mass. The sum $\sum_{i=1}^{n} \boldsymbol{H_{vOi}}$ of the tangential angular momentums of the $\boldsymbol{n}$ planets in the solar system revolving around the center O of mass is:

$$\sum_{i=1}^{n} \boldsymbol{H_{vOi}} = \sum_{i=1}^{n} \boldsymbol{R_{Oi}\, m_i v_{Oi}} \qquad (5.4.19)$$

Since the motion of the center O of mass of the solar system around the galactic center can be regarded as a uniform motion, the sum $\boldsymbol{H_O}$ of the tangential angular momentum of the planets and the Sun revolving around the center O of mass is a constant according to the formula for the conservation of tangential angular momentum, viz:

$$\boldsymbol{H_O = H_{vS}} + \sum_{i=1}^{n} \boldsymbol{H_{vOi} = R_S m_S v_S} + \sum_{i=1}^{n} \boldsymbol{R_{Oi}\, m_i v_{Oi} = C} \qquad (5.4.20)$$

The above formula is the conservation formula for the tangential angular momentum of the solar system rotating about the center O of mass. In the formula, C is a constant, $\boldsymbol{H_{vS}}$ is the tangential angular momentum of the sun rotating about the center O of mass of the solar system, and $\sum_{i=1}^{n} \boldsymbol{H_{vOi}}$ is the sum of the tangential angular momentums of the planets rotating about the center O of mass of the solar system.

5.4.4 Conservation formula for the sum of the tangential angular momentum of the synthetic Sun M_{SA} and Planet m_A

Assume that $\boldsymbol{M_{SA} = M - m_A}$ is the mass of the synthetic sun $\boldsymbol{M_{SA}}$. Assume that $\boldsymbol{m_A}$ is the mass of planet $\boldsymbol{m_A}$. Assume that $\boldsymbol{r_A}$ is the distance between planet $\boldsymbol{m_A}$ and the synthetic sun $\boldsymbol{M_{SA}}$. $\boldsymbol{v_A}$ is the tangential velocity of planet $\boldsymbol{m_A}$ as it orbits around the synthetic sun $\boldsymbol{M_{SA}}$. According to equation (5.4.17), the sum $\boldsymbol{H_{OA}}$ of the tangential angular momentum of the synthetic sun $\boldsymbol{M_{SA}}$ and planet $\boldsymbol{m_A}$ orbiting around the center O of mass of the solar system is:

$$\boldsymbol{H_{OA} = H_{vOA} + H_{vSA} = \frac{M_{SA}}{M} r_A m_A v_A} \qquad (5.4.21)$$

The $\boldsymbol{H_{vOA}}$ in the formula is the tangential angular momentum of the planet $\boldsymbol{m_A}$ rotating around the center O of mass; $\boldsymbol{H_{vSA}}$ is the tangential angular momentum of the synthetic sun $\boldsymbol{M_{SA}}$ rotating around the center O of mass of the solar system, and according to the law of conservation of tangential angular momentum, the sum $\boldsymbol{H_{OA}}$ of the tangential angular momentum is equal to the sum $\boldsymbol{H_O}$ of the tangential angular momentum of the solar system, that is:

$$\boldsymbol{H_O = H_{OA} = C} \qquad (5.4.22)$$

The **C** in the equation is a constant.

Assume that $\boldsymbol{M_{SB} = M - m_B}$ is the mass of the synthetic sun $\boldsymbol{M_{SB}}$. Assume that $\boldsymbol{m_B}$ is the mass of planet $\boldsymbol{m_B}$. Assume that $\boldsymbol{r_B}$ is the distance between planet $\boldsymbol{m_B}$ and the synthetic sun $\boldsymbol{M_{SB}}$. $\boldsymbol{v_B}$ is the tangential velocity of planet $\boldsymbol{m_B}$ as it orbits around the synthetic sun $\boldsymbol{M_{SB}}$. According to equation (5.4.17), the sum $\boldsymbol{H_{OB}}$ of the tangential angular momentum of the synthetic sun $\boldsymbol{M_{SB}}$ and planet $\boldsymbol{m_B}$ orbiting around the center O of mass of the solar system is:

$$\boldsymbol{H_{OB} = H_{vOB} + H_{vSB} = \frac{M_{SB}}{M} r_B m_B v_B} \qquad (5.4.23)$$

The $\boldsymbol{H_{vOB}}$ in the formula is the tangential angular momentum of the planet $\boldsymbol{m_B}$ rotating around the center O of mass; $\boldsymbol{H_{vSB}}$ is the tangential angular momentum of the synthetic sun $\boldsymbol{M_{SB}}$ rotating around the center O of mass of the solar system, and according to the law of conservation of tangential angular momentum, the sum H_{OB} of the tangential angular momentum is equal to the sum H_O of the tangential angular momentum of the solar system, that is:

$$H_O = H_{OB} = C \tag{5.4.24}$$

Based on the above equation and equation (5.4.22), the relationship equation can be obtained:

$$H_O = H_{OA} = H_{OB} =, \cdots, = H_{On} = C \tag{5.4.25}$$

The above formula is another expression formula of formula (5.4.20). According to the above formula and formula (5.4.21), the relation formula can be obtained:

$$H_O = \frac{M_{SA}}{M} r_A m_A v_A = \frac{M_{SB}}{M} r_B m_B v_B =, \cdots, = \frac{M_{Sn}}{M} r_n m_n v_n = C \tag{5.4.26}$$

The above formula is the conservation formula for the tangential angular momentum of the solar system. $M_{Sn} = M - m_n$ in the formula is the mass of the synthetic sun M_{Sn}; r_n is the distance between planet m_n and the synthetic sun M_{Sn}; and v_n is the tangential velocity of planet m_n as it revolves around the synthetic sun M_{Sn}.

5.4.5 Tangential angular momentum per unit mass of the solar system h_O

According to equation (5.3.4), the sum H_O of the tangential angular momentum of the Sun and Earth revolving around the center O of mass of the solar system is given by:

$$H_O = \frac{M}{M+m} rmv = J$$

Dividing the tangential angular momentum H_O by the total mass $(M+m)$ of the solar system gives the tangential angular momentum h_O per unit mass of the solar system:

$$h_O = \frac{H_O}{M+m} = \frac{M}{(M+m)^2} rmv \tag{5.4.27}$$

The h_O in the equation is a constant. According to equation (5.4.26), the tangential angular momentum H_{OA} and H_{On} of both planet m_A and planet m_n are equal to the tangential angular momentum H_O of the solar system.

Dividing the tangential angular momentum H_O in equation (5.4.26) by the total mass M of the solar system, gives the tangential angular momentum h_O per unit mass of the solar system, i.e:

$$h_O = \frac{H_O}{M} = \frac{M_{SA}}{M^2} r_A m_A v_A = \frac{M_{SB}}{M^2} r_B m_B v_B =, \cdots, = \frac{M_{Sn}}{M^2} r_n m_n v_n \tag{5.4.28}$$

The above equation can be expressed as:

$$h_O = \frac{H_O}{M} = \frac{H_{OA}}{M} = \frac{H_{OB}}{M} =, \cdots, = \frac{H_{On}}{M} \tag{5.4.29}$$

The above equation shows that the solar system has equal tangential angular momentum per unit mass.

5.4.6 Tangential angular momentum h_O possessed by a unit mass in cosmic space

In the Milky Way, there are countless stars orbiting the center of the galaxy. Assume that M' is the total mass of the galaxy; M'_n is the mass of the star n; and $M'_{Sn} = M' - m'_n$ is the mass of the synthetic sun M'_{Sn} of the galaxy. Assume that r'_n is the distance between the star m'_n and the synthetic sun M'_{Sn}; and v'_n is the tangential velocity of the star M'_n as it orbits the synthetic sun M'_{Sn}.

If the Milky Way is regarded as a solar system, and star A and star B are regarded as planets m_A and m_B, then according to equation (5.4.28) the Milky Way has a tangential angular momentum H'_O of :

$$H'_O = \frac{M'_A}{M'} r'_A m'_A v'_A = \frac{M'_B}{M'} r'_B m'_B v'_B =, \cdots, = \frac{M'_n}{M'} r'_n m'_n v'_n = C' \tag{5.4.30}$$

The C' in the formula is a constant. Divide the total mass M' of the galaxy by the above formula to obtain the relational formula:

$$h'_O = \frac{H'_O}{M'} = \frac{H'_{OA}}{M'} = \frac{H'_{OB}}{M'} =, \cdots, = \frac{H'_{On}}{M'} = \frac{C'}{M'} \tag{5.4.31}$$

Since H'_O is the tangential angular momentum of the galaxy and M' is the total mass of the galaxy, the tangential angular momentum h'_O is the tangential angular momentum per unit mass of the galaxy.

According to equation (5.4.29), the tangential angular momentum per unit mass of the Solar System is equal to

$\boldsymbol{h_O}$. Theoretically, the tangential angular momentum per unit mass at any point in cosmic space should be equal. In other words, the tangential angular momentum $\boldsymbol{h'_O}$ per unit mass of the Milky Way should be equal to the tangential angular momentum $\boldsymbol{h_O}$ per unit mass of the Solar System, i.e. $\boldsymbol{h'_O} = \boldsymbol{h_O}$.

If the tangential angular momentum $\boldsymbol{H'_{OA}}$, of star A is greater than the tangential angular momentum $\boldsymbol{H'_{OB}}$, of star B, then the energy $\boldsymbol{E_A}$, contained in the space of star A is greater than the energy $\boldsymbol{E_B}$, contained in the space of star B, i.e. $\boldsymbol{E_A} > \boldsymbol{E_B}$. Since energy flows from high to low, the kinetic energy $\boldsymbol{E_A}$, in the space of star A spreads into the space of star B until the tangential angular momentums per unit mass of stars A and B are equal, i.e. :

$$\boldsymbol{h_o} = \boldsymbol{h'_A} = \boldsymbol{h'_B} =, \cdots, = \boldsymbol{h'_n} = \boldsymbol{C} \qquad (5.4.32)$$

The physical meaning of the above equation is that the tangential angular momentum per unit mass of all stellar systems in cosmic space is equal. In other words, the tangential angular momentum $\boldsymbol{h_o}$ per unit mass possessed in the cosmic vacuum is equal.

Chapter 6 Derivation of the New Gravitational Formula

Introduction: This chapter is based on the centripetal force at perihelion and aphelion, and the formula for the conservation of tangential angular momentum. A new formula for gravitational is derived theoretically. And from this, the reasons for the anomalies in the orbits of cosmic probes are explained, and why the error in the measurement of the gravitational constant G is always too large.

6.1 Derive a new formula for gravitational based on the orbits of the planets around the center O of mass of the solar system

6.1.1 Derivation of a new gravitational formula

According to equation (2.5.13) and equation (2.5.17), the centripetal force F_{vO1} at the perihelion and the centripetal force F_{vO2} at the aphelion of the Earth's revolution around the center O of mass are:

$$\begin{cases} F_{vO1} = m\dfrac{v_{O1}^2}{R_{O1}} = m\dfrac{M}{M+m}\dfrac{v_1^2}{r_1} \\ F_{vO2} = m\dfrac{v_{O2}^2}{R_{O2}} = m\dfrac{M}{M+m}\dfrac{v_2^2}{r_2} \end{cases} \tag{6.1.1}$$

According to equation (3.1.7), the perigee tangential velocity v_{S1} of the Sun revolving around the center O of mass of the solar system is:

$$v_{S1} = \frac{m}{M+m}v_1$$

According to equation (2.1.13), the distance R_{S1} from the Sun's perigee to the center O of mass of the solar system is:

$$R_{S1} = \frac{m}{M+m}r_1$$

According to the centripetal force formula, the centripetal force F_{vS1} at the sun's perigee is:

$$F_{vS1} = M\frac{v_{S1}^2}{R_{S1}} = M\frac{m}{M+m}\frac{v_1^2}{r_1} \tag{6.1.2}$$

Comparing the above equation with equation (6.1.1), we can see that the centripetal force F_{vO1} at the Earth's perihelion is equal to the centripetal force F_{vS1} at the Sun's perigee, i.e:

$$F_{vO1} = F_{vS1} = m\frac{M}{M+m}\frac{v_1^2}{r_1} \tag{6.1.3}$$

Multiply the numerator and denominator of the above equation by r_1^2 to obtain the relational equation:

$$F_{vO1} = F_{vS1} = m\frac{M}{M+m}\frac{r_1^2v_1^2}{r_1^3} \tag{6.1.4}$$

The centripetal force F_{vO1} at the Earth's perihelion and the centripetal force F_{vS1} at the Sun's perigee are both equal to the universal gravitational force F_1 at perihelion, i.e:

$$F_{vO1} = F_{vS1} = \frac{r_1^2v_1^2}{M+m}\frac{Mm}{r_1^3} = K_{v1}\frac{Mm}{r_1^3} \tag{6.1.5}$$

Coefficient K_{v1} in the above equation is:

$$K_{v1} = \frac{r_1^2 v_1^2}{M+m} \tag{6.1.6}$$

Multiplying the square root of the coefficient K_{v1} by the Earth's mass m gives the relational equation:

$$m\sqrt{K_{v1}(M+m)} = r_1 m v_1 \tag{6.1.7}$$

According to equation (6.1.5), the centripetal force F_{vO2} at the Earth's apogee and the centripetal force F_{vS2} at the Sun's apogee are both equal to the gravitational force F_2 at the apogee, i.e:

$$F_{vO2} = F_{vS2} = \frac{r_2^2 v_2^2}{M+m}\frac{Mm}{r_2^3} = K_{v2}\frac{Mm}{r_2^3} \tag{6.1.8}$$

Coefficient K_{v2} in the above equation is:

$$K_{v2} = \frac{r_2^2 v_2^2}{M+m} \tag{6.1.9}$$

Multiplying the square root of the coefficient K_{v2} by the mass m of the Earth gives the relational equation:

$$m\sqrt{K_{v2}(M+m)} = r_2 m v_2 \tag{6.1.10}$$

According to equation (5.2.9), the conservation equation for the tangential angular momentum of the Earth's orbit around the Sun is:

$$H_v = rmv = r_1 m v_1 = r_2 m v_2 = C$$

Based on the above equation, the relationship equation can be obtained:

$$H_v = m\sqrt{K_{v1}(M+m)} = m\sqrt{K_{v2}(M+m)} = C \tag{6.1.11}$$

Since the mass M of the Sun, and the mass m of the Earth, are constants, the following equation can be derived from the above equation:

$$K = K_{v1} = K_{v2} \tag{6.1.12}$$

Based on the above equation, the relationship equation can be obtained:

$$K = \frac{r^2 v^2}{M+m} = \frac{r_1^2 v_1^2}{M+m} = \frac{r_2^2 v_2^2}{M+m} \tag{6.1.13}$$

The above formula is the new gravitational constant formula. The new gravitational constant K must be determined by physical experiments. The above formula is the constraint that the Earth's orbit around the Sun should satisfy.

Since both the centripetal force F_{vO1} at Earth's perihelion and the centripetal force F_{vS1} at the Sun's perigee are equal to the gravitational force F_1 at perihelion, the gravitational force F_1 at perihelion is:

$$F_1 = F_{vO1} = F_{vS1} = K\frac{Mm}{r_1^3} \tag{6.1.14}$$

According to the above equation, the gravitational force F_2 at the point of aphelion is:

$$F_2 = F_{vO2} = F_{vS2} = K\frac{Mm}{r_2^3} \tag{6.1.15}$$

Based on the gravitational force F_1 at perihelion and F_2 at aphelion, the gravitational force F at any point in the Earth's eggshell orbit is:

$$F = \frac{r^2 v^2}{M+m}\frac{Mm}{r^3} = K\frac{Mm}{r^3} \tag{6.1.16}$$

The above equation shows that the gravitational force F is inversely proportional to the cube of the distance r from the Earth to the Sun.

When the masses M and m are at rest, the gravitational force F is independent of the velocity v. The equation for the static gravitational force between the masses M and m at rest is:

$$F = K\frac{Mm}{r^3} \tag{6.1.17}$$

The formula above is the new formula for gravitational. The new gravitational constant K in the formula must be determined by physical experiments.

6.1.2 General relativity is a false theory of gravity

According to equation (6.1.13), the new gravitational constant K is:

$$K = \frac{r^2v^2}{M+m} = \frac{r_1^2v_1^2}{M+m} = \frac{r_2^2v_2^2}{M+m}$$

Strictly speaking, the above formula does not apply to the solar system, but only to the rotation of two points of mass around the center O of mass. The rotation of the Moon around the Earth satisfies the above formula. At this point, M is the mass of the Earth, m is the mass of the Moon, v is the tangential velocity of the Moon's revolution around the Earth, and r is the distance from the Earth to the Moon.

Since the center of mass of the Sun is very close to the center O of mass of the Solar System, the Earth's orbit around the Sun is approximately equal to the Earth's orbit around the center O of mass of the Solar System. Alternatively, the speed v and distance r of the Earth's orbit around the Sun approximately satisfy the above equation.

Multiplying the square root of the coefficient K by the mass of the Earth m gives the relational equation:

$$m\sqrt{K} = \frac{rmv}{\sqrt{M+m}} = \frac{r_1mv_1}{\sqrt{M+m}} = \frac{r_2mv_2}{\sqrt{M+m}}$$

Since the gravitational constant K, the masses M and m are all constants, the relation equation is obtained:

$$H_v = m\sqrt{K(M+m)} = rmv = r_1mv_1 = r_2mv_2 = C \qquad (6.1.18)$$

The above equation is the tangential conservation of angular momentum equation (5.2.9).

Note that due to the angle $\theta_1 = 90^0$ of the Earth's perihelion, the angular momentum $L_{v1} = r_1mv_1 \sin\theta_1$ at perihelion and the tangential angular momentum $H_{v1} = r_1mv_1$ at perihelion are equal, i.e:

$$L_{v1} = H_{v1} = r_1mv_1$$

Similarly, the angular momentum $L_{v2} = r_2mv_2 \sin\theta_2$ of the aphelion and the tangential angular momentum $H_{v2} = r_2mv_2$ of the aphelion are equal, i.e:

$$L_{v2} = H_{v2} = r_2mv_2$$

According to the law of conservation of angular momentum, the perihelion angular momentum $L_{v1} = r_1mv_1$ is equal to the aphelion angular momentum $L_{v2} = r_2mv_2$, i.e. $r_1mv_1 = r_2mv_2 = C$. Using this formula, we can derive the new formula for gravitational (6.1.17).

In summary, based on the model of the Sun and Earth revolving around the center O of mass of the solar system, only a new formula for gravitational can be derived, not Newton's formula for gravitational. From this it can be concluded that Newton's formula for universal gravitation is wrong.

Since Einstein's general theory of relativity, in the weak-field and low-speed case, is equivalent to Newton's formula for gravitational, general relativity is also a false theory of gravitational.

6.1.3 The conservation of angular momentum is false

Let θ be the angle between the tangential velocity v of the Earth and the distance r. Multiply equation (6.1.18) by $\sin\theta$. The law of conservation of angular momentum, $L_v = rmv\sin\theta = C$, gives the relational equation:

$$L_v = rmv\sin\theta = m\sqrt{K(M+m)}\sin\theta = C \qquad (6.1.19)$$

The trigonometric $\sin\theta$ in the formula is a variable, since the angle of entrainment θ varies over the range $90^0 \le \theta \le 180^0$. Since the gravitational constant K, the mass M of the Sun and the mass m of the Earth are constants, the angular momentum L_v is not equal to a constant, but to a variable, i.e:

$$L_v = rmv\sin\theta = m\sqrt{K(M+m)}\sin\theta \neq C \qquad (6.1.20)$$

Based on the above equation, it can be seen that the conservation of angular momentum law of classical theory is wrong.

6.1.4 Physical Implications of the Gravitational Constant K

Suppose that the solar system contains only the synthetic sun M_{SA} and the planet m_A. Both the synthetic sun M_{SA} and the planet m_A revolve around the center O of mass of the solar system.

Suppose $\boldsymbol{M}$ is the mass of the solar system; $\boldsymbol{m_A}$ is the mass of planet $\boldsymbol{m_A}$; and $\boldsymbol{M_{SA}} = \boldsymbol{M} - \boldsymbol{m_A}$ is the mass of the synthetic sun $\boldsymbol{M_{SA}}$. Suppose $\boldsymbol{v_A}$ is the tangential velocity of planet $\boldsymbol{m_A}$ orbiting the synthetic sun $\boldsymbol{M_{SA}}$; and $\boldsymbol{r_A}$ is the distance from planet $\boldsymbol{m_A}$ to the synthetic sun $\boldsymbol{M_{SA}}$. According to the new gravitational constant formula (6.1.13), the gravitational constant $\boldsymbol{K_A}$ of planet $\boldsymbol{m_A}$ is:

$$K_A = \frac{r_A^2 v_A^2}{M} \tag{6.1.21}$$

Suppose $\boldsymbol{m_B}$ is the mass of planet $\boldsymbol{m_B}$; $\boldsymbol{M_{SB}} = \boldsymbol{M} - \boldsymbol{m_B}$ is the mass of the synthetic sun $\boldsymbol{M_{SB}}$. Suppose $\boldsymbol{v_B}$ is the tangential velocity of planet $\boldsymbol{m_B}$ as it orbits the synthetic sun $\boldsymbol{M_{SB}}$; $\boldsymbol{r_B}$ is the distance from planet $\boldsymbol{m_B}$ to the synthetic sun $\boldsymbol{M_{SB}}$. According to the above equation, the gravitational constant $\boldsymbol{K_B}$ of planet $\boldsymbol{m_B}$ is:

$$K_B = \frac{r_B^2 v_B^2}{M} \tag{6.1.22}$$

According to equation (6.1.21) and equation (6.1.17), the new gravitational force $\boldsymbol{F_A}$ between the synthesized sun $\boldsymbol{M_{SA}}$ and the planet $\boldsymbol{m_A}$ is:

$$F_A = \frac{r_A^2 v_A^2}{M}\frac{(M - m_A)m_A}{r_A^3} = K_A\frac{(M - m_A)m_A}{r_A^3} \tag{6.1.23}$$

According to equation (6.1.22) and equation (6.1.17), the new gravitational force $\boldsymbol{F_B}$ between the synthesized sun $\boldsymbol{M_{SB}}$ and planet $\boldsymbol{m_B}$ is:

$$F_B = \frac{r_B^2 v_B^2}{M}\frac{(M - m_B)m_B}{r_B^3} = K_B\frac{(M - m_B)m_B}{r_B^3} \tag{6.1.24}$$

The $\boldsymbol{M}$ in the equation is the mass of the solar system.

Assume that $\boldsymbol{m_n}$ is the mass of planet $\boldsymbol{m_n}$; $\boldsymbol{M_{Sn}} = \boldsymbol{M} - \boldsymbol{m_n}$ is the mass of the synthetic sun $\boldsymbol{M_{Sn}}$. Assume that $\boldsymbol{v_n}$ is the tangential velocity of the revolution of planet $\boldsymbol{m_n}$ around the synthetic sun $\boldsymbol{M_{Sn}}$, and $\boldsymbol{r_n}$ is the distance from planet $\boldsymbol{m_n}$ to the synthetic sun $\boldsymbol{M_{Sn}}$.

Theoretically, the gravitational constant $\boldsymbol{K_A}$ in equation (6.1.21) and the gravitational constant $\boldsymbol{K_B}$ in equation (6.1.22) are both equal to the gravitational constant $\boldsymbol{K}$, i.e:

$$K = \frac{r_A^2 v_A^2}{M} = \frac{r_B^2 v_B^2}{M} =, \cdots, = \frac{r_n^2 v_n^2}{M} \tag{6.1.25}$$

The above formula is the formula for the gravitational constant of the solar system. It shows that the ratio $\frac{r^2 v^2}{M}$ is equal to the gravitational constant $\boldsymbol{K}$. The above formula applies not only to the Earth-Moon system and the solar system, but also to any galaxy in the universe.

In order to reveal the nature of the gravitational constant $\boldsymbol{K}$, the authors define the product $\boldsymbol{v_n r_n}$ of the tangential velocity $\boldsymbol{v_n}$ and the distance $\boldsymbol{r_n}$ as the velocity torque $\boldsymbol{\Omega_n}$ of the planet $\boldsymbol{m_n}$, namely:

$$\Omega_n = v_n r_n \tag{6.1.26}$$

Multiplying the above equation by the planetary mass $\boldsymbol{m_n}$ gives the tangential angular momentum $\boldsymbol{H_{vn}}$ of the planet $\boldsymbol{m_n}$ orbiting the synthetic sun $\boldsymbol{M_{Sn}}$, i.e:

$$H_{vn} = m_n \Omega_n = m_n v_n r_n \tag{6.1.27}$$

According to equation (6.1.25), the speed torque of planet $\boldsymbol{m_A}$ and planet $\boldsymbol{m_B}$ is:

$$\Omega_A = v_A r_A, \quad \Omega_B = v_B r_B \tag{6.1.28}$$

The gravitational constant equation (6.1.25) can be expressed as:

$$K = \frac{\Omega_A^2}{M} = \frac{\Omega_B^2}{M} =, \cdots, = \frac{\Omega_n^2}{M} \tag{6.1.29}$$

The above equation shows that the gravitational constant $\boldsymbol{K}$ is equal to the square of the velocity torque that the unit mass $\boldsymbol{1/M}$ of the system of mass points possesses.

6.1.5 The mass *m* of the Earth is recalculated on the basis of the new gravitational formula

If the new gravitational formula (6.1.17) is expressed in the form of Newton's gravitational formula, then the relational formula can be obtained:

$$F = K\frac{Mm}{r^3} = G\frac{Mm}{r^2} \qquad (6.1.30)$$

According to the above equation, Newton's gravitational constant G is:

$$G = K\frac{1}{r} \qquad (6.1.31)$$

The above equation shows that Newton's gravitational constant G is not a constant but a variable. Since Newton's gravitational constant G is inversely proportional to the distance r between two objects, the greater the distance r, the smaller the value of the gravitational constant G. The smaller the distance r, the greater the value of the gravitational constant G. The following two conclusions can be drawn from the above equation.

Conclusion 1. If the distance r between the Sun and the planets is greater, then the ratio of the Newtonian gravitational force F' to the actual gravitational force F will be smaller. From this it can be determined that if the cosmic probe is farther away from the sun, then astronomers will assume that the resistance the cosmic probe will experience at the edge of the solar system will be greater. This is the reason why spacecraft deviate from their theoretical orbits.

Conclusion 2, if the distance r between two microscopic particles is smaller, then the ratio of the Newtonian gravitational force F' to the actual gravitational force F will be larger. From this physicists believe that there is a strong nuclear force acting between microscopic particles inside the nucleus of an atom.

The current standard gravitational constant G_0 is:

$$G_0 = 6.674 \times 10^{-11} N \cdot m^2/kg^2$$

Assume that the distance between the centers of mass of the large and small spheres at which the gravitational constant is measured is $r_0 = 0.1m$; assume that the measured value G_0 is equal to the standard gravitational constant. Substituting the standard gravitational constant G_0 and the distance $r_0 = 0.1m$ into equation (6.1.31) gives the new gravitational constant K:

$$K = G_0 r_0 = 6.674 \times 10^{-12} N \cdot m^3/kg^2 \qquad (6.1.32)$$

According to the new formula for gravitational, the mass m of the Earth and the first cosmic velocity v can be estimated.

Suppose r is the distance from the center of the Earth to the ground; g is the gravitational acceleration of the ground; m is the mass of the Earth; and M is the mass of the object. The gravitational force gM of the object M on the ground is equal to the gravitational force F, i.e:

$$F = K\frac{Mm}{r^3} = gM \qquad (6.1.33)$$

According to the above equation, the mass of the Earth m is:

$$m = \frac{gr^3}{K} \qquad (6.1.34)$$

The new gravitational constant K is:

$$K = 6.674 \times 10^{-12} N \cdot m^3/kg^2$$

The gravitational acceleration $g = 9.78 \cdot m/s^2$ of the Earth's surface, and the average radius $r = 6.371 \times 10^6 \cdot m$ of the Earth. Substituting the data into the above equation gives the mass m of the Earth is:

$$m = \frac{9..78 \times (6.371 \times 10^6)^2}{6.674 \times 10^{-12}} = 3.789 \times 10^{32} \cdot kg \qquad (6.1.35)$$

The volume V of the earth is:

$$V = \frac{4}{3}\pi r^3 = \frac{4}{3} \times 3.14159 \times (6.371 \times 10^6)^3 m^3 = 1.083 \times 10^{27} \cdot cm^3 \qquad (6.1.36)$$

The average mass ρ contained in each cubic centimeter of the Earth's volume is:

$$\rho = \frac{m}{V} = \frac{3.789 \times 10^{32} kg}{1.083 \times 10^{27} cm^3} = 3.5 \times 10^5 \cdot kg/cm^3 \qquad (6.1.37)$$

The mass density ρ of the Earth's core is very large, as can be seen from the above equation. Due to the ultra-high temperature and pressure of the Earth's core, the atomic shells of the core are crushed so that the electrons outside the nucleus are compressed into the nucleus, turning the core into a high-density substance similar to a neutron star.

According to astronomical observations, neutron stars contain masses of up to 100 million tons per cubic centimeter. Obviously, the mass density of the Earth is much smaller than that of neutron stars.

The current theory is that the Earth has a mass $m = 5.965 \times 10^{24} \cdot kg$ and that the average mass density per cubic centimeter of the Earth ρ is:

$$\rho = \frac{m}{V} = \frac{5.965 \times 10^{27} g}{1.083 \times 10^{27} cm^3} = 5.51 \cdot g/cm^3 \qquad (6.1.38)$$

The average mass density ρ of the Earth clearly does not correspond to the high temperature and pressure nature of the Earth's core. The mass m of the Earth and the mass M of the Sun should be recalculated according to the new gravitational formula.

6.2 Mechanical energy $W_O = 0$ of motion of the Sun and Earth around the center O of mass of the solar system

6.2.1 Negative work E_K due to gravitational force

Suppose the distance r is changeable on path L_X. As the planet moves from point a to point b, the work A done by the new gravitational force F is given by equation (6.1.17):

$$A = -\int_{r_a}^{r_b} K\frac{Mm}{r^3}dr = \frac{KMm}{2}\left(\frac{1}{r_b^2} - \frac{1}{r_a^2}\right) \qquad (6.2.1)$$

When the planet is close to the Sun, at distance $r_b < r_a$, gravitational does positive work. Positive work done by gravitational can increase the velocity and kinetic energy of the planet. When the planet is far from the sun, at distance $r_b > r_a$, gravitational does negative work. Negative work done by gravitational decreases the velocity and kinetic energy of the planet.

The work A' done by Newton's gravitational force F' on path L_X is:

$$A' = -\int_{r_a}^{r_b} G\frac{M}{r^2}dr = GMm\left(\frac{1}{r_b} - \frac{1}{r_a}\right) \qquad (6.2.2)$$

Obviously, the work A done by the new gravitational force is not equal to the work A' done by Newtonian gravitational force.

As the planet moves from infinity to point b, i.e., $r_a = \infty$, the work $A_{\infty b}$ done by the new gravitational force F and the work $A'_{\infty b}$ done by the Newtonian gravitational force F' are:

$$A_{\infty b} = \frac{K}{2}\frac{Mm}{r^2}, \qquad A'_{\infty b} = G\frac{Mm}{r} \qquad (6.2.3)$$

The positive work done by gravitational force increases the velocity and kinetic energy of the planet.

If planet m moves from point a to infinity, i.e., when $r_b = \infty$, then the work E_K (gravitational energy) done by the new gravitational force F is:

$$E_K = \int_{r}^{\infty} K\frac{Mm}{r^3}dr = -\frac{K}{2}\frac{Mm}{r^2} \qquad (6.2.4)$$

The negative work E_K (gravitational energy) done by the new gravitational force F reduces the velocity and kinetic energy of the planet.

6.2.2 Mechanical energy W_O of the Sun and Earth in motion around the center O of mass of the solar system

When the earth moves from point a to point b, the work A done by the gravitational force F is always the same, whether the earth is moving around the sun or around the center O of mass.

According to equation (4.2.5), the sum E_O of the kinetic energies of the Earth and the Sun moving around the center O of mass of the solar system is given by:

$$E_O = \frac{1}{2}\frac{M}{M+m}mv^2$$

Adding the above equation to equation (6.2.4) gives the mechanical energy W_O for the motion of the Earth and Sun around the center O of mass of the solar system, namely:

$$W_O = E_O + E_K = \frac{1}{2}\frac{M}{M+m}mv^2 - \frac{K}{2}\frac{Mm}{r^2} \tag{6.2.5}$$

The mechanical energy W_v of the earth's motion around the sun is:

$$W_v = E_v + E_K = \frac{1}{2}mv^2 - \frac{K}{2}\frac{Mm}{r^2} \tag{6.2.6}$$

Comparing the above equation with equation (6.2.5), it is clear that the mechanical energy W_v of the Earth's orbit around the Sun is not equal to the mechanical energy W_O of the motion of the Earth and Sun around the center O of mass of the solar system, i.e., $W_v \neq W_O$.

The mechanical energy W_{vO} of the motion of the Earth around the center O of mass of the solar system is:

$$W_{vO} = E_{vO} + E_K = \frac{1}{2}mv_O^2 - \frac{K}{2}\frac{Mm}{r^2} \tag{6.2.7}$$

According to equation (3.1.3), the tangential velocity v_O of the ground around the center O of mass is given by:

$$v_O = \frac{M}{M+m}v$$

Substituting the above equation into equation (6.2.7) gives the mechanical energy W_{vO} of the Earth's motion around the center of mass O:

$$W_{vO} = \frac{1}{2}m\left(\frac{M}{M+m}\right)^2 v^2 - \frac{K}{2}\frac{Mm}{r^2} \tag{6.2.8}$$

By comparing the above equation with equation (6.2.5), the mechanical energy $W_{vO} \neq W_O$ can be determined.

6.2.3 The new gravitational force formula is consistent with the conservation of mechanical energy formula

According to equation (6.2.5), the mechanical energy W_{O1} of the Sun and Earth moving around the center O of mass of the solar system at the moment of perihelion is:

$$W_{O1} = E_{O1} + E_{K1} = \frac{1}{2}\frac{M}{M+m}mv_1^2 - \frac{K}{2}\frac{Mm}{r_1^2} \tag{6.2.9}$$

At the moment of aphelion, the mechanical energy W_{O2} of the Sun and Earth moving around the center O of mass of the solar system is:

$$W_{O2} = E_{O2} + E_{K2} = \frac{1}{2}\frac{M}{M+m}mv_2^2 - \frac{K}{2}\frac{Mm}{r_2^2} \tag{6.2.10}$$

According to the law of conservation of mechanical energy, the mechanical energy W_{O1} at perihelion is equal to the mechanical energy W_{O2} at aphelion, i.e:

$$W_{O1} - W_{O2} = \frac{1}{2}\frac{M}{M+m}m\left(v_1^2 - v_2^2\right) - \frac{KMm}{2}\left(\frac{1}{r_1^2} - \frac{1}{r_2^2}\right) = 0 \tag{6.2.11}$$

The difference value ΔE_O between the kinetic energy E_{O1} at perihelion and the kinetic energy E_{O2} at aphelion is:

$$\Delta E_O = E_{O1} - E_{O2} = \frac{1}{2}\frac{M}{M+m}m\left(v_1^2 - v_2^2\right) \tag{6.2.12}$$

Since the tangential velocity v_1 at perihelion and v_2 at aphelion are both perpendicular to the distance from the Earth to the Sun, the centripetal force equation can be used to verify that equation (6.2.11) is correct.

According to equation (6.1.13), the new gravitational constant K is:

$$K = \frac{r^2v^2}{M+m} = \frac{r_1^2v_1^2}{M+m} = \frac{r_2^2v_2^2}{M+m} \tag{6.2.13}$$

According to the above equation, the tangential velocity v_1 at perihelion is:

$$v_1^2 = \frac{K(M+m)}{r_1^2} \tag{6.2.14}$$

According to equation (6.2.13), the tangential velocity v_2 at the aphelion is:

$$v_2^2 = \frac{K(M+m)}{r_2^2} \tag{5.2.15}$$

Substituting the above equation and equation (6.2.14) into equation (6.2.11) gives the relationship equation:

$$W_{O1} - W_{O2} = \frac{1}{2}\frac{M}{M+m}m\left(\frac{K(M+m)}{r_1^2} - \frac{K(M+m)}{r_2^2}\right) - \frac{KMm}{2}\left(\frac{1}{r_1^2} - \frac{1}{r_2^2}\right) = 0$$

$$(6.2.16)$$

The above equation shows that the mechanical energy W_{O1} at perihelion is equal to the mechanical energy W_{O2} at aphelion, i.e. $W_{O1} = W_{O2}$. This confirms that the new gravitational formula (6.1.17) is consistent with the law of conservation of mechanical energy.

6.2.4 Mechanical energy $W_O = 0$ for the motion of the Sun and Earth around the center O of mass of the solar system

According to equation (6.2.9), at the moment of perihelion, the mechanical energy W_{O1} of the motion of the Sun and Earth around the center O of mass of the solar system is given by:

$$W_{O1} = \frac{1}{2}\frac{M}{M+m}mv_1^2 - \frac{K}{2}\frac{Mm}{r_1^2}$$

According to equation (6.2.14), the square v_1^2 of the tangential velocity at the perihelion is:

$$v_1^2 = \frac{K(M+m)}{r_1^2}$$

Substituting the above equation into equation (6.2.9) gives the relationship equation:

$$W_{O1} = \frac{1}{2}\frac{M}{M+m}m\frac{K(M+m)}{r_1^2} - \frac{K}{2}\frac{Mm}{r_1^2} = 0 \tag{6.2.17}$$

According to equation (6.2.10), the mechanical energy W_{O2} of the Sun and Earth moving around the center O of mass of the solar system at the moment of aphelion is:

$$W_{O2} = \frac{1}{2}\frac{M}{M+m}mv_2^2 - \frac{K}{2}\frac{Mm}{r_2^2}$$

According to equation (6.2.15), the tangential velocity v_2^2 at the aphelion is:

$$v_2^2 = \frac{K(M+m)}{r_2^2}$$

Substituting the above equation into equation (6.2.10) gives the relationship equation:

$$W_{O2} = \frac{1}{2}\frac{M}{M+m}m\frac{K(M+m)}{r_2^2} - \frac{K}{2}\frac{Mm}{r_2^2} = 0 \tag{6.2.18}$$

Based on the above equation and equation (6.2.17), it can be determined that the mechanical energy W_{O1} at perihelion is equal to the mechanical energy W_{O2} at aphelion, i.e:

$$W_{O1} = W_{O2} = 0$$

According to the law of conservation of mechanical energy, the mechanical energy W_O of the motion of the Sun and Earth around the center O of mass of the solar system is:

$$W_O = W_{O1} = W_{O2} = 0 \tag{6.2.19}$$

Substituting the above equation into equation (6.2.5) gives the relationship equation:

$$W_O = E_O + E_K = \frac{1}{2}\frac{M}{M+m}mv^2 - \frac{K}{2}\frac{Mm}{r^2} = 0 \tag{6.2.20}$$

Based on the above equation, the relationship equation can be obtained:

$$\frac{M}{M+m}mv^2 = K\frac{Mm}{r^2} \tag{6.2.21}$$

The above equation shows that the sum of the kinetic energy $\boldsymbol{E_O}$ of the Sun and Earth revolving around the center of mass O of the solar system is equal to the negative gravitational energy $-\boldsymbol{E_K}$.

Based on the above equation, the relationship equation can be obtained:

$$K = \frac{r^2v^2}{M+m}$$

The above equation is the new gravitational constant equation (6.1.13).

6.2.5 The law of conservation of angular momentum contradicts the law of conservation of mechanical energy

Multiply equation (6.2.21) by r^2m to obtain the relational equation:

$$\frac{M}{M+m}r^2m^2v^2 = KMm^2 \qquad (6.2.22)$$

Based on the above equation, the relationship equation can be obtained:

$$r^2m^2v^2 = Km^2(M+m) \qquad (6.2.23)$$

Assume that θ is the angle between the tangential velocity v of the Earth and the distance r. According to equation (2.5.2), the tangential velocity v of the Earth's orbit around the Sun can be decomposed into a transverse velocity v_φ and a radial velocity v_r, namely:

$$v = \sqrt{v_\varphi^2 + v_r^2}$$

Since the transverse velocity $v_\varphi = v\sin\theta$, radial velocity $v_r = v\cos\theta$, the above equation can be expressed as:

$$v = \sqrt{(v\sin\theta)^2 + (v\cos\theta)^2}$$

Substituting the above equation into equation (6.2.23) gives the relationship equation:

$$(r^2m^2v^2\sin^2\theta + r^2m^2v^2\cos^2\theta) = Km^2(M+m) \qquad (6.2.24)$$

According to the law of conservation of angular momentum $L_v = rmv\sin\theta = C$, the relation equation can be obtained:

$$r^2m^2v^2\sin^2\theta = C^2 \qquad (6.2.25)$$

Substituting the above equation into equation (6.2.24) gives the relationship equation:

$$r^2m^2v^2\cos^2\theta = Km^2(M+m) - C^2 \qquad (6.2.26)$$

Since the angle θ_1 at perihelion and θ_2 at aphelion are equal to 90^0, i.e., $\theta_1 = \theta_2 = 90^0$, the function $\sin 90^0 = 1$ and the function $\cos 90^0 = 0$. The above equation becomes:

$$Km^2(M+m) - C^2 = 0 \qquad (6.2.27)$$

Substituting the above equation into equation (6.2.26) gives the relationship equation:

$$r^2m^2v^2\cos^2\theta = 0 \qquad (6.2.28)$$

Since the tangential velocity $v \neq 0$, the distance $r \neq 0$, and the mass $m \neq 0$, the angle of entrainment $\theta = 90^0$. This result is clearly incorrect. It can be seen that the law of conservation of angular momentum contradicts the law of conservation of mechanical energy.

6.3 Newton's gravitational constant *G* is a variable, not a constant

6.3.1 Orbits of spacecraft do not conform to Newton's formula for gravitational

When NASA scientists analyzed the signals sent back by the Pioneer 10 spacecraft, they found that Pioneer 10's motion deviated from the orbit calculated by Newton's law of gravitational. Pioneer 10 should have been accelerating away from the solar system, but instead it was experiencing a deceleration anomaly. The farther it traveled from the Sun, the stronger the Sun's gravitational pull seemed to be. In December 1992, Pioneer 10's orbit changed extremely slightly. Then, in 1998, scientists noticed that Pioneer 10 was decelerating faster than expected, although the extra acceleration was very small. It was only about one ten billionth of the acceleration of Earth's gravitational.

This orbital anomaly was first discovered by Pioneer 10. However, when NASA scientists analyzed the orbital data of Pioneer 11, a sister spacecraft launched in April 1973, they found that similar deviations from the expected orbit had reappeared. Similar anomalies also occurred with the crashed Galileo probe to Jupiter and the Ulysses solar probe launched jointly by Europe and the United States. The Voyager 2 and Voyager 1 spacecraft, which were launched one after the other, also deviated significantly from their expected orbits. In particular, Voyager 1, which is currently the farthest away, is slowing down as it approaches the edge of the solar system, as if being held back by a pulling force.

Scientists initially thought the orbital anomaly was just a problem with the detector's internal equipment, but the same anomalies were observed in the Pioneer 11, Ulysses, and Galileo detectors, and the equipment was ruled out as the cause. The reason for the anomalous changes in the probes' orbits was not the prevailing gravitational pull of undiscovered stars, since Pioneer 10 and Pioneer 11 were 22 billion kilometers apart, so there would not have been such a large undiscovered star. Dr. John Anderson of NASA's Jet Propulsion Laboratory said: "It appears that the probes are not operating according to our known formula for universal gravitational." He said: "We've been working on this for years. Now we've come up with every reason we can think of."

After a thorough analysis and study of the spacecraft's orbit, scientists have concluded that an unexplained force is pulling the spacecraft. This is a negligibly small effect on the spacecraft itself, but it could mean that our understanding of gravitational may need to be corrected. They argue that the fact that Newton's formula for gravitational is no longer valid on the scale of the universe means that the "formula for gravitational" has limitations and is only valid under certain conditions.

6.3.2 Measurements of the universal gravitational constant *G* are always subject to large errors

In 1798, the British physicist Cavendish used a torsional equilibrium experiment to measure the gravitational force between two objects. From this, he determined the gravitational constant $\boldsymbol{G = 6.67 \times 10^{-11} \cdot Nm^2/kg^2}$..

The gravitational constant G is a very important physical constant used in many theoretical predictions and calculations. Unlike other physical constants, more than 200 years after the value of G was first determined, physicists still do not fully understand the gravitational constant.

In 1998, Bagley and Luther's team performed a torsional equilibrium experiment that measured $\boldsymbol{G = 6.674 \times 10^{-11} \cdot Nm^2/kg^2}$, which was enough to cast doubt on the previously measured G-values. The G-value results obtained by a very careful team differed from the previous experimental results by a staggering $\boldsymbol{0.15\%}$, while the previous experimental results had errors more than ten times smaller than this value. The results of this experiment shocked the scientific community at the time.

Although the gravitational constant G is the first known fundamental physical constant, its measurement accuracy is the worst of all physical constants. Over the past two centuries, scientists from different countries have measured more than two hundred values of G using different methods, but the accuracy of the measurement has improved by only about two orders of magnitude, and the values of G given by different experimental groups are not consistent within the margin of error.

At the end of 1999, the international committee CODATA decided to formally increase the error of the gravitational constant G from $\boldsymbol{0.0128\%}$ to $\boldsymbol{0.15\%}$. This unusual measure was developed to reflect the various discrepancies in the experiments described above. Of the 14 values of G included in CODATA-2014, 11 high-precision measurements were published after 2000.

In response to the current international inconsistency in high-precision G-value measurements, scientists have analyzed two possible reasons: first, there are still systematic errors that have not been properly assessed; and second, there may be unknown physical mechanisms.

6.3.3 Newton's gravitational constant *G* is a variable, not a constant

Through an experiment to measure the gravitational constant $\boldsymbol{G}$, the American scientist Long.D.R. discovered that the gravitational constant $\boldsymbol{G}$ is not a universally applicable constant in the universe.

In determining the value of the gravitational constant $\boldsymbol{G}$, the densities of the interacting spheres must be very uniform, otherwise the position of the center of mass would deviate from the center of the sphere, which would affect the accurate determination of the distance r. For this reason, Prof. Long.D.R replaced the spheres with tori. This substitution has two advantages.

One is that the gravitational field on the axis of the torus varies slowly with distance r within a certain range, and if the size and combination of the torus are properly matched, even a uniform gravitational field can be obtained over a larger range;

Second, on the axis of the torus, this gradually changing gravitational region is the region of stronger gravitational field. By placing a torsionally balanced sphere in this region, not only can the interacting center-of-mass distance r be more accurately determined, but the interacting force is also more pronounced.

Measurement experiments conducted by Prof. Long.D.R. showed two important results.

Result 1: The law that gravitational is inversely proportional to the square of the distance r is not strictly true in the laboratory context. The gravitational constant G varies slightly with the size of the ring. Note that this variation should be classified as a change in the gravitational constant due to a change in mass.

Result 2: The gravitational constant G is related to the distance r from the interacting center of mass. The formula for the gravitational constant $G(r)$ with respect to the distance r given by Prof. Lang is:

$$G(r) = G_0\left(1 + \varepsilon \ln\frac{r}{r_0}\right) \qquad (6.3.1)$$

The ε in the above equation is the correction constant and G_0 is the gravitational constant at distance r_0. From the above equation, we know that the gravitational constant $G(r)$ becomes larger as the distance r increases, or smaller as the distance r decreases. Note that we can derive the above equation using the new gravitational formula (6.1.17).

According to equation (6.1.31), the Newtonian gravitational constant G is

$$G = K\frac{1}{r} \qquad (6.3.2)$$

In an experiment to measure the gravitational constant G, suppose that r_0 is the distance between the centers of the large sphere M and the small sphere m; and suppose that G_0 is the gravitational constant determined by the experiment. Then the gravitational constant G_0 is:

$$G_0 = K\frac{1}{r_0} \qquad (6.3.3)$$

As the distance between the centers of the large sphere M and the small sphere m changes from r_0 to r, the value of the change in the gravitational constant $G(r)$ is:

$$\Delta G = K\int_{r_0}^{r}\frac{1}{r}dr = K\ln\frac{r}{r_0} \qquad (6.3.4)$$

As the distance between the centers of the large and small spheres changes from r_0 to r, the gravitational constant changes from G_0 to G, i.e:

$$G(r) = G_0 + \Delta G = G_0 + K\ln\frac{r}{r_0} \qquad (6.3.5)$$

Substituting equation (6.3.3) into the above equation gives the relationship equation:

$$G(r) = G_0 + G_0 r_0 \ln\frac{r}{r_0} = G_0\left(1 + r_0\ln\frac{r}{r_0}\right) \qquad (6.3.6)$$

Note that the distance r_0 in the formula is constant. If the distance $r_0 = \varepsilon$, then the above formula is equation (6.3.1). From this it can be seen that the gravitational constant $G = 6.67 \times 10^{-11} \cdot Nm^2/kg^2$ is not a universal physical constant in the universe.

6.3.4 Four results due to the gravitational constant G being a variable

According to equation (6.1.31), the Newtonian gravitational constant G is:

$$G = K\frac{1}{r}$$

The above equation shows that Newton's gravitational constant G is inversely proportional to the distance r.

Since the Moon's distance r from the Earth is much larger than the experimental distance r_0, i.e. $r \gg r_0$, in order for theoretical predictions of the Moon's motion around the Earth to agree with experimental observations, the gravitational constant G, which physicists use to calculate the Moon's orbit, must usually be smaller than the standard value of G_0, i.e. $G < G_0$. Otherwise, the theoretical orbit of the Moon would be very different from its actual orbit.

If the distance r between two objects is much larger than the experimental distance r_0, i.e. $r \gg r_0$ (e.g., the distance between the Earth and the Cosmic Probe), the value of Newton's gravitational constant G decreases as the distance r increases. However, physicists still use the standard gravitational constant G_0 to calculate the gravitational force F between two objects.

At this point, since the gravitational force F' calculated according to Newton's law of gravitational is much greater than the actual gravitational force F, physicists are bound to find that the spacecraft will experience tremendous drag at the edge of the solar system, as well as a slowing of the spacecraft's speed.

If the distance r between two objects is much smaller than the experimental distance r_0, i.e. $r \ll r_0$ (for example, the distance between elementary particles in the nucleus of an atom), then the value of Newton's gravitational constant G increases as the distance r decreases. However, physicists still use the standard gravitational constant G_0 to calculate the gravitational force F between two elementary particles.

At this point, since the gravitational force F' calculated according to Newton's law of universal gravitation is much smaller than the actual gravitational force F, physicists are bound to find that there is a huge suction force between the elementary particles in the nucleus. And this suction is what physicists call the strong nuclear force.

The gravitational constant G is inversely proportional to the distance r, which inevitably leads to four theoretical difficulties.

Difficulty 1: Measurements of the universal gravitational constant G are always subject to large errors.

Assuming that r_0 is the distance between two metal spheres A and B, the standard gravitational constant G_0, measured experimentally according to equation (6.1.31), is:

$$G_0 = K\frac{1}{r_0} \qquad (6.3.7)$$

According to the above equation, when the distance r_0 between the metal balls A and B changes, the value of the gravitational constant G_0 also changes.

The gravitational constant G has been accurately measured by scientists from many countries for more than 200 years. However, the gravitational constant G_0 measured by scientists from many countries inevitably has a large error because the experimental distances r_0 between the metal spheres A and B are not equal in multinational experiments.

The table below shows the errors in scientists' measurements of the gravitational constant G_0 over the past 200 years:

	Author	Year	G_0	Accuracy	%
1	Cavendish H.	1798	6.74	±0.05	+0.986
2	Richarz F. & Krigar-Menzel O.	1898	6.683	±0.011	+0.132
3	Richman etAl	1998	6.683	±0.011	+0.132
4	Luo etAl	1999	6.6699	±0.0007	-0.064
5	Fitzgerald &Armstrong	1999	6.6742	±0.0007	±0.01
6	Richman S.J. etAl	1999	6.6830	±0.0011	+0.132
7	Schurr， Noltting etAl	1999	6.6754	±0.0015	+0.018
8	Gundlach & Merkowitz	1999	6.67422	±0.00009	+0.0003
9	Quinn etAl	2000	6.67559	±0.00027	+0.021
10	Current data value	**2004**	**6.6742**	**±0.001**	**±0.0150**

Note: The data in the table is taken from the article "The Final Enlightenment of the Change in the Gravitational Constant (1)". (Http://lifterfans.bokee.com/)

Difficulty 2: If the real distance r between two objects is much larger than the experimental distance r_0, then the real gravitational force F is much smaller than Newton's gravitational force F'..

According to Newton's formula for gravitational, the gravitational force between the Earth and the spacecraft is F':

$$F' = G_0 \frac{Mm}{r^2} \qquad (6.3.8)$$

According to the new gravitational formula (6.1.17), the actual gravitational force F between the Earth and the spacecraft is:

$$F = K \frac{Mm}{r^3}$$

Substituting equation (6.3.7) into the above equation gives the relationship equation:

$$F = \frac{r_0}{r} G_0 \frac{Mm}{r^2} \qquad (6.3.9)$$

Substituting equation (6.3.8) into the above equation gives the relationship equation:

$$F = \frac{r_0}{r} F' \qquad (6.3.10)$$

The above equation shows that the greater the distance r from the spacecraft to the Earth, the greater the difference value between the gravitational force F' calculated according to Newton's law of universal gravitation and the actual gravitational force F. Assume an experimental distance $r_0 = 0.1m$ (r_0 is the experimental distance between metal ball A and metal ball B in the experiment). If the actual distance from the spacecraft to the Earth is $r = 5 \times 10^{12} m$ (500 million kilometers), the actual gravitational force F experienced by the spacecraft is:

$$F = \frac{r_0}{r} F' = \frac{0.1}{5 \times 10^{12}} F' = 2 \times 10^{-14} F' \qquad (6.3.11)$$

The above equation shows that Newton's gravitational force F' is much larger than the actual universal gravitational force F.

As the spacecraft moves away from the Sun, physicists believe that the spacecraft's orbit should change according to Newton's gravitational force F', but in reality the spacecraft's orbit changes according to the new gravitational force F. The farther the spacecraft is from the Sun, the greater the deviation $\Delta F = F' - F$ between Newton's gravitational force F' and the new gravitational force F. The increasing deviation ΔF corresponds to the increasing drag that the spacecraft experiences as it moves away from the Sun.

Because Newton's gravitational formula was wrong, physicists believed there was a mysterious force in the universe that prevented spacecraft from moving rapidly away from the sun. The Galileo spacecraft's orbit away from the sun deviated significantly from what the theory predicted, proving this point.

Difficulty 3: If the real distance r between two objects is much smaller than the experimental distance r_0, then the real gravitational force F is much larger than Newton's gravitational force F'.

Physicists believe that the gravitational constant G_0 applies to the gravitational force between any two objects. The Newtonian universal gravitational force F' that an electron in a hydrogen atom receives is:

$$F' = G_0 \frac{Mm}{r^2} \qquad (6.3.12)$$

According to the new gravitational formula, the gravitational force F on an electron in a hydrogen atom is:

$$F = K \frac{Mm}{r^3} = \frac{r_0}{r} G_0 \frac{Mm}{r^2} = \frac{r_0}{r} F' \qquad (6.3.13)$$

The above formula shows that the smaller the radius r of the electron's orbit of motion, the greater the difference value $\Delta F = F' - F$ between the Newtonian universal gravitational force F' and the actual gravitational force F. Assume that the distance between the metal sphere A and the metal sphere B is $r_0 = 0.1m$. The radius $r = 0.528 \times 10^{-10} m$ of the electron orbital in the ground state of the hydrogen atom,. Substituting the two data into the above equation yields the relational equation:

$$F = \frac{r_0}{r} F' = \frac{0.1}{0.528 \times 10^{-10}} F' = 1.894 \times 10^{9} F' \qquad (6.3.14)$$

The above equation shows that Newton's gravitational force F' is much smaller than the actual gravitational force F. From this it can be seen that the strong nuclear force F in the nucleus of an atom is the microscopic manifestation of gravitational.

In other words, scientists have determined from spectroscopic theory and scattering experiments that the radii of protons and neutrons are $\mathbf{0.87 \times 10^{-15}} \boldsymbol{m}$. The resulting minimum distance between protons and neutrons is $\mathbf{1.74 \times 10^{-15}} \boldsymbol{m}$. Assuming a distance $\boldsymbol{r} = \mathbf{2 \times 10^{-15}} \boldsymbol{m}$ between protons and neutrons, substituting this value into equation (6.3.14) yields the relational equation:

$$\boldsymbol{F} = \frac{\boldsymbol{r_0}}{\boldsymbol{r}}\boldsymbol{F'} = \frac{\mathbf{0.1}}{\mathbf{2 \times 10^{-15}}}\boldsymbol{F'} = \mathbf{5 \times 10^{13}}\boldsymbol{F'} \qquad (\mathbf{6.3.15})$$

The above equation shows that Newton's gravitational force $\boldsymbol{F'}$ is much smaller than the actual gravitational force $\boldsymbol{F}$. In other words, since Newton's gravitational force $\boldsymbol{F'}$ reduces the actual gravitational force $\boldsymbol{F}$ in the nucleus of an atom by a factor of $\mathbf{10^{13}}$, physicists believe that there is a strong nuclear force in the nucleus of an atom and that the gravitational force is a very small, weak force.

Difficulty 4: Leading to the Dark Energy and Dark Matter hypotheses.

To explain the accelerated expansion of the universe and the swirling motion of galaxies, physicists have developed the dark energy and dark matter hypothesis. Physicists believe that the accelerated expansion of the universe and the motion of galaxies are driven not by visible matter, but by dark energy and dark matter. However, physicists have not been able to observe any experimental evidence for the existence of dark energy and dark matter.

In summary, there is a large error in the measurement of the gravitational constant $\boldsymbol{G}$, and the actual orbit of the spacecraft does not conform to Newton's formula for gravitational. These two results are easy for physicists to discover. But so far, physicists have not found that the actual gravitational force $\boldsymbol{F}$ in the nucleus of an atom is much greater than Newton's universal gravitational force $\boldsymbol{F'}$. It is hoped that experimental physicists will discover and prove this result experimentally in the future.

6.3.5 The Casimir effect as a manifestation of gravitational in the microscopic realm

According to quantum mechanics, the universe is filled with zero-point energy particles of different wavelengths. Particles that can be removed are called real particles, while zero-point energy particles in a vacuum cannot be removed. Zero energy particles are not really particles; they are simply the lowest energy ground state, a variety of ever-present, rapidly changing, indivisible, energetic fields.

The well-known Casimir effect in quantum mechanics refers to the attractive pressure $\boldsymbol{F}$ between two parallel metal plates in a vacuum. This pressure $\boldsymbol{F}$ is caused by a smaller than normal number of virtual particles in the space between the plates. The peculiarity of this theory is that the Casimir force $\boldsymbol{F}$ normally causes only mutual attraction between objects, not mutual repulsion. This phenomenon was proposed by the Dutch physicist Hendrik Casimir in 1948, and the effect has since been discovered and named after him.

In 1948, the Dutch physicist Casimir proposed a program to detect the presence of Zero-Point Energy and theorized that Zero-Point Energy would take the form of particles. Zero-point energy has appeared and disappeared on a small scale.

Under normal circumstances, the vacuum is filled with zero-point energy particles of all wavelengths. But Casimir pointed out that if the tiny distance between two uncharged thin metal plates is $\boldsymbol{d}$, then only zero-point energy with a wavelength $\boldsymbol{\lambda}$ less than $\mathbf{2}\boldsymbol{d}$ can exist between the two plates.

In this case, the number of zero-point energy particles bound between the planes is much smaller than the number of zero-point energy particles outside, and therefore the energy of the fluctuations is smaller. The difference value between the inner and outer vacuum energy fluctuations creates an inward thrust $\boldsymbol{F}$ (force per unit area) on the plates, causing them to appear to be attracted to each other by the Casimir effect.

If two uncharged metal disks are close together, particles with longer wavelengths will be excluded from the metal disks. At this point the other wave particles outside the metal disks will create a thrust $\boldsymbol{F}$ that will bring the thin metal disks together. The closer the metal disks are the stronger the thrust $\boldsymbol{F}$. This phenomenon is known as the Casimir effect. it was first measured by physicists in 1996, and the actual measurements agree well with the theoretical calculations.

According to physicists, the Casimir effect is an important experimental result of quantum field theory because it confirms the existence of zero-point energy. Moreover, the most important significance of the Casimir force is that it is one of the macroscopic effects of quantum phenomena.

The authors of this book argue that there are three theoretical flaws in the physicists' explanation of the Casimir effect.

The first theoretical flaw is that physicists do not take into account that a smaller wavelength of zero-point energy, A, can enter the space between the two thin metal disks and thus displace the larger wavelength of zero-point energy, B. In this way, the number of zero-point energy particles in the space inside and outside the two thin metal disks does not change.

In other words, the positions of excluded zero-point energies of larger wavelengths can be filled by other energy particles of smaller wavelengths. Furthermore, physicists have not experimentally demonstrated that there is a difference value in the amount of zero point energy between the inside and outside of a thin metal disk. The zero-point energy difference value used by physicists to explain the Casimir effect is not theoretically convincing.

The second theoretical flaw is that physicists do not take into account that the energy of zero-point energy particles with smaller wavelengths is greater than the energy of zero-point energy particles with larger wavelengths.

According to quantum mechanics, the energy ε of a photon is:

$$\varepsilon = \frac{hc}{\lambda} \qquad (6.3.16)$$

The h in the formula is Planck's constant and λ is the wavelength of the photon. The above equation shows that the energy ε of a photon is inversely proportional to the wavelength λ.

Since the zero-point energy is considered to be the vacuum energy in the universe, the energy ε possessed by the zero-point energy should also satisfy the above formula, i.e., the shorter the wavelength λ of the zero-point energy, the greater the energy ε.

Since the zero-point energy particle A of smaller wavelength can enter the space inside the metal disk and replace the zero-point energy particle B of larger wavelength, the zero-point energy per unit area between the two metal disks should be greater than the zero-point energy per unit area outside the metal disks. Since the energy ε' per unit area inside the metal disk is greater than the energy ε per unit area outside the metal disk, the force F inside the metal disk should be a repulsive force rather than an attractive force.

The third theoretical flaw is that physicists do not consider the effects of gravitational.

Although there is no electromagnetic force between two neutral metal disks, there is a gravitational force F between the two disks. Inexplicably, physicists have not yet analyzed the Casimir effect in terms of gravitational. The authors suggest that this may be due to the incorrect Newtonian formula for gravitational.

The Casimir effect is very small and can only be detected when the distance r between two metal disks is small. However, the effect has a distinctly macroscopic nature. We know that 1 nanometer is equal to $10^{-9}m$ meter. If the distance between two thin metal disks is equal to 100 nanometers, which is the distance $r = 10^{-7}m$, the distance r is inserted into equation (6.3.13) to obtain the relational equation:

$$F = \frac{r_0}{r}F' = \frac{0.1}{10^{-7}}F' = 10^8 F' \qquad (6.3.17)$$

F' in the formula is the Newtonian gravitational force between the two metal disks, while F is the actual gravitational force between the two metal disks. The above formula shows that the Newtonian gravitational force F' is $10^8 m$ times smaller than the actual gravitational force F. Obviously, the actual gravitational force F in the Casimir effect has reached the macroscopic level. From this, it can be concluded that the Casimir effect is a manifestation of gravitational in the microscopic realm.

6.3.6 Newton's law of gravitational does not apply to microscopic particles moving at high speeds

The relationship between the new gravitational force F and the Newtonian gravitational force F' is given by equation (6.3.13):

$$F = \frac{r_0}{r}F'$$

The distance r_0 in the formula is the distance between two metal spheres in an experiment to measure the gravitational constant G. The distance r_0 is usually about $0.15\mathrm{m}$. The above formula shows that the smaller the radius r, the greater the deviation between Newton's gravitational force F' and the new gravitational force F.

According to atomic physics, the strong nuclear force is a short-range force with a range of action $1.5 \times 10^{-15} m$. Assuming an experimental distance $r_0 = 0.15m$, the relationship is obtained by substituting the two data into the above equation:

$$F = \frac{0.15}{1.5 \times 10^{-15}} F' = 1 \times 10^{14} F'$$

The above equation shows that Newton's microscopic universal gravitational force F' is much smaller than the actual universal gravitational force F. Based on the above equation, it can be determined that the strong nuclear force in the nucleus of an atom is the universal gravitational force.

Chapter 7 Sources of intrinsic mass m of an object

Introduction: The authors define the energy ε_h contained in Planck's constant h as an energy element. According to the mass-energy equivalence formula $\varepsilon = mc^2$, the energy ε and the mass m can be transformed into each other. Since the energy element ε_h possesses the mass m_h, and the mass m_h of the energy element ε_h is the smallest mass particle, the authors argue that the intrinsic mass m_c of a photon, as well as the intrinsic mass m of any object, consists of a certain amount of the mass m_h of the energy element.

7.1 Spiral speed of light conservation formula for energy element ε_h

7.1.1 Axiom of invariant speed of light for energy element ε_h

According to optics, photons have both fluctuating and particle nature. Diffraction and interference of light waves prove the fluctuating nature of photons. The photoelectric effect proves the particle nature of photons. The particle nature of photons is proven by the fact that photons can only carry quantized energy and cannot carry any value of energy.

The quantum nature of light waves was first discovered by the German scientist Planck. In 1901, Planck derived the formula for blackbody radiation consistent with electromagnetic wave radiation experiments, viz:

$$\rho = \frac{8\pi h\beta^3}{c^3}\frac{1}{e^{\frac{h\beta}{kT}} - 1}$$

In the formula, ρ is the energy density per unit frequency spacing; h is Planck's constant; β is the frequency of the radiated electromagnetic wave; k is Boltzmann's constant; and T is the temperature.

Planck found that the calculations of the formula would agree with the experimental results only if it was assumed that the emission and absorption of electromagnetic wave energy is not continuous, but is emitted and absorbed one by one in the form of particles. Physicists call such energy particles energons. The energy ε_c contained in each energon is equal to $h\beta$.

In 1905, Einstein further proposed that light waves are not continuous but are particles. Einstein called the energy particles contained in the light waves light quanta.In 1923, Compton used the concept of light quanta to successfully explain the Compton effect. The concept of light quanta was widely accepted and in 1926 light quanta were officially named photons by physicists.

According to optics, the energy ε_c of a photon is:

$$\varepsilon_c = h\beta \tag{7.1.1}$$

The β in the formula is the frequency of the light wave and h is Planck's constant. The energy ε_c of a photon is proportional to the frequency β of the photon. The speed of the photon's motion in the cosmic vacuum is the speed of light c.

The 26th International Conference on Weights and Measures (CGPM) voted that the exact value of Planck's constant h is:

$$h = 6.626 \times 10^{-34} J \cdot s = 4.135 \times 10^{-15} eV \cdot s$$

The $J \cdot s$ in the formula is joule-seconds; $eV \cdot s$ is electron volt-seconds. According to Equation (7.1.1), the energy ε_h contained within Planck's constant h is when the light wave frequency $\beta = 1$:

$$\varepsilon_h = h \times 1 = 6.626 \times 10^{-34} J = 4.135 \times 10^{-15} eV \tag{7.1.2}$$

Since the energy ε_h contained in Planck's constant h is the smallest energy particle that cannot be divided, this book defines the energy ε_h contained in Planck's constant h as an energy element, i.e:

$$\varepsilon_h = h \times 1 \tag{7.1.3}$$

The above equation is the definition of the energy element ε_h. In physics, the physical unit of Planck's constant h is $J \cdot s$ (energy × time). Although energy element ε_h and Planck's constant h have equal values, they have different physical units. The physical unit of energy element ε_h is J (joule).

Different photons have different light wave frequencies β. If the light wave frequency $\beta = 1$, the photon energy ε_c in equation (7.1.1) is equal to the energy element ε_h, i.e:

$$\varepsilon_c = h \times 1 = \varepsilon_h \tag{7.1.4}$$

According to the energy formula $\varepsilon_c = h\beta$ and the energy element formula $\varepsilon_h = h \times 1$ for photons, the energy ε_c of each photon is composed of the energy element ε_h, i.e:

$$\varepsilon_c = \sum_{i=1}^{v} \varepsilon_{hi} \tag{7.1.5}$$

It is clear that different photons contain different amounts β of the energy element ε_h.

In the vacuum of the universe, the speed of propagation of photons is always equal to the speed of light c. The authors define this property of the energy element ε_h as the axiom of invariance of the speed of light for energy elements. This leads to 8 theorems related to the speed c of light.

7.1.2 Difference between the new speed of light invariance theorem and the speed of light invariance principle of special relativity

The implication of the new speed of light invariance theorem is that the speed of propagation of light in a cosmic vacuum frame of reference is always equal to the speed of light c. The special relativity theory of the speed of light invariance principle implies that the speed of light in the vacuum of any inertial frame of reference is equal to the speed of light c.

To avoid misunderstanding by the reader, the following explanation illustrates the difference between the two.

(1) Implications of the principle of invariance of the speed of light for special relativity.

Special relativity states that the rights of different inertial frames of reference are equal and that there is no absolutely stationary frame of reference. The meaning of the special relativity principle of the invariance of the speed of light is that in the vacuum of any inertial frame of reference, the speed of propagation of light in any direction is equal to the speed of light c, and the speed of light c is independent of the speed of the inertial frame of reference. From this it can be seen that the "vacuum" in the special relativity principle of the invariance of the speed of light refers to the vacuum in different inertial frames of reference.

In other words, if two giant vacuum bottles A and B are moving in opposite directions. Special relativity tells us that the speed of light in each direction in the vacuum bottles A and B is equal to the speed of light c.

In the cosmic vacuum frame of reference (inertial system S), u is assumed to be the velocity of the terrestrial frame of reference (inertial system S'). If the vacuum bottle is at rest on the ground, then the vacuum in the bottle at that point is the vacuum of the terrestrial frame of reference. Since the ground is always moving, the vacuum in the vacuum bottle on the ground is now a vacuum in motion.

To avoid confusion and misunderstanding, let V be the vacuum in the cosmic vacuum frame of reference S. Let V' be the vacuum in the inertial frame of reference S'. Note that vacuum V' moves in vacuum V. Vacuum V' belongs to Einstein's inertial frame of reference. Obviously, both the Earth and the Sun are moving in the cosmic vacuum V. In addition, the Earth is moving in vacuum V' in the solar frame of reference.

According to special relativity, there is no absolutely stationary frame of reference, and all inertial frames of reference are equal in status. special relativity's principle of invariance of the speed of light means that in the vacuum of any inertial frame of reference, the speed of light propagating in all directions is equal to the speed of light c, and that the speed of light c is independent of the speed v of the frame of reference.

From this it can be seen that the vacuum in the principle of invariance of the speed of light includes not only the vacuum of the universe, but also the vacuum of any inertial frame of reference.

Since Einstein confused both the cosmic vacuum $\boldsymbol{V}$, which is at rest, and the inertial vacuum $\boldsymbol{V}'$, which is moving in the cosmic vacuum $\boldsymbol{V}$, as the same physical quantity, i.e., both $\boldsymbol{V}$ and $\boldsymbol{V}'$ are considered to be the cosmic vacuum $\boldsymbol{V}$. which led him to believe that the velocity of light in any inertial frame of reference is equal to the speed of light, $\boldsymbol{c}$. From this it can be ascertained that the principle of invariance of the speed of light in the special relativity and the speed vector calculation formula contradict each other.

(2) Implications of the new speed of light invariance theorem

The implication of the new invariant speed of light theorem is that in the cosmic vacuum frame of reference, light propagates through space in all directions at a speed equal to the speed of light $\boldsymbol{c}$. The speed of light $\boldsymbol{c}$ is independent of the state of motion of the light source.

In the vacuum $\boldsymbol{V}'$ of the inertial system, light usually travels at different speeds in all directions in space. In other words, the law of invariance of the speed of light applies only to the cosmic vacuum frame of reference; it does not apply to any other inertial system $\boldsymbol{S}'$ in motion..

Suppose the volume V of the vacuum bottle A is infinite. If the vacuum bottle A is stationary in the cosmic space frame of reference, then the volume V of the vacuum bottle A is the vacuum in the cosmic vacuum frame of reference. At this point, the speed of propagation of light in all directions in the vacuum bottle A is equal to the speed of light $\boldsymbol{c}$. Obviously, the Earth and the Sun move in the vacuum bottle A.

In the cosmic vacuum frame of reference $\boldsymbol{S}$, it is assumed that $\boldsymbol{v}$ is the velocity of motion in the inertial frame of reference $\boldsymbol{S}'$. According to Galileo's velocity transformation formula, the propagation velocities of light in the moving frame of reference $\boldsymbol{S}'$ are generally unequal. Obviously, in the moving frame of reference $\boldsymbol{S}'$, the speed of propagation of light along the direction of the velocity $\boldsymbol{v}$ is $\boldsymbol{c}' = \boldsymbol{c} - \boldsymbol{v}$, and the speed of propagation along the opposite direction of the velocity $\boldsymbol{v}$ is $\boldsymbol{c}' = \boldsymbol{c} + \boldsymbol{v}$.

7.1.3 Spiral speed of light conservation equation for the energy element ε_h

Suppose $\boldsymbol{v}$ is the displacement velocity of the object in the cosmic vacuum reference frame. Since the intrinsic mass $\boldsymbol{m}$ of the object consists of the intrinsic mass $\boldsymbol{m}_i$ of the elementary particles, and the intrinsic mass $\boldsymbol{m}_i$ consists of the mass $\boldsymbol{m}_h$, which is an energy element, the energy element $\boldsymbol{\varepsilon}_h$ contained in the intrinsic mass $\boldsymbol{m}_i$ should be displaced with a velocity equal to $\boldsymbol{v}$, i.e. the energy element $\boldsymbol{\varepsilon}_h$ is also displaced forward with the velocity $\boldsymbol{v}$.

Since the velocity of the energy element $\boldsymbol{\varepsilon}_h$ in the cosmic vacuum frame of reference is always equal to the velocity of light $\boldsymbol{c}$, and since the energy element $\boldsymbol{\varepsilon}_h$ can only move inside the intrinsic mass $\boldsymbol{m}_i$ of the elementary particles, the energy element $\boldsymbol{\varepsilon}_h$ inside the intrinsic mass $\boldsymbol{m}_i$ should have a circumferential velocity $\boldsymbol{v}_\omega$ perpendicular to the velocity $\boldsymbol{v}$, according to the velocity arithmetic formula. Alternatively, the energy element $\boldsymbol{\varepsilon}_h$ always rotates inside the intrinsic mass $\boldsymbol{m}_i$ with a circumferential velocity $\boldsymbol{v}_\omega$ around the center of intrinsic mass $\boldsymbol{m}_i$.

We know that electrons have a velocity of motion around the nucleus. In order to distinguish between the rotational speed of the energy element $\boldsymbol{\varepsilon}_h$ and the intrinsic mass $\boldsymbol{m}_i$, this book defines the circular speed $\boldsymbol{v}_\omega$ of the energy element $\boldsymbol{\varepsilon}_h$ rotating around the center O of the intrinsic mass $\boldsymbol{m}_i$ as the internal rotational speed of the energy element.

Since the energy element $\boldsymbol{\varepsilon}_h$ cannot move in a straight line within the intrinsic mass $\boldsymbol{m}_i$, but can only rotate in a spiral (i.e. circular motion + displacement motion), the straight-line speed of light $\boldsymbol{c}$ of the energy element $\boldsymbol{\varepsilon}_h$ changes to the spiral speed of light $\boldsymbol{c} = \sqrt{\boldsymbol{v}_\omega^2 + \boldsymbol{v}^2}$ within the intrinsic mass $\boldsymbol{m}_i$ at this point in time.

In other words, when the energy element $\boldsymbol{\varepsilon}_h$ has an internal rotational speed $\boldsymbol{v}_\omega \neq \boldsymbol{0}$ and a displacement speed $\boldsymbol{v} < \boldsymbol{c}$. The motion of the energy element $\boldsymbol{\varepsilon}_h$ is no longer a linear motion, but a spiral motion rotating around the center O of mass. This spiral motion is similar to the rotation of the moon around the earth.

We know that the Earth moves around the center O of mass of the solar system under the action of the gravitational force $\boldsymbol{F}$ and the centripetal force $\boldsymbol{f}$. Similarly, under the action of the universal gravitational force $\boldsymbol{F}$ and centripetal force $\boldsymbol{f}$, the energy elements $\boldsymbol{\varepsilon}_h$ contained within the elementary particles all revolve around the center O of mass of the elementary particles with Spiral speed of light $\boldsymbol{c} = \sqrt{\boldsymbol{v}_\omega^2 + \boldsymbol{v}^2}$

There is a type of binary stellar motion in the universe. That is, the motion of star A and star B around the center of both. If the two energy elements A and B are regarded as a binary stellar motion, then the motion of the two energy elements A and B around the center O of mass of the elementary particles is analogous to the motion of a binary star around its center.

Assume that $\boldsymbol{\delta}$ is the angle between the displacement velocity $\boldsymbol{v}$ of the elementary particle and the Spiral speed of light $\boldsymbol{c} = \sqrt{\boldsymbol{v}_\omega^2 + \boldsymbol{v}^2}$. The relationship between the Spiral speed of light $\boldsymbol{c}$, the displacement velocity $\boldsymbol{v}$, and the internal rotation velocity $\boldsymbol{v}_\omega$ is shown in Figure 7-1.

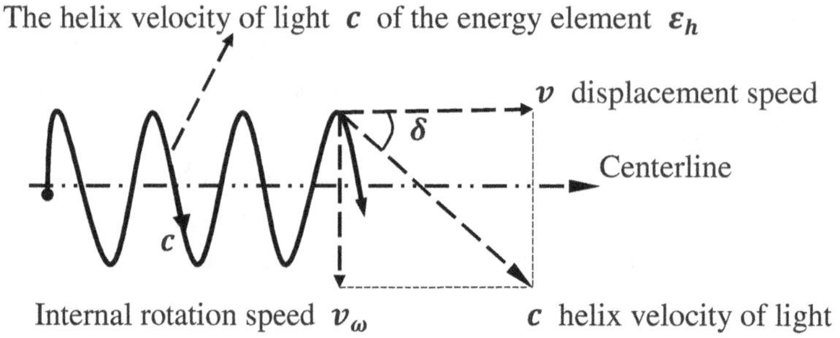

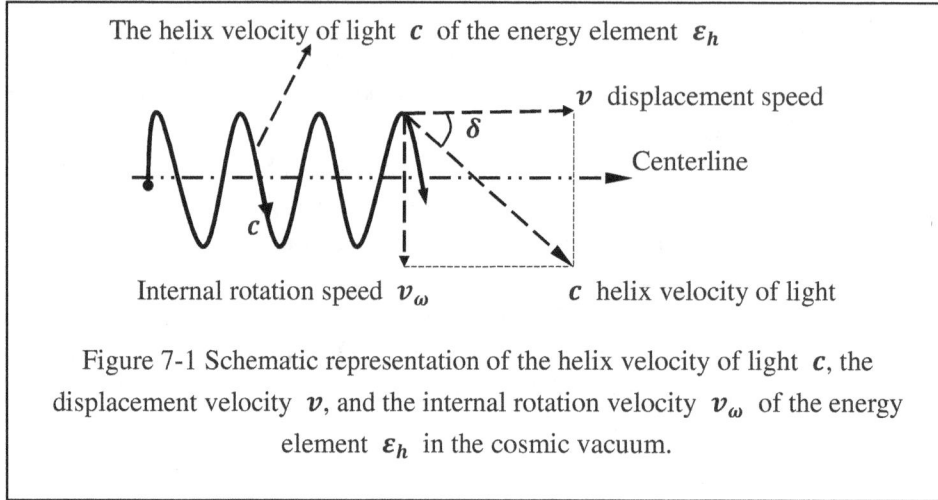

Figure 7-1 Schematic representation of the helix velocity of light $\boldsymbol{c}$, the displacement velocity $\boldsymbol{v}$, and the internal rotation velocity $\boldsymbol{v}_\omega$ of the energy element $\boldsymbol{\varepsilon}_h$ in the cosmic vacuum.

In the cosmic vacuum frame of reference, when the elementary particle $\boldsymbol{m}_i$ is at rest. According to the formula for the spiral speed of light of the energy element $\boldsymbol{c} = \sqrt{\boldsymbol{v}_\omega^2 + \boldsymbol{v}^2}$. The displacement velocity $\boldsymbol{v} = \boldsymbol{0}$ for the energy element $\boldsymbol{\varepsilon}_h$; the internal rotation velocity $\boldsymbol{v}_\omega = \boldsymbol{c}$; the clamping angle $\boldsymbol{\delta} = \boldsymbol{90^0}$. At this time, the trajectory of the energy element $\boldsymbol{\varepsilon}_h$ is a circular line.

In the cosmic vacuum frame of reference, when the displacement velocity $\boldsymbol{v}$ of the elementary particle $\boldsymbol{m}_i$ is equal to the speed of light $\boldsymbol{c}$, i.e. $\boldsymbol{v} = \boldsymbol{c}$. According to the formula $\boldsymbol{c} = \sqrt{\boldsymbol{v}_\omega^2 + \boldsymbol{v}^2}$ for the Spiral speed of light of the energy element. The internal rotational speed of energy element $\boldsymbol{\varepsilon}_h$ is $\boldsymbol{v}_\omega = \boldsymbol{0}$, and the clamp angle $\boldsymbol{\delta} = \boldsymbol{0}$. At this time, the trajectory of energy element $\boldsymbol{\varepsilon}_h$ is a straight line.

In the cosmic vacuum frame of reference, when the elementary particle $\boldsymbol{m}_i$ moves with the velocity $\boldsymbol{v}$. The motion path of the energy element $\boldsymbol{\varepsilon}_h$ inside the elementary particle $\boldsymbol{m}_i$ is a spiral, and the tangent speed of its path is always equal to the spiral speed of light $\boldsymbol{c}$. At this point, the spiral speed of light $\boldsymbol{c}$ can be decomposed into two perpendicular component speeds.

The velocity of the 1st component is the displacement velocity $\boldsymbol{v}$ of the energy element $\boldsymbol{\varepsilon}_h$, i.e:

$$\boldsymbol{v} = \boldsymbol{c} \cdot \cos \boldsymbol{\delta} \tag{7.1.6}$$

The displacement velocity $\boldsymbol{v}$ in equation is the velocity in the cosmic vacuum frame of reference. Note that the displacement velocity $\boldsymbol{v}$ is not only the displacement velocity of the elementary particle $\boldsymbol{m}_i$, but also the displacement velocity of the energy element $\boldsymbol{\varepsilon}_h$.

The 2nd component velocity is the internal rotational velocity $\boldsymbol{v}_\omega$ of the energy element $\boldsymbol{\varepsilon}_h$, viz:

$$\boldsymbol{v}_\omega = \boldsymbol{c} \cdot \sin \boldsymbol{\delta} \tag{7.1.7}$$

The internal rotation velocity $\boldsymbol{v}_\omega$ in equation is the velocity in the cosmic vacuum frame of reference. Note that the internal rotation velocity $\boldsymbol{v}_\omega$ is not the lateral displacement of the elementary particle $\boldsymbol{m}_i$. It should be the spin velocity of the elementary particle $\boldsymbol{m}_i$.

The internal rotation velocity $\boldsymbol{v}_\omega$ of the energy element $\boldsymbol{\varepsilon}_h$ is an invisible and imperceptible circular velocity hidden in the intrinsic mass $\boldsymbol{m}$ of the object. In other words, the internal rotation velocity $\boldsymbol{v}_\omega$ is the rotation velocity of the energy element $\boldsymbol{\varepsilon}_h$ and the spin velocity of the elementary particle $\boldsymbol{m}_i$. The internal rotational velocity $\boldsymbol{v}_\omega$ is neither the displacement velocity of the elementary particle $\boldsymbol{m}_i$ nor the spin velocity of the object.

The author believes that the internal rotation speed $v_{\omega} \neq 0$ of the energy element ε_h is the only reason for the formation of the intrinsic mass m of the object. When the internal rotation speed $v_{\omega} = 0$ of the energy element ε_h, the matter composed of the energy element ε_h at that time is a photon.

The spiral speed of light c of the energy element ε_h inside the elementary particle m_i is:

$$c = \sqrt{v_{\omega}^2 + v^2} \tag{7.1.8}$$

The above formula is the invariant formula for the spiral velocity of light for the energy element ε_h. Note that the spiral velocity of light c, the spin velocity v_{ω}, and the displacement velocity v are all velocities in the cosmic vacuum frame of reference. The invariant formula for the spiral speed of light is similar to the Pythagorean theorem in mathematics.

The Spiral speed of light invariant formula $c = \sqrt{v_{\omega}^2 + v^2}$ is a very important physical formula. It can be used to derive many important experimental laws in the classical electromagnetic field theory. From this point of view, the Spiral speed of light invariant formula is the cornerstone of electromagnetic field theory.

In the cosmic vacuum frame of reference S, assume that the inertial system S' moves with velocity u in the direction of the positive X-axis. If light is also displaced along the positive X-axis, then according to the Galilean velocity transformation formula, the propagation velocity of light in S' is c':

$$c' = c - u \geq 0 \tag{7.1.9}$$

In the inertial system S', let v'_{ω} be the internal rotation speed of the energy element ε_h; let v' be the displacement speed of the energy element ε_h along the direction of the axis of positive X'. According to the axiom of the invariant speed of light in the cosmic vacuum, the velocity v' takes the range of values:

$$u \leq v' \leq c - u \tag{7.1.10}$$

Since the internal rotational velocity v'_{ω} and the velocity u are perpendicular to each other, the internal rotational velocity $v'_{\omega} = v_{\omega}$. The resulting invariant formula for the Spiral speed of light for the inertial system S' is:

$$c' = c - u = \sqrt{v'^2_{\omega} + v'^2} = \sqrt{v_{\omega}^2 + v'^2} \tag{7.1.11}$$

The above formula is the formula for the invariant speed of light in the inertial system S'. The speed of light c' in the formula is the maximum speed along the positive X' axis. When the velocity $u = 0$ of the inertial system S', the above formula becomes the Spiral speed of light invariant formula for the cosmic vacuum frame of reference. At this point the speed of light $c' = c$ of the inertial system S'.

7.2 Sources of intrinsic mass *m* of an object

7.2.1 Derivation of a new mass-energy conversion equation

In the cosmic vacuum frame of reference, when an object is moving at velocity v. According to the mass-velocity formula of special relativity, the relationship between the rest mass m_0 of the object and the moving mass m is:

$$m = \frac{m_0}{\sqrt{1 - \frac{v^2}{c^2}}} \tag{7.2.1}$$

The mass-energy conversion formula for special relativity is:

$$E' = mc^2 = \frac{m_0 c^2}{\sqrt{1 - \frac{v^2}{c^2}}} \tag{7.2.2}$$

The speed of light c in the formula has nothing to do with the mass m of the object. The above formula shows that the value of mass energy $E' = mc^2$ increases with the increase in velocity v.

Since the above two formulas are not applicable to objects moving at the speed of light, the above two formulas are theoretically imperfect. To overcome the shortcomings of the above two formulas, this book derives a new mass-energy conversion formula based on the spiral speed of light formula $c = \sqrt{v_{\omega}^2 + v^2}$.

Let m_h be the mass of the energy element ε_h. In the cosmic vacuum, the energy element ε_h always travels at the speed of light c. According to the momentum formula $P = mv$, the momentum P_h of the energy element ε_h in the cosmic vacuum is:

$$P_h = m_h c \tag{7.2.3}$$

Substituting the spiral speed of light formula $c = \sqrt{v_\omega^2 + v^2}$ into the above formula gives the relational formula:

$$P_h = m_h c = m_h\sqrt{v_\omega^2 + v^2} \tag{7.2.4}$$

Squaring the above formula gives the relationship formula:

$$P_h^2 = m_h^2 c^2 = m_h^2(v_\omega^2 + v^2)$$

Dividing the above equation by the energy element mass m_h gives the relationship equation:

$$\frac{P_h^2}{m_h} = m_h c^2 = m_h v_\omega^2 + m_h v^2 \tag{7.2.5}$$

According to the kinetic energy formula $E = \frac{1}{2}mv^2$, $m_h v^2$ in the formula is energy. Therefore, both $m_h c^2$ and $m_h v_\omega^2$ in the formula are also energy.

Since m_h is the mass of the energy element, the energy $m_h c^2$ belongs to the energy possessed by the energy element ε_h, i.e:

$$\varepsilon_h = m_h c^2 = m_h v_\omega^2 + m_h v^2 \tag{7.2.6}$$

The above formula is the mass-energy conversion formula for the energy element ε_h. The above equation shows that the energy of the energy element ε_h can be decomposed into two component energies.

Since the internal rotation speed v_ω belongs to the tangential speed of the circle, this book defines the energy $m_h v_\omega^2$ in equation (7.2.6) as the internal rotation energy $\varepsilon_{h\omega}$ of the energy element, i.e:

$$\varepsilon_{h\omega} = m_h v_\omega^2 = m_h(c^2 - v^2) \tag{7.2.7}$$

The above formula is the internal rotational mass-energy conversion formula for an energy element. The above formula shows that the internal rotational energy $\varepsilon_{h\omega}$ of an energy element is proportional to the square of the internal rotational velocity v_ω.

Since the velocity v is the displacement velocity of the energy element ε_h, this book defines the energy $m_h v^2$ in equation (7.2.6) as the displacement energy ε_{hv} of the energy element, i.e:

$$\varepsilon_{hv} = m_h v^2 \tag{7.2.8}$$

The above equation is the displacement mass-energy conversion equation for an energy element. The above equation shows that the displacement energy ε_{hv} of the energy element is proportional to the square of the displacement velocity v.

Theoretically, since the intrinsic mass m_i of an elementary particle contains many energy elements ε_h, the intrinsic mass m_i is a collection of many energy element masses m_h, i.e:

$$m_i = \sum_i m_h \tag{7.2.9}$$

Multiplying the above equation by the square c^2 of the speed of light c gives the relational equation:

$$E_i = m_i c^2 = \sum_i m_h c^2 = \sum_i \varepsilon_h \tag{7.2.10}$$

Please note that the speed of light c in the formula is the Spiral speed of light of the energy element ε_h, not the displacement speed of the elementary particle m_i. Since the formula $\sum m_h c^2 = \sum \varepsilon_h$ is the set of energy, therefore $E_i = m_i c^2$ is the mass-energy conversion formula of elementary particle m_i.

Similarly, since the intrinsic mass m of an object contains the intrinsic masses m_i of many elementary particles, the intrinsic mass m of an object is a collection of many elementary particle intrinsic masses m_i, viz:

$$m = \sum_i^n \sum_i m_i \tag{7.2.11}$$

Multiplying the above equation by the square c^2 of the speed of light c gives the relational equation:

$$E = mc^2 = \sum_i^n \sum_i m_i c^2 \tag{7.2.12}$$

The above formula is the new mass-energy conversion formula. Note that the speed of light c in the formula is the Spiral speed of light of the energy element ε_h inside the intrinsic mass m_i of the elementary particle, not the displacement speed of the object.

Substituting spiral speed of light $c = \sqrt{v_\omega^2 + v^2}$ into equation (7.2.12), the new mass-energy conversion equation is:

$$E = mc^2 = mv_\omega^2 + mv^2 \tag{7.2.13}$$

According to equation (7.2.2), the mass-energy conversion formula for special relativity is:

$$E' = mc^2 = \frac{m_0 c^2}{\sqrt{1 - \frac{v^2}{c^2}}}$$

Compare the above formula with formula (7.2.13). The difference between the new mass-energy conversion formula and the mass-energy conversion formula for special relativity is as follows.

Difference 1: The new mass-energy conversion formula $E = mc^2$ holds for any object. Whereas special relativity's mass-energy conversion formula $E' = mc^2$ does not hold for objects moving at the speed of light.

Difference 2: The magnitude of the new mass-energy conversion formula $E = mc^2$ is independent of the change in velocity v, whereas the magnitude of the special relativity mass-energy conversion formula $E' = mc^2$ increases with increasing velocity v.

Difference 3: The speed of light c in the new mass-energy conversion formula $E = mc^2$ is the Spiral speed of light contained in the intrinsic mass m. And the speed of light c in the mass-energy conversion formula $E' = mc^2$ of special relativity has nothing to do with the intrinsic mass m.

Difference 4: The new mass-energy conversion formula $E = mc^2$ contains the spin energy of the object. And the mass-energy conversion formula $E' = mc^2$ of special relativity does not contain the spin energy of the object.

Difference 5: The new mass-energy conversion formula $E = mc^2$ satisfies the conservation of mass law. And the mass-energy conversion formula $E' = mc^2$ of special relativity does not satisfy the conservation of mass law.

Difference 6: The new mass-energy conversion formula $E = mc^2$ can be decomposed into two perpendicular components. The mass-energy conversion formula $E' = mc^2$ of special relativity cannot be decomposed into two perpendicular components.

It should be noted that since the new mass-energy conversion formula $E = mc^2$ is not derived using the mass-velocity formula of special relativity, the theoretical analyses and discussions that will unfold in the following part of this book have nothing to do with special relativity.

7.2.2 Mass m_h of the energy element ε_h

The mass m_h of the energy element ε_h is given by equation (7.2.6):

$$m_h = \frac{\varepsilon_h}{c^2} \tag{7.2.14}$$

The above formula is the formula for the mass of the energy element ε_h. The above equation shows that the mass m_h of the energy element is given by the energy element ε_h. According to equation (7.1.2), the energy element $\varepsilon_h = 6.626 \times 10^{-34} \cdot J$ and the speed of light in the vacuum of the universe $c = 3 \times 10^8 \cdot m/s$. Substituting the two data into the above equation, the mass m_h of the energy element can be obtained as:

$$m_h = \frac{\varepsilon_h}{c^2} = \frac{6.626 \times 10^{-34}}{(3 \times 10^8)^2} = 7.36 \times 10^{-51} \cdot kg \tag{7.2.15}$$

We can also use the energy ε_{me} of the electron to determine the mass of the energy element ε_h. The mass of an electron $m_e = 9.11 \times 10^{-31} \cdot kg$ and the energy of an electron $\varepsilon_{me} = 5.11 \times 10^5 \cdot eV$ (electron volts). The mass m_{eV} contained in an electron volt $1 \cdot eV$ unit is:

$$m_{eV} = \frac{m_e}{\varepsilon_{me}} = \frac{9.11 \times 10^{-31}}{5.11 \times 10^5} = 1.78 \times 10^{-36} \cdot kg$$

Since energy element ε_h has energy $\varepsilon_h = 4.135 \times 10^{-15} \cdot eV$ (electron volt), the mass m_h of energy element

$\boldsymbol{\varepsilon_h}$ is:

$$m_h = m_{eV}\varepsilon_h = 7.36 \times 10^{-51} \cdot kg \tag{7.2.16}$$

It is clear that the mass m_h in the above equation is equal to the mass m_h in equation (7.2.15).

The number n of energy elements ε_h contained in the electron mass m_e is:

$$n = \frac{m_e}{m_h} = \frac{9.11 \times 10^{-31}}{7.35 \times 10^{-51}} = 1.239 \times 10^{20} \tag{7.2.17}$$

Since the mass m_h of the energy element is the smallest mass particle, the mass m_c of the photon is:

$$m_c = m_h\beta = \frac{\varepsilon_h\beta}{c^2} = \frac{h\beta}{c^2} \tag{7.2.18}$$

Note that β in the equation has two meanings. The β in the product $\varepsilon_h\beta$ refers to the number of energy elements ε_h and the β in the product $h\beta$ refers to the frequency of light.

7.2.3 The intrinsic mass *m* of an object in modern physics has two sources

The intrinsic mass m of an object in modern physics has two sources

The first source of intrinsic mass m is the energy ε, which applies only to the mass m_c of a moving photon.

In quantum theory, photons are defined as gauge bosons. Photons are elementary particles that carry electromagnetic action, and photons are carriers of electromagnetic radiation. Unlike most elementary particles, the static mass m_0 of a photon is zero, otherwise the moving mass m_c of a photon is infinite.

The mass m_c of a photon is given by the photon energy E_c. Assume that m_c is the mass of the photon. According to equation (7.2.13), the mass-energy conversion equation for photons is:

$$E_c = m_c c^2 \tag{7.2.19}$$

According to the law of conservation of energy, the mass-energy $E_c = m_c c^2$ of a photon is equal to the energy $\varepsilon_c = h\beta$ of the photon, i.e:

$$\varepsilon_c = m_c c^2 = h\beta \tag{7.2.20}$$

According to the above equation, the mass m_c of the moving photon is:

$$m_c = \frac{h\beta}{c^2}$$

The second source of intrinsic mass m is the "Higgs field" (the source of non-photon mass).

Why do some elementary particles have mass, while others have zero mass? Physicists use the "Higgs field" to explain this. According to the Higgs mechanism, elementary particles gain mass by coupling with the Higgs field in the universe.

In 1964, the British physicist Higgs proved mathematically that if a "Higgs field" existed, then the particles that carry the force would have mass. Physicists believe that the Higgs field exists throughout the universe, and that the canonical boson, which has a "static mass of zero," can gain mass through "spontaneous symmetry breaking" in the Higgs field.

According to quantum field theory, everything is made up of quantum fields. Each elementary particle is a tiny oscillation of a quantum field. Just as photons are tiny vibrations of the electromagnetic field, quarks are tiny vibrations of the quark field, electrons are tiny vibrations of the electron field, gravitons are tiny vibrations of the gravitational field, and so on.

According to quantum mechanics, electrons cannot have mass because they are limited by "symmetry". Since the Higgs boson fills the entire universe like an electromagnetic field, the Higgs boson spontaneously breaks the "symmetry" of the electron. Since electrons are affected by the force of the "Higgs field", they no longer move at the speed of light c like photons. The electron thus gains mass.

It should be noted that since this book replaces the terms "Higgs field" and "spontaneous symmetry breaking" with the motion of the energy element ε_h around the center of the elementary particles, the mass m_h of the energy element is the source of the intrinsic mass m of all matter.

7.2.4 The internal rotational motion of the energy element ε_h is responsible for the formation of the intrinsic mass m of an object

According to equation (7.2.14), the mass m_h of the energy element is:

$$m_h = \frac{\varepsilon_h}{c^2}$$

The above formula is not only the definition of the mass m_h of the energy element, but also the source of the intrinsic mass m_i of the elementary particles.

According to the axiom of invariance of the speed of light, the energy element ε_h always moves at the speed of light c in the cosmic vacuum reference frame. If many energy elements ε_h are able to form masses m_i of elementary particles, the linear motion of the energy element ε_h must change.

Although the energy element ε_h cannot move in a straight line within the elementary particle intrinsic mass m_i. However, since the volume V_i of the elementary particle is much larger than the volume V_h of the energy element, a certain number of energy elements ε_h can rotate around the center of the elementary particle. The orbital speed of the energy element is equal to the spiral speed of light $c = \sqrt{v_\omega^2 + v^2}$, as shown in Figure 7-1.

In other words, since the energy element ε_h always moves at the speed of light, the only way for the energy element ε_h to become an elementary particle is to have an internal rotational speed, i.e., the energy element ε_h rotates around the center of the elementary particle. The internal rotational motion of the energy element ε_h is similar to the motion of the moon around the earth.

Since the mass-energy conversion equation $\varepsilon_c = m_c c^2$ of a photon and the mass-energy conversion equation $\varepsilon_i = m_i c^2$ of an elementary particle, the energy ε_i of an elementary particle should also consist of the energy $\varepsilon_h = m_h c^2$ of a certain number of energy elements. The difference between the mass-energy conversion equation $\varepsilon_c = m_c c^2$ of a photon and the mass-energy conversion equation $\varepsilon_i = m_i c^2$ of an elementary particle is that the energy element ε_h moves in a straight line in a photon, while it moves in a spiral in an elementary particle.

According to atomic physics, objects are composed of elementary particles such as electrons, quarks, and neutrinos. Since elementary particles are indivisible particles and the mass m_h of an energy element is the smallest mass particle, the intrinsic mass m_i of each elementary particle is composed of a certain number of masses m_h of energy elements. Its mathematical expression is:

$$m_i = \sum_i m_h = \sum_i \frac{\varepsilon_h}{c^2} \tag{7.2.21}$$

The above formula is the formula for the mass of an elementary particle. The m_i in the formula is the intrinsic mass of the elementary particle i. Multiplying the above formula by c^2, the square of the speed of light c, gives the relational formula:

$$\varepsilon_i = m_i c^2 = \sum_i m_h c^2 = \sum_i \varepsilon_h \tag{7.2.22}$$

The above formula is the mass-energy conversion equation of the elementary particle i. Note that the speed of light c in the formula is the spiral speed of light of the energy element ε_h in the elementary particle intrinsic mass m_i, not the speed of motion of the elementary particle i.

Since the masses of the elementary particles are all composed of the mass m_h of the energy element, and the intrinsic mass m of an object is all composed of a certain number of elementary particles, the mathematical expression for the intrinsic mass m of an object is:

$$m = \sum_i^n \sum_i m_i = \sum_{i=1}^n \sum_i \sum m_h = \sum_{i=1}^n \sum_i \sum \frac{\varepsilon_h}{c^2} \tag{7.2.23}$$

The above formula is the formula for the mass of an object. The speed of light c in the formula is the spiral speed of light of the energy element ε_h. The above formula shows that the intrinsic mass m of an object is made up of the mass m_h of a certain number of energy elements.

Since the intrinsic mass m is a scalar quantity, the magnitude of the intrinsic mass m is independent of the direction of motion of the energy element ε_h. Multiplying the above equation by c^2, the square of the speed of light c, gives the mass-energy conversion equation of the object, i.e:

$$E = mc^2 = \sum_{i=1}^{n}\sum_{i} m_i c^2 = \sum_{i=1}^{n}\sum_{i}\sum \varepsilon_h \tag{7.2.24}$$

The above equation shows that the mass-energy conversion equation $E = mc^2$ of an object comes from the set of energy elements ε_h. Note that the speed of light c in the equation is the spiral speed of light of the energy element ε_h, not the speed of motion of the object.

Since the authors replace the physical terms "Higgs field" and "spontaneous symmetry breaking" with the motion of the energy element ε_h around the center of the elementary particles, the mass m_h of the energy element is the source of the intrinsic mass m of all matter.

7.3 New mass velocity equation

7.3.1 Equation for the decomposition of the mass m_i of an elementary particle

Suppose m_i is the intrinsic mass of the elementary particle i. According to equation (7.2.13), the mass-energy conversion equation possessed by the elementary particle i is:

$$E_i = m_i c^2$$

The above equation can be modified to:

$$\frac{E_i}{c} = m_i c \tag{7.3.1}$$

According to equation (7.2.21) . The intrinsic mass m_i of the elementary particle i is composed of the mass m_h of a certain number of energy elements, i.e:

$$m_i = \sum_{i} m_h = \sum_{i} \frac{\varepsilon_h}{c^2}$$

If the elementary particle i is displaced in the cosmic vacuum frame of reference with the velocity v, the partial velocity of the energy element ε_h in the direction of the displacement of the elementary particle i is equal to the displacement velocity v. Substituting the spiral velocity of light $c = \sqrt{v_\omega^2 + v^2}$ of the energy element ε_h into equation (7.3.1), obtain the relation equation:

$$\frac{E_i}{c} = m_i c = m_i\sqrt{v_\omega^2 + v^2} \tag{7.3.2}$$

Dividing the above equation by the speed of light c, gives the relational equation:

$$m_i = \sqrt{\left(m_i\frac{v_\omega}{c}\right)^2 + \left(m_i\frac{v}{c}\right)^2} \tag{7.3.3}$$

The above formula is the mass-velocity formula for elementary particles.

When the displacement velocity of the elementary particle $v = 0$, the intrinsic mass m_i of the elementary particle is equal to the rest mass m_0. When the displacement velocity of the elementary particle $v \neq 0$, the intrinsic mass m_i of the elementary particle can be decomposed into two component masses.

If the displacement velocity of the elementary particle is $v = c$, the intrinsic mass of the elementary particle m_i is the mass of the photon.

7.3.2 New mass velocity equation

Since the velocity v_ω is the internal rotational velocity (i.e. circular velocity) of the energy element ε_h, in this book the first term $m_i v_\omega / c$ in the root notation of equation (7.3.3) is defined as the internal rotational mass $m_{i\omega}$ of the elementary particle i, i.e:

$$m_{i\omega} = m_i \frac{v_\omega}{c} = m_i \sqrt{1 - \frac{v^2}{c^2}} \tag{7.3.4}$$

The above formula is the formula for the internal rotating mass of elementary particles. The internal rotating mass $m_{i\omega}$ is the mass hidden inside the intrinsic mass m_i of the elementary particle, which is invisible and untouchable. The above equation shows that the internal rotating mass $m_{i\omega}$ of the elementary particle decreases with the increase of the displacement velocity v of the elementary particle.

When the displacement velocity of the elementary particle $v = 0$, the internal rotating mass $m_{i\omega}$ of the elementary particle is equal to the rest mass m_{i0} of the elementary particle, i.e. $m_{i\omega} = m_{i0}$. When the displacement velocity of the elementary particle $v = c$, the internal rotating mass $m_{i\omega} = 0$ of the elementary particle. At this time, the elementary particle is a photon.

Note that since the intrinsic mass m_i of elementary particles is an indivisible whole, the internal rotating mass $m_{i\omega}$ cannot stand alone when the displacement velocity $v \neq 0$.

Since the velocity v is the displacement velocity of the elementary particle, this book defines the second term $m_i v/c$ in the root notation of equation (7.3.3) as the displacement mass m_{iv} of the elementary particle i, i.e:

$$m_{iv} = m_i \frac{v}{c} \tag{7.3.5}$$

The above formula is the formula for the displaced mass of an elementary particle. The displaced mass m_{iv} is the invisible and untouchable mass hidden inside the intrinsic mass m_i of the elementary particle. The above formula shows that the displaced mass m_{iv} of the elementary particle is proportional to the displaced velocity v of the elementary particle.

The displaced mass of the elementary particle $m_{iv} = m_i$ when the displaced velocity of the elementary particle $v = c$. At this point, the displaced mass m_{iv} is the photon mass.

According to equation (7.2.23), the intrinsic mass m of a moving object is composed of the intrinsic masses m_i of the elementary particles, i.e:

$$m = \sum_i^n \sum_i m_i = \sum_i^n \sum_i \sum m_h \tag{7.3.6}$$

If an object moves with velocity v in the cosmic vacuum frame of reference, the intrinsic mass m of the moving object can be decomposed into two mass components according to equation (7.3.3), namely:

$$m = \sum_i^n \sum_i m_i = \sqrt{\left(m \frac{v_\omega}{c}\right)^2 + \left(m \frac{v}{c}\right)^2} \tag{7.3.7}$$

The above equation is the new mass-velocity equation. According to equation (7.3.4), the internal rotating mass m_ω is included in the intrinsic mass m of the moving object:

$$m_\omega = \sum_i^n \sum_i m_{i\omega} = m \frac{v_\omega}{c} = m \sqrt{1 - \frac{v^2}{c^2}} \tag{7.3.8}$$

The above equation shows that the internal rotational mass m_ω will decrease as the displacement velocity v increases. The internal rotational mass m_ω of an object is a mass that is hidden within the intrinsic mass m of the object and cannot be seen or felt.

According to equation (7.3.5), the displacement mass m_v contained in the intrinsic mass m of the moving object is:

$$m_v = \sum_i^n \sum_i m_{iv} = m \frac{v}{c} \tag{7.3.9}$$

The above equation shows that the displaced mass m_v is directly proportional to the displacement velocity v of the object. The displaced mass m_v of an object is a mass that is hidden within the intrinsic mass m of the object and cannot be seen or felt. The intrinsic mass m of a moving object can be expressed as:

$$m = \sqrt{m_\omega^2 + m_v^2} = \sqrt{\left(m\frac{v_\omega}{c}\right)^2 + \left(m\frac{v}{c}\right)^2} \tag{7.3.10}$$

The above equation is the new mass-velocity formula. The above equation shows that the intrinsic mass m of an object can be decomposed into an internal rotational mass m_ω and a displacement mass m_v.

Squaring the velocity of light $c = \sqrt{v_\omega^2 + v^2}$ of the spiral gives the relational equation:

$$c^2 = v_\omega^2 + v^2 \tag{7.3.11}$$

Multiplying the above equation by the intrinsic mass m gives the relational equation:

$$E_m = mc^2 = mv_\omega^2 + mv^2 \tag{7.3.12}$$

The speed of light c in the formula is the spiral speed of light of the energy element ε_h, not the speed of motion of the object. The above formula shows that the mass-energy conversion equation $E_m = mc^2$ of an object in the cosmic vacuum frame of reference contains two components.

Component 1 is: the displacement mass-energy $E_{mv} = mv^2$ resulting from the displacement velocity v of the object;

Component 2 is: the internal rotational mass-energy $E_{m\omega} = v_\omega^2$ produced by the internal rotational velocity v_ω of the object;

Note that the internal rotational mass-energy $E_{m\omega}$ changes in the opposite direction to the displacement mass-energy E_{mv}.

7.3.3 The assumption that the rest mass $m_{c0} = 0$ of a photon is wrong

The mass velocity formula for special relativity is:

$$m' = \frac{m_0}{\sqrt{1 - \frac{v^2}{c^2}}}$$

The m' in the formula is the moving mass. Note that the invariant in the above equation is the rest mass m_0. The above equation shows that the moving mass m' of the object increases as the velocity v increases. Since the speed of propagation of a photon is equal to the speed of light c, and the moving mass m_c of a photon, is not infinite, physicists believe that the rest mass $m_{c0} = 0$ of a photon.

According to the new mass-velocity formula (7.3.10), the intrinsic mass m of the object is:

$$m = \sqrt{m_\omega^2 + m_v^2} = \sqrt{\left(m\frac{v_\omega}{c}\right)^2 + \left(m\frac{v}{c}\right)^2}$$

Note that the invariant in the above equation is the intrinsic mass m of the object. Obviously, the rest mass m_0 at this point is the intrinsic mass m of the object as defined in this book.

The physical quantity that describes the change in mass of an object in special relativity is the moving mass m', while this book describes the change in mass of an object in two physical quantities, the internal rotating mass m_ω and the displaced mass m_v.

According to equation (7.3.8), the internal rotational mass m_ω contained in the object intrinsic mass m is:

$$m_\omega = m\sqrt{1 - \frac{v^2}{c^2}}$$

At the velocity $v = c$, the object's internal rotating mass $m_\omega = 0$, while the special relativity rest mass $m_0 = 0$. Obviously, the rest mass m_0 at this point, is again the internal rotating mass m_ω as defined in this book.

According to equation (7.3.9), the displacement mass m_v contained in the intrinsic mass m of the moving object is:

$$m_v = m\frac{v}{c}$$

The above equation shows that the displaced mass m_v of an object increases as its velocity v increases. Since the propagation velocity of a photon is equal to the speed of light c, the displaced mass m_{cv} of a photon is always equal to the mass $\boldsymbol{m_c}$ of the photon, i.e. $\boldsymbol{m_{cv}} = \boldsymbol{m_c}$.

Since this book defines three new concepts of intrinsic mass $\boldsymbol{m}$, internal rotational mass $\boldsymbol{m_\omega}$, and displacement mass $\boldsymbol{m_v}$, the new mass-velocity formula (7.3.10) avoids the error of the photon's rest mass $\boldsymbol{m_0} = \boldsymbol{0}$.

Since the intrinsic mass $\boldsymbol{m}$ of an object contains two different varying masses, the internal rotational mass $\boldsymbol{m_\omega}$ and the displacement mass $\boldsymbol{m_v}$, and modern physics uses only the moving mass $\boldsymbol{m}$ to describe these two different varying masses, this leads to the false assumption that the rest mass $\boldsymbol{m_{c0}} = \boldsymbol{0}$ of a photon.

Chapter 8 Deriving New External Force Formulas

Introduction: In Newtonian mechanics, the external force F is defined as the rate of change of momentum $P = mv$, i.e:

$$F = \frac{dP}{dt} = m\frac{dv}{dt} + v\frac{dm}{dt}$$

The Newtonian force F applies only to slow motion, not to fast motion. The authors theoretically derived a new formula for the external force based on the above formula and the spiral speed of light formula $c = \sqrt{v_\omega^2 + v^2}$, i.e:

$$F = \left(1 + \frac{v}{\sqrt{c^2 - v^2}}\right) ma$$

The above formula applies to both low and high speed motion.

8.1 The external force F_c can be decomposed into a displacement force F_v and an internal rotational force F_ω

8.1.1 Helical light speed momentum P_c, displacement momentum P_v, and internal rotation momentum P_ω of an object

According to equation (7.3.6), the intrinsic mass m of an object consists of the mass m_h of a certain number of energy elements, i.e:

$$m = \sum_i^n \sum_i m_i = \sum_i^n \sum_i \sum m_h$$

The mass m_h of the energy element is inside the intrinsic mass m_i of the elementary particle, always moving at the spiral speed of light. Multiplying the above equation by the speed of light c gives the relational equation:

$$P_c = mc = \sum_i^n \sum_i \sum m_h c \tag{8.1.1}$$

This book defines the product mc of mass and the speed of light, as the object's helical light-speed momentum P_c. The equation above shows that the object's helical light-speed momentum $P_c = mc$, is equal to the sum of the momentums of all the energy elements ε_h, within the intrinsic mass m.

According to the new mass-velocity formula (7.3.7), the intrinsic intrinsic mass m of a moving object can be decomposed into an internal rotational mass $m_\omega = m\frac{v_\omega}{c}$ and a displacement mass $m_v = m\frac{v}{c}$, namely:

$$m = \sqrt{\left(m\frac{v_\omega}{c}\right)^2 + \left(m\frac{v}{c}\right)^2}$$

Multiplying the above equation by the speed of light c gives the relational equation:

$$P_c = mc = \sqrt{(mv)^2 + (mv_\omega)^2} \tag{8.1.2}$$

The above formula is the momentum formula for the spiral velocity of light of an object. The speed of light c in the formula is the spiral speed of light of the energy element ε_h, not the speed of motion of the object.

According to Equation (8.1.2), the object's helical light speed momentum P_c can be decomposed into two components. Since v is the displacement velocity, the displacement momentum P_v of the object is:

$$P_v = mv \tag{8.1.3}$$

The above equation is the equation for the displacement momentum of an object.

Since v_ω is the internal rotational velocity, the internal rotational momentum P_ω of the object is:

$$P_\omega = mv_\omega \tag{8.1.4}$$

The above formula is the formula for the internal rotational momentum of the object. Substituting equation (8.1.3) and the above equation into equation (8.1.2) gives the formula for the spiral light-speed momentum P_c of the object:

$$P_c = mc = \sqrt{P_v^2 + P_\omega^2} \tag{8.1.5}$$

8.1.2 Force F in Newton's second law is displacement force F_v

In Newtonian mechanics, the differential expression for the Newtonian force F is:

$$F = \frac{dP}{dt} = m\frac{dv}{dt} + v\frac{dm}{dt} \tag{8.1.6}$$

The $P = mv$ in the formula is the momentum of the object. When the intrinsic intrinsic mass m is kept constant, the above formula becomes:

$$F = ma \tag{8.1.7}$$

The above formula is Newton's second law. The above formula only applies to objects moving at low speeds, not to objects moving at high speeds. In order to overcome the shortcomings of Newton's second law, this book launches a new analysis and exploration of the external force F by using the internal rotational velocity v_ω and displacement velocity v of the object.

Take the derivative of equation (8.1.5) with respect to time t when the intrinsic mass m of the object remains constant. Since the object's spiral light-speed momentum $P_c = mc$ is constant, the rate of change $\frac{dP_c}{dt} = 0$, i.e:

$$\frac{dP_c}{dt} = \frac{1}{\sqrt{P_v^2 + P_\omega^2}}\left(P_v\frac{dP_v}{dt} + P_\omega\frac{dP_\omega}{dt}\right) = 0 \tag{8.1.8}$$

Based on the above equation, the relationship equation can be obtained:

$$P_v\frac{dP_v}{dt} + P_\omega\frac{dP_\omega}{dt} = 0 \tag{8.1.9}$$

Take the derivative of equation (8.1.3) with respect to time t when the intrinsic mass m of the object is kept constant. The displacement force F_v of the object is:

$$F_v = \frac{dP_v}{dt} = m\frac{dv}{dt} = ma \tag{8.1.10}$$

The above equation is the equation for the displacement force of an object. The above equation shows that the magnitude of the displacement acceleration a is proportional to the displacement force $F_v = ma$ and inversely proportional to the intrinsic mass m of the object, and that the direction of the displacement acceleration a is in the same direction as the displacement force F_v.

Since the displacement acceleration $a = \frac{dv}{dt}$ is the rate of change of the displacement velocity v, the force F in Newton's second law is the displacement force F_v of the object.

8.1.3 Direction of internal rotational acceleration a_ω, opposite to direction of displacement acceleration a

Take the derivative of equation (8.1.4) with respect to time t when the intrinsic mass m of the object remains constant. The internal rotational force F_ω of the object is:

$$F_\omega = \frac{dP_\omega}{dt} = m\frac{dv_\omega}{dt} = ma_\omega \tag{8.1.11}$$

The above equation is the formula for the internal rotational force of an object. According to equation (7.1.8), the internal rotational speed v_ω of the object is:

$$v_{\omega\cdot} = \sqrt{c^2 - v^2} \tag{8.1.12}$$

Since the internal rotational velocity v_ω becomes smaller as the displacement velocity v increases, the internal rotational acceleration $\boldsymbol{a}_\omega$ becomes smaller as the external force $\boldsymbol{F}$ increases. The internal rotational acceleration $\boldsymbol{a}_\omega$ of the object can be obtained by taking the derivative of the internal rotational velocity $\boldsymbol{v}_\omega$ with respect to the time $\boldsymbol{t}$, which is:

$$\boldsymbol{a}_\omega = \frac{d\boldsymbol{v}_\omega}{dt} = \frac{d\boldsymbol{v}_\omega}{d\boldsymbol{v}}\frac{d\boldsymbol{v}}{dt} = -\frac{\boldsymbol{v}}{\sqrt{\boldsymbol{c}^2 - \boldsymbol{v}^2}}\frac{d\boldsymbol{v}}{dt} = -\frac{\boldsymbol{v}}{\sqrt{\boldsymbol{c}^2 - \boldsymbol{v}^2}}\boldsymbol{a} \tag{8.1.13}$$

The above equation is the formula for the internal rotational acceleration of an object. The above equation shows that the internal rotational acceleration $\boldsymbol{a}_\omega$ of the object is in the opposite direction of the displacement acceleration $\boldsymbol{a}$.

The internal rotational acceleration $\boldsymbol{a}_\omega = \boldsymbol{0}$ of the object when the displacement velocity $\boldsymbol{v} = \boldsymbol{0}$. When the displacement velocity $\boldsymbol{v} = \boldsymbol{c}$. Acceleration $\boldsymbol{a}_\omega = -\infty$ since the internal rotational velocity $\boldsymbol{v}_\omega = \sqrt{\boldsymbol{c}^2 - \boldsymbol{v}^2} = \boldsymbol{0}$.

8.1.4 Relationship Between Displacement Force F_v and Internal Rotational Force F_ω of an Object

Substituting equations (8.1.10) and (8.1.11) into equation (8.1.9) gives the relational equation:

$$\boldsymbol{P}_\omega \boldsymbol{F}_\omega = -\boldsymbol{P}_v \boldsymbol{F}_v \tag{8.1.14}$$

Substituting the object's displacement momentum equation (8.1.3) $\boldsymbol{P}_v = \boldsymbol{m}\boldsymbol{v}$ and the internal rotation momentum equation (8.1.4) $\boldsymbol{P}_\omega = \boldsymbol{m}\boldsymbol{v}_\omega$ into the above equation gives the relational equation:

$$\boldsymbol{F}_\omega = -\frac{\boldsymbol{v}}{\boldsymbol{v}_\omega}\boldsymbol{F}_v \tag{8.1.15}$$

The above equation shows that the direction of the internal rotational force $\boldsymbol{F}_\omega$ of the object is opposite to the direction of the displacement force $\boldsymbol{F}_v$. Substituting the internal rotational speed $\boldsymbol{v}_\omega . = \sqrt{\boldsymbol{c}^2 - \boldsymbol{v}^2}$ of the object into the above equation gives the relational equation:

$$\boldsymbol{F}_\omega = -\frac{\boldsymbol{v}}{\sqrt{\boldsymbol{c}^2 - \boldsymbol{v}^2}}\boldsymbol{F}_v = -\frac{\boldsymbol{v}}{\sqrt{\boldsymbol{c}^2 - \boldsymbol{v}^2}}\boldsymbol{m}\boldsymbol{a} \tag{8.1.16}$$

When the object's displacement velocity $\boldsymbol{v} = \boldsymbol{0}$, the object's internal rotational force $\boldsymbol{F}_\omega = \boldsymbol{0}$. When the displacement velocity $\boldsymbol{v} = \boldsymbol{c}$, the displacement velocity $\boldsymbol{v}$ is constant. At this time, since the displacement acceleration is $\boldsymbol{a} = \frac{d\boldsymbol{v}}{dt} = \boldsymbol{0}$, both the displacement force $\boldsymbol{F}_v = \boldsymbol{m}\boldsymbol{a}$ and the internal rotational force $\boldsymbol{F}_\omega$ are equal to 0, i.e., $\boldsymbol{F}_v = \boldsymbol{F}_\omega = \boldsymbol{0}$.

The force $\boldsymbol{F} = \boldsymbol{m}\boldsymbol{a}$ in Newton's second law is the external force on an object. Since the displacement force $\boldsymbol{F}_v$ decreases or increases the displacement velocity $\boldsymbol{v}$ of the object, and the internal rotation force $\boldsymbol{F}_\omega$ decreases or increases the internal rotation velocity $\boldsymbol{v}_\omega$ of the object, the external force $\boldsymbol{F}$ can be decomposed into two component forces.

The first component force is the displacement force $\boldsymbol{F}_v$. The displacement force $\boldsymbol{F}_v$ will decrease or increase the object's displacement velocity $\boldsymbol{v}$ and displacement acceleration $\boldsymbol{a}$. The displacement force $\boldsymbol{F}_v$ can change the object's state of motion and direction of motion. The displacement force $\boldsymbol{F}_v$ is the force $\boldsymbol{F}$ in Newton's second law.

The second component force is the internal rotational force $\boldsymbol{F}_\omega$. The internal rotational force $\boldsymbol{F}_\omega$ will decrease or increase the object's internal rotational velocity $\boldsymbol{v}_\omega$ and internal rotational acceleration $\boldsymbol{a}_\omega$. The internal rotational force $\boldsymbol{F}_\omega$ cannot change the state or direction of motion of the object. The internal rotational force $\boldsymbol{F}_\omega$ is the force that is hidden inside the object and cannot be seen.

8.2 Deriving a new formula for external forces

8.2.1 External forces in the cosmic vacuum frame of reference F

In the cosmic vacuum frame of reference, when an external force $\boldsymbol{F}$ acts on an object, the external force $\boldsymbol{F}$ is divided into a displacement force $\boldsymbol{F}_v$ and an internal rotational force $\boldsymbol{F}_\omega$. The displacement force $\boldsymbol{F}_v$ changes the displacement velocity $\boldsymbol{v}$ of the object, and the internal rotational force $\boldsymbol{F}_\omega$ changes the internal rotational velocity $\boldsymbol{v}_\omega$ of the object.

The displacement force $\boldsymbol{F}_v$ on the object is given by equation (8.1.10):

$$F_v = m\frac{dv}{dt} = ma$$

According to equation (8.1.11), the internal rotational force F_ω on the object is:

$$F_\omega = -m\frac{dv_\omega}{dt} = -m\frac{v}{\sqrt{c^2 - v^2}}a$$

Note that the internal rotational force F_ω decreases or increases the internal rotational velocity v_ω of the energy element ε_h. The internal rotational force F_ω is part of the external force F.

In the macroscopic world. Since the displacement velocity v of an object is much smaller than the speed of light c, the internal rotational force F_ω is very small. From an object subjected to a force F, one usually feels only the presence of the displacement force $F_v = ma$, and it is difficult to feel the presence of the internal rotational force F_ω.

In the microscopic world. One senses the presence of an internal rotational force F_ω from high-energy particles, yet one mistakenly sees it as a displacement force $F_v = ma$.

According to equation (8.1.15), the displacement force F_v and the internal rotation force F_ω change in opposite directions, i.e:

$$F_\omega = -\frac{v}{v_\omega}F_v$$

Since the direction of the external force F is the same as the direction of the displacement velocity v and the direction of the internal rotational force F_ω is opposite to the direction of the displacement force F_v, the external force F acting on the object should be:

$$F = F_v - F_\omega = \left(1 + \frac{v}{\sqrt{c^2 - v^2}}\right)ma \qquad (8.2.1)$$

The above equation is the formula for the external force in the cosmic vacuum frame of reference. Note that the external force F, the displacement velocity v, and the displacement acceleration a in the formula are all physical quantities in the cosmic vacuum frame of reference.

The above equation shows that the magnitude of the displacement acceleration a of an object is proportional to the external force F; it is inversely proportional to the intrinsic mass m of the object. The displacement acceleration a becomes smaller as the displacement velocity v increases. The direction of the displacement velocity v and the displacement acceleration a is the same as the direction of the external force F.

When the displacement acceleration of an object $a = 0$. Since the object is subjected to an external force $F = 0$. Therefore, the object is at rest, or the object is moving at a constant velocity v. These two states of motion of an object are Newton's first law (law of inertia). If the object is moved at a velocity v much less than the speed of light c, the above equation becomes:

$$F = ma \qquad (8.2.2)$$

The equation above is Newton's second law.

Theoretically, since the internal rotational velocity v_ω is perpendicular to the displacement velocity v, we can also assume that the internal rotational force F_ω is perpendicular to the displacement force F_v. Then the external force F acting on the object is:

$$F = \sqrt{F_v^2 + F_\omega^2} = \frac{c}{\sqrt{c^2 - v^2}}ma \qquad (8.2.3)$$

However, the above formula has a serious theoretical flaw, which is that the direction of the external force F does not coincide with the direction of the displacement of the object. Therefore, the correct new external force formula should be formula (8.2.1) instead of the above formula.

8.2.2 Acceleration Equations for Inertial Frames of reference

Assume that $S(x, y, z, t)$ are the coordinates of the S cosmic vacuum frame of reference; $S'(x', y', z', t')$ are the coordinates of the S' inertial frame of reference. Assume that the coordinate axes of the S vacuum frame of reference and the S' inertial frame of reference are parallel, with the x axis coinciding with the x' axis. Assume that the S' inertial frame of reference moves along the x axis at a velocity u in the S vacuum frame of reference, as shown in Figure 8-1.

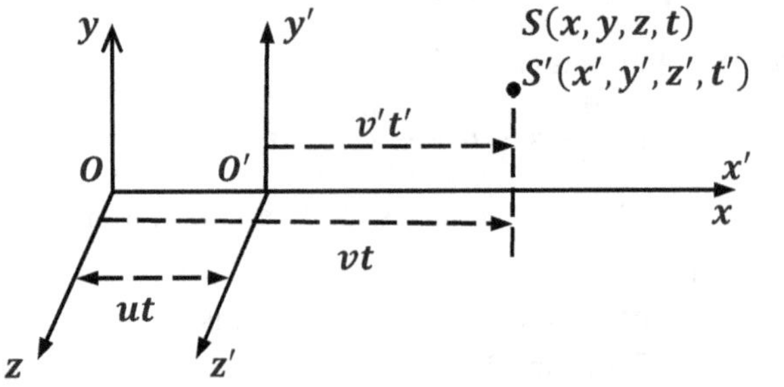

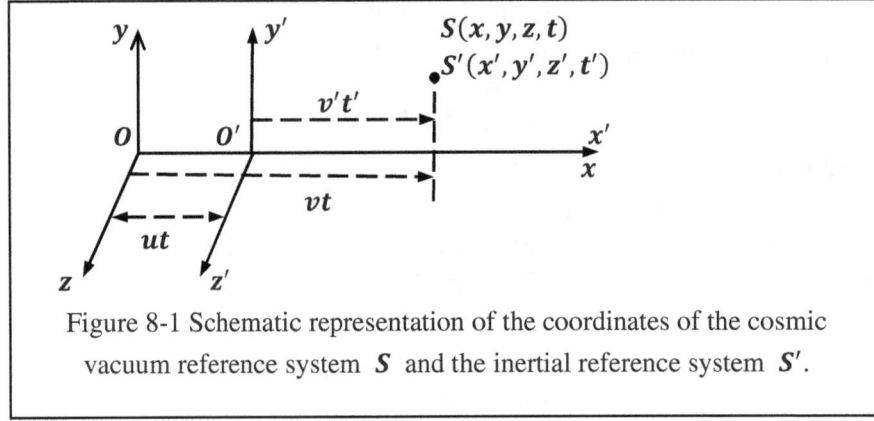

Figure 8-1 Schematic representation of the coordinates of the cosmic vacuum reference system S and the inertial reference system S'.

When the time of the object's motion $t = t' = 0$, the object is assumed to be at the coordinates $x = x' = 0$ of the x-axis. At a given time, the coordinates of the object's motion in the inertial system S and in the inertial system S' are x and x', respectively:

$$x = vt, \quad x' = v't' \tag{8.2.4}$$

Since the coordinates x and x' are the coordinates of the object, the two coordinates x and x' are the same point in space. According to the Galilean coordinate transformation formula, the coordinate transformation formula for the inertial system S and the inertial system S' is:

$$x' = x - ut, \quad y' = y, \quad z' = z, \quad t' = t \tag{8.2.5}$$

Since the coordinate x' contains the velocity u of the inertial system S', the coordinate x' depends on the choice of the inertial system. In other words, if the coordinates x and time t are constant, the coordinates x' have different values in different inertial systems.

Taking the derivative of the time transformation formula $t' = t$ with respect to time t' gives the relational formula $\frac{dt}{dt'} = 1$. Taking the derivative of the coordinate x' with respect to time t' gives the relational formula:

$$\frac{dx'}{dt'} = \frac{dx}{dt}\frac{dt}{dt'} - u\frac{dt}{dt'}$$

In the S frame of reference, the velocity of the object moving along the x axis is $v = \frac{dx}{dt}$; in the S' frame of reference, the velocity of the object moving along the x' axis is $v' = \frac{dx'}{dt'}$. The above formula becomes:

$$v' = v - u \tag{8.2.6}$$

The above formula is the Galilean velocity transformation formula. Since the above formula includes the velocity u of the motion of the inertial system S', the velocity v' is related to the choice of the inertial system. In other words, if the velocity v is constant, the velocity v' has different values in different inertial systems.

Taking the derivative of the above equation with respect to time t' gives the Galilean acceleration transformation equation, i.e:

$$a' = \frac{dv'}{dt'} = \frac{dv}{dt}\frac{dt}{dt'} \tag{8.2.7}$$

Since the above equation does not include the velocity of motion u of the inertial system, the acceleration a' in the inertial system is independent of the choice of inertial system. This leads to a corollary about acceleration, viz:

The acceleration a' in any inertial system is equal to the acceleration a in the cosmic frame of reference, i.e:

$$a' = a$$

This book refers to the above formula as the formula for acceleration in inertial systems. The above formula shows that the acceleration a' measured in any inertial system is the acceleration a in the cosmic vacuum frame of reference.

8.2.3 External force F' on an object in an inertial system S'

Substituting the velocity $\boldsymbol{v}$ from equation (8.2.6) into equation (8.2.1) yields the relationship equation:

$$\boldsymbol{F}=\left(\mathbf{1}+\frac{\boldsymbol{v}'+\boldsymbol{u}}{\sqrt{\boldsymbol{c}^2-(\boldsymbol{v}'+\boldsymbol{u})^2}}\right)\boldsymbol{ma}$$

Since the acceleration $\boldsymbol{a}'$ is equal to the acceleration $\boldsymbol{a}$, i.e. $\boldsymbol{a}'=\boldsymbol{a}$, the above equation can also be changed to:

$$\boldsymbol{F}'=\left(\mathbf{1}+\frac{\boldsymbol{v}'+\boldsymbol{u}}{\sqrt{\boldsymbol{c}^2-(\boldsymbol{v}'+\boldsymbol{u})^2}}\right)\boldsymbol{ma}' \qquad (\mathbf{8.2.8})$$

The above formula is the external force formula for the inertial system $\boldsymbol{S}'$. $\boldsymbol{F}'$ in the formula is the external force on the object in the inertial system $\boldsymbol{S}'$, $\boldsymbol{m}$ is the intrinsic mass of the object, $\boldsymbol{v}'$ is the velocity of the object in the inertial system $\boldsymbol{S}'$, $\boldsymbol{u}$ is the velocity of the inertial system $\boldsymbol{S}'$ in the cosmic vacuum frame of reference, and $\boldsymbol{c}$ is the (helical speed of light) in the cosmic vacuum frame of reference.

Based on the above two equations, the following three conclusions can be drawn.

Conclusion 1: The external force $\boldsymbol{F}$ of the cosmic vacuum frame of reference is equal to the external force $\boldsymbol{F}'$ of any inertial system, i.e.

$$\boldsymbol{F}=\boldsymbol{F}'$$

Conclusion 2: The mathematical expression for the external force formula is the same for all inertial systems except for the cosmic vacuum frame of reference.

Conclusion 3: The formula for the external force in the cosmic vacuum frame of reference is different from that of the inertial system. Since the former formula does not include the velocity $\boldsymbol{u}$ of the inertial system, the (relativity principle) in physics is wrong.

An object is at rest in an inertial system $\boldsymbol{S}'$ if its velocity $\boldsymbol{v}'=\mathbf{0}$ in the inertial system $\boldsymbol{S}'$. According to equation (8.2.8), the external force $\boldsymbol{F}_0'$ on the object is:

$$\boldsymbol{F}_0'=\left(\mathbf{1}+\frac{\boldsymbol{u}}{\sqrt{\boldsymbol{c}^2-\boldsymbol{u}^2}}\right)\boldsymbol{ma} \qquad (\mathbf{8.2.9})$$

However, when the object is at rest in the cosmic vacuum frame of reference $\boldsymbol{S}$, the external force $\boldsymbol{F}_0$ on the object is:

$$\boldsymbol{F}_0=\boldsymbol{ma}$$

The principle of relativity in physics refers to the fact that the mathematical formulations of a physical law remain unchanged in different inertial frames of reference. For example, Maxwell's equations hold in different inertial frames of reference. Since the above equations are different from equation (8.2.9), the principle of relativity in physics is false.

8.2.4 Inertial mass of an object m_F

Suppose $\boldsymbol{m}$ is the intrinsic mass of the object. In order to find the relationship between the intrinsic mass $\boldsymbol{m}$ and the external force $\boldsymbol{F}$, this book defines the inertial mass $\boldsymbol{m_F}$ of the object in the cosmic vacuum frame of reference as:

$$\boldsymbol{m_F}=\left(\mathbf{1}+\frac{\boldsymbol{v}}{\sqrt{\boldsymbol{c}^2-\boldsymbol{v}^2}}\right)\boldsymbol{m} \qquad (\mathbf{8.2.10})$$

The above formula shows that the inertial mass $\boldsymbol{m_F}$ of an object is proportional to the velocity $\boldsymbol{v}$. The new formula for the external force can be expressed as:

$$\boldsymbol{F}=\boldsymbol{m_F a}=\left(\mathbf{1}+\frac{\boldsymbol{v}}{\sqrt{\boldsymbol{c}^2-\boldsymbol{v}^2}}\right)\boldsymbol{ma} \qquad (\mathbf{8.2.11})$$

When the cosmic velocity of an object $\boldsymbol{v}=\mathbf{0}$, the external force $\boldsymbol{F}$ on the object is equal to the Newtonian force $\boldsymbol{F}'=\boldsymbol{ma}$, i.e. $\boldsymbol{F}=\boldsymbol{F}'=\boldsymbol{ma}$. When the cosmic velocity of an object $\boldsymbol{v}\neq\mathbf{0}$, the external force $\boldsymbol{F}$ on the object is not equal to the Newtonian force $\boldsymbol{F}'$, i.e. $\boldsymbol{F}\neq\boldsymbol{F}'=\boldsymbol{ma}$.

When the cosmic velocity $\boldsymbol{v}$ of an object approaches the speed of light $\boldsymbol{c}$, only a huge external force $\boldsymbol{F}$ can change the magnitude of the acceleration $\boldsymbol{a}$, because the inertial mass $\boldsymbol{m_F}$ of the object is so huge.

When the object's cosmic velocity $\boldsymbol{v}$ is equal to the speed of light $\boldsymbol{c}$, no enormous external force $\boldsymbol{F}$ can change the object's velocity because the object's inertial mass $\boldsymbol{m_F}$ is infinite.

From this it can be seen that the new external force formula (8.2.11) applies not only to macroscopic objects moving at low speeds, but also to microscopic particles moving at high speeds.

Substituting the velocity $v = v' + u$ from equation (8.2.6) into equation (8.2.10) gives the relational equation:

$$m_F = \left(1 + \frac{v' + u}{\sqrt{c^2 - (v' + u)^2}}\right) m \qquad (8.2.12)$$

The m_F in the formula is the inertial mass of the object; m is the intrinsic mass of the object; u is the velocity of the S' inertial system; and v' is the velocity of the object in the S' inertial system. The above equation shows that the inertial mass m_F of the object, is proportional to both the velocity u, and the velocity v'.

It should be noted that the inertial mass m_F is the mass m in motion in the classical theory. When the object is at rest in the S' inertial system. Since the velocity $v' = 0$. the static mass m_0' of the object in the S' inertial system is therefore:

$$m_0' = \left(1 + \frac{u}{\sqrt{c^2 - u^2}}\right) m \qquad (8.2.13)$$

The above equation shows that the static mass m_0' of an object in an inertial system S' will increase as the velocity u increases.

Since the static mass m_0' contains the velocity u of the inertial system S', the static mass m_0' of the object is related to the choice of the inertial system. Therefore, it can be stated that the static mass m_0' of the object in the inertial system does not satisfy the principle of relativity.

In summary, there are 5 types of mass of an object:

The first type of mass is: the intrinsic mass of the object m. The intrinsic mass m is independent of the object's state of motion.

The 2nd type of mass is: the displacement mass $m_v = m\frac{v}{c}$, whose mass m_v is related to the displacement velocity v of the object.

The 3rd type of mass is: the internal rotational mass $m_\omega = m\sqrt{1 - \frac{v^2}{c^2}}$, whose mass m_ω is related to the internal rotational velocity $v_\omega = \sqrt{c^2 - v^2}$ of the energy element ε_h.

The 4th type of mass is: the inertial mass $m_F = \left(1 + \frac{v'+u}{\sqrt{c^2-(v'+u)^2}}\right) m$, whose mass m_F is proportional both to the velocity u and to the velocity v'.

The 5th mass is: the static mass $m_0' = \left(1 + \frac{u}{\sqrt{c^2-u^2}}\right) m$, whose mass m_0' is proportional to the velocity u of the inertial system S'.

Since the photon's cosmic velocity $v = c$, the photon's internal rotational velocity $v_\omega = \sqrt{c^2 - v^2} = 0$. Therefore, the photon's internal rotational mass $m_\omega = m\sqrt{1 - \frac{v^2}{c^2}} = 0$. The experiment in which physicists verified that the photon's static mass $m_0 = 0$ was actually an experiment in which they verified that the photon's internal rotational mass $m_\omega = 0$.

Since the cosmic velocity of the photon $v = c$, the displaced mass m_v of the photon is equal to the intrinsic mass m of the photon, i.e. $m_v = m$. Since the inertial mass $m_F = \left(1 + \frac{v}{\sqrt{c^2-v^2}}\right) m$ of the photon is equal to infinity, the velocity of light c of the photon in the gravitational field remains constant.

Although a gravitational field cannot change the magnitude of the speed of light c, a gravitational field can change the direction of motion of a photon because it has a displacement force F_v, on the photon's displaced mass m_v, which bends the photon's trajectory. This is just like a magnetic field can only change the direction of motion of an electric charge, but not the speed of motion of the charge.

8.2.5 The relativity principle of physics is false

There are two types of relativity principles in physics, one is the relativity principle of mechanics and the other is the principle of narrow relativity.

The principle of relativity in mechanics states that the mathematical expression of the laws of mechanics is identical in all inertial systems, and that there is no inertial system superior to any other inertial system.

The principle of narrow relativity (the principle of narrow covariance) holds that all laws of physics are equivalent (equal-weighted) in all inertial frames of reference, that no inertial system has a superior position, and that there is no absolutely stationary frame of reference. Alternatively, all laws of physics are invariant in mathematical form under Lorentz transformations. Clearly the principle of narrow relativity is a generalization of the relativity principle of mechanics.

Physicists argue that no physical experiment performed within an inertial system can determine whether the inertial system itself is at relative rest or in uniform linear motion. Galileo demonstrated this idea on a large enclosed ship by saying that an observer could not determine whether the ship was moving or at rest. The derivation of the principle of relativity of mechanics is as follows:

The Galilean velocity transformation formula is $\boldsymbol{v} = \boldsymbol{v}' + \boldsymbol{u}$. Taking the derivative of the formula with respect to time $\boldsymbol{t}'$ gives the relational formula:

$$\frac{d\boldsymbol{v}}{dt}\frac{dt}{dt'} = \frac{d\boldsymbol{v}'}{dt'} = \boldsymbol{a}'$$

According to the axiom of time absoluteness, time $\boldsymbol{t}$ is equal to time $\boldsymbol{t}'$, i.e. $\boldsymbol{t} = \boldsymbol{t}'$. Since the time derivative $\frac{dt}{dt'} = \mathbf{1}$, the acceleration transformation formula is obtained:

$$\boldsymbol{a} = \boldsymbol{a}' \qquad (\mathbf{8.2.14})$$

Multiply the above equation by the mass $\boldsymbol{m}$ to obtain Newton's second law:

$$\boldsymbol{F} = \boldsymbol{ma} = \boldsymbol{ma}' = \boldsymbol{F}' \qquad (\mathbf{8.2.15})$$

The above formula shows that the mathematical expression of the external force $\boldsymbol{F}'$ is the same in all inertial systems. It is clear that the above equation is consistent with the principle of relativity in mechanics. However the above formula hides two problems that contradict special relativity.

Contradiction 1: Since equation (8.2.15) contains the axiom of temporal absoluteness, i.e., $\boldsymbol{t} = \boldsymbol{t}'$, equation (8.2.15) contradicts the Principle of Narrow Relativity.

Contradiction 2: Equation (8.2.15) hides the error that the velocity $\boldsymbol{v}$ of an object's motion can reach infinity. However the velocity $\boldsymbol{v}$ infinity is not in accordance with the principle of invariance of the speed of light in special relativity.

We can use equation (8.2.1) to show that equation (8.2.15) is false.

According to equation (8.2.1), the external force $\boldsymbol{F}$ on an object in a cosmic space frame of reference is:

$$\boldsymbol{F} = \left(\mathbf{1} + \frac{\boldsymbol{v}}{\sqrt{\boldsymbol{c}^2 - \boldsymbol{v}^2}}\right)\boldsymbol{ma}$$

According to the principle of relativity in mechanics, the external force $\boldsymbol{F}'$ on an object in an inertial system $\boldsymbol{S}'$ should be:

$$\boldsymbol{F}' = \left(\mathbf{1} + \frac{\boldsymbol{v}'}{\sqrt{\boldsymbol{c}^2 - \boldsymbol{v}'^2}}\right)\boldsymbol{ma} \qquad (\mathbf{8.2.16})$$

Since the velocity $\boldsymbol{v} = \boldsymbol{v}' + \boldsymbol{u}$ in the cosmic vacuum frame of reference $\boldsymbol{S}$, the external force $\boldsymbol{F}' \neq \boldsymbol{F}$. Since the inertial system $\boldsymbol{S}'$ is moving with velocity $\boldsymbol{u}$ in the cosmic vacuum frame of reference $\boldsymbol{S}$, the external force $\boldsymbol{F}'$ observed by an observer in the inertial system $\boldsymbol{S}'$ should be the external force in equation (8.2.8), viz:

$$\boldsymbol{F}' = \left(\mathbf{1} + \frac{\boldsymbol{v}' + \boldsymbol{u}}{\sqrt{\boldsymbol{c}^2 - (\boldsymbol{v}' + \boldsymbol{u})^2}}\right)\boldsymbol{ma} \qquad (\mathbf{8.2.17})$$

Note that the external force $\boldsymbol{F}' = \boldsymbol{F}$ in the above equation. Compare the above equation with equation (8.2.16). Since equation (8.2.16) does not contain the velocity $\boldsymbol{u}$ and the above equation does contain the velocity $\boldsymbol{u}$, the above equation is different from equation (8.2.16). From this we can conclude that the principle of relativity in mechanics is false.

To verify that the theory of relativity is wrong. The authors designed three new optical experiments based on the theory of light interference. The aim is to test whether the principle of relativity is consistent with objective facts.

Experiment 1; validation using a new laser interferometer having front and rear output apertures.

Experiment 2: Validation using a new laser interferometer having two output apertures in the same output mirror.

Experiment 3; validation using a new Michelson interferometer with long and short optical paths.

(For three new experimental programs in optics, see the discussion in Chapters 11 and 12.)

The procedure was as follows, with the new laser interferometer from Experiment 1 placed in a completely enclosed room. Rotate the laser interferometer in all directions in the room. If the laser interference fringes on the observation screen do not move (i.e., the number of movements of the interference fringes is equal to 0), then the room is stationary in the cosmic vacuum frame of reference.

If the phenomenon of moves appears in the laser interference fringes on the observation screen, then the maximum velocity of the laser interferometer (room) in the cosmic vacuum frame of reference can be calculated from the maximum number of moves in the interference fringes.

8.3 "Relativity of simultaneity" is independent of the state of motion of inertial systems

8.3.1 "Relativity of simultaneity" is independent of the state of motion of inertial systems

The idea of the relativity of simultaneity was the theoretical basis for Einstein's rejection of Newton's view of absolute time. Special relativity holds that event A and event B are simultaneous in one inertial reference system, but may not be simultaneous in other inertial systems. To illustrate this, Einstein subjectively imagined a famous train experiment. We can follow Einstein's lead and prove theoretically that Einstein's train experiment is theoretically wrong.

Suppose that points A and B on the platform of a train station are stationary in the cosmic space reference system; Suppose that there are three observation points on the station platform, E, M and D. Where point M is located at the midpoint of the line AB on the station platform; point E is located to the left of point M; and point D is located to the right of point M. Suppose that points A and B on the station platform are struck by lightning at the same time. This is shown in Figure 8-2.

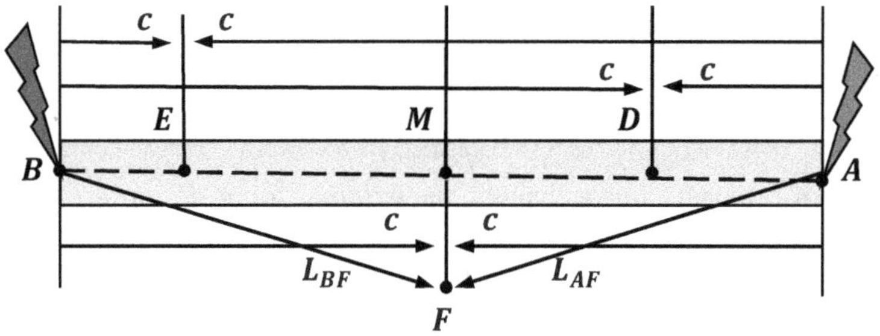

Figure 8-2 Schematic illustration of whether points A and B are struck by lightning at the same time, as observed by an observer at point E, point M, and point D.

According to the principle of invariance of the speed of light in the cosmic vacuum frame of reference, the speed of propagation of the electric flash of light is equal to the speed c of light, when points A and B are struck by lightning.

The distance L_{BE} is less than the distance L_{AE} since point E is located to the left of the midpoint M. As a result, the flash B from point B will first reach point E, while the flash A from point A will reach point E a little later.If the observer is watching from point E at the train station, then the observer will first see point B being struck by the flash, and only a little later will he see point A being struck by the flash.

The distance L_{BD} is greater than the distance L_{AD} since point D is located to the right of the midpoint M.

Therefore, the flash A from point A will first reach point D, while the flash B from point B will reach point D a little later. if the observer is watching from point D at the train station, then the observer will first see that point A has been struck by the lightning, and only a little later he will see that point B has been struck by the lightning.

Since point M is the midpoint of the distance L_{AB}, the distance $L_{BM} = L_{AM}$. therefore flash B and flash A will arrive at point M at the same time, and if the observer is watching from point M on the platform of the train station, the observer will see that point A and point B are struck by the lightning at the same time.

The relativity of simultaneity in special relativity means that if event A and event B occur at the same time in one inertial frame of reference, then event A and event B may not occur at the same time when observed in other inertial frames of reference.

Although points M and D are both observation points in the reference system of the train station, the observations of a stationary observer at point M are different from the observations of a stationary observer at point D.

Furthermore, at point D, whether the observer is observing at rest or at speed v (i.e., observing from a moving train), the observations are the same, i.e., point A is struck first by the lightning, and point B is struck by the lightning a little later. Obviously the observations of the stationary observer at point M are different from the observations of the moving observer at point D.

Since Einstein used a train frame of reference to transport an observer from point M to point D to observe, Einstein formulated the relativity principle of simultaneity based on the fact that the observations at point M and point D are different. This principle is obviously wrong, because in the cosmic vacuum reference system, the observation of a stationary observer at point M is different from the observation of a moving observer at point D.

The mistake Einstein made was to regard the observation of a moving observer at point D in the cosmic vacuum frame of reference as the observation of an inertial system.

In summary, according to the principle of relativity of simultaneity in special relativity, the relativity of simultaneity exists not only in different inertial systems, but also in the stationary reference system of the train station platform. From this it can be determined that the relativity of simultaneity is independent of the state of motion of the reference system.

There is another erroneous approach to the proof of the relativity principle regarding simultaneity.

Suppose plane A is the vertical plane of the tail of the flatbed train and plane B is on the vertical plane of the head of the flatbed train. Suppose point M is the midpoint of the flatbed train and point S is a light source located at the midpoint of the flatbed train. Suppose the train is moving to the right with speed v. Assume that at the instant the light source glows, the ground point E and the midpoint M of the train lie on the same vertical line. This is shown in Figure 8-3.

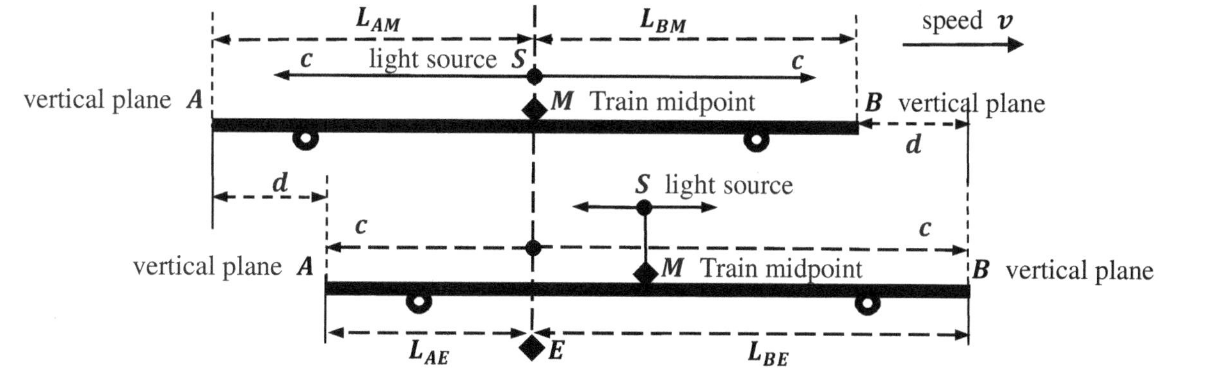

Figure 8-3 Schematic diagram of light source S emitting light at train midpoint M.

When the observer is observing at point M on the flat train. Since the distance $L_{AM} = L_{BM}$, the light ray c is arriving at both plane A and plane B according to the principle of special relativity that the speed of light is constant.

When the observer observes at point E on the ground. Since the distance $L_{AE} < L_{BE}$, the light ray c reaches the plane A first and a little later reaches the plane B. This leads to the conclusion that simultaneity is relative.

However, since point E on the ground and point M in the train are in the same space, and since the light observed at points E and M is the light emitted simultaneously by the light source S, while the observer's observations at points E and M are contradictory, it can be determined that the relativity principle of simultaneity is false.

8.3.2 The true meaning of "relativity of simultaneity"

Suppose a train passes through the platform with a speed of v. When lightning A and lightning B arrive at the same time at point M, the midpoint of the platform. If the observer inside the train is located to the left of point M (e.g., the observer is located at point E on the platform), he will see that point B is struck first by the lightning, and point A is struck by the lightning later. If the observer in the train is located to the right of point M (e.g., the observer is located at platform D), he will see that point A is struck by lightning first and point B is struck by lightning later.

If an observer inside the train happens to be at midpoint M on the platform, he will see points A and B struck by lightning at the same time. At this point, the observer at rest at midpoint M will also see points A and B struck by lightning at the same time. From this it can be determined that both the stationary observer and the moving observer at midpoint M will see points A and B struck by lightning at the same time. In other words, the "simultaneity" of the events seen by a stationary observer and a moving observer at midpoint M is the same.

Or, when an observer passes the midpoint M with speed v, at the location of point M, he can only see points A and B struck by lightning at the same time. He will never see point A struck by lightning first and point B struck by lightning later. It can be determined that the "simultaneity" of events A and B is independent of the state of motion of the inertial system.

Based on the lightning strikes observed by the observer at points E, M, and D on the station platform, it can be determined that the observer would see points A and B struck by lightning at the same time only if the observer is at the midpoint M. The observer does not see points A and B struck by lightning at the same time at points E and D.

Assume that the MF line is perpendicular to the AB line of the platform. At this point the distance from any point on the MF line to points A and B of the station is equal. Since point F is on the MF line, the distance L_{BF} is equal to the distance L_{AF}. If an observer observes a lightning strike at any point on the MF line, then point A and point B are always struck by lightning at the same time. This is shown in Figure 8-2.

Suppose that at time $t = 0$, point A and point B are struck by lightning at the same time. Since the distance L_{BM} is smaller than the distance L_{BF}, the observer sees point A and point B struck by lightning at the same time first at the position of point M, and later at the position of point F. The observer sees point A and B struck by lightning at the same time. Suppose t_M is the time when the observer sees point A and point B at point M being struck by lightning at the same time; t_F is the time when the observer sees point A and point B at point F being struck by lightning at the same time. The "simultaneity" observed by the observer at point M and point F has a time difference of Δt whose value is:

$$\Delta t = t_F - t_M \qquad (8.3.1)$$

If the distance L_{MF} is different, the time difference Δt is also different. The above formula is the real meaning of the concept of "relativity of simultaneity". The above formula shows that "Relativity of Simultaneity" can only occur on the vertical line of MF, not on the horizontal line of AB of the platform. However, special relativity applies the "relativity of simultaneity" to points M and D on the platform AB line.

In other words, when the lightning struck points A and B at the same time, Einstein placed an observer M at point M and then, using a fast-moving train, placed an observer D at point D. This is equivalent to observer M seeing that points A and B were struck by lightning at the same time, and observer D seeing that points A and B were not struck by lightning at the same time.

Einstein used the contradictory observations at points M and D to conclude the "relativity of simultaneity". Since there is no simultaneity between the observation at point M and the observation at point D, Einstein's use of a high-speed moving train to prove the conclusion of "relativity of simultaneity" is wrong.

While every point on the MF vertical line is a point of simultaneity, there is only one point of simultaneity on the

AB horizontal line of the station, and there cannot be multiple points of simultaneity. This is like a bullet can penetrate multiple vertical targets in the direction of motion, but the bullet can only hit one target in the horizontal direction; it is impossible for the bullet to hit multiple targets horizontally.

However, special relativity holds that there are multiple "simultaneity" points on the AB horizontal line. This view is equivalent to the idea that a single bullet can hit more than one horizontal target. It can be determined that the "relativity of simultaneity" view of special relativity does not correspond to objective facts.

8.3.3 The observer only observes lightning strike results characterized by simultaneity at the midpoint M on the AB line, and none of the lightning strike results observed at any other point on the AB line are characterized by simultaneity.

Assume that the train station platform is stationary in the cosmic vacuum reference system. Assume that the train station platform has three observation points E, M, and D on the line AB, where point M is at the midpoint of line AB, point E is to the left of midpoint M, and point D is to the right of midpoint M. This is shown in Figure 8-4.

Assume that the length $\boldsymbol{L}$ of the train, is equal to the length $\boldsymbol{L_{AB}}$ of the AB line, and assume that the train is moving horizontally to the right with speed $\boldsymbol{v}$. When the nose and tail of the train coincide with the distance $\boldsymbol{L_{AB}}$, points A and B on the train platform are struck by lightning at the same time.

When points A and B are struck by lightning at the same time, the observer can be anywhere in the train at this point. Suppose there are three observers in the train, $\boldsymbol{M'}$, $\boldsymbol{E'}$ and $\boldsymbol{D'}$ (the observers are represented by diamonds), and the distance between them is $\boldsymbol{M'M = E'E = D'D}$. This is shown in Figure 8-4.

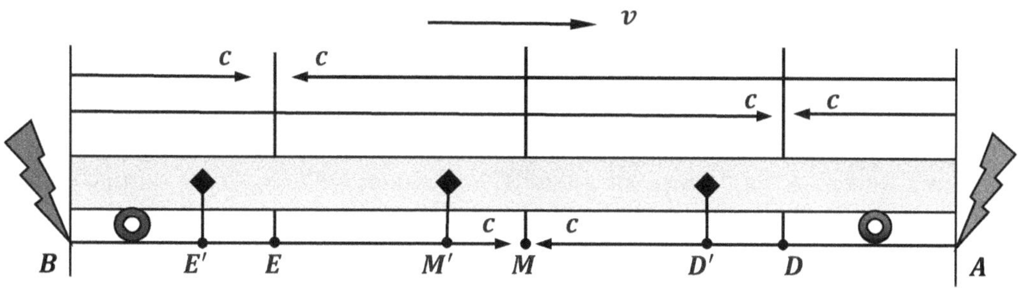

Figure 8-4: Schematic of an observer at points $\boldsymbol{M'}$, $\boldsymbol{E'}$, and $\boldsymbol{D'}$ on the train, observing whether points A and B on the ground are struck by lightning at the same time..

When the observer at point $\boldsymbol{M'}$ on the train moves to point M with the train. At this point, since point M is the midpoint, the observer at point $\boldsymbol{M'}$ on the train will see points A and B struck by lightning at the same time.

When the observer at point $\boldsymbol{E'}$ in the train moves with the train to point E. At this point, the observer at point $\boldsymbol{E'}$ in the train first sees that point B is struck by lightning, and then sees that point A is struck by lightning, since point E is located to the left of the center point M.

When the observer at point $\boldsymbol{D'}$ on the train moves with the train to point D. At this point, the observer at point $\boldsymbol{D'}$ on the train first sees that point A has been struck by lightning, and then sees that point B has been struck by lightning, since point D is located to the right of the center point M.

Based on the observations at points $\boldsymbol{M'}$, $\boldsymbol{E'}$ and $\boldsymbol{D'}$ on the train, it can be determined that observer $\boldsymbol{M'}$ can see points A and B being struck by lightning at the same time only if observer $\boldsymbol{M'}$ on the train moves to midpoint M. At both points E and D, the observer $\boldsymbol{M'}$ could not see that points A and B had been struck by lightning at the same time.

To summarize, only an observer located at the midpoint M on the ground would be able to see points A and B being struck by lightning at the same time; if the observer inside the train is not at point M, then he would not see points A and B being struck by lightning at the same time.

In addition, although the observer at point F on the MF vertical line also sees point A and point B being struck by lightning at the same time, the simultaneous striking of point A and point B by lightning seen at points M and F are not at the same moment. In other words, there is a time difference Δt between the simultaneity seen by the observer at point M and the simultaneity seen by the observer at point F.

8.4 Special relativity's coordinate transformation formula is wrong

8.4.1 Two contradictory effects of length contraction and expansion can be derived from special relativity

In the inertial system S, assume that the inertial system S' is moving with velocity u in the direction of the positive x-axis. Assume that the iron bar is moving with velocity u in the inertial system S, i.e., the iron bar is at rest in the inertial system S'. At this point the iron rod has a velocity $v' = 0$ in the inertial system S'. as shown in Figure 8-5.

According to special relativity, the equation for the positive transformation of coordinates between the inertial system S and the inertial system S' is:

$$x' = \frac{x - ut}{\sqrt{1 - \frac{u^2}{c^2}}}, \quad y' = y, \quad z' = z, \quad t' = \frac{t - \frac{u}{c^2}x}{\sqrt{1 - \frac{u^2}{c^2}}} \tag{8.4.1}$$

The equation for the inverse transformation of the coordinates between the inertial system S and the inertial system S' is:

$$x = \frac{x' + ut'}{\sqrt{1 - \frac{u^2}{c^2}}}, \quad y = y', \quad z = z', \quad t = \frac{t' + \frac{u}{c^2}x'}{\sqrt{1 - \frac{u^2}{c^2}}} \tag{8.4.2}$$

Assume that L is the length of motion of the iron rod in the inertial system S and L_0 is the intrinsic length of the iron rod in the inertial system S'. In the inertial system S, it is assumed that x_A is the starting point of the moving iron bar and x_B is the ending point of the moving iron bar. In the inertial system S', it is assumed that x'_A is the starting point of the intrinsic length L_0 of the iron bar and x'_B is the end point of the intrinsic length L_0 of the iron bar. This is shown in Figure 8-5.

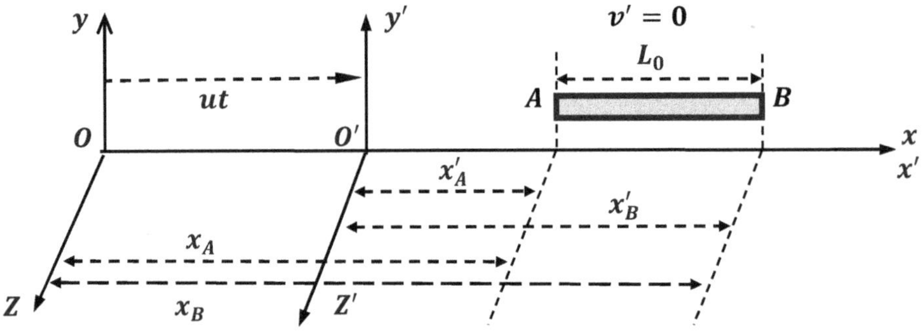

Figure 8-5 Schematic diagram of an iron bar AB moving with velocity u along the positive x-axis in an inertial system S and at rest in the inertial system S'.

In the inertial system S', if the observer is stationary with respect to the iron bar AB, then according to special relativity, the observer measures the stationary length L_0 of the iron bar AB as:

$$L_0 = x'_B - x'_A \tag{8.4.3}$$

In the inertial system S, if the observer is moving with respect to the iron bar AB, then the length L of the motion of the iron bar AB measured by the observer at moment t is:

$$L = x_B - x_A \tag{8.4.4}$$

The coordinate transformation formula of special relativity possesses a coordinate transformation formula for the forward direction (8.4.1) and a coordinate transformation formula for the reverse direction (8.4.2). We can derive the length contraction effect and the length expansion effect based on these two coordinate transformation formulas.

First, when the observer is at rest in the inertial system S. At the moment t, according to the coordinate transformation formula (8.4.1) in the positive direction, the coordinates x'_A of the starting point of the intrinsic length L_0 of the iron bar are:

$$x'_A = \frac{x_A - ut}{\sqrt{1 - \frac{u^2}{c^2}}} \tag{8.4.5}$$

At moment t, according to the coordinate transformation formula (8.4.1) in the positive direction, the terminal coordinate x'_B of the intrinsic length L_0 of the iron bar is:

$$x'_B = \frac{x_B - ut}{\sqrt{1 - \frac{u^2}{c^2}}} \tag{8.4.6}$$

Subtract equation (8.4.5) from the above equation to obtain the relational equation:

$$x'_B - x'_A = \frac{x_B - x_A}{\sqrt{1 - \frac{u^2}{c^2}}} \tag{8.4.7}$$

Since the intrinsic length of the iron bar $L_0 = x'_B - x'_A{}'$ belongs to the inertial system S', the length of motion of the iron bar $L = x_B - x_A$ belongs to the inertial system S, and thus the above equation can be expressed as:

$$L_0 = \frac{L}{\sqrt{1 - \frac{u^2}{c^2}}} \tag{8.4.8}$$

The above equation shows that the intrinsic length L_0 of the iron rod is contracted in the direction of motion, i.e., $L_0 > L$. The above equation is consistent with the relativistic length contraction effect.

In addition, when the observer is at rest in the inertial system S', at the moment t', according to the coordinate transformation formula (8.4.2) in the inverse direction, the coordinates x_A of the starting point of the iron bar's motion length L are:

$$x_A = \frac{x'_A + ut'}{\sqrt{1 - \frac{u^2}{c^2}}} \tag{8.4.9}$$

According to the coordinate transformation equation (8.4.2) in the inverse direction, the coordinates x_B of the end point of the length L of motion of the iron bar are:

$$x_B = \frac{x'_B + ut'}{\sqrt{1 - \frac{u^2}{c^2}}} \tag{8.4.10}$$

Subtracting equation (8.4.9) from the above equation gives the relational equation:

$$x_B - x_A = \frac{x'_B - x'_A}{\sqrt{1 - \frac{u^2}{c^2}}}$$

Since the length of motion of the iron bar $L = x_B - x_A$ belongs to the inertial system S, the intrinsic length of the iron bar $L_0 = x'_B - x'_A$ belongs to the inertial system S', so the above equation can be expressed as:

$$L = \frac{L_0}{\sqrt{1 - \frac{u^2}{c^2}}} \tag{8.4.11}$$

The above equation can be varied as:

$$L_0 = L\sqrt{1-\frac{u^2}{c^2}} \tag{8.4.12}$$

Since the coefficient $\sqrt{1-\frac{u^2}{c^2}} < 1$, the intrinsic length $L_0 < L$, i.e., the kinematic length L of the iron rod is increasing. However special relativity holds that the kinematic length L is shrinking, i.e., the intrinsic length $L_0 > L$ It can be determined from this that the above equation contradicts special relativity.

According to Equation (8.4.8), the intrinsic length L_0 of the iron bar AB is:

$$L_0 = \frac{L}{\sqrt{1-\frac{u^2}{c^2}}}$$

Compare the above equation with equation (8.4.12). Since the intrinsic length L_0 of the iron bar AB is both greater than the length of motion L (i.e., $L_0 > L$) and less than the length of motion L (i.e., $L_0 < L$), the formula for the coordinate transformation of special relativity is wrong.

8.4.2 Accelerated transformation formula for special relativity

According to special relativity, the velocity transformation formula for special relativity is:

$$v' = \frac{v-u}{1-\frac{vu}{c^2}} \tag{8.4.13}$$

Taking the derivative of the above equation with respect to time t' gives the relational equation:

$$\frac{dv'}{dt'} = \frac{dv'}{dt}\frac{dt}{dt'} = \frac{a\left(1-\frac{vu}{c^2}\right)-(v-u)\left(-a\frac{u}{c^2}\right)}{\left(1-\frac{vu}{c^2}\right)^2}\frac{dt}{dt'} \tag{8.4.14}$$

Simplify the above equation to obtain the relational equation:

$$a' = \frac{dv'}{dt'} = \frac{a\left(1-\frac{u^2}{c^2}\right)}{\left(1-\frac{vu}{c^2}\right)^2}\frac{dt}{dt'} \tag{8.4.15}$$

Take the derivative of the time t in formula (8.4.2) with respect to the time t' to obtain the relational formula:

$$\frac{dt}{dt'} = \frac{1}{\sqrt{1-\frac{u^2}{c^2}}}\left(1+\frac{v'u}{c^2}\right) > 1 \tag{8.4.16}$$

Since the velocity $u > 0$ and the velocity $v' \geq 0$, the derivative $\frac{dt}{dt'} > 1$. Substituting the above equation into equation (8.4.15) gives the relational equation:

$$a' = \frac{a\left(1-\frac{u^2}{c^2}\right)}{\left(1-\frac{vu}{c^2}\right)^2}\frac{1}{\sqrt{1-\frac{u^2}{c^2}}}\left(1+\frac{v'u}{c^2}\right)$$

Simplify the above equation to obtain the relational equation:

$$a' = \frac{a}{\left(1-\frac{vu}{c^2}\right)^2}\sqrt{1-\frac{u^2}{c^2}}\left(1+\frac{v'u}{c^2}\right)$$

Substituting the velocity v' from equation (8.4.13) into the above equation gives the relational equation:

$$a' = \frac{a}{\left(1-\frac{vu}{c^2}\right)^2}\sqrt{1-\frac{u^2}{c^2}}\left(1+\frac{u}{c^2}\frac{v-u}{1-\frac{vu}{c^2}}\right)$$

Simplify the above equation to obtain the relational equation:

$$a' = \frac{a}{\left(1 - \frac{vu}{c^2}\right)^3}\left(1 - \frac{u^2}{c^2}\right)^{\frac{3}{2}} \quad (8.4.17)$$

The above formula is the acceleration transformation formula of special relativity. When the object is at rest in the inertial system S', i.e., when the velocity $v = u$, the acceleration a' of the object in the inertial system S' is:

$$a' = a\sqrt{1 - \frac{u^2}{c^2}}$$

The above equation shows that the acceleration a' of an object in an inertial system S' is not equal to the acceleration a of the object in an inertial system S.

8.4.3 Special relativity holds that the external force F on an object in an inertial system S has different values in different inertial systems S'.

If an object moves with velocity v in an inertial system S. According to the mass-velocity formula of special relativity, the moving mass m of the object is:

$$m = \frac{m_0}{\sqrt{1 - \frac{v^2}{c^2}}}$$

Multiply the above equation by the acceleration a to obtain the relationship equation:

$$F = ma = \frac{m_0}{\sqrt{1 - \frac{v^2}{c^2}}}a \quad (8.4.18)$$

In the inertial system S, the external force on the object $F = m_0a$ when the object's velocity $v = 0$. In the inertial system S, the external force F on the object is equal to infinity when the object is moving at the speed of light c, i.e., when the velocity $v = c$.

Logically, Newton's second law $F = ma$ does not apply to high-speed motion, which means that the external force F is related to the velocity v. Physicists can consider equation (8.4.18) as a modified Newton's second law. However, physicists have so far refused to modify Newton's second law to equation (8.4.18) for fear that the coordinate transformation formula of special relativity will be rejected. This would lead to the collapse of the edifice of physics.

In the inertial system S', the velocity of motion of the object is v'. According to the mass-velocity formula, the object's moving mass m' is:

$$m' = \frac{m_0}{\sqrt{1 - \frac{v'^2}{c^2}}} \quad (8.4.29)$$

Multiplying the above equation by the acceleration a' gives the external force F' on the object in the inertial system S':

$$F' = m'a' = \frac{m_0}{\sqrt{1 - \frac{v'^2}{c^2}}}a' \quad (8.4.20)$$

Substituting the velocity v' from equation (8.4.13) and the acceleration a' from equation (8.4.17) into the above equation, we obtain the relation equation:

$$F' = m'a' = \frac{m_0}{\sqrt{1 - \frac{c^2(v-u)^2}{(c^2-vu)^2}}} \cdot \frac{a}{\left(1 - \frac{vu}{c^2}\right)^3}\left(1 - \frac{u^2}{c^2}\right)^{\frac{3}{2}} \quad (8.4.21)$$

If the speed of motion of the inertial system S' is $u = 0$, then the external force F' on the object in the inertial system S' is equal to the external force F on the object in the inertial system S, according to the above equation, i.e:

$$F' = F = \frac{m_0}{\sqrt{1 - \frac{v^2}{c^2}}} a \tag{8.4.22}$$

If the object moves in the inertial system S with velocity $v = u$, i.e. the object is at rest in the inertial system S', then the external force F' on the object in the inertial system S' is given by equation (8.4.21):

$$F' = \frac{m_0}{\left(1 - \frac{v^2}{c^2}\right)^{\frac{3}{2}}} a \neq \frac{m_0}{\sqrt{1 - \frac{v^2}{c^2}}} a = F \tag{8.4.23}$$

Since the above equation is not equal to equation (8.4.18), the external force F' in the inertial system S' is not equal to the external force F in the inertial system S, i.e. $F' \neq F$.

Theoretically, if the velocity of the inertial system S' is $u \neq 0$, then the external force F' in the inertial system S' is not equal to the external force F in the inertial system S, i.e:

$$F' = \frac{m_0}{\sqrt{1 - \frac{c^2(v-u)^2}{(c^2-vu)^2}}} \cdot \frac{a}{\left(1 - \frac{vu}{c^2}\right)^3} \left(1 - \frac{u^2}{c^2}\right)^{\frac{3}{2}} \neq \frac{m_0}{\sqrt{1 - \frac{v^2}{c^2}}} a = F \tag{8.4.24}$$

From the above equation, when an object on the ground is subjected to an external force F', the observer on the ground can calculate the magnitude of the external force F'; the observer on the train can also calculate the magnitude of the external force F''; in addition, the observer on the airliner can also calculate the magnitude of the external force F'''. Since the speed v'' of the train, is not equal to the speed v''' of the airplane, the external forces F', F'' and F''' are not equal, viz:

$$F' \neq F'' \neq F''' \tag{8.4.25}$$

The above equation shows that an external force F has different values in different inertial reference frames. Or, according to special relativity, we cannot determine which inertial frame's external force F is the real external force.

8.4.4 Using the mass-velocity formula to prove that the coordinate transformation formula of special relativity is wrong

Theoretically, the external force F on an object should be equal in magnitude in different inertial systems S'. For example, the gravitational force g on 1 kilogram of mass on the Earth is a Newtonian force of **9.8** kilograms. If observed and calculated in the lunar inertial system, then the gravitational force g' of a 1 kilogram mass on Earth should also be equal to **9.8** kilograms of Newtonian force. It should not be equal to **19.6** kilograms of Newtonian force, or **4.9** kilograms of Newtonian force.

In the cosmic vacuum reference system S, assume that v is the velocity of motion of the object. According to equation (8.2.1), the force F on the object in the cosmic vacuum reference system S is:

$$F = \left(1 + \frac{v}{\sqrt{c^2 - v^2}}\right) ma \tag{8.4.26}$$

The force F, velocity v and acceleration a in equation are all physical quantities in the cosmic space reference system. Note that the force F in the cosmospace reference system possesses uniqueness when the velocity v remains constant. That is, the force F has a value equal to F in all other inertial systems.

In the cosmic vacuum reference system S, assume that u is the velocity of motion of the inertial system S', and assume that the inertial system velocity u and the object velocity v are in the same direction. According to equation (8.2.17), the force F' on the object in the inertial system S' is:

$$F' = \left(1 + \frac{v' + u}{\sqrt{c^2 - (v' + u)^2}}\right) ma \tag{8.4.27}$$

Since the velocity v satisfies the Galilean velocity transformation formula, i.e., $v = v' + u$, and the acceleration a is equal in any inertial system, i.e., $a = a'$, the force F' of the inertial system S' is equal to the force F of the cosmic vacuum reference system, i.e., $F' = F$.

Since the force F is unique, its value should be the same in any inertial system. However, if the magnitude of the force $\boldsymbol{F'}$ is calculated according to the mass-velocity formula of special relativity, the force $\boldsymbol{F'}$ will have different values in different inertial systems $\boldsymbol{S'}$. At this point, one cannot determine which force $\boldsymbol{F'}$ is the true force. We can illustrate this with the following example.

Suppose the object is moving in the train inertial system $\boldsymbol{S''}$ with a velocity $\boldsymbol{v''}$, and if the observer is looking from the train inertial system $\boldsymbol{S''}$, then the force $\boldsymbol{F''}$ on the object in the train inertial system $\boldsymbol{S''}$ is, according to the mass-velocity formula:

$$\boldsymbol{F''} = \boldsymbol{m''a''} = \frac{\boldsymbol{m_0}}{\sqrt{1 - \frac{\boldsymbol{v''^2}}{c^2}}}\boldsymbol{a''} \tag{8.4.28}$$

The force $\boldsymbol{F''}$, velocity $\boldsymbol{v''}$, and acceleration $\boldsymbol{a''}$ in the formula are all physical quantities in the train inertial system $\boldsymbol{S''}$.

If the observer is looking from the ground inertial system $\boldsymbol{S'}$, then the force $\boldsymbol{F'}$ on the object in the ground inertial system $\boldsymbol{S'}$ is:

$$\boldsymbol{F'} = \boldsymbol{m'a'} = \frac{\boldsymbol{m_0}}{\sqrt{1 - \frac{\boldsymbol{v'^2}}{c^2}}}\boldsymbol{a'} \tag{8.4.29}$$

The force $\boldsymbol{F'}$, velocity $\boldsymbol{v'}$ and acceleration $\boldsymbol{a'}$ in the formula are all physical quantities in the terrestrial inertial system $\boldsymbol{S'}$.

If the observer is looking from the cosmic vacuum reference system $\boldsymbol{S}$, then the force $\boldsymbol{F}$ experienced by the object in the cosmic vacuum reference system $\boldsymbol{S}$ is:

$$\boldsymbol{F} = \boldsymbol{ma} = \frac{\boldsymbol{m_0}}{\sqrt{1 - \frac{\boldsymbol{v^2}}{c^2}}}\boldsymbol{a} \tag{8.4.30}$$

The force $\boldsymbol{F}$, velocity $\boldsymbol{v}$ and acceleration $\boldsymbol{a}$ in Eq. are all physical quantities in the cosmic vacuum reference system $\boldsymbol{S}$.

Since velocity $\boldsymbol{v} \neq \boldsymbol{v'} \neq \boldsymbol{v''}$ and acceleration $\boldsymbol{a} \neq \boldsymbol{a'} \neq \boldsymbol{a''}$, the force $\boldsymbol{F} \neq \boldsymbol{F'} \neq \boldsymbol{F''}$. Obviously, at this point in time, one cannot be sure which of the three forces, $\boldsymbol{F}$, $\boldsymbol{F'}$ and $\boldsymbol{F''}$, is the real force. If one believes that the force $\boldsymbol{F''}$ on the train is the real force, then the force $\boldsymbol{F''}$ on the train is not the real force when the train is traveling at high speed. This is because the mass-velocity formula only includes the velocity $\boldsymbol{v''}$ of the motion of the object on the train, not the velocity $\boldsymbol{u}$ of the train on the ground.

The fact that the objectively existing force $\boldsymbol{F}$ has different values in different inertial systems $\boldsymbol{S'}$ is a theoretical dilemma that cannot be overcome by the transformation formula of special relativity. The new force formula, however, has no such theoretical dilemma.

Since the mass velocity formula is derived from the velocity transformation formula of special relativity, the velocity transformation formula of special relativity must be theoretically wrong.

Chapter 9 Deriving a New Formula for Gravitational

Introduction: Based on the vortex field conservation formula and Stokes' theorem, the authors derive a new formula for gravitational.

9.1 Definition of an equiangular helical field

9.1.1 Conservation formula for an equiangular helical field

According to logarithm theory, the natural logarithm $\ln a$ $(a > 0)$ is a logarithm with a natural constant e as its base. Using e as the base of the logarithm makes it possible to simplify many mathematical formulas. The formula for the mathematical expression of the natural constant e is:

$$e = \lim_{x\to\infty}\left(1 + \frac{1}{x}\right)^x = 2.71828\cdots\cdots \qquad (9.1.1)$$

The physical meaning of the natural constant e is the limit of the growth of things per unit of time. If the independent variable x tends to positive infinity, then the limit of growth of things is equal to the constant e. If the independent variable x tends to negative infinity, then the limit of growth of things is also equal to the constant e.

In the objective world. The movements and changes of many things have a vortex structure, of which the equiangular spiral is a common vortex in nature. For example, the rotation of galaxies around the centers of star systems and the rotation of typhoons around the eye of the typhoon are all equiangular spirals. This is shown in Figure 9-1.

Figure 9-1 Schematic of an equiangular spiral field.

The creation and variation of an equiangular spiral field is related to a number of natural factors (e.g., mass m, electric charge q, etc.) that are usually the source of the equiangular spiral field.

Assume that the polar radius r is the independent variable of the equiangular vortex field; y is the dependent variable of the equiangular vortex field. Assume that the coefficient k is a physical quantity associated with the vortex source. (In a gravitational field, the coefficient k is the mass m; in an electromagnetic field, the coefficient k is the electric charge q). Let D_k be a constant associated with the vortex source k. (In the gravitational field, D_k is related to the gravitational constant G; in the electromagnetic field, the constant D_k is related to the vacuum permeability ε_0).

In the polar coordinate system (r, y). If the polar radius r takes values in the range $0 < r < +\infty$, then the variation between the polar radius r and the dependent variable y always satisfies the following exponential formula:

$$e^y \left(\frac{e}{r}\right)^{D_k k r} = 1 \qquad 0 < r < +\infty \tag{9.1.2}$$

The above formula is the conservation formula for an equiangular helical field. The k in the formula is the vortex source and D_k is the constant associated with the vortex source k. The dependent variable $y = 0$ when the polar radius $r = e$.

It should be noted that many of the laws of physics associated with equiangular helical fields (e.g., the gravitational force formula, the electric field force formula, etc.) can be derived using the conservation formula for equiangular helical fields.

9.1.2 Polar angles of equiangular helical fields

In the polar coordinate system (r, θ), the equation for the polar radius of an equiangular helical field is:

$$r = ae^{b\theta} \tag{9.1.3}$$

The a and b in the formula are constants, and θ is the polar angle of the equiangular helical field. Taking the natural logarithm of the above equation gives the relational equation:

$$\theta = \frac{1}{b} \ln\left(\frac{r}{a}\right) \tag{9.1.4}$$

The above equation is the formula for the polar angle of an equiangular helical field.

If the polar radius $r = a$, then the polar angle $\theta = 0$.

If the polar radius $r > a$, then the polar angle $\theta > 0$.

If the polar radius $r < a$, then the polar angle $\theta < 0$.

The polar angle coordinate line shown in Figure 9-2.

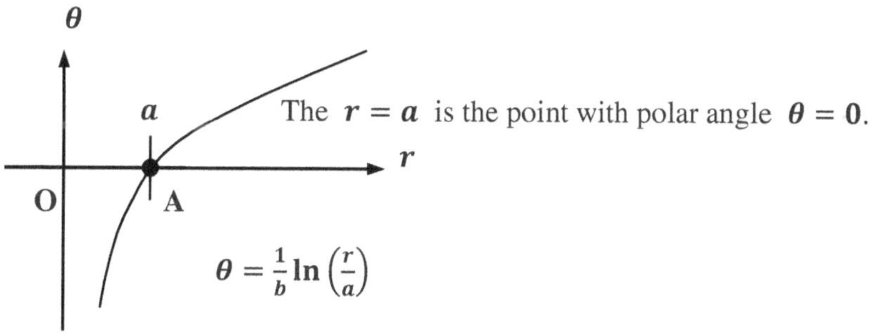

Figure 9-2 Schematic of the coordinate lines for polar angle θ.

Assume that point A is the starting point at the polar angle $\theta = 0$. Assume that the starting point A is on the r axis and that the distance from the starting point A to the center of the vortex O is $r = a$. This is shown in Figure 9-2.

It is clear that all polar angles θ rotated clockwise from the starting point A are positive and all polar angles θ rotated counterclockwise from the starting point A are negative.

If the points of intersection of the polar angle θ and the axis of r are A, B, C, and D, then the positive polar angles θ corresponding to the polar radius r are $\theta_A = 0$, $\theta_B = 360^0$, $\theta_C = 720^0$, and $\theta_D = 1080^0$, respectively, as shown in Figure 9-3.

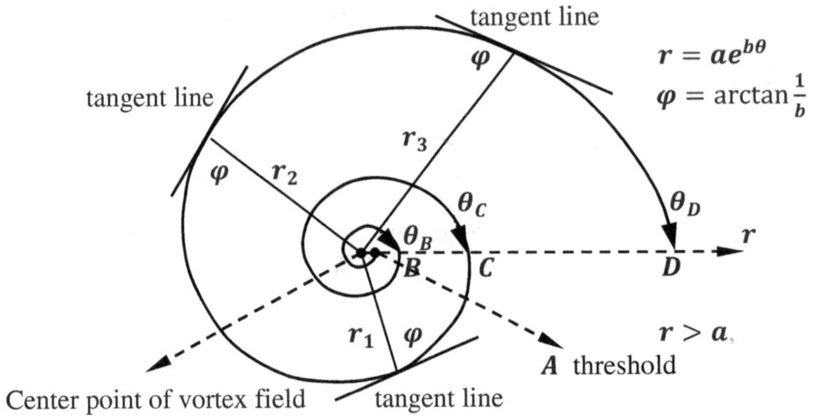

Figure 9-3 Schematic diagram of the clockwise increase of the positive polar angle $\boldsymbol{\theta}$. Point A is the starting point of the positive and negative polar angle $\boldsymbol{\theta}$. The starting point A has a polar radius $\boldsymbol{r = a}$. $\boldsymbol{\varphi}$ is the angle between the polar radius $\boldsymbol{r}$ and the tangent line. The positive polar angle $\boldsymbol{\theta}$ applies to macroscopic vortex fields. For example, the gravitational field of a star in the universe.

The angle $\boldsymbol{\varphi}$ between the polar radius $\boldsymbol{r}$ of any point on an equiangular spiral field and the tangent to that point is a constant angle. The term equiangular in an equiangular spiral field means that the pinch angles $\boldsymbol{\varphi}$ are all equal. This is shown in Figure 9-3.

According to astronomical observations. The Milky Way has the structure of an equiangular spiral field. The Milky Way's equiangular spiral field is a thin disk with four spiral arms. The inclination $\boldsymbol{\varphi}$ of the four spiral arms of the Milky Way is about **12°**.

In addition, according to Figure 9-2, all polar angles $\boldsymbol{\theta}$ rotated counterclockwise from the starting point A are negative polar angles. This is shown in Figure 9-4.

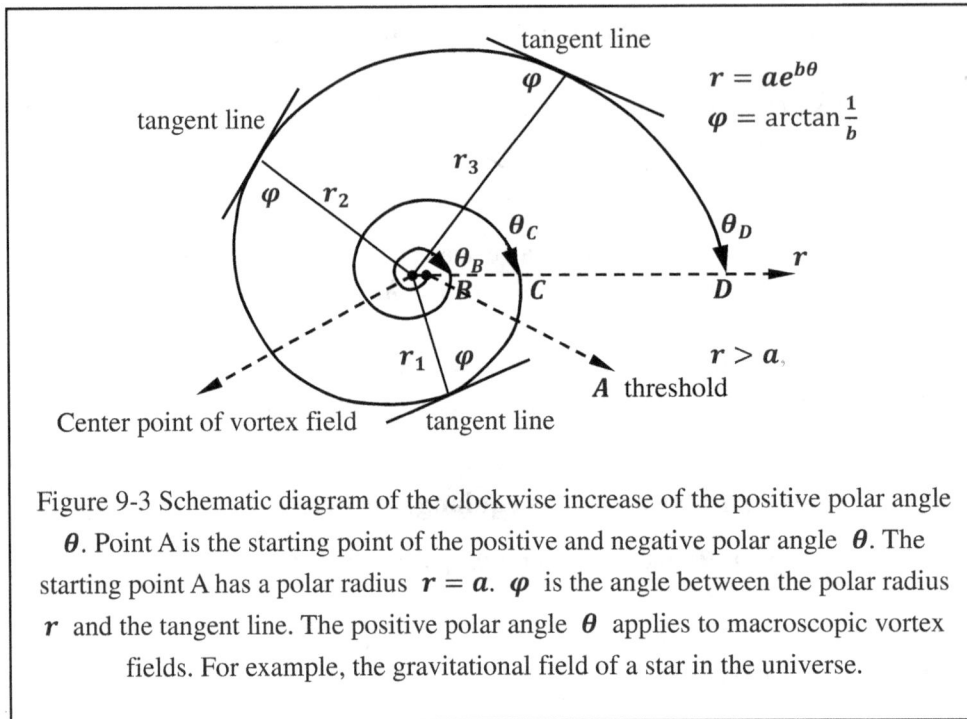

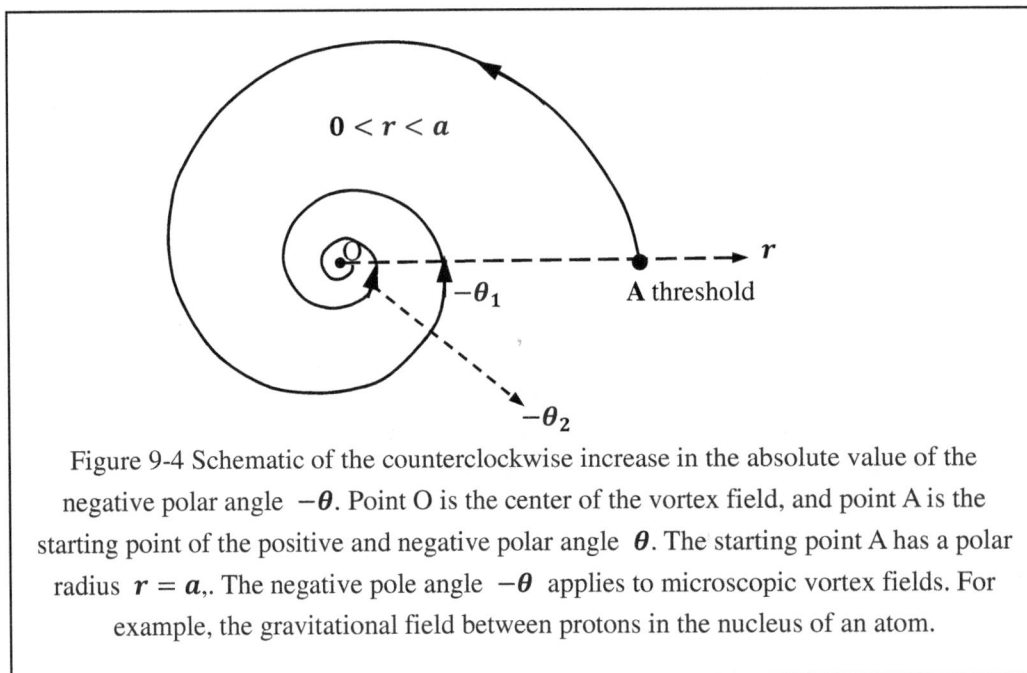

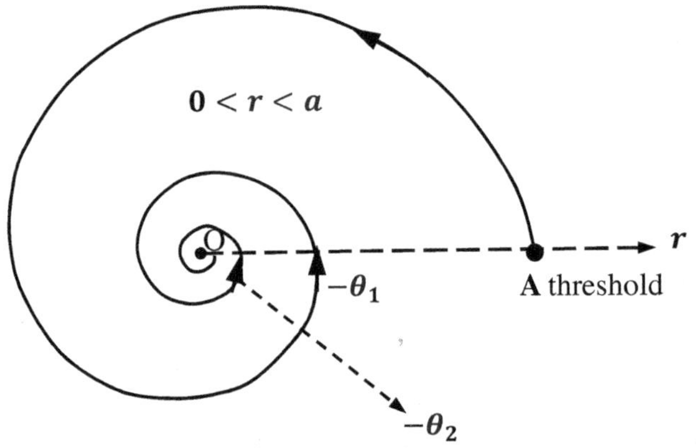

Figure 9-4 Schematic of the counterclockwise increase in the absolute value of the negative polar angle $\boldsymbol{-\theta}$. Point O is the center of the vortex field, and point A is the starting point of the positive and negative polar angle $\boldsymbol{\theta}$. The starting point A has a polar radius $\boldsymbol{r = a}$,. The negative pole angle $\boldsymbol{-\theta}$ applies to microscopic vortex fields. For example, the gravitational field between protons in the nucleus of an atom.

If constant $a = 1$; constant $1/b = D_k k$., the polar angle equation (9.1.4) becomes:

$$\theta = D_k k \cdot \ln r \qquad (9.1.5)$$

We can derive the above formula for the polar angle by using the conservation equation for an equiangular helical field.

9.1.3 Definition of an Equiangular Helix Field

Take the natural logarithm of equation (9.1.2). According to the logarithmic operation formula $\ln(ab) = \ln a + \ln b$, we can get the relation formula:

$$y + \ln\left(\frac{e}{r}\right)^{D_k kr} = 0 \qquad (9.1.6)$$

According to the logarithmic formula $\ln a^n = n\ln a$ and $\ln\left(\frac{a}{b}\right) = \ln a - \ln b$, the above formula becomes:

$$y = D_k kr(\ln r - 1) \qquad (9.1.7)$$

This book defines the dependent variable y as the vortex potential of the vortex field. The k in equation is the vortex source that generates the vortex field, such as mass m or electric charge q. D_k in equation is a constant associated with the vortex source k.

This book defines the set $\sum y$ of vortex potentials y in space as a vortex field. Since the space occupied by the vortex potential y is an infinite sphere, the vortex field $\sum y$ consisting of the vortex potential y is an infinite sphere.

It is important to note that the vortex field $\sum y$ as defined in this book is not a substance but a material property that the vortex source k has.

Theoretically, the development and variation of the vortex field $\sum y$ are unfolded according to equation (9.1.7). Different pole radii r have different vortex potentials y.

If polar radius $r = e$, then vortex potential $y = 0$.

If polar radius $r > e$, then vortex potential $y > 0$.

If polar radius $r<e$, then vortex potential $y < 0$.

The coordinate line for vortex potential y is shown in Figure 9-5.

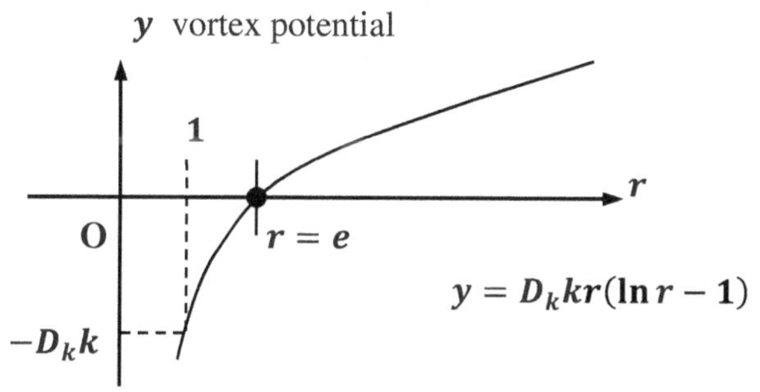

Figure 9-5 Schematic of the coordinate lines of the vortex potential y of a vortex field.

According to equation (9.1.7), if the polar radius r varies continuously to infinity in one direction. The curve formed by an infinite number of vortex potentials y, on the polar radius r is the vortex potential line.

9.1.4 Integration of the vortex potential y along the circular line L_R=2πR

Suppose $L = f(r)$ is a curve in the vortex field. Take the line element dl on the curve $L = f(r)$. According to equation (9.1.7), the product of the vortex potential y and the line element dl is:

$$ydl = D_k kr(\ln r - 1)dl \qquad (9.1.8)$$

Assume that the distance r from the vortex source k to each point on the circular line $L_R = 2\pi R$ is equal.

Since the vortex potential $\boldsymbol{y}$ is constant on the circular line $\boldsymbol{L_R}$, the integral $\oint \boldsymbol{ydl}$ of the vortex potential $\boldsymbol{y}$ along the circular line $\boldsymbol{L_R = 2\pi R}$ is:

$$\oint \boldsymbol{ydl} = \boldsymbol{yL_R} = \boldsymbol{2\pi R} \cdot \boldsymbol{D_k kr(\ln r - 1)} \qquad (9.1.9)$$

The vortex potential $\boldsymbol{y}$ is a vector since the loop integral $\oint \boldsymbol{ydl} \neq \boldsymbol{0}$ of the vortex potential $\boldsymbol{y}$. The direction of the vortex potential $\boldsymbol{y}$ is the same as the direction of the polar radius $\boldsymbol{r}$, or the direction is opposite.

Suppose the vortex source $\boldsymbol{k}$ is the center of the sphere; the distance $\boldsymbol{r}$ from point $\boldsymbol{k}$ to the circular line $\boldsymbol{L_R = 2\pi R}$ is the radius of the sphere. Suppose $\boldsymbol{\Omega}$ is the steradian angle possessed by the circular plane $\boldsymbol{P = \pi R^2}$; $\boldsymbol{\theta}$ is the angle between the radius $\boldsymbol{r}$ and the distance $\boldsymbol{Z}$. Suppose O is the center of the circular line $\boldsymbol{L_R = 2\pi R}$; $\boldsymbol{Z}$ is the perpendicular distance from point $\boldsymbol{k}$ to the center O. Suppose the surface $\boldsymbol{S}$ is bounded by the circular line $\boldsymbol{L_R = 2\pi R}$. This is shown in Figure 9-6.

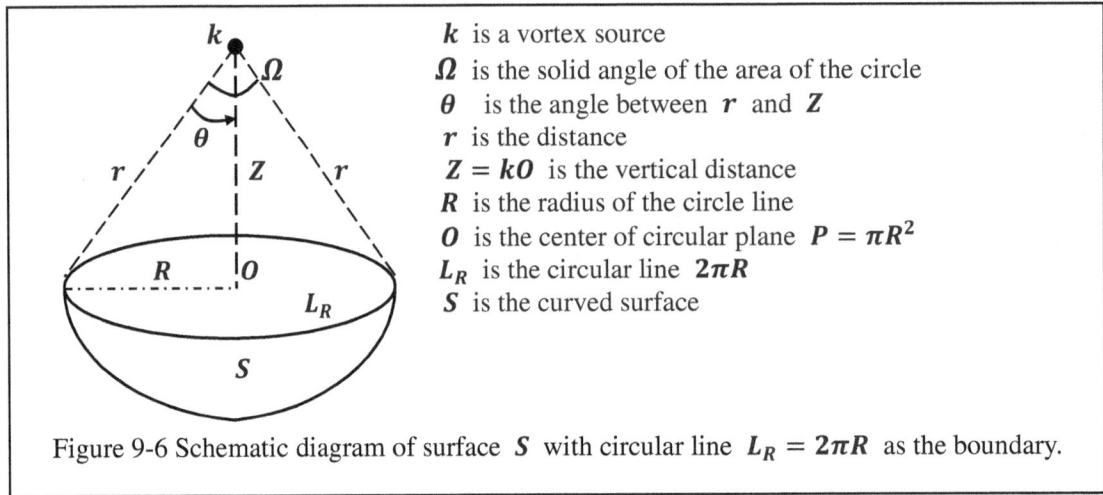

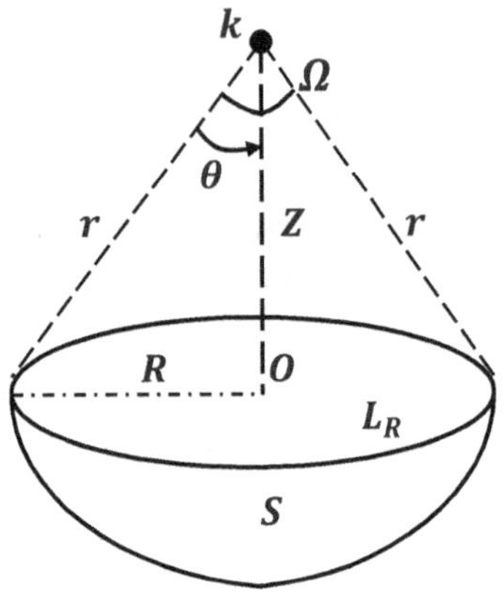

$\boldsymbol{k}$ is a vortex source
$\boldsymbol{\Omega}$ is the solid angle of the area of the circle
$\boldsymbol{\theta}$ is the angle between $\boldsymbol{r}$ and $\boldsymbol{Z}$
$\boldsymbol{r}$ is the distance
$\boldsymbol{Z = kO}$ is the vertical distance
$\boldsymbol{R}$ is the radius of the circle line
$\boldsymbol{O}$ is the center of circular plane $\boldsymbol{P = \pi R^2}$
$\boldsymbol{L_R}$ is the circular line $\boldsymbol{2\pi R}$
$\boldsymbol{S}$ is the curved surface

Figure 9-6 Schematic diagram of surface $\boldsymbol{S}$ with circular line $\boldsymbol{L_R = 2\pi R}$ as the boundary.

Note that there are an infinite number of surfaces $\boldsymbol{S}$ bounded by the circle line $\boldsymbol{L_R = 2\pi R}$.

When the distance $\boldsymbol{Z = 0}$, the vortex source $\boldsymbol{k}$ is inside the circular plane $\boldsymbol{P = \pi R^2}$. When the distance $\boldsymbol{Z \neq 0}$, the vortex source $\boldsymbol{k}$ is outside the circular plane $\boldsymbol{P}$. When the angle $\boldsymbol{\theta = 90^0}$. Since the radius of the circular line is $\boldsymbol{R = r}$, equation (9.1.9) becomes:

$$\oint \boldsymbol{ydl} = \boldsymbol{2\pi r} \cdot \boldsymbol{D_k kr(\ln r - 1)} \qquad (9.1.10)$$

At this point, the distance $\boldsymbol{Z = 0}$. The vortex source $\boldsymbol{k}$ is located at the center of the circular plane $\boldsymbol{P = \pi R^2}$.

9.1.5 Wang's circular crown S_W=$2\pi rR$

Suppose $\boldsymbol{S_\Omega}$ is the area of the spherical crown; $\boldsymbol{h}$ is the height of the crown. Suppose the spherical crown $\boldsymbol{S_\Omega = 2\pi hr}$ is bounded by the circular line $\boldsymbol{L_R = 2\pi R}$. This is shown in Figure 9-7.

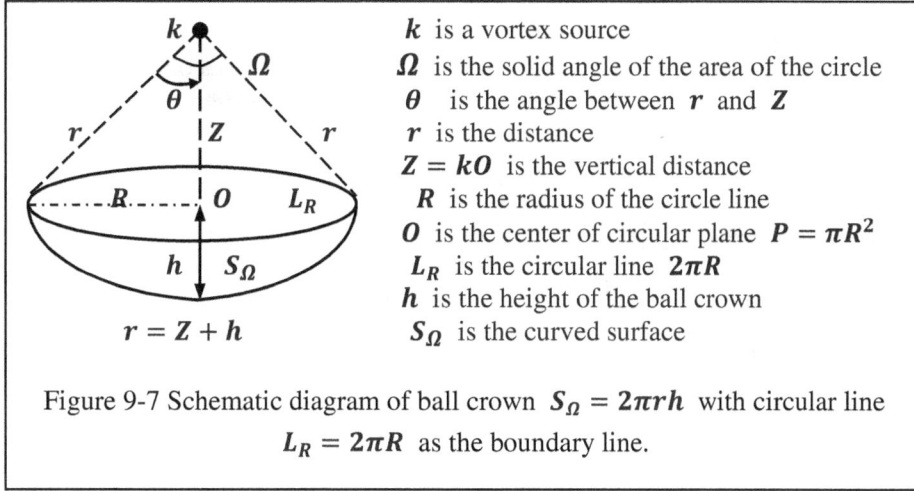

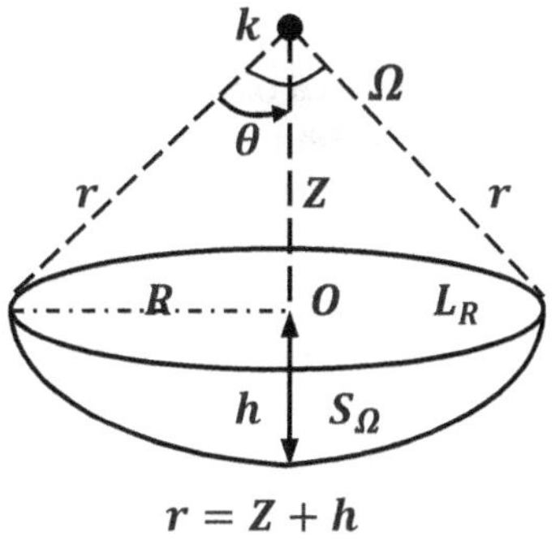

$\boldsymbol{k}$ is a vortex source
$\boldsymbol{\Omega}$ is the solid angle of the area of the circle
$\boldsymbol{\theta}$ is the angle between $\boldsymbol{r}$ and $\boldsymbol{Z}$
$\boldsymbol{r}$ is the distance
$\boldsymbol{Z} = \boldsymbol{kO}$ is the vertical distance
$\boldsymbol{R}$ is the radius of the circle line
$\boldsymbol{O}$ is the center of circular plane $\boldsymbol{P} = \boldsymbol{\pi R^2}$
$\boldsymbol{L_R}$ is the circular line $\boldsymbol{2\pi R}$
$\boldsymbol{h}$ is the height of the ball crown
$\boldsymbol{S_\Omega}$ is the curved surface

Figure 9-7 Schematic diagram of ball crown $\boldsymbol{S_\Omega} = \boldsymbol{2\pi rh}$ with circular line $\boldsymbol{L_R} = \boldsymbol{2\pi R}$ as the boundary line.

Since the distance $\boldsymbol{Z}$ is perpendicular to the circular plane $\boldsymbol{P} = \boldsymbol{\pi R^2}$, the distance $\boldsymbol{r}$ from the vortex source $\boldsymbol{k}$ to any point on the circular line $\boldsymbol{L_R} = \boldsymbol{2\pi R}$ is:

$$\boldsymbol{r} = \sqrt{\boldsymbol{R^2 + Z^2}} \qquad (\mathbf{9.1.11})$$

According to plane geometry, the spherical crown $\boldsymbol{S_\Omega}$ has a height $\boldsymbol{h}$ is:

$$\boldsymbol{h = r - Z = r(1 - \cos\theta)} \qquad (\mathbf{9.1.12})$$

The spherical crown area $\boldsymbol{S_\Omega}$ is:

$$\boldsymbol{S_\Omega} = \boldsymbol{2\pi rh} \qquad (\mathbf{9.1.13})$$

If the spherical crown $\boldsymbol{S_\Omega} = \boldsymbol{2\pi rh}$ is bordered by the circular line $\boldsymbol{L_R} = \boldsymbol{2\pi R}$, then the height $\boldsymbol{h}$ of the spherical crown is:

$$\boldsymbol{h = r - \sqrt{r^2 - R^2}} \qquad (\mathbf{9.1.14})$$

The area $\boldsymbol{S_\Omega}$ of the spherical crown with the circular line $\boldsymbol{L_R} = \boldsymbol{2\pi R}$ as the side line is:

$$\boldsymbol{S_\Omega} = \boldsymbol{2\pi rh} = \boldsymbol{2\pi r\left(r - \sqrt{r^2 - R^2}\right)} \qquad (\mathbf{9.1.15})$$

When the angle $\boldsymbol{\theta} = \mathbf{0}$, the radius of the circle line $\boldsymbol{R} = \mathbf{0}$, the area of the spherical crown $\boldsymbol{S_\Omega} = \mathbf{0}$. When the angle $\boldsymbol{\theta} = \mathbf{90^0}$, the radius of the circle line $\boldsymbol{R} = \boldsymbol{r}$, the area of the spherical crown $\boldsymbol{S_\Omega} = \boldsymbol{2\pi r^2}$. At this point, the area of the spherical crown $\boldsymbol{S_\Omega}$ is half the area of the sphere $\boldsymbol{S} = \boldsymbol{4\pi r^2}$.

This book mimics the spherical crown formula $\boldsymbol{S_\Omega} = \boldsymbol{2\pi rh}$ by defining the product $\boldsymbol{2\pi Rr}$ as Wang's circular crown $\boldsymbol{S_W}$, i.e:

$$\boldsymbol{S_W} = \boldsymbol{2\pi rR} \qquad (\mathbf{9.1.16})$$

The above formula is the Wang's circular crown formula.

Suppose that the spherical crown $\boldsymbol{S_\Omega} = \boldsymbol{2\pi rh}$ is bordered by the circular line $\boldsymbol{L_R} = \boldsymbol{2\pi R}$. Since the radius $\boldsymbol{R}$ of the circular line corresponds to the height $\boldsymbol{h}$ in the formula for the spherical crown $\boldsymbol{S_\Omega} = \boldsymbol{2\pi rh}$, the Wang's circular crown $\boldsymbol{S_W} = \boldsymbol{2\pi rR}$ is not a part on the sphere $\boldsymbol{S} = \boldsymbol{4\pi r^2}$. Wang's circular crown $\boldsymbol{S_W} = \boldsymbol{2\pi rR}$ is shown in Figure 9-8.

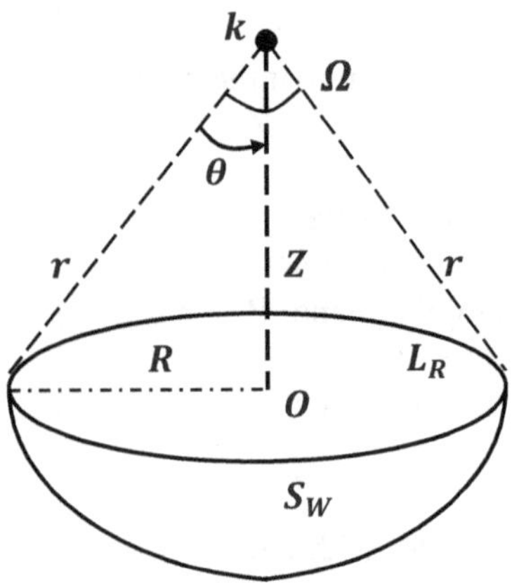

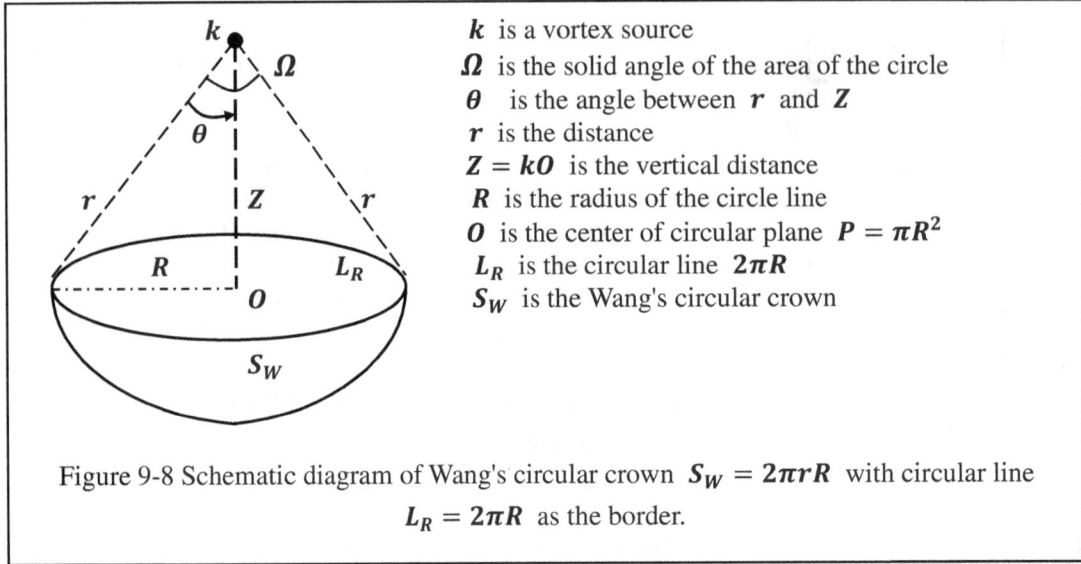

Figure 9-8 Schematic diagram of Wang's circular crown $S_W = 2\pi rR$ with circular line $L_R = 2\pi R$ as the border.

When the angle $\theta = 0$. Since radius $R = 0$, therefore Wang's circular crown $S_W = 0$. When angle $\theta = 90^0$. Since radius $R = r$, Wang's circular crown $S_W = 2\pi r^2$. At this time, Wang's circular crown S_W is equal to half of the sphere $S = 4\pi r^2$.

Theoretically, there are an infinite number of surfaces S bordered by the circular line $L_R = 2\pi R$. The spherical crown $S_\Omega = 2\pi rh$ is one of them. Wang's circular crown $S_W = 2\pi rR$ is also one of them.

According to Figure 9-7, the height h of the spherical crown $S_\Omega = 2\pi rh$ is:

$$h = r - \sqrt{r^2 - R^2} \tag{9.1.17}$$

According to Figure 9-7, the radius of the circular line $L_R = 2\pi R$ is $R = \sqrt{r^2 - Z^2}$; and the height of the spherical crown S_Ω is $h = r - Z$. When the distance $Z = 0$, Since the radius of the circular line $R = h = r$, Wang's circular crown S_W is equal to the spherical crown S_Ω, i.e:

$$S_W = S_\Omega = 2\pi r^2$$

when the distance $Z \neq 0$. Since the radius $R > h$, Wang's circular crown S_W is larger than the spherical crown S_Ω, i.e:

$$S_W > S_\Omega$$

9.2 Physical meaning of the positive and negative pole angles $\pm\beta$ of an equiangular helical field

9.2.1 Derivation of the Formula for the Polar Angle of an Equiangular Helical Field

Assuming equal distances r from the vortex source k to each point on the circular line $L_R = 2\pi R$, the integral $\oint ydl$ of the vortex potential y along the circular line $L_R = 2\pi R$ according to equation (9.1.9) is:

$$\oint ydl = 2\pi R \cdot D_k kr(\ln r - 1) \tag{9.2.1}$$

According to Stokes' theorem, the relation equation can be obtained:

$$\oint ydl = \iint \nabla \times y\, dS \tag{9.2.2}$$

The surface S in equation is bounded by the circle line $L_R = 2\pi R$. We denote the curl $\nabla \times y$ by β, i.e:

$$\beta = \nabla \times y \tag{9.2.3}$$

In this book, curl $\beta = \nabla \times y$ is defined as the polar angle of an equiangular helical field. Substituting the above equation into equation (9.2.2) gives the relation equation:

$$\oint y dl = \iint \beta\, dS \tag{9.2.4}$$

Substituting equation (9.2.1) into the above equation gives the relationship equation:

$$\iint \beta\, dS = 2\pi R \cdot D_k kr(\ln r - 1) \tag{9.2.5}$$

The above formula is the polar angle flux formula. The surface S in the formula is bounded by the circular line $L_R = 2\pi R$.

Theoretically: There are an infinite number of surfaces S bounded by the circular line $L_R = 2\pi R$. The spherical crown $S_\Omega = 2\pi hr$ is one of them, as is the Wang's circular crown $S_W = 2\pi Rr$.

When the surface S is a Wang's circular crown S_W, if the circular line $L_R = 2\pi R$ remains constant, then the Wang's circular crown S_W is just a function of the distance r. By taking the derivative of the Wang's circular crown S_W with respect to the distance r, we can obtain the relation equation:

$$dS_W = 2\pi R dr \tag{9.2.6}$$

Although the size of Wang's circular crown S_W varies with the distance r, the distance r at each point on the circular line $L_R = 2\pi R$ is always equal. Substituting the above equation into equation (9.2.5) yields the relation equation:

$$\iint \beta dS_W = \int \beta \cdot 2\pi R dr = 2\pi R \cdot D_k kr(\ln r - 1) \tag{9.2.7}$$

Simplifying the above equation by removing the constant $2\pi R$ gives the relational equation:

$$\int \beta dr = D_k kr(\ln r - 1) \tag{9.2.8}$$

Taking the derivative of the above equation with respect to the distance r gives the relational equation:

$$\beta = D_k k(\ln r - 1) + D_k kr \cdot \frac{1}{r}$$

Based on the above equation, the relationship equation can be obtained:

$$\beta = D_k k \cdot \ln r \tag{9.2.9}$$

The above equation is the formula for the polar angle of an equiangular vortex field. D_k in the formula is a constant related to the vortex source k.

9.2.2 Physical meaning of the positive and negative pole angles $\pm\beta$ of an equiangular helical field

According to equation (9.2.3), the polar angle β of an equiangular helical field is:

$$\beta = \nabla \times y$$

Since the polar angle β is equal to the curl $\nabla \times y$, the polar angle β is a vector. According to equation (9.2.9), the polar angle β of an equiangular helical field is:

$$\beta = D_k k \cdot \ln r$$

Since the polar radius r is included in equation the direction of the polar angle β has the following relationship with the direction of the polar radius r.

1), When the polar radius $r = 1$, the polar angle $\beta = 0$.

2), When the polar radius $r > 1$ the polar angle $\beta > 0$. In this case, the direction of the positive polar angle $+\beta$ is the same as that of the polar radius r, or the direction of the positive polar angle $+\beta$ is from the point $r = 1$ to infinity.

3), when the polar radius $r < 1$. Polar angle $\beta < 0$. At this time, the direction of the negative polar angle $-\beta$ is opposite to the direction of the polar radius r, or the direction of the negative polar angle $-\beta$ points from the point $r = 1$ to the center point O of the vortex field.

Since the vortex field with polar radius $r > 1$ is a macroscopic vortex field, the macroscopic positive pole angle $+\beta$ can be expressed as:

$$\beta = +D_k k \cdot \ln r, \quad r > 1 \tag{9.2.10}$$

The formula above is the positive polar angle formula. The plus sign " + " in the formula indicates that the positive polar angle $+\boldsymbol{\beta}$ increases in a clockwise direction. Since the gravitational force between the planets is suction, the equiangular helical field of the orthopole angle $+\boldsymbol{\beta}$ has suction properties.

Since the vortex field with polar radius $0 < r < 1$ is a microscopic vortex field, the microscopic negative polar angle $-\boldsymbol{\beta}$ can be expressed as:

$$\boldsymbol{\beta} = -D_k k \cdot \ln r, \qquad 0 < r < 1 \qquad (9.2.11)$$

The above formula is the negative pole angle formula. The negative sign " $-$ " in the formula indicates that the negative pole angle $-\boldsymbol{\beta}$ increases counterclockwise. Since the positive polar angle $+\boldsymbol{\beta}$ increasing clockwise possesses the property of suction, the negative polar angle $-\boldsymbol{\beta}$ increasing counterclockwise should possess the property of repulsion.

According to atomic physics, there is a strong nuclear force in the nucleus of an atom that binds protons or neutrons together in the nucleus. The strong nuclear force is a short-range gravitational force. The distance r at which the strong nuclear force acts is approximately in the range of $10^{-15}m < r < 10^{-10}m$.

During a nuclear reaction, a nucleus of large mass can split into two nuclei of smaller mass. This fact proves that when the distance is $0 < r < 10^{-15}m$, the gravitational force inside the nucleus is no longer a force of attraction, but a force of repulsion.

It is this repulsive force that prevents the gravitational force between protons and neutrons inside the nucleus from becoming infinite. Otherwise, a nucleus with a large mass would not split into two nuclei with smaller masses under the effect of the immense gravitational force.

In summary, any equiangular helical field should contain positive and negative pole angles $\pm\boldsymbol{\beta}$. This is shown in Figure 9-9.

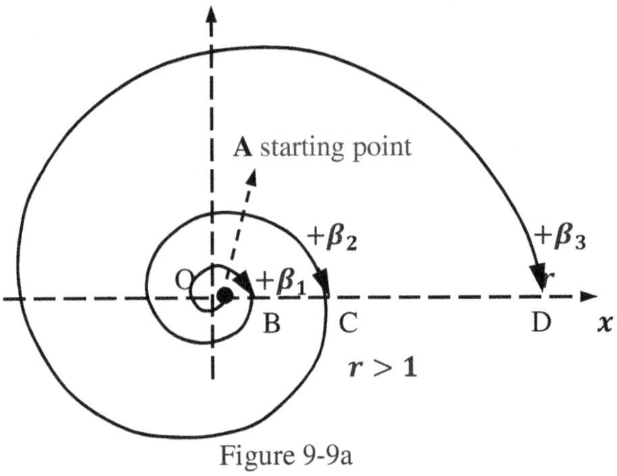

Figure 9-9a

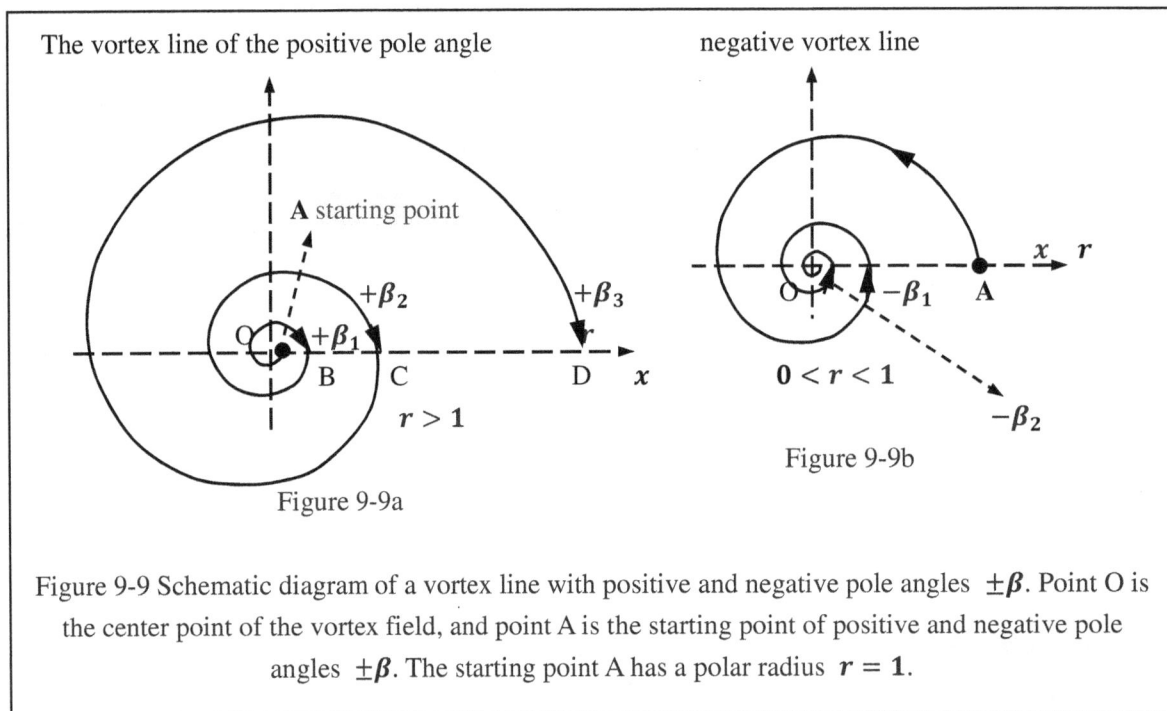

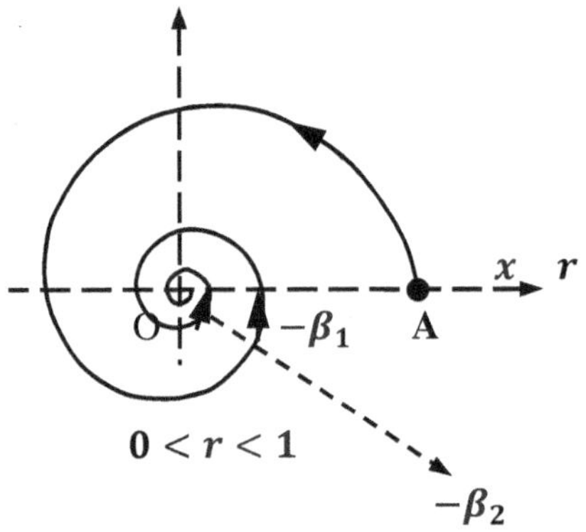

Figure 9-9b

Figure 9-9 Schematic diagram of a vortex line with positive and negative pole angles $\pm\boldsymbol{\beta}$. Point O is the center point of the vortex field, and point A is the starting point of positive and negative pole angles $\pm\boldsymbol{\beta}$. The starting point A has a polar radius $r = 1$.

When the polar radius $r > 1$, the polar angle between A, B, C, and D is positive polar angle $+\boldsymbol{\beta}$. This is shown in Figure 9-9a. An equiangular helical field with positive polar angle $+\boldsymbol{\beta}$ is a suction field. An equiangular helical field with positive pole angle $+\boldsymbol{\beta}$ applies to macroscopic vortex fields. For example, the gravitational field between the sun and a planet.

When the polar radius is $0 < r < 1$, the polar angle between O and A is the negative polar angle $-\boldsymbol{\beta}$. This is shown in Figure 9-9b. An equiangular helical field with negative polar angle $-\boldsymbol{\beta}$ is a repulsive force field. An equiangular helical field with negative pole angle $-\boldsymbol{\beta}$ applies to microscopic vortex fields. For example, the gravitational force between protons and neutrons in the nucleus of an atom.

In summary, the equation for the polar angle of an equiangular helical field should be expressed as:

$$\begin{cases} \beta = +D_k k \cdot \ln r, & r > 1 \\ \beta = 0, & r = 1 \\ \beta = -D_k k \cdot \ln r, & 0 < r < 1 \end{cases} \qquad (9.2.12)$$

The above formula is the formula for the polar angle of an equiangular helical field. The positive sign " + " in the formula indicates that the direction of the polar angle β is the same as the direction of the polar radius r, and the polar angle β increases clockwise. The negative sign " − " in the formula indicates that the direction of the polar angle β is opposite to the direction of the polar radius r, and the polar angle β increases counterclockwise.

A polar radius $r > 1$ in the formula means that the direction of the polar angle β is the same as the direction of the polar radius r. A polar radius of $0 < r < 1$ means that the direction of the polar angle β is opposite to the direction of the polar radius r. The polar angle coordinate lines of an equiangular helical field are shown in Figure 9-10.

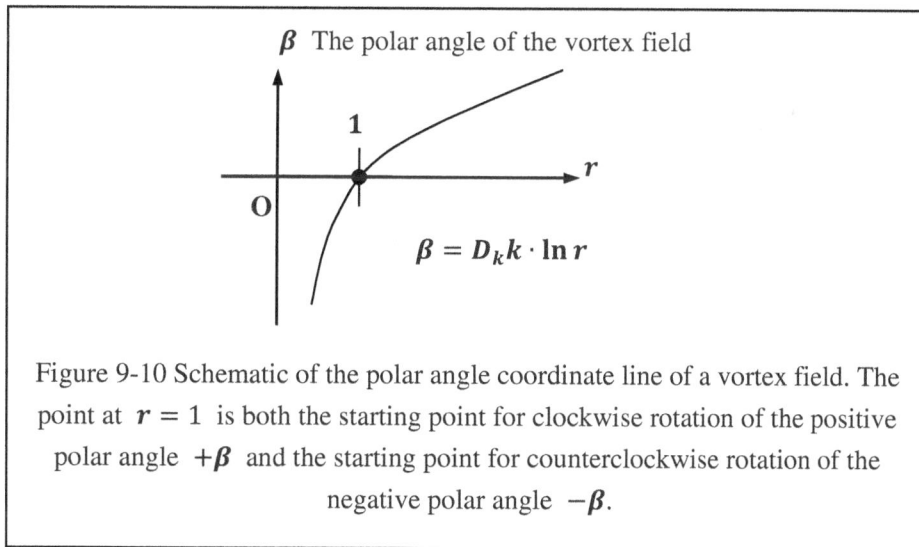

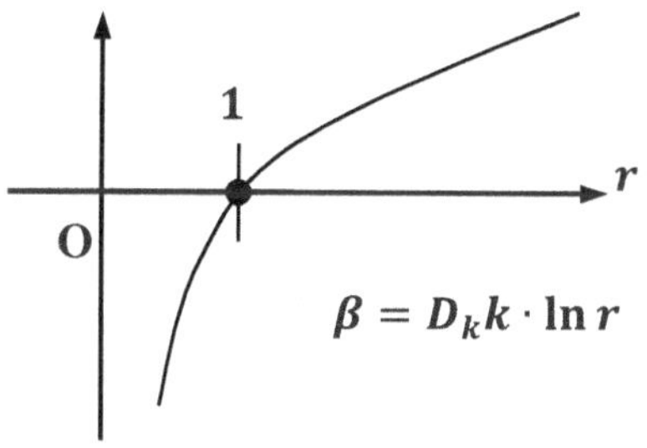

Figure 9-10 Schematic of the polar angle coordinate line of a vortex field. The point at $r = 1$ is both the starting point for clockwise rotation of the positive polar angle $+\beta$ and the starting point for counterclockwise rotation of the negative polar angle $-\beta$.

The derivative of equation (9.2.12) with respect to the polar radius r gives the relation equation:

$$\begin{cases} \dfrac{d\beta}{dr} = +D_k k \dfrac{1}{r} & r > 1 \\ \dfrac{d\beta}{dr} = 0, & r = 1 \\ \dfrac{d\beta}{dr} = -D_k k \dfrac{1}{r}, & 0 < r < 1 \end{cases} \qquad (9.2.13)$$

The $+D_k k \frac{1}{r}$ in equation denotes the rate of change of the positive polar angle β; the $-D_k k \frac{1}{r}$ in equation denotes the rate of change of the negative polar angle β.

Physicists believe that the polar angle β increases clockwise from the center point O of the vortex field. Because physicists do not realize that the polar angle β is divided into two cases, positive and negative, physicists do not realize the physical significance of the negative polar angle β (i.e., that the negative angle β possesses the property of repulsive force). This is where the infinity problem for gravitational arises.

9.2.3 Vortex fields of different substances have different minimum unit lengths l_m

If the polar radius r has a certain length. If the size of the polar radius r is measured in different unit lengths, then the polar radius r will have different values. For example, if the polar radius $r = 1cm$. If the unit length is $1m$, then the polar radius $r = 0.01m$; if the unit length is $1cm$, then the polar radius $r = 1cm$; and if the unit length is

$\boldsymbol{1mm}$, then the polar radius $\boldsymbol{r} = \mathbf{10}\boldsymbol{mm}$.

Although different polar radii $\boldsymbol{r}$ in the vortex field have different lengths, the smallest unit length $\boldsymbol{l_m}$ contained in the polar radius $\boldsymbol{r}$ is the same. Assuming that the polar radius $\boldsymbol{r}$ contains $\boldsymbol{n}$ smallest unit lengths $\boldsymbol{l_m}$, the polar radius $\boldsymbol{r}$ can be expressed as:

$$\boldsymbol{r} = \boldsymbol{n}(\boldsymbol{l_m}) \tag{9.2.14}$$

When $\boldsymbol{n} = \mathbf{1}$, the length of the polar radius $\boldsymbol{r} = \mathbf{1}$ is equal to the minimum unit length $\boldsymbol{l_m}$.

According to atomic physics, there is a strong nuclear force in the nucleus of an atom. The strong nuclear force is a short-range gravitational force. The distance $\boldsymbol{r}$ over which the strong nuclear force acts is approximately in the range $\mathbf{10^{-15}}\boldsymbol{m} < \boldsymbol{r} < \mathbf{10^{-10}}\boldsymbol{m}$. Assume that the smallest unit length of the gravitational vortex field $\boldsymbol{l_m} = \mathbf{10^{-15}}\boldsymbol{m}$.

When the distance $\boldsymbol{r} = \mathbf{10^{-15}}\boldsymbol{m}$, the strong nuclear force $\boldsymbol{F} = \mathbf{0}$.

When the distance $\boldsymbol{r} > \mathbf{10^{-15}}\boldsymbol{m}$, the strong nuclear force $\boldsymbol{F} > \mathbf{0}$. At this point, the strong force is a suction force. The suction force makes several positively charged $+\boldsymbol{q}$ protons and neutrons in the nucleus come close to each other to form the nucleus.

The strong nuclear force $\boldsymbol{F} < \mathbf{0}$ when the distance $\boldsymbol{r} < \mathbf{10^{-15}}\boldsymbol{m}$. In this case, the strong force is repulsive. The repulsive force prevents multiple positively charged $+\boldsymbol{q}$ protons in the nucleus from coming together to form a new particle.

Since photons are particles with mass, there is a gravitational force between photons. Since photons are not protons and neutrons, it is obviously wrong to use the unit length $\boldsymbol{l_m} = \mathbf{10^{-15}}\boldsymbol{m}$ as the smallest unit length of the gravitational field of photons.

In physics, the Planck length $\boldsymbol{l_p}$ is the smallest unit of length. Physicists believe that quantum effects begin to dominate in the microscopic realm at the scale of the Planck length $\boldsymbol{l_p}$. Physicists have derived the Planck length $\boldsymbol{l_p}$ based on quantum mechanics. The Planck length $\boldsymbol{l_p}$ is determined by the gravitational constant $\boldsymbol{G}$, the speed of light, and Planck's constant $\boldsymbol{h}$, i.e:

$$\boldsymbol{l_p} = \sqrt{\frac{\boldsymbol{hG}}{\boldsymbol{c}^3}} = 1.6 \times \mathbf{10^{-35}}\boldsymbol{m} \tag{9.2.15}$$

The Planck length $\boldsymbol{l_p}$ is approximately one part in **1022** of the diameter of a proton. Physicists believe that at this distance, gravitational and spacetime cease to exist and quantum effects dominate.

Assume that the Planck length $\boldsymbol{l_p}$ is the smallest unit length $\boldsymbol{l_m}$ in the gravitational field of a photon, i.e. $\boldsymbol{l_m} = \boldsymbol{l_p} = 1.6 \times \mathbf{10^{-35}}\boldsymbol{m}$. If the distance of the photons is $\boldsymbol{r} = \mathbf{1}(\mathbf{10^{-35}}\boldsymbol{m})$, the universal gravitational force between the photons $\boldsymbol{F} = \mathbf{0}$.

When the distance of photons $\boldsymbol{r} > 1(\mathbf{10^{-35}}\boldsymbol{m})$, the gravitational force between photons $\boldsymbol{F} > \mathbf{0}$. At this point, the gravitational force is a suction. The gravitational force makes the photons come close to each other to form a light wave that can travel a long distance.

The gravitational force $\boldsymbol{F} < \mathbf{0}$ between photons when their distance $\boldsymbol{r} < 1(\mathbf{10^{-35}}\boldsymbol{m})$. At this point, the gravitational force is repulsive. The repulsive force prevents the photons from coming together to form a new particle.

In summary, the vortex fields of different substances will have different minimum unit lengths $\boldsymbol{l_m}$.

9.2.4 Integration of the polar angle β along the circular line $L_R=2\pi R$

According to equation (9.2.9), the polar angle $\boldsymbol{\beta}$ of an equiangular helical field is:

$$\boldsymbol{\beta} = \boldsymbol{D_k k} \cdot \ln \boldsymbol{r}$$

Since the distance $\boldsymbol{r}$ varies continuously from the position $\boldsymbol{r} = 1(\boldsymbol{l_m})$ in one direction to infinity or converges to the vortex source $\boldsymbol{k}$, these innumerable distance points form a straight line. Since each point on the straight line has a polar angle $\boldsymbol{\beta}$, this book defines the straight line as the polar angle line of an equiangular spiral field. This polar angle line is the solid line that intersects the circular line in 9-11.

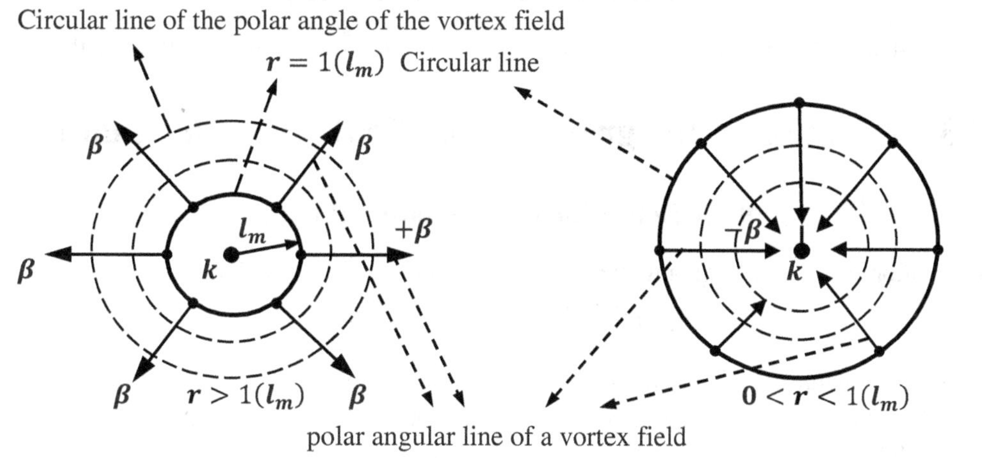

Figure 9-11 Schematic diagram of polar angle lines and polar circle lines for a vortex field. The arrows in the figure are the directions of the polar angle lines.

bviously, the polar angle $\boldsymbol{\beta}$ is not the same at every point on the polar angle line. Note that the direction of the polar angle line is the same as the direction of the polar angle $\boldsymbol{\beta}$.

The polar angle lines of an equiangular spiral field have the following properties:

(1), The vortex source $\boldsymbol{k}$ is the source of the polar angle line.

(2), The polar angles are not equal at each point on the polar angle lines.

(3), Polar angle lines are continuous disjoint curves.

(4), The directions of the polar angles all point to infinity or to the center of the vortex source.

The circular line of polar angles of an equiangular helical field.

When the distance $\boldsymbol{r}$ from the vortex source $\boldsymbol{k}$ to each point of the circular line $\boldsymbol{L_R} = \boldsymbol{2\pi R}$ is equal, the polar angles $\boldsymbol{\beta}$ are equal at each point on the circular line $\boldsymbol{L_R}$. In this book, a circular line with equal polar angles $\boldsymbol{\beta}$ is defined as a polar angle circular line. This polar angle circular line is the circular dashed line in Figure 9-11.

The circular line of the polar angle of an equiangular spiral field has the following properties:

(1), The source $\boldsymbol{k}$ of the vortex is the source of the circular line of the polar angle.

(2), The direction of the tangent at any point on the circular line of the polar angle is perpendicular to the polar radius $\boldsymbol{r}$ at that point.

(3), The polar angles at each point on the circular line of the polar angle are equal.

(4), The circular lines of polar angles are continuous, non-intersecting curves. Circular lines of equal polar angles can form a sphere of equal polar angles.

(5), The closer the circular line is to the vortex source $\boldsymbol{k}$, the smaller is the polar angle $\boldsymbol{\beta}$ at each point on the circular line. Conversely, the larger the polar angle $\boldsymbol{\beta}$ is at each point on the circular line.

Suppose $\boldsymbol{L} = \boldsymbol{f(r)}$ is a curve in an equiangular helical field. Take the line element $\boldsymbol{dl}$ on the curve $\boldsymbol{L} = \boldsymbol{f(r)}$. if the polar radius $\boldsymbol{r} > 1(\boldsymbol{l_m})$. According to equation (9.2.9), the product $\boldsymbol{\beta dl}$ of the polar angle $\boldsymbol{\beta}$ and the line element $\boldsymbol{dl}$ is:

$$\boldsymbol{\beta dl} = \boldsymbol{D_k k} \cdot \ln \boldsymbol{r\, dl}, \quad \boldsymbol{r} > 1(\boldsymbol{l_m}) \qquad (9.2.16)$$

Assume that each point on the circular line $\boldsymbol{L_R} = \boldsymbol{2\pi R}$ is an equal distance $\boldsymbol{r}$ from the vortex source $\boldsymbol{k}$. Since the polar radii $\boldsymbol{r}$ of each point on the circular line $\boldsymbol{L_R}$ are equal, the polar angle $\boldsymbol{\beta}$ on the circular line $\boldsymbol{L_R}$ is constant. The integral of the polar angle $\boldsymbol{\beta}$ along the circular line $\boldsymbol{L_R} = \boldsymbol{2\pi R}$ is:

$$\oint \boldsymbol{\beta dl} = \boldsymbol{\beta L_R} = \boldsymbol{2\pi R} \cdot \boldsymbol{D_k k} \cdot \ln \boldsymbol{r}, \quad \boldsymbol{r} > 1(\boldsymbol{l_m}) \qquad (9.2.17)$$

When the angle $\boldsymbol{\theta} = \boldsymbol{90^0}$ in Figure 9-8, the above equation becomes because the radius $\boldsymbol{R} = \boldsymbol{r}$:

$$\oint \beta dl = 2\pi D_k kr \cdot \ln r, \quad r > 1(l_m) \tag{9.2.18}$$

At this point the distance $Z = 0$ and the vortex source k is located in the center of the circular plane $P = \pi R^2$.

9.3 Derivation of the gravitational field strength equation

9.3.1 Definition of gravitational field

According to equation (9.1.2) , the equiangular vortex field conservation equation is:

$$e^y \left(\frac{e}{r}\right)^{D_k kr} = 1 \quad 0 < r < +\infty \tag{9.3.1}$$

The e in equation is a natural constant, the polar radius r is the independent variable of the vortex field, and y is the dependent variable of the vortex field. The k in equation is the vortex source, and D_k is the constant associated with the vortex source k. The dependent variable $y = 0$ when the polar radius $r = e$.

When the vortex source k is the mass m, i.e., the constant $k = m$ and the constant $D_k = D_m$. Substituting $k = m$ and $D_k = D_m$ into equation (9.3.1) yields the relational equation:

$$e^{y_m} \left(\frac{e}{r}\right)^{D_m mr} = 1, \quad 0 < r < +\infty \tag{9.3.2}$$

The above formula is a conservation formula for the vortex gravitational field. This book defines the dependent variable y_m in the formula as the gravitational field vortex potential. D_m in the formula is a constant associated with the gravitational field. When the polar radius $r = e$, the gravitational field vortex potential $y_m = 0$.

Taking the natural logarithm of equation (9.3.2) yields the relationship equation:

$$y_m = D_m mr(\ln r - 1) \tag{9.3.3}$$

The above equation is the gravitational field vortex potential equation. This book defines the set $\sum y_m$ of gravitational field vortex potentials y_m in space as a gravitational field.

Since the space occupied by the gravitational field vortex potential y_m is an infinite sphere, the gravitational field $\sum y_m$, which consists of the gravitational field vortex potential y_m, is an infinite sphere. Theoretically, the development and change of the gravitational field $\sum y_m$ are all unfolded according to equation (9.3.3).

It should be noted that the gravitational field $\sum y_m$ defined in this book is a material property that mass m has, not the cosmic vacuum, and the gravitational field $\sum y_m$ has nothing to do with the space-time bending of general relativity.

9.3.2 Formula for the polar angle of a gravitational field

Suppose O is the center of the circular line $L_R = 2\pi R$, and Z is the perpendicular distance from the mass m to the center O. Suppose Ω is the steradian angle possessed by the circular plane $P = \pi R^2$, and θ is the angle between the distance r and the perpendicular distance Z. Suppose the surface S is bounded by the circular line $L_R = 2\pi R$. This is shown in Figure 9-12.

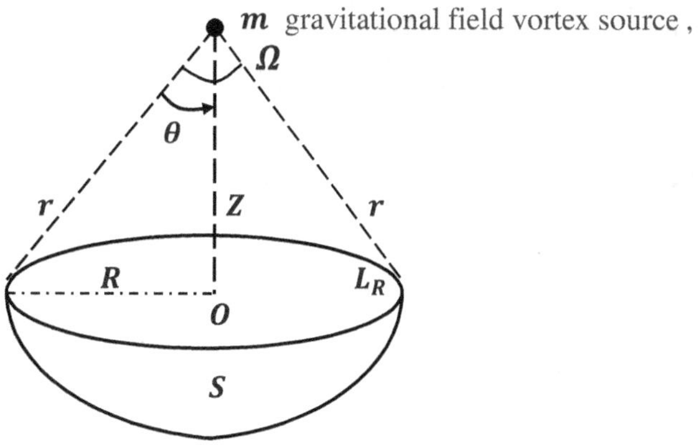

Figure 9-12 Schematic diagram of a gravitational field vortex source $\boldsymbol{m}$, and a surface $\boldsymbol{S}$ bounded by the circular line $\boldsymbol{L_R = 2\pi R}$.

Theoretically, there are an infinite number of surfaces $\boldsymbol{S}$ bounded by the circular line $\boldsymbol{L_R = 2\pi R}$.

Assume that the vortex source $\boldsymbol{k}$ is the mass $\boldsymbol{m}$ and that $\boldsymbol{D_k = D_m}$ is a constant associated with the gravitational field. According to equation (9.2.12), the mass $\boldsymbol{m}$ produces a polar angle $\boldsymbol{\beta}$ of :

$$\begin{cases} \boldsymbol{\beta = +D_m m \cdot \ln r}, & \boldsymbol{r > 1(l_m)} \\ \boldsymbol{\beta = 0}, & \boldsymbol{r = 1(l_m)} \\ \boldsymbol{\beta = -D_m m \cdot \ln r}, & \boldsymbol{0 < r < 1(l_m)} \end{cases} \quad (9.3.4)$$

The formula above is the formula for the polar angle of a gravitational field. The positive sign "+" in the formula indicates that the polar angle $\boldsymbol{\beta}$ rotates clockwise. At this point, the gravitational field is a positive equiangular spiral field. The negative sign "−" in the formula indicates that the polar angle $\boldsymbol{\beta}$ rotates counterclockwise. At this point, the gravitational field is a negative equiangular spiral field. The gravitational field contains positive and negative equiangular helical fields. This is shown in Figure 9-13.

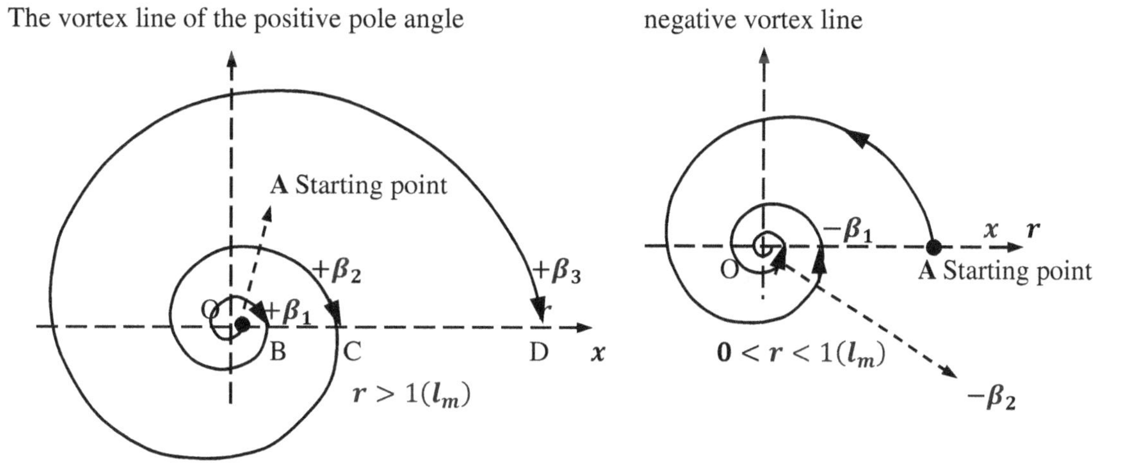

Figure 9-13 Schematic of positive and negative equiangular spirals line of a gravitational field. Point A is the starting point of the positive and negative equiangular spirals line.

A note about Figure 9-13:

Point O in the figure is the center of mass of the gravitational field. point A is the starting point of the positive and negative equi angular spiral line. The starting point A has a polar radius $\boldsymbol{r = 1(l_m)}$.

When the polar radius $r > 1(l_m)$ the equi angular spiral line between A, B, C, and D is rotating clockwise. The gravitational force F at this point is the suction force between macroscopic objects.

The equi angular spiral line between O and A is rotating counterclockwise when the polar radius $0 < r < 1(l_m)$. The gravitational force F at this point is the repulsive force within the microscopic nucleus.

9.3.3 Derivation of the gravitational field density equation

Assume that the distance r from the mass m to each point of the circular line $L_R = 2\pi R$ is equal. When the polar radius $r > 1(l_m)$, according to equation (9.3.4) . The integral $\oint \beta dl$ of the polar angle β along the circular line $L_R = 2\pi R$ is:

$$\oint \beta dl = 2\pi R \cdot D_m m \cdot \ln r, \quad r > 1(l_m) \tag{9.3.5}$$

According to Stokes' theorem, the relation equation can be obtained:

$$\oint \beta dl = \iint \nabla \times \beta\, dS \tag{9.3.6}$$

The surface S in equation is bounded by the circular line $L_R = 2\pi R$. We denote by J the curl $\nabla \times \beta$ of the polar angle β of the gravitational field, i.e:

$$J = \nabla \times \beta \tag{9.3.7}$$

In this book, Curl $J = \nabla \times \beta$ is defined as the gravitational field density. Equation (9.3.6) can be changed to:

$$\oint \beta dl = \iint J\, dS \tag{9.3.8}$$

Substituting equation (9.3.5) into the above equation gives the relationship equation:

$$\iint J\, dS = 2\pi R \cdot D_m m \cdot \ln r, \quad r > 1(l_m) \tag{9.3.9}$$

The above formula is the gravitational field density flux formula. The surface S in the formula is bounded by the circular line $L_R = 2\pi R$.

Theoretically: There are an infinite number of surfaces S bounded by the circular line $L_R = 2\pi R$. The spherical crown $S_\Omega = 2\pi hr$ is one of them, as is the Wang's circular crown $S_W = 2\pi Rr$.

When the surface S is a Wang's circular crown S_W, if the circular line $L_R = 2\pi R$ remains constant, then the Wang's circular crown S_W is just a function of the distance r. By taking the derivative of the Wang's circular crown S_W with respect to the distance r, we can obtain the relation equation:

$$dS_W = 2\pi R dr \tag{9.3.10}$$

Although the size of Wang's circular crown S_W varies with the distance r, the distance r at each point on the circular line $L_R = 2\pi R$ is always equal. Substituting the above equation into equation (9.3.9) yields the relation equation:

$$\iint J dS_W = \int J \cdot 2\pi R dr = 2\pi R \cdot D_m m \cdot \ln r, \quad r > 1(l_m) \tag{9.3.11}$$

Simplifying the above equation by removing the constant $2\pi R$ gives the relational equation:

$$\int J dr = D_m m \cdot \ln r, \quad r > 1(l_m) \tag{9.3.12}$$

Note that the above formula applies only to the case where the distance $r > 1(l_m)$. According to equation (9.3.4), mimicking the derivation process above, the relation formula can be obtained:

$$\begin{cases} \int J dr = +D_m m \cdot \ln r, & r > 1(l_m) \\ \int J dr = 0, & r = 1(l_m) \\ \int J dr = -D_m m \cdot \ln r, & 0 < r < 1(l_m) \end{cases} \tag{9.3.13}$$

The positive sign " + " in the formula indicates that the polar angle β of the gravitational field rotates clockwise. At this point, the gravitational field is a positive equiangular spiral field. The negative sign " − " in the formula indicates that the polar angle β of the gravitational field rotates counterclockwise. At this point, the gravitational field is a negative equiangular spiral field. Taking the derivative of the above equation with respect to the distance r gives the relational equation:

$$\begin{cases} J = +D_m m\dfrac{1}{r} & r > 1(l_m) \\ J = 0, & r = 1(l_m) \\ J = -D_m m\dfrac{1}{r}, & 0 < r < 1(l_m) \end{cases} \tag{9.3.14}$$

The above formula is the gravitational field density formula. The positive sign "+" in the formula indicates a positive gravitational field density, and the negative sign "−" indicates a negative gravitational field density.

When the polar radius $r = 1$, the gravitational field density $J = 0$. When the polar radius $r > 1(l_m)$, the gravitational field density $J > 0$. When the polar radius $0 < r < 1(l_m)$, the gravitational field density $J < 0$.

The derivative of equation (9.3.4) with respect to the polar radius r gives the relation equation:

$$\begin{cases} \dfrac{d\beta}{dr} = +D_m m\dfrac{1}{r} & r > 1(l_m) \\ \dfrac{d\beta}{dr} = 0, & r = 1(l_m) \\ \dfrac{d\beta}{dr} = -D_m m\dfrac{1}{r}, & 0 < r < 1(l_m) \end{cases}$$

Compare the above equation with equation (9.3.14). The relational equation can be obtained:

$$J = \frac{d\beta}{dr} \tag{9.3.15}$$

The above equation shows that the gravitational field density J is equal to the derivative of the polar angle β of the gravitational field with respect to the polar radius r.

9.3.4 Gravitational field density J is not gravitational potential U_W

In the classical theory, the Newtonian gravitational force F is:

$$F = G\frac{Mm}{r^2}$$

The work A done by the gravitational force F to move the mass M from point a to point b is:

$$A = \int_{r_a}^{r_b} F\,dl = \int_{r_a}^{r_b} G\frac{Mm}{r^2}dl = -GMm\left(\frac{1}{r_b} - \frac{1}{r_a}\right) \tag{9.3.16}$$

When the mass M moves from point a to infinity (i.e., point b), the gravitational potential energy W (i.e., the work done by the gravitational force F) possessed by the mass M is:

$$W = \int_{r}^{\infty} G\frac{Mm}{r^2}dl = GMm\frac{1}{r} \tag{9.3.17}$$

In classical theory, the gravitational potential U_W is defined as the ratio of the gravitational potential energy W to the mass M, i.e:

$$U_W = \frac{W}{M} \tag{9.3.18}$$

Substituting equation (9.3.17) into the above equation gives the relational equation:

$$U_W = G\frac{m}{r} \tag{9.3.19}$$

The above equation is the classical gravitational potential equation.

According to equation (9.3.14), when the polar radius $r > 1(l_m)$ the gravitational field density J is:

$$J = D_m\frac{m}{r}, \quad r > 1(l_m) \tag{9.3.20}$$

The D_m in the equation is a constant associated with the gravitational field. Compare the above equation with equation (9.3.19). Although the gravitational potential U_W is similar to the gravitational field density J, they have different physical meanings.

In other words, the gravitational potential U_W is derived from the work done by the gravitational force F, while the gravitational field density J is equal to the curl of the polar angle β of the gravitational field, viz:

$$J = \nabla \times \beta \tag{9.3.21}$$

Since the gravitational field density J is a vector quantity and the gravitational potential U_W is a scalar quantity, the gravitational field density J and the gravitational potential U_W are two different physical quantities.

9.3.5 Derivation of the new gravitational potential equation

When the distance $r > 1$, it is assumed that the distance r from the mass m to each point on the circular line $L_R = 2\pi R$ is equal. Since the gravitational field density J remains constant on the circular line $L_R = 2\pi R$, the integral $\oint Jdl$ of the gravitational field density J along the circular line $L_R = 2\pi R$ according to equation (9.3.14) is:

$$\oint Jdl = JL_R = 2\pi R \cdot D_m \frac{m}{r}, \quad r > 1(l_m) \tag{9.3.22}$$

The above equation is the loop integral equation for the gravitational field density. When the circular line $L_R = 2\pi R$ is the side line of the hemisphere, i.e. the circular line $L_R = 2\pi r$, the above formula becomes because the angle $\theta = 90^0$ and the radius $R = r$:

$$\oint Jdl = 2\pi D_m m \tag{9.3.23}$$

The above equation is the maximum formula for the loop integral of the gravitational field density. At this point, the mass m is in the center of the circular plane $P = \pi r^2$..

According to equation (9.3.22) and Stokes' theorem, the relation equation can be obtained:

$$\oint Jdl = \iint \nabla \times JdS = 2\pi R \cdot D_m \frac{m}{r}, \quad r > 1(l_m) \tag{9.3.24}$$

The surface S in equation is bounded by the circle line $L_R = 2\pi R$. We can denote Curl $\nabla \times J$ by the symbol U, i.e:

$$U = \nabla \times J \tag{9.3.25}$$

In this book, curl $U = \nabla \times J$ is defined as the gravitational potential of a gravitational field. Equation (9.3.24) can be changed to:

$$\oint Jdl = \iint U\,dS \tag{9.3.26}$$

Substituting equation (9.3.22) into the above equation gives the relational equation:

$$\iint U\,dS = 2\pi R \cdot D_m \frac{m}{r}, \quad r > 1(l_m) \tag{9.3.27}$$

The above formula is the gravitational potential flux formula. The surface S in the formula is bounded by the circular line $L_R = 2\pi R$.

Theoretically: There are an infinite number of surfaces S bounded by the circular line $L_R = 2\pi R$. The spherical crown $S_\Omega = 2\pi hr$ is one of them, as is the Wang's circular crown $S_W = 2\pi Rr$.

When the surface S is a Wang's circular crown S_W, if the circular line $L_R = 2\pi R$ remains constant, then the Wang's circular crown S_W is just a function of the distance r. By taking the derivative of the Wang's circular crown S_W with respect to the distance r, we can obtain the relation equation:

$$dS_W = 2\pi R dr$$

Although the size of Wang's circular crown S_W varies with the distance r, the distance r at each point on the circular line $L_R = 2\pi R$ is always equal. Substituting the above equation into equation (9.3.27) yields the relation equation:

$$\iint UdS_W = \int U \cdot 2\pi R dr = 2\pi R \cdot D_m \frac{m}{r}, \quad r > 1(l_m) \tag{9.3.28}$$

Simplifying the above equation by removing the constant $2\pi R$ gives the relational equation:

$$\int Udr = D_m \frac{m}{r}, \quad r > 1(l_m) \tag{9.3.29}$$

According to equation (9.3.14), mimicking the derivation process above, the relational equation can be obtained:

$$\begin{cases} \int U dr = +D_m \frac{m}{r}, & r > 1(l_m) \\ \int U dr = 0, & r = 1(l_m) \\ \int U dr = -D_m \frac{m}{r}, & 0 < r < 1(l_m) \end{cases} \quad (9.3.30)$$

Taking the derivative of the above equation with respect to the distance r gives the relational equation:

$$\begin{cases} U = -D_m \frac{m}{r^2}, & r > 1(l_m) \\ U = 0, & r = 1(l_m) \\ U = +D_m \frac{m}{r^2}, & 0 < r < 1(l_m) \end{cases} \quad (9.3.31)$$

The formula above is the new gravitational potential formula. D_m in the formula is a constant associated with the gravitational field. Since the new gravitational potential U belongs to the curl $U = \nabla \times J$, the new gravitational potential U is a vector, not a scalar.

It should be noted that the gravitational potential $U = \mp D_m \frac{m}{r^2}$, as defined in this book, is regarded in classical theory as the gravitational field strength $E = G\frac{m}{r^2}$. The reason for this discrepancy is that the Newtonian gravitational force F' is inversely proportional to the square of the distance r, i.e. $F' \propto 1/r^2$, whereas the new universal gravitational force F is inversely proportional to the cube of the distance r, i.e. $F \propto 1/r^3$.

9.3.6 Derivation of a new formula for the strength of the gravitational field

When the distance $r > 1(l_m)$, it is assumed that the distance r from the mass m to each point on the circular line $L_R = 2\pi R$ is equal. Since the gravitational potential U remains constant on the circular line $L_R = 2\pi R$, the integral $\oint U dl$ of the gravitational potential U along the circular line $L_R = 2\pi R$ according to equation (9.3.31) is:

$$\oint U dl = U L_R = -2\pi R \cdot D_m \frac{m}{r^2} \quad (9.3.32)$$

The above equation is the gravitational potential loop integral equation. When the circular line $L_R = 2\pi R$ is the boundary line of the hemisphere, that is, when the circular line $L_R = 2\pi r$. At this time, since the angle $\theta = 90^0$ and the radius $R = r$, the above equation becomes:

$$\oint U dl = -2\pi D_m \frac{m}{r} \quad (9.3.33)$$

The above equation is the formula for the maximum value of the gravitational potential loop integral. At this point, the mass m is the center of the circular plane $P = \pi r^2$. According to equation (9.3.32) and Stokes' theorem, the relational formula can be obtained:

$$\oint U dl = \iint \nabla \times U dS = -2\pi R \cdot D_m \frac{m}{r^2} \quad (9.3.34)$$

The surface S in equation is bounded by the circular line $L_R = 2\pi R$. We can denote the curl $\nabla \times U$ of the gravitational potential by the symbol E, i.e:

$$E = \nabla \times U \quad (9.3.35)$$

In this book, Curl $E = \nabla \times U$ is defined as the gravitational field strength. Equation (9.3.34) becomes:

$$\Phi_E = \oint U dl = \iint E\, dS \quad (9.3.36)$$

The above equation is the gravitational flux equation. Substituting equation (9.3.34) into the above equation gives the relational equation:

$$\Phi_E = \oint U dl = \iint E\, dS = -2\pi R \cdot D_m \frac{m}{r^2} \quad (9.3.37)$$

Theoretically: There are an infinite number of surfaces S bounded by the circular line $L_R = 2\pi R$. The spherical crown $S_{\Omega} = 2\pi hr$ is one of them, as is the Wang's circular crown $S_W = 2\pi Rr$.

When the surface S is a Wang's circular crown S_W, if the circular line $L_R = 2\pi R$ remains constant, then the Wang's circular crown S_W is just a function of the distance r. By taking the derivative of the Wang's circular crown S_W with respect to the distance r, we can obtain the relation equation:

$$dS_W = 2\pi R dr$$

Although the size of Wang's circular crown S_W varies with the distance r, the distance r at each point on the circular line $L_R = 2\pi R$ is always equal. Substituting the above equation into equation (9.3.37) yields the relation equation:

$$\iint E dS_W = \int E \cdot 2\pi R dr = -2\pi R \cdot D_m \frac{m}{r^2}, \quad r > 1(l_m) \tag{9.3.38}$$

Simplifying the above equation by removing the constant $2\pi R$ gives the relational equation:

$$\int E dr = -D_m \frac{m}{r^2}, \quad r > 1(l_m) \tag{9.3.39}$$

According to equation (9.3.31), mimicking the derivation process above, the relational equation can be obtained:

$$\begin{cases} \int E dr = -D_m \frac{m}{r^2}, & r > 1(l_m) \\ \int E dr = 0, & r = 1(l_m) \\ \int E dr = +D_m \frac{m}{r^2}, & 0 < r < 1(l_m) \end{cases} \tag{9.3.40}$$

Taking the derivative of the above equation with respect to the distance r gives the relational equation:

$$\begin{cases} E = +2D_m \frac{m}{r^3}, & r > 1(l_m) \\ E = 0, & r = 1(l_m) \\ E = -2D_m \frac{m}{r^3}, & 0 < r < 1(l_m) \end{cases} \tag{9.3.41}$$

The above formula is the new formula for the strength of the gravitational field. D_m in the formula is a constant associated with the gravitational field. l_m is the smallest unit length in the gravitational field. Since the gravitational field strength E belongs to Curl $E = \nabla \times U$, the gravitational field strength E is a vector, not a scalar.

When the distance r varies continuously in one direction from the position $r = 1(l_m)$ to infinity, or converges to the mass m, these innumerable distance points form a straight line. Since each point on the straight line has a gravitational field strength E, this book defines the straight line as a gravitational field strength line. This gravitational field strength line is the solid line that intersects the circular line in Figure 9-14.

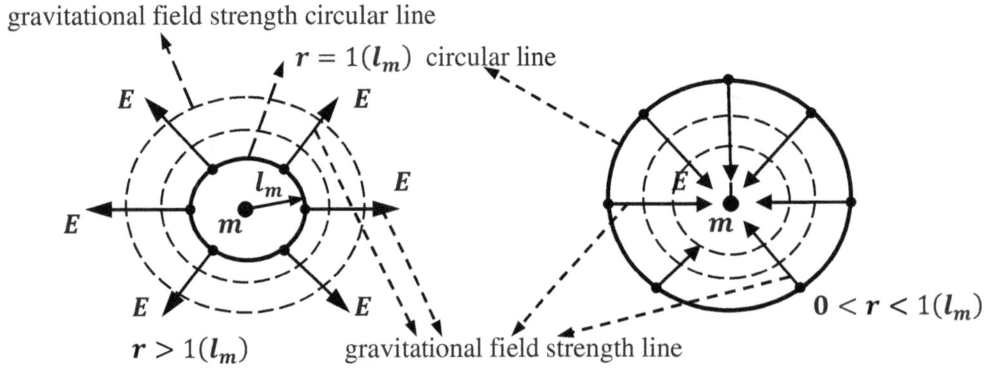

Figure 9-14 Schematic diagram of gravitational field strength lines and circular lines of force field strength. The arrows in the figure are the directions of the gravitational field strength lines.

Note that the direction of the gravitational field strength lines is the same as the direction of the gravitational field strength. The gravitational field intensity line has the following properties:

(1), The mass m is the source of the gravitational field intensity line.

(2), The gravitational field strength is not equal at every point on the gravitational field strength line.

(3), The gravitational field intensity lines are continuous, non-intersecting curves.

(4), The directions of the gravitational field strengths all point to infinity or to the center of mass.

When the distance r from the mass m to each point of the circular line $L_R = 2\pi R$ is equal, the gravitational field strength E is equal at each point on the circular line L_R. The circular line with equal gravitational field strength E is defined in this book as the gravitational field strength circular line. This gravitational field strength circular line is the circular dashed line in Figure 9-14.

The gravitational field strength circle line possesses the following properties:

(1), The mass m is the source of the gravitational field strength circular line.

(2), The direction of the tangent line at each point on the circular line is perpendicular to the polar radius r of that point.

(3), The gravitational field strength is equal at every point on the circular line.

(4), the gravitational field strength circle line is a continuous non-intersecting curve. Circular lines of equal values of gravitational field strength can form a sphere of equal gravitational field strength.

(5), the closer the circular line is to the mass m, the greater the gravitational field strength E at each point on the circular line. Conversely, the smaller the gravitational field strength E at each point on the circular line.

9.4 New formula for universal gravitation

9.4.1 Derivation of the new gravitational formula

According to equation (6.1.17), the new formula for gravitational is:

$$F = K\frac{Mm}{r^3}$$

The K in the equation is the new universal gravitational constant.

The gravitational field strength formula (9.3.41) multiplied by the mass M gives the relationship equation:

$$\begin{cases} F = EM = +2D_m\dfrac{Mm}{r^3}, & r > 1(l_m) \\ F = 0, & r = 1(l_m) \\ F = EM = -2D_m\dfrac{Mm}{r^3}, & 0 < r < 1(l_m) \end{cases} \tag{9.4.1}$$

when the distance $r > 1(l_m)$ Since the product EM is equal to the gravitational force F, the relational equation is obtained:

$$2D_m\frac{Mm}{r^3} = K\frac{Mm}{r^3}$$

Based on the above equation, the relationship equation can be obtained:

$$D_m = \frac{1}{2}K \tag{9.4.2}$$

Substituting the above equation into equation (9.4.1) gives the relationship equation:

$$\begin{cases} F = +K\dfrac{Mm}{r^3}, & r > 1(l_m) \\ F = 0, & r = 1(l_m) \\ F = -K\dfrac{Mm}{r^3}, & 0 < r < 1(l_m) \end{cases} \tag{9.4.3}$$

The above formula is the new universal gravitational formula. K in the formula is the new universal gravitational constant.

Substituting equation (9.4.2) into equation (9.3.41) gives the relationship equation:

$$\begin{cases} E = +K\dfrac{m}{r^3}, & r > 1(l_m) \\ E = 0, & r = 1(l_m) \\ E = -K\dfrac{m}{r^3}, & 0 < r < 1(l_m) \end{cases} \qquad (9.4.4)$$

The above formula is the new gravitational formula.

Substituting equation (9.4.2) into equation (9.3.31) gives the relationship equation:

$$\begin{cases} U = -\dfrac{1}{2}K\dfrac{m}{r^2}, & r > 1(l_m) \\ U = 0, & r = 1(l_m) \\ U = +\dfrac{1}{2}K\dfrac{m}{r^2}, & 0 < r < 1(l_m) \end{cases} \qquad (9.4.5)$$

The formula above is the new gravitational potential formula. K in the formula is the new universal gravitational constant.

9.4.2 Strong nuclear forces in atomic nuclei can be explained by the new formula for gravitational

According to atomic physics, the nucleus of an atom is made up of two particles, protons and neutrons. There is a strong nuclear force in the nucleus that holds the protons or neutrons together in the nucleus. Because the strong nuclear force is stronger than the Coulomb repulsion, the Coulomb repulsion between positively charged protons cannot split the nucleus. The strong nuclear force is a short-range gravitational force. It acts at a distance r in the range of about $10^{-15}m < r < 10^{-10}m$. The strong nuclear force is a short-range gravitational force.

Since the strong nuclear force has a minimum action distance $r = 10^{-15}m$, it can be assumed that the minimum unit length of the gravitational field $l_m = 10^{-15}m$.

When the polar radius $r = 1(l_m)$, the strong nuclear force $F = 0$.

When the polar radius $r > 1(l_m)$, the strong nuclear force $F > 0$, when the strong combined force is positive suction. The suction force causes several positively charged $+e$ protons and neutrons in the nucleus to come close to each other to form the nucleus.

When the polar radius $r < 1(l_m)$, the strong nuclear force $F < 0$. The strong combined force is now a negative repulsive force. The repulsive force prevents multiple positively charged $+e$ protons in the nucleus from coming together to form a new particle.

According to physics, there is a gravitational force F and an electric field force F_q between the electrons in an atom and the nucleus; both the gravitational force F and the electric field force F_q are suction forces. Since the electrons orbit the nucleus at high speed, according to the centripetal force formula, there should also be a centripetal force F_v between the high-speed electrons and the nucleus.

Suppose R_0 is the radius of an atom. When the distance r between the electron outside the nucleus and the nucleus is $10^{-10}m < r < R_0$, the sum of the gravitational force F and the electromagnetic field force F_q between the electron and the nucleus is equal to the centripetal force F_v, that is:

$$F + F_q = F_v \qquad (9.4.6)$$

The above formula makes several electrons with the nucleus to form an atom.

When the distance r between the electron and the nucleus is close to the atomic radius R_0, the sum of the gravitational force F and the electromagnetic force F_q between the electron and the nucleus is slightly less than the centripetal force F_v, i.e:

$$F + F_q < F_v \qquad (9.4.7)$$

The above equation allows the metal atom to have free electrons.

To summarize, when the distance between proton and neutron is $r < 10^{-15}m$, the gravitational force between proton and neutron is repulsive force $F_{(-)}$. When the distance between proton and neutron is $r > 10^{-15}m$, the gravitational force between proton and neutron is suction force $F_{(+)}$). At this time, the combined force F of the gravitational attraction $F_{(+)}$ and the electric field repulsion F_q of the positive electric charge reaches equilibrium, i.e:

$$F = F_{(+)} + F_q = 0 \qquad (9.4.8)$$

When the distance between the electron and the nucleus is $r > 10^{-10}m$, there is gravitational force F, centripetal force F_v, and electric field force F_q between the electron and the nucleus.

9.4.3 Reasons why light waves can travel long distances

According to astronomical observations, galaxies can be seen at distances of billions of light-years. The reason why light waves travel billions of light years without diffusion can be explained by the new formula for gravitational (9.4.3).

Assume that the Planck length l_p is the smallest unit length l_m in the gravitational field of the energy element ε_h, i.e. $l_m = l_p$. The universal gravitational force $F_h = 0$ between the energy elements A and B if the distance between them is $r = 1(l_p)$.

Suppose that energy elements A and B are contained in a photon. When the distance $r > 1(l_p)$ between the energy elements A and B, the gravitational force between the energy elements A and B is the suction force $F_{(+)}$ according to equation (9.4.3). The suction force $F_{(+)}$ allows energy elements A and B to be in close proximity to each other and to move in synchronization.

The gravitational force between energy elements A and B is the repulsive force $F_{(-)}$ when the distance between energy elements A and B is $r < 1(l_p)$. The repulsive force $F_{(-)}$ keeps the two energy elements of energy elements A and B away from each other, and they cannot come close enough to combine to form a new energy particle.

Since the energy elements ε_h contained in a photon are all adjacent at the top, bottom, left, and right, the energy elements ε_h in a photon always move forward in synchronization with each other. As a result, light waves can travel great distances in the universe. This is why we can see distant stars tens of billions of light-years away.

9.4.4 Reasons why the energy element ε_h can move in a circle around the center of an elementary particle

The reason for the ability of the energy element ε_h to move in a circular motion around the center of an elementary particle can be explained by the new gravitational formula (9.4.3).

According to astronomy, the phenomenon of binary motion exists in the universe, i.e. the motion of one star around another, with gravitational and centripetal forces between the two stars. Theoretically, two energy elements A and B move in a circle around the center of the elementary particles, which is similar to the motion of a binary star around the center O of mass of the system. Therefore, there is a gravitational force F and a centripetal force F_v between the energy elements A and B. Note that the gravitational force F and the centripetal force F_v are in opposite directions.

Since the velocity of the energy element ε_h is always equal to the speed of light c, the centripetal force F_v of the energy element ε_h moving around the center of the elementary particle is, according to the centripetal force formula:

$$F_v = m_h \frac{c^2}{R} = m_h \frac{2c^2}{r} \qquad (9.4.9)$$

The m_h in the formula is the mass of the energy element ε_h; R is the distance of the two energy elements A and B from the center of the elementary particle, i.e. $R = r/2$. The above formula shows that the centripetal force F_v increases as the distance r between A and B decreases.

Theoretically, there is always a special distance r_0 between two energy elements A and B, which makes the combined force $F = 0$. Alternatively, this special distance r_0 makes the gravitational force $F_{(+)}$ equal to the centripetal force F_v, i.e:

$$F_{(+)} = K\frac{m_h^2}{r_0^3} = m_h\frac{2c^2}{r_0} = F_v \qquad (9.4.10)$$

Note that half of the special distance r_0, i.e. $R = r_0/2$, is the radius of rotation of the energy element ε_h. Alternatively, the orbit of the energy element ε_h moving around the center of the elementary particle is the orbit in which the gravitational force $F_{(+)}$ is equal to the centripetal force F_v.

If the distance between the energy elements A and B is $r > r_0$, then the centripetal force F_v of the energy elements A and B moving around the center of the elementary particle is less than the gravitational force $F_{(+)}$. At this point, the energy elements A and B will approach each other.

On the contrary, if the distance $r < r_0$ between the energy elements A and B, then the centripetal force F_v of the energy elements A and B moving around the center of the elementary particle is greater than the gravitational force $F_{(+)}$. At this point, energy elements A and B will move away from each other.

9.4.5 New formula for calculating the Astronomical Unit of a planet to the Sun

The Astronomical Unit L_n of a planet to the Sun can be calculated using the Titius-Bode law. It was discovered by the German astronomer Johann Daniel Titius. It was later generalized by Johann Elert Bode, director of the Berlin Observatory, into an empirical formula, viz:

$$L_n = \frac{n+4}{10} \qquad (9.4.11)$$

In the formula, **n = 0, 3, 6, 12, 24, 48**, ... , (the latter number is twice the previous number). The Astronomical Unit of a planet to the Sun is as follows.

Mercury's Astronomical Unit $L_n = 0.4$.

Venus' Astronomical Unit $L_n = 0.7$.

Earth's Astronomical Unit $L_n = 1.0$.

Mars' Astronomical Unit $L_n = 1.6$.

Ceres's Astronomical Unit $L_n = 2.8$.

Jupiter's Astronomical Unit $L_n = 5.2$.

Saturn's Astronomical Unit $L_n = 10$.

Uranus's Astronomical Unit $L_n = 19.6$

Neptune's Astronomical Unit $L_n = 38.8$ (actual observed value is 30.2).

Pluto's Astronomical Unit $L_n = 77.2$ (actual observed value is 39.5).

Prof. Johann Daniel Titius predicted that there should be a planet at $L_n = 2.8$, and in 1801 the Italian astronomer Piazzi discovered Ceres at this distance. Ceres was too small, with a diameter of only 1,020 kilometers. Later it was known that there was an asteroid belt at Astronomical Unit $L_n = 2.8$.

Physicists have not yet figured out the physical meaning of the Titius-Bode law. The Titius-Bode law has been discarded by physicists because the actual observed values of the Astronomical Unit of Neptune and Pluto deviate greatly from the calculated values of the Titius-Bode law. We can use the vortex field theory to derive new Astronomical Unit formulas.

Substitute equation (9.4.2) into equation (9.3.4) to obtain the relationship equation:

$$\beta = +\frac{1}{2}Km \cdot \ln r \qquad (9.4.12)$$

Assume that $n = 1.2, 3, 4, \cdots$, is the sequential number of the planet. Canceling the natural logarithm of the above equation, the distance r_n from planet n to the Sun is:

$$r_n = e^{\frac{2\beta_n}{Km_n}} \qquad (9.4.13)$$

The K in equation is the new gravitational constant, m_n is the mass of planet n, and β_n is the polar angle of planet n in the solar vortex system. Assume that both the perihelion and aphelion of planet n are located on the horizontal axis of polar coordinates. Suppose $\beta_{n-1} = 2\pi \cdot n$ is the polar angle from the perihelion of planet n to the Sun and $\beta_{n-2} = \pi(2n+1)$ is the polar angle from the aphelion of planet n to the Sun. According to the above equation, the distance r_{n-1} from the perihelion of planet n to the Sun and the distance r_{n-2} from the aphelion of planet $\boldsymbol{n}$ to the Sun are respectively:

$$r_{n-1} = e^{\frac{4\pi \cdot n}{Km_n}}, \qquad r_{n-2} = e^{\frac{2\pi(2n+1)}{Km_n}} \tag{9.4.14}$$

According to equation (2.1.26), the mean distance $\boldsymbol{R_n}$ from planet $\boldsymbol{n}$ to the Sun is:

$$R_n = \frac{1}{2}(r_{n-1} + r_{n-2}) = \frac{1}{2} e^{\frac{4\pi \cdot n}{Km_n}} \left(1 + e^{\frac{2\pi}{Km_n}}\right) \tag{9.4.15}$$

If $\boldsymbol{n = 1}$, then $\boldsymbol{R_1}$ is the average distance from Mercury to the Sun. If $\boldsymbol{n = 2}$, then $\boldsymbol{R_2}$ is the average distance from Venus to the Sun. According to the above equation the average distance $\boldsymbol{R_3}$ from the Earth to the Sun is:

$$R_3 = \frac{1}{2}(r_{3-1} + r_{3-2}) = \frac{1}{2} e^{\frac{12\pi}{Km_3}} \left(1 + e^{\frac{2\pi}{Km_3}}\right) \tag{9.4.16}$$

Assume that the average distance $\boldsymbol{R_3}$ from the Earth to the Sun, is an Astronomical Unit. the Astronomical Unit $\boldsymbol{L_n}$ of planet $\boldsymbol{n}$ to the Sun is:

$$L_n = \frac{R_n}{R_3} = \frac{e^{\frac{4\pi \cdot n}{Km_n}} \left(1 + e^{\frac{2\pi}{Km_n}}\right)}{e^{\frac{12\pi}{Km_3}} \left(1 + e^{\frac{2\pi}{Km_3}}\right)} \tag{9.4.17}$$

Simplify the above equation to get the relationship equation:

$$L_n = e^{\frac{4\pi}{K}\left(\frac{n}{m_n} - \frac{3}{m_3}\right)} \frac{\left(1 + e^{\frac{2\pi}{Km_n}}\right)}{\left(1 + e^{\frac{2\pi}{Km_3}}\right)} \tag{9.4.18}$$

The above formula is the new Astronomical Unit formula. If $\boldsymbol{n = 3}$ in the formula, then the Astronomical Unit of the Earth to the Sun $\boldsymbol{L_3 = 1}$. Using the above formula, we can search for new planets beyond Pluto with $\boldsymbol{n = 11}$.

Chapter 10 Derivation of the wave equation for gravitational

Introduction: Based on the new gravitational field strength formula and Stokes' theorem, the wave equation for gravitational can be derived.

10.1 The dynamic gravitational force *F* can be decomposed into three gravitational components

10.1.1 The dynamic gravitational force *F* can be decomposed into three gravitational components

When the intrinsic masses m_1 and m_2 are at rest in the cosmic vacuum frame of reference, the gravitational force F_0 at rest between the resting masses m_1 and m_2 is given by equation (9.4.3) , i.e:

$$F_0 = K\frac{m_1 m_2}{r^3} \tag{10.1.1}$$

If the intrinsic mass m moves with velocity v in the cosmic vacuum frame of reference, then according to the new mass-velocity formula (7.3.10), the moving mass m can be decomposed into two mass components, one being the internal rotating mass m_ω and the other being the displacement mass m_v, viz:

$$m = \sqrt{m_\omega^2 + m_v^2} = \sqrt{\left(m\frac{v_\omega}{c}\right)^2 + \left(m\frac{v}{c}\right)^2} \tag{10.1.2}$$

When the intrinsic masses m_1 and m_2 move in the cosmic vacuum frame of reference. According to the above equation, the moving masses m_1 and m_2 can be expressed as:

$$m_1 = \sqrt{m_{\omega 1}^2 + m_{v1}^2}, \qquad m_2 = \sqrt{m_{\omega 2}^2 + m_{v2}^2} \tag{10.1.3}$$

Substituting the above equation into equation (10.1.1) gives the relationship equation:

$$F = K\frac{\sqrt{m_{\omega 1}^2 + m_{v1}^2}\sqrt{m_{\omega 2}^2 + m_{v2}^2}}{r^3} \tag{10.1.4}$$

Expanding the above equation gives the relational equation:

$$F = \sqrt{\left(\frac{K m_{\omega 1} m_{\omega 2}}{r^3}\right)^2 + \left(\frac{K m_{v1} m_{v2}}{r^3}\right)^2 + \left(\frac{K}{r^3}\right)^2 \left(m_{v1}^2 m_{\omega 2}^2 + m_{\omega 1}^2 m_{v2}^2\right)} \tag{10.1.5}$$

The above equation shows that the gravitational force F, which is dynamic between the moving masses m_1 and m_2, contains three gravitational components.

10.1.2 The internal rotating gravitational force F_ω between the moving masses m_1 and m_2

The first gravitational component included in the dynamic universal gravitational force F is the internal rotational gravitational force F_ω between the internal rotating masses $m_{\omega 1}$ and $m_{\omega 2}$, namely:

$$F_\omega = K\frac{m_{\omega 1} m_{\omega 2}}{r^3} \tag{10.1.6}$$

In the cosmic vacuum frame of reference, assume that the intrinsic mass m moves with velocity v. According to equation (7.3.8), the internal rotating mass m_ω contained in the moving mass m is:

$$m_\omega = m\sqrt{1-\frac{v^2}{c^2}} \qquad (10.1.7)$$

In the cosmic vacuum frame of reference, assume that the velocities of the intrinsic masses m_1 and m_2 are v_1 and v_2, respectively. According to the above equations, the internal rotating masses contained in the moving masses m_1 and m_2 are respectively:

$$m_{\omega 1} = m_1\sqrt{1-\frac{v_1^2}{c^2}}, \qquad m_{\omega 2} = m_2\sqrt{1-\frac{v_2^2}{c^2}} \qquad (10.1.8)$$

Substituting the above equation into equation (10.1.6). This book defines the gravitational force between the internal rotating masses $m_{\omega 1}$ and $m_{\omega 2}$ as the internal rotating gravitational force F_ω, i.e:

$$F_\omega = K\frac{m_1 m_2}{r^3}\sqrt{1-\frac{v_1^2}{c^2}}\sqrt{1-\frac{v_2^2}{c^2}} \qquad (10.1.9)$$

The above equation is the internal rotational gravitational equation. The above equation shows that the internal rotational gravitational force F_ω decreases as the velocity v_1 and v_2 increase.

When the cosmic velocity $v_1 = v_2 = 0$, the above equation becomes the equation for gravitational at rest. From this it can be seen that the internal rotational gravitational force F_ω is capable of changing the magnitude and direction of motion of the object's velocity v.

The internal rotational gravitational force $F_\omega = 0$ between the moving masses m_1 and m_2 if the velocity $v_1 = c$ or the velocity $v_2 = c$.

Since photons always move in the universe at the speed of light c, the internal rotational gravitational force F_ω between a photon and any massive object is always equal to 0. Similarly, the internal rotational gravitational force $F_\omega = 0$ between two photons.

If the cosmic velocities v_1 and v_2 are much smaller than the speed of light c, equation (10.1.9) becomes the static gravitational formula, i.e:

$$F_\omega = F_0 = K\frac{m_1 m_2}{r^3} \qquad (10.1.10)$$

The above equation shows that the internal rotational gravitational force F_ω and the static universal gravitational force F_0 are both forces of the same nature. The difference between the two is that the internal rotational gravitational force F_ω is a component of the dynamic universal gravitational force F. Both the static gravitational force F_0 and the internal rotational gravitational force F_ω are capable of changing the magnitude of the object's velocity v and direction of motion.

Since the speed of the Earth is much slower than the speed of light, the gravitational force between two stationary objects on the ground can be calculated using equation (10.1.10). Because of the Earth's rotation, there is a periodic variation in the magnitude of the gravitational force between two objects. In today's high-tech age, such small periodic variations in the gravitational force should be observable through physical experiments.

10.1.3 The displacement gravitation F_v between the moving masses m_1 and m_2

The second gravitational component included in the dynamic gravitational force F is the displaced gravitational force F_v between the displaced masses m_{v1} and m_{v2}, i.e:

$$F_v = K\frac{m_{v1} m_{v2}}{r^3} \qquad (10.1.11)$$

In the cosmic vacuum frame of reference, assume that intrinsic mass m moves with velocity v. According to equation (7.3.9), the displaced mass m_v contained in the moving mass m is

$$m_v = m\frac{v}{c} \qquad (10.1.12)$$

In the cosmic vacuum frame of reference, assume that the velocities of the intrinsic masses m_1 and m_2 are v_1 and v_2, respectively. According to the above equations, the displaced masses contained in the moving masses m_1 and m_2 are, respectively:

$$m_{v1} = m_1\frac{v_1}{c}, \qquad m_{v2} = m_2\frac{v_2}{c} \qquad (10.1.13)$$

Substitute the above equation into equation (10.1.11). In this book, the gravitational force between the displaced masses m_{v1} and m_{v2} is defined as the displaced gravitational force F_v, i.e:

$$F_v = K\frac{m_1v_1 \cdot m_2v_2}{c^2r^3} \qquad (10.1.14)$$

The above equation is the displacement gravitational equation.

According to electrodynamics, when charges q_1 and q_2 move in the same direction, the magnetic field force F_B between the moving electric charge q_1 and q_2 is:

$$F_B = \frac{\mu_0}{4\pi}\frac{q_1v_1 \cdot q_2v_2}{r^2}\sin\theta$$

The magnetic field force F_B changes only the direction of motion of the charge, not the magnitude of its velocity v.

Since the displacement gravitational force F_v and the magnetic field force F_B are both forces proportional to the displacement velocity v, the displacement gravitational force F_v, like the magnetic field force F_B, only changes the direction of the object's motion, not the magnitude of the object's velocity v.

The displacement gravitational force $F_v = 0$ when the velocity of the object is $v_1 = 0$ or the velocity of the object is $v_2 = 0$. The displacement gravitational force F_v is equal to the universal gravitational force F, i.e. $F_v = F$, when the velocity of the object is $v_1 = v_2 = c$.

Since photons always move at the speed of light c in the cosmic vacuum frame of reference, the displacement gravitational force F_v between photons is equal to the cosmic gravitational force F, i.e. $F_v = F$.

10.1.4 Displaced internal rotational gravitational $F_{v\omega}$ between displaced mass m_v and internal rotating mass m_ω

The third gravitational component included in the dynamic gravitational force F is the displaced internal rotational gravitational force $F_{v\omega}$ between the displaced mass m_v and the internal rotational mass m_ω, i.e:

$$F_{v\omega} = \sqrt{\left(K\frac{m_{v1}m_{\omega2}}{r^3}\right)^2 + \left(K\frac{m_{v2}m_{\omega1}}{r^3}\right)^2} \qquad (10.1.15)$$

The above equation is the displacement inertial rotation gravitational equation. According to the above equation, the displacement internal rotational gravitational force $F_{v\omega}$ between the moving masses m_1 and m_2 are respectively:

$$\begin{cases} F_{v1\omega2} = K\dfrac{m_{v1}m_{\omega2}}{r^3} = K\dfrac{m_1m_2}{r^3}\dfrac{v_1}{c}\sqrt{1-\dfrac{v_2^2}{c^2}} \\[2ex] F_{v2\omega1} = K\dfrac{m_{v2}m_{\omega1}}{r^3} = K\dfrac{m_1m_2}{r^3}\dfrac{v_2}{c}\sqrt{1-\dfrac{v_1^2}{c^2}} \end{cases} \qquad (10.1.16)$$

Displacement internal rotational gravitational $F_{v\omega}$ can change the velocity v and the direction of motion of an object. For photons, the displacement internal rotation gravitational force $F_{v\omega}$ can only change the direction of motion of the photon, not the speed of light.

Equation (10.1.15) can be expressed as:

$$F_{v\omega} = \sqrt{(F_{v1\omega2})^2 + (F_{v2\omega1})^2} \qquad (10.1.17)$$

Substituting the three gravitational components into equation (10.1.5) gives the relationship equation:

$$F = \sqrt{F_\omega^2 + F_v^2 + F_{v\omega}^2} \qquad (10.1.18)$$

In summary, the gravitational force between two objects is independent of the state of motion of the objects. Alternatively, the dynamic gravitational force F is always equal to the static gravitational force F_0, i.e:

$$F = F_0$$

10.1.5 The bending of light rays in the gravitational field can be explained by the displacement internal rotation gravitational force $F_{v\omega}$

When the cosmic velocities $v_1 = 0$ and $v_2 = 0$ for intrinsic masses m_1 and m_2, the gravitational force of displacement $F_v = 0$ and the gravitational force of internal rotation of displacement $F_{v\omega} = 0$ between intrinsic masses m_1 and m_2, since intrinsic masses m_1 and m_2 possess displaced masses $m_{v1} = 0$ and $m_{v2} = 0$.

Furthermore, since the intrinsic masses m_1 and m_2 have internal rotational masses $m_{\omega 1} = m_1$ and $m_{\omega 2} = m_2$, the internal rotational gravitational force F_ω between the intrinsic masses m_1 and m_2 is equal to the universal gravitational force F, i.e., $F_\omega = F$.

For the two photons, the internal rotational masses $m_{\omega 1} = 0$ and $m_{\omega 2} = 0$ of the two photons, since the photon velocities $v_1 = v_2 = c$. The internal rotational gravitational force $F_\omega = 0$ between the two photons0; and the displacement internal rotational gravitational force $F_{v\omega} = 0$.

In addition, since the two photons have the displaced masses $m_{v1} = m_1$ and $m_{v2} = m_2$, the displaced gravitational force between the two photons is $F_v = F$. It can be seen that there is only one displaced gravitational force F_v between the two photons, and that the displaced gravitational force F_v is equal to the gravitational force F between the two photon masses.

For the object mass m_1 and the photon mass m_2. In the cosmic vacuum frame of reference, if the object velocity $v_1 = 0$, then the displacement mass of the object $m_{v1} = 0$. Since the photon velocity $v_2 = c$, the photon has an internal rotational mass $m_{\omega 2} = 0$. When the photon passes near the object, the internal rotational gravitational force F_ω and the displacement gravitational force F_v between the object and the photon are both 0, that is, the:

$$F_\omega = K\frac{m_{\omega 1}m_{\omega 2}}{r^3} = 0, \qquad F_v = K\frac{m_{v1}m_{v2}}{r^3} = 0 \tag{10.1.19}$$

Furthermore, if the object velocity $v_1 \neq 0$ and the photon velocity $v_2 = c$, then according to equation (10.1.16) the displaced internal rotational gravitational forces $F_{v2\omega 1}$ and $F_{v1\omega 2}$ between the photon and the object are, respectively:

$$F_{v2\omega 1} = K\frac{m_1 m_2}{r^3}\sqrt{1 - \frac{v_1^2}{c^2}}, \qquad F_{v1\omega 2} = 0 \tag{10.1.20}$$

The displacement internal rotational gravitational force $F_{v\omega}$ of the between the object and the photon is:

$$F_{v\omega} = K\frac{m_1 m_2}{r^3}\sqrt{1 - \frac{v_1^2}{c^2}} \tag{10.1.21}$$

The above equation shows that when a photon passes near a massive object, the light will bend toward the object in the gravitational field due to the displacement internal rotational gravitational $F_{v\omega}$ acting on the photon. General relativity attributes this bending of light to spacetime bending.

10.2 Dynamic Gravitational Potential *U* and Dynamic Gravitational Field Strength *E*

10.2.1 The dynamic gravitational potential *U* can be decomposed into two components

According to equation (9.4.5), the gravitational potential U generated by the intrinsic mass m is:

$$U = -\frac{1}{2}K\frac{m}{r^2}, \quad r > 1(l_m) \tag{10.2.1}$$

The K in the equation is the new universal gravitational constant.

In the cosmic vacuum frame of reference, when the intrinsic mass m moves with velocity v. According to equation (10.1.2), the moving mass m can be decomposed into two components, one being the internal rotating mass m_ω and the other being the displaced mass m_v, i.e:

$$m = \sqrt{m_\omega^2 + m_v^2}$$

Substituting the above equation into equation (10.2.1) gives the relationship equation:

$$U = -\frac{K\,m}{2\,r^2} = -\sqrt{\left(-\frac{K\,m_\omega}{2\,r^2}\right)^2 + \left(-\frac{K\,m_v}{2\,r^2}\right)^2} = -\sqrt{U_\omega^2 + U_v^2}\,, \quad r > 1(l_m) \qquad (10.2.2)$$

The above equation shows that the dynamic gravitational potential U can be decomposed into two components, the internal rotational gravitational potential U_ω and the displacement gravitational potential U_v.

Since m_ω is the internal rotating mass, this book defines $-\frac{K\,m_\omega}{2\,r^2}$ in the root symbol of equation (10.2.2) as the internal rotating gravitational potential U_ω, i.e:

$$U_\omega = -\frac{K\,m_\omega}{2\,r^2}\,, \quad r > 1(l_m) \qquad (10.2.3)$$

According to equation (7.3.8), the internal rotating mass m_ω contained in the moving mass m is:

$$m_\omega = m\sqrt{1 - \frac{v^2}{c^2}} \qquad (10.2.4)$$

Substituting the above equation into equation (10.2.3) gives the relationship equation:

$$U_\omega = -\frac{K\,m}{2\,r^2}\sqrt{1 - \frac{v^2}{c^2}}\,, \quad r > 1(l_m) \qquad (10.2.5)$$

The above equation is the internal rotational gravitational potential equation. The above equation shows that the internal rotational gravitational potential U_ω decreases as the speed of motion v increases.

The internal rotational gravitational potential U_ω is equal to the static gravitational potential U_0 when the velocity $v = 0$ for the intrinsic mass m, i.e. $U_\omega = U_0$. From this it can be seen that both the internal rotational gravitational potential U_ω and the static gravitational potential U_0 are identical in nature. The difference between the two is that the internal rotational gravitational potential U_ω is moving while the static gravitational potential U_0 is stationary.

Since m_v is the displaced mass, this book defines $-\frac{K\,m_v}{2\,r^2}$ in the root symbol of equation (10.2.2) as the displaced gravitational potential U_v, i.e:

$$U_v = -\frac{K\,m_v}{2\,r^2}\,, \quad r > 1(l_m) \qquad (10.2.6)$$

According to equation (7.3.9), the displacement mass m_v contained in the moving mass m is:

$$m_v = m\frac{v}{c} \qquad (10.2.7)$$

Substituting the above equation into equation (10.2.6) gives the relationship equation:

$$U_v = -\frac{K\,mv}{2\,cr^2}\,, \quad r > 1(l_m) \qquad (10.2.8)$$

The above equation is the displacement gravitational potential equation. The above equation shows that the displacement gravitational potential U_v is proportional to the velocity v. When the velocity $v = 0$ for intrinsic mass m, the displacement gravitational potential $U_v = 0$.

10.2.2 Rotational gravitational field strength E_ω and displacement gravitational field strength E_v

According to equation (9.4.4), the gravitational field strength E produced by the intrinsic mass m is:

$$E = K\frac{m}{r^3}\,, \quad r > 1(l_m) \qquad (10.2.9)$$

When the intrinsic mass m moves with velocity v. Substituting the moving mass $m = \sqrt{m_\omega^2 + m_v^2}$ into the above equation gives the relational equation:

$$E = K\frac{m}{r^2} = \sqrt{\left(K\frac{m_\omega}{r^3}\right)^2 + \left(K\frac{m_v}{r^3}\right)^2} = \sqrt{E_\omega^2 + E_v^2}\,, \quad r > 1(l_m) \qquad (10.2.10)$$

Since m_ω is the internal rotating mass, this book defines $K\frac{m_\omega}{r^3}$ in the root symbol as the internal rotating gravitational field strength E_ω, i.e:

$$E_\omega = K\frac{m_\omega}{r^3}, \quad r > 1(l_m) \tag{10.2.11}$$

Substituting the internal rotating mass equation (10.2.4) into the above equation gives the relational equation:

$$E_\omega = K\frac{m}{r^3}\sqrt{1-\frac{v^2}{c^2}}, \quad r > 1(l_m) \tag{10.2.12}$$

The above equation is the formula for the internal rotating gravitational field strength. The above equation shows that the internal rotating gravitational field strength E_ω becomes smaller as the speed v increases.

Since m_v is the displaced mass, this book defines $K\frac{m_v}{r^3}$ in equation (10.2.10) as the displaced gravitational field strength E_v, i.e:

$$E_v = K\frac{m_v}{r^3}, \quad r > 1(l_m) \tag{10.2.13}$$

Substituting the displaced mass equation (10.2.7) into the above equation gives the relational equation:

$$E_v = K\frac{mv}{cr^3}, \quad r > 1(l_m) \tag{10.2.14}$$

The above equation is the displacement gravitational field strength equation. The above equation shows that the displacement gravitational field strength E_v is proportional to the intrinsic mass m and the velocity v.

10.2.3 Equation for the gravitational field strength flux Φ_E

Suppose that every point on the circular line $L_R = 2\pi R$ is at the same distance r from the intrinsic mass m. Since the gravitational potential U is constant on the circular line L_R, the integral $\oint Udl$ of the gravitational potential U along the circular line $L_R = 2\pi R$ is:

$$\oint Udl = UL_R = 2\pi RU, \quad r > 1(l_m) \tag{10.2.15}$$

Inserting equation (10.2.1) into the above equation gives the relational equation:

$$\oint Udl = -\pi RK\frac{m}{r^2}, \quad r > 1(l_m) \tag{10.2.16}$$

The above equation is the loop integral equation for the gravitational potential. According to Stokes' theorem, the relation equation can be obtained:

$$\oint Udl = \iint \nabla \times UdS = -\pi RK\frac{m}{r^2}, \quad r > 1(l_m) \tag{10.2.17}$$

According to equation (9.3.35), the gravitational field strength E is:

$$E = \nabla \times U \tag{10.2.18}$$

Substituting the above equation into equation (10.2.17) gives the relationship equation:

$$\Phi_E = \iint E\,dS = -\pi RK\frac{m}{r^2}, \quad r > 1(l_m) \tag{10.2.19}$$

The above equation is the gravitational field strength flux equation. The surface S in the formula is bordered by the circular line $L_R = 2\pi R$. When the intrinsic mass m is moving with velocity v. Substituting the moving mass $m = \sqrt{m_\omega^2 + m_v^2}$ into the above equation gives the relation equation:

$$\Phi_E = \iint E\,dS = -\sqrt{\left(-\pi RK\frac{m_\omega}{r^2}\right)^2 + \left(-\pi RK\frac{m_v}{r^2}\right)^2}, \quad r > 1(l_m) \tag{10.2.20}$$

Since m_ω is the internal rotating mass, this book defines $-\pi RK\frac{m_\omega}{r^2}$ in the root symbol as the internal rotating gravitational field strength flux $\Phi_{E\omega}$, i.e:

$$\Phi_{E\omega} = -\pi RK\frac{m_\omega}{r^2}, \quad r > 1(l_m) \tag{10.2.21}$$

Substituting the internal rotating mass equation (10.2.4) into the above equation gives the relational equation:

$$\boldsymbol{\Phi}_{E\omega} = -\pi RK\frac{m}{r^2}\sqrt{1-\frac{v^2}{c^2}}, \quad r > 1(l_m) \tag{10.2.22}$$

The above equation is the internal rotating gravitational field strength flux equation. The above equation shows that the internal rotating gravitational field strength flux $\boldsymbol{\Phi}_{E\omega}$ becomes smaller as the velocity $\boldsymbol{v}$ increases.

Since $\boldsymbol{m_v}$ is the displaced mass, this book defines $-\pi RK\frac{m_v}{r^2}$ in equation (10.2.20) as the displaced gravitational field strength flux $\boldsymbol{\Phi}_{Ev}$, i.e:

$$\boldsymbol{\Phi}_{Ev} = -\pi RK\frac{m_v}{r^2}, \quad r > 1(l_m) \tag{10.2.23}$$

Substituting the displaced mass $m_v = m\frac{v}{c}$ into the above equation gives the relational equation:

$$\boldsymbol{\Phi}_{Ev} = -\pi RK\frac{mv}{cr^2}, \quad r > 1(l_m) \tag{10.2.24}$$

The above equation is the displacement gravitational field strength flux equation. The above equation shows that the displacement gravitational field strength flux $\boldsymbol{\Phi}_{Ev}$ is proportional to the intrinsic mass $\boldsymbol{m}$ and velocity $\boldsymbol{v}$.

10.3 Line-mass kinematic potential of the gravitational field ε_{mLE}

10.3.1 Definition of mass motion potential ε_m

According to equation (9.4.4), the gravitational field strength $\boldsymbol{E}$ produced by the intrinsic mass $\boldsymbol{m}$ is:

$$E = K\frac{m}{r^3} \tag{10.3.1}$$

Suppose $\boldsymbol{m_1}$ is a mass moving in a gravitational field. According to equation (9.4.3), the gravitational force $\boldsymbol{F}$ between the intrinsic mass $\boldsymbol{m}$ and $\boldsymbol{m_1}$ is:

$$F = K\frac{m_1 m}{r^3} \tag{10.3.2}$$

Let $\boldsymbol{W}$ be the work done by the gravitational force $\boldsymbol{F}$ on the moving mass $\boldsymbol{m_1}$, namely:

$$W = \int_{r_a}^{r_b} Fdl = \int_{r_a}^{r_b} m_1 E dl \tag{10.3.3}$$

In electrodynamics, the electromotive force $\boldsymbol{\varepsilon}$ is defined as the ratio of the work $\boldsymbol{W}$ done by the non-static electric force $\boldsymbol{F}$ to the moving electric charge $\boldsymbol{q}$, i.e:

$$\varepsilon = \frac{W}{q} \tag{10.3.4}$$

This book imitates the definition of the electromotive force $\boldsymbol{\varepsilon}$ by defining the ratio $\frac{W}{m_1}$ of the work $\boldsymbol{W}$ done by the gravitational force $\boldsymbol{F}$ to the moving mass $\boldsymbol{m_1}$ as the mass motion potential $\boldsymbol{\varepsilon_m}$, viz:

$$\varepsilon_m = \frac{W}{m_1} = \int_{r_a}^{r_b} E dl \tag{10.3.5}$$

The above formula is the mass motion potential formula. When the gravitational work $\boldsymbol{W > 0}$, the direction of the mass motion potential $\boldsymbol{\varepsilon_m}$ is directed from intrinsic mass $\boldsymbol{m}$ to infinity. When the gravitational work $\boldsymbol{W < 0}$, the direction of mass motion potential $\boldsymbol{\varepsilon_m}$ is directed from infinity to intrinsic mass $\boldsymbol{m}$.

10.3.2 The integral of the time-varying gravitational field strength E_t over the curve L is equal to the mass-motion potential ε_m

When the intrinsic mass $\boldsymbol{m}$ moves with velocity $\boldsymbol{v}$ in the cosmic vacuum frame of reference, the gravitational field strength $\boldsymbol{E}$ produced by the intrinsic mass $\boldsymbol{m}$ is given by equation (10.2.10):

$$E = K\frac{m}{r^2} = \sqrt{\left(K\frac{m_\omega}{r^3}\right)^2 + \left(K\frac{m_v}{r^3}\right)^2} \quad (10.3.6)$$

Theoretically, the gravitational field strength E may or may not vary on the curve line L. In order to discuss accurately and to avoid misunderstandings, this book uses different symbols to indicate different gravitational field strengths. The meaning of the symbols is as follows:

(1) The symbol E denotes the gravitational field strength that remains constant on the curve L, (i.e., the gravitational field strength in which the intrinsic mass m, the velocity v, and the distance r all remain constant).

(2) The symbol E_t denotes the gravitational field strength that changes on the curve L, (i.e., the gravitational field strength at which one of the intrinsic mass m, the velocity v, and the distance r changes).

Suppose that each point on the circular line $L_R = 2\pi R$ has an equal distance r to the intrinsic mass m. When the intrinsic mass m, velocity v and distance r are constant on the circular line $L_R = 2\pi R$ and the polar radius $r > 1(l_m)$. The integral $\oint Edl$ of the gravitational field strength E on the circular line $L_R = 2\pi R$ is:

$$\oint Edl = EL_R = 2\pi R \cdot K\frac{m}{r^3} \quad (10.3.7)$$

If intrinsic mass m, velocity v, and distance r vary with time t, then the time-varying gravitational field strength E_t is expressed as:

$$E_t(m, r.v) = K\frac{m}{r^3} \quad (10.3.8)$$

The integral $\oint E_t dl$ of the time-varying gravitational field $E_t(m, r.v)$ on the circular line $L_R = 2\pi R$ is:

$$\oint E_t(m, r.v)dl = 0 \quad (10.3.9)$$

According to equation (10.3.2), the gravitational force F between intrinsic mass m and m_1 is:

$$F = m_1 E = K\frac{m_1 m}{r^3} \quad (10.3.10)$$

When the intrinsic mass m_1 moves from point a to point b, the gravitational field strength E belongs to the time-varying gravitational field strength E_t because of the change in distance r. The time-varying gravitational force $F_t = m_1 E_t$ does the work W on the curve L is:

$$W = \int_{r_a}^{r_b} m_1 E_t dl = -\frac{1}{2}Km_1 m\left(\frac{1}{r_b^2} - \frac{1}{r_a^2}\right) \quad (10.3.11)$$

Substituting the above equation into equation (10.3.5), the mass-motion potential ε_m produced by the time-varying gravitational field strength E_t on the curve L is:

$$\varepsilon_m = \frac{W}{m_1} = \int_{r_a}^{r_b} E_t dl = -\frac{1}{2}Km\left(\frac{1}{r_b^2} - \frac{1}{r_a^2}\right) \quad (10.3.12)$$

When the intrinsic mass m and velocity v are kept constant. If the distance $r_a \neq r_b$, then there exists a gravitational potential difference value ΔU between points a and b. At this time, the mass motion potential $\varepsilon_m \neq 0$. If the distance $r_a = r_b$, then there exists no gravitational potential difference value ΔU between points a and b. At this time, the mass motion potential $\varepsilon_m = 0$. From this it can be determined that the mass motion potential ε_m is essentially a line integral $\int E_t dl$ of the time-varying gravitational field strength E_t, that is:

$$\varepsilon_m = \int_{r_a}^{r_b} E_t dl, \quad r_a \neq r_b \quad (10.3.13)$$

10.3.3 Line-mass kinematic potential of a gravitational field ε_{mLE}

According to equation (10.3.6), the gravitational field strength E produced by the intrinsic mass m is:

$$E = K\frac{m}{r^3}$$

If intrinsic mass m, velocity v and distance r vary with time, different moments t_i correspond to different gravitational field strengths $E_i(m_i, r_i, v_i)$. In this book, the time-varying gravitational field strength corresponding to time $\boldsymbol{t}$ is denoted by $\boldsymbol{E_t(m, r.v)}$, i.e:

$$\boldsymbol{E_t(m, r.v) = K\frac{m}{r^3}} \quad \textbf{(10.3.14)}$$

At the moment $\boldsymbol{t}$, the time-varying gravitational field strength $\boldsymbol{E_t}$ can also be expressed theoretically as $\boldsymbol{E_t\left(\frac{dm}{dt}, \frac{dr}{dt}, \frac{dv}{dt}\right)}$, since there is still a rate of change $\boldsymbol{\left(\frac{dm}{dt}, \frac{dr}{dt}, \frac{dv}{dt}\right)}$ in the intrinsic mass $\boldsymbol{m}$, velocity $\boldsymbol{v}$, and distance $\boldsymbol{r}$. The above formula belongs to the gravitational field strength formula that does not include the rate of change $\boldsymbol{\left(\frac{dm}{dt}, \frac{dr}{dt}, \frac{dv}{dt}\right)}$.

Taking the derivative of equation (10.3.14) with respect to time $\boldsymbol{t}$ yields the relation equation:

$$\boldsymbol{\frac{\partial E_t}{\partial t} = \frac{\partial E_t}{\partial m}\frac{dm}{dt} + \frac{\partial E_t}{\partial v}\frac{dv}{dt} + \frac{\partial E_t}{\partial r}\frac{dr}{dt}} \quad \textbf{(10.3.15)}$$

Although the above formula includes the rate of change $\boldsymbol{\left(\frac{dm}{dt}, \frac{dr}{dt}, \frac{dv}{dt}\right)}$, the above formula is not a formula for the changing gravitational field strength, since the gravitational field strength $\boldsymbol{E_t \neq \frac{\partial E_t}{\partial t}}$.

Suppose points $\boldsymbol{a}$ and $\boldsymbol{b}$ are points on the curve $\boldsymbol{L}$. Integrating the time-varying gravitational field strength $\boldsymbol{E_t(m, r.v)}$ along the curve $\boldsymbol{L}$ gives the relationship equation:

$$\boldsymbol{\varepsilon_{mLE} = \int_{r_a}^{r_b} E_t(m, r.v)dl} \quad \textbf{(10.3.16)}$$

In this book, the line integral $\boldsymbol{\varepsilon_{mLE}}$ of the time-varying gravitational field strength $\boldsymbol{E_t(m, v, r)}$ is defined as the line-mass motion potential of the gravitational field. The above equation is the equation for the line-mass motion potential of the gravitational field.

When the intrinsic mass $\boldsymbol{m}$ and the velocity $\boldsymbol{v}$ are kept constant and the distance $\boldsymbol{r}$ is varied. According to equation (10.3.14), the relation equation can be obtained:

$$\boldsymbol{\varepsilon_{mLE} = \int_{r_a}^{r_b} E_t(m, r.v)dl = -\frac{1}{2}Km\left(\frac{1}{r_b^2} - \frac{1}{r_a^2}\right)} \quad \textbf{(10.3.17)}$$

If the distance $\boldsymbol{r_a \neq r_b}$, then the line-mass kinematic potential $\boldsymbol{\varepsilon_{mLE} \neq 0}$ of the gravitational field. If the distance $\boldsymbol{r_a = r_b}$, then the line-mass kinematic potential $\boldsymbol{\varepsilon_{mLE} = 0}$ of the gravitational field.

When the integral curve $\boldsymbol{L}$ starts from point $\boldsymbol{a}$ to point $\boldsymbol{b}$ and then returns from point $\boldsymbol{b}$ to point $\boldsymbol{a}$. Since the curve $\boldsymbol{L}$ is closed, the integral $\boldsymbol{\oint E_t dl}$ of the time-varying gravitational field strength $\boldsymbol{E_t(m, r.v)}$ on the closed curve $\boldsymbol{L}$ is:

$$\boldsymbol{\oint E_t(m, r.v)dl = \int_{r_a}^{r_b} E_t dl + \int_{r_b}^{r_a} E_t dl = 0} \quad \textbf{(10.3.18)}$$

The above equation is the loop integral equation for the strength of the time-varying gravitational field. Note that the above formula is the loop integral formula without the rate of change $\boldsymbol{\left(\frac{dm}{dt}, \frac{dr}{dt}, \frac{dv}{dt}\right)}$. Based on the above equation, physicists have determined that the gravitational field is a conserved field with a field source.

10.4 Derivation of the induction formula for gravitational fields

10.4.1 Mathematical expression for the strength of the time-varying gravitational field strength E_t of the

At the moment $\boldsymbol{t}$, the time-varying gravitational field strength $\boldsymbol{E_t}$ can also be expressed theoretically in terms of the rate of change $\boldsymbol{\left(\frac{dm}{dt}, \frac{dv}{dt}, \frac{dr}{dt}\right)}$, since there is still a rate of change $\boldsymbol{\left(\frac{dm}{dt}, \frac{dv}{dt}, \frac{dr}{dt}\right)}$ for the intrinsic mass $\boldsymbol{m}$, the velocity $\boldsymbol{v}$ and the distance $\boldsymbol{r}$, i.e:

$$\boldsymbol{E_t\left(\frac{dm}{dt}, \frac{dr}{dt}, \frac{dv}{dt}\right) = f\left(\frac{dm}{dt}, \frac{dr}{dt}, \frac{dv}{dt}\right)} \quad \textbf{(10.4.1)}$$

We use the gravitational potential $\boldsymbol{U}$ to derive the above equation. According to equation (10.2.1), the gravitational potential $\boldsymbol{U}$ generated by the intrinsic mass $\boldsymbol{m}$ is:

$$U = -\frac{1}{2}K\frac{m}{r^2} \tag{10.4.2}$$

If the intrinsic mass $\boldsymbol{m}$ moves with velocity $\boldsymbol{v}$ in the cosmic vacuum frame of reference, then according to the new mass-velocity formula (7.3.10), the moving mass $\boldsymbol{m}$ can be decomposed into two mass components, one being the internal rotating mass $\boldsymbol{m_\omega}$ and the other being the displacement mass $\boldsymbol{m_v}$, viz:

$$m = \sqrt{m_\omega^2 + m_v^2} = \sqrt{\left(m\frac{v_\omega}{c}\right)^2 + \left(m\frac{v}{c}\right)^2} \tag{10.4.3}$$

If intrinsic mass $\boldsymbol{m}$, distance $\boldsymbol{r}$, and velocity $\boldsymbol{v}$ vary with time $\boldsymbol{t}$, the derivative of equation (10.4.2) with respect to time $\boldsymbol{t}$ gives the relational equation:

$$\frac{\partial U}{\partial t} = \frac{\partial U}{\partial m}\frac{dm}{dt} + \frac{\partial U}{\partial r}\frac{dr}{dt} + \frac{\partial U}{\partial v}\frac{dv}{dt} = -\frac{1}{2}K\left(\frac{1}{r^2}\frac{dm}{dt} - 2\frac{m}{r^3}\frac{dr}{dt}\right) - \frac{\partial U}{\partial v}\frac{dv}{dt} \tag{10.4.4}$$

If velocity $\boldsymbol{v_r} = \frac{dr}{dt}$ and acceleration $\boldsymbol{a} = \frac{dv}{dt}$, then the above equation becomes:

$$\frac{\partial U}{\partial t} = -\frac{1}{2}K\left(\frac{1}{r^2}\frac{dm}{dt} - 2\frac{m}{r^3}v_r\right) + \frac{dU}{dv}a \tag{10.4.5}$$

When the intrinsic mass $\boldsymbol{m}$ and velocity $\boldsymbol{v}$ remain constant and the distance $\boldsymbol{r}$ changes. Since the rate of change of mass $\frac{dm}{dt} = \boldsymbol{0}$ and acceleration $\boldsymbol{a} = \boldsymbol{0}$, the above equation becomes:

$$\frac{\partial U}{\partial t} = K\frac{m}{r^3}\cdot v_r \tag{10.4.6}$$

Dividing the above equation by the velocity $\boldsymbol{v_r}$ gives the relational equation:

$$E_t = \frac{1}{v_r}\frac{\partial U}{\partial t} = K\frac{m}{r^3}$$

The above equation shows that the ratio of the rate of change $\frac{\partial U}{\partial t}$ of the gravitational potential, to the velocity $\boldsymbol{v_r}$, belongs to the time-varying gravitational field strength $\boldsymbol{E_t}$. Dividing equation (10.4.5) by the velocity $\boldsymbol{v_r}$, gives the relation equation:

$$E_t = \frac{1}{v_r}\frac{\partial U}{\partial t} = -\frac{1}{2}K\left(\frac{1}{v_r r^2}\frac{dm}{dt} - 2\frac{m}{r^3}\right) + \frac{a}{v_r}\frac{\partial U}{\partial v} \tag{10.4.7}$$

Theoretically, there is a serious error in the above formula. If the distance $\boldsymbol{r}$ is kept constant and the intrinsic mass $\boldsymbol{m}$ and the velocity $\boldsymbol{v}$ are changed, the above formula becomes because the rate of change $\frac{\partial U}{\partial v} \neq \boldsymbol{0}$ and the velocity $\boldsymbol{v_r} = \boldsymbol{0}$:

$$E_t = \frac{1}{v_r}\frac{\partial U}{\partial t} = \infty \tag{10.4.8}$$

When the distance $\boldsymbol{r}$ is kept constant and the intrinsic mass $\boldsymbol{m}$ and velocity $\boldsymbol{v}$ are varied, Equation (10.4.7) is wrong since the time-varying gravitational field strength $\boldsymbol{E_t}$ is not equal to infinity.

Since the ratio of the rate of change $\frac{\partial U}{\partial t}$ of the gravitational potential to the velocity belongs to the gravitational field strength $\boldsymbol{E_t}$, this velocity should be independent of the rate of change of the distance $\boldsymbol{v_r}$, and independent of the velocity $\boldsymbol{v}$. Otherwise there would be an error that the time-varying gravitational field strength $\boldsymbol{E_t}$ is equal to infinity.

Since the speed of light $\boldsymbol{c}$ is constant in the cosmic vacuum, the ratio of the rate of change $\frac{\partial U}{\partial t}$ to the speed of light $\boldsymbol{c}$ belongs to the time-varying gravitational field strength $\boldsymbol{E_t}$. Dividing equation (10.4.4) by the speed of light $\boldsymbol{c}$ gives the relational equation:

$$E_t = \frac{1}{c}\frac{\partial U}{\partial t} = \frac{1}{c}\frac{\partial U}{\partial m}\frac{dm}{dt} + \frac{1}{c}\frac{\partial U}{\partial r}\frac{dr}{dt} + \frac{1}{c}\frac{\partial U}{\partial v}\frac{dv}{dt} = -\frac{1}{2}K\left(\frac{1}{cr^2}\frac{dm}{dt} - 2\frac{v_r}{c}\frac{m}{r^3}\right) + \frac{a}{c}\frac{\partial U}{\partial v} \tag{10.4.9}$$

If the intrinsic mass $\boldsymbol{m}$ and velocity $\boldsymbol{v}$ are held constant and the distance $\boldsymbol{r}$ is varied, the above equation becomes the following since the rate of change $\frac{dm}{dt} = \boldsymbol{0}$ of mass and the acceleration $\boldsymbol{a} = \boldsymbol{0}$:

$$E_t = \frac{1}{c}\frac{\partial U}{\partial t} = K\frac{m}{r^2} \cdot \frac{v_r}{c} = \frac{v_r}{c}E \tag{10.4.10}$$

Since the ratio of the velocities $\frac{v_r}{c}$ has no physical unit, $\frac{1}{c}\frac{dU}{dt}$ in equation belongs to the gravitational field strength E_t. Since the gravitational field E_t generated by the intrinsic mass m at time t has uniqueness, $\frac{1}{c}\frac{\partial U}{\partial t}$ is equal to the time-varying gravitational field strength E_t, i.e:

$$E_t\left(\frac{dm}{dt}, \frac{dr}{dt}, \frac{dv}{dt}\right) = \frac{1}{c}\frac{\partial U}{\partial t} = -\frac{1}{2}K\left(\frac{1}{cr^2}\frac{dm}{dt} - 2\frac{v_r}{c}\frac{m}{r^3}\right) + \frac{a}{c}\frac{\partial U}{\partial v} \tag{10.4.11}$$

In summary, there are two formulas for expressing the strength of the time-varying gravitational field E_t when the intrinsic mass m, distance r and velocity v vary with time t.

One is a time-varying gravitational field strength formula that does not include the rate of change $\left(\frac{dm}{dt}, \frac{dr}{dt}, \frac{dv}{dt}\right)$, viz:

$$E_t(m, r.v) = K\frac{m}{r^3}$$

The loop integral $\oint E_t dl$ of the time-varying gravitational field strength $E_t(m, r.v)$ is:

$$\oint E_t(m, r.v)dl = 0 \tag{10.4.12}$$

The other is a time-varying gravitational field strength formula that includes the rate of change $\left(\frac{dm}{dt}, \frac{dr}{dt}, \frac{dv}{dt}\right)$, i.e:

$$E_t\left(\frac{dm}{dt}, \frac{dr}{dt}, \frac{dv}{dt}\right) = \frac{1}{c}\frac{\partial U}{\partial t} \tag{10.4.13}$$

The loop integral $\oint E_t dl$ of the time-varying gravitational field strength $E_t\left(\frac{dm}{dt}, \frac{dr}{dt}, \frac{dv}{dt}\right)$ is:

$$\oint E_t\left(\frac{dm}{dt}, \frac{dr}{dt}, \frac{dv}{dt}\right)dl = \oint \frac{1}{c}\frac{\partial U}{\partial t}dl \neq 0$$

If the gravitational field strength E remains constant on the closed curve L, then the loop integral $\oint Edl$ of the gravitational field strength E is:

$$\oint Edl = E\oint dl \tag{10.4.14}$$

Assuming that each point on the circular line $L_R = 2\pi R$ is equidistant r from the intrinsic mass m, the integral $\oint Edl$ of the gravitational field strength E along the circular line $L_R = 2\pi R$ is, since the gravitational field strength E is constant on the circular line L_R:

$$\oint Edl = EL_R = 2\pi RK\frac{m}{r^3} \tag{10.4.15}$$

Since the gravitational field strength E is constant on the circular line L_R, the loop integral $\oint Edl$ is not the mass motion potential ε_m. Compare equation (10.4.12) with the above equation. Although both formulas are integrals along the circular line $L_R = 2\pi R$, the gravitational field strength connotation in the two formulas is different. The gravitational field strength E in the above formula is a constant, while the gravitational field strength $E_t(m, r.v)$ in equation (10.4.12) is a variable.

Since physicists only consider the loop integral $\oint E_t(m, r.v)dl = 0$ for a change in distance r and do not consider the loop integral $\oint dl = EL_R$ for a constant distance r, physicists are wrong about the gravitational field being a conservative field.

10.4.2 Surface-mass kinematic potential of a gravitational field ε_{mSE}

Assume that each point on the circular line $L_R = 2\pi R$ is at an equal distance r from the intrinsic mass m. Assume that the surface S has the circular line $L_R = 2\pi R$ as its boundary. According to the gravitational field strength flux equation (10.2.19) , the gravitational field strength flux Φ_E through the surface S is:

$$\Phi_E = \iint E\,dS = -\pi RK\frac{m}{r^2} \tag{10.4.16}$$

Note that the gravitational field strength E in the formula is constant on the circular line $L_R = 2\pi R$.

When the intrinsic mass $\boldsymbol{m}$, distance $\boldsymbol{r}$, and velocity $\boldsymbol{v}$ vary, taking the derivative of the above equation with respect to time $\boldsymbol{t}$ gives the relational equation:

$$\frac{\partial \Phi_E}{\partial t} = \iint \frac{\partial E}{\partial t} dS = \frac{\partial \Phi_E}{\partial m}\frac{dm}{dt} + \frac{\partial \Phi_E}{\partial r}\frac{dr}{dt} + \frac{\partial \Phi_E}{\partial v}\frac{dv}{dt} \tag{10.4.17}$$

Note that since the intrinsic mass $\boldsymbol{m}$, distance $\boldsymbol{r}$, and velocity $\boldsymbol{v}$ have changed, the $\boldsymbol{E}$ in the formula at this point should be the time-varying gravitational field strength $\boldsymbol{E_t(m,r.v)}$. Alternatively, the gravitational field strength $\boldsymbol{E}$ in the above equation should be replaced by the time-varying gravitational field strength $\boldsymbol{E_t}$, i.e:

$$\frac{\partial \Phi_E}{\partial t} = \iint \frac{\partial E_t}{\partial t} dS = \frac{\partial \Phi_E}{\partial m}\frac{dm}{dt} + \frac{\partial \Phi_E}{\partial r}\frac{dr}{dt} + \frac{\partial \Phi_E}{\partial v}\frac{dv}{dt} = -\pi RK\left(\frac{1}{r^2}\frac{dm}{dt} - 2\frac{m}{r^3}v_r\right) + \frac{\partial \Phi_E}{\partial v}a \tag{10.4.18}$$

The above formula is the formula for the rate of change of the gravitational field strength flux. The formula has velocity $\boldsymbol{v_r} = \frac{dr}{dt}$ and acceleration $\boldsymbol{a} = \frac{dv}{dt}$. Dividing the above formula by the speed of light $\boldsymbol{c}$ gives the relational formula:

$$\varepsilon_{mSE} = \frac{1}{c}\frac{\partial \Phi_E}{\partial t} = \frac{1}{c}\iint \frac{\partial E_t}{\partial t} dS = -\frac{\pi RK}{c}\left(\frac{1}{r^2}\frac{dm}{dt} - 2\frac{m}{r^3}v_r\right) + \frac{\partial \Phi_E}{\partial v}\frac{dv}{dt} \tag{10.4.19}$$

This book defines $\boldsymbol{\varepsilon_{mSE}}$ as the surface-mass motion potential of a gravitational field. The above equation is the formula for the surface-mass kinematic potential of the gravitational field. $\boldsymbol{E_t(m,r,v)}$ in the formula is the time-varying gravitational field strength.

When the gravitational field strength flux $\boldsymbol{\Phi_E}$ increases, the surface-mass motion potential $\boldsymbol{\varepsilon_{mSE} > 0}$ of the time-varying gravitational field because the gravitational field strength flux variation $\boldsymbol{\Delta\Phi_E > 0}$.

When the gravitational field strength flux $\boldsymbol{\Phi_E}$ decreases, the surface-mass motion potential $\boldsymbol{\varepsilon_{mSE} < 0}$ of the time-varying gravitational field because the gravitational field strength flux variation $\boldsymbol{\Delta\Phi_E < 0}$.

When intrinsic mass $\boldsymbol{m}$ and velocity $\boldsymbol{v}$ remain constant and distance $\boldsymbol{r}$ changes. Since the rate of change of mass $\frac{dm}{dt} = \boldsymbol{0}$ and acceleration $\boldsymbol{a = 0}$, equation (10.4.19) becomes:

$$\varepsilon_{mSE} = 2\pi Km\frac{v_r}{c}\frac{R}{r^3} \tag{10.4.20}$$

According to equation (10.3.17), the line-mass kinematic potential $\boldsymbol{\varepsilon_{mLE}}$ (gravitational potential difference value $\boldsymbol{\Delta U}$) generated by the intrinsic mass $\boldsymbol{m}$ is:

$$\varepsilon_{mLE} = \int_{r_a}^{r_b} E_t(m,r)dl = -\frac{1}{2}Km\left(\frac{1}{r_b^2} - \frac{1}{r_a^2}\right)$$

Compare equation (10.4.20) with the equation above. Since the velocity ratio $\frac{v_r}{c}$ has no physical units, $\boldsymbol{2\pi Km\frac{v_r}{c}}$ and $\frac{1}{2}\boldsymbol{Km}$ have the same physical units.

Since $\frac{R}{r^3}$ and $\left(\frac{1}{r_b^2} - \frac{1}{r_a^2}\right)$ have the same physical units, the surface mass kinematic potential $\boldsymbol{\varepsilon_{mSE}} = \frac{1}{c}\iint \frac{\partial E_t}{\partial t}$ belongs to the mass kinematic potential $\boldsymbol{\varepsilon_m}$ (gravitational potential difference value $\boldsymbol{\Delta U}$).

10.4.3 Derivation of the Gravitational Field Induction Formula

Assuming that each point on the circular line $\boldsymbol{L_R = 2\pi R}$ is equidistant $\boldsymbol{r}$ from the intrinsic mass $\boldsymbol{m}$, the integral $\boldsymbol{\oint Edl}$ of the gravitational field strength $\boldsymbol{E}$ along the circular line $\boldsymbol{L_R = 2\pi R}$ is, since the gravitational field strength $\boldsymbol{E}$ is constant on the circular line $\boldsymbol{L_R}$:

$$\oint Edl = EL_R = 2\pi RK\frac{m}{r^3} \tag{10.4.21}$$

Note that $\boldsymbol{E}$ in equation is not the time-varying gravitational field strength $\boldsymbol{E_t}$. Assume that the surface $\boldsymbol{S}$ is bounded by the closed circular line $\boldsymbol{L_R = 2\pi R}$. When the intrinsic mass $\boldsymbol{m}$ and velocity $\boldsymbol{v}$ are kept constant and the distance $\boldsymbol{r}$ is varied, the surface-mass kinematic potential $\boldsymbol{\varepsilon_{mSE}}$ of the time-varying gravitational field is, according to equation (10.4.20):

$$\varepsilon_{mSE} = 2\pi Km\frac{v_r}{c}\frac{R}{r^3}$$

Comparing the above equation with equation (10.4.21), we can see that the surface-mass kinematic potential $\boldsymbol{\varepsilon_{mSE}}$ is not equal to the integral $\oint \boldsymbol{Edl}$ of the gravitational field strength $\boldsymbol{E}$ on the circular line $\boldsymbol{L_R = 2\pi}$, i.e:

$$\boldsymbol{\varepsilon_{mSE} \neq \oint Edl} \qquad (10.4.22)$$

If the time-varying gravitational field strength $\boldsymbol{E_t}$ does not contain the rate of change $\left(\frac{dm}{dt}, \frac{dr}{dt}, \frac{dv}{dt}\right)$, then the surface-mass kinematic potential $\boldsymbol{\varepsilon_{mSE}}$ is not equal to the time-varying gravitational field strength $\boldsymbol{E_t(m, r, v)}$, integral $\oint \boldsymbol{E_t dl}$ on the closed circular line $\boldsymbol{L_R = 2\pi R}$, i.e:

$$\boldsymbol{\varepsilon_{mSE} \neq \oint E_t(m, r, v)dl = 0} \qquad (10.4.23)$$

Theoretically, there should be a time-varying gravitational field strength $\boldsymbol{E_t}\left(\frac{dm}{dt}, \frac{dr}{dt}, \frac{dv}{dt}\right)$ on the circular line $\boldsymbol{L_R = 2\pi R}$ such that the integral $\oint \boldsymbol{E_t}\left(\frac{dm}{dt}, \frac{dr}{dt}, \frac{dv}{dt}\right)\boldsymbol{dl \neq 0}$. We can derive this time-varying gravitational field from Stokes' theorem.

According to equation (10.2.1), the gravitational potential $\boldsymbol{U}$ generated by the intrinsic mass $\boldsymbol{m}$ is:

$$\boldsymbol{U = -\frac{1}{2}K\frac{m}{r^2}}$$

Suppose that the gravitational potential $\boldsymbol{U}$ integrates along the closed curve $\boldsymbol{L}$, and suppose that the surface $\boldsymbol{S}$ has the closed curve $\boldsymbol{L}$ as its boundary. According to Stokes' theorem, the relation equation can be obtained:

$$\boldsymbol{\oint Udl = \iint \nabla \times U\, dS} \qquad (10.4.24)$$

Note that the gravitational potential $\boldsymbol{U}$ is constant on the closed curve $\boldsymbol{L}$. Otherwise the integral $\oint \boldsymbol{Udl = 0}$. According to equation (10.2.18) the gravitational field strength $\boldsymbol{E}$ is equal to the curl $\boldsymbol{\nabla \times U}$ of the gravitational potential $\boldsymbol{U}$, i.e:

$$\boldsymbol{E = \nabla \times U} \qquad (10.4.25)$$

Inserting the above equation into equation (10.4.24) gives the relational equation:

$$\boldsymbol{\Phi_E = \oint Udl = \iint E\, dS} \qquad (10.4.26)$$

The derivative of the above equation with respect to time $\boldsymbol{t}$ gives the relational equation:

$$\boldsymbol{\frac{\partial \Phi_E}{\partial t} = \oint \frac{\partial U}{\partial t}dl = \iint \frac{\partial E}{\partial t}dS}$$

Note that since the intrinsic mass $\boldsymbol{m}$, distance $\boldsymbol{r}$, and velocity $\boldsymbol{v}$ have changed, $\boldsymbol{E}$ in equation should be the time-varying gravitational field strength $\boldsymbol{E_t(m, r. v)}$. Alternatively, the gravitational field strength $\boldsymbol{E}$ in equation above should be replaced by the time-varying gravitational field strength $\boldsymbol{E_t}$, viz:

$$\boldsymbol{\frac{\partial \Phi_E}{\partial t} = \oint \frac{\partial U}{\partial t}dl = \iint \frac{\partial E_t}{\partial t}dS} \qquad (10.4.27)$$

Dividing the above equation by the speed of light $\boldsymbol{c}$, gives the relational equation:

$$\boldsymbol{\varepsilon_{mSE} = \frac{1}{c}\frac{\partial \Phi_E}{\partial t} = \oint \frac{1}{c}\frac{\partial U}{\partial t}dl = \iint \frac{1}{c}\frac{\partial E_t}{\partial t}dS} \qquad (10.4.28)$$

The above equation shows that the surface-mass kinematic potential $\boldsymbol{\varepsilon_{mSE}}$ is equal to the integral $\oint \frac{1}{c}\frac{\partial U}{\partial t}\boldsymbol{dl}$ of the rate of change $\frac{1}{c}\frac{\partial U}{\partial t}$ on the closed curve $\boldsymbol{L}$. According to equation (10.4.13), the time-varying gravitational field $\boldsymbol{E_t}$ is:

$$\boldsymbol{E_t\left(\frac{dm}{dt}, \frac{dr}{dt}, \frac{dv}{dt}\right) = \frac{1}{c}\frac{\partial U}{\partial t}}$$

Substituting the above equation into equation (10.4.28) gives the relationship equation:

$$\boldsymbol{\varepsilon_{mSE} = \oint E_t dl = \iint \frac{1}{c}\frac{\partial E_t}{\partial t}dS} \qquad (10.4.29)$$

The above equation is the time-varying gravitational field induction equation. Note that the time-varying gravitational field strength $\boldsymbol{E_t}\left(\frac{dm}{dt}, \frac{dr}{dt}, \frac{dv}{dt}\right)$ on the left side of the equation and the time-varying gravitational field strength $\boldsymbol{E}_t(\boldsymbol{m}, \boldsymbol{r}, \boldsymbol{v})$ on the right side of the equation.

According to electromagnetic theory, Faraday's law of electromagnetic induction in Maxwell's equations is:

$$\oint \boldsymbol{E}d\boldsymbol{l} = -\iint \frac{\partial \boldsymbol{B}}{\partial t} d\boldsymbol{S}$$

The $\boldsymbol{E}$ in the formula is the time-varying electromagnetic field strength and $\boldsymbol{B}$ is the time-varying magnetic flux density.Compare equation (10.4.29) with the above formula. It is clear that the time-varying gravitational field induction formula is analogous to Faraday's law of electromagnetic induction.

10.5 Derivation of the gravitational wave equation

10.5.1 Divergence ∇·E_t = 0 of the time-varying gravitational field E_t

According to equation (10.2.19), the gravitational field strength flux $\boldsymbol{\Phi}_E$ through the surface $\boldsymbol{S}$ is:

$$\boldsymbol{\Phi}_E = \iint \boldsymbol{E}_t d\boldsymbol{S} \tag{10.5.1}$$

Suppose the point $\boldsymbol{P}$ is the center of the sphere $\boldsymbol{S} = 4\pi r^2$. Suppose the interior of the sphere $\boldsymbol{S}$ contains no intrinsic mass $\boldsymbol{m}$. As the sphere $\boldsymbol{S}$ shrinks toward the center point $\boldsymbol{P}$, the flux $\boldsymbol{\Phi}_E$ through the sphere $\boldsymbol{S} = 4\pi r^2$ is always equal to 0, because the lines of the gravitational field strength into and out of the sphere $\boldsymbol{S}$ are always equal, i.e:

$$\boldsymbol{\Phi}_E = \oiint \boldsymbol{E}_t d\boldsymbol{S} = 0 \tag{10.5.2}$$

Note that the above equation holds if the intrinsic mass $\boldsymbol{m}$ is outside the sphere $\boldsymbol{S} = 4\pi r^2$.

According to the definition of divergence, the divergence $\nabla \cdot \boldsymbol{E}_t$ of the gravitational field strength $\boldsymbol{E}_t$ at point $\boldsymbol{P}$ in space is:

$$\nabla \cdot \boldsymbol{E}_t = \lim_{\Delta V \to 0} \frac{\oiint \boldsymbol{E}_t d\boldsymbol{S}}{\Delta V} \tag{10.5.3}$$

Substituting equation (10.5.2) into the above equation gives the relationship equation:

$$\nabla \cdot \boldsymbol{E}_t = 0 \tag{10.5.4}$$

The above equation shows that the divergence $\nabla \cdot \boldsymbol{E}_t$ of the gravitational field strength $\boldsymbol{E}_t$ at any point in space is always equal to zero.

10.5.2 Wave equation for the strength of the gravitational field

Using equation (10.4.29) and Stokes' theorem, the relation equation can be obtained:

$$\oint \boldsymbol{E}_t d\boldsymbol{l} = \iint \nabla \times \boldsymbol{E}_t d\boldsymbol{S} = -\iint \frac{1}{c}\frac{\partial \boldsymbol{E}_t}{\partial t} d\boldsymbol{S} \tag{10.5.5}$$

Based on the above equation, the relationship equation can be obtained:

$$\nabla \times \boldsymbol{E}_t = -\frac{1}{c}\frac{\partial \boldsymbol{E}_t}{\partial t} \tag{10.3.6}$$

Taking Curl from the above equation gives the relationship equation:

$$\nabla \times (\nabla \times \boldsymbol{E}_t) = -\nabla \times \left(\frac{1}{c}\frac{\partial \boldsymbol{E}_t}{\partial t}\right) \tag{10.5.7}$$

According to curl theory, the curl $\nabla \times (\nabla \times \boldsymbol{E}_t)$ on the left side of the equal sign in the above equation is:

$$\nabla \times (\nabla \times \boldsymbol{E}_t) = \nabla(\nabla \cdot \boldsymbol{E}_t) - \nabla^2 \boldsymbol{E}_t \tag{10.5.8}$$

The $\nabla(\nabla \cdot \boldsymbol{E}_t)$ in equation is the gradient of the divergence $\nabla \cdot \boldsymbol{E}_t$ at point $\boldsymbol{P}$. According to equation (10.5.4), the divergence $\nabla \cdot \boldsymbol{E}_t$ of the gravitational field strength $\boldsymbol{E}_t$ at any point in space is always equal to 0, i.e:

$$\nabla \cdot \boldsymbol{E}_t = 0$$

Since the gradient $\nabla \boldsymbol{C} = \boldsymbol{0}$ for constant $\boldsymbol{C}$, the gradient $\nabla(\nabla \cdot \boldsymbol{E}_t)$ of the divergence $\nabla \cdot \boldsymbol{E}_t$ is equal to 0, i.e:

$$\nabla(\nabla \cdot \boldsymbol{E}_t) = \boldsymbol{0} \tag{10.5.9}$$

Substituting the above equation into equation (10.5.8) gives the relationship equation:

$$\nabla \times (\nabla \times \boldsymbol{E}_t) = -\nabla^2 \boldsymbol{E}_t \tag{10.5.10}$$

The curl $\nabla \times \left(\frac{1}{c}\frac{\partial \boldsymbol{E}_t}{\partial t}\right)$ on the right side of the equal sign in equation (10.5.7) is:

$$\nabla \times \left(\frac{1}{c}\frac{\partial \boldsymbol{E}_t}{\partial t}\right) = \frac{1}{c}\frac{\partial}{\partial t}\nabla \times \boldsymbol{E}_t \tag{10.5.11}$$

Substituting the above equation and equation (10.5.10) into equation (10.5.7) gives the relationship equation:

$$\nabla^2 \boldsymbol{E}_t - \frac{1}{c^2}\frac{\partial^2 \boldsymbol{E}_t}{\partial t^2} = 0 \tag{10.5.12}$$

The above equation is the gravitational wave equation. The above equation shows that the propagation speed of the gravitational field strength is the speed of light c. Multiplying the above equation by the intrinsic mass m gives the relational equation:

$$\nabla^2 (m\boldsymbol{E}_t) - \frac{1}{c^2}\frac{\partial^2}{\partial t^2}(m\boldsymbol{E}_t) = 0 \tag{10.5.13}$$

Since the gravitational force $\boldsymbol{F} = m\boldsymbol{E}_t$, the above equation becomes:

$$\nabla^2 \boldsymbol{F} - \frac{1}{c^2}\frac{\partial^2 \boldsymbol{F}}{\partial t^2} = 0 \tag{10.5.14}$$

The equation above is the gravitational wave equation. The above equation shows that a gravitational wave travels at the speed of light c.

10.5.3 Fluctuations in gravitational *F* and fluctuations in curved spacetime are two different concepts

In general relativity, gravitational waves are ripples in space and time in the universe. Einstein predicted, based on general relativity, that violent celestial activity would cause the surrounding time and space to fluctuate together. This is where gravitational waves come from.

General relativity teaches that in the case of a non-spherical symmetric distribution of matter. Gravitational waves are generated when there is a change in the motion of matter or in the mass distribution of a material system.

If the gravitational wave passes vertically through the plane of the space-time circle, the space-time circle will be distorted due to the curvature of space-time. The space inside the circle will be stretched in one direction according to the frequency of the gravitational wave and correspondingly compressed in the direction perpendicular to it.

Although general relativity predicts the existence of gravitational waves, gravitational waves in general relativity refer to the perturbation of cosmic space-time caused by the accelerated motion of objects. In other words, gravitational waves in general relativity are the ripples of curved spacetime. Gravitational waves in general relativity do not belong to the gravitational waves of formula (10.5.14).

We can describe the difference between the two in terms of water ripples. If the cosmic vacuum V is regarded as a horizontal surface, then, according to general relativity, a circle of ripples excited by a falling rock is a space-time bending wave (ripple).

If we follow the wave equation in equation (10.5.14), then the ripples caused by the falling rock are the gravitational force $\boldsymbol{F}$, which is constantly fluctuating outward. Obviously, the fluctuation of the gravitational force of the universe and the fluctuation of curved space-time are two different concepts.

The gravitational force $\boldsymbol{F}$ of the Sun on the Earth is very large. It takes about 8 minutes for the Sun's gravitational force $\boldsymbol{F}$ to reach the Earth's position. However, physicists never use the Sun's gravitational force $\boldsymbol{F}$ on the Earth to determine the propagation speed of gravitational waves. This fact is enough to prove that gravitational waves from the Sun to the Earth are completely different from the gravitational waves of general relativity.

The scientific community has been searching for the existence of gravitational waves. The most common way to detect gravitational waves is to observe the motion of massive objects. This is because their motion emits more gravitational wave radiation. Examples include supermassive black holes formed by mergers of galaxies beyond the Milky Way, the rotation of spinning pulsars, and the gravitational collapse of supernovae.

Scientists typically use laser interferometers to detect them. According to physicists, gravitational waves themselves are very small, thirty or forty orders of magnitude smaller than the electromagnetic force. It is estimated that the change in a gravitational wave caused by a differential laser beam is only the size of a proton. They are therefore extremely difficult to detect, even in a vacuum and in an isolated oscillating environment.

In summary, gravitational waves in physics are ripples in the curvature of time and space, not the gravitational waves of equation (10.5.14).

10.5.4 Scientists have experimentally determined that a changing gravitational field propagates at the speed of light *c*

On December 26, 2012, the team of Keyun Tang from the Institute of Geology and Geophysics of the Chinese Academy of Sciences (IGGP) held a press conference to announce that they had successfully obtained the world's first observational evidence that "the gravitational field of the universe propagates at the speed of light". The paper was published in the latest issue of the Chinese Science Bulletin.

The abstract of Prof. K. Y. Tang's paper is as follows Since the true position of the Sun cannot be directly observed, the apparent position is used to approximate the true position in the actual gravitational tidal formula of the Earth. It is found that this approximation is equivalent to hiding the assumption that the gravitational force of the universe propagates at the speed of light. In this paper, the viscoelastic phase lag of the Earth is corrected from the gravitational tide observations at Shiquanhe Station in Tibet and Wushi Station in Xinjiang; the theoretical value of the lunar universal gravitational tide is subtracted from the gravitational tide observations, and the quasi-solar is extracted. Cosmic gravitational wave observational data; derive and solve the cosmic gravitational wave propagation equation, and obtain the cosmic gravitational wave propagation speed of about 0.93-1.05 times the speed of light, with a relative error of 5%. This is the first strong observational evidence that the speed of gravitational is equal to the speed of light.

In September 2010, after learning about Prof. Tang's work, Chinese scientist Prof. Luo Mingjin quickly thought of using long-series data to separate the phase difference between solar and lunar waves to determine the propagation speed of gravitational.

Using three years of long series data from the superconducting gravimeter at the Wuhan Station, which has high observational accuracy, Prof. Luo analyzed the phase lag data of the solar semidiurnal S2 wave and the lunar semidiurnal M2 wave obtained by harmonic analysis, and analyzed them in the frequency domain, confirming the gravitational effect on each other. The propagation speed of the effect is equal to the speed of light.

Chapter 11 Interference experiments with one-way light paths prove that the special relativity principle of constant speed of light is not consistent with objective facts

Introduction: The Michelson-Morley experiment is a closed two-way light path interference experiment, not a one-way light path interference experiment. No one in history has ever verified that the special theory of relativity's principle of the invariance of the speed of light is consistent with objective facts using an interference experiment with a one-way light path.

Current lasers are single aperture output light beams. The authors have designed two new types of laser interferometers based on the working principle of the lasers.

The first is a front and back hole laser interferometer, that is, the laser uses the front hole and the back hole to output beams of the same wavelength for optical interference.

The second is a double-hole laser interferometer, i.e., two identical laser tubes are used to output beams of the same wavelength for optical interference.

The use of two new laser interferometer can be directly verified special relativity theory of light speed invariance principle does not conform to the objective facts.

11.1 Theoretical basis for unidirectional optical path interference experiments

11.1.1 Structure of helium-neon lasers

The helium-neon laser was invented by scientists at Bell Laboratories in the United States in 1961. The helium-neon laser has the advantages of simple structure, high energy conversion efficiency, low cost, simple operation, good laser quality and stable operation.

The laser light output from a He-Ne laser has good monochromaticity and directionality, and is high-quality coherent light. The red light output from a He-Ne laser has a wavelength of **632.8** nanometers.

The key component of a He-Ne laser is the laser tube. The laser tube is made of hard glass and is generally a layered sleeve construction. The inner layer is the discharge tube and the outer layer is the gas storage tube. The helium-neon laser consists of three parts: the discharge tube, the resonant cavity, and the laser power supply.

The discharge tube consists of a capillary glass discharge tube, a gas storage tube, and an electrode. the resonant cavity consists of a fully reflecting mirror and a partially reflecting mirror. the resonant cavity can be divided into internal and external cavity types. The structure of the internal cavity He-Ne laser is shown in Figure 11-1.

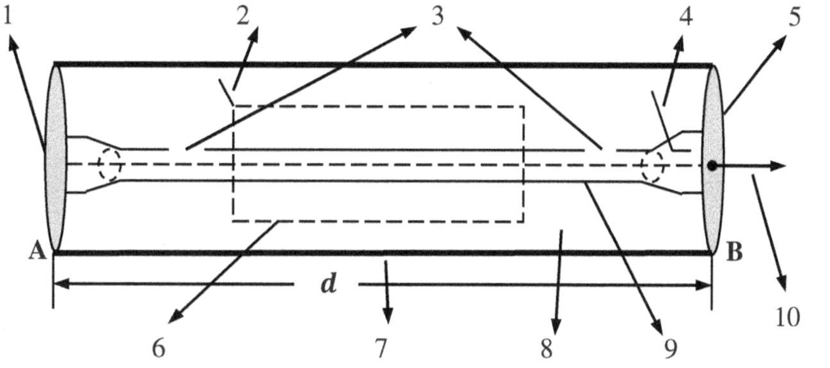

Figure 11-1 Schematic diagram of the structure of an intracavity helium-neon laser.

Figure 11-1 Numbering Instructions;
1. fully reflective mirror;
2. metal rod cathode;
3. capillary glass tube vent;
4. metal rod anode;
5. flat partially reflecting mirror;
6. Hollow metal tube cathode;
7. round glass tube for gas storage;
8. helium-neon gas mixture; 9. capillary glass discharge tube
9. capillary glass discharge tube
10. the output laser.
d is the length of the resonant cavity.

Illustration of Figure 11-1.

The center of the helium-neon laser is a capillary glass discharge tube about 1.5 mm in diameter, and the outer sheath of the capillary glass discharge tube is a gas storage glass tube about 45 mm in diameter. The anode of the laser is a tungsten rod and the cathode is a hollow metal tube. The discharge tube has a planar reflecting mirror affixed to the A end and a planar partially reflecting mirror affixed to the B end, which are perpendicular to the axis of the discharge tube and parallel to each other to form a resonant cavity.

The discharge tube is the core device for generating laser light. The discharge tube consists of three parts: a capillary glass discharge tube, a gas storage tube, and an electrode. It is shown in Figure 11-2.

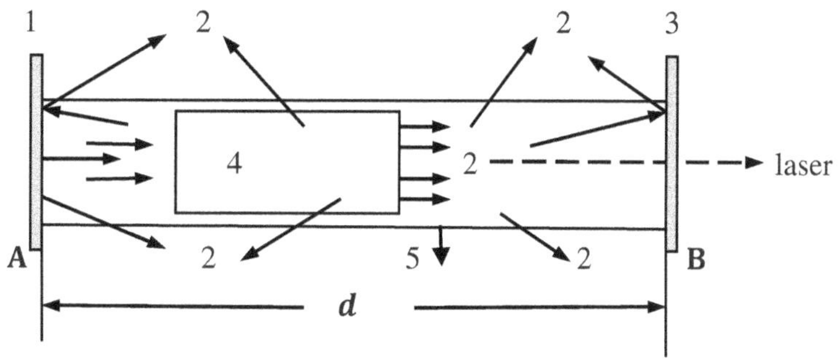

Figure 11-2 Schematic diagram of a capillary glass discharge tube and parallel plane resonant cavity.

Figure 11-2 Numbering Instructions;
1. fully reflecting mirror;
2. photons;
3. planar partially reflecting mirror;
4. helium-neon gas mixture;
5. capillary glass discharge tube;
d is the length of the resonant cavity.

The discharge tube is filled with helium and neon gases of greater than **99.99%** purity. Anodes and cathodes are attached to the ends of the capillary glass discharge tube to make the gases conductive. The structure of the capillary glass discharge tube is determined by the energy level structure of the neon atom. Capillary glass discharge tubes not only require strict straightness, but also have special requirements for their inner diameter.

11.1.2 Structure of parallel-plane resonant cavities

According to the operating principle of the He-Ne laser, the optical resonant cavity consists of two flat reflecting mirrors, A and B, which are perpendicular to the axis of the capillary glass discharge tube.

The function of the optical resonant cavity is to preferentially amplify light of a certain frequency and in the same direction, while suppressing light of other frequencies and directions. This is shown in Figure 11-2.

All photons that do not move along the axis of the resonant cavity rapidly exit the cavity and are no longer in contact with the active medium. Photons moving along the axis of the resonant cavity continue to travel through the resonant cavity, oscillating back and forth as they are reflected by the two plane mirrors.

As the photon travels, it continuously encounters excited particles, producing stimulated radiation. The number of photons traveling along the axis of the resonant cavity continues to increase, creating an intense light in the resonant cavity that travels in the same direction, frequency, and phase, which is called a laser.

To direct the laser light out of the resonant cavity, the flat mirror B is made into a partially transmitting mirror. The transmitted light becomes the output laser light, while the partially reflected light remains in the resonant cavity to further excite new photons. The optical resonant cavity works as follows

1). Provide feedback energy.

2). Select the direction and frequency of the light wave.

A parallel-plane resonant cavity is a coaxial resonant cavity consisting of two parallel-plane mirrors A and B, as shown in Figure 11-2.

Parallel plane cavity is the first optical resonant cavity proposed in the history of laser technology development. In a parallel plane cavity, a light beam parallel to the axis of the resonant cavity, after being reflected by a parallel plane mirror, its propagation direction is still parallel to the axis. The light beam goes back and forth an infinite number of times and does not escape from the cavity.

In other words, within the resonant cavity of the capillary glass discharge tube, photons whose direction of motion is not parallel to the axis of the discharge tube will quickly escape the resonant cavity and no longer come into contact with the helium-neon gas mixture inside the discharge tube.

Any photon whose direction of motion is parallel to the axis of the capillary glass discharge tube will oscillate continuously between the two plane mirrors A and B. The photon will then move in the direction of the axis of the capillary glass discharge tube. The photon movement continuously encounters excited particles to produce excited radiation, the number of photons moving along the axis of the discharge tube is increasing, the formation of the same direction of propagation, frequency and phase of light in the capillary glass discharge tube.

According to laser theory, if the length of the capillary glass discharge tube is d (the length of the resonant cavity), then the relationship between the output laser wavelength λ and the length d is:

$$d = n\frac{\lambda}{2} \tag{11.1.1}$$

The n in the equation is a positive integer. The above equation shows that the length d of the resonant cavity is an integer multiple of the half-wavelength $\lambda/2$. In other words, light waves that do not satisfy the above equation are canceled out in the resonant cavity. Finally, only light waves with wavelength $\lambda = 2d/n$ remain. At this point, the laser light exists as a standing wave inside the capillary glass discharge tube. The standing wave inside the resonant cavity is shown in Figure 11-3.

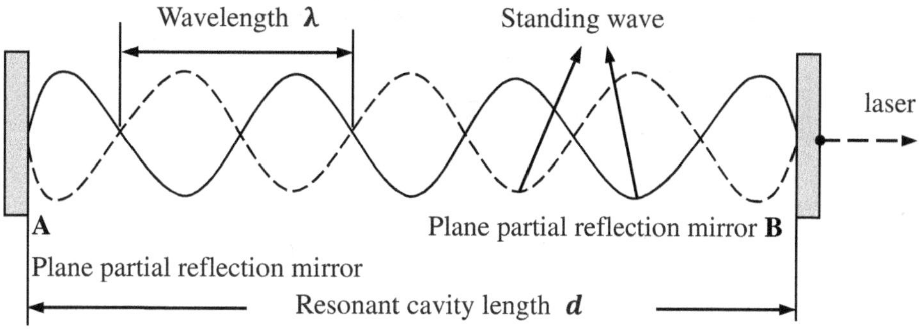

Figure 11-3 Schematic of laser standing waves in a parallel plane Resonant cavity.

11.1.3 Conditions under which two columns of light waves A and B overlap to produce interference fringes

Light waves propagate in a medium in the form of sinusoidal waves. When two columns of light waves of the same frequency overlap, if the phase of the two columns of light waves is different, there will be a phenomenon of light intensity decrease or light intensity increase.

For example, a beam of monochromatic light is divided into two columns of light waves by a beam splitter, and then they overlap in a certain spatial area. It will be found that the light intensity in the overlapping region is not uniformly distributed. Its brightness and darkness vary with its position in space. The brightest spot exceeds the sum of the light intensities of the original two light waves, while the darkest spot may have a light intensity of zero. This redistribution of light intensity is known as optical interference fringing.

There are three prerequisites for the overlapping of two columns of coherent light waves A and B to produce interference fringes.

1). The frequencies υ of the two columns of light waves A and B must be equal;

2). The directions of oscillation of the two columns of light waves A and B are the same;

3). The two columns of light waves A and B have a fixed phase difference.

Laser light is a high-quality coherent light because it has the same frequency, phase, and direction of oscillation. This property of lasers has been used to make laser interferometers with very high precision.

11.2 Structure and optical path of front and rear aperture laser interferometers

11.2.1 Structure of front and rear aperture lasers

Current lasers are single-aperture output lasers. People usually use a beamsplitter to decompose the laser into laser 1 and laser 2, and then use these two rays to conduct interference experiments.

In order to verify the principle of invariant speed of light, the authors designed a new type of double-aperture laser in two types, a front and back-aperture laser with one capillary glass discharge tube, and a double-aperture laser with two capillary glass discharge tubes.

When a helium-neon gas mixture is used to generate laser light, the front and rear aperture lasers consist of a capillary glass discharge tube, two identical flat partially reflecting mirrors A and B, and a power supply. The structures of the front and rear aperture lasers are shown in Figure 11-4.

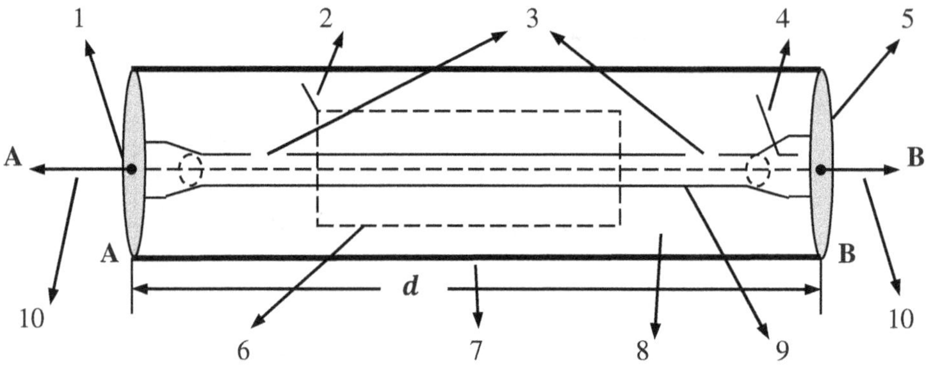

Figure 11- 4 Structural diagram of he-ne laser with front and rear holes.

Figure 11-4 Numbered Description.
1. Planar partially reflecting mirror;
2. cathode;
3. capillary tube vent;
4. anode;
5. planar partially reflecting mirror;
6. hollow cylindrical metal plate cathode;
7. round glass tube for gas storage;
8. helium-neon gas mixture;
9. capillary glass discharge tube;
10. the output laser.

d is the length of the resonant cavity.

Note that by replacing the planar partially reflecting mirror A in Figure 11-4 with a planar fully reflecting mirror, Figure 11-4 becomes the single-aperture laser in Figure 11-1.

Obviously, the two lasers output from the A and B ends of the laser in Figure 11-4 are lasers of the same frequency, with a fixed phase difference and opposite propagation directions.

11.2.2 Optical paths of the front and rear aperture laser interferometers

The front and rear aperture laser interferometer consists of the following parts.

1). Front and back aperture lasers;.
2). Planar reflector mirror C;
3). Planar Reflector mirror D; .
4). Planar reflecting mirror E;.
5). Transmissive reflector mirror R..
6). Observation screen P;.
7). Stand and swivel device.

Assume that the four points C, D, R, and E are the vertices of a planar rectangle. Assume that the length $\boldsymbol{d}$ is the distance between the laser output apertures A and B (i.e. $\boldsymbol{d} = \boldsymbol{AB}$), the length $\boldsymbol{k}$ is the distance between point A and point C and between point B and point E (i.e. $\boldsymbol{k} = \boldsymbol{AC} = \boldsymbol{BE}$), and the length $\boldsymbol{h}$ is the distance between point C and point D and between point E and point R (i.e. $\boldsymbol{h} = \boldsymbol{CD} = \boldsymbol{ER}$). Front-to-rear aperture laser interferometer optical path. This is shown in Figure 11-5.

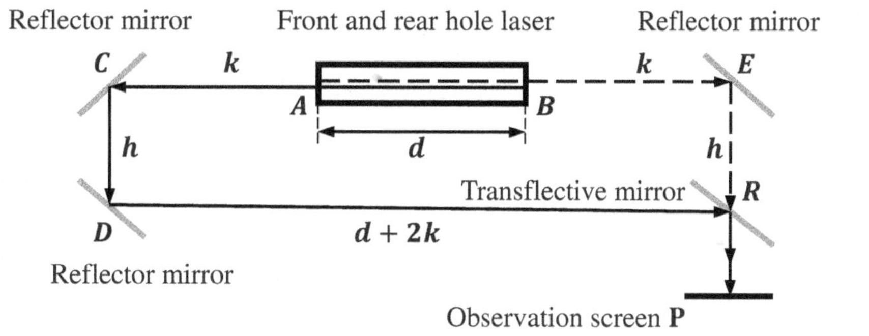

Figure 11-5. The structure and optical path diagram of the front and rear hole laser interferometer.

The optical path of the light beam A in Figure 11-5 from the output port B to the observation screen P is Starting point B → A → C → D → R → P.

The optical path of the light beam B from the output port A to the observation screen P is:
Starting point A → B → E → R → P.

Beam A and light beam B arrive at the transreflector mirror R and then propagate simultaneously toward the observation screen P. At this time, light beam A and light beam B are coherent light since light beam A and light beam B have the same frequency and direction of vibration and the phase difference between them is constant. It is thus determined that the unidirectional light beam A and the unidirectional light beam B produce interference fringes on the observation screen P.

It should be noted that the interference fringes produced by the Michelson interferometer belong to the interference fringes of the closed bidirectional optical path, while the interference fringes produced by the front and rear aperture laser interferometers belong to the interference fringes of the unidirectional optical path. No scientist in history has ever performed an interference experiment with a unidirectional light path.

11.2.3 Definitions of the cosmic vacuum frame of reference, the Newtonian ground frame of reference, and the Einsteinian ground frame of reference

In order to analyze and discuss the relationship between the principle of invariance of the speed of light and the Michelson-Morley experiment, it is necessary to clearly define the following three frames of reference with different meanings.

1). Definition of the cosmic vacuum frame of reference.

The cosmic vacuum contains many celestial bodies and matter. This book defines the cosmic vacuum that does not contain celestial bodies and matter as the cosmic vacuum frame of reference. The axiom of invariance of the speed of light defined in this book applies only to the cosmic vacuum frame of reference and not to any other moving inertial system.

2). Definition of Newtonian ground frame of reference.

According to Galileo's velocity transformation formula $\boldsymbol{v}' = \boldsymbol{v} \pm \boldsymbol{u}$, light has different speeds of motion in different $\boldsymbol{S}'$ inertial systems. Since all celestial bodies are moving in the cosmic vacuum frame of reference, the ground frame of reference is also moving in the cosmic vacuum frame of reference.

In the cosmic vacuum frame of reference, assume that $\boldsymbol{v}$ is the velocity of the ground frame of reference. According to Galileo's velocity transformation formula, the propagation speed of light in the direction of velocity $\boldsymbol{v}$ in the ground frame of reference is $\boldsymbol{c}' = \boldsymbol{c} - \boldsymbol{v}$, and the propagation speed in the direction opposite to velocity $\boldsymbol{v}$ is $\boldsymbol{c}' = \boldsymbol{c} + \boldsymbol{v}$.

In Newtonian mechanics, since the propagation velocities of light in all directions in the ground frame of reference are generally not equal, the author defines the ground frame of reference that conforms to Galileo's velocity transformation formula as the Newtonian ground frame of reference.

3). Definition of Einsteinian ground frame of reference.

The principle of invariance of the speed of light in special relativity means that in any inertial frame of reference, the speed of propagation of light in all directions is equal to the speed of light c, which does not change with the motion of the observer and the light source.

According to the principle of invariance of the speed of light in special relativity, in the ground frame of reference, the propagation speed of light in all directions is equal to the speed of light c. The author defines the ground frame of reference in which the propagation speed of light in all directions is equal to the speed of light $\boldsymbol{c}$ as the Einstein ground frame of reference.

Obviously, in the Einstein ground frame of reference. The speed of propagation of light in each direction $\boldsymbol{c}'$ is equal to the speed of light $\boldsymbol{c}$, i.e., $\boldsymbol{c}' = \boldsymbol{c}$.

11.3 Experimental results predicted by the Newtonian ground frame of reference

11.3.1 Three methods for calculating the number of interference fringe moves

Let $\boldsymbol{L}$ be the intrinsic length of the object. If the observer is stationary with respect to the object, then according to the length contraction effect in special relativity, the length of the stationary object as measured by the observer is $\boldsymbol{L}$.

Assuming that the velocity $\boldsymbol{v}$ is the relative velocity between the object and the observer, the length $\boldsymbol{L}'$ of the moving object as measured by the observer is:

$$\boldsymbol{L}' = \boldsymbol{L}\sqrt{1 - \frac{\boldsymbol{v}^2}{\boldsymbol{c}^2}}$$

Special relativity tells us that if the object is at rest and the observer is moving at velocity $\boldsymbol{v}$, then the length $\boldsymbol{L}$ of the object as measured by the observer will also contract.

In the cosmic vacuum frame of reference, the ground is assumed to be moving at velocity $\boldsymbol{v}$. Since the interferometer is at rest on the ground, the velocity $\boldsymbol{v}$ is also the velocity of the interferometer. The light interference experiment uses the following 3 equations to calculate the optical path length.

Equation 1 is Galileo's velocity transformation formula, i.e. $\boldsymbol{c}' = \boldsymbol{c} \pm \boldsymbol{v}$.

Equation 2 is the principle of invariance of the speed of light in special relativity, i.e. $\boldsymbol{c}' = \boldsymbol{c}$.

Equation 3 is the length contraction formula, i.e. $\boldsymbol{L}' = \boldsymbol{L}\sqrt{1 - \frac{\boldsymbol{v}^2}{\boldsymbol{c}^2}}$.

Theoretically, there are three ways to calculate the number $\boldsymbol{n}$ of moves of the interference fringes.

Method one is to calculate the number of moves $\boldsymbol{n}$ of the interference fringes using the Galilean velocity transformation formula $\boldsymbol{c}' = \boldsymbol{c} \pm \boldsymbol{v}$. For example, the Michelson-Morley experiment used this method to calculate the number of moves $\boldsymbol{n}$ of the interference fringes, which was calculated to be $\boldsymbol{n} = \mathbf{0.37}$.

Method 2 is to calculate the number of moves $\boldsymbol{n}'$ of the interference fringes using the Galilean velocity transformation formula $\boldsymbol{c}' = \boldsymbol{c} \pm \boldsymbol{v}$ and the length contraction formula $\boldsymbol{L}' = \boldsymbol{L}\sqrt{1 - \frac{\boldsymbol{v}^2}{\boldsymbol{c}^2}}$. For example, Lorenz used this method to calculate the number of moves $\boldsymbol{n}'$ of the interference fringe. The result of the calculation was $\boldsymbol{n}' = \mathbf{0}$, so Lorenz explained the zero result of the Michelson-Morley experiment.

Method 3 is to calculate the number of moves $\boldsymbol{n}'''$ of the interference fringes using the principle of invariance of the speed of light in special relativity, i.e., the formula $\boldsymbol{c}' = \boldsymbol{c}$.

Logically, special relativity should use Method 3 to calculate the number of moves $\boldsymbol{n}'''$ of interference fringes. However, since the establishment of special relativity, no physicist has used Method 3 to calculate the number of moves $\boldsymbol{n}'''$ of interference fringes.

The authors calculated the number of moves n, n', and n''' of the interference fringes using each of the three methods above. We can use the experimental results of the front and back hole laser interferometer to verify which theoretical calculation is correct, and thus determine which formula is correct.

11.3.2 Optical path difference value $\Delta L_{A\leftarrow}$ when the initial direction of the light beam A is opposite to the direction of the velocity v

Note that the Galilean velocity transformation formula $c' = c \pm v$ is used below to calculate the number of moves n of the interference fringes.

In a cosmic vacuum frame of reference, assume that the ground is moving to the right with velocity v. Since the front and rear aperture laser interferometers are stationary on the ground, the interferometers are also moving to the right with velocity v. Assume that C, D, and E are planar reflectors and R is a transreflector.

If the length of the resonant cavity of the laser tube $d = AB$ is parallel to the direction of the velocity v, then light beam A and light beam B are also parallel to the direction of the velocity v. At this point, the optical path length $k = AC = BE$ is parallel to the direction of the velocity v, and the optical path length $h = CD = ER$ is perpendicular to the velocity v. This is shown in Figure 11-6.

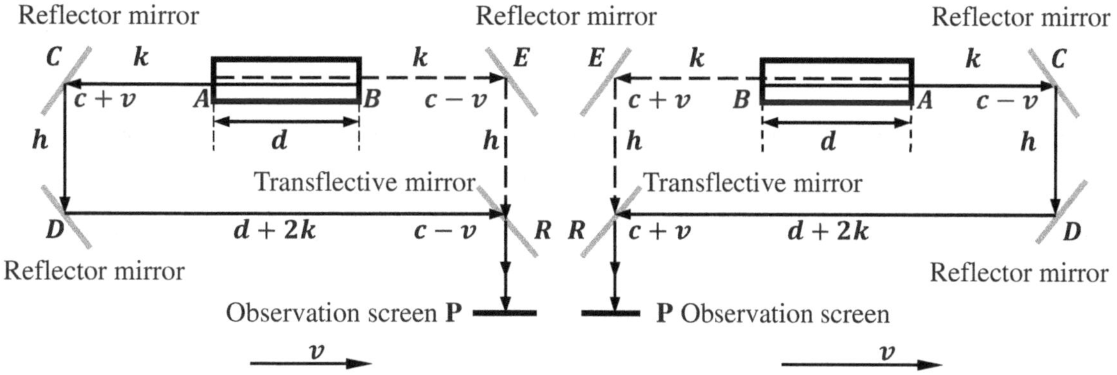

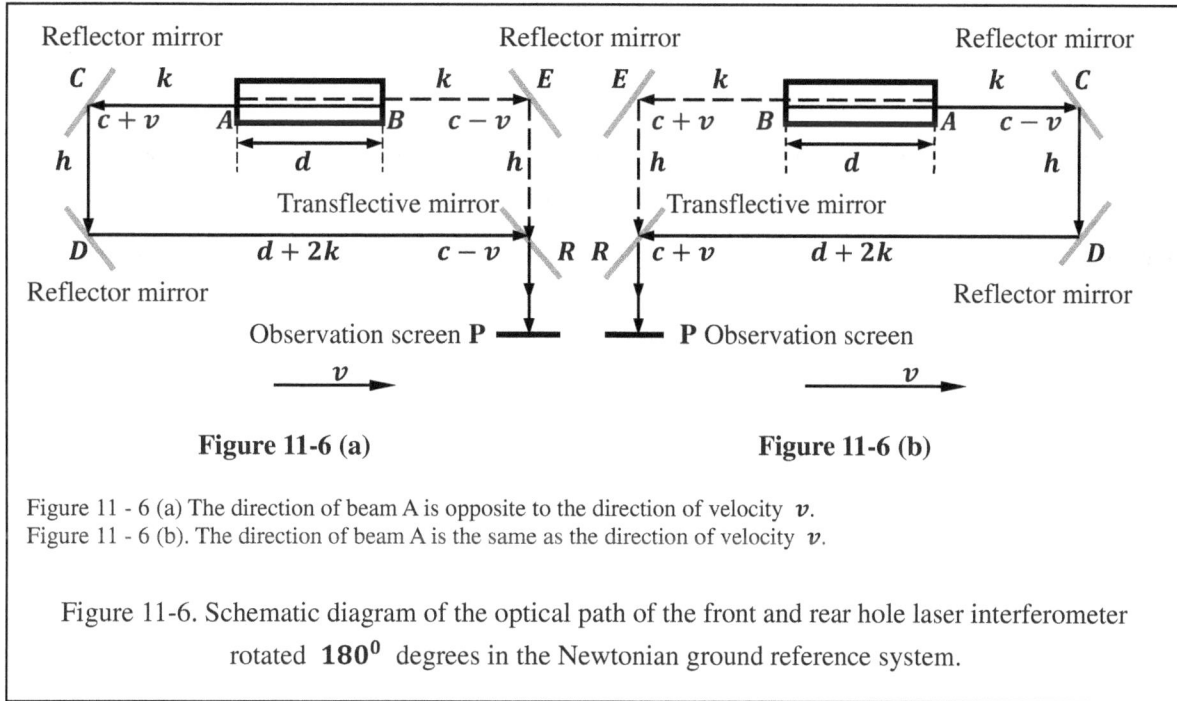

Figure 11-6 (a) **Figure 11-6 (b)**

Figure 11 - 6 (a) The direction of beam A is opposite to the direction of velocity v.
Figure 11 - 6 (b). The direction of beam A is the same as the direction of velocity v.

Figure 11-6. Schematic diagram of the optical path of the front and rear hole laser interferometer rotated $\mathbf{180^0}$ degrees in the Newtonian ground reference system.

In both the cosmic vacuum frame of reference and the Newtonian ground frame of reference, both light beam A and light beam B propagate vertically upward and downward with a speed of $\sqrt{c^2 - v^2}$. However, in Einstein's ground frame of reference, both light beam A and light beam B propagate vertically upward and downward at the speed c of light.

Since the observer is stationary with respect to the interferometer, there is no length contraction effect on the optical path in the interference experiment according to special relativity, i.e:

$$d' = d, \qquad k' = k, \qquad h' = h$$

In a Newtonian ground frame of reference, if the initial direction of light beam A is opposite to the direction of velocity v, then the initial direction of light beam B is the same as the direction of velocity v.

The propagation speed of light beam A on route **B→A→C** is $c'_A = c + v$, and on route **D→R** is $c'_A = c - v$. The propagation speed of light beam B on route **A→B→E** is $c'_B = c - v$. This is shown in Figure 11-6(a).

In the Newtonian ground frame of reference, if the initial direction of the light beam A is opposite to the velocity

$\boldsymbol{v}$, then the time $t_{A\leftarrow}$ taken for the light beam A to reach the transreflector mirror R is:

$$t_{A\leftarrow} = \frac{d+k}{c+v} + \frac{h}{\sqrt{c^2-v^2}} + \frac{d+2k}{c-v} \tag{11.3.1}$$

The optical path length $L_{A\leftarrow}$ of light beam A reaching the transreflector mirror R is:

$$L_{A\leftarrow} = ct_{A\leftarrow} = c\left(\frac{d+k}{c+v} + \frac{h}{\sqrt{c^2-v^2}} + \frac{d+2k}{c-v}\right) \tag{11.3.2}$$

If the initial direction of the light beam B is in the same direction as the velocity v, then the time $t_{B\rightarrow}$ taken by the light beam B to reach the transreflector mirror R is:

$$t_{B\rightarrow} = \frac{d+k}{c-v} + \frac{h}{\sqrt{c^2-v^2}} \tag{11.3.3}$$

The optical path length $L_{B\rightarrow}$ of light beam B reaching the transreflector mirror R is:

$$L_{B\rightarrow} = ct_{B\rightarrow} = c\left(\frac{d+k}{c-v} + \frac{h}{\sqrt{c^2-v^2}}\right) \tag{11.3.4}$$

In the Newtonian ground frame of reference, the optical path length difference value $\Delta L_{A\leftarrow}$ between light beam A and light beam B when they reach the transreflector mirror R is:

$$\Delta L_{A\leftarrow} = L_{A\leftarrow} - L_{B\rightarrow} = c\left(\frac{d+k}{c+v} + \frac{k}{c-v}\right) \tag{11.3.5}$$

Note that the optical path length difference value $\Delta L_{A\leftarrow}$ has nothing to do with the principle of invariance of the speed of light and the length contraction effect in special relativity.

11.3.3 Optical path length difference value $\Delta L_{A\rightarrow}$ when the direction of the light beam A is the same as the direction of the velocity v

If the interferometer is rotated horizontally by 180^0, the initial direction of light beam A is the same as the direction of velocity v, and the initial direction of light beam B is opposite to the direction of velocity v.

In the Newtonian inertial frame of reference, light beam A propagates with velocity $c'_A = c - v$ on the B→A→C path and with velocity $c'_A = c + v$ on the D→R path. light beam B propagates with velocity $c'_B = c + v$ on the A→B→E path. This is shown in Figure 11-6(b).

In the Newtonian ground frame of reference, if the initial direction of the light beam A is the same as the direction of the velocity v, then the time $t_{A\rightarrow}$ for the light beam A to reach the transreflector mirror R is:

$$t_{A\rightarrow} = \frac{d+k}{c-v} + \frac{h}{\sqrt{c^2-v^2}} + \frac{d+2k}{c+v} \tag{11.3.6}$$

The optical path length $L_{A\rightarrow}$ of the light beam A reaching the transreflector mirror R is:

$$L_{A\rightarrow} = ct_{A\rightarrow} = c\left(\frac{d+k}{c-v} + \frac{h}{\sqrt{c^2-v^2}} + \frac{d+2k}{c+v}\right) \tag{11.3.7}$$

If the initial direction of the light beam B is opposite to the velocity v, then the time $t_{B\leftarrow}$ for the light beam B to reach the transmission mirror R is:

$$t_{B\leftarrow} = \frac{d+k}{c+v} + \frac{h}{\sqrt{c^2-v^2}} \tag{11.3.8}$$

The optical path length $L_{B\leftarrow}$ of the light beam B reaching the transmission mirror R is:

$$L_{B\leftarrow} = ct_{B\leftarrow} = c\left(\frac{d+k}{c+v} + \frac{h}{\sqrt{c^2-v^2}}\right) \tag{11.3.9}$$

In the Newtonian ground frame of reference, if the initial direction of light beam A is the same as the velocity v, then the optical path length difference value $\Delta L_{A\rightarrow}$ between light beam A and light beam B when they reach the transreflector mirror R is:

$$\Delta L_{A\rightarrow} = L_{A\rightarrow} - L_{B\leftarrow} = c\left(\frac{d+k}{c-v} + \frac{k}{c+v}\right) \tag{11.3.10}$$

Note that the optical path length difference value $\Delta L_{A\rightarrow}$ is independent of the light speed invariance principle and the length contraction effect.

11.3.4 In the Newtonian ground frame of reference, the largest interference striped movement value *n*

In the Newtonian ground frame of reference, if the interferometer is rotated horizontally 180^0, then the difference value $\Delta L_{\leftrightarrow}$ between the optical path length difference value $\Delta L_{A\rightarrow}$ of equation (11.3.10) and the optical path length difference value $\Delta L_{A\leftarrow}$ of equation (11.3.5) is:

$$\Delta L_{\leftrightarrow} = \Delta L_{A\rightarrow} - \Delta L_{A\leftarrow} = c\left(\frac{d+k}{c-v} + \frac{k}{c+v}\right) - c\left(\frac{d+k}{c+v} + \frac{k}{c-v}\right)$$

Since the velocity v is much smaller than the speed of light c, i.e., $v \ll c$, the above equation can be simplified to:

$$\Delta L_{\leftrightarrow} = \frac{cd}{c-v} - \frac{cd}{c+v} = \frac{2cvd}{c^2 - v^2} \approx \frac{2vd}{c} \tag{11.3.11}$$

Note that the value $\Delta L_{\leftrightarrow}$ of the change in the optical path length difference value is independent of the principle of invariance of the speed of light and the effect of length contraction.

If the value $\Delta L_{\leftrightarrow}$ of the change in the optical path length difference value increases, then the circular stripe on the observation screen P will gradually increase in size, and new stripes will continuously appear from the center of the circle.

On the contrary, if the value $\Delta L_{\leftrightarrow}$ of the change in optical path length difference value decreases, the circular stripe on the observation screen P will gradually shrink and continuously disappear from the center of the circle.

In the Newtonian ground frame of reference, if the front and rear aperture laser interferometers are rotated horizontally by 180^0, then the maximum number of moves n of the interference fringes is:

$$n = \frac{\Delta L_{\leftrightarrow}}{\lambda} = \frac{2vd}{c\lambda} \tag{11.3.12}$$

Note that the maximum value n is independent of the principle of invariance of the speed of light and the length contraction effect.

11.3.5 Length contraction effects are negligible in interference experiments with one-way optical paths

Note that the number of moves n' of interference fringes is calculated below using the Galilean velocity conversion formula $c' = c \pm v$ and the length contraction formula $L' = L\sqrt{1 - \frac{v^2}{c^2}}$.

To explain why the results of the Michelson-Morley experiment were always equal to 0, in 1895 Lorentz formulated the hypothesis that the length L of an object contracts in the direction of motion. The formula for this is:

$$L' = L\sqrt{1 - \frac{v^2}{c^2}} \tag{11.3.13}$$

Note that the velocity v in the formula is the velocity of the object moving in the cosmic vacuum, not the relative velocity between the object and the observer. If the velocity of the object $v = 0$, then there is no contraction effect on the length L of the object, regardless of whether the observer is at rest or in motion.

Since the Lorentz length contraction formula is not derived from the principle of invariance of the speed of light, the Lorentz length contraction formula can be used for the Newtonian ground frame of reference. From the Lorentz length contraction formula, the length d' of the length d in Figure 11-6 after contraction in the direction of motion is:

$$d' = d\sqrt{1 - \frac{v^2}{c^2}} \tag{11.3.14}$$

If Lorentz's equation for length contraction is correct, then the value $\Delta L'_{\leftrightarrow}$ of the change in the optical path length difference value after contraction of the resonant cavity length d according to equation (11.3.11) should be:

$$\Delta L'_{\leftrightarrow} = \frac{cd'}{c-v} - \frac{cd'}{c+v} = \frac{2cvd'}{c^2 - v^2} \tag{11.3.15}$$

Substituting the contracted length $\boldsymbol{d'}$ into the above equation, since the velocity $\boldsymbol{v}$ is much smaller than the speed of light $\boldsymbol{c}$, i.e. $\boldsymbol{v} \ll \boldsymbol{c}$, the above equation can be simplified to:

$$\Delta L'_{\leftrightarrow} = \frac{2cvd'}{c^2 - v^2} = \frac{2vd}{\sqrt{c^2 - v^2}} \approx \frac{2vd}{c} \qquad (11.3.16)$$

Note that the difference value $\Delta L'_{\leftrightarrow}$ is independent of the principle of invariance of the speed of light, although the length contraction effect is included in the difference value $\Delta L'_{\leftrightarrow}$. Compare the above equation with equation (11.3.11). In the one-way optical path interference experiment, since the difference value $\Delta L'_{\leftrightarrow} = \Delta L_{\leftrightarrow}$, the effect of length contraction effect on the difference value $\Delta L_{\leftrightarrow}$ is negligible.

The maximum number of moves $\boldsymbol{n'}$ of interference fringes in the Newtonian ground frame of reference is:

$$n' = \frac{\Delta L'_{\leftrightarrow}}{\lambda} = \frac{2vd}{c\lambda}$$

Note that the maximum number of moves $\boldsymbol{n'}$ is independent of the principle of invariance of the speed of light, although it includes the effect of length contraction. The above equation is the same as equation (11.3.12).

11.3.6 Experimental results predicted by the Newtonian ground frame of reference

Assume that the distance between the two planar partially reflecting mirrors A and B in the resonant cavity of the laser is $\boldsymbol{d = 0.1m}$. If the interference experiment uses a HeNe laser as the light source, the laser wavelength $\boldsymbol{\lambda = 6.328 \times 10^{-7} m}$.

Theoretically, the velocity $\boldsymbol{v}$ in equation (11.3.12) is the velocity of the ground frame of reference in the cosmic vacuum frame of reference. However, the magnitude of the velocity $\boldsymbol{v}$ has not been determined by physicists so far because physicists deny the existence of Newtonian absolute spacetime.

According to the available astronomical observations, the velocity $\boldsymbol{v}$ of the ground frame of reference in the cosmic vacuum frame of reference can take the following three different values.

The first ground speed $\boldsymbol{v}$ is the speed of the Earth relative to the cosmic background radiation (CMB).

According to the Doppler effect, the energy of light increases as you move toward the light source. Moving away from the source, the energy of the light decreases. Because the cosmic background radiation is absolutely static. Therefore scientists determine the speed of the Earth's motion through the universe by calculating the magnitude of the redshift value of the background radiation.

According to astronomers' calculations, the Earth's velocity of motion with respect to the CMB is $\boldsymbol{v \approx 627km/s}$. From this, it can be determined that the velocity $\boldsymbol{v}$ of motion of the ground frame of reference with respect to the CMB, is:

$$v = 627km/s \approx 20.9 \times 10^{-4} c$$

Substitute the velocity $\boldsymbol{v = 20.9 \times 10^{-4} c}$ of the Earth's motion relative to the CMB, into equation (11.3.12). In the Newtonian ground frame of reference, if the front and rear aperture laser interferometers are rotated $\boldsymbol{180^0}$, then the maximum number of moves $\boldsymbol{n_1}$ of the interference fringes is:

$$n_1 = \frac{2vd}{c\lambda} = \frac{2 \times 0.1 \times 20.9 \times 10^{-4}}{6.328 \times 10^{-7}} = 660.5 \qquad (11.3.17)$$

In the Newtonian ground frame of reference, if the interferometer is rotated $\boldsymbol{90^0}$, then the maximum number of moves $\boldsymbol{n_{1\perp}}$ of the interference fringes is:

$$n_{1\perp} = \frac{1}{2} n_1 = 330.2 \qquad (11.3.18)$$

Since the Sun orbits around the center of the Milky Way galaxy and the Earth has its own rotational velocity, the cosmic velocity $\boldsymbol{v}$ at any point on the Earth will vary periodically on a day-to-day basis. To minimize the error due to the rotation of the interferometer, the interferometer can be fixed and the number of moves of the interference fringes can be observed by the diurnal variation.

If the interference fringes move towards the center of the screen during 12 hours, then after 12 hours the interference fringes will move outward from the center of the screen. From equation (11.3.17), the number of moves of the interference fringe per hour is $\boldsymbol{n_1/12 = 55}$.

The second ground velocity $\boldsymbol{v}$ is the speed of the solar system's revolution around the galactic center. According to astronomers' observational calculations, the speed $\boldsymbol{v_O} = \mathbf{250}\boldsymbol{km/s}$ of the solar system's revolution around the galactic center. If the Earth's rotation and revolution speeds are ignored, the Earth's average speed $\boldsymbol{v}$ should be equal to the speed $\boldsymbol{v_O}$, i.e. $\boldsymbol{v} = \boldsymbol{v_O} \approx \mathbf{8 \times 10^{-4}}\boldsymbol{c}$.

Substitute the velocity $\boldsymbol{v} = \mathbf{8 \times 10^{-4}}\boldsymbol{c}$ of the Sun's revolution around the galactic center, into equation (11.3.12). In the Newtonian ground frame of reference, if the interferometer is rotated $\mathbf{180^0}$, then the maximum number of moves $\boldsymbol{n_2}$ of the interference fringes is:

$$n_2 = \frac{2vd}{c\lambda} = \frac{2 \times 0.1 \times 8 \times 10^{-4}}{6.328 \times 10^{-7}} = 252.8 \qquad (11.3.19)$$

If the interferometer is rotated $\mathbf{90^0}$, then the maximum number of moves $\boldsymbol{n_{2\perp}}$ of the interference fringes is:

$$n_{2\perp} = \frac{1}{2} n_2 = 126.4 \qquad (11.3.20)$$

If the interference fringes move towards the center of the screen during 12 hours, then after 12 hours the interference fringes will move outward from the center of the screen. The maximum number of moves $\boldsymbol{n_{2\perp}/12 = 21}$ of the interference fringes can be obtained from the above equation.

The third ground speed $\boldsymbol{v}$ is the speed of the Earth's revolution around the Sun. According to astronomers' calculations, the average speed of the Earth's revolution around the Sun is $\mathbf{30}$ kilometers per second. The earth's velocity $\boldsymbol{v}$ is one ten-thousandth of the speed of light $\boldsymbol{c}$, i.e., $\boldsymbol{v} = \mathbf{1 \times 10^{-4}}\boldsymbol{c}$. Substitute the earth's velocity $\boldsymbol{v}$ into equation (11.3.12). In the Newtonian ground frame of reference, if the front and rear aperture laser interferometers are rotated $\mathbf{180^0}$, then the maximum number of moves $\boldsymbol{n_3}$ of the interference fringes is:

$$n_3 = \frac{2vd}{c\lambda} = \frac{2 \times 0.1 \times 10^{-4}}{6.328 \times 10^{-7}} = 31.6 \qquad (11.3.21)$$

Since the velocity $\boldsymbol{v}$ of the ground frame of reference in the cosmic vacuum frame of reference cannot have more than one value at the same time, only one of the calculations $\boldsymbol{n}$, $\boldsymbol{n_2}$, and $\boldsymbol{n_3}$ is correct. In other words, only one of the three calculations is the velocity in the cosmic vacuum frame of reference, and the other two calculations are wrong.

Since the cosmic background radiation (CMB) is absolutely stationary, the maximum number of moves $\boldsymbol{n}$ of the interference fringes in the Newtonian ground frame of reference should be approximated by the following values:

$$n = \frac{2vd}{c\lambda} = \frac{2 \times 0.1 \times 20.9 \times 10^{-4}}{6.328 \times 10^{-7}} = 660.5$$

Note that the prerequisites for obtaining the above calculations are that the distance between the reflecting mirrors in the A and B planes portions of the resonant cavities of the front and rear aperture lasers is $\boldsymbol{d} = \mathbf{0.1}\boldsymbol{m}$, and that the interferometric experiments are performed using a helium-neon laser as the light source.

11.4 Experimental results of Einsteinian ground reference frame predictions

Note that the number of moves $\boldsymbol{n'''}$ of interference fringes is calculated below using the formula $\boldsymbol{c'} = \boldsymbol{c}$ for the invariance principle of the speed of light in special relativity.

11.4.1 Optical path difference value $\Delta L_{A\leftarrow}'''$ when the initial direction of the light beam A is opposite to the direction of the velocity v

In a cosmic vacuum frame of reference, assume that the ground is moving to the right with velocity $\boldsymbol{v}$. Since the front and rear aperture laser interferometers are stationary on the ground, the interferometers are also moving to the right with velocity $\boldsymbol{v}$. Assume that C, D, and E are planar reflectors and R is a transreflector.

If the length of the resonant cavity of the laser tube $\boldsymbol{d} = \boldsymbol{AB}$ is parallel to the direction of the velocity $\boldsymbol{v}$, then light beam A and light beam B are also parallel to the direction of the velocity $\boldsymbol{v}$. At this point, the optical path length $\boldsymbol{k} = \boldsymbol{AC} = \boldsymbol{BE}$ is parallel to the direction of the velocity $\boldsymbol{v}$, and the optical path length $\boldsymbol{h} = \boldsymbol{CD} = \boldsymbol{ER}$ is perpendicular to the velocity $\boldsymbol{v}$. This is shown in Figure 11-7.

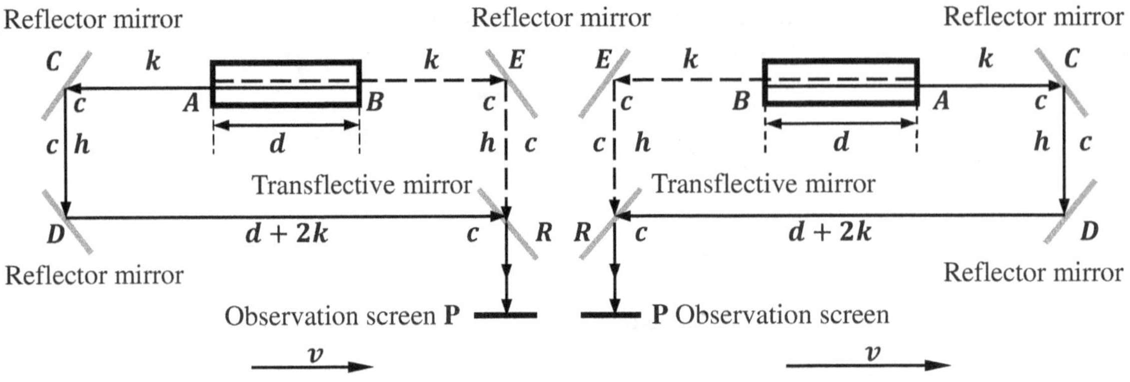

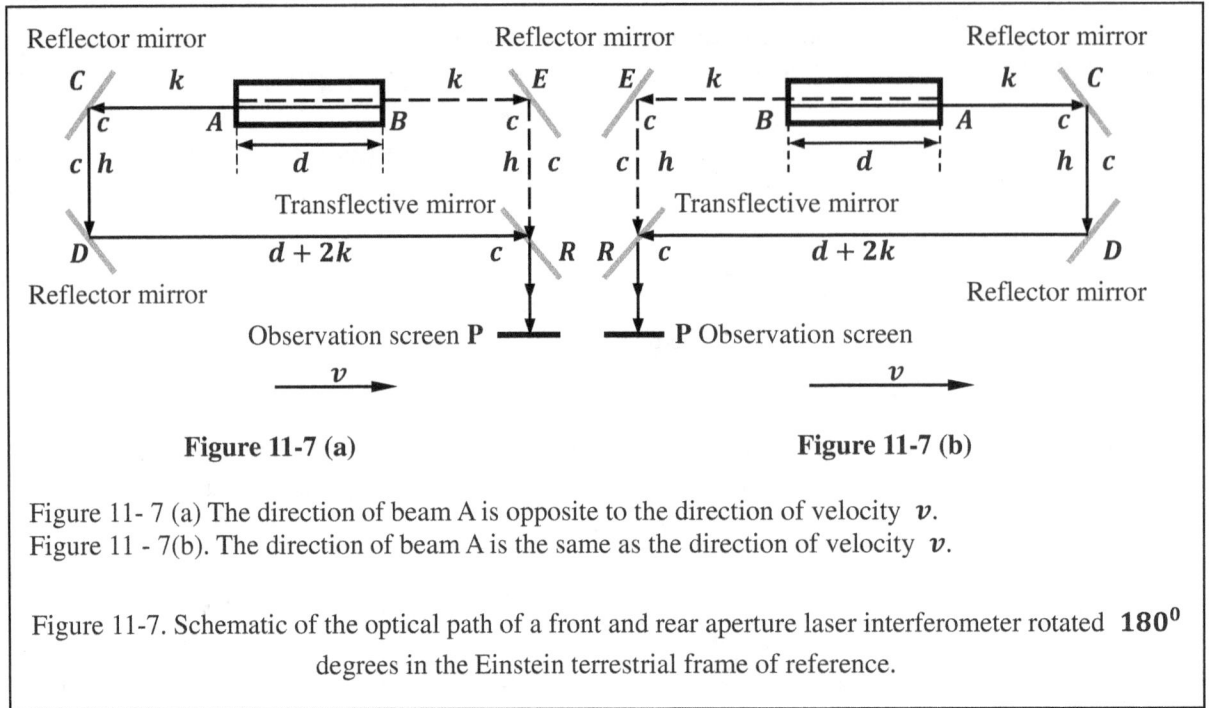

Figure 11-7 (a) **Figure 11-7 (b)**

Figure 11- 7 (a) The direction of beam A is opposite to the direction of velocity $\boldsymbol{v}$.
Figure 11 - 7(b). The direction of beam A is the same as the direction of velocity $\boldsymbol{v}$.

Figure 11-7. Schematic of the optical path of a front and rear aperture laser interferometer rotated $\mathbf{180^0}$ degrees in the Einstein terrestrial frame of reference.

In both the cosmic vacuum frame of reference and the Newtonian ground frame of reference, both light beam A and light beam B propagate vertically upward and downward with a speed of $\sqrt{c^2 - v^2}$. However, in Einstein's ground frame of reference, both light beam A and light beam B propagate vertically upward and downward at the speed $\boldsymbol{c}$ of light.

Since the observer is stationary with respect to the interferometer, there is no length contraction effect on the optical path in the interference experiment according to special relativity, i.e:

$$d' = d, \qquad k' = k, \qquad h' = h$$

In Einstein's ground frame of reference, the speed of propagation of the light beam A and the speed of light B on each route is equal to the speed of light $\boldsymbol{c}$. If the initial direction of the light beam A is opposite to the direction of the speed v, then the time $t'''_{A\leftarrow}$ for the light beam A to reach the transmissive mirror sub R is:

$$t'''_{A\leftarrow} = \frac{d+k}{c} + \frac{h}{c} + \frac{d+2k}{c} \tag{11.4.1}$$

The optical path length $L'''_{A\leftarrow}$ of light beam A reaching the transreflector mirror R is:

$$L'''_{A\leftarrow} = ct'''_{A\leftarrow} = 2d + 3k + h \tag{11.4.2}$$

In Einstein's ground frame of reference, the time $t'''_{B\rightarrow}$ for the light beam B to reach the transmissive mirror R is:

$$t'''_{B\rightarrow} = \frac{d+k}{c} + \frac{h}{c} \tag{11.4.3}$$

The optical path length $L'''_{B\rightarrow}$ of the light beam B reaching the transreflector mirror R is:

$$L'''_{B\rightarrow} = ct'''_{B\rightarrow} = d + k + h \tag{11.4.4}$$

In the Einsteinian ground frame of reference, if the initial direction of light beam A is opposite to the direction of velocity v, then the optical path length difference value $\Delta L'''_{A\leftarrow}$ between light beam A and light beam B arriving at the transreflector mirror R is:

$$\Delta L'''_{A\leftarrow} = L'''_{A\leftarrow} - L'''_{B\rightarrow} = d + 2k \tag{11.4.5}$$

Note that the optical path length difference value $\Delta L'''_{A\leftarrow}$ contains the light speed invariance principle.

11.4.2 Maximum number of moves of interference fringes n''' in a Einsteinian ground frame of reference

If the front and rear aperture laser interferometers are rotated by $\mathbf{180^0}$, the initial direction of light beam A is the same as the direction of velocity $\boldsymbol{v}$, and the initial direction of light beam B is opposite to the direction of velocity $\boldsymbol{v}$.

In the Einsteinian ground frame of reference, if the initial direction of the light beam A is the same as the direction of the velocity $\boldsymbol{v}$, then the time $t'''_{A\rightarrow}$ for the light beam A to reach the transmissive mirror R is:

$$t'''_{A\rightarrow} = \frac{d+k}{c} + \frac{h}{c} + \frac{d+2k}{c} \qquad (11.4.6)$$

The optical path length $L'''_{A\rightarrow}$ of light beam A reaching the transreflector mirror R is:

$$L'''_{A\rightarrow} = ct'''_{A\rightarrow} = 2d + 3k + h \qquad (11.4.7)$$

In the Einsteinian ground frame of reference, if the initial direction of the light beam B is opposite to the velocity $\boldsymbol{v}$, then the time $t'''_{B\leftarrow}$ for the light beam B to reach the transmissive mirror sub R is:

$$t'''_{B\leftarrow} = \frac{d+k}{c} + \frac{h}{c} \qquad (11.4.8)$$

The optical path length $L'''_{B\leftarrow}$ of the light beam B reaching the transreflector mirror R is:

$$L'''_{B\leftarrow} = ct'''_{B\leftarrow} = d + k + h \qquad (11.4.9)$$

In the Einsteinian ground frame of reference, if the initial direction of light beam A is the same as the direction of velocity $\boldsymbol{v}$, then the optical path length difference value $\Delta L'''_{A\rightarrow}$ between light beam A and light beam B when they arrive at the transreflector mirror R is:

$$\Delta L'''_{A\rightarrow} = L'''_{A\rightarrow} - L'''_{B\leftarrow} = d + 2k \qquad (11.4.10)$$

When the interferometer is rotated horizontally by 180^0, is the difference value $\Delta L'''_{\leftrightarrow}$ between the optical path length difference value $\Delta L'''_{A\rightarrow}$ of the above equation and the optical path length difference value $\Delta L'''_{A\leftarrow}$ of equation (11.4.5).

$$\Delta L'''_{\leftrightarrow} = \Delta L'''_{A\rightarrow} - \Delta L'''_{A\leftarrow} = (d+2k) - (d+2k) = 0 \qquad (11.4.11)$$

In the Einsteinian ground frame of reference, if the interferometer is rotated 180^0, then the maximum number of moves n''' of interference fringes is:

$$n''' = \frac{\Delta L'''_{\leftrightarrow}}{\lambda} = 0 \qquad (11.4.12)$$

Comparing the above equation with equation (11.3.17), if the interferometer is rotated by 180^0, then the maximum number of moves of the interference fringes in Einstein's ground frame of reference is $n''' = 0$, whereas in Newton's ground frame of reference, the maximum number of moves of the interference fringes is $n_1 = 660.5$.

11.4.3 If the interferometer is rotated 90^0, then the maximum number of movements of the interference fringes is $n_\perp'''$

If the interferometer is rotated 90^0, then the direction of light beam A and light beam B are both perpendicular to the direction of velocity $\boldsymbol{v}$, as shown in Figure 11-8.

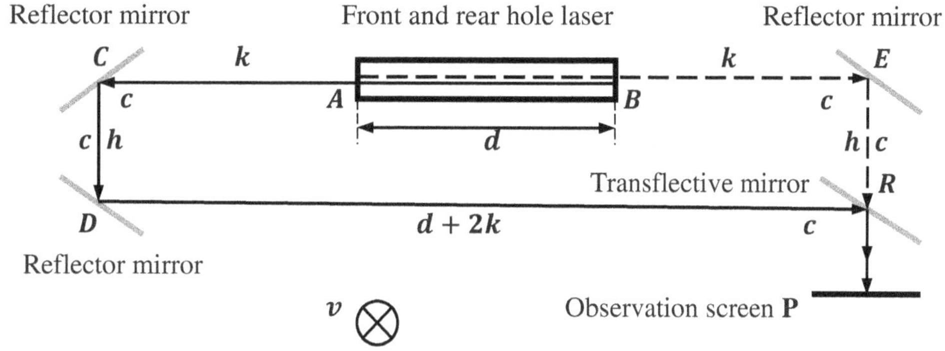

Figure 11-8. Schematic of the optical paths of beam A and beam B perpendicular to the velocity v in the Einstein ground reference system.

Since the observer is stationary with respect to the interferometer, there is no length contraction effect on the optical path in the interference experiment according to special relativity, i.e.

In Einstein's ground frame of reference, the time $t'''_{A\downarrow}$ for the light beam A to reach the transmissive mirror R is:

$$t'''_{A\downarrow} = \frac{d+k}{c} + \frac{h}{c} + \frac{d+2k}{c} \tag{11.4.13}$$

The optical path length $L'''_{A\downarrow}$ of light beam A arriving at the transreflector mirror R is:

$$L'''_{A\downarrow} = ct'''_{A\downarrow} = 2d + 3k + h \tag{11.4.14}$$

In the Einsteinian ground frame of reference, the time $t'''_{B\downarrow}$ for the light beam B to reach the transmissive mirrors R is:

$$t'''_{B\downarrow} = \frac{d+k}{c} + \frac{h}{c} \tag{11.4.15}$$

The optical path length $L'''_{B\downarrow}$ of light beam B arriving at the transreflector mirror R is:

$$L'''_{B\downarrow} = ct'''_{B\downarrow} = d + k + h \tag{11.4.16}$$

In the Einsteinian ground frame of reference, if light beam A and light beam B are perpendicular to the velocity $\boldsymbol{v}$, then the difference value in optical path length $\Delta L'''_{AB\downarrow}$ between light beam A and light beam B as they arrive at the transmissive mirror R is:

$$\Delta L'''_{AB\downarrow} = L'''_{A\downarrow} - L'''_{B\downarrow} = d + 2k \tag{11.4.17}$$

If the interferometer is rotated 90^0, then the difference value $\Delta L'''_{\perp}$ between the above optical path length difference value $\Delta L'''_{AB\downarrow}$ and the optical path length difference value $\Delta L'''_{A\rightarrow}$ of equation (11.4.10) is:

$$\Delta L'''_{\perp} = \Delta L'''_{AB\downarrow} - \Delta L'''_{A\rightarrow} = (d+2k) - (d+2k) = 0 \tag{11.4.18}$$

If the interferometer is rotated 90^0, then the maximum number of shifts of the interference fringes $n'''_{\perp}$ is:

$$n'''_{\perp} = \frac{\Delta L'''_{\perp}}{\lambda} = 0 \tag{11.4.19}$$

To summarize, rotating the front and rear aperture laser interferometers in all directions in space, the principle of invariance of the speed of light in special relativity is correct if the number of moves $n'''_{\perp} = 0$ of the interference fringes observed on the observing screen P.

On the contrary, the principle of invariance of the speed of light in special relativity is false if the maximum number of moves $n = 660.5$ of the interference fringes, is observed on the observation screen P.

11.5 Verification of the principle of invariance of the speed of light with a two-aperture laser interferometer

11.5.1 Structure and optical path of a double-aperture laser interferometer

The principle of invariance of the speed of light in special relativity can also be verified with a double-aperture laser interferometer. The double-aperture laser is the core component of the interferometer. The double-aperture laser consists of two glass capillary discharge tubes, planar partially reflecting mirrors and planar fully reflecting mirrors, a helium-neon gas mixture, and a power supply. Its structure is shown in Figure 11-9.

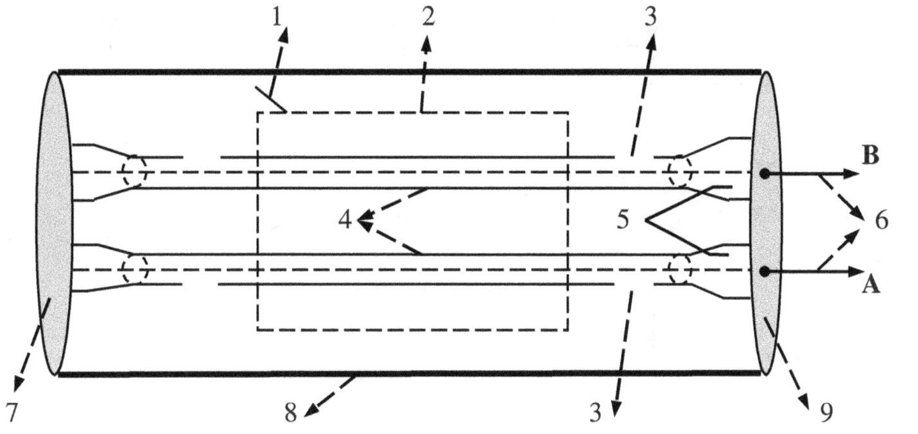

Figure 11-9 Schematic diagram of the structure of a two-hole laser.

Figure 11-9 Numbering instructions.

1. cathode.
2. Hollow metal tube cathode;
3. helium-neon gas mixture;
4. capillary glass discharge tube;
5. anode tungsten rod;
6. an output laser;
7. a planar totally reflective mirror;
8. gas storage round tube;
9. planar partially reflecting mirror.

Illustration of Figure 11-9.

1) A double-aperture laser contains two capillary glass discharge tubes of identical dimensions.

2) Two capillary glass discharge tubes share a hollow metal tube cathode. The two anodes are tungsten rods of the same size. Since the cathode and anode of the two glass discharge tubes are identical, the voltage experienced by the two glass discharge tubes is equal.

3) The center axes of the two capillary glass discharge tubes are parallel.

4) The reflecting surface of the planar total reflection mirror is parallel to the planar partial reflection mirror. The two reflecting surfaces are strictly perpendicular to the center axis of the capillary glass discharge tubes.

5) Since the two capillary glass discharge tubes are located in the same gas storage tube, the proportion of helium-neon gas mixture in the two capillary glass discharge tubes is exactly equal.

6) A, B two output holes output laser frequency $\boldsymbol{v}$ completely equal. This technical index is the key index to ensure the success of the new optical experiment. Theoretically speaking:Since the resonant cavity length L of the two capillary glass discharge tubes in the double-aperture laser is equal, the laser frequency $\boldsymbol{v}$ in the two capillary glass discharge tubes should be equal.

The optical path of a dual-aperture laser interferometer is shown in Figure 11-10.

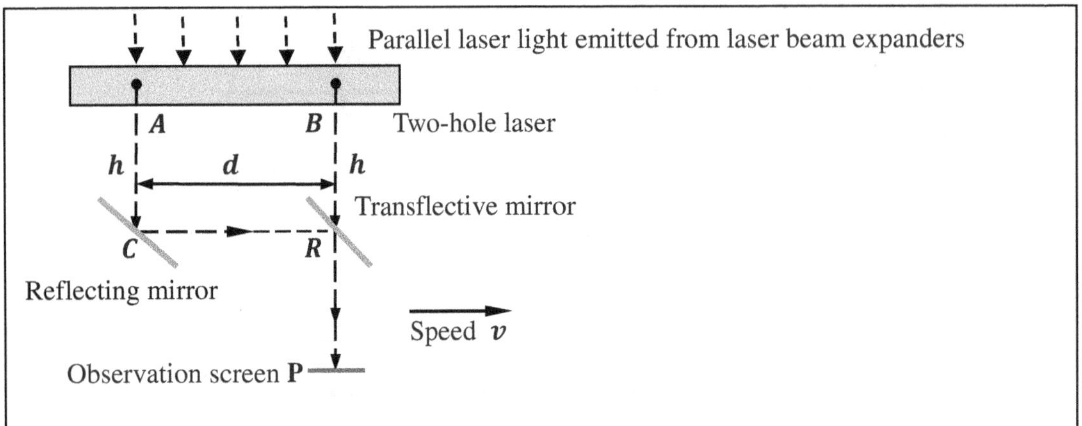

Figure 11-10 Schematic diagram of the structure and optical path of the two-hole laser interferometer.

A and B in Figure 11-10 are laser output holes. The optical path of light beam A to the observation screen P is A→C→R→P, and the optical path of light beam B to the observation screen P is B→R→P.

In Figure 11-10, light path C→R is perpendicular to light path B→R. d is the distance between holes A and B, and h is the length of light path A→C, which is also the length of light path B→R. light beam A is reflected at point C to point R, where it is reflected back to the viewing screen P. light beam B reaches the viewing screen P directly through the transreflector mirror R.

Beam A and light beam B arrive at the transreflector mirror R and then propagate simultaneously toward the observation screen P. At this time, light beam A and light beam B are coherent light since light beam A and light beam B have the same frequency and direction of vibration and the phase difference between them is constant. It is thus determined that the unidirectional light beam A and the unidirectional light beam B produce interference fringes on the observation screen P.

It should be noted that. Since the laser beam expander is capable of expanding the diameter of the laser beam, laser **A** and laser **B** in Figure 11-10 may also be parallel lasers output from the laser beam expander. As shown in Figure 11-11.

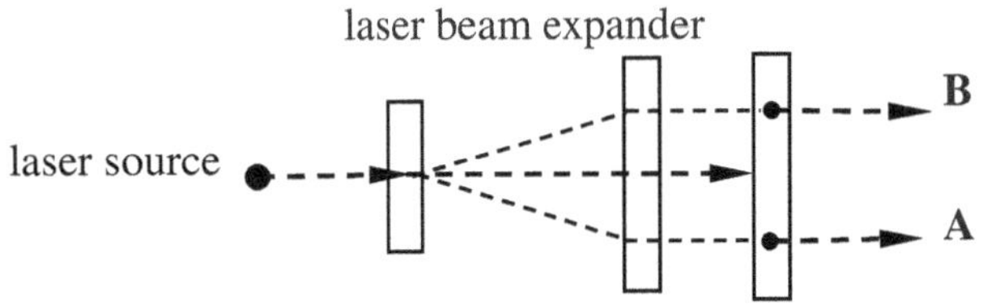

Figure 11-11 Schematic diagram of parallel laser A and laser B output from a laser beam expander mirror.

11.5.2 Optical path length difference value $\Delta L_{A\rightarrow}$ when the direction of light beam A is the same as the direction of velocity v

Note that the Galilean velocity transformation formula $c' = c \pm v$ is used below to calculate the number n of moves of the interference fringes.

In the cosmic vacuum frame of reference, assume that the ground is moving to the right with velocity v. Since the double-aperture laser interferometer is stationary on the ground, the interferometer is also moving to the right with velocity v in the cosmic vacuum.

If light beam A is parallel to the velocity $\boldsymbol{v}$ after reflection at point C, then the optical path length $\boldsymbol{d} = \boldsymbol{CR}$ is parallel to the velocity $\boldsymbol{v}$ and the optical path length $\boldsymbol{h} = \boldsymbol{AC} = \boldsymbol{BR}$ is perpendicular to the velocity $\boldsymbol{v}$. This is shown in Figure 11-10.

In the Newtonian ground frame of reference, if the direction of the light beam A is the same as the direction of the velocity v, then the propagation velocity $c'_{A\rightarrow}$ of the light beam A on the path C→R is:

$$c'_{A\rightarrow} = c - v$$

The time $t_{A\rightarrow}$ for light beam A to reach the transreflector mirror R is:

$$t_{A\rightarrow} = \frac{h}{\sqrt{c^2 - v^2}} + \frac{d}{c - v} \qquad (11.5.1)$$

The optical path length $L_{A\rightarrow}$ of the rightward propagating light beam A when it reaches the transreflecting mirror R is:

$$L_{A\rightarrow} = ct_{A\rightarrow} = c\left(\frac{h}{\sqrt{c^2 - v^2}} + \frac{d}{c - v}\right) \qquad (11.5.2)$$

The time $t_{B\downarrow}$ for the light beam B perpendicular to the velocity $\boldsymbol{v}$ to reach the transreflector mirror R is:

$$t_{B\downarrow} = \frac{h}{\sqrt{c^2 - v^2}} \qquad (11.5.3)$$

The optical path length $L_{B\downarrow}$ of light beam B reaching the transreflector mirror R is:

$$L_{B\downarrow} = ct_{B\downarrow} = \frac{ch}{\sqrt{c^2 - v^2}} \qquad (11.5.4)$$

In a Newtonian ground frame of reference, if the direction of light beam A is the same as the direction of velocity $\boldsymbol{v}$, then the optical path length difference value $\Delta L_{A\rightarrow}$ between light beam A and light beam B when they reach the transreflector mirror R is:

$$\Delta L_{A\rightarrow} = L_{A\rightarrow} - L_{B\downarrow} = \frac{cd}{c - v} \qquad (11.5.5)$$

Note that the light path length difference value $\Delta L_{A\rightarrow}$ does not contain the light speed invariance principle.

11.5.3 Optical path length difference value $\Delta L_{A\leftarrow}$ when the direction of light beam A is opposite to the direction of velocity *v*

If the interferometer is rotated $\mathbf{180^0}$, then the direction of the light beam A is opposite to the velocity $\boldsymbol{v}$. In the Newtonian ground frame of reference, the propagation velocity $c'_{A\leftarrow}$ of light beam A on the path C→R is:

$$c'_{A\leftarrow} = c + v$$

The time $t_{A\leftarrow}$ for light beam A to reach the transreflector mirror R is:

$$t_{A\leftarrow} = \frac{h}{\sqrt{c^2 - v^2}} + \frac{d}{c + v} \qquad (11.5.6)$$

The optical path length $L_{A\leftarrow}$ of the leftward propagating light beam A when it reaches the transreflecting mirror R is:

$$L_{A\leftarrow} = ct_{A\leftarrow} = c\left(\frac{h}{\sqrt{c^2 - v^2}} + \frac{d}{c + v}\right) \qquad (11.5.7)$$

The time $t_{B\downarrow}$ for the light beam B perpendicular to the velocity $\boldsymbol{v}$ to reach the transreflector mirror R is:

$$t_{B\downarrow} = \frac{h}{\sqrt{c^2 - v^2}} \qquad (11.5.8)$$

The optical path length $L_{B\downarrow}$ of light beam B reaching the transreflector mirror R is:

$$L_{B\downarrow} = ct_{B\downarrow} = \frac{ch}{\sqrt{c^2 - v^2}} \qquad (11.5.9)$$

In the Newtonian ground frame of reference, if the direction of light beam A is opposite to the direction of velocity $\boldsymbol{v}$, then the optical path length difference value $\Delta L_{A\leftarrow}$ between light beam A and light beam B when they reach the transreflector mirror R is:

$$\Delta L_{A\leftarrow} = L_{A\leftarrow} - L_{B\downarrow} = \frac{cd}{c+v} \qquad (11.5.10)$$

Note that the light path length difference value $\Delta L_{A\leftarrow}$ does not contain the light speed invariance principle and the length contraction effect.

11.5.4 Experimental results predicted by the Newtonian ground frame of reference

In the Newtonian ground frame of reference, if the interferometer is rotated horizontally 180^0, then the difference value $\Delta L_{\leftrightarrow}$ between the optical path length difference value $\Delta L_{A\rightarrow}$ of equation (11.5.5) and the optical path length difference value $\Delta L_{A\leftarrow}$ of equation (11.5.10) is:

$$\Delta L_{\leftrightarrow} = \Delta L_{A\rightarrow} - \Delta L_{A\leftarrow} = \frac{cd}{c-v} - \frac{cd}{c+v} = \frac{2cvd}{c^2-v^2} \approx \frac{2vd}{c} \qquad (11.5.11)$$

Note that the difference value $\Delta L_{\leftrightarrow}$ does not include the principle of invariance of the speed of light. In the Newtonian ground frame of reference, the maximum number of moves n of the interference fringes is:

$$n = \frac{\Delta L_{\leftrightarrow}}{\lambda} = \frac{2vd}{c\lambda} \qquad (11.5.12)$$

Note that the maximum value n does not include the light-speed invariance principle.

Assuming that the distance between output aperture A and output aperture B is $d = 0.01m$. If a HeNe laser is used as the light source for the interference experiment, the wavelength of the laser $\lambda = 6.328 \times 10^{-7}m$. According to astronomers' calculations, the velocity of the Earth with respect to the Cosmic Background Radiation (CMB) is $v \approx 627 \cdot km/s = 20.9 \times 10^{-4}c$.

In the Newtonian ground frame of reference, if the interferometer is rotated 180^0, then according to equation (11.5.12), the maximum number of moves of the interference fringes n_1 is:

$$n_1 = \frac{2vd}{c\lambda} = \frac{2 \times 0.01 \times 20.9 \times 10^{-4}c}{6.328 \times 10^{-7}c} = 66.0 \qquad (11.5.13)$$

If the interferometer is rotated 90^0, the maximum number of moves $n_{1\perp}$ of the interference fringes is:

$$n_{1\perp} = \frac{1}{2}n_1 = 33 \qquad (11.5.14)$$

In order to minimize the error generated by the rotation of the double-aperture interferometer, the interferometer can be fixed and the number of moves of the interference fringes can be observed by diurnal variation. If the interference fringes move towards the center of the screen during 12 hours, then after 12 hours the interference fringes will move outward from the center of the screen. From equation (11.5.13) the maximum number of moves per hour of the interference fringe $n_1/12 = 5.5$.

According to astronomers' observational calculations, the speed $v_O = 250km/s$ of the solar system's revolution around the galactic center. If the Earth's rotation and revolution speeds are ignored, the Earth's average speed v should be equal to the speed v_O, i.e. $v = v_O \approx 8 \times 10^{-4}c$.

Substitute the velocity $v = 8 \times 10^{-4}c$ of the Sun's revolution around the galactic center, into equation (11.5.12). In the Newtonian ground frame of reference, if the interferometer is rotated 180^0, then the maximum number of moves n_2 of the interference fringes is:

$$n_2 = \frac{2vd}{c\lambda} = \frac{2 \times 0.01 \times 8 \times 10^{-4}}{6.328 \times 10^{-7}} = 25.3 \qquad (11.5.15)$$

If the interferometer is rotated 90^0, then the maximum number of moves $n_{2\perp}$ of the interference fringes is:

$$n_{2\perp} = \frac{1}{2}n_2 = 12.6$$

According to astronomers' calculations, the average speed of the Earth's revolution around the Sun is **30** kilometers per second. The earth's velocity v is one ten-thousandth of the speed of light c, i.e., $v = 1 \times 10^{-4}c$. Substitute the earth's velocity v into equation (11.5.12). In the Newtonian ground frame of reference, if the front and rear aperture laser interferometers are rotated 180^0, then the maximum number of moves n_3 of the interference fringes is:

$$n_3 = \frac{2vd}{c\lambda} = \frac{2 \times 0.01 \times 10^{-4}}{6.328 \times 10^{-7}} = 3.16 \qquad (11.5.16)$$

Since the velocity v of the ground frame of reference in the cosmic vacuum frame of reference cannot have more than one value at the same time, only one of the calculations n, n_2, and n_3 is correct. In other words, only one of the three calculations is the velocity in the cosmic vacuum frame of reference, and the other two calculations are wrong.

Since the cosmic background radiation (CMB) is absolutely stationary, the maximum number of moves n of the interference fringes in the Newtonian ground frame of reference should be approximated by the following values:

$$n = \frac{2vd}{c\lambda} = \frac{2 \times 0.01 \times 20.9 \times 10^{-4}}{6.328 \times 10^{-7}} = 66$$

Note that the prerequisites for obtaining the above calculations are that the distance between output aperture A and output aperture B is $d = 0.01m$ and that the interference experiments use a He-Ne laser as the light source.

11.5.5 Experimental results of Einsteinian ground reference frame predictions

In Einstein's ground frame of reference, the optical path length $L'''_{A\rightarrow}$ of the light beam A as it propagates from the reflector C to the right to the transreflector R is:

$$L'''_{A\rightarrow} = c\frac{d}{c} = d$$

If the double-aperture interferometer is rotated 180^0, then the optical path length $L'''_{A\leftarrow}$ of the light beam A propagating leftward from the reflector C to reach the transreflector R is:

$$L'''_{A\leftarrow} = c\frac{d}{c} = d$$

The difference value $\Delta L'''_{A\leftrightarrow}$ between the optical path length $L'''_{A\rightarrow}$ and the optical path length $L'''_{A\leftarrow}$ is:

$$\Delta L'''_{A\leftrightarrow} = L'''_{A\rightarrow} - L'''_{A\leftarrow} = d - d = 0 \qquad (11.5.17)$$

The maximum number of moves n''' of interference fringes in the Einsteinian ground reference frame is:

$$n''' = \frac{\Delta L'''_{A\leftrightarrow}}{\lambda} = 0 \qquad (11.5.18)$$

Comparing the above results with those of equation (11.5.13), the maximum number of moves of the interference fringes in the Newtonian ground frame of reference is $n_1 = 66$.

If the double-aperture interferometer is rotated 90^0, then the light beam A is perpendicular to the velocity v. At this point, the optical path length $L'''_{A\downarrow}$ of the light beam A from the reflector C to the transreflector R is:

$$L'''_{A\downarrow} = c\frac{d}{c} = d$$

If the double-aperture interferometer is rotated 90^0, then the difference value $\Delta L'''_{\perp}$ between the optical path length $L'''_{A\downarrow}$ and the optical path length $L'''_{A\rightarrow}$ is:

$$\Delta L'''_{\perp} = L'''_{A\downarrow} - L'''_{A\rightarrow} = d - d = 0 \qquad (11.5.19)$$

Note that the difference value $\Delta L'''_{\perp}$ incorporates the principle of invariance of the speed of light. In the Einsteinian ground frame of reference, if the interferometer is rotated by 90^0, then the maximum number of shifts of the interference fringes $n'''_{\perp}$ is:

$$n'''_{\perp} = \frac{\Delta L'''_{\perp}}{\lambda} = 0 \qquad (11.5.21)$$

To summarize, rotate the two-aperture interferometer in all directions in space, and the principle of invariance of the speed of light in special relativity is correct if the number of moves $n'''_{\perp} = 0$ of the interference fringes observed on the observing screen P.

On the contrary, the principle of invariance of the speed of light in special relativity is false if the maximum number of moves $n = 66.05$ of the interference fringes, is observed on the observation screen P.

Chapter 12 Verification of the speed-of-light invariance principle for special relativity with long and short optical path Michelson interferometers

Introduction:Since the establishment of special relativity more than 100 years ago, no one has verified the principle of the invariance of the speed of light in special relativity with the Michelson interferometer with long and short light paths. Since the experimental results predicted by the Newtonian ground reference frame and the Einstein ground reference frame are different, we can use that experiment to verify that the principle of invariance of the speed of light is wrong.

12.1 The zero result of the Michelson-Morley experiment is independent of the velocity *v*

Note that the number of moves n of the interference fringes is calculated according to the Galilean velocity transformation formula $c' = c \pm v$.

12.1.1 Lengths of transverse and longitudinal optical paths for the Michaelison-Morley experiment

The Michelson interferometer is a precision optical instrument invented by the American physicist Michelson to prove the existence of "Ether". It consists of a light source S, a beam splitter O, two mutually perpendicular plane reflecting mirrors m_1 and m_2, and an observation screen P.

The Michelson-Morley experiment is an interference experiment in which the length of the longitudinal optical path is equal to the length of the transverse optical path, and the experimental optical path is shown in Figure 12-1.

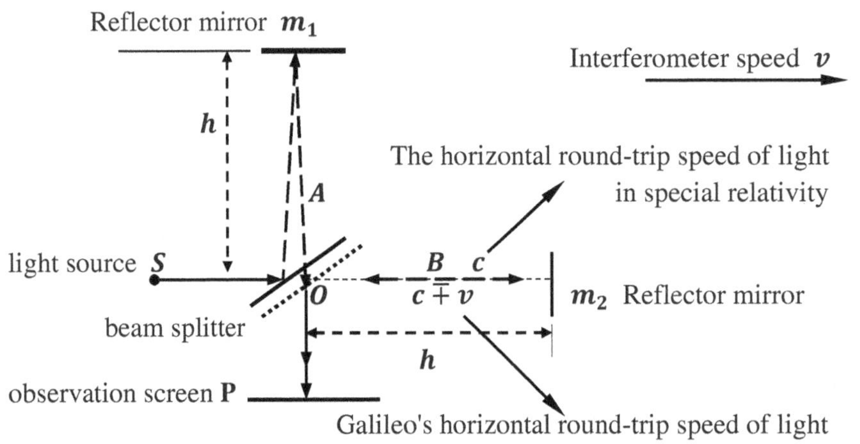

Figure 12-1 Optical path diagram of the Michelson-Morley experiment.

The experimental procedure is that the light beam S is split into a longitudinal light beam A and a transverse light beam B at point O of the beam splitter. The longitudinal light beam A propagates up and down from point O of the beam splitter, returns to point O and propagates toward the observation screen P. The transverse light beam B propagates back and forth horizontally from the point O of the beam splitter, returns to the point O and propagates toward the observation screen P.

The longitudinal light beam A and the transverse light beam B produce interference fringes on the observation screen P. The rotating interferometer observes the number $\boldsymbol{n}$ of shifts of the interference fringes on the observation screen P to determine whether Newtonian absolute spacetime exists.

However, since the experimental result is always equal to 0, the Michelson-Morley experiment cannot prove that Newtonian absolute spacetime exists objectively. We can prove that Newtonian absolute spacetime exists objectively by using interference experiments with long and short light paths.

In the cosmic vacuum frame of reference, the ground is assumed to be moving to the right at velocity $\boldsymbol{v}$. Since the Michelson interferometer is at rest on the ground, velocity $\boldsymbol{v}$ is also the velocity of the interferometer in the cosmic vacuum.

In both the cosmic vacuum frame of reference and the Newtonian ground frame of reference, both light beam A propagate vertically upward and downward with a speed of $\sqrt{c^2 - v^2}$. However, in Einstein's ground frame of reference, both light beam A propagate vertically upward and downward at the speed $\boldsymbol{c}$ of light.

Assume that $\boldsymbol{h}$ is the distance from point O of the beam splitter to the plane reflector mirrors $\boldsymbol{m_1}$ and $\boldsymbol{m_2}$, and that $\boldsymbol{t_{A\uparrow}}$ is the time required for the longitudinal light beam A to reach the plane mirror $\boldsymbol{m_1}$ from point O upward. The vertical distance $\boldsymbol{h}$ between the point O of the beam splitter and the plane mirror $\boldsymbol{m_1}$ is:

$$h^2 = (ct_{A\uparrow})^2 - (vt_{A\uparrow})^2 = t_{A\uparrow}^2(c^2 - v^2) \qquad (12.1.1)$$

In the Newtonian ground frame of reference, the time $\boldsymbol{t_{A\uparrow}}$ required for the longitudinal light beam A to travel upward from point O to point $\boldsymbol{m_1}$ is:

$$t_{A\uparrow} = \frac{h}{\sqrt{c^2 - v^2}} \qquad (12.1.2)$$

The time required for light beam A to return from point $\boldsymbol{m_1}$ to point O is also equal to $\boldsymbol{t_{A\uparrow}}$. The time $\boldsymbol{t_{A\updownarrow}}$ required for light beam A to travel upward from point O to point $\boldsymbol{m_1}$ and return to point O is:

$$t_{A\updownarrow} = 2t_{A\uparrow} = \frac{2h}{\sqrt{c^2 - v^2}} \qquad (12.1.3)$$

In the Newtonian ground frame of reference, the optical path length $\boldsymbol{L_{A\updownarrow}}$ possessed by light beam A as it travels upward from point O to point $\boldsymbol{m_1}$ and back to point O is:

$$L_{A\updownarrow} = ct_{A\updownarrow} = \frac{2ch}{\sqrt{c^2 - v^2}} \qquad (12.1.4)$$

In the classical theory, $\boldsymbol{L_{A\updownarrow}}$ is the length of the longitudinal optical path in the Michelson-Morley experiment.

Since the Michelson interferometer is moving to the right in the universe with velocity $\boldsymbol{v}$, the light beam B propagates to the right in the Newtonian ground frame of reference with velocity $\boldsymbol{c - v}$ from point O. The time $\boldsymbol{t_{B\rightarrow}}$ required for the light beam B to reach the plane mirror $\boldsymbol{m_2}$ from point O to the right is:

$$t_{B\rightarrow} = \frac{h}{c - v}$$

The velocity of light beam B returning to point O from point $\boldsymbol{m_2}$ of the plane mirror is $\boldsymbol{c + v}$. The time $\boldsymbol{t_{B\leftarrow}}$ required for light beam B to return to point O from point $\boldsymbol{m_2}$ is:

$$t_{B\leftarrow} = \frac{h}{c + v}$$

The time $\boldsymbol{t_{B\leftrightarrow}}$ required for light beam B to reach point $\boldsymbol{m_2}$ from point O to the right and return to point O from point $\boldsymbol{m_2}$ is:

$$t_{B\leftrightarrow} = t_{B\rightarrow} + t_{B\leftarrow} = \frac{h}{c - v} + \frac{h}{c + v} = \frac{2ch}{c^2 - v^2} \qquad (12.1.5)$$

In the Newtonian ground frame of reference, the optical path length $\boldsymbol{L_{B\leftrightarrow}}$ possessed by light beam B as it travels from point O to the right to point m_2 and then from point m_2 back to point O is:

$$L_{B\leftrightarrow} = ct_{B\leftrightarrow} = \frac{2c^2h}{c^2 - v^2} \qquad (12.1.6)$$

In the classical theory, $L_{B\leftrightarrow}$ is the length of the transverse optical path in the Michelson-Morley experiment.

12.1.2 Maximum number of moves of interference fringes in the Newtonian ground frame of reference *n*

In the Michelson-Morley experiment, if the light beam B is parallel to the velocity v, then the difference $\Delta L_{B\leftrightarrow}$ between the optical path length $L_{B\leftrightarrow}$ in the transverse direction and $L_{A\updownarrow}$ in the longitudinal direction is:

$$\Delta L_{B\leftrightarrow} = L_{B\leftrightarrow} - L_{A\updownarrow} = \frac{2c^2h}{c^2 - v^2} - \frac{2ch}{\sqrt{c^2 - v^2}} \qquad (12.1.7)$$

If the Michelson interferometer is rotated by 90^0, then the two light paths A and B exchange their states with respect to the direction of the velocity v, i.e., light path A becomes a transverse light path and light path B becomes a longitudinal light path.

When the transverse light beam A reaches the point m_1 from the point O to the right and then returns to the point O from the point m_1, the optical path length $L_{A\leftrightarrow}$ possessed by the transverse light beam A is:

$$L_{A\leftrightarrow} = \frac{2c^2h}{c^2 - v^2} \qquad (12.1.8)$$

The optical path length $L_{B\updownarrow}$ possessed by light beam B when the longitudinal light beam B travels upward from point O to point m_2 and then returns from point m_2 to point O is:

$$L_{B\updownarrow} = \frac{2ch}{\sqrt{c^2 - v^2}} \qquad (12.1.9)$$

The difference $\Delta L_{B\updownarrow}$ between the optical path length $L_{B\updownarrow}$ in the longitudinal direction and $L_{A\leftrightarrow}$ in the transverse direction is:

$$\Delta L_{B\updownarrow} = L_{B\updownarrow} - L_{A\leftrightarrow} = \frac{2ch}{\sqrt{c^2 - v^2}} - \frac{2c^2h}{c^2 - v^2} \qquad (12.1.10)$$

Since light beam A and light beam B exchange states with respect to the direction of velocity v, the difference $\Delta L_{B\leftrightarrow}$ between the transverse length difference $\Delta L_{B\updownarrow}$ and the longitudinal length difference $\Delta L_{\updownarrow\leftrightarrow}$ is:

$$\Delta L_{\updownarrow\leftrightarrow} = \Delta L_{B\leftrightarrow} - \Delta L_{B\updownarrow} \qquad (12.1.11)$$

Substituting equations (12.1.7) and (12.1.10) into the above equation yields the relationship:

$$\Delta L_{\updownarrow\leftrightarrow} = \left(\frac{2c^2h}{c^2 - v^2} - \frac{2ch}{\sqrt{c^2 - v^2}}\right) - \left(\frac{2ch}{\sqrt{c^2 - v^2}} - \frac{2c^2h}{c^2 - v^2}\right)$$

Simplify the above formula to get the relation:

$$\Delta L_{\updownarrow\leftrightarrow} = 4h\frac{c^2\sqrt{c^2 - v^2} - c^3 + cv^2}{(c^2 - v^2)\sqrt{c^2 - v^2}} \qquad (12.1.12)$$

Since the velocity v is much smaller than the speed of light c, i.e., $v \ll c$, the above equation can be simplified to:

$$\Delta L_{\updownarrow\leftrightarrow} = \frac{4hv^2}{c^2} \qquad (12.1.13)$$

In the Newtonian ground frame of reference, the maximum number of moves n of interference fringes that should be observed on the viewing screen is:

$$n = \frac{\Delta L_{\updownarrow\leftrightarrow}}{\lambda} = \frac{4hv^2}{c^2\lambda} \qquad (12.1.14)$$

Note that the maximum number of moves n does not include the light speed invariance principle.

12.1.3 Zero result of the Michelson-Morley experiment independent of ground velocity *v*

According to Michelson-Morley experiment, the wavelength of sodium light $\lambda = 5.9 \times 10^{-7}m$, and the distance from plane reflectors m_1 and m_2 to the beam splitter $h = 5.5m$.

Theoretically, the velocity v in equation (12.1.14) should be the velocity of the ground moving through the universe. Three different values can be used for the velocity of the ground.

The first ground velocity v_1 is the speed of motion of the earth relative to the cosmic background radiation.

According to astronomers' calculations, the velocity of the Earth's motion with respect to the cosmic background radiation is $627 \cdot km/s$, i.e., the ground speed $v_1 = 627 \cdot km/s \approx 20.9 \times 10^{-4}c$.

Substitute the sodium wavelength $\lambda = 5.9 \times 10^{-7}m$, the distance $h = 5.5m$, and the cosmic background velocity v_1 into equation (12.1.14). In the Newtonian ground frame of reference, the maximum number of moves n_1 of interference fringes that should be observed is:

$$n_1 = \frac{4hv_1^2}{c^2\lambda} = \frac{4 \times 5.5 \times (20.9 \times 10^{-4}c.)^2}{5.9 \times 10^{-7}c^2} = 162.87 \qquad (12.1.15)$$

The second ground speed v_2 is the speed of the Sun's revolution around the center of the Milky Way.

According to astronomers' observations and calculations, the speed of the Sun's revolution around the center of the Milky Way is $v_O = 250 \cdot km/s$. If the Earth's rotation and revolution speeds are not taken into account, the Earth's average speed v should be equal to the Sun's revolution around the center of the Milky Way v_O, i.e., $v_2 = v_O \approx 8 \times 10^{-4}c$.

Substitute the velocity v_2 into equation (12.1.14). The maximum number of moves n_2 of interference fringes that should be observed in the Newtonian ground frame of reference is:

$$n_2 = \frac{4hv_1^2}{c^2\lambda} = \frac{4 \times 5.5 \times (8 \times 10^{-4}c.)^2}{5.9 \times 10^{-7}c^2} = 23.86 \qquad (12.1.16)$$

The third ground speed v_3 is the average speed of the earth's revolution around the sun.

According to astronomers' observational calculations, the average speed v of the Earth's revolution around the Sun is one ten-thousandth of the speed of light c, i.e., $v_3 = 1 \times 10^{-4}c$. Substituting the speed v_3 into equation (12.1.14), The maximum number of moves n_3 of interference fringes that should be observed in the Newtonian ground frame of reference is:

$$n_3 = \frac{4hv_1^2}{c^2\lambda} = \frac{4 \times 5.5 \times (1 \times 10^{-4}c.)^2}{5.9 \times 10^{-7}c^2} = 0.37 \qquad (12.1.17)$$

In the Michelson-Morley experiment, the velocity v_3 was considered to be the velocity of the Earth relative to the Ether.

The accuracy of the Michelson interferometer is **0.01%** and the number of moves of the interference fringes that can be observed is **0.01**. However, no moves of the interference fringes were observed in the experiments. it can be determined that the number of moves of the interference fringes is always equal to 0 no matter which of the ground velocities v is equal to v_1, v_2, or v_3, i.e:

$$n_1 = n_2 = n_3 = 0$$

From the above equation, the zero result of the Michelson-Morley experiment is independent of the ground speed v. In other words, the result of the Michelson-Morley experiment is always equal to zero, regardless of the speed at which the interferometer is moving.

Since the light paths in the Michelson-Morley experiment were all closed two-way light paths, the Michelson-Morley experiment did not prove that the speed of light, c, in one-way light paths is equal in all directions on the ground.

12.2 The null result of the Michelson-Morley experiment equals the prediction of Einstein's ground reference frame

12.2.1 Special relativity distorts Lorentz's length contraction equation

In 1892, the Dutch physicist Lorenz formulated the length contraction hypothesis, which states that the length L

of a moving object, contracts in the direction of motion. Its formula is:

$$L' = L\sqrt{1-\frac{v^2}{c^2}} \qquad (12.2.1)$$

The velocity v in the formula is the velocity of motion of the object, not the relative velocity between the observer and the object. The above formula was subjectively assumed by Lorentz, not derived from the coordinate transformation formula of special relativity. Lorentz used the above formula to successfully explain the results of the experiment.

Since the interferometer is in motion in the cosmic vacuum, the length h of the transverse optical path of the interferometer is reduced in the direction of motion, and its reduced length h' is:

$$h' = h\sqrt{1-\frac{v^2}{c^2}} \qquad (12.2.2)$$

According to the above equation and the Galilean velocity transformation formula $c' = c \pm v$, in the Newtonian ground frame of reference, the optical path length $L'_{B\leftrightarrow}$ for the transverse light beam B to reach point m_2 from point O to the right, and to return to point O from point m_2, is:

$$L'_{B\leftrightarrow} = c\frac{h'}{c-v} + c\frac{h'}{c+v} = \frac{2c^2h}{c^2-v^2}\sqrt{1-\frac{v^2}{c^2}} = \frac{2ch}{\sqrt{c^2-v^2}} \qquad (12.2.3)$$

Note that the optical path length $L'_{B\leftrightarrow}$ contains the length contraction effect and does not contain the principle of invariance of the speed of light.

According to equation (12.1.4), in the Newtonian ground reference system, the optical path length $L_{A\updownarrow}$ of the light beam A traveling upward from point O to point m_1, and returning from point m_1 to point O, is:

$$L_{A\updownarrow} = \frac{2ch}{\sqrt{c^2-v^2}}$$

The difference $\Delta L'_{\updownarrow\leftrightarrow}$ between the length $L'_{B\leftrightarrow}$ of the transverse optical path, and the length $L_{A\updownarrow}$ of the longitudinal optical path, is:

$$\Delta L'_{\updownarrow\leftrightarrow} = L'_{B\leftrightarrow} - L_{A\updownarrow} = \frac{2ch}{\sqrt{c^2-v^2}} - \frac{2ch}{\sqrt{c^2-v^2}} = 0 \qquad (12.2.5)$$

If the Michelson interferometer is rotated by 90^0, the two light paths A and B exchange their states with respect to the direction of the velocity v, i.e., light path A becomes a transverse light path and light path B becomes a longitudinal light path.

Since light beam A becomes transverse, the optical path length difference $\Delta L'_{\updownarrow\leftrightarrow}$ is increased by a factor of two, i.e:

$$2\Delta L'_{\updownarrow\leftrightarrow} = 0$$

In the Newtonian ground frame of reference, the number of moves n' of interference fringes is:

$$n' = \frac{2\Delta L'_{\updownarrow\leftrightarrow}}{\lambda} = 0 \qquad (12.2.6)$$

Note that the number of moves of the interference fringes $n' = 0$ contains the length contraction effect and does not contain the principle of invariance of the speed of light. Since the calculated result n' equals the experimental result, the speed of light propagation in all directions on the ground conforms to the Galilean velocity transformation formula $c' = c \pm v$.

That is, under Lorentz's assumption of length contraction, the null result of the Michelson-Morley experiment proves that the cosmic vacuum frame of reference is an absolutely stationary frame of reference.

It should be noted that the velocity v in Lorentz's length contraction formula is the velocity of the object's motion, whereas the velocity v in the special relativity length contraction formula is the relative velocity between the observer and the object.

Alternatively, when an object and an observer are moving at the same time with velocity v, Lorentz argues that the intrinsic length L_0 of the object produces a length contraction effect, while Einstein argues that the intrinsic length

L_0 of the object does not produce a length contraction effect. It follows that special relativity distorts Lorentz's length contraction formula.

12.2.2 The null result of the Michelson-Morley experiment equals the prediction of Einstein's ground reference frame

In the cosmic vacuum frame of reference, assume that the ground frame of reference is moving with velocity $\boldsymbol{v}$. Since the Michelson interferometer is at rest on the ground, the velocity $\boldsymbol{v}$ of the ground frame of reference is the velocity of the Michelson interferometer.

Since the observer is stationary with respect to the interferometer, there is no length contraction effect on the optical path of the interference experiment in Einstein's ground frame of reference.

In Einstein's ground frame of reference, the length $L_{A\updownarrow}$ of the longitudinal optical path of light beam A from point O up to point m_1 and back to point O from point m_1 is:

$$L_{A\updownarrow} = c\left(\frac{h}{c}+\frac{h}{c}\right) = 2h \quad (12.2.7)$$

The length $L'''_{B\leftrightarrow}$ of the transverse optical path of light beam B from point O to the right to point m_2 and back to point O from point m_2 is:

$$L'''_{B\leftrightarrow} = c\left(\frac{h}{c}+\frac{h}{c}\right) = 2h \quad (12.2.8)$$

In Einstein's ground frame of reference, the difference $\Delta L'''_{\updownarrow\leftrightarrow}$ between the longitudinal length $L_{A\updownarrow}$ and the transverse length $L'''_{B\leftrightarrow}$ is:

$$\Delta L'''_{\updownarrow\leftrightarrow} = L_{A\updownarrow} - L'''_{B\leftrightarrow} = 2h - 2h = 0 \quad (12.2.9)$$

If the Michelson interferometer is rotated by 90^0, the two light paths A and B exchange their states with respect to the direction of the velocity $\boldsymbol{v}$, i.e., light path A becomes a transverse light path and light path B becomes a longitudinal light path.

Since light beam A becomes transverse, the optical path length difference $\Delta L'''_{\updownarrow\leftrightarrow}$ is increased by a factor of two, i.e:

$$2\Delta L'''_{\updownarrow\leftrightarrow} = 0 \quad (12.2.10)$$

The maximum number of moves n''' of interference fringes in Einstein's ground reference frame is:

$$n''' = \frac{2\Delta L'''_{\updownarrow\leftrightarrow}}{\lambda} = 0 \quad (12.2.11)$$

Since the above zero result is consistent with the zero result of the Michelson-Morley experiment, physicists believe that the zero result of the Michelson-Morley experiment proves the principle of invariance of the speed of light.

It should be noted that the Michelson-Morley experiment is a round-trip two-way light path interference experiment, not a one-way light path interference experiment.

12.3 Verification of the principle of invariance of the speed of light with a long and short optical path interferometer

12.3.1 Difference $\Delta L_{B\leftrightarrow}$ in length between long optical path B and short optical path A when long optical path B is parallel to the velocity v

In the Michelson-Morley experiment, the lengths of both light path A and light path B are equal to h. In order to verify that the principle of invariance of the speed of light in special relativity is wrong, the authors changed the two equal-length light paths of the Michelson-Interferometer into a short light path A and a long light path B.

The Michelson-Morley experiment can be divided into two types.

Type 1 is an experiment in which both light path A and light path B are of equal length.

Type 2 is an experiment with short optical path A and long optical path B.

Rotate the interferometer with long and short optical paths by 90^0 and observe whether the interference fringes are moves or not. if the number of moves of the interference fringes $n = 0$, then the principle of invariance of the speed of light is correct. On the contrary, if the number of moves of the interference fringes $n \neq 0$, then the principle of invariance of the speed of light is wrong.

Assume that h is the length of the short optical path A and $(h + d)$ is the length of the long optical path B, where d is the newly added length of the interferometer. Assume that m_1 is the plane reflecting mirror of the short light path A and m_2 is the plane reflecting mirror of the long light path B. Assume that the short optical path A is perpendicular to the velocity v and the long optical path B is parallel to the velocity v. As shown in Figure 12-2.

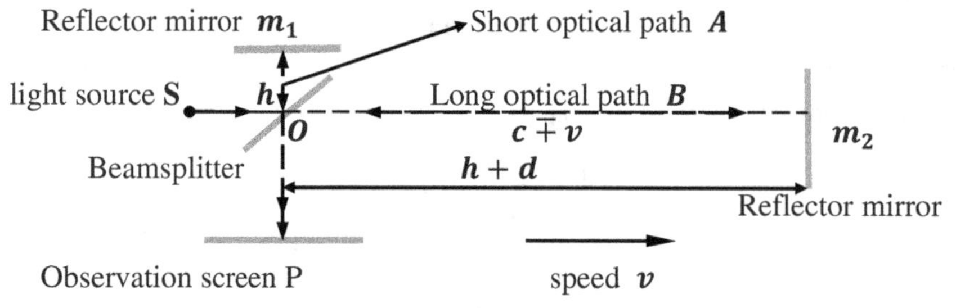

Figure 12-2 In the Newtonian ground reference frame, the optical path diagram of the long and short optical path interferometer.

In the Newtonian ground frame of reference, the transverse length $L_{B\leftrightarrow}$ of a long light beam B traveling from point O to the right to point m_2 and returning from point m_2 to point O is:

$$L_{B\leftrightarrow} = c\frac{h+d}{c-v} + c\frac{h+d}{c+v} = \frac{2c^2(h+d)}{c^2 - v^2} \tag{12.3.1}$$

The longitudinal length $L_{A\updownarrow}$ of the short light beam A from point O up to point m_1 and back to point O from point m_1 is:

$$L_{A\updownarrow} = \frac{2ch}{\sqrt{c^2 - v^2}} \tag{12.3.2}$$

In the Newtonian ground frame of reference, the difference $\Delta L_{B\leftrightarrow}$ between the transverse length $L_{B\leftrightarrow}$ and the longitudinal length $L_{A\updownarrow}$ is:

$$\Delta L_{B\leftrightarrow} = L_{B\leftrightarrow} - L_{A\updownarrow} = \frac{2c^2(h+d)}{c^2 - v^2} - \frac{2ch}{\sqrt{c^2 - v^2}} \tag{12.3.3}$$

If the length $d = 0$, the above equation becomes equation (12.1.7), i.e:

$$\Delta L_{B\leftrightarrow} = \frac{2c^2 h}{c^2 - v^2} - \frac{2ch}{\sqrt{c^2 - v^2}} \tag{12.3.4}$$

12.3.2 When the long optical path B is perpendicular to the velocity *v*, the length difference $\Delta L_{B\updownarrow}$ between the long optical path B and the short optical path A

If the long optical path B is perpendicular to the velocity v, then the positions of the long optical path B and the short optical path A, as shown in Figure 12-3.

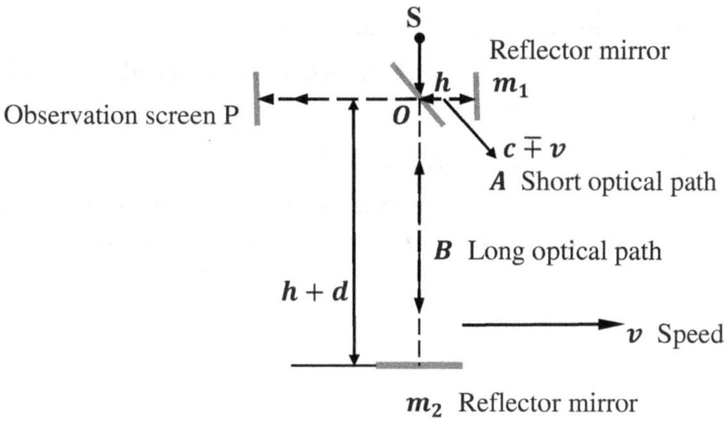

Figure 12-3 Schematic diagram of the optical path with the long and short optical paths rotated 90^0 degrees in the Newtonian ground reference system.

In the Newtonian ground frame of reference, the transverse length $L_{A\leftrightarrow}$ of light beam A from point O to the right to point m_1 and back to point O from point m_1 is:

$$L_{A\leftrightarrow} = c\frac{h}{c - v} + c\frac{h}{c + v} = \frac{2c^2h}{c^2 - v^2} \quad (12.3.5)$$

The longitudinal length $L_{B\updownarrow}$ of light beam B from point O down to point m_2 and back to point O from point m_2 is:

$$L_{B\updownarrow} = \frac{2c(h + d)}{\sqrt{c^2 - v^2}} \quad (12.3.6)$$

In the Newtonian ground frame of reference, the difference $\Delta L_{B\updownarrow}$ between the longitudinal length $L_{B\updownarrow}$ and the transverse length $L_{A\leftrightarrow}$ is:

$$\Delta L_{B\updownarrow} = L_{B\updownarrow} - L_{A\leftrightarrow} = \frac{2c(h + d)}{\sqrt{c^2 - v^2}} - \frac{2c^2h}{c^2 - v^2} \quad (12.3.7)$$

If the length $d = 0$, the above equation becomes equation (12.1.10), i.e:

$$\Delta L_{B\updownarrow} = \frac{2ch}{\sqrt{c^2 - v^2}} - \frac{2c^2h}{c^2 - v^2} \quad (12.3.8)$$

12.3.3 Experimental results predicted by the Newtonian ground reference system

According to Figure 12-2, if the Michelson interferometer is rotated by 90^0, the short optical path A and the long optical path B are exchanged with respect to the direction of the velocity v, i.e., the short optical path A becomes transverse optical path, and the long optical path B becomes longitudinal optical path. As shown in Figure 12-3.

If the transverse optical path B becomes the longitudinal optical path, then the difference $\Delta L_{\updownarrow\leftrightarrow}$ between the length difference $\Delta L_{B\leftrightarrow}$ of equation (12.3.3) and the length difference $\Delta L_{B\updownarrow}$ of equation (12.3.7) is:

$$\Delta L_{\updownarrow\leftrightarrow} = \Delta L_{B\leftrightarrow} - \Delta L_{B\updownarrow}$$

Substituting equations (12.3.3) and (12.3.7) into the above equation yields the relationship:

$$\Delta L_{\updownarrow\leftrightarrow} = \left(\frac{2c^2(h + d)}{c^2 - v^2} - \frac{2ch}{\sqrt{c^2 - v^2}}\right) - \left(\frac{2c(h + d)}{\sqrt{c^2 - v^2}} - \frac{2c^2h}{c^2 - v^2}\right)$$

Since the velocity v is much smaller than the speed of light c, i.e., $v \ll c$, the above equation can be simplified to:

$$\Delta L_{\updownarrow\leftrightarrow} = \frac{2c^2(2h + d)}{c^2 - v^2} - \frac{2c(2h + d)}{\sqrt{c^2 - v^2}} \approx \frac{2(2h + d)v^2}{c^2} \quad (12.3.9)$$

The above equation shows that the difference $\Delta L_{\updownarrow\leftrightarrow}$ contains two components.

The first component is the difference $\frac{4hv^2}{c^2}$ caused by the length h.

Since the length h is included in the long optical path B, the smallest length h should be chosen for the experiment in order to improve the accuracy of the experiment.

The second component is the difference $\frac{2v^2d}{c^2}$ due to the length d.

Long and short optical path interferometers use this component to verify the principle of invariance of the speed of light. If the length h is much smaller than the length d, i.e., $h \ll d$, then equation (12.3.9) becomes.

$$\Delta L_{\updownarrow\leftrightarrow} \approx \frac{2v^2d}{c^2} \tag{12.3.10}$$

In order to improve the accuracy of the experiment, a larger length d should be chosen whenever possible.

The maximum number of moves n of interference fringes that should be observed in the Newtonian ground frame of reference is:

$$n = \frac{\Delta L_{\updownarrow\leftrightarrow}}{\lambda} = \frac{2(2h+d)v^2}{\lambda c^2} \tag{12.3.11}$$

In summary, the Michelson-Morley experiment can be divided into two types.

The first type of experiment is an experiment in which the lengths of light path A and light path B are equal, and the result of the experiment is always equal to zero.

Since the Michelson-Morley experiment is an interference experiment with two-way light paths, the equality of the lengths of the two light paths A and B is a defect of this experiment. It is this flaw that makes the zero result of the Michelson-Morley experiment inconsistent with Newton's absolute view of space-time and consistent with Einstein's view of space-time.

The second type of experiment is the experiment with short light path A and long light path B.

If special relativity is correct, then the experimental result for the long and short light paths should be equal to 0, i.e., the number of moves of the interference fringes $n = 0$.

If Newtonian absolute spacetime is is correct, then the experimental results for the long and short light paths should not be equal to 0, i.e., the number of moves of the interference fringes $n \neq 0$.

Assume that the length of the short optical path A is $h = 0.05m$; the length of the long optical path B is $d = 2m$; and the laser wavelength $\lambda = 6.328 \times 10^{-7}m$. Relative to the cosmic background radiation (CMB), the velocity of the earth $v_1 = 20.9 \times 10^{-4}c$. According to Equation (12.3.11), if the interferometer is rotated by 90^0, then the maximum number of moves of the interference fringes n_1 is:

$$n_1 = \frac{2(2h+d)v^2}{\lambda c^2} = \frac{2 \times 2.1 \times (20.9 \times 10^{-4}c)^2}{6.238 \times 10^{-7}c^2} = 28.99 \tag{12.3.12}$$

The average speed of the Sun's motion around the galactic center $v_2 = 8 \times 10^{-4}c$. If the interferometer is rotated 90^0, then the maximum number of moves of the interference fringes n_2 is:

$$n_2 = \frac{2(2h+d)v^2}{\lambda c^2} = \frac{2 \times 2.1 \times (8 \times 10^{-4}c)^2}{6.238 \times 10^{-7}c^2} = 4.24 \tag{12.3.13}$$

The average speed of the Earth's revolution around the Sun is $v = 1 \times 10^{-4}c$. If the interferometer is rotated 90^0, then the maximum number of moves of the interference fringes n_3 is:

$$n_3 = \frac{2(2h+d)v^2}{\lambda c^2} = \frac{2 \times 2.1 \times (1 \times 10^{-4}c)^2}{6.238 \times 10^{-7}c^2} = 0.06 \tag{12.3.14}$$

To summarize, in the Newtonian ground frame of reference, if the interferometer rotates 90^0, then the number of moves of the interference fringes $n \neq 0$.

12.3.4 Length difference $\Delta L_{B\leftrightarrow}'''$ between the long optical path B and the short optical path A in the Einstein ground frame of reference

In Einstein's ground frame of reference, if the long optical path B is parallel to the velocity v, then the optical path of the interferometer is shown in Figure 12-4.

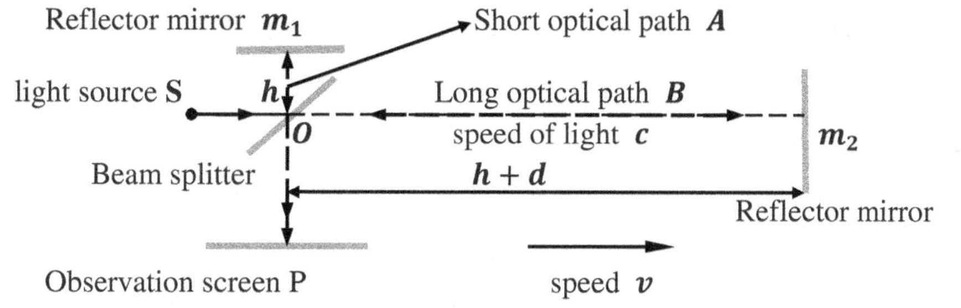

Figure 12-4 In the Einstein's ground reference frame, the optical path diagram of the long and short optical path interferometer.

Since the observer is stationary with respect to the interferometer, there is no length contraction effect on the optical path in the interference experiment according to special relativity.

In Einstein's ground frame of reference, the longitudinal length $L_{A\updownarrow}$ of the light beam A from point O up to point m_1 and back to point O from point m_1 is:

$$L_{A\updownarrow} = c\left(\frac{h}{c} + \frac{h}{c}\right) = 2h \qquad (12.3.15)$$

The transverse length $L'''_{B\leftrightarrow}$ of light beam B from point O to the right to point m_2 and back to point O from point m_2 is:

$$L'''_{B\leftrightarrow} = c\left(\frac{h+d}{c} + \frac{h+d}{c}\right) = 2(h+d) \qquad (12.3.16)$$

In the Einstein ground frame of reference, the difference $\Delta L'''_{B\leftrightarrow}$ between the transverse length $L'''_{B\leftrightarrow}$ and the longitudinal length $L_{A\updownarrow}$ is:

$$\Delta L'''_{B\leftrightarrow} = L'''_{B\leftrightarrow} - L_{A\updownarrow} = 2d \qquad (12.3.17)$$

12.3.5 Experimental results predicted by Einstein's ground reference frame

If the Michelson interferometer is rotated by 90^0, then the short optical path A and the long optical path B exchange their states with respect to the direction of the velocity v, i.e., the short optical path A becomes the transverse optical path, and the long optical path B becomes the longitudinal optical path. As shown in Figure 12-5.

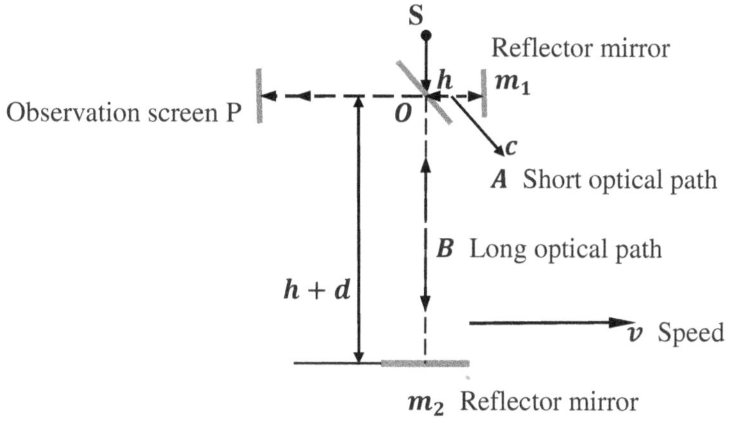

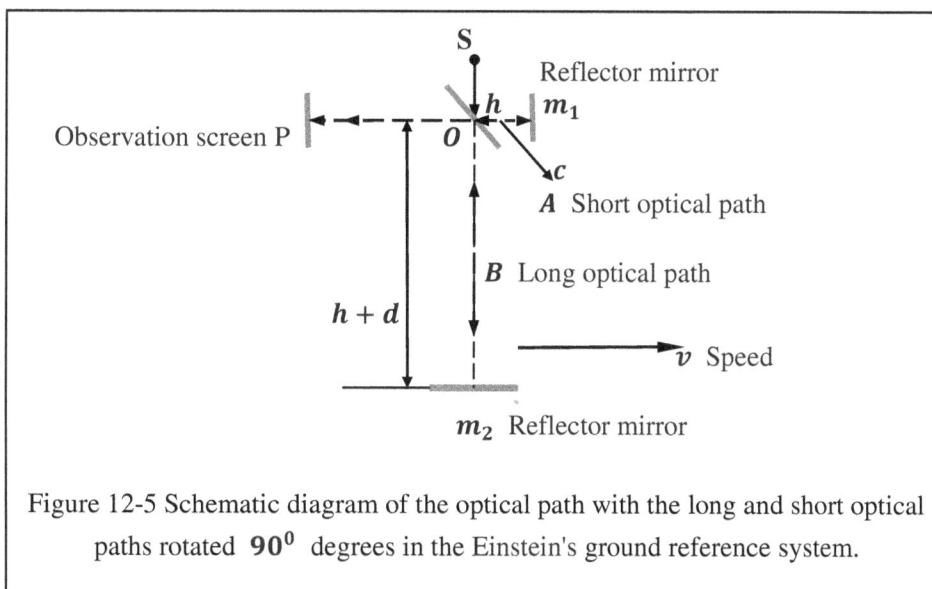

Figure 12-5 Schematic diagram of the optical path with the long and short optical paths rotated 90^0 degrees in the Einstein's ground reference system.

In Einstein's ground frame of reference, The longitudinal length $L_{B\updownarrow}$ of the long optical path B when light beam B travels down from point O to point m_2 and then returns from point m_2 to point O is:

$$L_{B\updownarrow} = c\left(\frac{h+d}{c} + \frac{h+d}{c}\right) = 2(h+d) \qquad (12.3.18)$$

The transverse length $L'''_{A\leftrightarrow}$ of the short optical path A as it travels from point O to the right to point m_1 and then returns from point m_1 to point O is:

$$L'''_{A\leftrightarrow} = c\left(\frac{h}{c} + \frac{h}{c}\right) = 2h \qquad (12.3.19)$$

The optical path length difference $\Delta L'''_{B\updownarrow}$ between the longitudinal length $L_{B\updownarrow}$ and the transverse length $L'''_{A\leftrightarrow}$ is:

$$\Delta L'''_{B\updownarrow} = L_{B\updownarrow} - L'''_{A\leftrightarrow} = 2d \qquad (12.3.20)$$

According to Figure 12-4, if the Michelson interferometer is rotated by 90^0, the short optical path A and the long optical path B are exchanged with respect to the direction of the velocity v, i.e., the short optical path A becomes transverse optical path, and the long optical path B becomes longitudinal optical path. As shown in Figure 12-5.

If the transverse optical path B becomes the longitudinal optical path, then the difference $\Delta L_{\updownarrow\leftrightarrow}$ between the length difference $\Delta L'''_{B\updownarrow}$ of equation (12.3.20) and the length difference $\Delta L'''_{B\leftrightarrow}$ of equation (12.3.17) is:

$$\Delta L'''_{\updownarrow\leftrightarrow} = \Delta L'''_{B\updownarrow} - \Delta L'''_{B\leftrightarrow}$$

Substituting equations (12.3.17) and (12.3.20) into the above equation yields the relationship:

$$\Delta L'''_{\updownarrow\leftrightarrow} = 2d - 2d = 0 \qquad (12.3.21)$$

In Einstein's ground frame of reference, if the interferometer is rotated 90^0, then the maximum number of moves of the interference fringes $n'''_{\perp}$ is:

$$n'''_{\perp} = \frac{\Delta L'''_{\updownarrow\leftrightarrow}}{\lambda} = 0 \qquad (12.3.22)$$

The above equation shows that the number of moves $n'''_{\perp}$ of the interference fringes of the long and short light paths is always equal to 0, in Einstein's ground frame of reference.

Comparing equation (12.3.22) with equation (12.3.12), the principle of invariance of the speed of light in special relativity is correct if the number of moves of the interference fringes $n'''_{\perp} = 0$.

Conversely, if the number of moves of the interference fringes $n = 28.99$, the principle of the invariance of the speed of light is false.

Chapter 13 Mechanisms of Electron Formation

Introduction: According to the high-energy γ-photons passing through the lead plate into positive and negative electrons, the collision of positive and negative electrons into two low-energy γ-photon energy planes, as well as the axiom of invariance of the speed of light of the energy element ε_h, it is possible to determine that the electron is a hollow circular energy tube possessing the property of spin.

13.1 Universal gravitational force F between photon masses m_{c1} and m_{c2}

13.1.1 The internal rotational gravitational force F_ω between the photon masses m_{c1} and m_{c2} is equal to 0, i.e. F_ω=0

In the cosmic vacuum frame of reference, assume that the masses m_1 and m_2 have velocities v_1 and v_2, respectively, and that the dynamic gravitational force F between the moving masses m_1 and m_2 is given by equation (10.1.5):

$$F = \sqrt{\left(\frac{Km_{\omega1}m_{\omega2}}{r^3}\right)^2 + \left(\frac{Km_{v1}m_{v2}}{r^3}\right)^2 + \left(\frac{K}{r^3}\right)^2 \left(m_{v1}^2 m_{\omega2}^2 + m_{\omega1}^2 m_{v2}^2\right)} \qquad (13.1.1)$$

The above equation shows that the gravitational force F, which is dynamic between the moving masses m_1 and m_2, contains three gravitational components.

According to equation (10.1.6), the internal rotating gravitational force F_ω between the internal rotating masses $m_{\omega1}$ and $m_{\omega2}$ is:

$$F_\omega = K\frac{m_{\omega1}m_{\omega2}}{r^3} \qquad (13.1.2)$$

According to equation (10.1.8), the internal rotating masses contained in the moving masses m_1 and m_2 are respectively:

$$m_{\omega1} = m_1\sqrt{1-\frac{v_1^2}{c^2}}, \qquad m_{\omega2} = m_2\sqrt{1-\frac{v_2^2}{c^2}} \qquad (13.1.3)$$

Substitute the above equation into equation (13.1.2). The internal rotating gravitational force F_ω between the internal rotating masses $m_{\omega1}$ and $m_{\omega2}$ is:

$$F_\omega = K\frac{m_1m_2}{r^3}\sqrt{1-\frac{v_1^2}{c^2}}\sqrt{1-\frac{v_2^2}{c^2}} \qquad (13.1.4)$$

The above equation shows that the internal rotational gravitational force F_ω decreases with increasing cosmic velocities v_1 and v_2.

If the moving mass m_1 is the photon mass m_c and the moving mass m_2 is the object mass. Since the velocity $v_1 = c$ for the photon mass m_c and $v_2 < c$ for the object, the internal rotational gravitational force F_ω between the photon mass m_c and the object mass m_2 is:

$$F_\omega = K\frac{m_cm_2}{r^3}\sqrt{1-\frac{c^2}{c^2}}\sqrt{1-\frac{v_2^2}{c^2}} = 0 \qquad (13.1.5)$$

In addition the internal rotational gravitational force $F_\omega = 0$ between the photon masses m_{c1} and m_{c2}, i.e:

$$F_\omega = K\frac{m_{c1}m_{c2}}{r^3}\sqrt{1-\frac{c^2}{c^2}}\sqrt{1-\frac{c^2}{c^2}} = 0 \tag{13.1.6}$$

13.1.2 Displacement gravitational force F_v between photon masses m_{c1} and m_{c2}

According to equation (10.1.11), the displacement gravitational force F_v between the displaced masses m_{v1} and m_{v2} is:

$$F_v = K\frac{m_{v1}m_{v2}}{r^3} \tag{13.1.7}$$

According to equation (10.1.13), the displaced masses contained in the moving masses m_1 and m_2, respectively, are:

$$m_{v1} = m_1\frac{v_1}{c}, \qquad m_{v2} = m_2\frac{v_2}{c} \tag{13.1.8}$$

Substitute the above equation into equation (13.1.7). The moving gravitational force F_v between the displaced masses m_{v1} and m_{v2} is:

$$F_v = K\frac{m_1v_1 \cdot m_2v_2}{c^2r^3} \tag{13.1.9}$$

The above equation shows that the displacement gravitational force F_v is proportional to the velocities v_1 and v_2.

If the moving mass m_1 is the photon mass m_c and the moving mass m_2 is the object mass. Since the velocity $v_1 = c$ for the photon mass m_c and $v_2 < c$ for the object, the displacement gravitational force F_v between the photon mass m_c and the object mass m_2 is:

$$F_v = K\frac{m_c \cdot m_2v_2}{cr^3} \tag{13.1.10}$$

The above equation shows that the direction of propagation of light changes when it passes near a massive object.

In addition, the displaced gravitational force F_v between the two photon masses m_{c1} and m_{c2} is:

$$F_v = K\frac{m_{c1}m_{c2}}{r^3} \tag{13.1.11}$$

The above equation shows that there is a gravitational force between two photons.

13.1.3 Displacement-internal rotating gravitational force $F_{v\omega}$ between displaced mass m_v and internal rotating mass m_ω

According to equation (10.1.15), the displacement-internal rotation gravitational force $F_{v\omega}$ between the displaced mass m_v and the internal rotating mass m_ω is:

$$F_{v\omega} = \sqrt{\left(K\frac{m_{v1}^2m_{\omega2}^2}{r^3}\right)^2 + \left(K\frac{m_{v2}^2m_{\omega1}^2}{r^3}\right)^2} \tag{13.1.12}$$

According to equation (10.1.16), the displacement-internal rotational gravitational force $F_{v\omega}$ between the moving masses m_1 and m_2, respectively, is:

$$\begin{cases} F_{v1\omega2} = K\dfrac{m_{v1}m_{\omega2}}{r^3} = K\dfrac{m_1m_2}{r^3}\dfrac{v_1}{c}\sqrt{1-\dfrac{v_2^2}{c^2}} \\ F_{v2\omega1} = K\dfrac{m_{v2}m_{\omega1}}{r^3} = K\dfrac{m_1m_2}{r^3}\dfrac{v_2}{c}\sqrt{1-\dfrac{v_1^2}{c^2}} \end{cases} \tag{13.1.13}$$

If the moving mass m_1 is the photon mass m_c and the moving mass m_2 is the object mass. Since the velocity $v_1 = c$ for the photon mass m_c and $v_2 < c$ for the object, the gravitational force $F = F_{v1\omega2}$ between the photon mass m_c and the object mass m_2 is:

$$F = F_{v1\omega 2} = K\frac{m_c m_2}{r^3}\sqrt{1-\frac{v_2^2}{c^2}} \qquad (13.1.14)$$

The above equation shows that the direction of propagation of light changes when it passes near a massive object.

13.1.4 Universal gravitational force F_c between photon masses m_{cA} and m_{cB}

According to equation (9.4.3), the new gravitational force is:

$$\begin{cases} F = +K\dfrac{Mm}{r^3}, & r > 1(l_m) \\ F = 0, & r = 1(l_m) \\ F = -K\dfrac{Mm}{r^3}, & 0 < r < 1(l_m) \end{cases}$$

The l_m in the formula is the smallest unit length in the gravitational field. According to the above equation, the gravitational force F_c between the photon masses m_{cA} and m_{cB} is:

$$\begin{cases} F_c = +K\dfrac{m_{cA}m_{cB}}{r^3}, & r > 1(l_m) \\ F_c = 0, & r = 1(l_m) \\ F_c = -K\dfrac{m_{cA}m_{cB}}{r^3}, & 0 < r < 1(l_m) \end{cases} \qquad (13.1.15)$$

when the distance $r > 1(l_m)$. Since the gravitational force $F > 0$, the gravitational force F_c between the photon masses m_{cA} and m_{cB} is the suction force $F_{c(+)}$, i.e:

$$F_{c(+)} = K\frac{m_{cA}m_{cB}}{r^3} \qquad (13.1.16)$$

When the distance $r = 1(l_m)$. Since the gravitational force $F = 0$, the gravitational force F_c between the photon masses m_{cA} and m_{cB} is:

$$F_c = 0$$

If the distance $r < l_m$. Since the gravitational force $F < 0$, the gravitational force F_c between the photon masses m_{cA} and m_{cB} is the repulsive force $F_{c(-)}$, i.e:

$$F_{c(-)} = -K\frac{m_{cA}m_{cB}}{r^3} \qquad (13.1.17)$$

Since there is a suction force $F_{c(+)}$ and a repulsion force $F_{c(-)}$ between the photon masses m_{cA} and m_{cB}, two photons moving in the same direction will only be adjacent to each other and will not overlap each other under the action of the gravitational force F_c.

13.1.5 The gravitational force F_c of a photon possesses particle properties

According to the mass-energy conversion equation $\varepsilon_c = m_c c^2$ for photons and the energy equation $\varepsilon_c = h\beta$ for photons, the mass m_c of a photon is:

$$m_c = \frac{h\beta}{c^2} \qquad (13.1.18)$$

The β in equation is the frequency of the light wave. Assume that β_A and β_B are the frequencies of photons A and B, respectively. The masses m_{cA} and m_{cB} of the photons are respectively:

$$m_{cA} = \frac{h\beta_A}{c^2}, \qquad m_{cB} = \frac{h\beta_B}{c^2} \qquad (13.1.19)$$

Substituting the photon masses m_{cA} and m_{cB} into equation (13.1.16) gives the relation equation:

$$F_c = K\frac{h^2\beta_A\beta_B}{c^4 r^3} \qquad (13.1.20)$$

The above equation is the gravitational formula for a photon. Since the change in frequency β, the value $\Delta\beta$ has a particle nature. Therefore, the change in gravitational force F_c between photons A and B has a particle nature.

In the cosmic vacuum reference system, the equation for the speed of light is:

$$\lambda\beta = c \tag{13.1.21}$$

When the wavelength λ and frequency β of the light are variables, differentiating the above equation gives the relational equation:

$$\lambda d\beta + \beta d\lambda = 0 \tag{13.1.22}$$

Based on the above equation, the relationship equation can be obtained:

$$d\lambda = -\frac{\lambda}{\beta}d\beta = -\frac{c}{\beta^2}d\beta \tag{13.1.23}$$

The above equation shows the relationship between the wavelength change value $\Delta\lambda$ and the frequency change value $\Delta\beta$. Since the frequency change value $\Delta\beta$ has a particle nature, the wavelength change value $\Delta\lambda$ also has a particle nature. That is, the wavelength change value $\Delta\lambda$ can only be an integer change, not a fractional change. From this point of view, the photon wavelength λ change value $\Delta\lambda$ has particle nature.

13.2 Universal gravitational force F_h between two energy elements ε_h

13.2.1 The mass m_h of an energy element can be decomposed into an internal rotational mass $m_{h\omega}$ and a displacement mass m_{hv}

In the cosmic vacuum reference system, when the elementary particle m_i moves with velocity v. The orbit of the energy element ε_h inside the elementary particle m_i is a helix, and the tangent velocity of its orbit is always equal to the helical velocity of light c. According to equation (7.1.8), the helical velocity of light c of the energy element ε_h is:

$$c = \sqrt{v_\omega^2 + v^2} \tag{13.2.1}$$

Note that the velocity v is not only the displacement velocity of the elementary particle m_i, but also the displacement velocity of the energy element ε_h. The velocity v_ω in the formula is the internal rotation velocity of the energy element ε_h and also the spin velocity of the elementary particle m_i. The internal rotation velocity v_ω of the energy element ε_h is an invisible and imperceptible circular velocity hidden in the mass m of the object.

This book argues that the internal rotation speed $v_\omega \neq 0$ of the energy element ε_h is the only condition for the formation of the mass m of the object. When the internal rotation speed $v_\omega = 0$ of the energy element ε_h, the matter composed of the energy element ε_h at that time is a photon.

The spiral speed of light invariance theorem is a very important physical theorem. It can be used to derive many important experimental laws in classical electromagnetic field theory. It also explains the formation of electric charge and the reason why electric charge has an attractive or repulsive force. In this respect, the spiral theorem of the invariance of the speed of light is the cornerstone of electromagnetic field theory.

According to equation (7.3.7), the intrinsic mass m of the object is:

$$m = \sqrt{\left(m\frac{v_\omega}{c}\right)^2 + \left(m\frac{v}{c}\right)^2} \tag{13.2.2}$$

According to the above equation, the mass m_h of the energy element ε_h is:

$$m_h = \sqrt{\left(m_h\frac{v_\omega}{c}\right)^2 + \left(m_h\frac{v}{c}\right)^2} = \sqrt{(m_{h\omega})^2 + (m_{hv})^2} \tag{13.2.3}$$

The above equation shows that the mass m_h of the energy element can be decomposed into an internal rotational mass $m_{h\omega}$ and a displacement mass m_{hv}.

According to equation (7.3.8), the internal rotational mass $m_{h\omega}$ contained in the energy element mass m_h is:

$$m_{h\omega} = m_h\frac{v_\omega}{c} = m_h\sqrt{1 - \frac{v^2}{c^2}} \tag{13.2.4}$$

The above equation shows that the internal rotating mass $m_{h\omega}$ of the energy element ε_h decreases as the displacement velocity v increases.

According to equation (7.3.9), the displacement mass m_{hv} contained in the energy element mass m_h is:

$$m_{hv} = m_h \frac{v}{c} \tag{13.2.5}$$

The above equation shows that the displacement mass m_{hv} of the energy element ε_h is proportional to the displacement velocity v.

13.2.2 Universal gravitational force F_h between two energy elements ε_h

In the cosmic vacuum frame of reference, assume that the displacement velocities of the two energy elements ε_h are v_1 and v_2. According to equation (13.1.1), the universal gravitational force F_h between the two energy elements ε_h is:

$$F_h = \sqrt{\left(\frac{K m_{h\omega1} m_{h\omega2}}{r^3}\right)^2 + \left(\frac{K m_{hv1} m_{hv2}}{r^3}\right)^2 + \left(\frac{K}{r^3}\right)^2 \left(m_{hv1}^2 m_{h\omega2}^2 + m_{h\omega1}^2 m_{hv2}^2\right)} \tag{13.2.6}$$

According to equation (13.1.4), the internal rotational gravitational force F_ω between the internal rotating masses $m_{\omega1}$ and $m_{\omega2}$ is:

$$F_\omega = K \frac{m_1 m_2}{r^3} \sqrt{1 - \frac{v_1^2}{c^2}} \sqrt{1 - \frac{v_2^2}{c^2}}$$

According to the above equation, the internal rotational gravitational force $F_{h\omega}$ between the internal rotating masses $m_{h\omega1}$ and $m_{h\omega2}$ of the two energy elements is:

$$F_{h\omega} = K \frac{m_{h\omega1} m_{h\omega2}}{r^3} = K \frac{m_h^2}{r^3} \sqrt{1 - \frac{v_1^2}{c^2}} \sqrt{1 - \frac{v_2^2}{c^2}} \tag{13.2.7}$$

When the displacement velocity of the two energy elements is equal, i.e., $v_1 = v_2 = v$. According to the above equation, the internal rotational gravitational force $F_{h\omega}$ between the two energy element masses m_h is:

$$F_{h\omega} = K \frac{m_h^2}{r^3} \left(1 - \frac{v^2}{c^2}\right) \tag{13.2.8}$$

The above equation shows that the internal rotational gravitational force $F_{h\omega}$ between the energy elements ε_h decreases as the cosmic velocity v increases.

When m_1 is the energy element mass m_h and m_2 is the object mass. According to equation (13.1.4), the internal rotational gravitational force F_ω between the energy element mass m_h and the object mass m_2 is:

$$F_\omega = K \frac{m_h m_2}{r^3} \sqrt{1 - \frac{v_1^2}{c^2}} \sqrt{1 - \frac{v_2^2}{c^2}} \tag{13.2.9}$$

In the cosmic vacuum frame of reference, when the energy element ε_h moves in a straight line. Since the velocity $v_1 = c$ of the energy element ε_h and the velocity $v_2 < c$ of the object, the internal rotational gravitational force F_ω between the mass m_h of the energy element and the mass m_2 of the object is:

$$F_\omega = K \frac{m_h m_2}{r^3} \sqrt{1 - \frac{c^2}{c^2}} \sqrt{1 - \frac{v_2^2}{c^2}} = 0 \tag{13.2.10}$$

The above equation shows that when the energy element ε_h moves in a straight line. The internal rotational gravitational force F_ω between the energy element mass m_h and the object mass m_2 is always equal to zero.

In the cosmic vacuum frame of reference, when the energy element ε_h moves in a straight line. Since the energy element ε_h has a velocity $v = c$, the internal rotational gravitational force $F_{h\omega} = 0$ between two linearly moving energy elements ε_h, viz:

$$F_{h\omega} = K \frac{m_h m_h}{r^3} \sqrt{1 - \frac{c^2}{c^2}} \sqrt{1 - \frac{c^2}{c^2}} = 0 \tag{13.2.11}$$

According to equation (13.1.9). the displacement gravitational force F_v between the displaced masses m_{v1} and

m_{v2} is:

$$F_v = K\frac{m_1 v_1 \cdot m_2 v_2}{c^2 r^3}$$

According to the above equation, the displacement gravitational force F_{hv} between the displaced masses m_{hv1} and m_{hv2} of the two energy elements is:

$$F_{hv} = K\frac{m_h^2 v_1 v_2}{c^2 r^3} \tag{13.2.12}$$

When the displacement velocities of the two energy elements ε_h are equal, i.e., $v_1 = v_2 = v$. According to the above equation, the displacement gravitational force F_{hv} between two energy element masses m_h is:

$$F_{hv} = K\frac{m_h^2 v^2}{c^2 r^3} \tag{13.2.13}$$

When the moving mass m_1 is the energy element mass m_h and the moving mass m_2 is the object mass. Since the energy element ε_h has velocity $v_1 = c$ and the object velocity $v_2 < c$, the displacement gravitational force F_v between the energy element mass m_h and the object mass m_2 is:

$$F_v = K\frac{m_h \cdot m_2 v_2}{c r^3} \tag{13.2.14}$$

The displacement gravitational force F_{hv} between two linearly moving energy element masses m_h is:

$$F_{hv} = K\frac{m_h^2}{r^3} \tag{13.2.15}$$

According to equation (13.1.13), the displacement-internal rotational gravitational force $F_{v\omega}$ between the moving masses m_1 and m_2, respectively, is:

$$\begin{cases} F_{v1\omega2} = K\dfrac{m_{v1} m_{\omega2}}{r^3} = K\dfrac{m_1 m_2}{r^3}\dfrac{v_1}{c}\sqrt{1-\dfrac{v_2^2}{c^2}} \\ F_{v2\omega1} = K\dfrac{m_{v2} m_{\omega1}}{r^3} = K\dfrac{m_1 m_2}{r^3}\dfrac{v_2}{c}\sqrt{1-\dfrac{v_1^2}{c^2}} \end{cases}$$

When the moving mass m_1 is the energy element mass m_h and the moving mass m_2 is the object mass. Since the energy element ε_h velocity $v_1 = c$ and the object velocity $v_2 < c$, the displacement-internal rotational gravitational force $F_{v1\omega2}$ between the energy element mass m_h and the object mass m_2 is:

$$F_{v1\omega2} = K\frac{m_h m_2}{r^3}\sqrt{1-\frac{v_2^2}{c^2}} \tag{13.2.16}$$

The above equation shows that the direction of propagation of the energy element ε_h changes when the energy element ε_h, moving in a straight line, passes through the neighborhood of a massive object.

13.2.3 The gravitational force F_h between energy elements ε_h has a particle nature

According to equation (7.2.14), the mass m_h of the energy element ε_h is:

$$m_h = \frac{\varepsilon_h}{c^2} \tag{13.2.17}$$

According to equation (13.2.15), the gravitational force F_h between two energy elements ε_h is:

$$F_h = K\frac{m_h^2}{r^3} = K\frac{\varepsilon_h^2}{c^4 r^3} \tag{13.2.18}$$

Since the change value $\Delta\varepsilon_h$ of the energy element ε_h has a particle nature, the change value ΔF_h of the gravitational force between the energy elements ε_h also has a particle nature.

Alternatively, the changing value ΔF_h of the gravitational force can only change in the form of an integer, not in the form of a fraction. From this it can be determined that the gravitational force F_h of the energy element has a particle nature.

According to equation (9.4.3), the new formula for the gravitational force is:

$$\begin{cases} F = +K\dfrac{Mm}{r^3}, & r > 1(l_m) \\ F = 0, & r = 1(l_m) \\ F = -K\dfrac{Mm}{r^3}, & 0 < r < 1(l_m) \end{cases} \quad (13.2.19)$$

The l_m in the formula is the smallest unit length in the gravitational field. According to the above equation, the universal gravitational force F_h between two energy elements ε_h is:

$$\begin{cases} F_h = +K\dfrac{\varepsilon_h^2}{c^4 r^3}, & r > 1(l_m) \\ F_h = 0, & r = 1(l_m) \\ F_h = -K\dfrac{\varepsilon_h^2}{c^4 r^3}, & 0 < r < 1(l_m) \end{cases} \quad (13.2.20)$$

When the distance $r > 1(l_m)$. Since the gravitational force $F > 0$, the gravitational force F_h between the two energy elements is the suction force $F_{h(+)}$, i.e:

$$F_{h(+)} = K\frac{\varepsilon_h^2}{c^4 r^3}, \quad r > 1(l_m) \quad (13.2.21)$$

When the distance $r = 1(l_m)$. Since the gravitational force $F = 0$, the gravitational force F_h between the two energy elements is:

$$F_h = 0, \quad r = 1(l_m)$$

When the distance $r < l_m$. Since the gravitational force $F < 0$, the gravitational force F_h between the two energy elements is repulsive $F_{h(-)}$, i.e:

$$F_{h(-)} = -K\frac{\varepsilon_h^2}{c^4 r^3}, \quad r < 1(l_m) \quad (13.2.22)$$

Since there are suction and repulsion forces between energy elements ε_h, two energy elements ε_h moving in the same direction will only be adjacent to each other under the action of gravitational force F_h and will not overlap together.

Since photons are composed of energy elements ε_h, the energy elements ε_h contained in the photon energy $\varepsilon_c = h\beta$ will not be concentrated in a single point. Nor will it be concentrated in a straight line. Since a light wave is a plane wave, the energy elements ε_h contained in the photon energy ε_c should be distributed in a plane.

13.3 A photon is a plane composed of a certain number of energy elements ε_h

13.3.1 Number of energy elements ε_h contained in the energy ε_c of a photon n=β·1s

The quantum nature of light waves was first discovered by the German scientist Planck, who in 1901 derived the formula for blackbody radiation consistent with radiation experiments, viz:

$$\rho_v = \frac{8\pi h\beta^3}{c^3}\frac{1}{e^{\frac{h\beta}{kT}} - 1}$$

In the equation ρ_v is the energy density per unit frequency spacing; h is Planck's constant; β is the frequency of the electromagnetic wave; k is Boltzmann's constant; and T is the temperature.

Planck found that the calculations of the formula could be compatible with the experimental results only if it was assumed that the emission and absorption of the energy of an electromagnetic wave is not continuous, but is emitted and absorbed one by one in the form of particles. The energy contained in each energy particle of an electromagnetic wave is $\varepsilon_c = h\beta$.

In 1905, Einstein further proposed that light waves are not continuous but particles. Einstein called the energy particles contained in light waves light quanta. In 1923, the American physicist Compton successfully explained the Compton effect with the concept of light quanta. The concept of light quanta was widely accepted and applied, and in 1926 light quanta were officially named photons by physicists.

According to the fluctuation theory. The frequency β of a light wave is the number of times the light wave makes a periodic change per unit of time. The formula for the frequency of a light wave is:

$$\beta = \frac{1}{T} = \frac{c}{\lambda} \tag{13.3.1}$$

The period T in the formula is the time it takes for a light wave to complete one oscillation; the wavelength λ is the distance between two adjacent wave crests.

Suppose l_0 is the unit length that measures the size of the wavelength λ. Although different photons have different wavelengths λ, the unit length l_0 contained in the wavelength λ is the same. Assume that the coefficient k is the number of unit lengths l_0 contained in the wavelength λ. The wavelength λ can be expressed as:

$$\lambda = kl_0 \tag{13.3.2}$$

Since the wavelength of red light $\lambda = 650 \times 10^{-9}m$, the coefficient of red light $k = 650$ and the unit length $l_0 = 10^{-9}m$. Since the wavelength λ contains k unit lengths l_0, the frequency β can be expressed as:

$$\beta = \frac{c}{\lambda} = \frac{c}{kl_0} \tag{13.3.3}$$

Since both the speed of light c and the unit length l_0 are constants, the frequency β is inversely proportional to the coefficient k. In other words, the smaller the value of the coefficient k, the greater the frequency β. According to physics, the energy ε_c of a photon is:

$$\varepsilon_c = h\beta$$

Substituting equation (13.3.3) into the above equation gives the relationship equation:

$$\varepsilon_c = \frac{hc}{\lambda} = \frac{hc}{k} \cdot \frac{1}{l_0} \tag{13.3.4}$$

For any light wave. Since h, c, and l_0 are constants, the energy $\varepsilon_c\lambda = hc$ contained in the wavelength λ of any light wave is equal. The above equation shows that the energy hc contained in the wavelength λ is uniformly divided into k parts. The smaller the value of the coefficient k, the greater the energy ε_c of the photon. Conversely, the larger the value of the coefficient k, the smaller the energy ε_c of the photon.

Since the energy hc contained in the wavelength λ of a light wave is equally divided into k parts, the energy ε_c of a photon is essentially the energy possessed per unit length l_0. According to the above equation, the energy of photon A and photon B are respectively:

$$\varepsilon_{cA} = \frac{hc}{k_A} \cdot \frac{1}{l_0}, \qquad \varepsilon_{cB} = \frac{hc}{k_B} \cdot \frac{1}{l_0}$$

The ratio of the energies of photon A and photon B is:

$$\frac{\varepsilon_{cA}}{\varepsilon_{cB}} = \frac{k_B}{k_A} \tag{13.3.5}$$

Since Planck's constant $h = 6.626 \times 10^{-34}J \cdot s$ and the energy element $\varepsilon_h = 6.626 \times 10^{-34}J$, the ratio $\frac{h}{\varepsilon_h} = 1s$ of the two. The number n of energy elements ε_h contained in the energy ε_c of a photon is:

$$n = \frac{\varepsilon_c}{\varepsilon_h} = \frac{h\beta}{\varepsilon_h} = \beta \cdot 1s$$

Substituting equation (13.3.3) into the above equation gives the relationship equation:

$$n = \beta \cdot 1s = \frac{1}{k} \cdot \frac{c \cdot 1s}{l_0} \tag{13.3.6}$$

Since ε_c is the energy contained in a unit length l_0 of a light wave, the quantity n is also the number of energy elements ε_h contained in a unit length l_0 of a light wave. From this it can be seen that the frequency β of a light wave has three meanings.

Implication 1 is the number of periodic changes made by the light wave in $1s$ time.

Implication 2 is the number of energy elements ε_h contained in the energy ε_c of a photon $n = \beta \cdot 1s$.

Implication 3 is that the number of energy elements ε_h contained in the unit length l_0 of the light wave is $n = \frac{c \cdot}{kl_0}$.

13.3.2 A photon is an energy plane consisting of energy elements ε_h

According to optical theory, light waves are transverse waves, which have the property of plane-polarized light. Light being plane-polarized light can be verified by using two equally polarized lenses, A and B. The experimental procedure to verify that a light wave is a plane wave is shown in Figure 13-1.

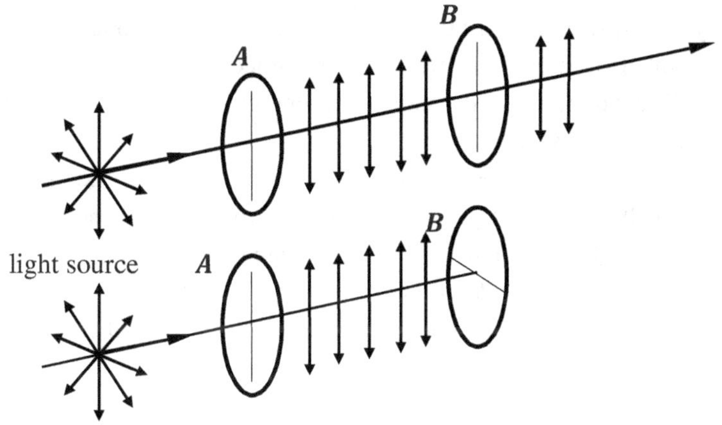

Figure 13-1 Schematic of an experiment to verify that light waves are plane waves. A and B are two identical optical polarizers.

When polarized lens A is illuminated with natural light (such as daylight or sunlight), the light passing through lens A is plane-polarized light. If lens A is fixed and lens B is rotated slowly, the intensity of the light passing through lens B changes periodically as lens B is rotated. When lens B is rotated $\mathbf{90°}$, the intensity of the light passing through lens B changes from brightest to darkest.

Since the direction of oscillation of light transmitted through lens A is not symmetrical with respect to the direction of propagation of light, photons are plane light. Figuratively speaking, a photon is like a page.

According to equation (13.3.6), the number n of energy elements ε_h contained in the energy ε_c of a photon is:

$$n = \beta \cdot 1s = \frac{1}{k} \cdot \frac{c \cdot 1s}{l_0}$$

Assume that the size of the wavelength λ is measured by the unit length $l_0 = 1 \times 10^{-9} m$. Since the speed of light $c = 3 \times 10^8 \cdot m/s$ and the wavelength $\lambda = 650 \times 10^{-9} m$ for the red wave, the number n of energy elements ε_h contained in the energy ε_c of the red photon is:

$$n = \frac{1}{k} \cdot \frac{c \cdot 1s}{l_0} = \frac{1}{650} \cdot \frac{3 \times 10^8}{10^{-9}} = 4.615 \times 10^{14} \qquad (13.3.7)$$

The coefficient $k = 650$ in the equation

According to high-energy physics, a high-energy γ light beam is a light wave with a wavelength $\lambda < 0.01 \times 10^{-9} m$. If the wavelength $\lambda = 0.001 \times 10^{-9} m$ of the γ beam, then the number n_γ of the energy element ε_h contained in the energy ε_c of the high-energy γ photon is:

$$n_\gamma = \frac{1}{k} \cdot \frac{c \cdot 1s}{l_0} = \frac{1}{0.001} \cdot \frac{3 \times 10^8}{10^{-9} m} = 3 \times 10^{20} \qquad (13.3.8)$$

The coefficient $k = 0.001$ in the equation.

According to equation (7.2.17), the number n_e of energy elements ε_h contained in the electron mass m_e is:

$$n_e = \frac{m_e}{m_h} = 1.239 \times 10^{20} \qquad (13.3.9)$$

It is clear that the number n_e of energy elements in an electron, is less than the number n_γ of energy elements in a γ photon, i.e. $n_e < n_\gamma$.

According to equation (13.3.4), the energy ε_c of the photon is:

$$\varepsilon_c = \frac{hc}{\lambda} = \frac{hc}{k} \cdot \frac{1}{l_0}$$

Since ε_c is the energy contained in a wavelength λ, the energy $\varepsilon_c \lambda = hc$ contained in any wavelength λ is equal. Since the energy hc is uniformly divided into k parts, the energy ε_c of a photon is the energy contained within a unit length l_0.

Since a photon is composed of energy elements ε_h and the shape of a photon is planar, the energy elements ε_h contained within the photon's energy ε_c should be distributed in a certain plane rather than over the unit length l_0.

Theoretically, we can use the number of energy elements ε_h contained within a unit length l_0 to represent the magnitude of the photon energy ε_c. One can also use the number of energy elements ε_h contained within the plane S_ε to express the magnitude of the photon energy ε_c.

Assume that the unit length l_0 is the transverse length of the photon energy plane (i.e., the length in the direction of the photon's motion). Assume that b is the longitudinal length of the photon energy plane (i.e., the length perpendicular to the photon in the direction of motion). In this book, the photon energy plane S_ε is defined as:

$$S_\varepsilon = l_0 \cdot b \qquad (13.3.10)$$

According to equation (13.3.6), the number n of energy elements ε_h contained in the energy plane S_ε of a photon is:

$$n = \frac{c \cdot 1s}{\lambda} = \frac{1}{k} \cdot \frac{c \cdot 1s}{l_0}$$

13.3.3 Number of energy elements ε_h contained in a transverse length l_0 and a longitudinal length b

Suppose l_m is the smallest unit length between energy elements ε_h. when the distance $r > 1(l_m)$. According to equation (13.2.21), the gravitational force F_h between two energy elements is the suction force $F_{h(+)}$, i.e:

$$F_{h(+)} = K\frac{\varepsilon_h^2}{c^4 r^3}, \quad r > 1(l_m) \qquad (13.3.11)$$

Under the action of the suction force $F_{h(+)}$. Two energy elements ε_h moving in the same direction gradually approach each other and cannot move away from each other.

When the distance $r < 1(l_m)$. According to equation (13.2.22), the repulsive force $F_{h(-)}$ between two energy elements ε_h is:

$$F_{h(-)} = -K\frac{\varepsilon_h^2}{c^4 r^3}, \quad r < 1(l_m) \qquad (13.3.12)$$

Under the action of the repulsive force $F_{h(-)}$. Two energy elements ε_h moving in the same direction gradually move away from each other and cannot merge into a new larger energy particle.

When the distance $r = 1(l_m)$ the gravitational force F_h between the two energy elements is:

$$F_h = 0, \quad r = 1(l_m) \qquad (13.3.13)$$

At this point, the gravitational force F_h between the two energy elements ε_h reaches a state of equilibrium, since the distance r between the two energy elements ε_h neither increases nor decreases.

Since photons are composed of energy elements, the energy elements ε_h in each photon are in gravitational equilibrium with each other, i.e. $F_h = 0$.

Since the minimum unit length $l_m \neq 0$ between energy elements ε_h, only a certain number of energy elements ε_h can be accommodated on a unit length l_0. The number n_{l0} of energy elements ε_h that can be accommodated on a transverse unit length l_0 is:

$$n_{l0} = \frac{l_0}{l_m} \qquad (13.3.14)$$

Note that n_{l0} in equation is also the number of longitudinal queues owned by the energy plane S_ε.

When the number of energy elements ε_h contained in a photon is very large, the excess energy elements ε_h can only be arranged on the longitudinal length b, since there is no room for n energy elements ε_h on the transverse length l_0. It is this reason that causes the photon energy ε_c to exhibit a planar $S_\varepsilon = l_0 \cdot b$ distribution rather than a line distribution.

The number n_b of energy elements ε_h that can be accommodated on a longitudinal length b is:

$$n_b = \frac{n}{n_{l0}} = \frac{1}{k} \cdot \frac{c \cdot 1s}{l_0^2} l_m \qquad (13.3.15)$$

Note that n_b in equation is also the number of transverse queues owned by the energy plane S_ε. The size of the longitudinal length b is:

$$b = n_b l_m = \frac{1}{k} \cdot \frac{c \cdot 1s}{l_0^2} l_m^2 \qquad (13.3.16)$$

Obviously, the shorter the wavelength $\lambda = kl_0$ of the light wave and the smaller the coefficient k, the greater the longitudinal length b of the photon. The energy plane of the photon, S_ε, is shown in Figure 13-2.

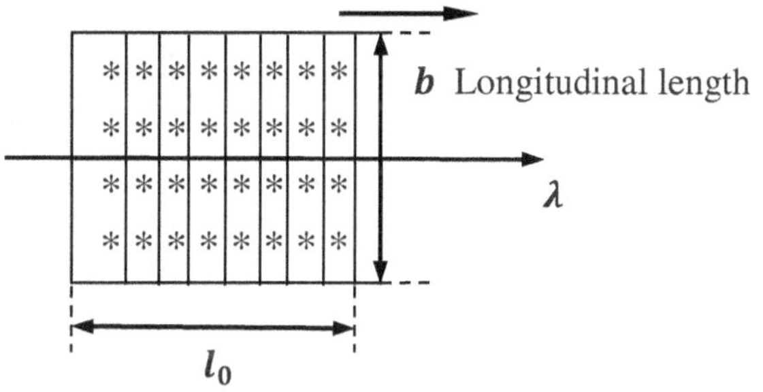

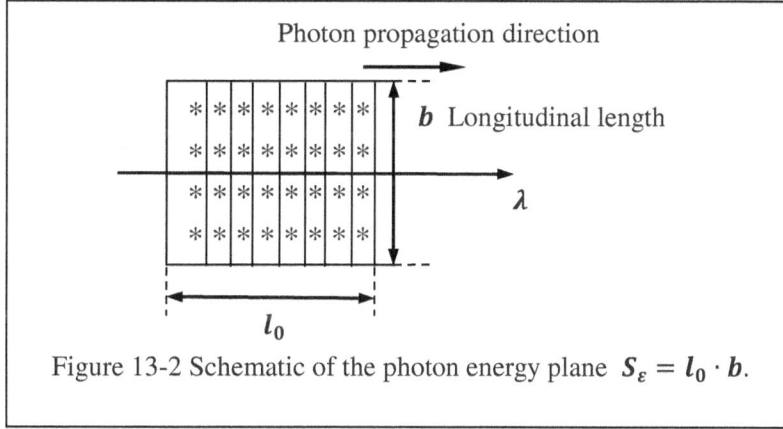

Figure 13-2 Schematic of the photon energy plane $S_\varepsilon = l_0 \cdot b$.

Note: Figure 13-2 contains 4 transverse queues (**4** unit lengths l_0) and **8** longitudinal queues. The asterisk * in the figure is the energy element ε_h and b is the longitudinal length of the photon energy plane S_ε.

Although the photon energy ε_c is distributed on the energy plane $S_\varepsilon = l_0 \cdot b$, the longitudinal length b is a variable since the unit length l_0 remains constant while the photon energy ε_c can take different values.

In other words. Since the photon energy ε_c is inversely proportional to the wavelength λ, and the unit length l_0 is a fixed number, the smaller the wavelength λ, the larger the number of energy elements ε_h possessed on the longitudinal length b is: At this point, the value of the longitudinal length b is larger.

Conversely, the greater the wavelength λ, the smaller the number of energy elements ε_h possessed on the longitudinal length b. The smaller the value of the longitudinal length at this point. From this point of view, the number of energy elements ε_h possessed on the energy plane $S_\varepsilon = l_0 \cdot b$ is proportional to the longitudinal length b.

13.3.4 The energy of a photon $\varepsilon_c = h\beta$ is a plane with neat vertical and horizontal alignments

In a military parade, there is an orderly formation of soldiers marching through the square. If the energy plane of the photons $S_\varepsilon = l_0 \cdot b$ is considered as a queue of soldiers and the energy element ε_h as a soldier, then according to equation (13.3.14) the number of soldiers n_{l0} contained in a lateral queue (i.e. within a unit length l_0) is:

$$n_{l0} = \frac{l_0}{l_m}$$

According to equation (13.3.15), the number n_b of soldiers contained in a longitudinal queue (i.e. within a longitudinal length b) is:

$$n_b = \frac{n}{n_{l0}} = \frac{nl_m}{l_0}$$

The number n of soldiers contained in the photon energy plane $S_\varepsilon = l_0 \cdot b$ is:

$$n = n_{l0} \cdot n_b \qquad (13.3.17)$$

It is clear that the energy elements ε_h in the photon energy plane $S_\varepsilon = l_0 \cdot b$ are all moving forward at the speed of light c in step. as shown in Figure 13-3.

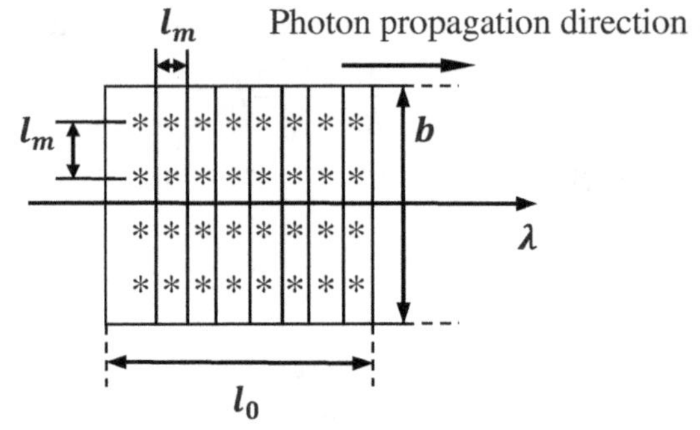

Figure 13-3 Schematic diagram of the horizontal queue of energy elements contained within a unit length l_0.

The photon energy plane $S_\varepsilon = l_0 \cdot b$ in Figure 13-3 contains four transverse queues and eight longitudinal queues. A transverse queue contains 8 energy elements ε_h (8 asterisks *).

According to equation (13.3.3), the frequency β of the light wave is:

$$\beta = \frac{c}{kl_0}$$

According to equation (13.3.16), the longitudinal length b is:

$$b = \frac{1}{k} \cdot \frac{c \cdot 1s}{l_0^2} l_m^2 = \frac{(\beta \cdot 1s) l_m^2}{l_0} \qquad (13.3.18)$$

Substituting the above equation in the photon energy plane $S_\varepsilon = l_0 \cdot b$ gives the relationship equation:

$$S_\varepsilon = l_0 \cdot b = (\beta \cdot 1s) l_m^2 \qquad (13.3.19)$$

Since the photon energy $\varepsilon_c = h\beta$ is distributed over the rectangular area $S_\varepsilon = l_0 \cdot b$, the energy density σ on the photon energy plane S_ε is:

$$\sigma = \frac{\varepsilon_c}{S_\varepsilon} = \frac{h\beta}{(\beta \cdot 1s) l_m^2} = \frac{\varepsilon_h}{l_m^2} \qquad (13.3.20)$$

The above equation is the energy density formula for the photon plane. Since the energy element ε_h and the minimum unit length l_m are constants, the energy density σ of any photon energy plane is equal. Alternatively, the energy density σ of a photonic plane is a constant. The above formula shows that each minimum unit area squared l_m^2 contains only one energy element ε_h.

It should be noted that the energy density σ of the photon energy plane and the photon energy density per unit area are two different concepts. The former refers to the density of the energy element ε_h inside the photon, while the latter refers to the density of photons per unit area.

13.3.5 Difference between the energy plane of low-energy photons and that of high-energy photons

Since the transverse side length of the photon energy plane is always equal to the unit length l_0, an increase in the photon energy plane S_c is an increase in the longitudinal side length b, with the unit length l_0 remaining constant.

According to equation (13.3.18), the longitudinal length b is:

$$b = \frac{(\beta \cdot 1s) l_m^2}{l_0}$$

As the photon frequency β keeps decreasing. Since the unit length l_0 always remains constant, the longitudinal side length b becomes constantly smaller. It can be determined that the longitudinal side length b is smaller in the energy plane of low-energy photons. This is shown in Figure 13-4.

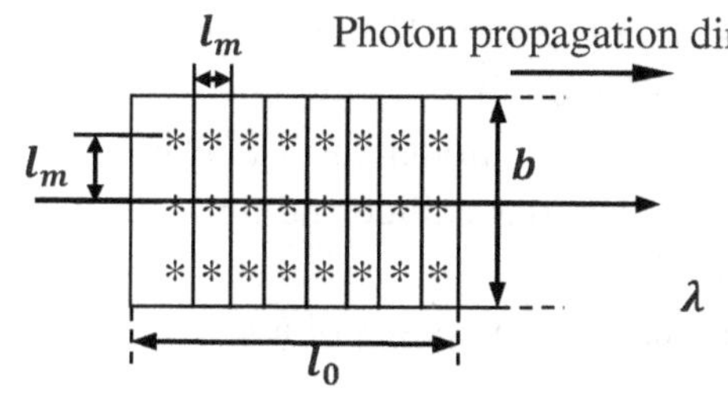

Figure 13-4 Schematic diagram of low-energy photon energy planes.

Conversely, when the photon frequency $\boldsymbol{\beta}$ is increasing, the longitudinal side length $\boldsymbol{b}$ becomes continuously larger because the unit length $\boldsymbol{l_0}$ always remains constant. From this it can be determined that the longitudinal side length $\boldsymbol{b}$ of the high-energy photon energy plane is large. This is shown in Figure 13-5.

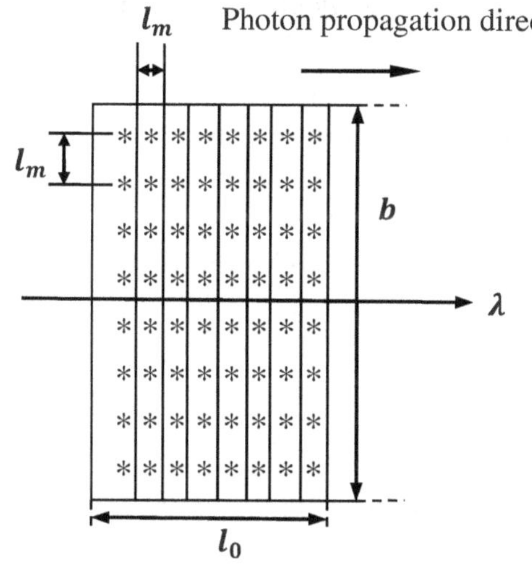

Figure 13-5 Schematic of high-energy photon energy planes.

If photons are compared to a thin page of a book, then the intensity of light is like the thickness of a book. Obviously, the more pages the book has, the thicker the book is, and by the same token, the more photons there are, the greater the light intensity.

13.4 Energy planes of low-energy γ photons can be bent and rolled into hollow circular energy tubes

13.4.1 The energy plane of one high-energy γ photon can be fissioned into the energy planes of two low-energy γ photons

According to high-energy physics experiments, when a high-energy $\boldsymbol{\gamma}$ photon with an energy $\boldsymbol{\varepsilon}$ greater than $\boldsymbol{1022 \cdot keV}$ collides with the nucleus of a lead plate atom, the high-energy $\boldsymbol{\gamma}$ photon can be split into two positive and negative electrons with an energy of $\boldsymbol{511 \cdot keV}$, and the reaction equation is:

$$\gamma \rightarrow e^{+} + e^{-} \tag{13.4.1}$$

The remaining energy of the high-energy $\boldsymbol{\gamma}$ photon is converted into the kinetic energy of the positron and the negative electron.

When the positive and negative electrons collide, the positive and negative electrons disappear, and the mass of the positive and negative electrons is transformed into two low-energy γ photon energy planes of equal energy ($\mathbf{511 \cdot keV}$), the same frequency, and opposite directions of propagation. The reaction equation is:

$$e^{+} + e^{-} \rightarrow 2\gamma \qquad (13.4.2)$$

Based on the fact that the high-energy γ photon passes through the lead plate and turns into positive and negative electrons, and the fact that the positive and negative electrons collide and turn into two low-energy γ photon energy planes, the following two conclusions can be deduced.

Conclusion 1, both positive and negative electrons are composed of a low-energy γ photon with energy equal to $\mathbf{511 \cdot keV}$.

Since electrons are not low-energy γ photons, the energy plane of a γ photon is not an electron. However, since the collision of positive and negative electrons transforms into the energy planes of two γ photons that propagate in opposite directions, the interior of the positive and negative electrons should contain curved closed energy curved surfaces.

Obviously, this curved closed energycurved surface should be a hollow circular energy tube. Since the positive and negative electrons are hollow circular energy tubes, it is only when the positive and negative electrons collide that the positive and negative electrons can be transformed into the energy plane of two γ photons.

Conclusion 2, the lead nucleus tears the energy plane of the high-energy γ photon into two energy planes of the low-energy γ photon and rolls these two low-energy planes into two hollow circular energy tubes.

These two hollow circular energy tubes constitute the positive and negative electrons. When the positive and negative electrons collide, the two hollow circular energy tubes (positive and negative electrons) are transformed into two γ photon energy planes that propagate in opposite directions.

Theoretically, the energy element ε_h contained within the electron is not a spherical energy particle stacked haphazardly. Or rather, the electron is not a spherical energy particle.

Otherwise, it would be difficult to explain why, when positive and negative electrons collide, they do not emit three or four other low-energy photons, but always emit two energy planes of low-energy γ photons of equal energy ($\mathbf{511 \cdot keV}$), the same frequency, and opposite propagation directions.

13.4.2 Lead nuclei cut the energy plane of a high-energy γ photon into the energy planes of two low-energy γ photons

Theoretically: the surface of the lead atomic nucleus is not a smooth sphere, but angular and uneven. When the energy plane of a photon collides with a point B on the nucleus of a lead plate, assume that the collision point B is a very tiny bump.

When the low-energy plane $S_\varepsilon = l_0 \cdot b$ collides with point B of the lead nucleus. Since the length b of the low-energy plane S_ε is very small, the low-energy plane S_ε is not cut into two energy planes by the B point. The low-energy plane S_ε is usually either all reflected or all refracted at the B point. This is shown in Figure 13-6.

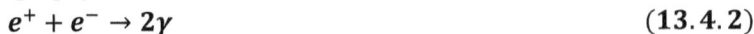

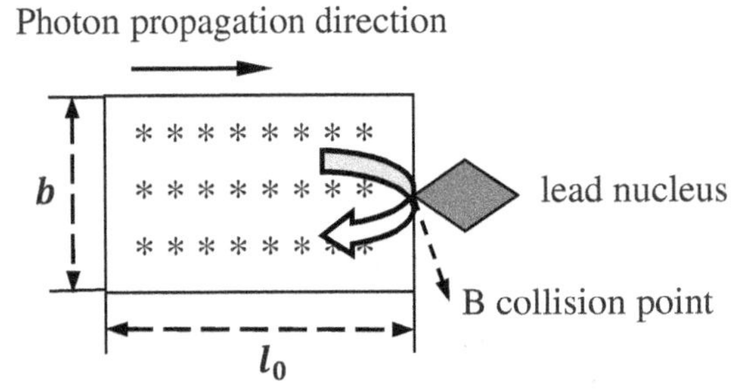

Figure 13-6 Schematic of collisional rebound or refraction of a low-energy photon energy plane with a lead plate nucleus.

The collision of the low-energy plane $S_\varepsilon = l_0 \cdot b$ with the B cut point is like letting a short plank of wood collide with the blade of a knife, at which point the knife can hardly cut the short plank in half, and the short plank usually bounces back from the knife.

When the high-energy plane $S_\varepsilon = l_0 \cdot b$ collides with the B point of the lead nucleus. Since the length b of the high-energy plane S_ε is large, the B collision point is equivalent to a sharp knife blade, which splits the energy plane of the high-energy γ photon in two and turns it into two low-energy γ photons, as shown in Figure 13-7.

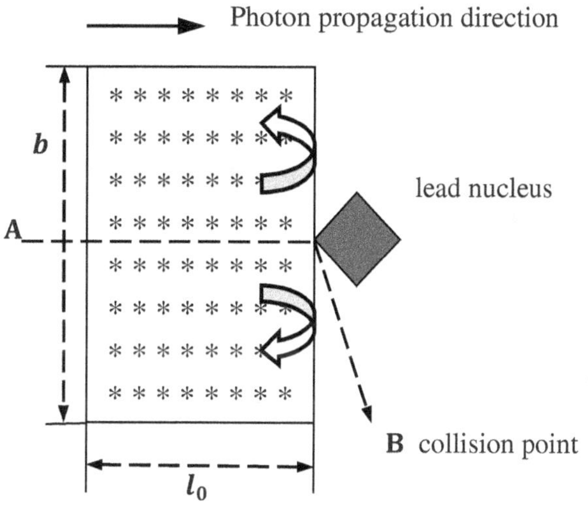

Figure 13-7 Schematic diagram of the energy plane $S_\varepsilon = l_0 \cdot b$ of a high-energy γ photon, cut by a lead nucleus along the dashed line AB.

The horizontal line AB in the figure is the splitting line of the high-energy γ photon energy plane.

The collision of the high-energy plane $S_\varepsilon = l_0 \cdot b$ with the B cutting point is like letting the midpoint of a slender plank collide with the blade of a knife, at which point the knife easily cuts the slender plank into two short planks.

It should be noted that the two low-energy γ photons produced by the cut turn out to be positive and negative electrons, rather than the energy plane of the γ photons, $S_\varepsilon = l_0 \cdot b$. Otherwise, two γ photons with different directions of propagation should have appeared, not positive and negative electrons.

13.4.3 The electron is a hollow circular energy tube

When positive and negative electrons collide, the positive and negative electrons disappear and the mass of the positive and negative electrons is transformed into two planes of γ photon energy of equal energy ($\mathbf{511} \cdot \boldsymbol{keV}$) and opposite direction of propagation. This fact shows that the electrons contain γ photon energy within them.

According to physics, the rest mass of a photon $m_0 = \mathbf{0}$. If the γ photon is at rest inside an electron, why is the mass m_e of the resting electron not equal to 0, and why does the γ photon at rest inside the electron suddenly have the speed of light c when the positive and negative electrons disappear?

Modern physics considers the photon to be a particle of energy. When a high-energy γ photon collides with the nucleus of a lead atom, it becomes a positive or negative electron. Since the displacement velocity v of the electron is less than the speed of light c, physicists cannot explain how the velocity of the γ photon inside the electron suddenly becomes less than the speed of light, or why the speed of light c of the γ photon inside the electron suddenly becomes 0.

Since photons are all composed of the energy element ε_h, the collision of a γ photon with the nucleus of a lead atom belongs to the elastic collision of the energy element ε_h with the nucleus. Since the lead nucleus cannot incorporate all of the energy element ε_h in the energy plane of the γ photon into the electron in a single instant, it does not make logical sense for the classical theory to view the electron as an energy particle.

Since the photon energy plane $S_\varepsilon = l_0 \cdot b$ always moves forward at the speed of light c, and the electron displacement velocity v is smaller than the speed of light c, the γ photon contained in the electron is clearly not the energy plane S_ε.

Since the energy plane $\boldsymbol{S}_\varepsilon$ of the high-energy $\boldsymbol{\gamma}$ photon is cut into two low-energy $\boldsymbol{\gamma}$ photons by the lead nucleus, the energy plane $\boldsymbol{S}_\varepsilon = \boldsymbol{l}_0 \cdot \boldsymbol{b}$ of the new low-energy $\boldsymbol{\gamma}$ photon will be bent during the cutting and collision rebound process by the law of conservation of momentum and the law of conservation of angular momentum during the cutting process, and thus become a curved closed-energy curved surface. the closed-energy curved surface is similar to the seawater surfing schematic in Figure 13-8.

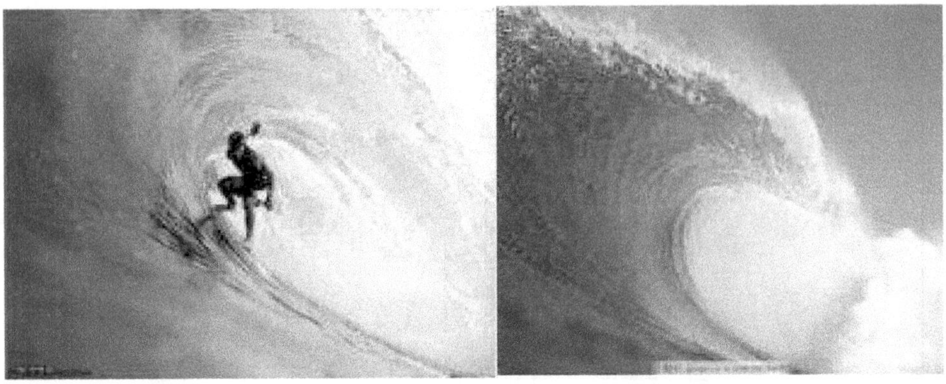

Figure 13-8 Curved surface formed by seawater.

We know that when the underground reef blocks the advance of the waves, the curved surface of seawater consisting of water will appear on the sea surface. as shown in Figure 13-8.

If we regard the energy element $\boldsymbol{\varepsilon}_h$ as a water molecule, the energy plane of low-energy $\boldsymbol{\gamma}$ photons as ocean waves, and the atomic nucleus as an underground reef, then the process of the energy plane of low-energy $\boldsymbol{\gamma}$ photons turning into a closed energy curved surface is similar to the process of the formation of seawater curved surface.

Further, the photon energy planes cut on both sides of the horizontal line AB in Figure 13-7, under the action of the lead nuclei, curl one inward to form a hollow circular energy tube and the other outward to form a hollow circular energy tube. Since these two photon hollow circular energy tubes rotate in opposite directions, these two photon hollow circular energy tubes form a positron and a negatron.

Since the high-energy $\boldsymbol{\gamma}$ photons are cut by the lead nucleus into two hollow circular energy tubes with opposite rotational directions, this book refers to the hollow circular energy tubes as electron circular tubes. The electron circular tube is similar to several bearings arranged in parallel, and the balls in the bearings correspond to the longitudinally arranged energy element $\boldsymbol{\varepsilon}_h$. This is shown in Figure 13-9.

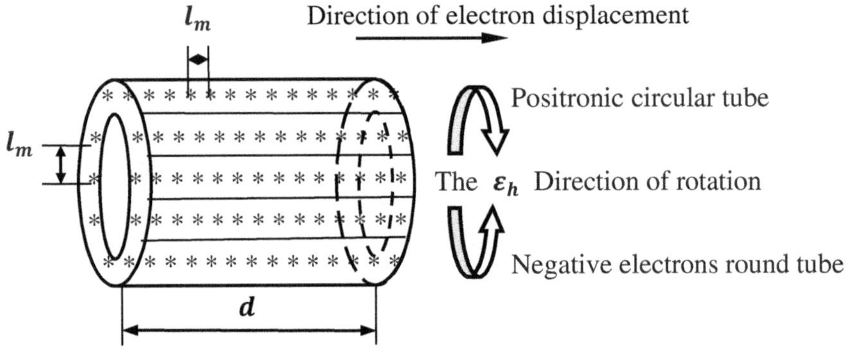

Figure 13-9 Schematic diagram of an energy circular tube (electron circular tube) formed by the energy element $\boldsymbol{\varepsilon}_h$ moving at the speed of light $\boldsymbol{c}$ in a spiral.

It should be noted that the energy element $\boldsymbol{\varepsilon}_h$ is all in rotational motion about the centerline of the tube at the speed of light c in the electron circular tube, as shown in Figure 13-10.

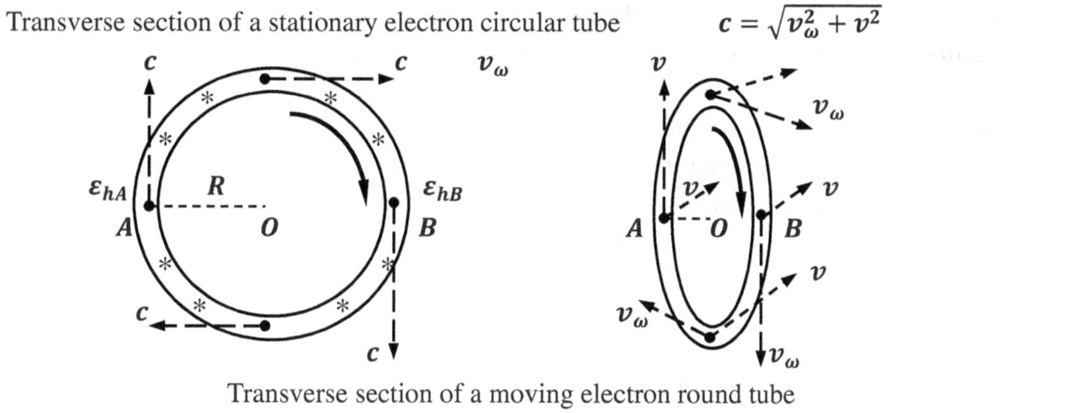

Figure 13-10 Schematic of transverse sections of a stationary and moving electron circular tube. R is the distance from the energy element ε_h to point O at the center of the tube.

The rotational motion of the energy element ε_h in the electron circular tube is analogous to the rotational motion of the Earth around the Sun.

In summary, a high-energy γ photon energy plane, cut by the nucleus into two hollow circular energy tubes (electron round tube). At this time, the electron round tube is not static, but around the center line of the electron round tube to do spiral motion (displacement + vertical rotation).

Since electrons have the property of spin and the hollow circular energy tube rotates perpendicularly to the displacement line with the velocity v_ω, the perpendicular rotation of the hollow circular energy tube around the displacement line is the spin of the electron.

Physicists consider the electron to be a spherical particle, while this book considers the electron to be a hollow circular tube of energy (electron round tube). Using the electron round tube theory can explain why two γ photons in two opposite directions are emitted when positive and negative electrons collide and disappear, and why electrons have spin properties and other related theoretical issues.

For two electron round tubes moving in the same direction, if one electron round tube spins clockwise and the other electron tube spins counterclockwise, then the two electron round tubes with opposite spin directions form a positive electron and a negative electron.

When the collision of positive and negative electrons disappears, because the spin direction of the electron circular tube contained in the positive and negative electrons is opposite, the disappearance of the positive and negative electrons must become two γ photon energy planes with opposite propagation directions.

If the electron round tube of positive and negative electrons does not rotate in the opposite direction, then it is difficult to explain why, when the positive and negative electrons collide and disappear, the positive and negative electrons do not change into three or four other low-energy photons, but always into two low-energy γ photons of the same energy, the same frequency, and opposite directions of propagation.

13.4.4 The energy element ε_h always moves inside the electron at the spiral speed of light c

Since the speed of the positive and negative electrons before the collision disappears is small, the speed of light c of the two γ photons radiated when the collision of the positive and negative electrons disappears is not the result of the action of other external forces, but rather, the hollow circular energy tubes of the positive and negative electrons turn into energy planes of the γ photons.

Since the energy element ε_h always moves at the speed of light c, the energy element ε_h in the electron circular tube always rotates around the center line of the electron circular tube at the helical speed of light c (displacement speed + circular line speed). Otherwise, the energy element ε_h in the electron tube becomes invisible in an instant.

When the electron is at rest in the cosmic vacuum frame of reference, the trajectory of the rotational motion of the energy element ε_h in the electron circular tube is a circular line. At this time, the direction of rotation of the energy element ε_h is perpendicular to the centerline of the electron circular tube.

When the electron moves with speed v in the cosmic vacuum frame of reference, the trajectory of the rotational motion of the energy element ε_h is a helix. At this time, the direction of rotation of the energy element ε_h (the direction of the speed of light) is not perpendicular to the direction of displacement of the electron.

Assuming that δ is the angle between the direction of light speed and the direction of electron displacement for the energy element ε_h, the angle δ becomes smaller as the electron displacement velocity v increases.

When the electron is at rest, i.e., when the electron displacement velocity $v = 0$, the angle $\delta = 90^0$. When the electron displacement velocity v is equal to the speed of light c, i.e., when the displacement velocity $v = c$, the angle $\delta = 0$. This is shown in Figure 13-11.

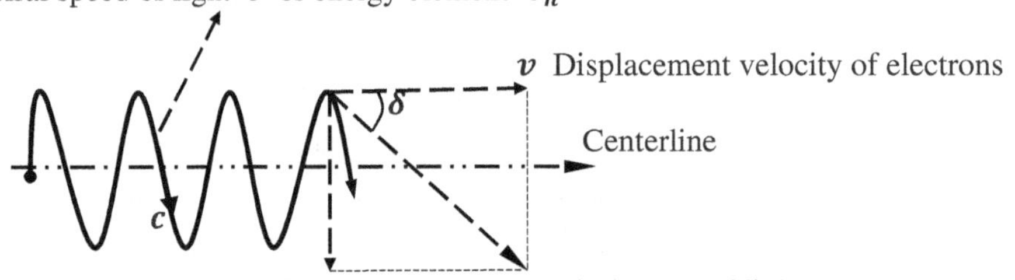

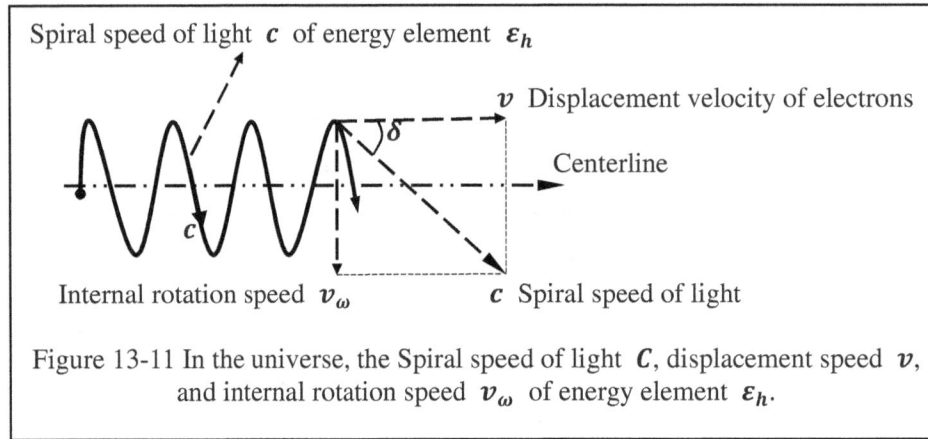

Figure 13-11 In the universe, the Spiral speed of light C, displacement speed v, and internal rotation speed v_ω of energy element ε_h.

The spiral speed of light c of the energy element ε_h can be decomposed into two mutually perpendicular partial speeds.

One of them is: the velocity v of the energy element ε_h in the direction of the electron displacement. note that the velocity v is at the same time the displacement velocity of the electron circular tube, viz:

$$v = c \cdot \cos\delta \qquad (13.4.3)$$

The second is: the internal rotational velocity v_ω of the energy element ε_h in the direction perpendicular to the direction of the electron displacement. note that the internal rotational velocity v_ω of the energy element is also the spin velocity of the electron circular tube, i.e:

$$v_\omega = c \cdot \sin\delta \qquad (13.4.4)$$

The spiral speed of light c of the energy element ε_h is:

$$c = \sqrt{v_\omega^2 + v^2} \qquad (13.4.5)$$

The above equation is the equation for the conservation of the speed of light for the energy element. Note that the speed c of light, the spin speed v_ω, and the displacement speed v, all three are speeds in the cosmic vacuum frame of reference. The above formula is similar to Pythagoras theorem in math.

When the γ photon's rotational velocity $v_\omega \neq 0$ and displacement velocity $v < c$ within the electron, the energy element ε_h in the γ photon is no longer in a straight-line motion form, but in a spiral motion form, and the hollow, circular energy tube of the γ photon is the electron.

The principle of conservation of the spiral speed of light is a very important physical theorem, and many important experimental laws in the classical electromagnetic field theory can be analytically derived using the above formula. The spiral speed of light conservation formula is the cornerstone of electromagnetic field theory.

13.4.5 Spiral speed of light conservation formula in the S' inertial system

Since the speed c of light, belongs to the speed of the cosmic vacuum frame of reference, both the velocity v_ω, and the velocity v, in equation (13.4.5) also belong to the speed of the cosmic vacuum frame of reference.

Suppose that the inertial system S' in the cosmic vacuum frame of reference S moves in the direction of the positive X-axis with the velocity u, i.e. with the velocity $u \geq 0$. If the photon is also displaced in the direction of the positive X-axis in the cosmic vacuum frame of reference S, then the velocity c' of propagation of the photon in S' is:

$$c' = c - u \geq 0 \tag{13.4.6}$$

Assume that the electron circular tube is displaced along the $+X'$ axis with a velocity $v' - u \geq 0$ and that the electron circular tube has a spin velocity v'_ω in the inertial system S'. Since the electron spin velocity v'_ω and the velocity u are perpendicular to each other, the spin velocity $v'_\omega = v_\omega$. the spiral speed of light c' in the inertial system S' is:

$$c' = c - u = \sqrt{v'^2_\omega + v'^2} = \sqrt{v^2_\omega + v'^2} \tag{13.4.7}$$

The above formula is the formula for the conservation of the spiral velocity of light in the S' inertial system. The spiral velocity of light c' in the above formula is the maximum velocity propagating along the $+X'$ axis in the S' inertial system.

When the inertial system S' velocity $u = 0$, the above formula becomes the formula for the conservation of the spiral velocity of light in the cosmic vacuum. At this point, the spiral velocity of light $c' = c$.

13.5 Derivation of the formula for the energy element directional force $f_{\varepsilon h}$

13.5.1 The energy element ε_h contains within it the rotational energy element $\varepsilon_{h\omega}$ and the displacement energy element ε_{hv}

Multiplying equation (13.4.5) by the energy element ε_h gives the relation equation:

$$\varepsilon_h c = \sqrt{(\varepsilon_h v_\omega)^2 + (\varepsilon_h v)^2} \tag{13.5.1}$$

Dividing the above equation by the speed of light c, yields the relational equation that:

$$\varepsilon_h = \sqrt{\left(\varepsilon_h \frac{v_\omega}{c}\right)^2 + \left(\varepsilon_h \frac{v}{c}\right)^2} = \sqrt{(\varepsilon_{h\omega})^2 + (\varepsilon_{hv})^2} \tag{13.5.2}$$

The above formula shows that the energy element ε_h contains two components of energy within it. Since the velocity v_ω is the internal rotation speed of the energy element ε_h, this book defines the first component $\varepsilon_{h\omega}$ contained within the root sign of the formula as the rotational energy element of the energy element ε_h, i.e:

$$\varepsilon_{h\omega} = \varepsilon_h \frac{v_\omega}{c} \tag{13.5.3}$$

According to the spiral speed of light formula $c = \sqrt{v^2_\omega + v^2}$, the internal rotation speed v_ω of the energy element ε_h is:

$$v_\omega = \sqrt{c^2 - v^2} \tag{13.5.4}$$

Substituting the above equation into equation (13.5.3) gives the relationship equation:

$$\varepsilon_{h\omega} = \varepsilon_h \sqrt{1 - \frac{v^2}{c^2}} \tag{13.5.5}$$

The above equation is the rotational energy element equation. The above equation shows that the rotational energy element $\varepsilon_{h\omega}$ becomes smaller as the electron displacement velocity v increases.

Since the velocity v is the displacement velocity of the energy element ε_h (i.e., the displacement velocity of the electron), this book defines the second component ε_{hv} contained within the root sign of equation (13.5.2) as the displacement energy element of the energy element ε_h, i.e:

$$\varepsilon_{hv} = \varepsilon_h \frac{v}{c} \tag{13.5.6}$$

The above equation is the displacement to energy element equation. The above equation shows that the displacement energy element ε_{hv} is proportional to the displacement velocity v of the electron.

According to equation (13.5.2), the energy elements ε_{hA} and ε_{hB} are, respectively:

$$
\begin{cases}
\varepsilon_{hA} = \sqrt{\left(\frac{\varepsilon_{hA} v_{\omega A}}{c}\right)^2 + \left(\frac{\varepsilon_{hA} v_A}{c}\right)^2} = \sqrt{(\varepsilon_{h\omega A})^2 + (\varepsilon_{hvA})^2} \\
\varepsilon_{hB} = \sqrt{\left(\frac{\varepsilon_{hB} v_{\omega B}}{c}\right)^2 + \left(\frac{\varepsilon_{hB} v_B}{c}\right)^2} = \sqrt{(\varepsilon_{h\omega B})^2 + (\varepsilon_{hvB})^2}
\end{cases} \quad (13.5.7)
$$

13.5.2 The directional force $f_{\varepsilon h}$ of the energy element ε_h contains within it the rotational directional force $f_{\varepsilon h\omega}$ and the displacement directional force $f_{\varepsilon hv}$

According to the spiral speed of light formula $c = \sqrt{v_\omega^2 + v^2}$ for the energy element ε_h, it can be seen that there are two directions of motion contained within the energy element ε_h, one in the direction of the rotational speed v_ω and the other in the direction of the displacement speed v.

Suppose r is the distance between the energy elements ε_{hA} and ε_{hB}. Assume that v_A is the displacement velocity of the energy element ε_{hA}, v_B is the displacement velocity of the energy element ε_{hB}, β is the angle between the displacement velocities v_A and v_B, and θ is the angle between the distance r and the displacement velocity v_A. This is shown in Figure 13-12.

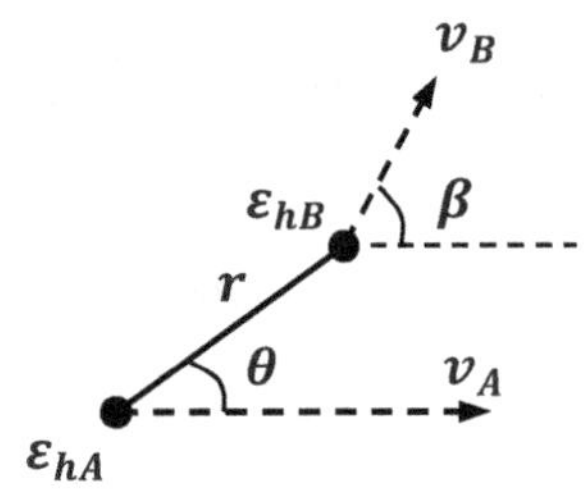

Figure 13-12 Schematic of the displacement direction of the energy elements ε_{hA} and ε_{hB}.

If the angle $\beta = 0$ between the displacement velocities v_A and v_B, then the energy elements ε_{hA} and ε_{hB} are moving in the same direction.

Since the electron contains only the energy element ε_h, the directional force $f_{\varepsilon h}$ of the energy element ε_h is the source of the electric field force F_E and the magnetic field force F_B.

Since the electromagnetic field force F is inversely proportional to the square of the distance r, the directional force $f_{\varepsilon h}$ between the energy elements ε_{hA} and ε_{hB} can be expressed as:

$$
f_{\varepsilon h} = T\frac{\varepsilon_{hA} \cdot \varepsilon_{hB}}{r^2}\sin\theta\cos\beta \quad (13.5.8)
$$

The above equation is a mathematical expression for the axiom of the directional force $f_{\varepsilon h}$ of the energy element. T in the formula is the directional force constant.

Substituting equation (13.5.7) into the above equation gives the relationship equation:

$$
f_{\varepsilon h} = T\frac{\sqrt{\left(\frac{\varepsilon_{hA} v_{\omega A}}{c}\right)^2 + \left(\frac{\varepsilon_{hA} v_A}{c}\right)^2} \cdot \sqrt{\left(\frac{\varepsilon_{hB} v_{\omega B}}{c}\right)^2 + \left(\frac{\varepsilon_{hB} v_B}{c}\right)^2}}{r^2}\sin\theta\cos\beta \quad (13.5.9)
$$

Since the directional force between the rotational energy element $\varepsilon_{h\omega}$ and the displacement energy element ε_{hv} is equal to zero, the above equation becomes:

$$
f_{\varepsilon h} = \sqrt{\left(\frac{T}{c^2}\frac{\varepsilon_{hA} v_{\omega A} \cdot \varepsilon_{hB} v_{\omega B}}{r^2}\sin\theta\cos\beta\right)^2 + \left(\frac{T}{c^2}\frac{\varepsilon_{hA} v_A \cdot \varepsilon_{hB} v_B}{r^2}\sin\theta\cos\beta\right)^2} \quad (13.5.10)
$$

The above equation shows that the directional force $f_{\varepsilon h}$ of the energy element ε_h contains two component directional forces within it.

Since the velocity v_ω is the internal rotation speed of the energy element ε_h, this book defines the first component $f_{\varepsilon h\omega}$ contained within the root sign of the above equation as the rotational directional force of the energy element ε_h, i.e:

$$f_{\varepsilon h\omega} = \frac{T}{c^2}\frac{\varepsilon_{hA}v_{\omega A}\cdot\varepsilon_{hB}v_{\omega B}}{r^2}\sin\theta\cos\beta \qquad (13.5.11)$$

According to equation (13.5.4) , the rotational velocities $v_{\omega A}$ and $v_{\omega B}$ of the energy elements ε_{hA} and ε_{hB}, respectively, are:

$$v_{\omega A} = \sqrt{c^2 - v_A^2}\,, \qquad v_{\omega B} = \sqrt{c^2 - v_B^2} \qquad (13.5.12)$$

Substituting the above equation into equation (13.5.11) yields the relationship equation:

$$f_{\varepsilon h\omega} = T\frac{\varepsilon_{hA}\cdot\varepsilon_{hB}}{r^2}\sqrt{1-\frac{v_A^2}{c^2}}\sqrt{1-\frac{v_B^2}{c^2}}\sin\theta\cos\beta \qquad (13.5.13)$$

The above equation is the formula for the rotational directional force of the energy element. The above equation shows that the rotational directional force $f_{\varepsilon h\omega}$ becomes smaller as the electron displacement velocities v_A and v_B increase.

Since the velocity v is the displacement velocity of the energy element ε_h (i.e., the displacement velocity of the electron), this book defines the second component $f_{\varepsilon hv}$ contained within the root sign of equation (13.5.10) as the displacement-directional force of the energy element ε_h, i.e:

$$f_{\varepsilon hv} = \frac{T}{c^2}\frac{\varepsilon_{hA}v_A\cdot\varepsilon_{hB}v_B}{r^2}\sin\theta\cos\beta \qquad (13.5.14)$$

The above equation is the displacement direction force equation for the energy element. The above equation shows that the displacement direction force $f_{\varepsilon hv}$ is proportional to the displacement velocities v_A and v_B of the electron.

If the rotational directional force $f_{\varepsilon h\omega}$ is greater than the displacement directional force $f_{\varepsilon hv}$, i.e., $f_{\varepsilon h\omega} > f_{\varepsilon hv}$, then the directional force $f_{\varepsilon h}$ is characterized by the electric field force F_E.

If the rotational directional force $f_{\varepsilon h\omega}$ is smaller than the displacement directional force $f_{\varepsilon hv}$, i.e., $f_{\varepsilon h\omega} < f_{\varepsilon hv}$, then the directional force $f_{\varepsilon h}$ is characterized by the magnetic field force F_B.

13.5.3 Causes of stabilized rotation of electron round tubes

The energy elements ε_h inside the electron circular tube are in two states of rotational motion with respect to each other.

The first state of motion is the rotational motion of the energy element ε_{hA} in the same direction as the neighboring energy element ε_h.

In Figure 13-10, the energy element ε_{hA} at point A rotates in the same direction as the neighboring energy element ε_h, i.e., upward at the same time.

Suppose r is the distance between the energy element ε_{hA} and the neighboring energy element ε_h, and suppose l_m is the minimum unit length in the gravitational field. If the distance $r = 1(l_m)$, then the gravitational force F_{mh} between the energy element ε_{hA} and the neighboring energy element ε_h at point A is equal to 0, i.e:

$$F_{mh} = 0 \qquad (13.5.15)$$

Assume that v is the displacement velocity of energy element ε_{hA} and neighboring energy element ε_h (i.e., the displacement velocity of electrons), β is the angle between the displacement velocity v of the energy element ε_{hA} and the displacement velocity v of the neighboring energy element ε_h, and θ is the angle between the distance r and velocity v.

According to the directional force equation (13.5.8), the directional force $f_{\varepsilon h}$ between an energy element ε_{hA} and a neighboring energy element ε_h is:

$$f_{\varepsilon h} = T\frac{\varepsilon_{hA}\cdot\varepsilon_h}{r^2}\sin\theta\cos\beta \qquad (13.5.16)$$

The T in the equation is the directional force constant. Since the angle of pinch $\beta = 0$, the above equation becomes:

$$f_{\varepsilon h} = T\frac{\varepsilon_{hA} \cdot \varepsilon_h}{r^2}\sin\theta \qquad (13.5.17)$$

Since the energy element ε_{hA} at point A moves in the same direction as the neighboring energy element ε_h, there is a suction force $f_{\varepsilon h(+)}$ and a gravitational force F_{mh} between the energy element ε_{hA} at point A and the neighboring energy element ε_h, and the combined force F of the two is:

$$F = F_{mh} + f_{\varepsilon h(+)} \qquad (13.5.18)$$

If the distance $r > l_m$, then the gravitational force F_{mh} between energy elements ε_h is the suction force $F_{mh(+)}$. Since the combined force $F > 0$, the energy element ε_{hA} at point A is constantly close to the neighboring energy element ε_h.

If the distance $r < l_m$, then the universal gravitational force F_{mh} between the energy elements ε_h is the repulsive force $F_{mh(-)}$. At this point the repulsive force $F_{mh(-)}$ is greater than the directional force $f_{\varepsilon h(+)}$, i.e:

$$F_{mh(-)} > f_{\varepsilon h(+)}$$

Since the combined force $F < 0$, the energy element ε_{hA} at point A will keep moving away from the neighboring energy elementε_h.

If the distance $r = r_m < l_m$, then the gravitational force F_{mh} between the energy element ε_{hA} at point A and the neighboring energy element ε_h is the repulsive force $F_{mh(-)}$. At this point the repulsive force $F_{mh(-)}$ is canceled by the suction force $f_{\varepsilon h(+)}$, i.e:

$$F = F_{mh(-)} + f_{\varepsilon h(+)} = 0 \qquad (13.5.19)$$

Since the combined force $F = 0$, the distance r between the energy element ε_{hA} at point A and the neighboring energy element ε_h remains constant and rotates synchronously.

In summary, the combined force $F = 0$ between adjacent energy elements ε_h is the factor that determines the motion of adjacent energy elements ε_h in the same direction.

The second kind state of motion is that the energy element ε_h exhibits motion in the opposite direction.

In Figure 13-10, the energy elements ε_{hA} at point A and ε_{hB} at point B at both ends of the electron circular tube diameter $2R$, show motion in opposite directions, i.e., ε_{hA} moves upward and ε_{hB} moves downward.

There are three different forces between the energy elements ε_{hA} and ε_{hB}.

The first force is the gravitational force F_{mh} between the energy elements ε_{hA} and ε_{hB}.

In Figure 13-10, the distance between the energy elements ε_{hA} and ε_{hB} is $2R$. According to equation (13.2.21), the gravitational force F_{mh} between the energy elements ε_{hA} and ε_{hB} is:

$$F_{mh} = K\frac{\varepsilon_h^2}{c^4(2R)^3}, \quad 2R > l_m \qquad (13.5.20)$$

The K in the formula is the gravitational constant, R is the radius of the electron circular tube, and l_m is the minimum unit length between the energy elements ε_h. Since the distance $2R > l_m$, the gravitational force F_{mh} between the energy elements ε_{hA} and ε_{hB} is the suction force.

The 2nd force is the centripetal force F_v of the energy element ε_h moving around the centerline of the electron circular tube at the spiral speed of light c.

When the electron circular tube is moving with velocity v in vacuum. The circular linear velocity v_ω of the energy element ε_h perpendicular to the direction of velocity is:

$$v_\omega = \sqrt{c^2 - v^2} \qquad (13.5.21)$$

Since the energy element ε_h moves around the centerline of the electron circular tube at the spiral speed of light c, according to the centripetal force equation, the centripetal force $F_{v\omega}$ of the energy element ε_{hA} (or the energy element ε_{hB}) is:

$$F_{v\omega} = m_h\frac{v_\omega^2}{R} = m_h\frac{c^2 - v^2}{R} \qquad (13.5.22)$$

The v in the equation is the displacement velocity of the electron tube.

The third force is the directional force $f_{\varepsilon h}$ caused by the direction of motion of the energy element ε_h.

In Figure 13-10, the directional force $f_{\varepsilon h}$ between the energy elements ε_{hA} and ε_{hB} is the repulsive force

$f_{\varepsilon h(-)}$ since the energy elements ε_{hA} and ε_{hB} are moving in opposite directions.

Since the centripetal force $F_{v\omega}$ is also a repulsive force, the repulsive force $F_{(-)}$ between the energy elements ε_{hA} and ε_{hB} is:

$$F_{(-)} = f_{\varepsilon h} + F_{v\omega} \tag{13.5.23}$$

The combined force F between the energy elements ε_{hA} and ε_{hB} is:

$$F = F_{mh} + F_{(-)} \tag{13.5.24}$$

If the radius R of the electron circular tube neither increases nor decreases, then the gravitational force F_{mh} between the energy elements ε_{hA} and ε_{hB} should be equal to the repulsive force $F_{(-)}$, i.e:

$$F_{mh} = -F_{(-)} \tag{13.5.25}$$

If the force of gravitational $F_{mh} > F_{(-)}$, then the energy elements ε_{hA} and ε_{hB} will keep moving closer together . So that the radius R of the electron round tube will keep shrinking, which will lead to the collapse and disappearance of the electron round tube.

Conversely, if the force of gravitational $F_{mh} < F_{(-)}$, then the energy elements ε_{hA} and ε_{hB} will keep moving away . So that the radius R of the electron circular tube will keep increasing, which will lead to the expansion and disappearance of the electron circular tube.

When the gravitational force $F_{mh} = F_{(-)}$, the radius R of the electron tube neither increases nor decreases. At this point, the electron tube maintains an equilibrium rotational motion.

In summary, the combined force F between the energy elements ε_h in an electron circular tube is always equal to 0, i.e:

$$F = F_{mh} + F_{(-)} = 0 \tag{13.5.26}$$

13.5.4 Radius *R* and length *d* of an electron circular tube

According to physical experiments, after the collision of positive and negative electrons, the positive and negative electrons become two low-energy γ photons in opposite directions. According to the photoelectric effect, electrons can absorb photons of other frequencies.

When an electron is at rest in the cosmic vacuum frame of reference, the stationary electron circular tube contains only γ photons and no photons of other frequencies due to the displacement velocity of the electron $v = 0$. From this it can be determined that γ photons are the skeleton of the electron circular tube.

Since the energy element ε_h is always rotating around the centerline of the electron circular tube at the helical speed of light c inside the tube, the electron circular tube has both spin and displacement motions. The high-speed circular motion of the energy element ε_h makes the electron look like a solid particle.

When the displacement directions of electron circular tubes A and B electron circular tubes are the same, if the rotational directions of electron circular tubes A and B are opposite, then electron circular tubes A and B are electrons of the opposite sex. If the direction of rotation of electron circular tubes A and B is the same, then electron circular tubes A and B are electrons of the same sex.

When the displacements of electron circular tubes A and B electron circular tubes are in opposite directions, if the rotations of electron circular tubes A and B are in opposite directions, then electron circular tubes A and B are electrons of the same sex. If electron circular tubes A and B rotate in the same direction, then electron circular tubes A and B are opposite sex electrons.

According to high-energy physics, the fact that electrons jumping in different orbits absorb or radiate photons of different frequencies suggests that the electrons contain photons of different frequencies within them. Since the γ photons only form the skeleton of the electron round tube, there are some other frequencies of photons attached to the electron skeleton, and it is these frequencies of photons that make the stationary electron round tube obtain the displacement velocity v.

Since γ photons move at the helical speed of light c inside the electron circular tube, these photons of different frequencies also move at the helical speed of light c inside the electron circular tube. as shown earlier in Figure 13-9.

When the energy plane of the high-energy γ photon collides with the B point of the nucleus, the B point is equivalent to a sharp knife blade, splitting the energy plane of the high-energy γ photon in two into two electron round tubes, as shown earlier in Figure 13-7.

The horizontal line AB in Figure 13-7 is the splitting line of the high-energy γ photon energy plane. Since the electron round tubes is rolled up along the AB cut line, the circumference of the electron round tubes is equal to the transverse unit length l_0. The radius R of the electron round tubes is:

$$R = \frac{l_0}{2\pi} \tag{13.5.27}$$

According to the photon energy formula, the wavelength λ of the photon is:

$$\lambda = \frac{hc}{\varepsilon_c} \tag{13.5.28}$$

Planck's constant in the formula $h = 4.136 \times 10^{-15} \cdot eV \cdot s$; the speed of light $c = 3 \times 10^8 \cdot m/s$.

According to atomic physics, positive and negative electrons can only be produced when a light beam with energy $\varepsilon_c > 1022 \cdot keV$ (electron volts) irradiates a lead plate. Assuming that the energy $\varepsilon_c = 1200 \cdot keV$ for γ photons, the wavelength λ of γ photons is:

$$\lambda = \frac{hc}{\varepsilon_c} = \frac{4.136 \times 10^{-15} \times 3 \times 10^8}{1.2 \times 10^6} = 1.034 \times 10^{-12} m \tag{13.5.29}$$

The frequency β of the γ photon is:

$$\beta = \frac{c}{\lambda} = \frac{3 \times 10^8}{1.034 \times 10^{-12}} = 2.901 \times 10^{20} Hz \tag{13.5.30}$$

If the unit length of the transverse direction $l_0 = 1 \times 10^{-12} m$, then the radius R of the electron circular tube is:

$$R = \frac{l_0}{2\pi} = \frac{1 \times 10^{-12}}{2\pi} = 1.59 \times 10^{-13} \tag{13.5.31}$$

According to equation (13.3.18), the longitudinal length b of the energy plane of high-energy γ photons is:

$$b = (\beta \cdot 1s)\frac{l_m^2}{l_0} \tag{13.5.32}$$

Since the longitudinal length b is cut in half, the length d of the γ electron round tube (i.e. the stationary electron round tube) is equal to half of the longitudinal length b, i.e:

$$d = \frac{b}{2} = (\beta \cdot 1s)\frac{l_m^2}{2l_0} \tag{13.5.33}$$

The length d in the formula is the length of the stationary electron round tube. Since the electron round tube absorbs photon energy and then leaps to other orbitals, the lengths that the electron round tubes in different orbitals have are different.

Assume that the longitudinal length b is equal to the transverse unit length l_0. If the unit length $l_0 = 1 \times 10^{-12} m$, then the length d of the electron circular tube is:

$$d = 5 \times 10^{-13} \mathrm{m}$$

In 1964, scientists from Harvard University and Cornell University proved that the radius r of an electron is in the range of $10^{-13} m$ to $10^{-14} m$ through electron experiments.

Assume that the longitudinal length b is equal to the transverse unit length l_0. According to equation (13.5.32), the minimum unit length l_m between two energy elements ε_h is:

$$l_m = \sqrt{\frac{l_0 b}{\beta}} = \sqrt{\frac{10^{-12} \times 10^{-12}}{2.901 \times 10^{20}}} = 5.87 \times 10^{-23} m \tag{13.5.34}$$

when the radius r is equal to the minimum unit length l_m. Since radius $r = 1$, the gravitational force F_h between two energy elements ε_h is:

$$F_h = 0$$

Chapter 14 Momentum P_e and mass energy ε_e of electrons

Introduction: The momentum of an electron $\boldsymbol{P_e} = \boldsymbol{m_e v}$. Since the mass $\boldsymbol{m_e}$ of an electron can be expressed in terms of the energy $\boldsymbol{\varepsilon} = \boldsymbol{hv}$ of a photon, the momentum $\boldsymbol{P_e}$ of an electron can also be expressed in terms of the energy $\boldsymbol{\varepsilon} = \boldsymbol{hv}$ of a photon.

14.1 The moving mass m_e of an electron can be decomposed into a displacement mass m_{ev} and a spin mass $m_{e\omega}$

14.1.1 Moving mass m_e and displacement velocity v of an electron

When a positron and a negatron collide, the positron and the negatron become two $\boldsymbol{\gamma}$ photons that propagate in opposite directions. Assuming that $\boldsymbol{\gamma}$ is the frequency of the $\boldsymbol{\gamma}$ photon, the mass $\boldsymbol{m_\gamma}$ of the $\boldsymbol{\gamma}$ photon is according to equation (13.1.18):

$$\boldsymbol{m_\gamma = \frac{h\gamma}{c^2}} \tag{14.1.1}$$

Since the displacement velocity $\boldsymbol{v}$ of the electron is not included in equation, and the $\boldsymbol{\gamma}$ photon is always rotating at the speed of light $\boldsymbol{c}$ inside the electron circular tube, the mass $\boldsymbol{m_\gamma}$ of the $\boldsymbol{\gamma}$ photon does not change with the change of the electron's displacement velocity $\boldsymbol{v}$.

According to atomic physics, when an atom contains multiple electrons, the multiple electrons are distributed in orbitals at different energy levels. This is shown in Figure 14-1.

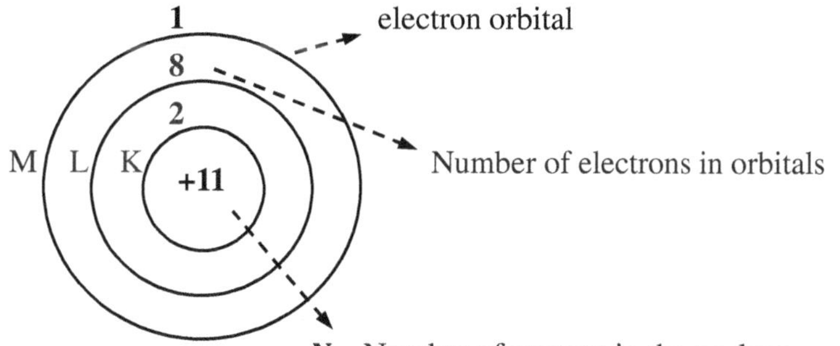

Figure 14-1 Schematic of the layered arrangement of electrons outside the nucleus of an N_a atom.

According to high-energy physics, electrons jumping in different orbits absorb or radiate photons of different frequencies. The electron absorbs the most external energy when it leaps from the ground state to the ionized state. This fact indicates that the electron circular tube contains photons of different frequencies.

According to atomic physics, the electron motion orbits in an atom are divided into several levels, and each electron usually has a stable motion orbit. Assuming that $\boldsymbol{\delta_i}$ is the $\boldsymbol{\delta_i}$ photon frequency, the electron absorbs the $\boldsymbol{\delta_i}$ photon and transmigrates to the stable orbit of the $\boldsymbol{i}$ level.

According to the photon energy formula, the energy of $\boldsymbol{\delta_i}$ photon $\boldsymbol{\varepsilon_\delta} = \boldsymbol{h\delta_i}$. Assuming that $\boldsymbol{m_{\delta i}}$ is the mass of

δ_i photon according to equation (13.1.18), the mass of δ_i photon $m_{\delta i}$ is:

$$m_{\delta i} = \frac{h\delta_i}{c^2} \tag{14.1.2}$$

Assume that m_e is the moving mass of the electron. When the electron is at rest in the cosmic vacuum, the moving mass m_e of the electron at this time is equal to the rest mass m_{e0} of the electron, i.e:

$$m_e = m_{e0} \tag{14.1.3}$$

Since the stationary electron circular tube contains only γ photons, the mass m_{e0} of the stationary electrons, is equal to the mass m_γ of the γ photons, i.e:

$$m_{e0} = m_\gamma = \frac{h\gamma}{c^2} \tag{14.1.4}$$

To summarize, when an electron is at rest in the cosmic vacuum, the mass m_e of the electron in motion, is equal to the mass m_{e0} of the electron at rest, ie:

$$m_e = m_{e0} = m_\gamma = \frac{h\gamma}{c^2} \tag{14.1.5}$$

When a stationary electron absorbs the energy $\varepsilon_\delta = h\delta_i$ of a δ_i photon, the electron leaps into a stabilized motion on the i orbit. Since the energy $\varepsilon_\delta = h\delta_i$ of the foreign photon absorbed by the electron is attached to the electron circular tube, the mass m_e of the electron moving in that stable orbit is:

$$m_e = m_\gamma + m_{\delta i} = \frac{h(\gamma + \delta\ \ )}{c^2} \tag{14.1.6}$$

The length of the electron circular tube increases after the δ_i photon absorbed by the resting electron in the presence of factors such as gravitational. This is shown in Figure 14-2.

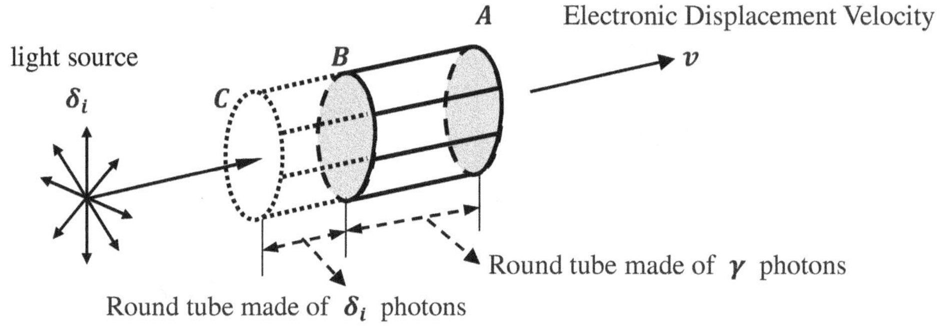

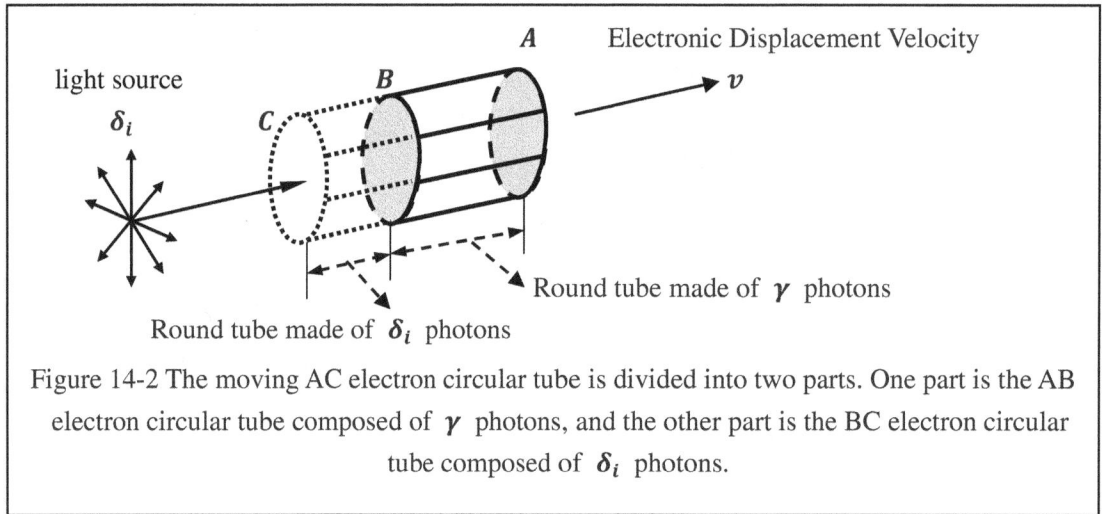

Figure 14-2 The moving AC electron circular tube is divided into two parts. One part is the AB electron circular tube composed of γ photons, and the other part is the BC electron circular tube composed of δ_i photons.

Since electrons have different energies in different orbitals, electrons in different orbitals have different moving masses m_e. When electrons are in the ground state, they have the smallest moving mass m_e. When the electron is in the ionized state, the electron has a relatively large moving mass m_e.

Assuming that β is the frequency of β photon, according to the photon energy formula, the energy of β photon $\varepsilon_\beta = h\beta$. After the electron absorbs the energy of one β photon, the electron leaps to the n layer of the orbit to stabilize the motion.

Since the energy absorbed by the electron $\varepsilon_\beta = h\beta$ is attached to the electron circular tube, the mass m_e of the electron moving in that stable orbit is:

$$m_e = m_\gamma + m_{\delta i} + m_\beta = \frac{h(\gamma + \delta_i + \beta)}{c^2} \tag{14.1.7}$$

The above equation is the equation for the moving mass of an electron. The above formula shows that the electron's moving mass m_e is a variable. When the photon energy $\varepsilon = h(\delta_i + \beta)$ absorbed by the electron changes, the electron's moving mass m_e will change accordingly.

If the photon energy absorbed by the electron $\varepsilon = h(\delta_i + \beta) = 0$, then the electron is at rest in the cosmic vacuum frame of reference, when the electron's moving mass m_e is equal to the rest mass m_{e0}, viz:

$$m_e = m_{e0} = m_\gamma = \frac{h\gamma}{c^2} \qquad (14.1.8)$$

14.1.2 Displacement velocity *v* of an electron

Suppose v is the displacement velocity of the electron in the cosmic vacuum frame of reference and m_e is the mass of the moving electron. According to the kinetic energy formula in mechanics, the kinetic energy E_e of the moving electron in the cosmic vacuum frame of reference is:

$$E_e = \frac{1}{2} m_e v^2 \qquad (14.1.9)$$

Substituting the mass m_e of the moving electron in equation (14.1.7) into the above equation gives the relational equation:

$$E_e = \frac{1}{2}\frac{h(\gamma + \delta_i + \beta)}{c^2} v^2 \qquad (14.1.10)$$

According to the law of conservation of energy, the kinetic energy E_e of the moving electron should be equal to the energy $h(\delta_i + \beta)$ of the photon absorbed by the stationary electron, i.e:

$$E_e = E_{\delta i} + E_\beta = h(\delta_i + \beta) \qquad (14.1.11)$$

Substituting the above equation into equation (14.1.10) gives the relationship equation:

$$E_e = \frac{1}{2}\frac{h(\gamma + \delta_i + \beta)}{c^2} v^2 = h(\delta_i + \beta) \qquad (14.1.12)$$

Simplifying the above equation gives the relationship equation:

$$v = c\sqrt{\frac{2(\delta_i + \beta)}{\gamma + \delta_i + \beta}} \qquad (14.1.13)$$

The above equation is the electron displacement velocity equation. The above equation shows that when an electron absorbs a β photon in a stable orbit, the displacement velocity v of the electron increases as the frequency β of the absorbed photon increases.

When the electron retreats from the new orbit to the original stable orbit, the electron radiates β photons, at which point the electron displacement velocity v changes to:

$$v = c\sqrt{\frac{2\delta_i}{\gamma + \delta_i}} \qquad (14.1.14)$$

When an electron radiates δ_i photons outward in a stable orbit, the electron's displacement velocity $v = 0$ in the cosmic vacuum frame of reference because of the photon energy $\varepsilon = 0$ absorbed by the electron. At this time, the electron's spin speed $v_\omega = c$.

14.1.3 Displacement mass m_{ev} of the electron and spin mass $m_{e\omega}$ of the electron

Assuming that v is the velocity of an electron in the cosmic vacuum frame of reference, it follows from equation (13.4.5) that the spiral speed c of light for the energy element ε_h is:

$$c = \sqrt{v^2 + v_\omega^2}$$

Note that the spiral speed c of light, the spin speed v_ω and the displacement speed v are all velocities in the cosmic vacuum frame of reference. Multiplying equation (14.1.7) by the speed c of light gives the relational equation:

$$m_e c = \frac{h(\gamma + \delta_i + \beta)}{c} \qquad (14.1.15)$$

Substituting the spiral speed of light $c = \sqrt{v^2 + v_\omega^2}$ into the left-hand side of the equal sign gives the relational

equation:

$$m_e\sqrt{v^2+v_\omega^2}=\frac{h(\gamma+\delta_i+\beta)}{c} \tag{14.1.16}$$

Dividing both sides of the above equation by the speed of light c, gives the relationship equation:

$$m_e=\sqrt{\left(m_e\frac{v}{c}\right)^2+\left(m_e\frac{v_\omega}{c}\right)^2}=\frac{h(\gamma+\delta_i+\beta)}{c^2} \tag{14.1.17}$$

The above equation is the mass-velocity equation for the electron. Although the electron's moving mass m_e becomes larger as the energy of the photon absorbed by the electron $\varepsilon=h(\delta_i+\beta)$ increases, the electron's moving mass m_e will never equal infinity.

According to equation (14.1.17), the electron's moving mass m_e can be decomposed into two component masses.

Since the velocity v is the displacement velocity of the electron, this book defines the 1st partial mass $m_e v/c$, within the root symbol of equation (14.1.17) as the electron's displacement mass m_{ev}, viz:

$$m_{ev}=m_e\frac{v}{c} \tag{14.1.18}$$

Since the electron's moving mass m_e is an indivisible whole, the electron's displaced mass m_{ev} cannot stand alone. The above equation shows that the magnitude of the electron displacement mass m_{ev} is proportional to the electron displacement velocity v.

The displaced mass of the electron $m_{ev}=0$ when the electron's displacement velocity $v=0$. The displaced mass of the electron $m_{ev}=m_e$ when the electron's displacement velocity $v=c$.

Since the velocity v_ω is the spin velocity of the electron, this book defines the 2nd partial mass $m_e v_\omega/c$, within the root symbol of equation (14.1.17), as the spin mass $m_{e\omega}$ of the electron, i.e:

$$m_{e\omega}=m_e\frac{v_\omega}{c}=m_e\sqrt{1-\frac{v^2}{c^2}} \tag{14.1.19}$$

Since the electron's moving mass m_e is an indivisible whole, the electron's spin mass $m_{e\omega}$ cannot exist separately. The above equation shows that the spin mass $m_{e\omega}$ of the electron becomes smaller as the electron displacement velocity v increases.

An important corollary can be obtained from equation (14.1.19): any physical quantity that is proportional to the electron spin mass $m_{e\omega}$ is correspondingly reduced in value by a factor of k, i.e:

$$k=\sqrt{1-\frac{v^2}{c^2}} \tag{14.1.20}$$

The above equation shows that the transformation factor k is a variable, not a constant.

14.1.4 When the electron displacement velocity v=0, the electron's spin mass $m_{e\omega}$ is equal to the electron's rest mass m_{e0}

According to equation (14.1.19), the moving mass m_e of the electron is:

$$m_e=\frac{m_{e\omega}}{\sqrt{1-\frac{v^2}{c^2}}} \tag{14.1.21}$$

The $m_{e\omega}$ in equation is the spin mass of the electron.

According to the mass-velocity formula of special relativity, the electron's moving mass m_e is:

$$m_e=\frac{m_{e0}}{\sqrt{1-\frac{v^2}{c^2}}} \tag{14.1.22}$$

The m_{e0} in the formula is the rest mass of the electron. Compare the above formula with formula (14.1.21). Since the mass m_e in both formulas is the moving mass of the electron, the rest mass m_{e0} in the above formula is equivalent to the spin mass $m_{e\omega}$ in formula (14.1.21).

Since there is no concept of the spin mass $m_{e\omega}$ of the electron in modern physics, but only the concept of the rest mass m_{e0} of the electron, there must exist the problem of the electron moving mass m_e equal to infinity in modern physics.

Obviously, when the electron displacement velocity $v = c$, since the rest mass of the electron $m_{e0} \neq 0$, so according to the mass-velocity formula of special relativity, at this time, the moving mass of the electron m_e is equal to infinity, that is:

$$m_e = \infty \tag{14.1.23}$$

It is important to note that the problem of a moving mass m_e equal to infinity does not exist in the new physics. It is clear that when the electron displacement velocity $v = c$, according to equation (14.1.19), the spin mass $m_{e\omega}$ of the electron is equal to 0, i.e:

$$m_{e\omega} = 0$$

According to equation (14.1.18), at this point the electron's moving mass m_e is equal to the displaced mass m_{ev}, i.e:

$$m_e = m_{ev} \tag{14.1.24}$$

When the electron displacement velocity $v = 0$, according to equation (14.1.19), the spin mass $m_{e\omega}$ of the electron is equal to the rest mass m, i.e:

$$m_{e\omega} = m_e = m_{e0}$$

At this point, the moving mass m_e of special relativity is equal to the rest mass m_{e0}, i.e:

$$m_e = m_{e0}$$

14.2 Mass-energy ε_e of the electron

14.2.1 Mass-energy conversion equation for electrons

According to Equation (14.1.7), the moving mass m_e of the electron is:

$$m_e = \frac{h(\gamma + \delta_i + \beta)}{c^2}$$

When the photon energy absorbed by the electron $h(\delta_i + \beta) = 0$. Since the displacement velocity of the electron $v = 0$, the rest mass m_{e0} of the electron is:

$$m_{e0} = m_\gamma = \frac{h\gamma}{c^2} \tag{14.2.1}$$

According to the mass-energy conversion equation $\varepsilon = mc^2$, the mass energy ε_{e0} of the static electron is:

$$\varepsilon_{e0} = m_\gamma c^2 = h\gamma \tag{14.2.2}$$

The mass-energy ε_e of the moving electron is:

$$\varepsilon_e = m_e c^2 \tag{14.2.3}$$

Substituting the moving mass m_e of the electron in equation (14.1.7) into the above equation gives the relation equation:

$$\varepsilon_e = m_e c^2 = h(\gamma + \delta_i + \beta) \tag{14.2.4}$$

The above formula is the mass-energy formula for a moving electron. When an electron radiates a β photon, the electron's mass energy ε_e changes. Since the frequency γ of a low-energy γ photon is a constant, the mass energy ε_e of an electron is proportional to the frequency of light absorbed by the electron $(\delta_i + \beta)$.

When the photon frequency $\delta_i + \beta = \gamma$ absorbed by the electron, the mass m_e of the electron is two times the mass m_γ of the low-energy γ photon, i.e. $m_e = 2m_\gamma$. At this time, since the displacement velocity v of the electron is equal to the speed of light c, i.e. $v = c$, the kinetic energy E_e of the electron moving at the speed of light is:

$$E_e = \frac{1}{2} m_e c^2 = m_\gamma c^2 \tag{14.2.5}$$

The above equation shows that when the electron displacement velocity v is equal to the speed of light c, the kinetic energy of the electron E_e is equal to the energy ε_γ of the γ photon, and the mass energy of the electron ε_e is equal to twice the energy ε_γ of the γ photon, viz:

$$\varepsilon_e = m_e c^2 = 2m_\gamma c^2$$

14.2.2 The mass energy ε_e of an electron can be decomposed into the spin mass energy $\varepsilon_{e\omega}$ and the displacement mass energy ε_{ev}

According to equation (7.2.21), the mass m_e of the energy element ε_h contained in the electron is:

$$m_e = \sum m_h = \sum \frac{\varepsilon_h}{c^2}$$

Since the energy element ε_h is always moving around the centerline of the electron circular tube at the spiral speed of light c, according to the mass-energy conversion equation $\varepsilon = mc^2$, the mass-energy conversion equation for the electron is:

$$\varepsilon_e = \sum m_h c^2 = \sum \varepsilon_h = m_e c^2 \qquad (14.2.6)$$

The speed of light c in equation is the spiral speed of light of the energy element ε_h, not the displacement speed v of the electron. Substituting the electron mass m_e from equation (14.1.7) into the above equation gives the relational equation:

$$\varepsilon_e = m_e c^2 = h(\gamma + \delta_i + \beta) \qquad (14.2.7)$$

The above equation shows that the mass energy ε_e of an electron is a variable. When an electron absorbs a β photon in a stable orbit, or radiates a β photon, the electron's mass energy ε_e changes.

If the electron radiates a photon of energy $h(\delta_i + \beta)$, then the moving electron becomes a static electron. At this point the mass energy ε_{e0} of the static electron is:

$$\varepsilon_{e0} = m_{e0} c^2 = h\gamma \qquad (14.2.8)$$

Substituting the spiral speed of light formula $c = \sqrt{v_\omega^2 + v^2}$ into equation (14.2.6) gives the relational formula:

$$\varepsilon_e = m_e c^2 = m_e v_\omega^2 + m_e v^2 \qquad (14.2.9)$$

The above formula is the mass-energy equivalence formula for the electron. Note that the speed of light c in the formula is the spiral speed of light of the energy element ε_h, not the displacement speed v of the electron. The above formula shows that the mass energy ε_e of the electron can be decomposed into two component mass energies.

14.2.3 Spin mass energy of the electron $\varepsilon_{e\omega}$

Since the velocity v_ω is the spin velocity of the electron, this book defines the first component mass energy $m_e v_\omega^2$ on the right-hand side of the equal sign of Equation (14.2.9) as the spin mass energy $\varepsilon_{e\omega}$ of the electron, viz:

$$\varepsilon_{e\omega} = m_e v_\omega^2 = m_e(c^2 - v^2) \qquad (14.2.10)$$

The above formula is the electron spin mass energy formula. The above formula shows that the spin mass energy $\varepsilon_{e\omega}$ of an electron is proportional to the square of the electron's spin speed v_ω.

According to the spiral speed of light formula $c = \sqrt{v_\omega^2 + v^2}$, the spin speed v_ω is:

$$v_\omega = \sqrt{c^2 - v^2}$$

Substituting the velocity v from equation (14.1.13) into the above equation gives the relationship equation:

$$v_\omega = \sqrt{c^2 - c^2 \frac{2(\delta_i + \beta)}{\gamma + \delta_i + \beta}}$$

Simplifying the above equation gives the relationship equation:

$$v_\omega = c\sqrt{\frac{\gamma - (\delta_i + \beta)}{\gamma + \delta_i + \beta}} \qquad (14.2.11)$$

The above formula is the formula for the spin speed v_ω of the electron. Substituting the mass m_e from equation (14.1.7) and the spin speed v_ω from the above equation into equation (14.2.10) gives the relational equation:

$$\varepsilon_{e\omega}=\frac{h(\gamma+\delta_i+\beta)}{c^2}\cdot c^2\frac{\gamma-(\delta_i+\beta)}{\gamma+\delta_i+\beta}$$

Simplifying the above equation gives the relationship equation:

$$\varepsilon_{e\omega}=m_e v_\omega^2=h(\gamma-\delta_i-\beta) \tag{14.2.12}$$

The above equation shows that the larger the photon energy $h(\delta_i+\beta)$ absorbed by the electron, the smaller the electron's spin mass energy $\varepsilon_{e\omega}$.

The spin velocity of the electron $v_\omega=\sqrt{c^2-v^2}$ becomes smaller as the velocity v increases. When the photon energy absorbed by the electron, $h(\delta_i+\beta)=0$, the electron velocity $v=0$ and the electric field E of the static electron reaches its maximum value, while the magnetic field $B=0$ for the static electron. From this it can be determined that the electron spin velocity v_ω is the only cause of the electric field E.

14.2.4 Displaced mass energy of an electron ε_{ev}

Since the velocity v is the displacement velocity of the electron, this book defines the second component mass-energy $m_e v^2$ to the right of the equal sign in equation (14.2.9) as the electron displacement mass-energy ε_{ev}, viz:

$$\varepsilon_{ev}=m_e v^2 \tag{14.2.13}$$

The above equation is the equation for the displacement mass energy of an electron. The above equation shows that the displaced mass energy ε_{ev} of the electron is proportional to the square of the displaced velocity v. Substituting the mass m_e from equation (14.1.7) and the velocity v from equation (14.1.13) into the above equation gives the relational equation:

$$\varepsilon_{ev}=\frac{h(\gamma+\delta_i+\beta)}{c^2}\cdot c^2\frac{2(\delta_i+\beta)}{\gamma+\delta_i+\beta}$$

Simplifying the above equation gives the relationship equation:

$$\varepsilon_{ev}=m_e v^2=2h(\delta_i+\beta) \tag{14.2.14}$$

The above equation shows that the greater the photon energy $h(\delta_i+\beta)$ absorbed by the electron, the greater the displaced mass energy ε_{ev} of the electron. It should be noted that the energy ε_{ev} in the above equation is the displaced mass energy of the electron, not the electron's site-shift energy E_e, i.e:

$$\varepsilon_{ev}\neq E_e=\frac{1}{2}m_e v^2$$

When the photon energy $h(\delta_i+\beta)\neq 0$ absorbed by the electron, the displacement velocity $v\neq 0$ of the electron. The electric field E of the electron becomes smaller as the displacement velocity v increases, while the magnetic field B of the electron becomes larger as the displacement velocity v increases. From this it can be determined that the displacement velocity v of the electron is the only cause of the magnetic field B.

14.2.5 Kinetic energy of an electron E_e

According to classical kinematics, the kinetic energy E_e possessed by a moving electron is:

$$E_e=\frac{1}{2}m_e v^2 \tag{14.2.15}$$

Substituting the mass m_e from equation (14.1.7) and the velocity v from equation (14.1.13) into the above equation gives the relationship equation:

$$E_e=\frac{1}{2}\frac{h(\gamma+\delta_i+\beta)}{c^2}\cdot c^2\frac{2(\delta_i+\beta)}{\gamma+\delta_i+\beta}$$

Simplifying the above equation gives the relationship equation:

$$E_e=h(\delta_i+\beta) \tag{14.2.16}$$

The above equation is the kinetic energy equation for an electron. The above equation shows that the electron kinetic energy E_e is equal to the photon energy $h(\delta_i+\beta)$ absorbed by the electron.

The electron kinetic energy $E_e=0$ when the electron velocity $v=0$. When the electron velocity $v=c$. According to equation (14.1.13), the relation equation can be obtained:

$$c = c\sqrt{\frac{2(\delta_i + \beta)}{\gamma + \delta_i + \beta}} \qquad (14.2.17)$$

Simplifying the above equation gives the relationship equation:

$$\delta_i + \beta = \gamma \qquad (14.2.18)$$

Substituting the above equation into equation (14.2.16) gives the relationship equation:

$$E_e = h(\delta_i + \beta) = h\gamma \qquad (14.2.19)$$

The above equation shows that when the energy of the photon absorbed by the electron $\varepsilon = h(\delta_i + \beta)$ is equal to the energy of the γ photon $\varepsilon_\gamma = h\gamma$, the electron moves with a velocity $v = c$.

According to equation (14.1.7) , the mass m_e of the electron is:

$$m_e = \frac{h(\gamma + \delta_i + \beta)}{c^2}$$

Substituting equation (14.2.18) into the above equation gives the relationship equation:

$$m_e = \frac{2h\gamma}{c^2} \qquad (14.2.20)$$

According to equation (14.1.1) the mass m_γ of the γ photon is:

$$m_\gamma = \frac{h\gamma}{c^2}$$

Substituting the above equation into equation (14.2.20) gives the relationship equation:

$$m_e = 2m_\gamma \qquad (14.2.21)$$

The above equation shows that when the electron velocity $v = c$, the mass m_e of the electron is twice the mass m_γ of the low-energy γ photon.

The mass-energy of a low-energy γ photon $\varepsilon_\gamma = m_\gamma c^2 = 511 \cdot kvV$. The mass-energy of an electron $\varepsilon_e = 2\varepsilon_\gamma = 1022 \cdot kvV$ when the electron's velocity $v = c$. According to the mass-energy conversion formula $\varepsilon = mc^2$, the electron's mass m_e will not equal infinity at this time.

When the electron's velocity $v = c$, special relativity holds that the electron's mass m_e is equal to infinity.

14.3 Spin momentum $P_{me\omega}$ and displacement momentum P_{mev} of electrons

14.3.1 Definition of the electron's helical light speed momentum P_{mec}

According to Equation (7.2.21) , the mass of the electron m_e is equal to the mass of the energetic element contained in the electron, i.e.:

$$m_e = \sum m_h = \sum \frac{\varepsilon_h}{c^2}$$

Since the energy element ε_h always moves around the centerline of the electron circular tube at the helical speed of light c, according to the momentum formula $P = mv$, this book defines the product of the mass $\sum m_h$ and the speed of light c as the electron's helical light speed momentum P_{mec}, i.e:

$$P_{mec} = \sum m_h c = m_e c \qquad (14.3.1)$$

Note that the helical light speed momentum $P_{mec} = m_e c$ of an electron, and the momentum $P_{mev} = mv$ of an electron, are two different physical quantities. The momentum $P_{mev} = mv$ of the electron in classical theory, is defined in this book as the displacement momentum of the electron.

According to Equation (14.1.7) , the mass m_e of the electron is:

$$m_e = \frac{h(\gamma + \delta_i + \beta)}{c^2} \qquad (14.3.2)$$

Substituting the above equation into equation (14.3.1) gives the relationship equation:

$$P_{mec} = m_e c = \frac{h(\gamma + \delta_i + \beta)}{c} \qquad (14..3)$$

The above equation shows that the electron's helical light speed momentum P_{mec} is a variable. When an electron absorbs an β photon in a stable orbit, or radiates an β photon, the electron's helical light speed momentum P_{mec} changes.

When the electron radiates the absorbed photon energy $h(\delta_i + \beta)$, the electron becomes a static electron. Since the energy of the static electron is equal to the energy of the γ photon, the helical light speed momentum P_{mec0} of the static electron is:

$$P_{mec0} = m_{e0}c = \frac{h\gamma}{c} \qquad (14.3.4)$$

Since the energy element ε_h always moves in the electron at the helical speed of light c, substituting the helical speed of light formula $c = \sqrt{v_\omega^2 + v^2}$ into Equation (14.3.1) gives the relational formula:

$$P_{mec} = m_e c = \sqrt{(m_e v)^2 + (m_e v_\omega)^2} \qquad (14.3.5)$$

The above equation is the formula for the helical light speed momentum of an electron. The above equation shows that the helical light speed momentum P_{mec} of the electron can be decomposed into two component momentums.

14.3.2 Displacement momentum P_{mev} of an electron

Since the velocity v is the displacement velocity of the electron, this book defines the first component momentum $m_e v$ inside the root sign of Equation (14.3.5) as the electron's displacement momentum P_{mev}, i.e:

$$P_{mev} = m_e v \qquad (14.3.6)$$

Substituting the mass m_e from equation (14.1.7) and the velocity v from equation (14.1.13) into the above equation gives the relationship equation:

$$P_{mev} = m_e v = \frac{h(\gamma + \delta_i + \beta)}{c^2} \cdot c\sqrt{\frac{2(\delta_i + \beta)}{\gamma + \delta_i + \beta}}$$

Simplifying the above equation gives the relationship equation:

$$P_{mev} = m_e v = \frac{h}{c}\sqrt{2(\delta_i + \beta)(\gamma + \delta_i + \beta)} \qquad (14.3.7)$$

The above equation is the equation for the displacement momentum of the electron. The above equation shows that the greater the photon energy $h(\delta_i + \beta)$ absorbed by the electron, the greater the displacement momentum P_{mev} of the electron.

According to equation (14.1.13), the displacement velocity v of the electron is:

$$v = c\sqrt{\frac{2(\delta_i + \beta)}{\gamma + \delta_i + \beta}} \qquad (14.3.8)$$

When the photon energy absorbed by the electron $h(\delta_i + \beta) = 0$. Since the velocity $v = 0$ of motion of the electron, the displaced momentum P_{mev} of the electron is:

$$P_{mev} = 0 \qquad (14.3.9)$$

When the photon energy $h(\delta_i + \beta) = h\gamma$ absorbed by the electron. Since the displacement velocity $v = c$ of the electron, the displacement momentum P_{mev} of the electron is:

$$P_{mev} = m_e c = 2\frac{h\gamma}{c} \qquad (14.3.10)$$

14.3.3 Spin momentum of the electron $P_{me\omega}$

Since the velocity v_ω is the spin velocity of the electron, this book defines the second component momentum $m_e v_\omega$ within the root sign of equation (14.3.5) as the electron's spin momentum $P_{me\omega}$, i.e:

$$P_{me\omega} = m_e v_\omega = m_e\sqrt{c^2 - v^2} \qquad (14.3.11)$$

The above equation is the formula for the spin momentum of an electron. The above equation shows that the spin momentum of the electron, $P_{me\omega}$, is proportional to the electron's spin velocity v_ω. Substituting the electron mass m_e from equation (14.1.7) and the velocity v_ω from equation (14.2.11) into the above equation gives the relational

equation:

$$P_{me\omega} = m_e v_\omega = \frac{h(\gamma + \delta_i + \beta)}{c^2} \cdot c\sqrt{\frac{\gamma - (\delta_i + \beta)}{\gamma + \delta_i + \beta}}$$

Simplifying the above equation gives the relationship equation:

$$P_{me\omega} = m_e v_\omega = \frac{h\sqrt{\gamma^2 - (\delta_i + \beta)^2}}{c} \tag{14.3.12}$$

The above equation shows that the larger the energy $h(\delta_i + \beta)$ of the foreign photon absorbed by the electron, the smaller the electron's spin momentum $P_{me\omega}$.

Chapter 15 The electric charge e of an electron can be decomposed into a rotational electric charge e_ω and a magnetic electric charge e_v

Introduction: Physicists consider the electron to be a spherical particle, while this book considers the electron to be a hollow circular tube of energy (electron circular tube). The electron circular tube theory can explain why a moving electron has an electric field $\boldsymbol{E}$ and a magnetic field $\boldsymbol{B}$.

15.1 The electric field force F_E and the magnetic field force F_B are both directional forces $f_{\varepsilon h}$ between energy elements ε_h

15.1.1 Electric charge $\pm e$ for both positive and negative electrons is generated by an electron circular tube

Based on the analytical discussion in Chapter 13, it is clear that the electron is a hollow circular tube of energy (electron circular tube). The structure of an electron circular tube is shown in Figure 15-1.

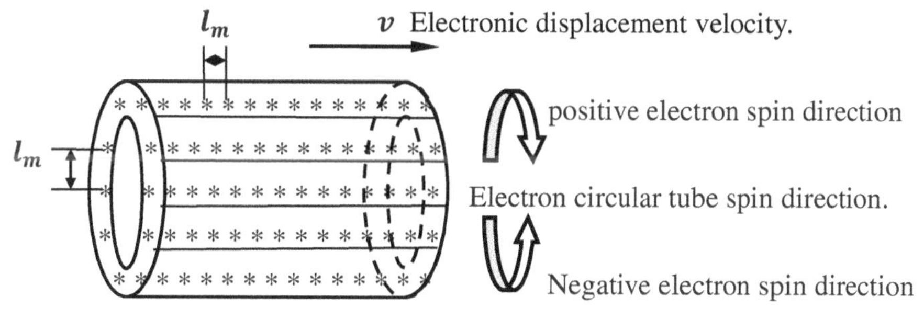

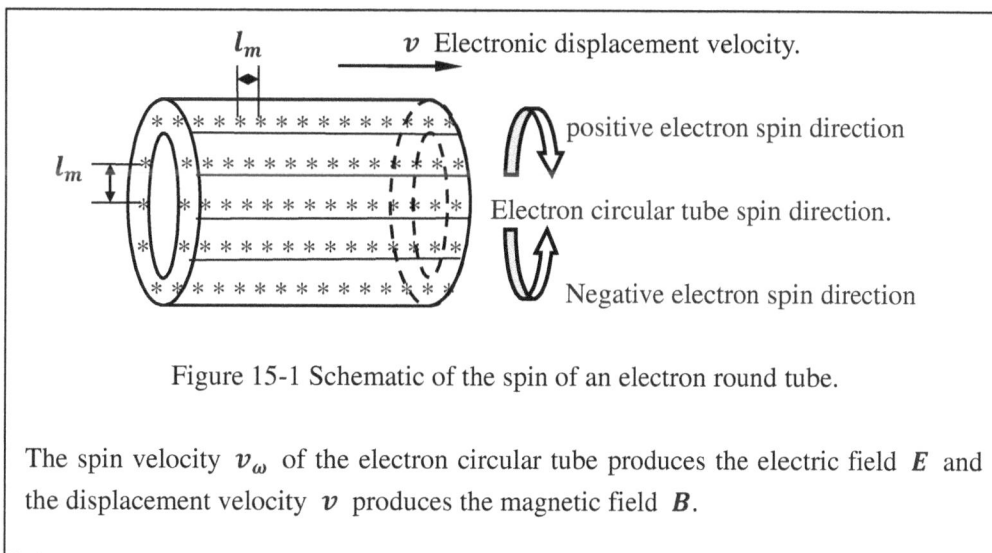

Figure 15-1 Schematic of the spin of an electron round tube.

The spin velocity v_ω of the electron circular tube produces the electric field $\boldsymbol{E}$ and the displacement velocity v produces the magnetic field $\boldsymbol{B}$.

The energy element ε_h is all rotating around the centerline of the electron circular tube at the spiral speed of light c. The transverse section of the electron circular tube is shown in Figure 15-2.

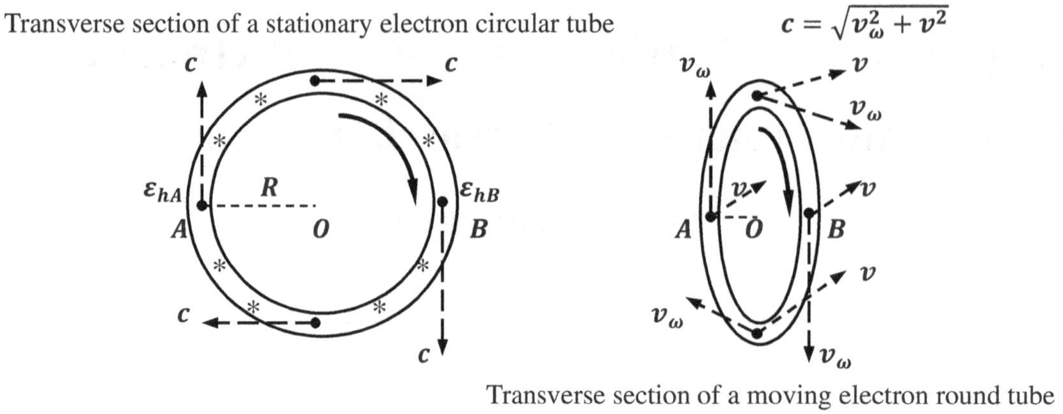

Figure 15-2 Schematic of transverse sections of a stationary and moving electron circular tube. ***R*** is the distance from the energy element ε_h to point O at the center of the tube.

When the positron and the negatron collide and disappear, the electric charge $\pm e$ of both the positron and the negatron disappears, and at this time the positron and the negatron round tubes turn out to be two γ photons propagating in opposite directions. From this, it can be determined that the electric charge $\pm e$ of both positive and negative electrons are generated by the electron round tube.

15.1.2 The same direction of rotation of the spin velocity $v_{\omega A}$ of electron e_A and the spin velocity $v_{\omega B}$ of electron e_B is responsible for the repulsive force between electrons e_A and e_B

According to the spiral speed of light formula $c = \sqrt{v_\omega^2 + v^2}$ for the energy element ε_h, it can be seen that there are two directions of motion contained within the energy element ε_h, one in the direction of the internal rotation speed v_ω (i.e., in the direction of the electron spin velocity v_ω) and the other in the direction of the displacement velocity v (i.e., in the direction of the electron displacement velocity v).

Suppose r is the distance between the energy elements ε_{hA} and ε_{hB}, and suppose v_A is the displacement velocity of the energy element ε_{hA} and v_B is the displacement velocity of the energy element ε_{hB}. Assume that β is the angle between the displacement velocity v_A and the displacement velocity v_B, and θ is the angle between the distance r and the displacement velocity v_A. As shown in Figure 15-3.

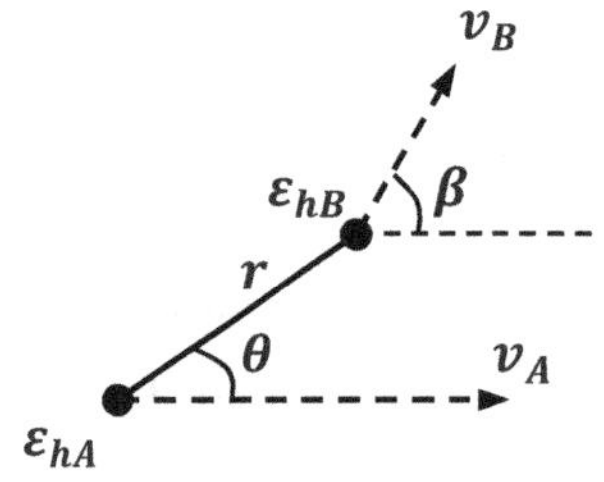

Figure 15-3 Schematic of the displacement direction of the energy elements ε_{hA} and ε_{hB}.

According to the directional force equation (13.5.8), the directional force $f_{\varepsilon h}$ between the energy elements ε_{hA} and ε_{hB} is:

$$f_{\varepsilon h} = T\frac{\varepsilon_{hA} \cdot \varepsilon_{hB}}{r^2}\sin\theta\cos\beta$$

The T in equation is the directional force constant. If the pinch angle $0 \le \beta < 90^0$, then the directional force

$f_{\varepsilon h}$ is the suction force $f_{\varepsilon h(+)}$. If the pinch angle $90^0 < \beta \le 180^0$, then the directional force $f_{\varepsilon h}$ is the repulsive force $f_{\varepsilon h(-)}$.

If the static electron circular tubes e_A and e_B are both rotating clockwise, then electrons e_A and e_B are the same kind of electrons. Assume that the energy element ε_{hA} in electron e_A and ε_{hB} in electron e_B are adjacent to each other. Since energy elements ε_{hA} and ε_{hB} rotate in opposite directions (i.e., energy element ε_{hA} rotates downward and energy element ε_{hB} rotates upward), energy elements ε_{hA} and ε_{hB} are moving in opposite directions, i.e., the angle $\beta = 180^0$ of the direction of motion of energy elements ε_{hA} and ε_{hB} is shown in Figure 15-4.

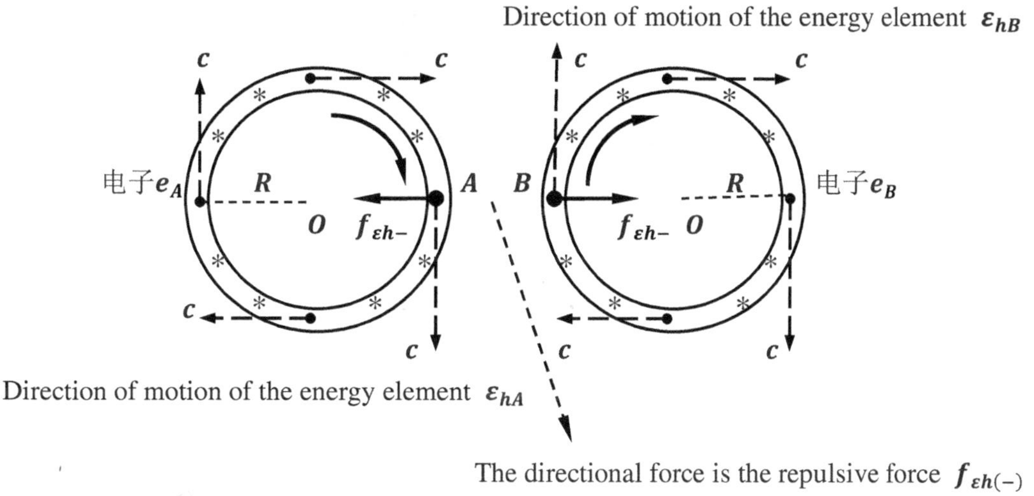

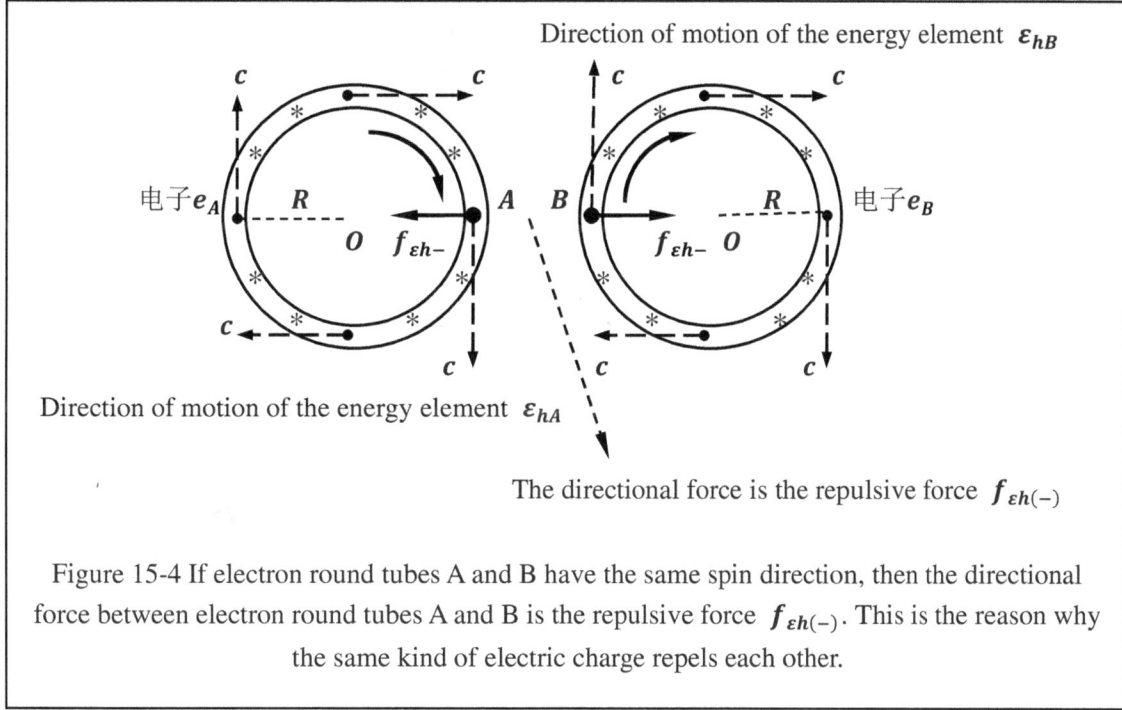

Figure 15-4 If electron round tubes A and B have the same spin direction, then the directional force between electron round tubes A and B is the repulsive force $f_{\varepsilon h(-)}$. This is the reason why the same kind of electric charge repels each other.

Since the energy elements ε_{hA} and ε_{hB} move in opposite directions, the directional force $f_{\varepsilon h}$ between the energy elements ε_{hA} and ε_{hB} is a repulsive force $f_{\varepsilon h(-)}$ according to the axiom of directional force of the energy element ε_h (i.e., Axiom 5).

Since the directional force $f_{\varepsilon h}$ of the energy element ε_{hA} on the A semicircular plane and ε_{hB} on the B semicircular plane is repulsive, the rotational electric field force F_ω between electrons of the same species e_A and e_B is repulsive.

It should be noted that there are two kinds of forces between the electron round tube e_A and e_B. One force is the gravitational force F_{me} between the electron round tube masses m_{eA} and m_{eB}, and the other force is the directional force $f_{\varepsilon h}$ caused by the direction of motion of the energy element ε_h. The force F between the electron round tubes e_A and e_B is equal to the sum of both the gravitational force F_{me} and the directional force $f_{\varepsilon h}$, viz:

$$F = F_{me} + f_{\varepsilon h} \qquad (15.1.1)$$

Since the energy elements ε_{hA} and ε_{hB} move in opposite directions, i.e., the angle $\beta = 180^0$, the directional force $f_{\varepsilon h}$ between the electron circular tubes e_A and e_B, which are rotating in the same direction, is the repulsive force $f_{\varepsilon h(-)}$.

When two negative electrons are outside the nucleus. Since the distance r between the two negative electrons is relatively large, the gravitational force F_{me} between the two negative electrons is less than the directional force $f_{\varepsilon h(-)}$. At this time the force between the two negative electrons is the repulsive force $F_{(-)}$, i.e:

$$F_{(-)} = F_{me} + f_{\varepsilon h(-)} < 0 \qquad (15.1.2)$$

This conclusion is clearly consistent with the objective fact that electrons of the same sex repel each other.

When two positrons are inside the nucleus. Since the distance r between the two positrons is very small, the gravitational force F_{me} between the two positrons is greater than the directional force $f_{\varepsilon h(-)}$, so the force between the two positrons is the suction force $F_{(+)}$, viz:

$$F_{(+)} = F_{me} + f_{\varepsilon h(-)} > 0 \tag{15.1.3}$$

This conclusion is clearly consistent with the objective fact that positrons exist within the nucleus of an atom.

15.1.3 The opposite direction of rotation of the spin velocity $v_{\omega A}$ of electron e_A and the spin velocity $v_{\omega B}$ of electron e_B is responsible for the repulsive force between electrons e_A and e_B

When the static electron circular tube e_A rotates counterclockwise and the static electron circular tube e_B rotates clockwise, electron e_A is a positive electron and electron e_B is a negative electron.

Assume that the energy element ε_{hA} in electron e_A and ε_{hB} in electron e_B are adjacent to each other. Since the energy elements ε_{hA} and ε_{hB} have the same direction of rotation (i.e., the energy elements ε_{hA} and ε_{hB} both rotate upward), the energy elements ε_{hA} and ε_{hB} have the same direction of motion, i.e., the angle $\beta = 0$ between the directions of motion of the energy elements ε_{hA} and ε_{hB}. This is shown in Figure 15-5.

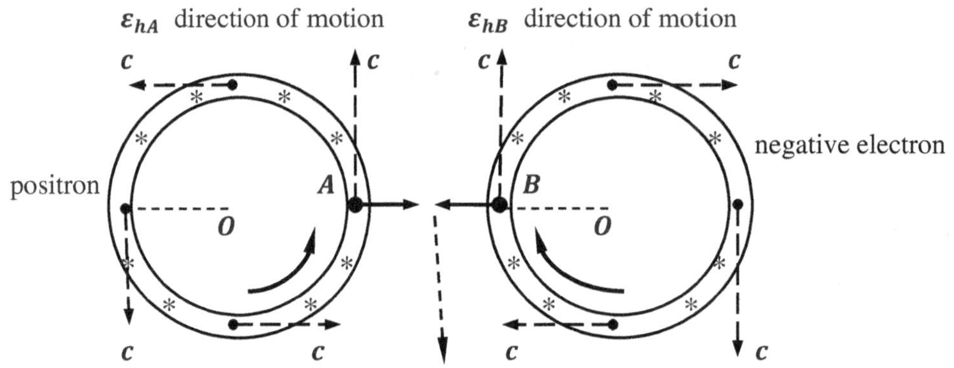

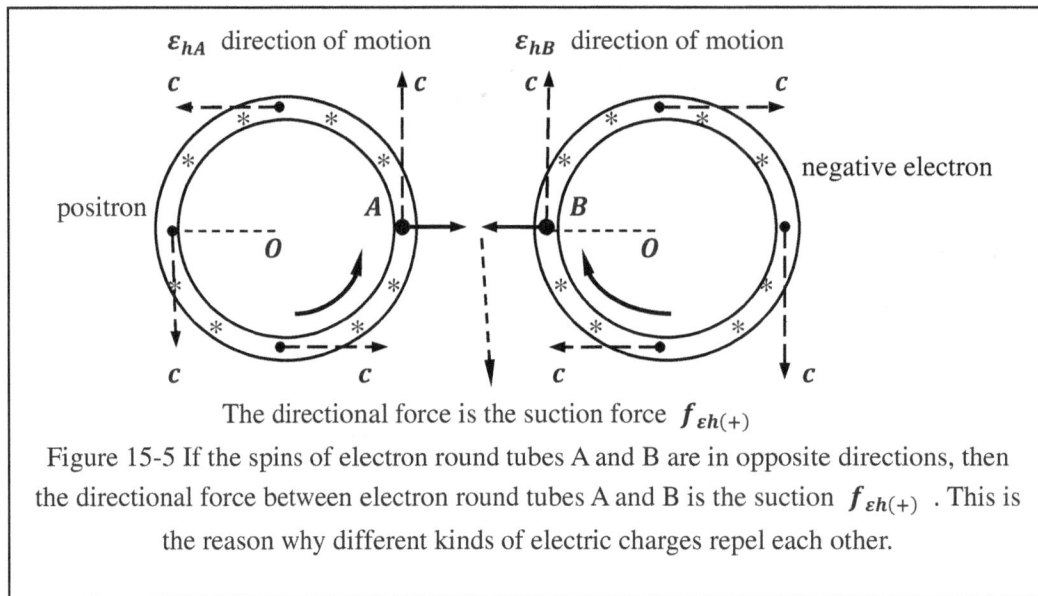

The directional force is the suction force $f_{\varepsilon h(+)}$

Figure 15-5 If the spins of electron round tubes A and B are in opposite directions, then the directional force between electron round tubes A and B is the suction $f_{\varepsilon h(+)}$. This is the reason why different kinds of electric charges repel each other.

Since the energy elements ε_{hA} and ε_{hB} move in the same direction, according to the axiom of directional force of the energy element ε_h (i.e., Axiom 5), the directional force $f_{\varepsilon h}$ between the energy elements ε_{hA} and ε_{hB} is the suction force $f_{\varepsilon h(+)}$.

Since the directional force $f_{\varepsilon h}$ of the energy element ε_{hA} on the A semicircular plane and ε_{hB} on the B semicircular plane is suction, the rotational electric field force F_{ω} between the positive and negative electrons e_A and e_B is suction.

When the distance r between the positive and negative electrons is large, the force between the circular tube of positive and negative electrons is equal to the sum of the gravitational force F_{me} and the directional force $f_{\varepsilon h(+)}$, viz:

$$F_{(+)} = F_{me} + f_{\varepsilon h(+)} > 0 \tag{15.1.4}$$

This conclusion is clearly consistent with the objective fact that positive and negative electrons are attracted to each other.

When the distance r between the positive and negative electron circular tubes e_A and e_B is very small, the positive and negative electron circular tubes e_A and e_B are converted into two γ-photons with opposite directions, i.e:

$$e_+ + e_- \rightarrow 2\gamma \tag{15.1.5}$$

This conclusion is clearly consistent with the objective fact that the collision of positive and negative electrons converts into two γ photons.

In summary, the essence of the electrostatic field force $\boldsymbol{F_0}$ is the directional force $\boldsymbol{f_{\varepsilon h}}$ between the energy elements $\boldsymbol{\varepsilon_h}$.

15.1.4 Conversion factor η (charge-to-mass ratio) between electric charge e and mass m_e of an electron

According to equation (14.1.7) the mass $\boldsymbol{m_e}$ of the electron is:

$$\boldsymbol{m_e = \frac{h(\gamma + \delta_i + \beta)}{c^2}} \qquad (15.1.6)$$

When the photon energy $\boldsymbol{\varepsilon = h(\delta_i + \beta) = 0}$ absorbed by the electron, the velocity $\boldsymbol{v = 0}$ of the electron and the mass $\boldsymbol{m_{e0}}$ of the stationary electron is:

$$\boldsymbol{m_{e0} = \frac{h\gamma}{c^2}} \qquad (15.1.7)$$

A static electron has a electric charge of static electric charge $\boldsymbol{e_0}$.

When the electron absorbs photon energy $\boldsymbol{\varepsilon = h(\delta_i + \beta)}$, the mass $\boldsymbol{m_e}$ of the electron becomes larger as the absorbed photon energy $\boldsymbol{\varepsilon}$ increases.

Assuming that $\boldsymbol{\eta}$ is the conversion factor (charge-to-mass ratio) between the electron mass $\boldsymbol{m_e}$ and the electron electric charge $\boldsymbol{e}$, viz:

$$\boldsymbol{e = \eta m_e} \qquad (15.1.8)$$

The above equation is the charge-to-mass conversion equation for electrons, and $\boldsymbol{\eta}$ in the equation is the charge-to-mass conversion factor (charge-to-mass ratio). Substituting the electron mass $\boldsymbol{m_e}$ in equation (15.1.6) into the above equation, we can get the relationship equation:

$$\boldsymbol{e = \eta\frac{h(\gamma + \delta_i + \beta)}{c^2}} \qquad (15.1.9)$$

The above equation is the charge-photon energy conversion equation for the electron. Since the motion velocity $\boldsymbol{v}$ of the electron is proportional to the photon energy absorbed by the electron $\boldsymbol{\varepsilon = h(\delta_i + \beta)}$, if the photon energy absorbed by the electron $\boldsymbol{\varepsilon = h(\delta_i + \beta) = 0}$, the electron is at rest in the cosmic vacuum frame of reference, and according to the above formula, the static electric charge of the static electron $\boldsymbol{e_0}$ is:

$$\boldsymbol{e_0 = \eta\frac{h\gamma}{c^2}} \qquad (15.1.10)$$

The above equation shows that the static electric charge $\boldsymbol{e_0}$ of the electron is composed of the mass of the $\boldsymbol{\gamma}$ photon $\boldsymbol{m_\gamma = h\gamma/c^2}$. Since photons have wave-particle duality, the electric charge $\boldsymbol{e}$ of the electron also has wave-particle duality.

If the photon energy absorbed by the stationary electron $\boldsymbol{\varepsilon = h(\delta_i + \beta)}$, then the mass $\boldsymbol{m_e}$ of the electron becomes larger and the electric charge $\boldsymbol{e}$ of the moving electron follows.

Although the electric charge $\boldsymbol{e}$ of the moving electron becomes larger as the absorbed photon energy $\boldsymbol{\varepsilon = h(\delta_i + \beta)}$ increases, according to the axiom of the invariance of the speed of light, the maximum photon energy absorbed by the stationary electron $\boldsymbol{\varepsilon = h(\delta_i + \beta) = h\gamma}$. From this, we obtain the relational equation:

$$\boldsymbol{\delta_i + \beta = \gamma} \qquad (15.1.11)$$

Substituting the above equation into equation (15.1.9) gives the relationship equation:

$$\boldsymbol{e = 2\eta\frac{h\gamma}{c^2} = 2e_0} \qquad (15.1.12)$$

The above equation shows that the maximum value of the motion electric charge $\boldsymbol{e}$ is twice the static electric charge $\boldsymbol{e_0}$, the

If the photon energy absorbed by the static electron $\boldsymbol{\varepsilon = h(\delta_i + \beta) = 0}$, then one gets equation:

$$\boldsymbol{\eta = \frac{e_0}{m_{e0}}} \qquad (15.1.13)$$

The above equation shows that the charge-to-mass ratio $\boldsymbol{\eta}$ is equal to the ratio of the static electric charge $\boldsymbol{e_0}$ of the electron to the static mass $\boldsymbol{m_{e0}}$. From this it can be determined that the charge-to-mass ratio $\boldsymbol{\eta}$ is a constant, independent of the photon energy absorbed by the electron $\boldsymbol{\varepsilon = h(\delta_i + \beta)}$.

According to the physical experiment:

Static mass of electron $m_{e0} = 9.10953 \times 10^{-31} \cdot kg$,

Charge of static electron $e_0 = 1.602189 \times 10^{-19} \cdot C$

The charge-to-mass ratio η of the electron is:

$$\eta = \frac{e_0}{m_{e0}} = \frac{1.602189 \times 10^{-19}}{9.10953 \times 10^{-31}} = 1.7588 \times 10^{11} \cdot C/kg \qquad (15.1.14)$$

Obviously, the value of the charge-to-mass ratio η, is what physicists have specified as the charge-to-mass ratio.

In classical theory, the static electric charge e_0 is the electric charge at rest in the inertial system S' Since there are an infinite number of inertial systems S', there can be an infinite number of electrostatic charges e_0 for electrons.

According to the axiom of cosmic vacuum reference system, there is only one cosmic vacuum reference system, while there are countless inertial systems S'. In order to distinguish the static electric charge in the cosmic vacuum frame of reference from that in other inertial systems S', this book defines the static electric charge e_0 in the cosmic vacuum frame of reference as the static electric charge of an electron.

Since the standard electric charge of an electron specified by a physicist is unique, the static electric charge e_0 of an electron defined in this book is the standard electric charge of an electron specified by a physicist, i.e., the static electric charge $e_0 = 1.602 \times 10^{-19} \cdot C$.

15.2 The electric charge *e* of a moving electron can be decomposed into two component charges, the rotational electric charge e_ω and the magnetic electric charge e_v

15.2.1 The electric charge *e* of a moving electron can be decomposed into two component charges

The electron is one of the fundamental particles that make up matter, and the electron has the property of rotation. An important property that the electron electric charge e has is its quantum nature.

When a charged body is at rest, the electric charge q of any charged body can only be an integer multiple of the electric charge e of the electron. The electric charge of a proton has the same absolute value as the electric charge of an electron, except that the electric charge of a proton is positive. There is no electric charge in nature that exists separately from matter.

According to the mass-velocity equation for electrons (14.1.17), the electron mass m_e can be decomposed into two component masses, viz:

$$m_e = \sqrt{\left(m_e \frac{v_\omega}{c}\right)^2 + \left(m_e \frac{v}{c}\right)^2} \qquad (15.2.1)$$

The above equation is the mass-velocity equation for the electron. The $m_{e\omega} = m_e \frac{v_\omega}{c}$ in the formula is the spin mass of the electron and $m_{ev} = m_e \frac{v}{c}$ is the displacement mass of the electron. Substituting the above equation into the charge-mass conversion equation (15.1.8) gives the relational equation:

$$e = \eta m_e = \sqrt{\left(\eta m_e \frac{v_\omega}{c}\right)^2 + \left(\eta m_e \frac{v}{c}\right)^2} \qquad (15.2.2)$$

The above equation can also be expressed as:

$$e = \sqrt{\left(e \frac{v_\omega}{c}\right)^2 + \left(e \frac{v}{c}\right)^2} \qquad (15.2.3)$$

The above equation is the charge-velocity equation for an electron. The above equation shows that the electric charge e of a moving electron can be decomposed into two component charges.

15.2.2 Rotational electric charge e_ω (spin electric charge) of a moving electron

According to Equation (15.2.3), the electric charge $\boldsymbol{e}$ of the moving electron can be decomposed into two component charges.

Since the velocity $\boldsymbol{v}_\omega$ is the spin velocity of the electron, this book defines the first component electric charge $\boldsymbol{e}\frac{\boldsymbol{v}_\omega}{c}$ inside the root sign of equation (15.2.3) as the rotational electric charge $\boldsymbol{e}_\omega$ (spin electric charge) of the moving electron, i.e:

$$\boldsymbol{e}_\omega = \boldsymbol{e}\frac{\boldsymbol{v}_\omega}{\boldsymbol{c}} \qquad (\mathbf{15.2.4})$$

The above equation shows that the rotational electric charge $\boldsymbol{e}_\omega$ (spin electric charge) of an electron, is proportional to the electron's spin speed $\boldsymbol{v}_\omega$. According to the spiral speed of light equation $\boldsymbol{c} = \sqrt{\mathbf{v}_\omega^2 + \boldsymbol{v}^2}$, the electron's spin speed $\boldsymbol{v}_\omega$ is:

$$\boldsymbol{v}_\omega = \sqrt{\boldsymbol{c}^2 - \boldsymbol{v}^2} \qquad (\mathbf{15.2.5})$$

Substituting the above equation into equation (15.2.4) gives the relationship equation:

$$\boldsymbol{e}_\omega = \boldsymbol{e}\frac{\boldsymbol{v}_\omega}{\boldsymbol{c}} = \boldsymbol{e}\sqrt{\mathbf{1} - \frac{\boldsymbol{v}^2}{\boldsymbol{c}^2}} \qquad (\mathbf{15.2.6})$$

The above equation is the equation for the rotational electric charge of an electron (spin-charge equation). As the displacement velocity $\boldsymbol{v}$ of the electron increases, the rotational electric charge $\boldsymbol{e}_\omega$ of the electron becomes smaller. An important corollary can be obtained from the above equation that the physical quantity proportional to the rotational electric charge $\boldsymbol{e}_\omega$ is correspondingly reduced in value by a factor of $\sqrt{\mathbf{1} - \boldsymbol{v}^2/\boldsymbol{c}^2}$.

15.2.3 Derivation of the rotational electric charge equation for electrons

According to electromagnetic field theory, the electric field force between two moving electrons is not equal to the Coulomb force. If the rotational electric charge $\boldsymbol{e}_\omega$ varies with the spin speed $\boldsymbol{v}_\omega$, then there should be a certain variation relationship between the rotational electric charge $\boldsymbol{e}_\omega$ and the spin speed $\boldsymbol{v}_\omega$. Assuming that the rotational electric charge $\boldsymbol{e}_\omega$ is proportional to the spin velocity $\boldsymbol{v}_\omega$, i.e:

$$\boldsymbol{e}_\omega = \boldsymbol{g}\boldsymbol{v}_\omega \qquad (\mathbf{15.2.7})$$

The $\boldsymbol{g}$ in the equation is the coefficient to be solved, and the relationship equation can be obtained by substituting equation (15.2.5) into the above equation:

$$\boldsymbol{e}_\omega = \boldsymbol{g}\sqrt{\boldsymbol{c}^2 - \boldsymbol{v}^2} \qquad (\mathbf{15.2.8})$$

When the electron is at rest in the cosmic vacuum, the rotational electric charge $\boldsymbol{e}_\omega = \boldsymbol{e} = \boldsymbol{gc}$ of the electron since the displacement velocity $\boldsymbol{v} = \mathbf{0}$ of the electron, so the coefficient $\boldsymbol{g}$ is:

$$\boldsymbol{g} = \frac{\boldsymbol{e}}{\boldsymbol{c}} \qquad (\mathbf{15.2.9})$$

Substituting the above equation into equation (15.2.8), the rotational electric charge $\boldsymbol{e}_\omega$ is:

$$\boldsymbol{e}_\omega = \boldsymbol{e}\sqrt{\mathbf{1} - \frac{\boldsymbol{v}^2}{\boldsymbol{c}^2}}$$

The above equation derived using equation (15.2.8) is the same as equation (15.2.6). There are a number of current physics experiments involving rotating charges $\boldsymbol{e}_\omega$ (e.g., high-energy physics experiments).

When the displacement velocity $\boldsymbol{v} = \mathbf{0}$ of the electron, the rotational electric charge $\boldsymbol{e}_\omega$ of the electron is equal to the static electric charge $\boldsymbol{e}_0$, i.e:

$$\boldsymbol{e}_\omega = \boldsymbol{e}_0$$

The above equation shows that the rotational electric charge $\boldsymbol{e}_\omega$ of a moving electron and the static electric charge $\boldsymbol{e}_0$ are charges of the same nature, i.e., they are both charges that can be accelerated by the electric field $\boldsymbol{E}$.

According to equation (15.2.6), when the displacement velocity $\boldsymbol{v}$ of the electron increases. Since the rotational electric charge $\boldsymbol{e}_\omega$ of the electron will become smaller, the force $\boldsymbol{F}$ of the electric field $\boldsymbol{E}$ on the moving electron will also become smaller.

When the displacement velocity $\boldsymbol{v}$ of the electron reaches the speed of light. The rotational electric charge of the electron $\boldsymbol{e_\omega = 0}$ because the displacement velocity $\boldsymbol{v = c}$ of the $\boldsymbol{\gamma}$ photon within the electron. At this point the electron is subjected to an electric field force $\boldsymbol{F = 0}$ in the electric field $\boldsymbol{E}$.

15.2.4 Relationship between the rotational electric charge e_ω of an electron, and the photon energy $\varepsilon = h(\delta_i + v)$ absorbed by the electron

When the displacement velocity $\boldsymbol{v}$ of the electron increases, according to (15.2.5), the spin velocity $\boldsymbol{e_\omega}$ of the electron becomes smaller. Assuming that the photon energy $\boldsymbol{\varepsilon = h(\delta_i + \beta)}$ absorbed by the electron, the displacement velocity $\boldsymbol{v}$ of the electron according to equation (14.1.13) is:

$$\boldsymbol{v = c\sqrt{\frac{2(\delta_i + \beta)}{\gamma + \delta_i + \beta}}}$$

Substituting the above equation into equation (15.2.6) gives the relationship equation:

$$\boldsymbol{e_\omega = e\sqrt{1 - \frac{v^2}{c^2}} = e\sqrt{1 - \frac{2(\delta_i + \beta)}{\gamma + \delta_i + \beta}}} \qquad (15.2.10)$$

Simplifying the above equation gives the relationship equation:

$$\boldsymbol{e_\omega = e\sqrt{\frac{\gamma - (\delta_i + \beta)}{\gamma + \delta_i + \beta}}} \qquad (15.2.11)$$

The above equation shows that the rotational electric charge $\boldsymbol{e_\omega}$ of the electron, becomes smaller as the photon energy $\boldsymbol{\varepsilon = h(\delta_i + \beta)}$ absorbed by the electron, increases.

When the photon energy absorbed by the electron $\boldsymbol{\varepsilon = h(\delta_i + \beta) = 0}$, the rotational electric charge $\boldsymbol{e_\omega}$ of the electron is equal to the static electric charge $\boldsymbol{e_0}$ of the electron, i.e. $\boldsymbol{e_\omega = e_0}$.

When the photon energy absorbed by the electron $\boldsymbol{\varepsilon = h\gamma}$, i.e., $\boldsymbol{\delta_i + \beta = \gamma}$, the rotational electric charge $\boldsymbol{e_\omega = 0}$ of the electron.

Substituting equation (15.1.9) electric charge $\boldsymbol{e}$ into equation (15.2.11) gives the relationship equation:

$$\boldsymbol{e_\omega = \eta\frac{h(\gamma + \delta_i + \beta)}{c^2}\sqrt{\frac{\gamma - (\delta_i + \beta)}{\gamma + \delta_i + \beta}}} \qquad (15.2.12)$$

Simplifying the above equation gives the relationship equation:

$$\boldsymbol{e_\omega = \eta\frac{h\sqrt{\gamma^2 - (\delta_i + \beta)^2}}{c^2}} \qquad (15.2.13)$$

The above equation shows that the rotational electric charge $\boldsymbol{e_\omega}$ of the electron becomes smaller as the photon energy $\boldsymbol{\varepsilon = h(\delta_i + \beta)}$ absorbed by the electron increases. When the photon energy absorbed by the electron $\boldsymbol{\varepsilon = 0}$, the rotational electric charge $\boldsymbol{e_\omega}$ of the electron is:

$$\boldsymbol{e_\omega = \eta\frac{h\gamma}{c^2} = e_0} \qquad (15.2.14)$$

The above equation is equation (15.1.10). In other words, when the photon energy absorbed by the electron $\boldsymbol{\varepsilon = h(\delta_i + \beta) = 0}$, the rotational electric charge $\boldsymbol{e_\omega}$ of the electron, is equal to the static electric charge $\boldsymbol{e_0}$ of the electron, because of the displacement velocity $\boldsymbol{v = 0}$ of the electron, i.e:

$$\boldsymbol{e_\omega = e_0}$$

The above equation shows that the rotating electric charge $\boldsymbol{e_\omega}$ and the static electric charge $\boldsymbol{e_0}$ are charges of the same nature. Since the static electric charge $\boldsymbol{e_0}$ is capable of generating an electric field $\boldsymbol{E_0}$, the rotating electric charge $\boldsymbol{e_\omega}$ must also be capable of generating an electric field $\boldsymbol{E_\omega}$. The difference between the two is that the former generates a static electric field $\boldsymbol{E_0}$, while the latter generates a moving electric field $\boldsymbol{E_\omega}$.

15.2.5 Magnetic electric charge e_v of a moving electron

According to equation (15.2.3), the electric charge $\boldsymbol{e}$ of the moving electron can be decomposed into two component charges. Since the velocity $\boldsymbol{v}$ is the displacement velocity of the electron, this book defines the second component electric charge $\boldsymbol{ev/c}$ within the root sign of equation (15.2.3) as the magnetic electric charge $\boldsymbol{e_v}$ of the moving electron, i.e:

$$\boldsymbol{e_v = e\frac{v}{c}} \qquad (\mathbf{15.2.15})$$

The above equation is the formula for the magnetic electric charge of an electron. The magnetic electric charge $\boldsymbol{e_v}$ of the electron is the magnetic electric charge that physicists have been searching for a long time. The above formula shows that the magnetic electric charge $\boldsymbol{e_v}$ of an electron is directly proportional to the displacement velocity $\boldsymbol{v}$. The greater the displacement velocity $\boldsymbol{v}$ of the electron, the greater the magnetic electric charge $\boldsymbol{e_v}$, and the stronger the magnetic field $\boldsymbol{B}$ produced by the moving electron.

When the displacement velocity $\boldsymbol{v = 0}$ of the electron. Since the magnetic electric charge $\boldsymbol{e_v = 0}$, the static electron cannot produce the magnetic field $\boldsymbol{B}$.

Substituting the rotational electric charge $\boldsymbol{e_\omega}$ and the magnetic electric charge $\boldsymbol{e_v}$ into equation (15.2.3) yields the relation equation:

$$\boldsymbol{e = \sqrt{e_\omega^2 + e_v^2}} \qquad (\mathbf{15.2.16})$$

When the electron displacement velocity $\boldsymbol{v = 0}$, the rotational electric charge $\boldsymbol{e_\omega}$ of the electron is equal to the static electric charge $\boldsymbol{e_0}$, i.e., $\boldsymbol{e_\omega = e_0}$, and the magnetic electric charge $\boldsymbol{e_v = 0}$ of the electron.

When the electron displacement velocity $\boldsymbol{v}$ increases, the spin velocity $\boldsymbol{v_\omega}$ and rotational electric charge $\boldsymbol{e_\omega}$ of the moving electrons become smaller, while the magnetic electric charge $\boldsymbol{e_v}$ and magnetic flux density $\boldsymbol{B}$ increase.

To summarize, in the cosmic vacuum frame of reference, the moving electron contains within it two charges of different nature.

One is the rotational electric charge $\boldsymbol{e_\omega}$ capable of generating the electric field $\boldsymbol{E}$, whose magnitude becomes smaller as the displacement velocity $\boldsymbol{v}$ of the electron increases. When the displacement velocity $\boldsymbol{v = 0}$ of the electron, the rotational electric charge $\boldsymbol{e_\omega}$ of the electron is equal to the static electric charge $\boldsymbol{e_0}$, i.e. $\boldsymbol{e_\omega = e_0}$.

The other is a magnetic electric charge $\boldsymbol{e_v}$ capable of generating a magnetic field $\boldsymbol{B}$. The magnitude of the magnetic electric charge $\boldsymbol{e_v}$ becomes larger as the displacement velocity $\boldsymbol{v}$ of the electron increases. When the displacement velocity $\boldsymbol{v = 0}$ of the electron, the magnetic electric charge $\boldsymbol{e_v = 0}$ of the electron.

Since the electric charge $\boldsymbol{q}$ is a collection of electron charges $\boldsymbol{e}$, i.e:

$$\boldsymbol{q = \sum e}$$

When the electric charge $\boldsymbol{q}$ is moving with velocity $\boldsymbol{v}$ in the cosmic vacuum, the relation equation can be obtained by substituting equation (15.2.16) electric charge $\boldsymbol{e}$ into the above equation:

$$\boldsymbol{q = \sum \sqrt{e_\omega^2 + e_v^2} = \sqrt{q_\omega^2 + q_v^2}} \qquad (\mathbf{15.2.17})$$

The above equation is the charge-velocity equation. The above equation shows that the moving electric charge $\boldsymbol{q}$ can be decomposed into two component charges.

When the electric charge displacement velocity $\boldsymbol{v = 0}$, the rotational electric charge $\boldsymbol{q_\omega}$ contained in the electric charge $\boldsymbol{q}$ is equal to the static electric charge $\boldsymbol{q_0}$, i.e:

$$\boldsymbol{q_\omega = q_0}$$

When the electric charge displacement velocity $\boldsymbol{v = 0}$, the moving electric charge $\boldsymbol{q}$ produces a magnetic electric charge $\boldsymbol{q_v = 0}$.

If the displacement velocity $\boldsymbol{v}$ of the electric charge $\boldsymbol{q}$ increases, then the rotational electric charge $\boldsymbol{q_\omega}$ contained in the moving electric charge $\boldsymbol{q}$ becomes smaller, while the magnetic electric charge $\boldsymbol{q_v}$ and the magnetic flux density $\boldsymbol{B}$ contained in the moving electric charge $\boldsymbol{q}$ become larger.

15.2.6 Relation between the magnetic electric charge e_v of an electron, and the photon energy $\varepsilon=h(\delta_i+\beta)$ absorbed by the electron

Assuming that the photon energy absorbed by the electron is $\boldsymbol{\varepsilon} = \boldsymbol{h}(\boldsymbol{\delta}_i + \boldsymbol{\beta})$, and according to equation (14.1.13), the displacement velocity $\boldsymbol{v}$ of the electron is:

$$v = c\sqrt{\frac{2(\delta_i + \beta)}{\gamma + \delta_i + \beta}} \quad (15.2.18)$$

Substituting the above equation into equation (15.2.15) gives the relationship equation:

$$e_v = e\frac{v}{c} = e\sqrt{\frac{2(\delta_i + \beta)}{\gamma + \delta_i + \beta}} \quad (15.2.19)$$

Substituting equation (15.1.9) electric charge $\boldsymbol{e}$ into the above equation gives the relational equation:

$$e_v = \eta\frac{h(\gamma + \delta_i + \beta)}{c^2}\sqrt{\frac{2(\delta_i + \beta)}{\gamma + \delta_i + \beta}} \quad (15.2.20)$$

Simplifying the above equation gives the relationship equation:

$$e_v = \eta\frac{h\sqrt{2(\gamma + \delta_i + \beta)(\delta_i + \beta)}}{c^2} \quad (15.2.21)$$

The above equation shows that the magnetic electric charge $\boldsymbol{e}_v$ of the moving electron becomes larger as the photon energy absorbed by the electron $\boldsymbol{\varepsilon} = \boldsymbol{h}(\boldsymbol{\delta}_i + \boldsymbol{\beta})$ increases. When the photon energy absorbed by the electron $\boldsymbol{\varepsilon} = \boldsymbol{h}(\boldsymbol{\delta}_i + \boldsymbol{\beta}) = \boldsymbol{0}$, the displacement velocity of the electron $\boldsymbol{v} = \boldsymbol{0}$ and the magnetic electric charge of the electron $\boldsymbol{e}_v = \boldsymbol{0}$.

15.3 Rotational electric charge e_ω' and magnetic electric charge e_v' in inertial system S'

15.3.1 Charge-velocity equation for an electron in an inertial system S'

According to equation (15.2.3), the charge-velocity equation for an electron in the cosmic vacuum frame of reference is:

$$e = \sqrt{\left(e\frac{v_\omega}{c}\right)^2 + \left(e\frac{v}{c}\right)^2}$$

Since the speed of light $\boldsymbol{c}$, the spin speed $\boldsymbol{v}_\omega$ and the displacement speed $\boldsymbol{v}$ all belong to the speed in the cosmic vacuum frame of reference, the electric charge $\boldsymbol{e}$ belongs to the electric charge in the cosmic vacuum frame of reference.

Suppose that the inertial system $\boldsymbol{S}'$ is moving in the cosmic vacuum frame of reference with velocity $\boldsymbol{u}$ in the direction of the $+\boldsymbol{X}$ axis. According to Galileo's velocity transformation formula, the maximum velocity $\boldsymbol{c}'$ of a photon traveling in the inertial system $\boldsymbol{S}'$ in the direction of the $+\boldsymbol{X}'$ axis is:

$$c' = c - u \quad (15.3.1)$$

Assume that $\boldsymbol{v}'$ is the displacement velocity of the electron along the axis of $+\boldsymbol{X}'$, $\boldsymbol{v}'_\omega$ is the spin velocity of the electron, and the spiral speed of light $\boldsymbol{c}'$ of the electron in the inertial system $\boldsymbol{S}'$ is:

$$c' = \sqrt{v_\omega'^2 + v'^2} = c - u \quad (15.3.2)$$

When the displacement velocity of the electron $\boldsymbol{v}' = \boldsymbol{0}$, the spin velocity of the electron $\boldsymbol{v}'_\omega = \boldsymbol{c}' = \boldsymbol{c} - \boldsymbol{u}$.

When the displacement velocity of the electron $\boldsymbol{v}' = \boldsymbol{c}' = \boldsymbol{c} - \boldsymbol{u}$, the spin velocity of the electron $\boldsymbol{v}'_\omega = \boldsymbol{0}$.

When the inertial system $\boldsymbol{S}'$ is moving with speed $\boldsymbol{u} = \boldsymbol{c}$, the electron's spin velocity $\boldsymbol{v}'_\omega = \boldsymbol{v} = \boldsymbol{0}$.

Assume that $\boldsymbol{e}'$ is the electric charge of the electron moving along the axis of $+\boldsymbol{X}'$. According to equation

(15.2.3), the charge-velocity equation for an electron in an inertial system $\boldsymbol{S'}$ is:

$$e' = \sqrt{\left(e'\frac{v'_{\omega}}{c'}\right)^2 + \left(e'\frac{v'}{c'}\right)^2} \qquad (15.3.3)$$

The above equation is the charge-velocity equation for an electron in an inertial system $\boldsymbol{S'}$. Note that the displacement velocity $\boldsymbol{v'}$ and the velocity $\boldsymbol{u}$ are both in the same direction, and the electron's spin velocity $\boldsymbol{v'_{\omega}}$ is perpendicular to the velocity $\boldsymbol{u}$.

When the velocity of the inertial system $\boldsymbol{S'}$ is $\boldsymbol{u} = \boldsymbol{0}$, the displacement velocity of the electron $\boldsymbol{v'} = \boldsymbol{v}$, at which point equation (15.3.3) becomes the charge-velocity equation for the electron in the cosmic vacuum frame of reference.

Since the velocity $\boldsymbol{u}$ of the Earth's motion in the cosmic vacuum is much smaller than the speed of light $\boldsymbol{c}$, the Earth's inertial system can be considered to be the cosmic vacuum reference system.

15.3.2 The electric charge *e* in the charge-to-mass ratio equation is the rotational electric charge of the electron e_{ω}

According to the charge-velocity equation (15.2.3), the electric charge $\boldsymbol{e}$ of the moving electron is:

$$e = e\sqrt{\left(\frac{v_{\omega}}{c}\right)^2 + \left(\frac{v}{c}\right)^2} \qquad (15.3.4)$$

According to the mass-velocity equation (14.1.17) , the mass $\boldsymbol{m_e}$ of the moving electron is:

$$m_e = m_e\sqrt{\left(\frac{v_{\omega}}{c}\right)^2 + \left(\frac{v}{c}\right)^2} \qquad (15.3.5)$$

Dividing the electric charge $\boldsymbol{e}$ by the mass $\boldsymbol{m_e}$, the charge-to-mass ratio $\boldsymbol{\eta}$ of the moving electron is:

$$\eta = \frac{e}{m_e} \qquad (15.3.6)$$

The above equation shows that the charge-to-mass ratio $\boldsymbol{\eta}$ is a constant independent of the displacement velocity $\boldsymbol{v}$ of the electron.

Based on the above equation, the relationship equation can be obtained:

$$\eta = \frac{e}{m_e} = \frac{e_0}{m_{e0}} \qquad (15.3.7)$$

The above equation shows that the charge-to-mass ratio $\boldsymbol{\eta}$ is equal to the ratio of the static electric charge $\boldsymbol{e_0}$ of the electron to the static mass $\boldsymbol{m_{e0}}$ of the electron.

In classical theory, physicists define the charge-to-mass ratio $\boldsymbol{k}$ of an electron as:

$$k = \frac{e}{m_e} \qquad (15.3.8)$$

The $\boldsymbol{e}$ in the formula is the electric charge of the electron and $\boldsymbol{m_e}$ is the moving mass of the electron. According to physical experiments that measure the charge-to-mass ratio of electrons, the charge-to-mass ratio $\boldsymbol{k}$ of an electron changes as the displacement velocity $\boldsymbol{v}$ of the electron increases.

In special relativity, the electric charge $\boldsymbol{e}$ of an electron is a constant, and the electron's moving mass $\boldsymbol{m_e}$ increases as the electron displacement velocity $\boldsymbol{v}$ increases. Assuming that $\boldsymbol{m_{e0}}$ is the static mass of the electron and $\boldsymbol{v}$ is the velocity of the electron's motion, the mass-velocity equation for the electron is:

$$m_e = \frac{m_{e0}}{\sqrt{1 - \frac{v^2}{c^2}}} \qquad (15.3.9)$$

Substituting the above equation into equation (15.3.8), the equation for the charge-to-mass ratio in special relativity is:

$$k = \frac{e}{m_e} = \frac{e}{m_{e0}}\sqrt{1 - \frac{v^2}{c^2}} \qquad (15.3.10)$$

Since the static mass m_{e0} of the electron is independent of the change in the electron's velocity v, physicists believed that the charge-to-mass ratio k of the electron becomes smaller as the electron's velocity v increases. However, new physics considers the charge-to-mass ratio η to be a constant, and η is independent of the electron's speed v of motion.

According to Equation (15.2.6), the rotational electric charge e_ω of the moving electron is:

$$e_\omega = e\frac{v_\omega}{c} = e\sqrt{1-\frac{v^2}{c^2}} \qquad (15.3.11)$$

The above equation shows that the rotational electric charge e_ω of the electron becomes smaller as the displacement velocity v increases. Comparing the above equation with equation (15.3.10), it can be determined that k is equal to the ratio of the electron's rotational electric charge e_ω, to its static mass m_{e0}, i.e:

$$k = \frac{e_\omega}{m_{e0}} = \frac{e}{m_{e0}}\sqrt{1-\frac{v^2}{c^2}} \qquad (15.3.12)$$

The above equation is the formula for the rotational electric charge e_ω-velocity ratio of an electron. Although the above formula has the same mathematical form as formula (15.3.10), the theoretical implications of the two are completely opposite.

The difference between the two is that the rotational electric charge e_ω in the equation $k = \frac{e_\omega}{m_{e0}}$ is a function of the velocity v, while the moving mass m_e in the equation $k = \frac{e}{m_e}$ is a function of the velocity v.

Since the rotating electric charge $e_\omega = e\sqrt{1-\frac{v^2}{c^2}}$ in the moving electron and the static electric charge e_0 are charges of the same nature, the rotating electric charge e_ω is subjected to the electric field force F in the electric field, which causes the moving electron to acquire an acceleration a.

When the displacement velocity v of the electron increases. Since the rotational electric charge e_ω in the moving electron becomes smaller as the velocity v increases, the electric field force F on the moving electron becomes smaller.

When the electron displacement velocity v is equal to the speed of light c. Since the rotational electric charge $e_\omega = 0$ in the moving electron, so the electric field force on the moving electron $F = 0$.

Chapter 16 electric charge energy ε_e=ec^2 and electric charge momentum P_e=ec for electrons

Introduction: The mass-energy conversion formula for an object is $\boldsymbol{\varepsilon = mc^2}$. This chapter mimics the formula $\boldsymbol{\varepsilon = mc^2}$ and analyzes and discusses the electric charge energy of electrons $\boldsymbol{\varepsilon_e = ec^2}$.

16.1 Electric charge energy ε_e=ec^2 of an electron

16.1.1 The electric charge energy ε_e=ec^2 of an electron can be decomposed into two component electric charge energies

In physics, the mass-energy conversion equation for an object is:

$$\boldsymbol{\varepsilon = mc^2} \qquad (16.1.1)$$

We mimic the mass-energy conversion formula by defining the product $\boldsymbol{ec^2}$ of the electron's electric charge $\boldsymbol{e}$ and the speed of light $\boldsymbol{c}$ squared, as the electron's electric charge energy $\boldsymbol{\varepsilon_e}$, i.e:

$$\boldsymbol{\varepsilon_e = ec^2} \qquad (16.1.2)$$

When an electron is moving with velocity $\boldsymbol{v}$ in the cosmic vacuum reference center, according to equation (15.2.16) , the electric charge $\boldsymbol{e}$ of the moving electron can be decomposed into a rotational electric charge $\boldsymbol{e_\omega}$ and a magnetic electric charge $\boldsymbol{e_v}$, viz:

$$\boldsymbol{e = \sqrt{e_\omega^2 + e_v^2}}$$

Substituting the above equation into equation (16.1.2) gives the relationship equation:

$$\boldsymbol{\varepsilon_e = ec^2 = \sqrt{(e_\omega c^2)^2 + (e_v c^2)^2}} \qquad (16.1.3)$$

The above equation is the equation for the electric charge energy of an electron. The above equation shows that the electric charge energy $\boldsymbol{\varepsilon_e}$ of an electron can be decomposed into two component electric charge energies.

16.1.2 Rotational electric charge energy of an electron $\varepsilon_{e\omega}$ (spin charge energy)

Since the electric charge $\boldsymbol{e_\omega}$ is the rotational electric charge of the electron, this book defines the first component electric charge energy $\boldsymbol{e_\omega c^2}$, within the root sign of Equation (16.1.3) as the rotational electric charge energy $\boldsymbol{\varepsilon_{e\omega}}$ (spin charge energy) of the electron, i.e:

$$\boldsymbol{\varepsilon_{e\omega} = e_\omega c^2} \qquad (16.1.4)$$

The above formula is the rotational charge energy formula (spin charge energy formula) for electrons. According to equation (15.2.6) the rotational electric charge $\boldsymbol{e_\omega}$ of the electron is:

$$\boldsymbol{e_\omega = e\sqrt{1 - \frac{v^2}{c^2}}}$$

Substituting the above equation into equation (16.1.4) gives the relationship equation:

$$\boldsymbol{\varepsilon_{e\omega} = e_\omega c^2 = c^2 e\sqrt{1 - \frac{v^2}{c^2}} = \varepsilon_e\sqrt{1 - \frac{v^2}{c^2}}} \qquad (16.1.5)$$

The above equation shows that the rotational electric charge energy $\boldsymbol{\varepsilon_{e\omega}}$ (spin charge energy) becomes smaller as the displacement velocity $\boldsymbol{v}$ of the electron increases. When the displacement velocity of the electron $\boldsymbol{v = 0}$, the rotational electric charge energy $\varepsilon_{e\omega} = \varepsilon_e$.

16.1.3 Magnetic electric charge energy ε_{ev} of an electron

Since the electric charge e_v is the magnetic electric charge of the electron, this book defines the second component electric charge energy $e_v c^2$, inside the root sign of equation (16.1.3) as the magnetic electric charge energy ε_{ev} of the electron, i.e:

$$\varepsilon_{ev} = e_v c^2 \tag{16.1.6}$$

The above equation is the magnetic electric charge energy equation for an electron. According to equation (15.2.15) the magnetic electric charge e_v of an electron is:

$$e_v = e\frac{v}{c}$$

Substituting the above equation into equation (16.1.6) gives the relationship equation:

$$\varepsilon_{ev} = e_v c^2 = c^2 e\frac{v}{c} = \varepsilon_e \frac{v}{c} \tag{16.1.7}$$

The above equation shows that the magnetic electric charge energy ε_{ev} of the electron is proportional to the displacement velocity v. When the displacement velocity $v = 0$ of the electron, the magnetic electric charge energy $\varepsilon_{ev} = 0$.

The electric charge energy ε_e of an electron can be expressed as:

$$\varepsilon_e = \sqrt{\varepsilon_{e\omega}^2 + \varepsilon_{ev}^2} \tag{16.1.8}$$

16.2 Electric charge momentum $P_e = ec$ of an electron

16.2.1 Magnetic electric charge momentum P_{ev} of an electron

When an electron moves with velocity v in the cosmic vacuum frame of reference, according to classical mechanics the momentum P_m of the electron is:

$$P_m = m_e v \tag{16.2.1}$$

According to equation (15.1.8), the conversion formula for the electric charge e and mass m_e of an electron is:

$$e = \eta m_e \tag{16.2.2}$$

The charge-to-mass ratio $\eta = 1.7588 \times 10^{11} \cdot C/kg$ in equation Multiplying equation (16.2.1) by the charge-to-mass ratio η gives the relational equation:

$$P_{ev} = \eta P_m = ev \tag{16.2.3}$$

This book defines the product ev of the electron's electric charge e, and displacement velocity v, as the electron's magnetic electric charge momentum P_{ev}. The above equation is the magnetic electric charge momentum equation for the electron. The above equation shows that the product ηP_m of the electron's momentum P_m and the charge-to-mass ratio η is equal to the electron's magnetic electric charge momentum P_{ev}.

According to Equation (14.1.13) and equation (15.1.9) , the displacement velocity v and electric charge e of the electron are, respectively:

$$v = c\sqrt{\frac{2(\delta_i + \beta)}{\gamma + \delta_i + \beta}}, \qquad e = \eta \frac{h(\gamma + \delta_i + \beta)}{c^2}$$

Substituting the above equation into equation (16.2.3), the magnetic electric charge momentum P_{ev} of the electron is:

$$P_{ev} = ev = \eta \frac{h(\gamma + \delta_i + \beta)}{c^2} \cdot c\sqrt{\frac{2(\delta_i + \beta)}{\gamma + \delta_i + \beta}}$$

Simplifying the above equation gives the relationship equation:

$$P_{ev} = \frac{\eta h}{c}\sqrt{2(\gamma + \delta_i + \beta)(\delta_i + \beta)} \tag{16.2.4}$$

The above equation shows that the magnetic electric charge momentum $\boldsymbol{P_{ev}}$ becomes larger as the photon energy absorbed by the electron $\boldsymbol{\varepsilon = h(\delta_i + \beta)}$ increases. When the photon energy absorbed by the electron $\boldsymbol{\varepsilon = h(\delta_i + \beta) = 0}$, the magnetic electric charge momentum $\boldsymbol{P_{e0}}$ of the static electron is:

$$\boldsymbol{P_{e0} = 0} \qquad (16.2.5)$$

When the photon energy $\boldsymbol{\varepsilon = h(\delta_i + \beta) = h}$ absorbed by the electron. Since $\boldsymbol{\delta_i + \beta = \gamma}$, the displacement velocity $\boldsymbol{v}$ of the electron is equal to the speed of light $\boldsymbol{c}$. The magnetic electric charge momentum $\boldsymbol{P_{ec}}$ of an electron moving at the speed of light $\boldsymbol{c}$ is:

$$\boldsymbol{P_{ec} = \eta\frac{2h\gamma}{c}} \qquad (16.2.6)$$

The above equation shows that the magnetic electric charge momentum $\boldsymbol{P_{ec}}$ of an electron moving at the speed of light $\boldsymbol{c}$ is not equal to infinity. According to equation (16.2.3), the momentum $\boldsymbol{P_m}$ of an electron is:

$$\boldsymbol{P_m = \frac{P_{ec}}{\eta} = \frac{e}{\eta}c = mc} \qquad (16.2.7)$$

Since the magnetic electric charge momentum $\boldsymbol{P_{ec}}$ is not equal to infinity, the momentum of the electron $\boldsymbol{P_m}$ is not equal to infinity, viz:

$$\boldsymbol{P_m \neq \infty}$$

However in special relativity the momentum $\boldsymbol{P_m = \infty}$ of an electron moving at the speed of light $\boldsymbol{c}$.

16.2.2 Electric charge momentum of an electron P_e

When an electron moves with velocity $\boldsymbol{v}$ in the cosmic vacuum frame of reference, the variation of the electron's spin velocity $\boldsymbol{v_\omega}$ and displacement velocity $\boldsymbol{v}$ is bounded by the spiral speed of light formula, i.e:

$$\boldsymbol{c = \sqrt{v_\omega^2 + v^2}} \qquad (16.2.8)$$

Multiplying the above equation by the electric charge $\boldsymbol{e}$ of the electron gives the relational equation:

$$\boldsymbol{P_e = ec = \sqrt{(ev_\omega)^2 + (ev)^2}} \qquad (16.2.9)$$

This book defines $\boldsymbol{ec}$, the product of the electric charge $\boldsymbol{e}$ of an electron and the speed of light $\boldsymbol{c}$, as the electric charge momentum of a moving electron. The above equation is the electric charge momentum formula for the electron. The above equation shows that the electric charge momentum $\boldsymbol{P_e}$ of a moving electron can be decomposed into two component electric charge momentums.

According to equation (15.1.9), the electric charge $\boldsymbol{e}$ of the moving electron is:

$$\boldsymbol{e = \eta\frac{h(\gamma + \delta_i + \beta)}{c^2}}$$

Substituting the above equation into equation (16.2.9) gives the relationship equation:

$$\boldsymbol{P_e = ec = \eta\frac{h(\gamma + \delta_i + \beta)}{c}} \qquad (16.2.10)$$

The above equation shows that the electric charge momentum $\boldsymbol{P_e}$ becomes larger as the photon energy absorbed by the electron $\boldsymbol{\varepsilon = h(\delta_i + \beta)}$ increases. When the photon energy absorbed by the electron $\boldsymbol{\varepsilon = h(\delta_i + \beta) = 0}$, the electric charge momentum $\boldsymbol{P_e}$ of the static electron is:

$$\boldsymbol{P_e = \eta\frac{h\gamma}{c}} \qquad (16.2.11)$$

The electric charge momentum $\boldsymbol{P_e}$ in equation is at this point equal to the momentum of the $\boldsymbol{\gamma}$ photon contained within the static electron.

16.2.3 Spin-charge momentum of an electron $P_{e\omega}$

Since the velocity $\boldsymbol{v_\omega}$ is the spin velocity of the electron, this book defines the first component electric charge momentum $\boldsymbol{ev_\omega}$ inside the root sign of equation (16.2.9) as the spin electric charge momentum $\boldsymbol{P_{e\omega}}$ of the moving electron, i.e:

$$\boldsymbol{P_{e\omega} = ev_\omega} \qquad (16.2.12)$$

The above equation is the equation for the spin electric charge momentum of a moving electron. According to equation (16.2.8) , the spin velocity v_ω of the electron is:

$$v_\omega = \sqrt{c^2 - v^2} \tag{16.2.13}$$

Substituting the above equation into equation (16.2.12) gives the relationship equation:

$$P_{e\omega} = e\sqrt{c^2 - v^2} \tag{16.2.14}$$

The above equation shows that the spin electric charge momentum $P_{e\omega}$ of the electron becomes smaller as the electron displacement velocity v increases. When the displacement velocity $v = 0$ of the electron, the spin electric charge momentum $P_{e\omega} = P_e$ of the electron.

The electric charge momentum P_e of a moving electron can be expressed as:

$$P_e = ec = \sqrt{(ev_\omega)^2 + (ev)^2} = \sqrt{P_{e\omega}^2 + P_{ev}^2} \tag{16.2.15}$$

When the displacement velocity $v = 0$ of the electron. Since the spin velocity $v_\omega = c$ of the electron, the spin electric charge momentum $P_{e\omega} = P_e = ec$ contained in the moving electron and the magnetic electric charge momentum $P_{ev} = 0$.

When the displacement velocity $v = c$ of the electron. Since the spin velocity $v_\omega = 0$ of the electron, the spin electric charge momentum $P_{e\omega} = 0$ of the moving electron and the magnetic electric charge momentum $P_{ev} = P_e = ec$.

16.3 The essence of the current density J of a conductor is the magnetic electric charge momentum per unit volume $P_{\rho v}=nev$

16.3.1 Free electric charge density per unit volume of a conductor ρ

Assume that n is the free electron density per unit volume of the conductor. The free electric charge density ρ per unit volume of the conductor is:

$$\rho = ne \tag{16.3.1}$$

When a conductor is at rest in the cosmic vacuum and free electrons are moving in the conductor with velocity v. According to equation (15.2.16), the electric charge e of the moving electron can be decomposed into a rotational electric charge e_ω and a magnetic electric charge e_v, i.e:

$$e = \sqrt{e_\omega^2 + e_v^2}$$

Substituting the above equation into equation (16.3.1), the free electric charge density ρ per unit volume of the conductor is:

$$\rho = ne = \sqrt{(ne_\omega)^2 + (ne_v)^2} \tag{16.3.2}$$

The above equation is the formula for the free electric charge density per unit volume of a conductor. The above equation shows that the free electric charge density ρ per unit volume of a conductor, can be decomposed into two component electric charge densities.

16.3.2 Spin electric charge density ρ_ω per unit volume of a conductor

Since the velocity v_ω is the spin velocity of the electron, this book defines the first component electric charge density ne_ω within the root sign of equation (16.3.2), as the spin electric charge density ρ_ω per unit volume of the conductor, viz:

$$\rho_\omega = ne_\omega \tag{16.3.3}$$

The above equation is the formula for the spin electric charge density per unit volume of a conductor. According to equation (15.2.6), the spin electric charge e_ω is:

$$e_\omega = e\frac{v_\omega}{c} = e\sqrt{1 - \frac{v^2}{c^2}}$$

Substituting the above equation into equation (16.3.3) gives the relationship equation:

$$\rho_\omega = ne\sqrt{1-\frac{v^2}{c^2}} = \rho\sqrt{1-\frac{v^2}{c^2}} \tag{16.3.4}$$

The above equation is the formula for the spin electric charge density per unit volume of a conductor. The above equation shows that the spin electric charge density ρ_ω per unit volume of the conductor, becomes smaller as the displacement velocity v of the free electrons increases.

When the displacement velocity $v = 0$ of free electrons, the spin electric charge density ρ_ω per unit volume is equal to the stationary free electric charge density ρ_0, i.e. $\rho_\omega = \rho_0$. From this, it can be determined that the spin electric charge density ρ_ω and the stationary free electric charge density ρ_0 are both electric charge densities with the same properties, i.e., both of them are capable of generating electric fields.

16.3.3 Magnetic electric charge density ρ_v per unit volume of a conductor

Since the velocity v is the displacement velocity of the electron, this book defines the second component electric charge momentum ne_v, within the root sign of equation (16.3.2) as the magnetic electric charge density ρ_v per unit volume of the conductor, viz:

$$\rho_v = ne_v \tag{16.3.5}$$

The above equation is the formula for the magnetic electric charge density per unit volume of a conductor. The above equation shows that the magnetic electric charge density ρ_v per unit volume of the conductor becomes larger as the displacement velocity v of the electron increases. The magnetic electric charge density $\rho_v = \rho$ per unit volume when the displacement velocity $v = c$ of the electron.

According to equation (15.2.15), the magnetic electric charge e_v is:

$$e_v = e\frac{v}{c}$$

Substituting the above equation into equation (16.3.5) gives the relationship equation:

$$\rho_v = ne\frac{v}{c} = \rho\frac{v}{c} \tag{16.3.6}$$

The above equation is the formula for the magnetic electric charge density per unit volume of a conductor. The above equation shows that the magnetic electric charge density ρ_v per unit volume of a conductor, is proportional to the displacement velocity v, of the free electrons.

When the displacement velocity $v = 0$ of the free electron, the magnetic electric charge density $\rho_v = 0$ and the static electron produces no magnetic field. If the electron displacement velocity v is greater, then the magnetic electric charge density ρ_v per unit volume is greater and the magnetic field B produced by the moving electrons is stronger. It can be determined that the magnetic electric charge density ρ_v produces the magnetic field B.

In summary, the electric charge density ρ per unit volume can be expressed as:

$$\rho = \sqrt{\rho_\omega^2 + \rho_v^2} \tag{16.3.7}$$

The spin electric charge density ρ_ω in equation produces the electric field E and the magnetic electric charge density ρ_v produces the magnetic field B.

16.3.4 Spin electric charge momentum $P_{\rho\omega}$ and magnetic electric charge momentum $P_{\rho v}$ per unit volume of a conductor

According to equation (16.2.9), the electric charge momentum P_e of the moving electron is:

$$P_e = ec = \sqrt{(ev_\omega)^2 + (ev)^2}$$

Multiplying the above equation by the free electron density n gives the relational equation:

$$P_\rho = nP_e = nec = \sqrt{(nev_\omega)^2 + (nev)^2} \tag{16.3.8}$$

This book defines $P_\rho = nec$ as the free electric charge momentum per unit volume of a conductor. The above equation is the formula for the free electric charge momentum per unit volume of a conductor. The above equation shows that the free electric charge momentum P_ρ per unit volume of a conductor, can be decomposed into two component electric charge momentums.

Since the velocity $\boldsymbol{v}_\omega$ is the spin velocity of the electron, this book defines the first component electric charge momentum $\boldsymbol{nev}_\omega$ inside the root sign of equation (16.3.8) as the spin electric charge momentum $\boldsymbol{P}_{\rho\omega}$ per unit volume of the conductor, viz:

$$\boldsymbol{P}_{\rho\omega} = \boldsymbol{nev}_\omega \qquad (16.3.9)$$

The above equation is the equation for the spin electric charge momentum per unit volume of the conductor. According to equation (16.2.13), the spin velocity $\boldsymbol{v}_\omega$ of the electron is:

$$\boldsymbol{v}_\omega = \sqrt{\boldsymbol{c}^2 - \boldsymbol{v}^2}$$

Substituting the above equation into equation (16.3.9) gives the relationship equation:

$$\boldsymbol{P}_{\rho\omega} = \boldsymbol{ne}\sqrt{\boldsymbol{c}^2 - \boldsymbol{v}^2} \qquad (16.3.10)$$

The above equation is the formula for the spin electric charge momentum per unit volume of a conductor. The above equation shows that the spin electric charge momentum $\boldsymbol{P}_{\rho\omega}$ per unit volume of the conductor becomes smaller as the displacement velocity $\boldsymbol{v}$ of the free electrons increases.

Since the velocity $\boldsymbol{v}$ is the displacement velocity of the electron, this book defines the second component electric charge momentum $\boldsymbol{nev}$ within the root sign of equation (16.3.8) as the magnetic electric charge momentum $\boldsymbol{P}_{\rho v}$ per unit volume of the conductor, viz:

$$\boldsymbol{P}_{\rho v} = \boldsymbol{nev} \qquad (16.3.11)$$

The above equation is the formula for the magnetic electric charge momentum per unit volume of a conductor. The above equation shows that the magnetic electric charge momentum $\boldsymbol{P}_{\rho v}$ per unit volume of a conductor, is proportional to the displacement velocity $\boldsymbol{v}$ of the electrons. When the electron displacement velocity $\boldsymbol{v} = \boldsymbol{c}$, the magnetic electric charge momentum $\boldsymbol{P}_{\rho v}$ per unit volume of a conductor is:

$$\boldsymbol{P}_{\rho v} = \boldsymbol{nP}_e$$

The $\boldsymbol{P}_e$ in the equation is the electric charge momentum of the moving electron.

16.3.5 current density *J* is essentially the magnetic electric charge momentum $P_{\rho v}$ per unit volume of the conductor

According to electromagnetic theory, electric current $\boldsymbol{I}$ is the result of the directional moves of free electrons in a conductor. Physicists describe the strength and direction of electric current at each point of a conductor in terms of current density $\boldsymbol{J}$. The formula for current density $\boldsymbol{J}$ is:

$$\boldsymbol{J} = \frac{\boldsymbol{I}}{\boldsymbol{S}} \qquad (16.3.12)$$

The $\boldsymbol{S}$ in the formula is the conductor cross section. electric current $\boldsymbol{I}$ equation is:

$$\boldsymbol{I} = \boldsymbol{nevS} \qquad (16.3.13)$$

Substituting the above equation into equation (16.3.12) gives the relationship equation:

$$\boldsymbol{J} = \boldsymbol{nev} \qquad (16.3.14)$$

Comparing the above equation with equation (16.3.11) gives the relationship equation:

$$\boldsymbol{J} = \boldsymbol{P}_{\rho v} = \boldsymbol{nev} \qquad (16.3.15)$$

The above equation shows that the current density $\boldsymbol{J}$ of a conductor is essentially the magnetic electric charge momentum $\boldsymbol{P}_{\rho v}$ per unit volume of the conductor.

16.3.6 Electric current *I'* in inertial system *S'*

Suppose $\boldsymbol{n}$ is the density of free electrons per unit volume of the wire, and $\boldsymbol{S}$ is the cross-section of the wire. When the wire is stationary in the cosmic vacuum, the free electrons in the wire pass through the cross-section $\boldsymbol{S}$ of the wire with a speed $\boldsymbol{v}$. According to electromagnetic field theory, the electric current $\boldsymbol{I}$ across the cross-section $\boldsymbol{S}$ of the wire is:

$$\boldsymbol{I} = \boldsymbol{nevS}$$

Note that the wire in equation is at rest in the cosmic vacuum frame of reference. If the inertial system $\boldsymbol{S}'$

(terrestrial frame of reference) is moving with velocity $\boldsymbol{u}$ in the cosmic vacuum frame of reference, then the current-carrying wire, which is at rest in the inertial system $\boldsymbol{S'}$ (terrestrial frame of reference), is moving with velocity $\boldsymbol{u}$ in the cosmic vacuum frame of reference.

When the wire is at rest in the inertial system $\boldsymbol{S'}$ (terrestrial frame of reference) and the free electrons are moving with velocity $\boldsymbol{v}$ in the cosmic vacuum frame of reference. If the direction of motion of the free electrons is the same as the direction of motion of the inertial system $\boldsymbol{S'}$, then the displacement velocity $\boldsymbol{v'}$ of the free electrons in the wire is:

$$\boldsymbol{v' = v - u} \qquad (16.3.16)$$

The electric current $\boldsymbol{I'}$ in a stationary wire in the ground reference frame is:

$$\boldsymbol{I' = nev'S = ne(v - u)S} \qquad (16.3.17)$$

The above equation shows that when the wire and the free electrons are both at rest relative to each other. Since the velocity $\boldsymbol{v' = 0}$ of the free electrons in the wire, the electric current $\boldsymbol{I' = 0}$ in the wire.

Note that the electric current $\boldsymbol{I'}$ in the above equation is the electric current when the wire is at rest in the inertial system $\boldsymbol{S'}$. If the cosmic velocity $\boldsymbol{u}$ of the inertial system $\boldsymbol{S'}$ is equal to the cosmic velocity $\boldsymbol{v}$ of the free electrons, i.e. $\boldsymbol{u = v}$, then the electric current $\boldsymbol{I' = 0}$ in the wire.

Theoretically, the electric current $\boldsymbol{I}$ in the present electromagnetic field theory belongs essentially to the electric current $\boldsymbol{I'}$ in the inertial system $\boldsymbol{S'}$ and not to the electric current $\boldsymbol{I}$ in the cosmic vacuum frame of reference.

16.4 Physicists incorrectly define the rate of change of electric charge *dq/dt* as electric current *I*

16.4.1 Electric charge momentum P_S contained in free electrons in the cross-section *S* of the conductor

Assume that $\boldsymbol{S}$ is the cross-section of the wire, $\boldsymbol{n}$ is the density of free electrons per unit volume of the wire, and $\boldsymbol{L}$ is the length of the wire. The volume $\boldsymbol{V}$ of the wire is:

$$\boldsymbol{V = SL} \qquad (16.4.1)$$

The number of free electrons $\boldsymbol{nV}$ contained in the volume $\boldsymbol{V}$ of the conductor is:

$$\boldsymbol{nV = nSL} \qquad (16.4.2)$$

Divide the above equation by the length $\boldsymbol{L}$ to get the relationship equation:

$$\boldsymbol{n\frac{V}{L} = nS} \qquad (16.4.3)$$

The $\boldsymbol{nS}$ in equation is the number of free electrons contained in the cross-section $\boldsymbol{S}$ of the wire.

According to equation (15.2.16), the electric charge $\boldsymbol{e}$ of a moving electron can be decomposed into a rotational electric charge $\boldsymbol{e_\omega}$ and a magnetic electric charge $\boldsymbol{e_v}$, viz:

$$\boldsymbol{e = \sqrt{e_\omega^2 + e_v^2}}$$

Multiplying the above equation by the number of free electrons $\boldsymbol{nS}$ in the cross-section $\boldsymbol{S}$ gives the relational equation:

$$\boldsymbol{q_S = enS = \sqrt{(e_\omega nS)^2 + (e_v nS)^2}} \qquad (16.4.4)$$

The $\boldsymbol{q_S}$ in equation is the amount of free electric charge contained in the cross-section $\boldsymbol{S}$ of the wire. According to equation (15.2.6) and equation (15.2.15), the rotational charge $\boldsymbol{e_\omega}$ and the magnetic electric charge $\boldsymbol{e_v}$ are, respectively:

$$\boldsymbol{e_\omega = e\sqrt{1 - \frac{v^2}{c^2}}}, \qquad \boldsymbol{e_v = e\frac{v}{c}} \qquad (16.4.5)$$

Substituting the above equation into equation (16.4.4) gives the relationship equation:

$$q_S = enS = \sqrt{\left(e\sqrt{1-\frac{v^2}{c^2}}nS\right)^2 + \left(e\frac{v}{c}nS\right)^2} \tag{16.4.6}$$

According to equation (16.2.9), the electric charge momentum P_e of the moving electron is:

$$P_e = ec = \sqrt{(ev_\omega)^2 + (ev)^2} \tag{16.4.7}$$

Multiplying the above equation by the number of free electrons nS in the cross-section S gives the relational equation:

$$P_S = P_e nS = \sqrt{\left(e\sqrt{c^2 - v^2}nS\right)^2 + (evnS)^2} \tag{16.4.8}$$

The above equation is the formula for the free electric charge momentum in the cross section of the wire. $P_S = P_e nS$ in the formula is the free electric charge momentum contained in the cross section S of the wire.

Multiplying equation (16.4.6) by the speed of light c gives the relation equation:

$$cq_S = cenS = \sqrt{\left(e\sqrt{c^2 - v^2}nS\right)^2 + (evnS)^2} \tag{16.4.9}$$

Comparing the above equation with equation (16.4.8), it can be determined that the product $cq_S = cenS$ is equal to the free electric charge momentum P_S contained in the cross section S of the wire, i.e:

$$P_S = cq_S = \sqrt{\left(e\sqrt{c^2 - v^2}nS\right)^2 + (evnS)^2} \tag{16.4.10}$$

The above equation shows that the free electric charge momentum P_S in the cross-section S of the wire can be decomposed into two component electric charge momentums.

16.4.2 Spin electric charge momentum $P_{S\omega}$ and magnetic electric charge momentum P_{Sv} contained in the free electrons in the cross-section S of the conductor

Since the velocity v_ω is the spin velocity of the electron, this book defines the first component electric charge momentum $nSe\sqrt{c^2 - v^2}$ inside the root sign of equation (16.4.10), as the spin electric charge momentum $P_{S\omega}$ of the cross-section S of the wire, viz:

$$P_{S\omega} = enS\sqrt{c^2 - v^2} \tag{16.4.11}$$

The above equation is the equation for the spin electric charge momentum in the cross section of the wire. The above equation shows that the spin electric charge momentum $P_{S\omega}$ in the cross-section S of the wire becomes smaller as the displacement velocity v of the free electron increases.

The spin electric charge momentum $P_{S\omega}$ in the cross section S of the wire is when the displacement velocity $v = 0$ of the free electron:

$$P_{S\omega} = P_S$$

Since the velocity v is the displacement velocity of the free electron, this book defines the second component electric charge momentum $evnS$, within the root sign of equation (16.4.10), as the magnetic electric charge momentum P_{Sv}, of the cross-section S of the wire, viz:

$$P_{Sv} = evnS \tag{16.4.12}$$

The above equation is the equation for the magnetic electric charge momentum in the cross-section of a wire. The above equation shows that the magnetic electric charge momentum P_{Sv} in the cross section S of the wire is proportional to the displacement velocity v.

When the displacement velocity $v = c$ of the free electron, the magnetic electric charge momentum P_{Sv} in the cross-section S of the wire is:

$$P_{Sv} = P_S$$

In summary, the free electric charge momentum P_S in the cross-section S of the wire can be expressed as:

$$P_S = cq_S = \sqrt{P_{S\omega}^2 + P_{Sv}^2} \tag{16.4.13}$$

When the displacement velocity $v = 0$ of the free electron. Since the spin velocity $v_\omega = c$ of the electron, the spin electric charge momentum $\boldsymbol{P_{S\omega}} = \boldsymbol{ecnS}$ in the free electric charge $\boldsymbol{q_S}$ and the magnetic electric charge momentum $\boldsymbol{P_{Sv}} = \boldsymbol{0}$.

When the displacement velocity $\boldsymbol{v} = \boldsymbol{c}$ of the electron. Since the spin velocity $\boldsymbol{v_\omega} = \boldsymbol{0}$ of the electron, the spin electric charge momentum $\boldsymbol{P_{S\omega}} = \boldsymbol{0}$ in the free electric charge $\boldsymbol{q_S}$ and the magnetic electric charge momentum $\boldsymbol{P_{Sv}} = \boldsymbol{ecnS}$.

16.4.3 Electric current *I* is essentially the magnetic electric charge momentum P_{Sv}=*evnS* of the free electrons contained in the cross-section *S* of the wire

According to equation (16.4.6), the free electric charge $\boldsymbol{q_S}$ contained in the cross-section $\boldsymbol{S}$ of the wire is:

$$\boldsymbol{q_S} = \boldsymbol{enS}$$

Assume that $\boldsymbol{v}$ is the displacement velocity of the free electron. Multiplying the above equation by the electron velocity $\boldsymbol{v}$ gives the relational equation:

$$\boldsymbol{I} = \boldsymbol{q_S v} = \boldsymbol{enSv} \qquad (16.4.14)$$

The above equation is the classical electric current equation. The direction of displacement of the free electric charge $\boldsymbol{q_S}$ is perpendicular to the cross-section $\boldsymbol{S}$ of the wire . The free electric charge $\boldsymbol{q_S}$ in the formula is not the electric charge contained in the sphere $\boldsymbol{S} = \boldsymbol{4\pi r^2}$, nor is it the electric charge that passes through the crown $\boldsymbol{S_\Omega} = \boldsymbol{2\pi hr}$ of the sphere.

According to equation (16.4.12), the product $\boldsymbol{q_S v}$ of the free electric charge $\boldsymbol{q_S}$ in the cross-section $\boldsymbol{S}$ of the wire and the velocity $\boldsymbol{v}$ is the magnetic electric charge momentum $\boldsymbol{P_{Sv}}$, i.e:

$$\boldsymbol{P_{Sv}} = \boldsymbol{q_S v} = \boldsymbol{enSv}$$

Comparing the above equation with equation (16.4.14) gives the relationship equation:

$$\boldsymbol{I} = \boldsymbol{P_{Sv}} = \boldsymbol{enSv} \qquad (16.4.15)$$

The above equation shows that electric current $\boldsymbol{I}$ is essentially the magnetic electric charge momentum $\boldsymbol{P_{Sv}}$ in the cross-section $\boldsymbol{S}$ of the wire.

According to equation (16.4.13), the free electric charge momentum $\boldsymbol{P_S}$ in the cross-section $\boldsymbol{S}$ of the wire can be decomposed into spin electric charge momentum $\boldsymbol{P_{S\omega}}$ and magnetic electric charge momentum $\boldsymbol{P_{Sv}}$, i.e:

$$\boldsymbol{P_S} = \boldsymbol{c q_S} = \sqrt{\boldsymbol{P_{S\omega}^2} + \boldsymbol{P_{Sv}^2}}$$

Since the magnetic electric charge momentum $\boldsymbol{P_{Sv}}$ is equal to the classical conduction electric current $\boldsymbol{I}$ (i.e. $\boldsymbol{P_{Sv}} = \boldsymbol{I}$), the spin electric charge momentum $\boldsymbol{P_{S\omega}}$ should be equal to the spin electric current $\boldsymbol{I_\omega}$, i.e:

$$\boldsymbol{P_{S\omega}} = \boldsymbol{I_\omega} \qquad (16.4.16)$$

Similarly, the free electric charge momentum $\boldsymbol{P_S}$ in the cross-section $\boldsymbol{S}$ of the wire should be equal to the total electric current $\boldsymbol{I_S}$ in the wire, viz:

$$\boldsymbol{I_S} = \boldsymbol{P_S} = \boldsymbol{c q_S} = \boldsymbol{cenS} \qquad (16.4.17)$$

In summary, the total electric current $\boldsymbol{I_S}$ in the conductor cross-section $\boldsymbol{S}$ can be expressed as:

$$\boldsymbol{I_S} = \sqrt{\boldsymbol{I_\omega^2} + \boldsymbol{I^2}} \qquad (16.4.18)$$

The above equation is the equation for the total electric current in the cross-section $\boldsymbol{S}$ of the wire. $\boldsymbol{I}$ in the formula is the classical conduction electric current , the conduction electric current $\boldsymbol{I}$ is able to generate the magnetic field $\boldsymbol{B}$, and the spin electric current $\boldsymbol{I_\omega}$ is able to generate the electric field $\boldsymbol{E}$.

Since the total electric current $\boldsymbol{I_S}$ in the wire is the sum of the vectors of spin electric current $\boldsymbol{I_\omega}$ and conduction electric current $\boldsymbol{I}$, the free electrons in the wire spiral along the surface of the wire.

When the displacement velocity $\boldsymbol{v} = \boldsymbol{0}$ of free electrons. Since the spin velocity $\boldsymbol{v_\omega} = \boldsymbol{c}$ of free electrons, the spin electric current $\boldsymbol{I_\omega} = \boldsymbol{P_{S\omega}} = \boldsymbol{ecnS}$ in the cross section $\boldsymbol{S}$ of the wire and the conduction electric current $\boldsymbol{I} = \boldsymbol{P_{Sv}} = \boldsymbol{0}$.

When the displacement velocity $\boldsymbol{v} = \boldsymbol{c}$ of free electrons. Since the spin velocity $\boldsymbol{v_\omega} = \boldsymbol{0}$ of free electrons, the spin electric current $\boldsymbol{I_\omega} = \boldsymbol{P_{S\omega}} = \boldsymbol{0}$ in the cross section $\boldsymbol{S}$ of the wire and the conduction electric current $\boldsymbol{I} = \boldsymbol{P_{Sv}} = \boldsymbol{ecnS}$.

Since the higher the voltage the greater the velocity $\boldsymbol{v}$ of the electrons, the shorter the time $\boldsymbol{t}$ for the electric current I to travel from point a to point b, and thus the less energy the electrons radiate outward. This is the reason why high voltage transmission is able to conserve electrical energy.

16.4.4 Physicists are wrong to define the rate of change of electric charge *dQ/dt* as the conduction electric current *I=enSv*

Assume that S is the cross-section of the wire and n is the density of free electrons contained per unit volume of the wire. Assume L is the length of the wire and l' is the moves distance of the electrons. As shown in Figure 16-1.

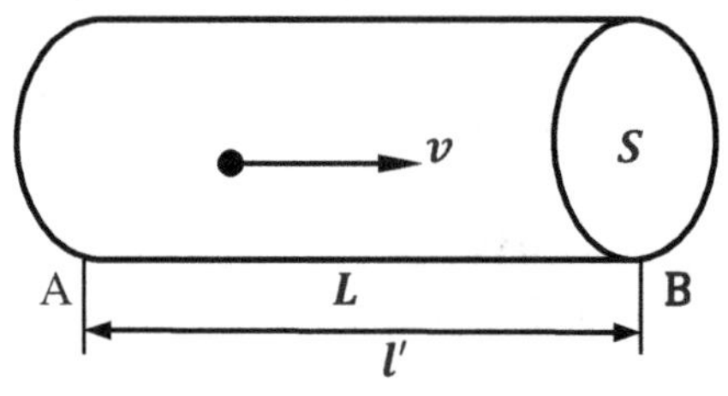

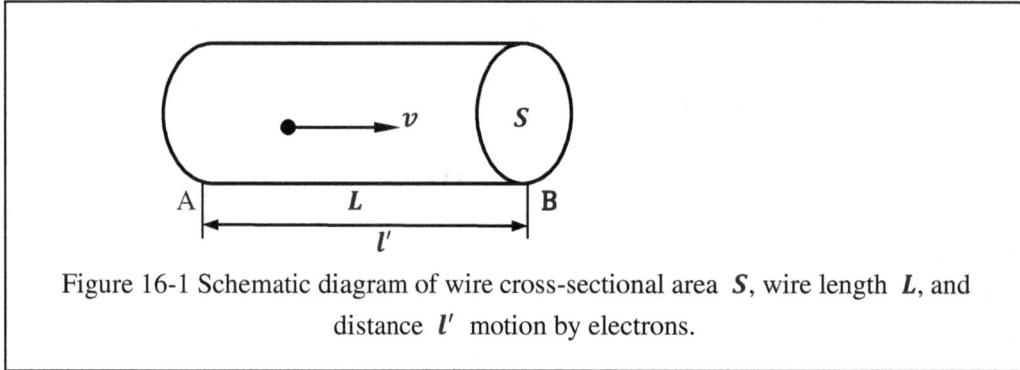

Figure 16-1 Schematic diagram of wire cross-sectional area S, wire length L, and distance l' motion by electrons.

The volume V of the conductor is:

$$V = SL \tag{16.4.19}$$

The number of free electrons nV contained in the volume V is:

$$nV = nSL \tag{16.4.20}$$

The free electric charge Q_V contained in the volume V is:

$$Q_V = enV = enSL \tag{16.4.21}$$

When the length L of the wire changes and the cross section S of the wire remains constant. The derivative of the above equation with respect to time t yields the relationship equation:

$$\frac{dQ_V}{dt} = enS\frac{dL}{dt} = enSv_L \tag{16.4.22}$$

Since the free electric charge Q_V in the volume V has only magnitude and no direction, the rate of change of electric charge $\frac{dQ_V}{dt}$ is a scalar. The rate $v_L = \frac{dL}{dt}$ in the equation is the rate of change of the length of the wire, not the displacement velocity v of the free electrons.

In order to derive the displacement current equation $\iint \frac{\partial D}{\partial t} dS$, and thus the electromagnetic wave equation, physicists define the rate of change $\frac{dQ}{dt}$ of charge, as the conductionelectric current I, i.e:

$$I = \frac{dQ}{dt} \tag{16.4.23}$$

The Q in the equation is the free electric charge in the wire through the cross-section S. Note that physicists, at this point, consider both electric charge Q_V and electric charge Q to be the same charge, i.e. $Q = Q_V$. Since the free electric charge Q has only magnitude and no direction, the physicist defines electric current I as scalar. The physicist derives the above equation as follows.

The free electric charge Q_V contained in the volume V is:

$$Q_V = enV = enSL$$

The free electric charge q_S contained in the cross-section S of the conductor is:

$$q_S = \frac{Q_V}{L} = enS \tag{16.4.24}$$

Suppose v is the displacement velocity of the free electron. If the free electron moves from point A to point B in time dt, then the free electric charge Q through the cross section of A is:

$$Q = q_S l' = enSl' \tag{16.4.25}$$

When the moves distance $\boldsymbol{l'}$ of the electron varies and the cross section $\boldsymbol{S}$ of the wire remains constant. The derivative of the above equation with respect to time $\boldsymbol{t}$ gives the relation equation:

$$\frac{dQ}{dt} = q_S \frac{dl'}{dt} = enSv = I \tag{16.4.26}$$

The above equation is the classical electric current formula for conduction. It should be noted that in deriving the electric current formula, physicists made the mistake of secretly switching physical quantities. In other words, physicists sneakily switched the wire length $\boldsymbol{L}$ in equation (16.4.21) to the moves distance $\boldsymbol{l'}$ of the electron.

Since the rate of change of electric charge $\frac{dQ}{dt}$ in equation (16.4.25) is a scalar and $\boldsymbol{enSv}$ is a vector, it is wrong for physicists to define the scalar $\frac{dQ}{dt}$ as the vector $\boldsymbol{enSv}$. The correct equation for the rate of change of electric charge $\frac{dQ}{dt}$ should be equation (16.4.22).

Chapter 17 Derivation of Coulomb's Law

Introduction: Coulomb's law is a law obtained through physical experimentation, and this book derives Coulomb's law using the equation for conservation of the vortex field.

17.1 Positive and negative vortex fields in electric fields

17.1.1 Definition of electric field

According to equation (9.1.2) , the equiangular vortex field conservation equation is:

$$e^{y}\left(\frac{e}{r}\right)^{D_k kr} = 1 \qquad 0 < r < +\infty \tag{17.1.1}$$

The e in equation is a natural constant, the polar radius r is the independent variable of the vortex field, and y is the dependent variable of the vortex field. The k is the vortex source, and D_k is the constant associated with the vortex source k. The dependent variable $y = 0$ when the polar radius $r = e$.

When the vortex source k is the electric charge q, i.e., the constant $k = q$ and the constant $D_k = D_q$. Substituting $k = q$ and $D_k = D_q$ into equation (17.1.1) yields the relational equation:

$$e^{y_q}\left(\frac{e}{r}\right)^{D_q qr} = 1, \quad 0 < r < +\infty \tag{17.1.2}$$

The above equation is the equation for the conservation of the vortex electric field. This book defines the dependent variable y_q in the formula as the electric field vortex potential. D_q in the formula is a constant associated with the electric field. When the pole radius $r = e$, the electric field vortex potential $y_q = 0$.

Taking the natural logarithm of equation (17.1.2) yields the relationship equation:

$$y_q = D_q qr(\ln r - 1) \tag{17.1.3}$$

The above equation is the electric field vortex potential equation. This book defines the set $\sum y_q$ of electric field vortex potentials y_q in space as an electric field.

Since the space occupied by the electric field vortex potential y_q is an infinite sphere, the electric field $\sum y_q$ consisting of the electric field vortex potential y_q is an infinite sphere. Theoretically, the development of the electric field $\sum y_q$ are unfolded according to equation (17.1.3).

It is important to note that the electric field $\sum y_q$ as defined in this book is not a substance, but rather a material property possessed by the electric charge q.

Physicists consider the electric field to be a special substance that exists in the space around a charged body. Although this special substance is not made of molecular atoms, it exists objectively. Physicists are wrong in this view.

The reason the electric field $\sum y_q$ has a force on the electric charge placed in it is because the nature of the electric charge force is a directional force between the energy elements ε_h. Since the directional force is a material property of the energy element ε_h, the electric field is not a special substance but a material property of the energy element ε_h.

17.1.2 Positive and negative vortex fields in electric fields

According to equation (9.2.12), the polar angle β of the equiangular vortex field is:

$$\begin{cases} \beta = +D_k k \cdot \ln r, & r > 1 \\ \beta = 0, & r = 1 \\ \beta = -D_k k \cdot \ln r, & 0 < r < 1 \end{cases}$$

Substituting the constant $k = q$ and the constant $D_k = D_q$ into the above equation, the polar angle β of the vortex field generated by the electric charge q is:

$$\begin{cases} \beta = +D_q q \cdot \ln r, & r > 1(l_m) \\ \beta = 0, & r = 1(l_m) \\ \beta = -D_q q \cdot \ln r, & 0 < r < 1(l_m) \end{cases} \tag{17.1.4}$$

The above equation is the formula for the polar angle of an electric field. The l_m in equation is the smallest unit length of the equiangular helical field. The positive sign " + " in the formula means that the polar angle β increases in clockwise rotation. The negative sign " − " in the formula means that the polar angle β increases in counterclockwise rotation.

A polar radius $r > 1(l_m)$ in the formula means that the direction of the polar angle β is the same as the direction of the polar radius r. A polar radius of $0 < r < 1(l_m)$ means that the direction of the polar angle β is opposite to the direction of the polar radius r. The polar angle coordinate lines of the electric field are shown in Figure 17-1.

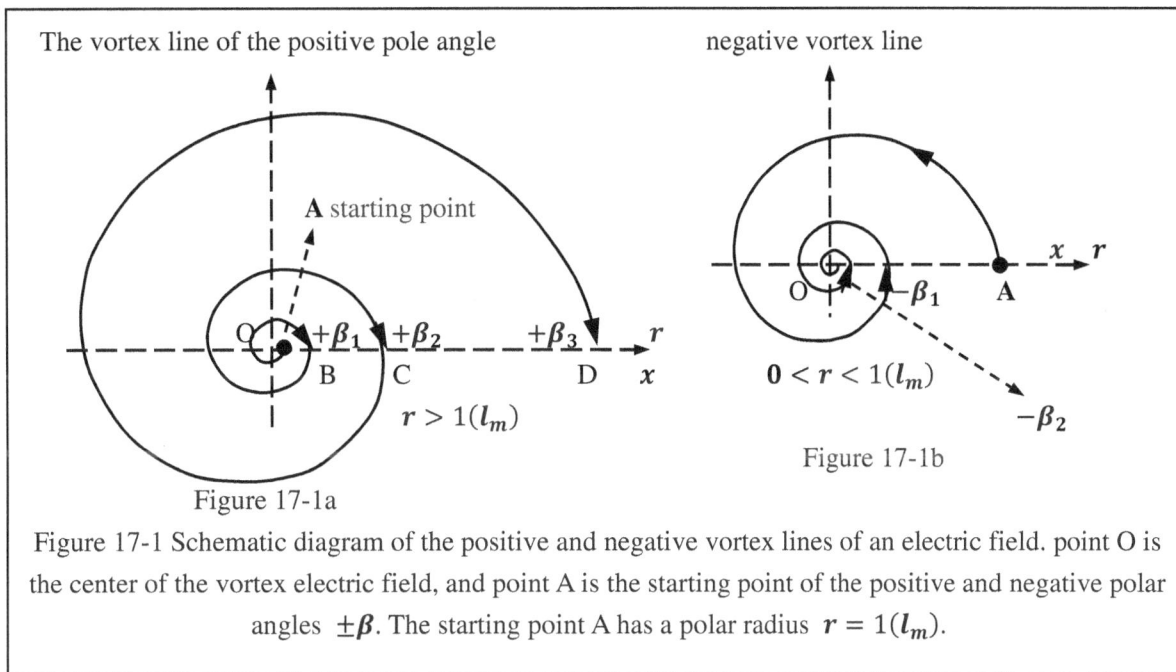

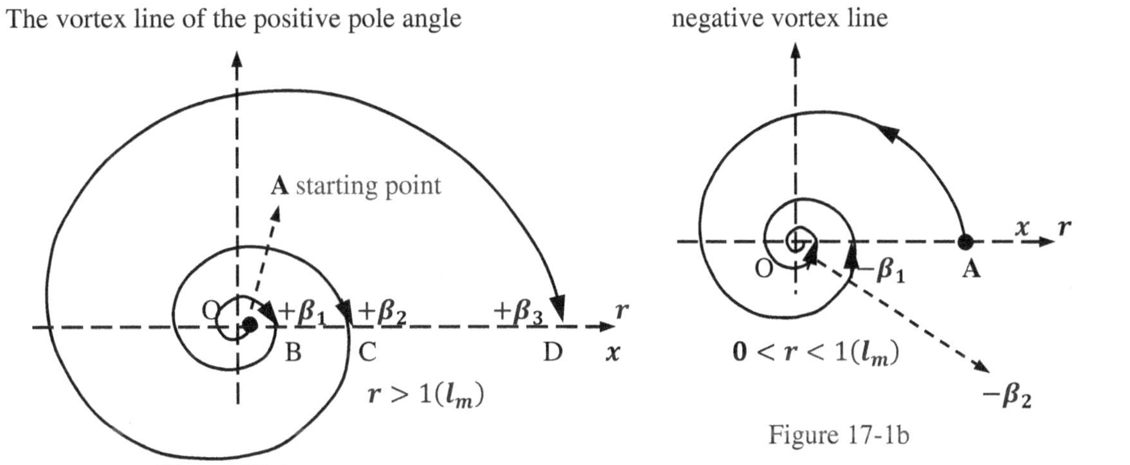

Figure 17-1a

Figure 17-1b

Figure 17-1 Schematic diagram of the positive and negative vortex lines of an electric field. point O is the center of the vortex electric field, and point A is the starting point of the positive and negative polar angles $\pm\beta$. The starting point A has a polar radius $r = 1(l_m)$.

A note about Figure 17-1:

Point O in the figure is the center of charge. point A is the starting point of the positive and negative equi angular spiral line. The starting point A has a polar radius $r = 1(l_m)$.

When the pole radius $r > 1(l_m)$, the equi angular spiral line between A, B, C and D is rotating clockwise, when the electric field force F between positive and negative charges is the suction force between macroscopic electric fields.

When the polar radius $0 < r < 1(l_m)$ the equi angular spiral line between O and A is rotating counterclockwise. At this time, if the positive and negative charges do not cancel each other into γ photons that propagate in opposite directions, then the electric field force F between the positive and negative charges is the repulsive force in the microscopic field.

According to the mathematical theory, the formula for the polar angle θ of an equiangular vortex field is:

$$\theta = \frac{1}{b}\ln\left(\frac{r}{a}\right)$$

Comparing the above equation with equation (17.1.4), if the coefficients $a = 1$ and $\frac{1}{b} = \pm D_q q$, then the electric field pole angle β is the pole angle θ.

Taking the derivative of equation (17.1.4) with respect to the polar radius r yields the relation equation:

$$\begin{cases} \dfrac{d\beta}{dr} = +D_q q\dfrac{1}{r} & r > 1(l_m) \\ \dfrac{d\beta}{dr} = 0, & r = 1(l_m) \\ \dfrac{d\beta}{dr} = -D_q q\dfrac{1}{r}, & 0 < r < 1(l_m) \end{cases} \qquad (17.1.5)$$

The $+D_q q\frac{1}{r}$ in equation denotes the rate of change of the positive polar angle β; the $-D_q q\frac{1}{r}$ in equation denotes the rate of change of the negative polar angle β.

17.1.3 Integration of the electrostatic field polar angle β_0 along the circular line $L_R=2\pi R$

Assume that the electric charge q_0 is at rest in the cosmic vacuum frame of reference. Suppose O is the center of the circular line $L_R = 2\pi R$, and Z is the perpendicular distance of the static electric charge q_0 to the center O. Suppose Ω is the steradian angle possessed by the circular plane $P = \pi R^2$, and θ is the angle between the radius r and the perpendicular distance Z. as shown in Figure 17-2.

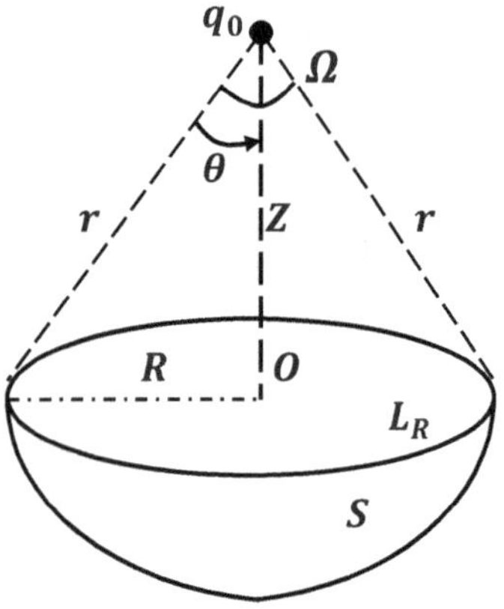

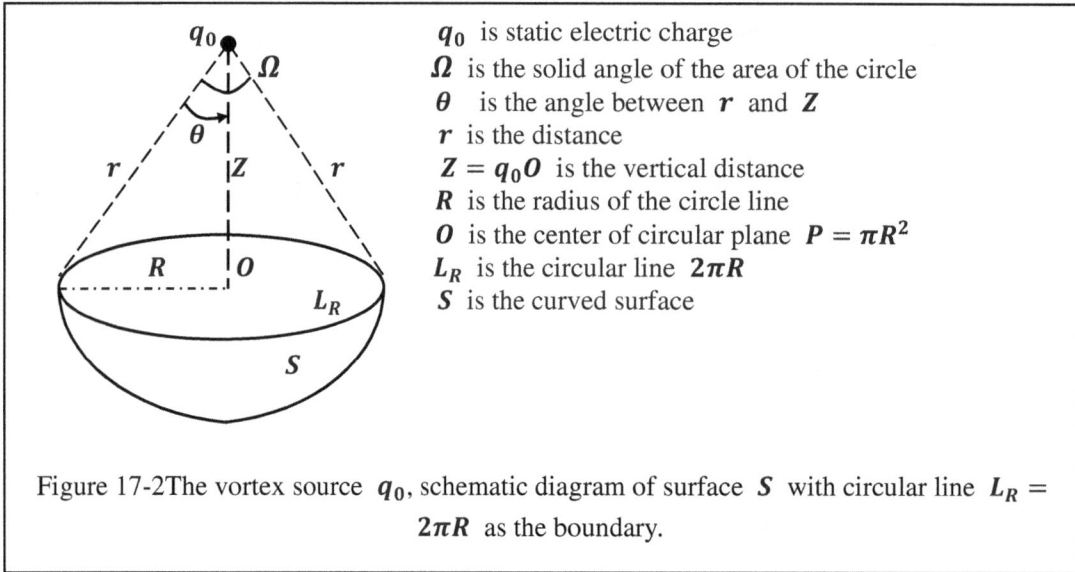

q_0 is static electric charge
Ω is the solid angle of the area of the circle
θ is the angle between r and Z
r is the distance
$Z = q_0 O$ is the vertical distance
R is the radius of the circle line
O is the center of circular plane $P = \pi R^2$
L_R is the circular line $2\pi R$
S is the curved surface

Figure 17-2The vortex source q_0, schematic diagram of surface S with circular line $L_R = 2\pi R$ as the boundary.

If the static electric charge q_0 is within the circular plane $P = \pi R^2$, then the distance $Z = 0$. If the static electric charge q_0 is outside the circular plane P, then the distance $Z \neq 0$.

Note that the surface S in Figure 17-2 is a surface with the circle line $L_R = 2\pi R$ as a sideline. Theoretically: there are an infinite number of surfaces S with the circle line L_R as a sideline.

According to equation (9.1.15) , the spherical crown S_Ω with the circular line $L_R = 2\pi R$ as the sideline is:

$$S_\Omega = 2\pi hr = 2\pi r\left(r - \sqrt{r^2 - R^2}\right) \qquad (17.1.6)$$

According to equation (9.1.16) , the Wang's circular crown S_W with the circle line $L_R = 2\pi R$ as the edge is:

$$S_W = 2\pi Rr \qquad (17.1.7)$$

Note that the Wang's circular crown $S_W = 2\pi Rr$ and the spherical crown $S_\Omega = 2\pi hr$ are two different surfaces.

According to equation (17.1.4), the polar angle β_0 of the electrostatic field is:

$$\beta_0 = +D_q q_0 \cdot \ln r\,, \qquad r > 1(l_m) \tag{17.1.8}$$

When the polar radius r varies continuously from the position $r = 1(l_m)$ in one direction to infinity, or converges to the static electric charge q_0, these numerous points of polar radius r form a straight line. Since each point on the straight line has an electrostatic field polar angle β_0, this book defines the straight line as an electrostatic field polar angle line. The electric field polar angle line is the solid line that intersects the circular line in Figure 17-3.

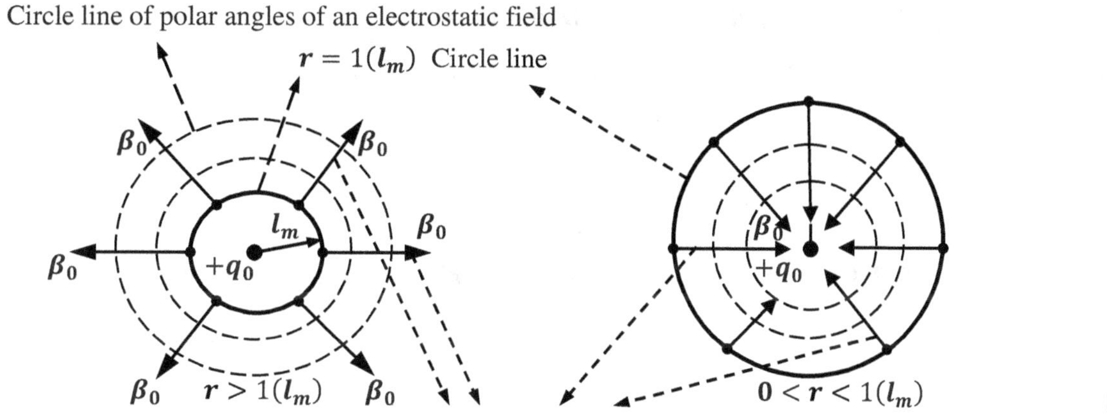

Figure 17-3 Schematic diagram of positive electrostatic field pole angle lines and pole angle circle lines. The arrows in the figure show the direction of the positive polar angle lines.

The electrostatic field polar angle line possesses the following properties:

(1) The static electric charge q_0 is the source of the electric field polar angle line.

(2) The electric field polar angles are not equal at every point on the electrostatic field polar angle line.

(3) The electrostatic field polar angle lines are continuous non-intersecting curves.

(4) The directions of the electrostatic field polar angles all point to infinity, or to the center of charge.

When the distance r from the static electric charge q_0 to each point of the circular line $L_R = 2\pi R$ is equal, the electrostatic field polar angles β_0 are equal at each point on the circular line L_R. In this book, the circular line with equal electrostatic field polar angles β_0 is defined as the electrostatic field polar angle circular line. This electrostatic field polar angle circular line is the circular dashed line in Figure 17-3.

The electric field pole angle circular line possesses the following properties:

(1) The static electric charge q_0 is the source of the circular line of the extreme angle of the electric field.

(2) The direction of the tangent at each point on the circular line is perpendicular to the radius r of that point.

(3) The electric field polar angles are equal at every point on the circular line.

(4) Electrostatic field polar angle circular lines are continuous non-intersecting curves. Electric field polar angle circular lines of equal values can form an equal electric field polar angle spherical surface.

(5) When the polar radius $r > 1(l_m)$. The closer the circular line is to the static electric charge q_0, the smaller the electric field polar angle β_0 is at each point on the circular line. Conversely, the greater the electrostatic field pole angle β_0 at points on the circular line.

Assume that $L = f(r)$ is a curve in an electrostatic field, and take the line element dl on the curve $L = f(r)$. according to equation (17.1.8), the product $\beta_0 dl$ of the electrostatic field pole angle β_0 and the line element dl, is:

$$\beta_0 dl = D_q q_0 \cdot \ln r\, dl\,, \qquad r > 1(l_m) \tag{17.1.9}$$

The electrostatic field polar angle β_0 is constant on the circular line L_R if each point on the circular line $L_R = 2\pi R$ is equidistant r from the static electric charge q_0. The integral of the electrostatic field polar angle β_0 along the circular line L_R is:

$$\oint \boldsymbol{\beta_0 dl} = \boldsymbol{\beta_0 L_R} = \boldsymbol{2\pi R \cdot D_q q_0 \cdot \ln r}, \quad r > 1(l_m) \tag{17.1.10}$$

According to Figure 17-2. the circular line $\boldsymbol{L_R = 2\pi R}$ encloses the circular plane $\boldsymbol{P = \pi R^2}$. if the static electric charge $\boldsymbol{q_0}$ is inside the circular plane $\boldsymbol{P}$, then the distance $\boldsymbol{Z = 0}$. If the electrostatic charge $\boldsymbol{q_0}$ is outside of the circular plane $\boldsymbol{P}$, then the distance $\boldsymbol{Z \neq 0}$.

When the angle $\boldsymbol{\theta = 90^0}$ in Figure 17-2. Since the radius $\boldsymbol{R = r}$, equation (17.1.10) becomes:

$$\oint \boldsymbol{\beta_0 dl} = \boldsymbol{2\pi r \cdot D_q q_0 \cdot \ln r}, \quad r > 1(l_m) \tag{17.1.11}$$

At this point the distance $\boldsymbol{Z = 0}$ and the static electric charge $\boldsymbol{q_0}$ is located at the center of the circular plane $\boldsymbol{P = \pi R^2}$.

17.2 Electrostatic potential φ_0

17.2.1 Derivation of the static electric potential equation

Assume that the polar radius $\boldsymbol{r}$ remains constant on the circular line $\boldsymbol{L_R = 2\pi R}$. According to equation (17.1.8), the integral of the electrostatic field polar angle $\boldsymbol{\beta_0}$ over the circular line $\boldsymbol{L_R = 2\pi R}$ is:

$$\oint \boldsymbol{\beta_0 dl} = \boldsymbol{2\pi R \cdot D_q q_0 \cdot \ln r}, \quad r > 1(l_m) \tag{17.2.1}$$

According to Stokes' formula, the relationship equation can be obtained:

$$\oint \boldsymbol{\beta_0 dl} = \iint \boldsymbol{\nabla \times \beta_0 \, dS} \tag{17.2.2}$$

The surface $\boldsymbol{S}$ in equation is bounded by the circular line $\boldsymbol{L_R = 2\pi R}$. We denote by $\boldsymbol{\varphi_0}$ the Curl $\boldsymbol{\nabla \times \beta_0}$ of the electrostatic field polar angle $\boldsymbol{\beta_0}$, i.e:

$$\boldsymbol{\varphi_0 = \nabla \times \beta_0} \tag{17.2.3}$$

This book defines Curl $\boldsymbol{\varphi_0 = \nabla \times \beta_0}$ as the static electric potential . equation (17.2.2) becomes:

$$\oint \boldsymbol{\beta_0 dl} = \iint \boldsymbol{\varphi_0 \, dS} \tag{17.2.4}$$

Substituting equation (17.2.1) into the above equation gives the relationship equation:

$$\iint \boldsymbol{\varphi_0 \, dS} = \boldsymbol{2\pi R \cdot D_q q_0 \cdot \ln r}, \quad r > 1(l_m) \tag{17.2.5}$$

The above equation is the static electric potential flux equation, in which the surface $\boldsymbol{S}$ is bounded by the circular line $\boldsymbol{L_R = 2\pi R}$. There are countless surfaces $\boldsymbol{S}$ bordered by the circular line $\boldsymbol{L_R = 2\pi R}$, the spherical crown $\boldsymbol{S_\Omega = 2\pi hr}$ is one of them, and Wang's circular crown $\boldsymbol{S_W = 2\pi Rr}$ is also one of them.

When the circular line $\boldsymbol{L_R = 2\pi R}$ remains constant. If the surface $\boldsymbol{S}$ in equation (17.2.5) is Wang's circular crown $\boldsymbol{S_W = 2\pi Rr}$, then Wang's circular crown $\boldsymbol{S_W}$ is a function of the radius $\boldsymbol{r}$. Differentiating Wang's circular crown $\boldsymbol{S_W}$ with respect to radius $\boldsymbol{r}$ yields the relation equation:

$$\boldsymbol{dS_W = 2\pi R dr} \tag{17.2.6}$$

Although the magnitude of Wang's circular crown $\boldsymbol{S_W = 2\pi Rr}$ changes as the radius $\boldsymbol{r}$ changes, the polar radius $\boldsymbol{r}$ at each point on the circular line $\boldsymbol{L_R = 2\pi R}$ is always equal. Substituting the above equation into equation (17.2.5) yields the relation equation:

$$\iint \boldsymbol{\varphi_0 dS_W} = \int \boldsymbol{\varphi_0 \cdot 2\pi R dr} = \boldsymbol{2\pi R \cdot D_q q_0 \cdot \ln r}, \quad r > 1(l_m) \tag{17.2.7}$$

Simplifying the above equation by removing the constant $\boldsymbol{2\pi R}$ gives the relational equation:

$$\int \boldsymbol{\varphi_0 dr} = \boldsymbol{D_q q_0 \cdot \ln r}, \quad r > 1(l_m) \tag{17.2.8}$$

According to equation (17.1.4), mimicking the derivation process above, the relationship equation can be obtained:

$$
\begin{cases}
\int \varphi_0 dr = +D_q q_0 \cdot \ln r & r > 1(l_m) \\
\int \varphi_0 dr = 0, & r = 1(l_m) \\
\int \varphi_0 dr = -D_q q_0 \cdot \ln r & 0 < r < 1(l_m)
\end{cases}
\quad (17.2.9)
$$

The $+D_q q_0 \cdot \ln r$ in the formula indicates clockwise electric field, and $-D_q q_0 \cdot \ln r$ indicates counterclockwise electric field. By taking the derivative of the above equation, we can get the relation equation:

$$
\begin{cases}
\varphi_0 = +D_q q_0 \frac{1}{r}, & r > 1(l_m) \\
\varphi_0 = 0, & r = 1(l_m) \\
\varphi_0 = -D_q q_0 \frac{1}{r}, & 0 < r < 1(l_m)
\end{cases}
\quad (17.2.10)
$$

The above equation is the static electric potential equation. The $+D_q q_0 \frac{1}{r}$ in the equation represents the static electric potential of clockwise electric field, and the $-D_q q_0 \frac{1}{r}$ in the equation represents the static electric potential of counterclockwise electric field.

Note that when radius $r = 1(l_m)$, radius r is equal to the minimum unit length l_m of the electric field. The vortex fields of different things will have different minimum unit lengths l_m.

Compare equation (17.2.10) with equation (17.1.5). The relationship equation can be obtained:

$$
\varphi_0 = \frac{d\beta_0}{dr} \quad (17.2.11)
$$

The above equation shows that the electrostatic potential φ_0 is equal to the derivative of the electrostatic field pole angle β_0 with respect to the pole radius r.

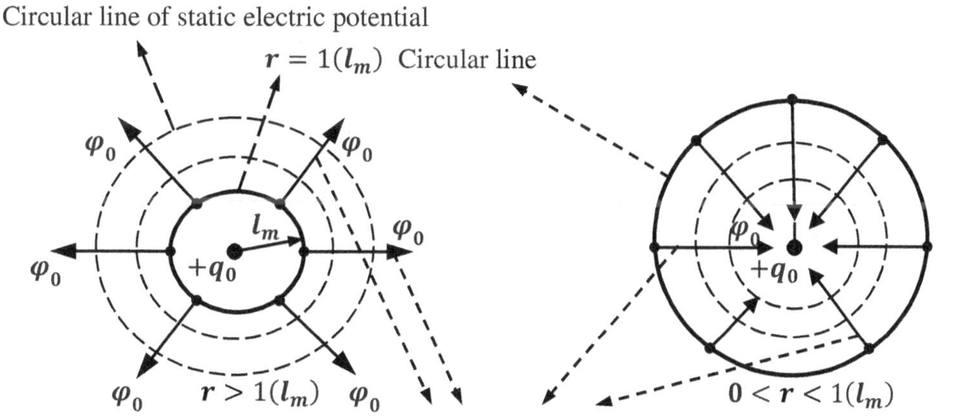

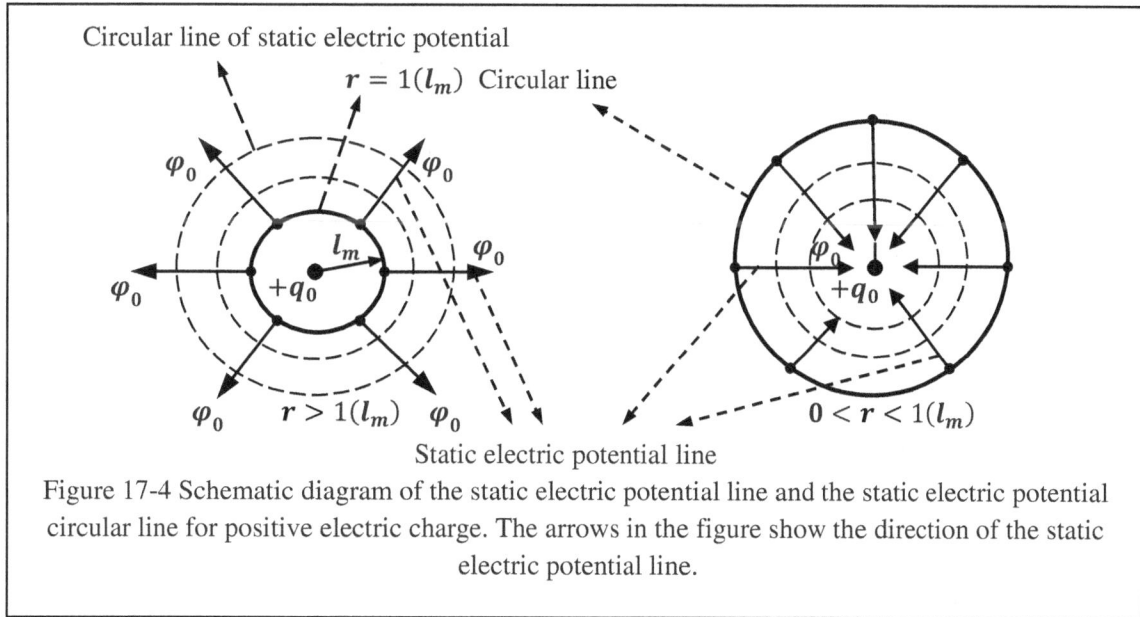

Figure 17-4 Schematic diagram of the static electric potential line and the static electric potential circular line for positive electric charge. The arrows in the figure show the direction of the static electric potential line.

According to equation (17.2.10), when the static electric charge q_0 is positive electric charge $+q_0$. If the polar radius $r > 1(l_m)$, then the static electric potential φ_0 is positive, and the direction of static electric potential φ_0 points to infinity.

If the polar radius $r < 1(l_m)$, then the static electric potential φ_0 is negative, and the direction of the static electric potential φ_0 points to the electric charge $+q_0$.

When the radius r varies continuously from the position $r = 1(l_m)$ to infinity in one direction, or converges to the static electric charge q_0, these innumerable radius r points form a straight line. Since each point on the line has static electric potential φ_0, this book defines the line as a static electric potential line. This static electric potential line is the solid line that intersects the circular line in Figure 17-4.

When the distance r from the static electric charge q_0 to each point of the circular line $L_R = 2\pi R$ is equal, the static electric potential φ_0 at each point on the circular line L_R is equal. In this book, a circular line with equal static electric potential φ_0 is defined as a static electric potential circular line. This static electric potential circular line is the circular dashed line in Figure 17-4.

Note that the static electric potential as defined by physicists is a scalar quantity, whereas the static electric potential as defined by the book is a vector quantity.

17.2.2 Integration of the static electric potential φ_0 along the circular line L_R=$2\pi R$

Suppose $L = f(r)$ is a curve in an electrostatic field. Take the line element dl on the curve $L = f(r)$. the product of the electrostatic potential φ_0 and the line element dl is:

$$\varphi_0 dl = D_q \frac{q_0}{r} dl \tag{17.2.12}$$

The static electric potential φ_0 is constant on the circular line L_R if each point on the circular line $L_R = 2\pi R$ is equidistant r from the static electric charge q_0. The integral of the static electric potential φ_0 along the circular line L_R is:

$$\oint \varphi_0 dl = \varphi_0 L_R = 2\pi R \cdot D_q \frac{q_0}{r} \tag{17.2.13}$$

When the angle $\theta = 90^0$ in Figure 17-4. Since the radius $R = r$, the above equation becomes:

$$\oint \varphi_0 dl = 2\pi D_q q_0 \tag{17.2.14}$$

The above equation is the formula for the maximum value of the circular line integral of the static electric potential φ_0. At this point the distance $Z = 0$ and the static electric charge q_0 is located at the center of the circular line plane $P = \pi R^2$.

17.2.3 Static electric potential φ_0 is a vector, not a scalar

According to the electric field theory, the electric field strength E of the electric charge Q is:

$$E = K\frac{Q}{r^2} \tag{17.2.15}$$

In the formula K is a constant and r is the distance from electric charge q to electric charge Q. The electric potential energy E_P of electric charge q is equal to the work W done by the electric field force $F = qE$ when the positive electric charge q is moved from point A to infinity, i.e:

$$E_P = W = \int_r^{\infty} F\,dl = \int_r^{\infty} K\frac{Qq}{r^2} dl = K\frac{Qq}{r} \tag{17.2.16}$$

Physicists define the electric potential φ as the ratio of the electric potential energy E_P of a electric charge q to the electric charge q it carries, i. e:

$$\varphi = \frac{E_P}{q} = K\frac{Q}{r} \tag{17.2.17}$$

In order to derive the wave equation of an electromagnetic wave, physicists propose the Gauge Theory, and the electric potential φ must be a scalar in the Gauge Theory. Otherwise, if the electric potential φ is a vector, then physicists cannot derive the wave equation of an electromagnetic wave.

According to equation (17.2.3) the static electric potential φ_0 is:

$$\varphi_0 = \nabla \times \beta_0$$

Since Curl $\nabla \times \beta_0$ is a vector, the static electric potential φ_0 defined in this book is a vector, not a scalar.

17.3 Electrostatic field strength E_0

17.3.1 Derivation of the formula for the strength of an electrostatic field

Assume that the electric charge $\boldsymbol{q_0}$ is at an equal distance $\boldsymbol{r}$ to each point on the circular line $\boldsymbol{L_R} = \boldsymbol{2\pi R}$. When the static electric potential $\boldsymbol{\varphi_0}$ is integrated along the circular line $\boldsymbol{L_R}$, the relation equation can be obtained according to equation (17.2.13) and Stokes' theorem:

$$\oint \boldsymbol{\varphi_0 dl} = \iint \boldsymbol{\nabla} \times \boldsymbol{\varphi_0\, dS} = \boldsymbol{2\pi R \cdot D_q \frac{q_0}{r}} \qquad (17.3.1)$$

The surface $\boldsymbol{S}$ in equation is bounded by the circular line $\boldsymbol{L_R} = \boldsymbol{2\pi R}$. We can use the symbol $\boldsymbol{E_0}$ to denote the Curl $\boldsymbol{\nabla} \times \boldsymbol{\varphi_0}$ of the static electric potential $\boldsymbol{\varphi_0}$, i.e:

$$\boldsymbol{E_0} = \boldsymbol{\nabla} \times \boldsymbol{\varphi_0} \qquad (17.3.2)$$

This book defines Curl $\boldsymbol{E_0} = \boldsymbol{\nabla} \times \boldsymbol{\varphi_0}$ as the electrostatic field strength. Substituting the above equation into equation (17.3.1) yields the relation equation:

$$\oint \boldsymbol{\varphi_0 dl} = \iint \boldsymbol{E_0\, dS} = \boldsymbol{2\pi R \cdot D_q \frac{q_0}{r}} \qquad (17.3.3)$$

The surface $\boldsymbol{S}$ in equation is bordered by the circular line $\boldsymbol{L_R} = \boldsymbol{2\pi R}$. When the circular line $\boldsymbol{L_R} = \boldsymbol{2\pi R}$ remains constant. If the surface $\boldsymbol{S}$ in equation is Wang's circular crown $\boldsymbol{S_W} = \boldsymbol{2\pi Rr}$, then Wang's circular crown $\boldsymbol{S_W}$ is a function of the radius $\boldsymbol{r}$. Differentiating Wang's circular crown $\boldsymbol{S_W}$ with respect to radius $\boldsymbol{r}$ yields the relation equation:

$$\boldsymbol{dS_w} = \boldsymbol{2\pi Rdr}$$

Although the size of Wang's circular crown $\boldsymbol{S_w} = \boldsymbol{2\pi Rr}$ varies with the radius $\boldsymbol{r}$, the radius $\boldsymbol{r}$ at each point on the circular line $\boldsymbol{L_R} = \boldsymbol{2\pi R}$ is equal. Substituting the above equation into equation (17.3.3) yields the relational equation:

$$\iint \boldsymbol{E_0 dS_W} = \int \boldsymbol{E_0 \cdot 2\pi Rdr} = \boldsymbol{2\pi R \cdot D_q \frac{q_0}{r}} \qquad (17.3.4)$$

Simplifying the above equation and eliminating the constant $\boldsymbol{2\pi R}$ yields the relational equation:

$$\int \boldsymbol{E_0 dr} = \boldsymbol{D_q \frac{q_0}{r}} \qquad (17.3.5)$$

Take the derivative of the above equation with respect to the radius $\boldsymbol{r}$ to obtain the relational equation:

$$\boldsymbol{E_0} = -\boldsymbol{D_q \frac{q_0}{r^2}} \qquad (17.3.6)$$

The above equation is the electrostatic field strength equation, where $\boldsymbol{D_q}$ is the coefficient to be determined. Since the electrostatic field strength $\boldsymbol{E_0}$ belongs to Curl $\boldsymbol{E_0} = \boldsymbol{\nabla} \times \boldsymbol{\varphi_0}$, the electrostatic field strength $\boldsymbol{E_0}$ is a vector.

According to equation (17.2.10), the static electric potential $\boldsymbol{\varphi_0}$ is:

$$\begin{cases} \boldsymbol{\varphi_0} = +\boldsymbol{D_q \frac{q_0}{r}}, & \boldsymbol{r} > 1(\boldsymbol{l_m}) \\ \boldsymbol{\varphi_0} = \boldsymbol{0}, & \boldsymbol{r} = 1(\boldsymbol{l_m}) \\ \boldsymbol{\varphi_0} = -\boldsymbol{D_q \frac{q_0}{r}}, & 0 < \boldsymbol{r} < 1(\boldsymbol{l_m}) \end{cases} \qquad (17.3.7)$$

Integrating the above equation along the circular line $\boldsymbol{L_R} = \boldsymbol{2\pi R}$ and then mimicking the derivation process of equation (17.3.6) yields the relational equation:

$$\begin{cases} \boldsymbol{E_0} = -\boldsymbol{D_q \frac{q_0}{r^2}}, & \boldsymbol{r} > 1(\boldsymbol{l_m}) \\ \boldsymbol{E_0} = \boldsymbol{0}, & \boldsymbol{r} = 1(\boldsymbol{l_m}) \\ \boldsymbol{E_0} = +\boldsymbol{D_q \frac{q_0}{r^2}}, & 0 < \boldsymbol{r} < 1(\boldsymbol{l_m}) \end{cases} \qquad (17.3.8)$$

The above equation is the formula for the strength of the electrostatic field. Note that when radius $\boldsymbol{r} = 1(\boldsymbol{l_m})$, radius $\boldsymbol{r}$ is equal to the minimum unit length $\boldsymbol{l_m}$ of the electrostatic field.

If radius $r = 1(l_m)$, then the electrostatic field strength $E_0 = 0$ of electric charge $+q_0$. If radius $r > 1(l_m)$, then the electrostatic field strength $E_0 < 0$ of electric charge $+q_0$. If radius $0 < r < 1(l_m)$, then the electrostatic field strength $E_0 > 0$ of electric charge $+q_0$.

17.3.2 Electrostatic field lines and electrostatic field circles lines

When the static electric charge q_0 is positive electric charge $+q_0$. Since the electrostatic field strength $E_0 < 0$, the direction of the electrostatic field strength E_0 is opposite to the vector radius r, i.e., the direction of the electrostatic field strength E_0 is pointing to the positive electric charge $+q_0$.

When the radius r varies continuously from the position $r = l_m$ in one direction to infinity, or converges to a electric charge q_0, these innumerable points on the radius r form a straight line. Since each point on the straight line has an electrostatic field strength E_0, this book defines the straight line as an electrostatic field strength line. The electrostatic field strength line is the solid line that intersects the circular line in Figure 17-5.

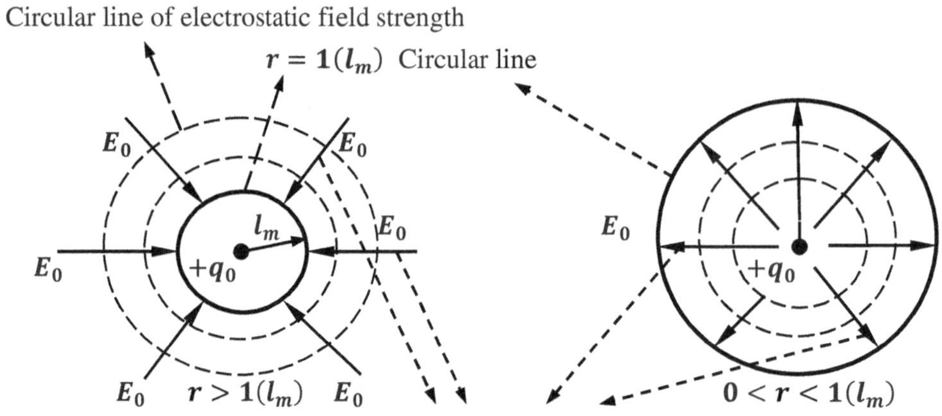

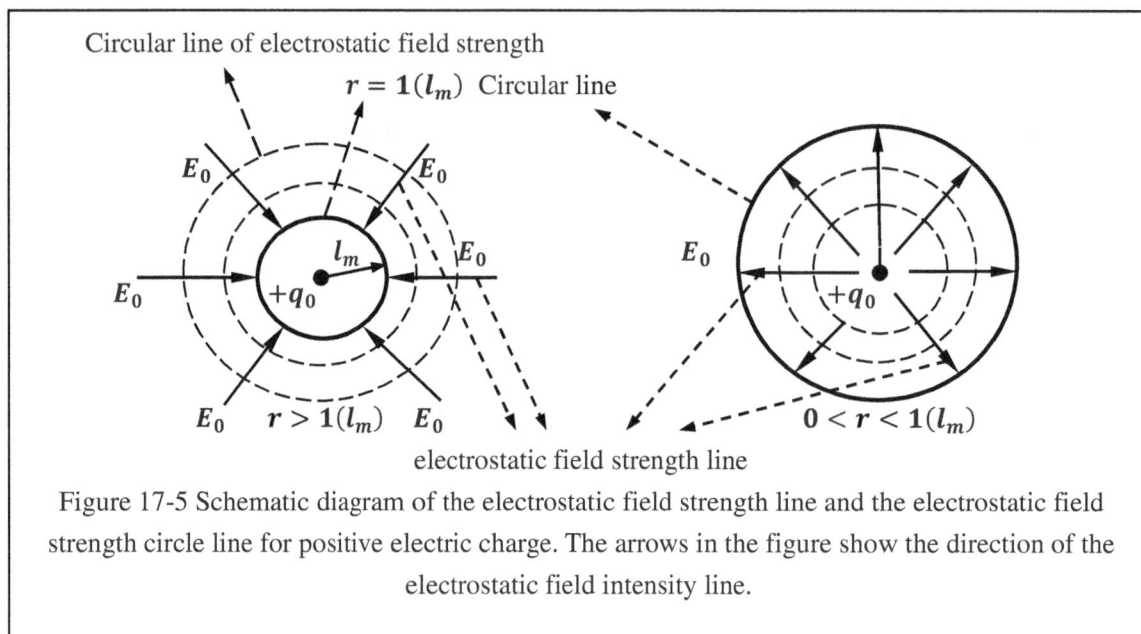

Figure 17-5 Schematic diagram of the electrostatic field strength line and the electrostatic field strength circle line for positive electric charge. The arrows in the figure show the direction of the electrostatic field intensity line.

The electrostatic field lines of an electric charge possess the following properties:

(1) The static electric charge q_0 is the source of the electrostatic field lines.

(2) The electrostatic field strength is not equal at every point on the electrostatic field lines.

(3) The electrostatic field lines are not closed. When the radius $r > 1(l_m)$, the directions of the electrostatic field strengths for static electric charge $+q_0$ are all toward electric charge $+q_0$. The directions of the electrostatic field strengths for static electric charge $-q_0$ are all toward infinity.

(4) The electrostatic field lines are continuous nonintersecting curves.

(5) Every electrostatic field line is a vortex line.

The circular line of the electrostatic field of a charge:

When the distance r from the static electric charge q_0 to each point of the circular line $L_R = 2\pi R$ is equal, the electrostatic field strength E_0 is equal at each point on the circular line L_R. In this book, the circular line with equal electrostatic field strength E_0 is defined as the circular line of electrostatic field strength. This electrostatic field strength circular line is the circular dashed line in Figure 17-5.

The electrostatic field circular line has the following properties:

(1), The static electric charge q_0 is the source of the electrostatic field circular line.

(2), The direction of the tangent line at each point on the electrostatic field circle line is perpendicular to the radius r of the point.

(3), The strength of the electrostatic field at each point on the circular line of the electrostatic field is equal.

(4), the electrostatic field circle line is a continuous non-intersecting curve. Numerically equal electrostatic field circular lines can form an equivalent electrostatic field sphere.

(5), when the radius $r > 1(l_m)$, the closer the circular line is to the electric charge q_0, the greater the electrostatic field strength E_0 at the points on that line. Conversely, the smaller it is:

17.3.3 Integration of the electrostatic field strength E_0 along the circular line $L_R=2\pi R$

Assume that $L = f(r)$ is a curve in the electric field, and take the line element dl on the curve $L = f(r)$. according to equation (17.3.8) , the product of the electrostatic field strength E_0 and the line element dl is:

$$E_0 dl = -D_q \frac{q_0}{r^2} dl, \qquad r > 1(l_m)$$

The integral $\int E_0 dl$ of the electrostatic field strength E_0 along the curve $L = f(r)$ is of two types.

The first type is an integral whose radius r is a variable.

Assume that the radius r is variable on the curve $L_X = f(r)$. Assume that the curve L_X starts at a and ends at b. According to equation (17.3.8), the integral of the electrostatic field strength E_0 along the curve L_X is:

$$\int E_0 dl = \int_{r_a}^{r_b} E_0 dl = -D_q q_0 \int_{r_a}^{r_b} \frac{1}{r^2} dl = D_q q_0 \left(\frac{1}{r_b} - \frac{1}{r_a}\right) \qquad (17.3.9)$$

When the start point a and the end point b of the curve $L_X = f(r)$ are the same point, the above equation becomes:

$$\int_{r_a}^{r_b} E_0 dl = 0 \qquad (17.3.10)$$

If the curve $L_X = f(r)$ goes from the starting point a to the point b and then returns from the point b to the starting point a, then the curve L_X is a closed curve. The integral $\oint E_0 dl$ of the electrostatic field strength E_0 on the closed curve L_X is:

$$\oint E_0 dl = \int_{r_a}^{r_b} E_0 dl + \int_{r_b}^{r_a} E_0 dl = 0 \qquad (17.3.11)$$

The above equation shows that the integral $\oint E_0 dl$ of the electrostatic field strength E_0 over the closed curve L_X is always equal to 0 if the radius r varies over the closed circular line L_X.

The 2nd type is an integral whose radius r remains constant on the circular line $L_R = 2\pi R$.

The electrostatic field strength E_0 is constant on the circular line L_R if the distances r from the static electric charge q_0 to each point on the circular line $L_R = 2\pi R$ are equal. According to Equation (17.3.8), the integral of the electrostatic field strength E_0 along the circular line L_R is:

$$\oint E_0 dl = E_0 L_R = -2\pi R \cdot D_q \frac{q_0}{r^2}, \qquad r > 1(l_m) \qquad (17.3.12)$$

When the angle $\theta = 90^0$ in Figure 17-2. Since the radius $R = r$, the above equation changes to:

$$\oint E_0 dl = -2\pi D_q \frac{q_0}{r} \qquad (17.3.13)$$

At this point the distance $Z = 0$ and the static electric charge q_0 is located in the center of the circular plane $P = \pi R^2$.

It is important to point out that since physicists only consider loop integrals with varying radius r and do not consider loop integrals with constant radius r, physicists are wrong about the electrostatic field strength E_0 loop integrals always being equal to zero.

17.4 Derivation of Coulomb's law

17.4.1 Derivation of Coulomb's law

According to Coulomb's law, the electric field force between charges of the same sex is the repulsive force $F_0 < 0$ and the electric field force between charges of opposite sex is the suction force $F_0 > 0$. We can derive the formula for the electric field force based on this property of the electric field force.

According to equation (17.3.6), the electrostatic field strength E_0 generated by the static electric charge q_0 is:

$$E_0 = -D_q \frac{q_0}{r^2}$$

The D_q in the formula is a constant. Assuming that r is the distance from static electric charge Q_0 to static electric charge q_0, multiply the above equation by static electric charge Q_0. If we use the symbol F_0 to denote the product $Q_0 E_0$, then we get the relation equation:

$$F_0 = Q_0 E_0 = -D_q \frac{Q_0 q_0}{r^2}, \qquad r > 1(l_m) \tag{17.4.1}$$

Assume the constant $D_q > 0$. If the charges q_0 and Q_0 are homosexual charges, then the product $F_0 < 0$. If the charges q_0 and Q_0 are anisotropic charges, then the product $F_0 > 0$.

According to Coulomb's law, the electrostatic field force F_0 between the electrostatic charges Q_0 and q_0 is:

$$F_0 = \frac{1}{4\pi\varepsilon_0} \frac{Q_0 q_0}{r^2}$$

The formula ε_0 is the vacuum dielectric constant. Comparing the above equation with equation (17.4.1), the constant D_q is equal to the Coulomb constant $\frac{1}{4\pi\varepsilon_0}$, i.e:

$$D_q = \frac{1}{4\pi\varepsilon_0} \tag{17.4.2}$$

Substituting the above equation into equation (17.4.1) gives the relationship equation:

$$F_0 = -\frac{1}{4\pi\varepsilon_0} \frac{Q_0 q_0}{r^2}, \qquad r > 1(l_m) \tag{17.4.3}$$

The above formula is the new electric field force formula. Note that the new electric field force formula differs from Coulomb's law by a negative sign.

Substituting equation (17.4.2) into equation (17.3.8) , the electrostatic field strength E_0 is:

$$\begin{cases} E_0 = -\dfrac{1}{4\pi\varepsilon_0} \dfrac{q_0}{r^2}, & r > 1(l_m) \\ E_0 = 0, & r = 1(l_m) \\ E_0 = +\dfrac{1}{4\pi\varepsilon_0} \dfrac{q_0}{r^2}, & 0 < r < 1(l_m) \end{cases} \tag{17.4.4}$$

The above formula is the new electrostatic field strength formula. Multiplying the above equation by the electric charge Q_0 gives the relational equation:

$$\begin{cases} F_0 = -\dfrac{1}{4\pi\varepsilon_0} \dfrac{Q_0 q_0}{r^2}, & r > 1(l_m) \\ F_0 = 0, & r = 1(l_m) \\ F_0 = +\dfrac{1}{4\pi\varepsilon_0} \dfrac{Q_0 q_0}{r^2}, & 0 < r < 1(l_m) \end{cases} \tag{17.4.5}$$

The above equation is the new electrostatic field force equation. Note that if radius $r = 1(l_m)$, then radius r is equal to the minimum unit length l_m of the electric field.

If radius $r = 1(l_m)$, then the electrostatic field force $F_0 = 0$. If radius $r > 1(l_m)$, then the electrostatic field force F_0 between same-sex charges is $F_0 < 0$. At this point, the electric field force F_0 is the repulsive force $F_{(-)}$. If radius $r < 1(l_m)$, then the electrostatic field force F_0 between same-sex charges is $F_0 > 0$, at which point the electric field force F_0 is the suction force $F_{(+)}$.

When electric charge Q_0 and electric charge q_0 are positive and negative charges. Since the collision of positive and negative electrons turns into two γ photons that propagate in opposite directions, there exists only a suction force $F_{(+)}$ between the positive and negative charges, and a repulsive force $F_{(-)}$ that does not exist.

17.4.2 If the distance r is held constant on the circular line L_R=2πR, then the electrostatic field circuital theorem $\oint E_0\,dl=0$ is incorrect

According to electrodynamics, the electric field force F_0 on a electric charge q in an electrostatic field E_0 is:

$$F_0 = qE_0 = \frac{1}{4\pi\varepsilon_0}\frac{qq_0}{r^2} \tag{17.4.6}$$

When the electric charge q moves along the curve $L = f(r)$ from point a to point b, the work W done by the electric field force F_0 is:

$$W = \int_{r_a}^{r_b} F_0 dl = -\frac{qq_0}{4\pi\varepsilon_0}\left(\frac{1}{r_b} - \frac{1}{r_a}\right) \tag{17.4.7}$$

The voltage U between point a and point b is:

$$U = \frac{W}{q} = \int_{r_a}^{r_b} E_0 dl = -\frac{q_0}{4\pi\varepsilon_0}\left(\frac{1}{r_b} - \frac{1}{r_a}\right) \tag{17.4.8}$$

Suppose that the radius r is changeable on the curve $L = f(r)$. When the curve $L = f(r)$ reaches point b from point a and then returns to point a from point b. Since the curve $L = f(r)$ is closed, the voltage U of the electrostatic field on the closed curve L is:

$$U = \oint E_0 dl = \int_{r_a}^{r_b} E_0 dl + \int_{r_b}^{r_a} E_0 dl = 0 \tag{17.4.9}$$

Physicists get the relationship equation based on the above equation:

$$\oint E_0 dl = 0 \tag{17.4.10}$$

The above formula is the electrostatic field circuital theorem. the way physicists derive the above formula does not conform to the rules of mathematical integration when the strength E_0 of the electrostatic field, is held constant on the curve L. We may wish to perform the following proof.

The integral $\int E_0 dl$ of the electrostatic field strength E_0 along the curve $L = f(r)$ is of two types.

The first type is an integral whose radius r is a variable.

Suppose that the radius r is changeable on the curve $L_X = f(r)$. When the curve $L_X = f(r)$ arrives at point b from point a and then returns to point a from point b. Since the curve $L_X = f(r)$ is closed, the integral of the electrostatic field strength E_0 along the closed curve $L_X = f(r)$ according to Equation (17.4.10) is:

$$\oint E_0 dl = 0$$

The above equation shows that the integral $\oint E_0 dl$ of the electrostatic field strength E_0 over the closed curve L_X is always equal to 0 if the radius r varies over the closed circular line L_X.

The 2nd type is an integral whose radius r remains constant on the circular line $L_R = 2\pi R$.

The electrostatic field strength E_0 is constant on the circular line L_R if the distance r from the static electric charge q_0 to each point on the circular line $L_R = 2\pi R$ is equal. According to equation (17.4.4), the integral of the electrostatic field strength E_0 along the circular line $L_R = 2\pi R$ is:

$$\oint E_0 dl = E_0 L_R = -\frac{R}{2\varepsilon_0}\frac{q_0}{r^2}, \qquad r > 1(l_m) \tag{17.4.11}$$

The above equation shows that the circuital integral of the electrostatic field E_0 is not equal to 0. When the angle $\theta = 90^0$ in Figure 17-2. Since the radius $R = r$, the above equation becomes:

$$\oint E_0 dl = -\frac{1}{2\varepsilon_0}\frac{q_0}{r}$$

At this point the distance $Z = 0$ and the static electric charge q_0 is located in the center of the circular plane $P = \pi R^2$.

To summarize, since physicists only consider loop integrals with changing radius r and not those with constant radius r, the loop law for electrostatic fields is wrong.

17.4.3 Inconsistency between the rules for calculating the electric field circuital theorem $\oint E_0\, dl=0$ and the rules for calculating the magnetic field circuital theorem $\oint Bdl=\mu_0 I$

According to electromagnetic field theory, both the electrostatic field strength $\boldsymbol{E_0}$ and the magnetic magnetic flux density $\boldsymbol{B}$ are inversely proportional to the square of the distance $\boldsymbol{r}$. However, the electrostatic field circuital theorem $\oint \boldsymbol{E_0 dl} = \boldsymbol{0}$, and the magnetic field circuital theorem $\oint \boldsymbol{Bdl} = \boldsymbol{\mu_0 I}$. The difference in the results of the integration of the two is due to the inconsistency of the rules of calculation of the two.

In other words, a varying distance $\boldsymbol{r}$ is a computational rule for the electrostatic field circuital theorem $\oint \boldsymbol{E_0 dl} = \boldsymbol{0}$. The distance $\boldsymbol{r}$ remaining constant is a rule of calculation for the magnetic field circuital theorem $\oint \boldsymbol{Bdl} = \boldsymbol{\mu_0 I}1$.

According to the magnetic field circuital theorem, the integral of the magnetic flux density $\boldsymbol{B}$ along the closed curve $\boldsymbol{L}$ is:

$$\oint \boldsymbol{Bdl} = \boldsymbol{\mu_0 I} \tag{17.4.12}$$

The above equation shows that the magnetic field circuital theorem $\oint \boldsymbol{Bdl} \neq \boldsymbol{0}$. However, if the distance $\boldsymbol{r}$ is a variable, then it can be deduced that the integral $\oint \boldsymbol{Bdl} = \boldsymbol{0}$ of the magnetic flux density $\boldsymbol{B}$ along the closed curve $\boldsymbol{L}$.

According to the theory of magnetic fields, a moving electric charge $\boldsymbol{q}$ produces a magnetic flux density $\boldsymbol{B}$ is:

$$\boldsymbol{B} = \frac{\boldsymbol{\mu_0}}{\boldsymbol{4\pi}}\frac{\boldsymbol{qv}}{\boldsymbol{r^2}}\sin\boldsymbol{\theta}$$

The $\boldsymbol{\theta}$ in the formula is the angle between the distance $\boldsymbol{r}$ and the electric charge displacement velocity $\boldsymbol{v}$. When the distance $\boldsymbol{r}$ can be varied arbitrarily on the curve $\boldsymbol{L_X} = \boldsymbol{f(r)}$, if the starting point of the curve $\boldsymbol{L_X} = \boldsymbol{f(r)}$ is $\boldsymbol{a}$ and the ending point is $\boldsymbol{b}$, then the integral of the magnetic flux density $\boldsymbol{B}$ on the curve $\boldsymbol{L_X} = \boldsymbol{f(r)}$ is:

$$\int_{r_a}^{r_b} \boldsymbol{Bdl} = \frac{\boldsymbol{\mu_0 qv}\sin\boldsymbol{\theta}}{\boldsymbol{4\pi}}\int_{r_a}^{r_b}\frac{\boldsymbol{1}}{\boldsymbol{r^2}}\boldsymbol{dr} = -\frac{\boldsymbol{\mu_0 qv}\sin\boldsymbol{\theta}}{\boldsymbol{4\pi}}\left(\frac{\boldsymbol{1}}{\boldsymbol{r_a}} - \frac{\boldsymbol{1}}{\boldsymbol{r_b}}\right) \tag{17.4.13}$$

If curve $\boldsymbol{L_X} = \boldsymbol{f(r)}$ starts at point $\boldsymbol{a}$ and reaches point $\boldsymbol{b}$ and then returns from point $\boldsymbol{b}$ to point $\boldsymbol{a}$, then curve $\boldsymbol{L_X}$ is a closed curve. The integral of the magnetic flux density $\boldsymbol{B}$ over the closed curve $\boldsymbol{L_X}$ is:

$$\oint \boldsymbol{Bdl} = \int_{r_a}^{r_b}\boldsymbol{Bdl} + \int_{r_b}^{r_a}\boldsymbol{Bdl} = \boldsymbol{0} \tag{17.4.14}$$

Comparing the above equation with equation (17.4.9), it can be determined that the reason for the magnetic field loop integral $\oint \boldsymbol{Bdl} = \boldsymbol{0}$ is the same as the reason for the electrostatic field circuital theorem $\oint \boldsymbol{E_0 dl} = \boldsymbol{0}$, i.e., distance $\boldsymbol{r}$ is the variable.

According to equation (17.4.14) . it can be determined that the magnetic field $\boldsymbol{B}$ is a conservative field (i.e., the work $\boldsymbol{W}$ done by the magnetic force is independent of the route $\boldsymbol{L}$ along which the magnetic force does its work), that the magnetic field $\boldsymbol{B}$ is a spinless field, and that the magnetic induction line is not a closed curve. This conclusion is clearly inconsistent with the physicist's view that magnetic field $\boldsymbol{B}$ is not a conservative field.

In summary, when the distance $\boldsymbol{r}$ is a variable, integrate both the electrostatic field strength $\boldsymbol{E_0}$ and the magnetic flux density $\boldsymbol{B}$ along the closed curve $\boldsymbol{L_X} = \boldsymbol{f(r)}$. One can obtain the relation equation:

$$\oint \boldsymbol{E_0 dl} = \boldsymbol{0}, \quad \oint \boldsymbol{Bdl} = \boldsymbol{0} \tag{17.4.15}$$

If the distance $\boldsymbol{r}$ from the electric charge $\boldsymbol{q}$ to each point on the circular line $\boldsymbol{L_R} = \boldsymbol{2\pi R}$ is equal, then the integrals of the electrostatic field strength $\boldsymbol{E_0}$ and the magnetic flux density $\boldsymbol{B}$, both along the circular line $\boldsymbol{L_R} = \boldsymbol{2\pi R}$, are, respectively:

$$\oint \boldsymbol{E_0 dl} = \boldsymbol{E_0 L_R}, \quad \oint \boldsymbol{Bdl} = \boldsymbol{BL_R} \tag{17.4.16}$$

The above equation shows that if the distance $\boldsymbol{r}$ is kept constant, then neither the loop integral $\oint \boldsymbol{E_0 dl}$ of the electrostatic field nor the loop integral $\oint \boldsymbol{Bdl}$ of the magnetic field is equal to zero.

Chapter 18 Electromagnetic potential φ generated by a moving electric charge q

Introduction: This chapter derives the electromagnetic potential φ generated by the motion electric charge q from the electric field vortex potential formula and Stokes' theorem.

18.1 Polar angle β of the electromagnetic field produced by a moving electric charge q

18.1.1 Definition of the electromagnetic field vortex potential y_q

Suppose O is the center of the circular line $L_R = 2\pi R$, and Z is the perpendicular distance from the electric charge q to the center O. Assume that the electric charge q is moving down the distance Z with velocity v in the cosmic vacuum frame of reference. Suppose Ω is the steradian angle possessed by the circular plane $P = \pi R^2$, and θ is the angle between the radius r and the perpendicular distance Z. Assume that the surface S is bordered by the circular line $L_R = 2\pi R$. This is shown in Figure 18-1.

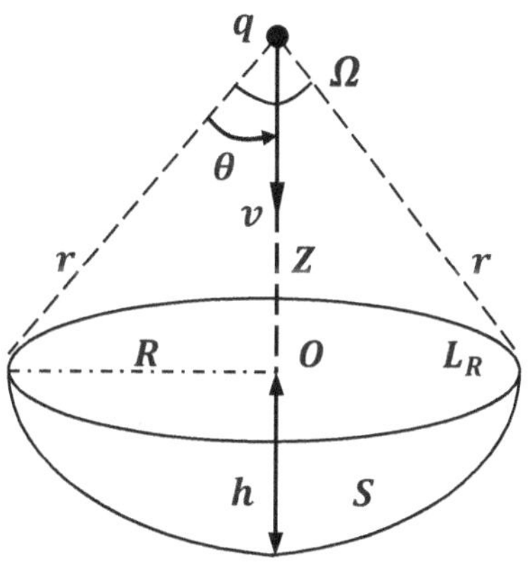

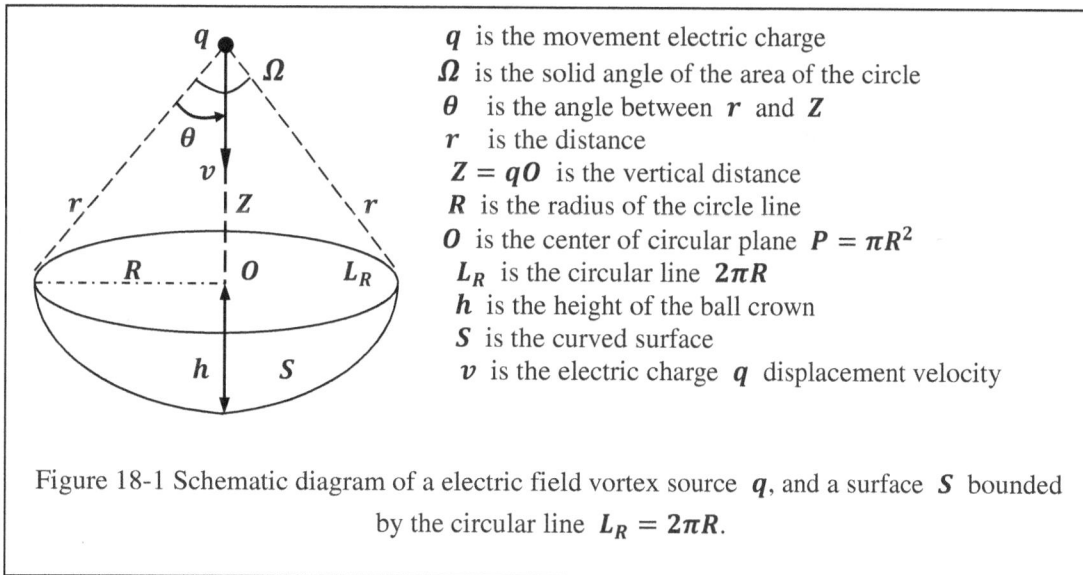

q is the movement electric charge
Ω is the solid angle of the area of the circle
θ is the angle between r and Z
r is the distance
$Z = qO$ is the vertical distance
R is the radius of the circle line
O is the center of circular plane $P = \pi R^2$
L_R is the circular line $2\pi R$
h is the height of the ball crown
S is the curved surface
v is the electric charge q displacement velocity

Figure 18-1 Schematic diagram of a electric field vortex source q, and a surface S bounded by the circular line $L_R = 2\pi R$.

According to the electric field vortex potential equation (17.1.3), the electrostatic field vortex potential y_{q0} generated by the static electric charge q_0 is:

$$y_{q0} = D_q q_0 r(\ln r - 1) \tag{18.1.1}$$

According to equation (17.4.2), the Coulomb constant D_q is:

$$D_q = \frac{1}{4\pi\varepsilon_0}$$

The ε_0 in equation is the vacuum dielectric constant. Substituting the Coulomb constant D_q into equation

(18.1.1) , the electrostatic field vortex potential y_{q0} generated by the static electric charge q_0 is:

$$y_{q0} = \frac{1}{4\pi\varepsilon_0} q_0 r(\ln r - 1) \tag{18.1.2}$$

Since the static electric charge q_0 does not generate the magnetic field B, the electrostatic field vortex potential y_{q0} contains only the electric field E, not the magnetic field B.

According to Figure 18-1, the function sinθ is:

$$\sin\theta = \frac{R}{r} = \frac{R}{\sqrt{R^2 + Z^2}} \tag{18.1.3}$$

If a electric charge q is moving with velocity v in the cosmic vacuum frame of reference, then according to the classical electromagnetic theory, the magnetic flux density B produced by the moving electric charge q is:

$$B = \frac{\mu_0}{4\pi}\frac{qv}{r^2}\sin\theta \tag{18.1.4}$$

In the formula μ_0 is the vacuum permeability, and θ is the angle between the radius r and the velocity v. The above equation shows that the magnetic flux density B has a maximum value in the direction perpendicular to the velocity v.

Since the moving electric charge q has a magnetic field B and an electric field E, the vortex potential y_q generated by the moving electric charge q contains the magnetic field B and the electric field E. In this book, the vortex potential y_q generated by the moving electric charge q is defined as the electromagnetic field vortex potential.

Since the magnetic field B, generated by the moving electric charge q, has a maximum in the direction perpendicular to the velocity v, the electromagnetic fields vortex potential y_q generated by the moving electric charge q is:

$$y_q = \frac{\sin\theta}{4\pi\varepsilon_0} qr(\ln r - 1) \tag{18.1.5}$$

The above formula is the electromagnetic field vortex potential formula. Since the moving electric charge q has a magnetic field B and electric field E, and the magnetic field B and electric field E are perpendicular to each other, so the direction of the electromagnetic field vortex potential y_q is not parallel to the direction of the magnetic field B, nor parallel to the direction of the electric field E, but between the direction of the electric field and the direction of the magnetic field between the two.

When the radius r and the magnitude of the electric charge q are kept constant, the electromagnetic field vortex potential y_q generated by the moving electric charge q is different in different directions in space.

If the effect of the pinch angle θ on the electromagnetic field vortex potential y_q is not considered, then the values of the electromagnetic field vortex potential y_q are equal in all directions. The pinch angle θ only affects the magnitude of the electromagnetic field vortex potential y_q in that direction, not the direction of the electromagnetic field vortex potential y_q.

If the angle $\theta = 0$, then the radius r lies in the line of the direction of motion of electric charge q. Since the function $\sin 0^0 = 0$, the electromagnetic field vortex potential $y_q = 0$.

If the angle of pinch $\theta = 90^0$, then the radius r is perpendicular to the direction of the velocity v. Since the function $\sin 90^0 = 1$, the electromagnetic field vortex potential y_q has a maximum value.

18.1.2 Integration of the electromagnetic field vortex potential y_q along the circular line $L_R=2\pi R$

Assume that $L = f(r)$ is a curve in the electromagnetic field. Take the line element dl on the curve $L = f(r)$. according to equation (18.1.5), the product of the electromagnetic field vortex potential y_q and the line element dl is:

$$y_q dl = \frac{\sin\theta}{4\pi\varepsilon_0} qr(\ln r - 1)dl \tag{18.1.6}$$

Assume that the distance r from the moving electric charge q to each point on the circular line $L_R = 2\pi R$ is equal. Since the electromagnetic field vortex potential y_q is constant on the circular line L_R, the integral of the electromagnetic field vortex potential y_q along the circular line $L_R = 2\pi R$ is:

$$\oint y_q dl = y_q L_R = 2\pi R \cdot \frac{q\sin\theta}{4\pi\varepsilon_0} r(\ln r - 1) \tag{18.1.7}$$

According to Figure 18-1. if the moving electric charge q is in the center of the plane $P = \pi R^2$, then the distance $Z = 0$. If the moving electric charge q is outside the circular plane $P = \pi R^2$, then the distance $Z \neq 0$.

When the angle $\theta = 90^0$. Since radius $R = r$, equation (18.1.7) becomes:

$$\oint y_q dl = \frac{q}{2\varepsilon_0} r^2(\ln r - 1) \tag{18.1.8}$$

At this point the distance $Z = 0$ and the moving electric charge q is located in the center of the circular plane $P = \pi R^2$.

18.1.3 Definition of electromagnetic field pole angle β

According to equation (18.1.7) and Stokes' theorem, the relation equation can be obtained:

$$\oint y_q dl = \iint \nabla \times y_q dS = 2\pi R \cdot \frac{q\sin\theta}{4\pi\varepsilon_0} r(\ln r - 1) \tag{18.1.9}$$

The Curl $\nabla \times y_q$ in equation is the Curl of the electromagnetic field vortex potential y_q , and the surface S is bounded by the circular line $L_R = 2\pi R$. We denote the Curl $\nabla \times y_q$ by β, i.e:

$$\beta = \nabla \times y_q \tag{18.1.10}$$

This book refers to Curl β as the electromagnetic field pole angle. Substituting the above equation into equation (18.1.9) gives the relational equation:

$$\iint \beta dS = 2\pi R \cdot \frac{q\sin\theta}{4\pi\varepsilon_0} r(\ln r - 1) \tag{18.1.11}$$

The above equation is the electromagnetic field pole angle flux equation. The surface S of the above equation is bordered by the circular line $L_R = 2\pi R$.

When the circular line $L_R = 2\pi R$ remains constant. If the surface S in equation (18.1.11) is Wang's circular crown $S_W = 2\pi R r$, then Wang's circular crown S_W is a function of the radius r. Differentiating Wang's circular crown S_W with respect to radius r yields the relation equation:

$$dS_W = 2\pi R dr \tag{18.1.12}$$

Although the size of Wang's circular crown S_W changes with the change of the radius r, the radius r of each point on the ring line $L_R = 2\pi R$ is always the same. Substituting the above formula into formula (18.1.11), the relational formula can be obtained:

$$\iint \beta dS_W = \int \beta \cdot 2\pi R dr = 2\pi R \cdot \frac{q\sin\theta}{4\pi\varepsilon_0} r(\ln r - 1) \tag{18.1.13}$$

Simplifying the above equation and eliminating the constant $2\pi R$ yields the relational equation:

$$\int \beta dr = \frac{q\sin\theta}{4\pi\varepsilon_0} r(\ln r - 1) \tag{18.1.14}$$

Take the derivative of the above formula with respect to the radius r, and get the relational formula:

$$\beta = \frac{q\sin\theta}{4\pi\varepsilon_0} \cdot \ln r \tag{18.1.15}$$

The β in the formula is the polar angle of the electromagnetic field possessed by radius r. Depending on the range of values of radius r, the above equation can be expressed as:

$$\begin{cases} \beta = +\dfrac{q\sin\theta}{4\pi\varepsilon_0} \cdot \ln r, & r > 1(l_m) \\ \beta = 0, & r = 1(l_m) \\ \beta = -\dfrac{q\sin\theta}{4\pi\varepsilon_0} \cdot \ln r, & 0 < r < 1(l_m) \end{cases} \tag{18.1.16}$$

The above formula is the electromagnetic field polar angle formula. $\ln r$ in the formula indicates that the direction of the polar angle $\boldsymbol{\beta}$ is the same as the direction of the vector radius $\boldsymbol{r}$. $\ln r < 0$ indicates that the direction of the polar angle $\boldsymbol{\beta}$ is opposite to the direction of the vector radius $\boldsymbol{r}$.

The positive sign " + " in the formula indicates that the polar angle $\boldsymbol{\beta}$ rotates clockwise, and the negative sign " − " indicates that the polar angle $\boldsymbol{\beta}$ rotates counterclockwise. Positive and negative equiangular helical fields contained in the electromagnetic field. As shown earlier in Figure 17-1.

18.1.4 Integration of the electromagnetic field polar angle β along the circular line $L_R=2\pi R$

Suppose $\boldsymbol{L} = \boldsymbol{f}(\boldsymbol{r})$ is a curve in the electromagnetic field. Take the line element $\boldsymbol{dl}$ on the curve $\boldsymbol{L} = \boldsymbol{f}(\boldsymbol{r})$. according to equation (18.1.16), the product of the electromagnetic field polar angle $\boldsymbol{\beta}$ and the line element $\boldsymbol{dl}$ is:

$$\boldsymbol{\beta dl} = +\frac{\boldsymbol{q\sin\theta}}{\boldsymbol{4\pi\varepsilon_0}}\cdot \ln r\, dl, \qquad r > 1(l_m) \qquad (\mathbf{18.1.17})$$

Assume that the distance $\boldsymbol{r}$ from the moving electric charge $\boldsymbol{q}$ to each point on the circular line $\boldsymbol{L_R} = \boldsymbol{2\pi R}$ is equal. Since the electromagnetic field polar angle $\boldsymbol{\beta}$ is constant on the circular line $\boldsymbol{L_R}$, the integral of the electromagnetic field polar angle $\boldsymbol{\beta}$ along the circular line $\boldsymbol{L_R} = \boldsymbol{2\pi R}$ is:

$$\oint \boldsymbol{\beta dl} = \boldsymbol{\beta L_R} = +2\pi R\cdot\frac{\boldsymbol{q\sin\theta}}{\boldsymbol{4\pi\varepsilon_0}}\cdot \ln r \qquad r > 1(l_m) \qquad (\mathbf{18.1.18})$$

The above equation is the loop integral equation for the electromagnetic field pole angle $\boldsymbol{\beta}$. When the angle of pinch $\boldsymbol{\theta} = \mathbf{90^0}$, the above equation becomes since the radius $\boldsymbol{R} = \boldsymbol{r}$:

$$\oint \boldsymbol{\beta dl} = +\frac{q}{2\varepsilon_0}r\cdot \ln r, \qquad r > 1(l_m) \qquad (\mathbf{18.1.19})$$

The above equation is the formula for the maximum value of the loop integral of the electromagnetic field pole angle. At this point the distance is $\boldsymbol{Z} = \boldsymbol{0}$ and the moving electric charge $\boldsymbol{q}$ is at the center of the circular plane $\boldsymbol{P} = \boldsymbol{\pi R^2}$.

18.2 Electromagnetic potential φ

18.2.1 Derivation of the electromagnetic potential equation

According to equation (18.1.18) and Stokes' theorem, the relationship equation is obtained:

$$\oint \boldsymbol{\beta dl} = \iint \nabla\times\boldsymbol{\beta}\, \boldsymbol{dS} = +2\pi R\cdot\frac{\boldsymbol{q\sin\theta}}{\boldsymbol{4\pi\varepsilon_0}}\cdot \ln r\,, \qquad r > 1(l_m) \qquad (\mathbf{18.2.1})$$

The surface $\boldsymbol{S}$ in equation is bounded by the circular line $\boldsymbol{L_R} = \boldsymbol{2\pi R}$. We denote by $\boldsymbol{\varphi}$ the Curl $\nabla\times\boldsymbol{\beta}$ of the polar angle $\boldsymbol{\beta}$ of the electromagnetic field, i.e:

$$\varphi = \nabla\times\boldsymbol{\beta} \qquad (\mathbf{18.2.2})$$

This book defines Curl $\boldsymbol{\varphi} = \nabla\times\boldsymbol{\beta}$ as the electromagnetic potential, and equation (18.2.1) becomes:

$$\oint \boldsymbol{\beta dl} = \iint \varphi\, \boldsymbol{dS} = +2\pi R\cdot\frac{\boldsymbol{q\sin\theta}}{\boldsymbol{4\pi\varepsilon_0}}\cdot \ln r\,, \qquad r > 1(l_m) \qquad (\mathbf{18.2.3})$$

The above equation is both a loop integral equation for the electromagnetic field pole angle $\boldsymbol{\beta}$ and a flux equation for the electromagnetic potential $\boldsymbol{\varphi}$.

When the circular line $\boldsymbol{L_R} = \boldsymbol{2\pi R}$ remains constant. If the surface $\boldsymbol{S}$ in equation (18.2.3) is Wang's circular crown $\boldsymbol{S_W} = \boldsymbol{2\pi Rr}$, then Wang's circular crown $\boldsymbol{S_W}$ is a function of the radius $\boldsymbol{r}$. Differentiating Wang's circular crown $\boldsymbol{S_W}$ with respect to radius $\boldsymbol{r}$ yields the relation equation:

$$\boldsymbol{dS_w} = \boldsymbol{2\pi Rdr}$$

Although Wang's circular crown $\boldsymbol{S_w}$ varies with radius $\boldsymbol{r}$, the radius $\boldsymbol{r}$ at each point on the circular line $\boldsymbol{L_R} = \boldsymbol{2\pi R}$ is always equal. Substituting the above equation into equation (18.2.3) yields the relational equation:

$$\iint \varphi\, \boldsymbol{dS_W} = \int \varphi\cdot 2\pi Rdr = +2\pi R\cdot\frac{\boldsymbol{q\sin\theta}}{\boldsymbol{4\pi\varepsilon_0}}\cdot \ln r\,, \qquad r > 1(l_m) \qquad (\mathbf{18.2.4})$$

Simplifying the above equation and eliminating the constant $\boldsymbol{2\pi R}$ yields the relational equation:

$$\int \varphi dr = +\frac{q\sin\theta}{4\pi\varepsilon_0}\cdot \ln r, \quad r > 1(l_m) \tag{18.2.5}$$

Take the derivative of the above equation with respect to the radius r to obtain the relational equation:

$$\varphi = +\frac{\sin\theta}{4\pi\varepsilon_0}\cdot\frac{q}{r}, \quad r > 1(l_m) \tag{18.2.6}$$

Depending on the range of values of the radius r, the relationship equation is obtained:

$$\begin{cases} \varphi = +\dfrac{\sin\theta}{4\pi\varepsilon_0}\cdot\dfrac{q}{r}, & r > 1(l_m) \\ \varphi = 0, & r = 1(l_m) \\ \varphi = -\dfrac{\sin\theta}{4\pi\varepsilon_0}\cdot\dfrac{q}{r}, & 0 < r < 1(l_m) \end{cases} \tag{18.2.7}$$

The above equation is the electromagnetic potential equation. Note that if radius $r = 1(l_m)$, then radius r is equal to the minimum unit length l_m of the electromagnetic field, i.e. $r = l_m$.

When the electric charge q is positive $+q$ and the radius $r > 1(l_m)$ Since the electromagnetic potential $\varphi > 0$, the electromagnetic potential φ is:

$$\varphi = +\frac{1}{4\pi\varepsilon_0}\frac{q\sin\theta}{r}, \quad r > 1(l_m) \tag{18.2.8}$$

If the radius r varies continuously from the position $r = 1(l_m)$ to infinity in one direction, or if the radius r converges to the electric charge q, then these radius points form a straight line. Since each point on the straight line has an electromagnetic potential φ, this book defines the line as an electromagnetic potential line. This electromagnetic potential line is the solid line that intersects the circular line in Figure 18-2.

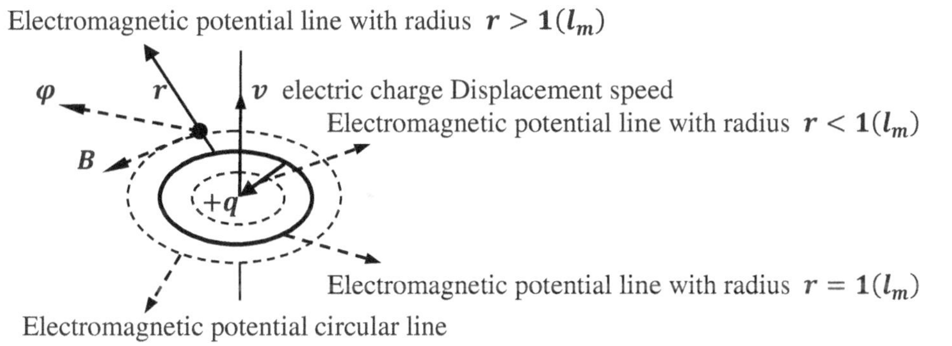

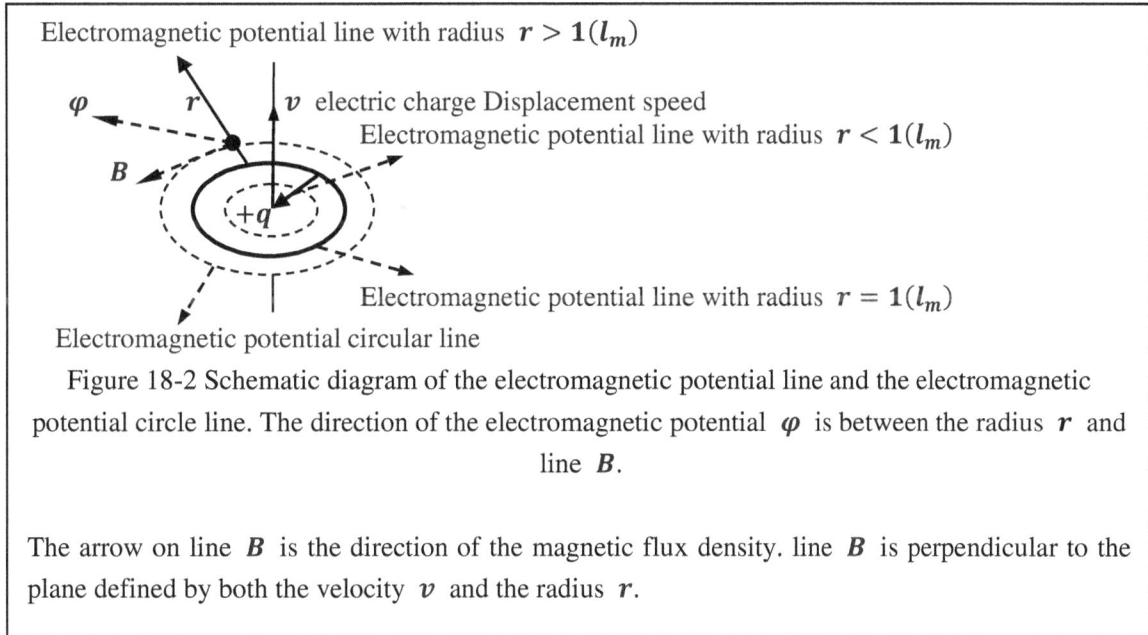

Figure 18-2 Schematic diagram of the electromagnetic potential line and the electromagnetic potential circle line. The direction of the electromagnetic potential φ is between the radius r and line B.

The arrow on line B is the direction of the magnetic flux density. line B is perpendicular to the plane defined by both the velocity v and the radius r.

When the distance r from the electric charge q to each point of the circular line $L_R = 2\pi R$ is equal, the electromagnetic potential φ at each point on the circular line L_R is equal. In this book, the circular line with equal electromagnetic potential φ is defined as the electromagnetic potential circular line. This electromagnetic potential circular line is the circular dashed line in Figure 18-2.

18.2.2 Integration of the electromagnetic potential φ along the circular line $L_R=2\pi R$

Assume that $L = f(r)$ is a curve in the electric field, and take the line element dl on the curve $L = f(r)$. according to equation (18.2.8) , the product of the electromagnetic potential φ and the line element dl is:

$$\varphi dl = +\frac{1}{4\pi\varepsilon_0}\frac{q\sin\theta}{r}dl$$

Integrate the above equation to get the relationship equation:

$$\int \varphi dl = \int \frac{1}{4\pi\varepsilon_0}\frac{q\sin\theta}{r}dl \tag{18.2.9}$$

The above equation is the electromagnetic potential line flow equation. The integral $\int \varphi dl$ is of two types according to the variation of the radius r.

The first type is an integral whose radius r is a variable.

Assume that the radius r is variable on the curve $L_X = f(r)$. Assume that the curve L_X starts at a and ends at b. According to equation (18.2.9), the integral of the electromagnetic potential φ along the curve $L_X = f(r)$ is:

$$\int \varphi dl = \int_{r_a}^{r_b} \varphi dl = \frac{q\sin\theta}{4\pi\varepsilon_0}\int_{r_a}^{r_b}\frac{1}{r}dl = \frac{q\sin\theta}{4\pi\varepsilon_0}(\ln r_b - \ln r_a) \tag{18.2.10}$$

According to equation (18.1.15), the electromagnetic field polar angle β generated by a electric charge q at point a and point b in space are, respectively:

$$\beta_a = \frac{q\sin\theta}{4\pi\varepsilon_0}\ln r_a\text{,}\quad \beta_b = \frac{q\sin\theta}{4\pi\varepsilon_0}\ln r_b$$

Substituting the above equation into equation (18.2.10) yields the relationship equation:

$$\int \varphi dl = \beta_b - \beta_a \tag{18.2.11}$$

The above equation shows that if the electromagnetic potential φ is integrated along the route L_X, then the integral is equal to the electromagnetic field pole angle difference $\Delta\beta$.

The curve L_X is a closed curve if the curve $L_X = f(r)$ goes from the starting point a to the point b and then returns from the point b to the starting point a. The integral $\oint \varphi dl$ of the electromagnetic potential φ on the closed curve L_X is:

$$\oint \varphi dl = \int_{r_a}^{r_b} \varphi dl + \int_{r_b}^{r_a} \varphi dl = 0 \tag{18.2.12}$$

The above equation shows that the integral $\oint \varphi dl$ of the electromagnetic potential φ over the closed curve L_X is always equal to 0 if the radius r varies over the closed circular line L_X.

The 2nd type is an integral whose radius r remains constant on the circular line $L_R = 2\pi R$.

The electromagnetic potential φ is constant on the circular line L_R if the distance r from the moving electric charge q to each point on the circular line $L_R = 2\pi R$ is equal. According to Equation (18.2.8), the integral of the electromagnetic potential φ along the circular line $L_R = 2\pi R$ is:

$$\oint \varphi dl = \varphi L_R = 2\pi R\cdot\frac{1}{4\pi\varepsilon_0}\frac{q\sin\theta}{r} \tag{18.2.13}$$

If then the angle $\theta = 90^0$ in Figure 18-1, then since the radius $R = r$, the above equation becomes:

$$\oint \varphi dl = \frac{q}{2\varepsilon_0} \tag{18.2.14}$$

At this point the distance $Z = 0$ and the static electric charge q_0 is located in the center of the circular plane $P = \pi R^2$.

18.3 Rotational electric potential φ_ω

18.3.1 The electromagnetic potential φ can be decomposed into a rotational electric potential φ_ω and a Magnetic-electric potential φ_v

If the electron is moving with velocity v in the cosmic vacuum frame of reference, then according to the electric charge-velocity equation (15.2.3) the electric charge e of the electron is:

$$e=\sqrt{\left(e\frac{v_\omega}{c}\right)^2+\left(e\frac{v}{c}\right)^2} \tag{18.3.1}$$

Since electric charge q is a collection of electron charges e, i.e., $q=\sum e$, if electric charge q is moving with velocity v in the cosmic vacuum, then the moving electric charge q is:

$$q=\sqrt{\left(q\frac{v_\omega}{c}\right)^2+\left(q\frac{v}{c}\right)^2}=\sqrt{q_\omega^2+q_v^2} \tag{18.3.2}$$

The above equation is the charge-velocity equation for a moving electric charge. The above equation shows that the moving electric charge q can be decomposed into two component charges $q\frac{v_\omega}{c}$ and $q\frac{v}{c}$.

Since the velocity v_ω is the spin velocity of the electron, this book defines the first component $q\frac{v_\omega}{c}$ within the root sign of equation (18.3.2) as the rotational electric charge q_ω of the moving electric charge, i.e:

$$q_\omega=q\frac{v_\omega}{c}=q\sqrt{1-\frac{v^2}{c^2}} \tag{18.3.3}$$

The above equation is the rotational electric charge equation. The above equation shows that the magnitude of the rotational electric charge q_ω becomes smaller as the electron velocity v increases.

Since the velocity v is the displacement velocity of the charge, this book defines the second component $q\frac{v}{c}$ within the root sign of equation (18.3.2) as the magnetic electric charge q_v of the moving electric charge, i.e:

$$q_v=q\frac{v}{c} \tag{18.3.4}$$

The above equation is the magnetic electric charge equation. magnetic electric charge q_v is the magnetic charge that physicists have been searching for a long time.

When the velocity $v=0$ of the moving electric charge. Since magnetic electric charge $q_v=0$, the static electric charge q_0 does not produce the magnetic field B. The greater the electric charge displacement velocity v, the greater the magnetic electric charge q_v, and the stronger the magnetic field B produced by the moving electric charge.

According to equation (18.2.7), the electromagnetic potential φ is:

$$\varphi=+\frac{1}{4\pi\varepsilon_0}\frac{q\sin\theta}{r}, \qquad r>1(l_m)$$

Substituting $q=\sqrt{q_\omega^2+q_v^2}$ from equation (18.3.2) into the above equation gives the relationship equation:

$$\varphi=+\frac{1}{4\pi\varepsilon_0}\sqrt{\left(\frac{q_\omega\sin\theta}{r}\right)^2+\left(\frac{q_v\sin\theta}{r}\right)^2}=\sqrt{\varphi_\omega^2+\varphi_v^2}, \qquad r>1(l_m) \tag{18.3.5}$$

The above equation shows that the electromagnetic potential φ generated by a moving electric charge q can be decomposed into a rotational electric potential φ_ω and a Magnetic-electric potential φ_v.

18.3.2 Derivation of the formula for the rotational electric potential φ_ω

Since the electric charge q_ω is the rotational electric charge contained in the moving electric charge q, this book defines the first component electric potential φ_ω inside the root sign of equation (18.3.5) as the rotational electric potential , viz:

$$\varphi_\omega=+\frac{1}{4\pi\varepsilon_0}\frac{q_\omega}{r}\sin\theta, \qquad r>1(l_m) \tag{18.3.6}$$

According to equation (18.3.3), the rotating electric charge q_ω is:

$$q_\omega=q\sqrt{1-\frac{v^2}{c^2}}$$

Substituting the above equation into equation (18.3.6) gives the relationship equation:

$$\varphi_\omega = +\frac{1}{4\pi\varepsilon_0}\frac{q}{r}\sqrt{1-\frac{v^2}{c^2}}\sin\theta, \quad r > 1(l_m) \qquad (18.3.7)$$

The above equation is the rotational electric potential equation. The above equation shows that the rotational electric potential φ_ω decreases with the increase of the electric charge motion speed v.

If the angle $\theta = 0$, then the radius r coincides with the direction of motion of the electric charge q. Since the function $\sin 0^0 = 0$, the rotation electric potential $\varphi_\omega = 0$.

If the angle $\theta = 90^0$, then the radius r is perpendicular to the velocity v, at which point the rotational electric potential φ_ω has the maximum value, i.e:

$$\varphi_\omega = +\frac{1}{4\pi\varepsilon_0}\frac{q}{r}\sqrt{1-\frac{v^2}{c^2}}, \quad r > 1(l_m) \qquad (18.3.8)$$

The above equation is the maximum rotational electric potential equation. If the electric charge velocity $v = 0$, then the above equation becomes static electric potential φ_0, i.e., $\varphi_\omega = \varphi_0$.

From this, it can be determined that the rotational electric potential φ_ω and the static electric potential φ_0 are electric potentials of the same nature. The difference between the two is that the rotational electric potential φ_ω is moving, while the static electric potential φ_0 is stationary.

Since the transformation factor $\sqrt{1-\frac{v^2}{c^2}}$ is scalar and the rotating electric field strength φ_ω contains only the radius r, the direction of the rotating electric field strength φ_ω is either in the same direction as the vector radius r or in the opposite direction.

According to equation (18.1.3), the function $\sin\theta$ is:

$$\sin\theta = \frac{R}{r} = \frac{R}{\sqrt{R^2+Z^2}}$$

Substituting the above equation into equation (18.3.7) yields the relationship equation:

$$\varphi_\omega = +\frac{1}{4\pi\varepsilon_0}\frac{qR}{R^2+Z^2}\sqrt{1-\frac{v^2}{c^2}}, \quad r > 1(l_m) \qquad (18.3.9)$$

The above equation is another expression of the rotational electric potential equation. If the circular line $L_R = 2\pi R$ is the boundary line of half a sphere, i.e., the circular line $L_R = 2\pi r$, then the above equation becomes equation (18.3.8) because the radius $R = \mathrm{r}$ and distance $Z = 0$.

If the radius r takes values in different ranges, then according to equation (18.3.7), the rotational electric potential φ_ω is:

$$\begin{cases} \varphi_\omega = +\frac{1}{4\pi\varepsilon_0}\frac{q}{r}\sqrt{1-\frac{v^2}{c^2}}\sin\theta, & r > 1(l_m) \\ \varphi_\omega = 0, & r = 1(l_m) \\ \varphi_\omega = -\frac{1}{4\pi\varepsilon_0}\frac{q}{r}\sqrt{1-\frac{v^2}{c^2}}\sin\theta, & 0 < r < 1(l_m) \end{cases} \qquad (18.3.10)$$

The above formula is the formula for the rotational electric potential.

If the radius $r = 1(l_m)$, then the rotational electric potential $\varphi_\omega = 0$. If radius $r > 1(l_m)$, then the rotational electric potential $\varphi_\omega > 0$ for positive electric charge. If radius $r < 1(l_m)$, then rotational electric potential $\varphi_\omega < 0$ for positive electric charge.

18.3.3 Integration $\oint \varphi_\omega dl$ of the rotated electric potential φ_ω along the circular line $L_R=2\pi R$

when the distance r from the electric charge q to each point of the circular line $L\ \ = 2\pi R$ is equal. Since the rotational electric potential φ_ω is equal at every point on the circular line L_R, the integral $\oint \varphi_\omega dl$ of the rotational electric potential φ_ω along the circular line L_R is:

$$\oint \varphi_\omega dl = \varphi_\omega L_R = +\frac{1}{4\pi\varepsilon_0}\frac{q\sin\theta}{r}\sqrt{1-\frac{v^2}{c^2}}\cdot 2\pi R, \qquad r > 1(l_m) \qquad (18.3.11)$$

Simplifying the above equation gives the relationship equation:

$$\oint \varphi_\omega dl = +\frac{1}{2\varepsilon_0}\frac{Rq\sin\theta}{r}\sqrt{1-\frac{v^2}{c^2}}, \qquad r > 1(l_m) \qquad (18.3.12)$$

The above equation is the equation for the circular line integral of the rotational electric potential. The above equation shows that the circular line integral $\oint \varphi_\omega dl$ is proportional to the motion electric charge q and inversely proportional to the radius r. The circular line integral $\oint \varphi_\omega dl$ becomes smaller as the velocity v increases.

Substituting equation (18.1.3) the function $\sin\theta$ into equation (18.3.12) yields the relationship equation:

$$\oint \varphi_\omega dl = +\frac{1}{2\varepsilon_0}\frac{R^2 q}{(R^2+Z^2)}\sqrt{1-\frac{v^2}{c^2}}, \qquad r > 1(l_m) \qquad (18.3.13)$$

The above formula is an alternative expression for the integral formula for the circular line of the rotational electric potential. When the circular line $L_R = 2\pi R$ is the edge of half a sphere, i.e., circular line $L_R = 2\pi r$. Since the angle $\theta = 90^0$, the distance $Z = 0$, the radius $R = r$, the above formula becomes:

$$\oint \varphi_\omega dl = +\frac{q}{2\varepsilon_0}\sqrt{1-\frac{v^2}{c^2}}, \qquad r > 1(l_m) \qquad (18.3.14)$$

The above equation is the maximum rotational electric potential circular line integral equation. In this case, the motion electric charge q is located at the center of the circular line $L_R = 2\pi r$ plane. The above equation shows that the maximum rotational electric potential circular line integral $\oint \varphi_\omega dl$ is independent of the radius R and radius r.

18.4 Magnetic-electric potential φ_v

18.4.1 Derivation of the formula for the Magnetic-electric potential φ_v

Since the electric charge q_v is the magnetic electric charge contained in the motion electric charge q , this book defines the second component within the root sign of equation (18.3.5), the electric potential φ_v, as the Magnetic-electric potential , i.e:

$$\varphi_v = +\frac{1}{4\pi\varepsilon_0}\frac{q_v}{r}\sin\theta, \qquad r > 1(l_m) \qquad (18.4.1)$$

According to equation (18.3.4), the magnetic electric charge q_v contained in the motion electric charge q is:

$$q_v = q\frac{v}{c} \qquad (18.4.2)$$

Substituting the above equation into equation (18.4.1) gives the relationship equation:

$$\varphi_v = +\frac{1}{4\pi\varepsilon_0}\frac{qv}{cr}\sin\theta = \varphi\frac{v}{c}, \qquad r > 1(l_m) \qquad (18.4.3)$$

The above equation is the Magnetic-electric potential equation. Since the formula contains two vectors, velocity v and radius r, the direction of Magnetic-electric potential φ_v is perpendicular to the plane defined by both velocity v and radius r. The angle θ in the formula only affects the magnitude of the Magnetic-electric potential φ_v in that direction, not the direction of the Magnetic-electric potential φ_v.

When the angle $\theta = 0$. Since the radius r coincides with the direction line of motion of electric charge q, the Magnetic-electric potential $\varphi_v = 0$.

When angle $\theta = 90^0$. Since the radius r is perpendicular to the direction of motion of electric charge q, the Magnetic-electric potential φ_v has the maximum value, i.e:

$$\varphi_v = +\frac{1}{4\pi\varepsilon_0}\frac{qv}{cr}, \qquad r > 1(l_m) \tag{18.4.4}$$

The above equation shows that the Magnetic-electric potential φ_v is proportional to the electric charge q and velocity v, and inversely proportional to the radius r. When the electric charge velocity $v = 0$, the Magnetic-electric potential $\varphi_v = 0$.

Dividing equation (18.4.3) by the speed of light c gives the relation equation:

$$\varphi_B = \frac{1}{c}\varphi_v = +\frac{1}{4\pi\varepsilon_0}\frac{qv}{c^2 r}\sin\theta, \qquad r > 1(l_m) \tag{18.4.5}$$

This book defines $\varphi_B = \frac{1}{c}\varphi_v$ as the magnetic potential of a moving electric charge. According to electromagnetism, the product of the vacuum permittivity ε_0 and the vacuum permeability μ_0 is:

$$\varepsilon_0\mu_0 = \frac{1}{c^2} \tag{18.4.6}$$

Substituting the above equation into equation (18.4.5) gives the relationship equation:

$$\varphi_B = +\frac{1}{c}\varphi_v = +\frac{\mu_0}{4\pi}\frac{qv\sin\theta}{r}, \qquad r > 1(l_m) \tag{18.4.7}$$

The above equation is the magnetic potential equation. Since the above formula contains two vectors, velocity v and radius r, the direction of the magnetic potential φ_B is perpendicular to the plane defined by both the velocity v and radius r.

In summary, the electromagnetic potential φ can be expressed as:

$$\varphi = \sqrt{\varphi_\omega^2 + \varphi_v^2} \tag{18.4.8}$$

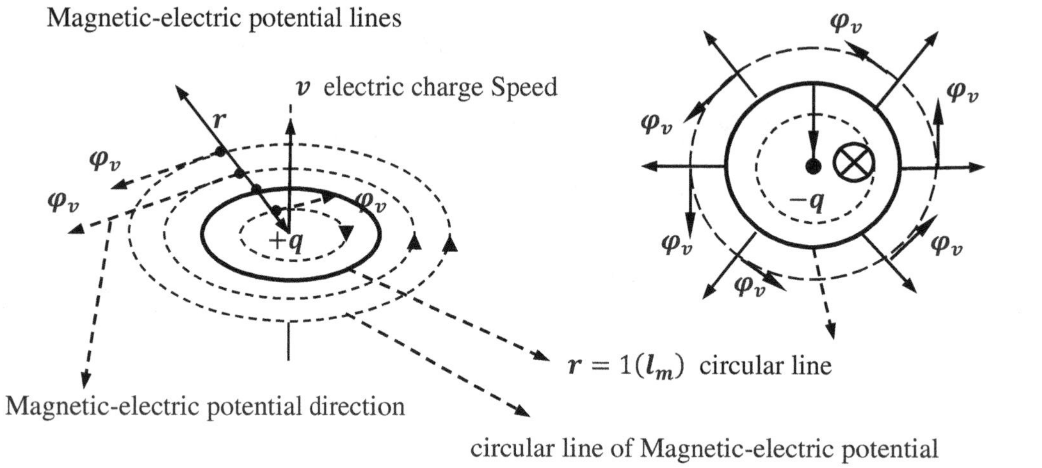

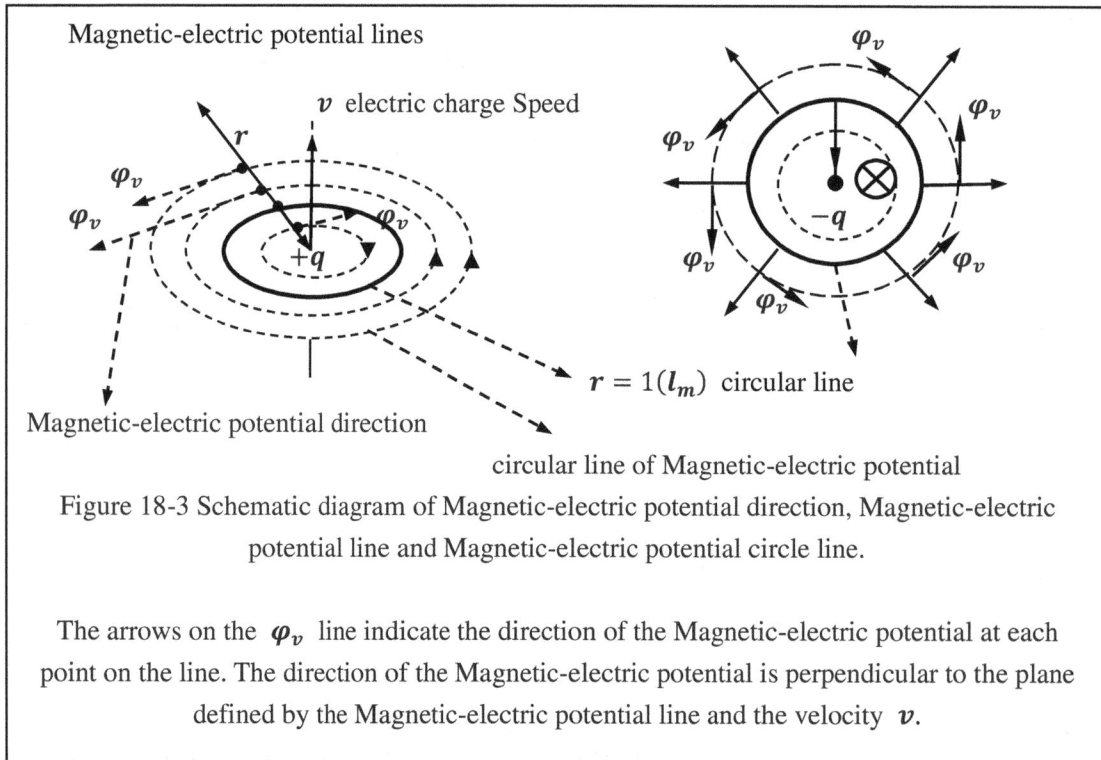

Figure 18-3 Schematic diagram of Magnetic-electric potential direction, Magnetic-electric potential line and Magnetic-electric potential circle line.

The arrows on the φ_v line indicate the direction of the Magnetic-electric potential at each point on the line. The direction of the Magnetic-electric potential is perpendicular to the plane defined by the Magnetic-electric potential line and the velocity v.

Note that the Magnetic-electric potential direction is not the same as the direction of the Magnetic-electric potential line. The Magnetic-electric potential direction is perpendicular to the plane defined by both the radius r and the velocity v, while the Magnetic-electric potential line is oriented in the direction of the vector radius r. The direction of the Magnetic-electric potential φ_v can be determined by the right-hand rule.

When the distance r from the electric charge q to each point of the circular line $L_R = 2\pi R$ is equal, the Magnetic-electric potential φ_v is equal at each point on the circular line L_R. In this book, a circular line with equal

Magnetic-electric potential $\boldsymbol{\varphi_v}$ is defined as a Magnetic-electric potential circular line. This Magnetic-electric potential circular line is the circular dashed line in Figure 18-3.

18.4.2 Integration $\oint \varphi_v\, dl$ of Magnetic-electric potential φ_v along circular line $L_R=2\pi R$

When the distance $\boldsymbol{r}$ from the electric charge $\boldsymbol{q}$ to each point of the circular line $\boldsymbol{L_R = 2\pi R}$ is equal. Since the Magnetic-electric potential $\boldsymbol{\varphi_v}$ is equal at every point on the circular line $\boldsymbol{L_R}$, the integral $\oint \boldsymbol{\varphi_v dl}$ of the Magnetic-electric potential $\boldsymbol{\varphi_v}$ over the circular line $\boldsymbol{L_R}$ is:

$$\oint \boldsymbol{\varphi_v dl} = \boldsymbol{\varphi_v L_R} = +\frac{\mathbf{1}}{\boldsymbol{4\pi\varepsilon_0}}\frac{\boldsymbol{qv}\sin\boldsymbol{\theta}}{\boldsymbol{cr}}\cdot \boldsymbol{2\pi R}, \quad \boldsymbol{r} > 1(\boldsymbol{l_m}) \qquad (\mathbf{18.4.9})$$

Simplify the above equation to get the relationship equation:

$$\oint \boldsymbol{\varphi_v dl} = +\frac{\mathbf{1}}{\boldsymbol{2\varepsilon_0}}\frac{\boldsymbol{Rqv}\sin\boldsymbol{\theta}}{\boldsymbol{cr}}, \quad \boldsymbol{r} > 1(\boldsymbol{l_m}) \qquad (\mathbf{18.4.10})$$

The above equation is the equation of the circular line integral of Magnetic-electric potential. The above equation shows that the circular line integral $\oint \boldsymbol{\varphi_v dl}$ of the Magnetic-electric potential is proportional to the electric charge $\boldsymbol{q}$ and the velocity $\boldsymbol{v}$, and inversely proportional to the radius $\boldsymbol{r}$.

Substituting equation (18.1.3) the function $\sin\boldsymbol{\theta}$ into equation (18.4.10) gives the relationship equation:

$$\oint \boldsymbol{\varphi_v dl} = +\frac{\mathbf{1}}{\boldsymbol{2\varepsilon_0}}\frac{\boldsymbol{R^2qv}}{\boldsymbol{c(R^2+Z^2)}}, \quad \boldsymbol{r} > 1(\boldsymbol{l_m}) \qquad (\mathbf{18.4.11})$$

The above formula is an alternative expression for the Magnetic-electric potential circular line integral formula. When the circular line $\boldsymbol{L_R = 2\pi R}$ is the edge line of half a sphere, i.e., circular line $\boldsymbol{L_R = 2\pi r}$. Since the angle $\boldsymbol{\theta} = \mathbf{90^0}$, distance $\boldsymbol{Z} = \mathbf{0}$, radius $\boldsymbol{R} = \boldsymbol{r}$, so the above formula becomes:

$$\oint \boldsymbol{\varphi_v dl} = +\frac{\boldsymbol{qv}}{\boldsymbol{2\varepsilon_0 c}}, \quad \boldsymbol{r} > 1(\boldsymbol{l_m}) \qquad (\mathbf{18.4.12})$$

The above equation is the maximum Magnetic-electric potential circular line integral equation. At this time, the motion electric charge $\boldsymbol{q}$ is located in the center of the circular line $\boldsymbol{L_R = 2\pi r}$ plane. The above equation shows that the maximum Magnetic-electric potential circular line integral $\oint \boldsymbol{\varphi_\omega dl}$ is independent of radius $\boldsymbol{R}$ and radius $\boldsymbol{r}$.

Chapter 19 Derivation of the magnetic flux density formula for the motion electric charge *q*

Introduction: This chapter derives a formula for the magnetic flux density generated by the motion electric charge q based on the magnetic electric potential formula and Stokes' theorem.

19.1 The integral $\oint Edl$ of the electromagnetic field strength E along the circular line $L_R=2\pi R$

19.1.1 Derivation of the equation for the strength of the electromagnetic field

Suppose that electric charge q is moving with velocity v in the cosmic vacuum frame of reference. According to the electric charge - velocity equation (18.3.2), the moving electric charge q can be decomposed into the rotational electric charge q_ω and the magnetic electric charge q_v, i.e:

$$q = \sqrt{\left(q\frac{v_\omega}{c}\right)^2 + \left(q\frac{v}{c}\right)^2} = \sqrt{q_\omega^2 + q_v^2}$$

If the distance r from the motion electric charge q to each point on the circular line $L_R = 2\pi R$ is equal, then the electromagnetic potential φ is constant on the circular line L_R. When the electromagnetic potential φ is integrated along the circular line $L_R = 2\pi R$, the relation equation can be obtained according to equation (18.2.13) and Stokes' theorem:

$$\oint \varphi dl = \iint \nabla \times \varphi\, dS = 2\pi R \cdot \frac{1}{4\pi\varepsilon_0}\frac{q}{r}\sin\theta \tag{19.1.1}$$

The surface S in equation is bounded by the circular line $L_R = 2\pi R$. We can denote the Curl $\nabla \times \varphi$ of the electromagnetic potential φ by the symbol E, i.e:

$$E = \nabla \times \varphi \tag{19.1.2}$$

Since the rotating electric charge q_ω produces an electric field and the magnetic electric charge q_v produces a Magnetic-electric field, this book defines Curl $E = \nabla \times \varphi$ as the electromagnetic field strength. Substituting the above equation into equation (19.1.1) yields the relational equation:

$$\iint E\, dS = +2\pi R \cdot \frac{1}{4\pi\varepsilon_0}\frac{q}{r}\sin\theta \tag{19.1.3}$$

The surface S in equation has the circle line $L_R = 2\pi R$ as its boundary. when the radius R of the circular line is kept constant. If the surface S is Wang's circular crown $S_w = 2\pi Rr$, then Wang's circular crown S_w is just a function of the radius r. Differentiating Wang's circular crown $S_w = 2\pi Rr$ with respect to radius r yields the relation equation:

$$dS_w = 2\pi R dr$$

Substituting the above equation into equation (19.1.3) yields the relationship equation:

$$\iint E dS_W = \int E \cdot 2\pi R dr = +2\pi R \cdot \frac{\sin\theta\, q}{4\pi\varepsilon_0\, r} \tag{19.1.4}$$

Simplifying the above equation and eliminating the constant $2\pi R$ yields the relational equation:

$$\int Edr = +\frac{\sin\theta\, q}{4\pi\varepsilon_0\, r} \tag{19.1.5}$$

Take the derivative of the above equation with respect to the radius r to obtain the relational equation:

$$E = -\frac{\sin\theta\ q}{4\pi\varepsilon_0\, r^2} \tag{19.1.6}$$

The above equation shows that the electromagnetic field strength $E < 0$ for positive electric charge $+q$ and $E > 0$ for negative electric charge $-q$. Depending on the range of values of the radius r, the electromagnetic field strength E can be expressed as:

$$\begin{cases} E = -\dfrac{1}{4\pi\varepsilon_0}\dfrac{q}{r^2}\sin\theta\ , & r > 1(l_m) \\ E = 0, & r = 1(l_m) \\ E = +\dfrac{1}{4\pi\varepsilon_0}\dfrac{q}{r^2}\sin\theta\ , & 0 < r < 1(l_m) \end{cases} \tag{19.1.7}$$

The above formula is the electromagnetic field strength formula. If radius $r = 1(l_m)$, then the electromagnetic field strength $E = 0$. If radius $r > 1(l_m)$, then the electromagnetic field strength $E < 0$ for electric charge $+q$. If radius $r < 1(l_m)$, then the electromagnetic field strength $E > 0$ for electric charge $+q$.

When the radius r varies continuously from the position $r = 1(l_m)$ to infinity in one direction, or converges to electric charge q, these numerous points of radius r form a straight line. Since each point on the straight line has an electromagnetic field strength E, this book defines the straight line as an electromagnetic field strength line. This line of electromagnetic field strength is the solid line that intersects the circular line in Figure 19-1.

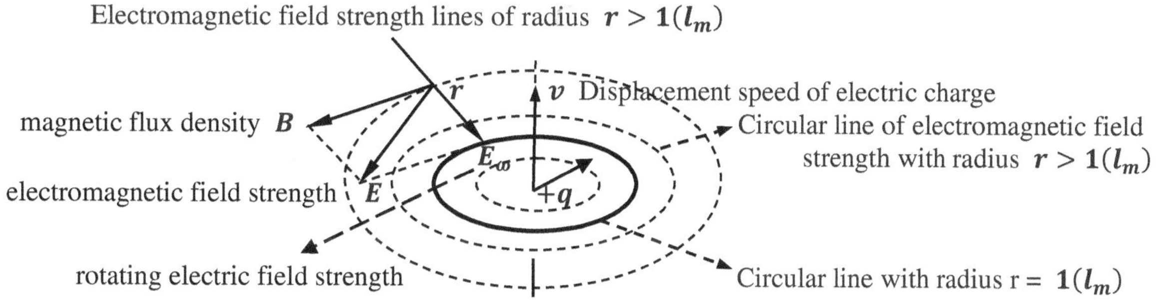

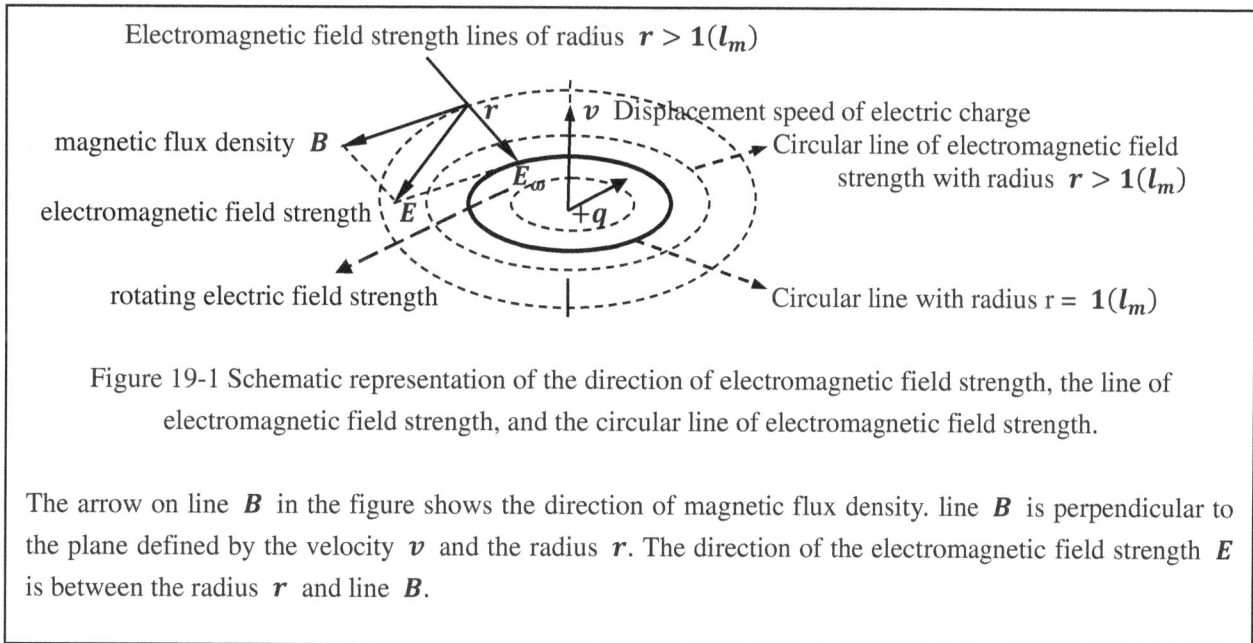

Figure 19-1 Schematic representation of the direction of electromagnetic field strength, the line of electromagnetic field strength, and the circular line of electromagnetic field strength.

The arrow on line B in the figure shows the direction of magnetic flux density. line B is perpendicular to the plane defined by the velocity v and the radius r. The direction of the electromagnetic field strength E is between the radius r and line B.

Note that the direction of the electromagnetic field strength E is different from the direction of the electromagnetic field strength lines. The direction of the electromagnetic field strength E is between the radius r and the B line. The direction of the line of electromagnetic field strength is the direction of the vector radius r.

If the distance r from the electric charge q to each point of the circular line $L_R = 2\pi R$ is equal, then the electromagnetic field strength E is equal at each point on the circular line L_R. This book defines a circular line with equal electromagnetic field strength E as an electromagnetic field strength circular line. This electromagnetic field strength circular line is the circular dashed line in Figure 19-1.

19.1.2 Integration $\oint Edl$ of the electromagnetic field strength E along the circular line $L_R=2\pi R$

Assume that $L = f(r)$ is a curve in the electromagnetic field, and take the line element dl on the curve $L = f(r)$. according to Equation (19.1.7) , the product of the electromagnetic field strength E and the line element dl is:

$$Edl = -\frac{1}{4\pi\varepsilon_0}\frac{q\sin\theta}{r^2}dl, \qquad r > 1(l_m)$$

Integrating the above equation along the curve $L = f(r)$ yields the relation equation:

$$\int Edl = -\int \frac{1}{4\pi\varepsilon_0}\frac{q\sin\theta}{r^2}dl \tag{19.1.8}$$

The integral $\int Edl$ of the electromagnetic field strength E is of two types.

The first type is an integral whose radius r is a variable.

Assume that the radius r is variable on the curve $L_X = f(r)$. Assume that the curve L_X starts at a and ends at b. According to equation (19.1.8), the integral of the electromagnetic field strength E along the curve $L_X = f(r)$ is:

$$\int Edl = -\frac{q\sin\theta}{4\pi\varepsilon_0}\int_{r_a}^{r_b}\frac{1}{r^2}dl = \frac{q\sin\theta}{4\pi\varepsilon_0}\left(\frac{1}{r_b} - \frac{1}{r_a}\right) \tag{19.1.9}$$

The above equation is the equation for the line flow of electromagnetic field strength for a change in radius r. According to equation (18.2.8), the motion electric potential φ produced by electric charge q at cosmic vacuum point a and point b are, respectively:

$$\varphi_a = \frac{\sin\theta}{4\pi\varepsilon_0}\frac{q}{r_a}, \qquad \varphi_b = \frac{\sin\theta}{4\pi\varepsilon_0}\frac{q}{r_b}$$

Substituting the above equation into equation (19.1.9) gives the relationship equation:

$$\int Edl = \varphi_b - \varphi_a \tag{19.1.10}$$

The above equation shows that when the electromagnetic field strength E is integrated along the route L_X, its integral $\int Edl$ is equal to the electromagnetic potential difference $\Delta\varphi = \varphi_b - \varphi_a$.

If the curve $L_X = f(r)$ goes from the starting point a to the point b and returns from the point b to the starting point a. The curve L_X is a closed curve. The integral $\oint Edl$ of the electromagnetic field strength E on the closed curve L_X is:

$$\oint Edl = \int_{r_a}^{r_b} Edl + \int_{r_b}^{r_a} Edl = 0 \tag{19.1.11}$$

The above equation shows that the integral $\oint Edl$ of the electromagnetic field strength E over the closed curve L_X is always equal to 0, if the radius r varies over the closed circular line L_X.

The 2nd type is an integral whose radius r remains constant on the circular line $L_R = 2\pi R$.

If the distance r from the electric charge q to each point on the circular line $L_R = 2\pi R$ is equal, then the electromagnetic field strength E is constant on the circular line L_R. According to Equation (19.1.7), the integral $\oint Edl$ of the electromagnetic field strength E along the circular line $L_R = 2\pi R$ is:

$$\oint Edl = EL_R = -2\pi R\cdot\frac{1}{4\pi\varepsilon_0}\frac{q}{r^2}\sin\theta, \qquad r > 1(l_m) \tag{19.1.12}$$

When the angle $\theta = 90^0$ in Figure 18-1, the above equation becomes since the radius $R = r$:

$$\oint Edl = -\frac{q}{2\varepsilon_0 r} \tag{19.1.13}$$

At this point the distance $Z = 0$ and the motion electric charge q is located in the center of the circular plane $P = \pi R^2$.

19.2 Rotating electric field strength E_ω produced by moving electric charge q

19.2.1 The electromagnetic field strength E can be decomposed into two component electric field strengths

According to equation (19.1.6), the electromagnetic field strength equation:

$$E = -\frac{1}{4\pi\varepsilon_0}\frac{q}{r^2}\sin\theta\,, \qquad r > 1(l_m) \tag{19.2.1}$$

The θ in equation is the angle between the electric charge velocity v and the radius r. According to the electric charge-velocity equation (18.3.2), the motion electric charge q contains the rotational electric charge q_ω and the magnetic electric charge q_v, i.e:

$$q = \sqrt{\left(q\frac{v_\omega}{c}\right)^2 + \left(q\frac{v}{c}\right)^2} = \sqrt{q_\omega^2 + q_v^2} \tag{19.2.2}$$

In the classical theory, the electric field strength E' is:

$$E' = \frac{1}{4\pi\varepsilon_0}\frac{q}{r^2} \tag{19.2.3}$$

Although the electric charge q in the equation can move, the electric field strength E' does not include the electric charge velocity v because physicists do not decompose the electric charge q, into the rotating electric charge q_ω and magnetic electric charge q_v.

Substituting equation (19.2.2) into equation (19.2.1) yields the relationship equation:

$$E = -\sqrt{\left(-\frac{\sin\theta\, q_\omega}{4\pi\varepsilon_0\, r^2}\right)^2 + \left(+\frac{\sin\theta\, q_v}{4\pi\varepsilon_0\, r^2}\right)^2} = -\sqrt{(E_\omega)^2 + (E_v)^2}\,, \qquad r > 1(l_m)$$

$$(19.2.4)$$

The above equation shows that the electromagnetic field strength E can be decomposed into two component electric field strengths. According to equation (18.3.3), the rotating electric charge q_ω contained in the moving electric charge q is:

$$q_\omega = q\sqrt{1 - \frac{v^2}{c^2}} \tag{19.2.5}$$

Since the electric charge q has positive electric charge $+q$ and negative electric charge $-q$, the rotational electric charge q_ω also has positive rotational electric charge $+q_\omega$ and negative rotational electric charge $-q_\omega$.

According to equation (18.3.4), the magnetic electric charge q_v contained in the motion electric charge q is:

$$q_v = q\frac{v}{c} \tag{19.2.6}$$

Note that the magnetic electric charge q_v is also differentiated between a positive magnetic electric charge $+q_v$ and a negative magnetic electric charge $-q_v$. There are two factors that affect the positive or negative sign of the magnetic electric charge q_v, one is the positive or negative sign of the moving electric charge q, and the other is the direction of the charge velocity v. For example, wires with the same direction of electric current I attract each other, and wires with opposite direction of electric current I repel each other.

When the velocity $v = 0$ of motion of the electric charge. At this point, equation (19.2.4) changes to because of the rotating electric charge $q_\omega = q$ and magnetic electric charge $q_v = 0$:

$$E = -\frac{1}{4\pi\varepsilon_0}\frac{q}{r^2}\,, \qquad r > 1(l_m) \tag{19.2.7}$$

According to equation (17.4.4), the electrostatic field strength E_0 generated by the static electric charge q_0 is:

$$E_0 = -\frac{1}{4\pi\varepsilon_0}\frac{q_0}{r^2} \tag{19.2.8}$$

Compare the above equation with equation (19.2.7). It can be determined that the electromagnetic field strength E

is equal to the electrostatic field strength $\boldsymbol{E}_0$, i.e., $\boldsymbol{E} = \boldsymbol{E}_0$, when the velocity $\boldsymbol{v} = \boldsymbol{0}$ of the electric charge $\boldsymbol{q}$.

19.2.2 Definition of rotating electric field strength E_ω

Since the transformation factor $\sqrt{1 - \frac{v^2}{c^2}}$ in equation (19.2.5) is a dimensionless coefficient, the rotating electric charge $\boldsymbol{q}_\omega$ and the static electric charge $\boldsymbol{q}_0$ are electric charges of the same nature . That is, both belong to the class of electric charges that produce an electric field strength .

Since the electric charge $\boldsymbol{q}_\omega$ is the electric charge that generates the electric field strength , this book defines the first component electric field strength $\boldsymbol{E}_\omega$ inside the root sign of equation (19.2.4) as the rotating electric field strength, i.e:

$$\boldsymbol{E}_\omega = -\frac{1}{4\pi\varepsilon_0}\frac{q_\omega}{r^2}\sin\theta\,, \qquad r > 1(l_m) \qquad (19.2.9)$$

The above equation shows that the rotational electric field strength $\boldsymbol{E}_\omega < \boldsymbol{0}$ for positively rotating electric charge $+\boldsymbol{q}_\omega$ and $\boldsymbol{E}_\omega > \boldsymbol{0}$ for negatively rotating electric charge $-\boldsymbol{q}_\omega$. Substituting the rotating electric charge $\boldsymbol{q}_\omega$ in equation (19.2.5) into the above equation, we can get the relation equation:

$$\boldsymbol{E}_\omega = -\frac{1}{4\pi\varepsilon_0}\frac{q}{r^2}\sqrt{1 - \frac{v^2}{c^2}}\sin\theta = E\sqrt{1 - \frac{v^2}{c^2}}\,, \qquad r > 1(l_m) \qquad (19.2.10)$$

The above formula is the rotating electric field strength formula. The $\boldsymbol{E}_\omega$ in the formula is the rotating electric field strength in the cosmic vacuum frame of reference. The above formula shows that positive electric charge $+\boldsymbol{q}$ produces rotating electric field strength $\boldsymbol{E}_\omega < \boldsymbol{0}$ and negative electric charge $-\boldsymbol{q}$ produces rotating electric field strength $\boldsymbol{E}_\omega > \boldsymbol{0}$. The rotating electric field strength $\boldsymbol{E}_\omega$ decreases as the electric charge velocity $\boldsymbol{v}$ increases.

Since the transformation factor $\sqrt{1 - \frac{v^2}{c^2}}\sin\theta$ in equation (19.2.10) is a dimensionless coefficient, the rotating electric field $\boldsymbol{E}_\omega$ is an electric field strength. In addition, since the electric field force between homosexual electric charges is the repulsive force $\boldsymbol{F}_{(-)}$ and the electric field force between anisotropic electric charges is the suction force $\boldsymbol{F}_{(+)}$, Equation (19.2.10) takes a negative sign.

Since the rotating electric field strength $\boldsymbol{E}_\omega$ contains the vector radius $\boldsymbol{r}$, the direction of the rotating electric field strength $\boldsymbol{E}_\omega$ is either in the same direction as the direction of the vector radius $\boldsymbol{r}$ or in the opposite direction to the direction of the vector radius $\boldsymbol{r}$.

If the electric charge moves with velocity $\boldsymbol{v} = \boldsymbol{0}$, then the rotating electric field strength $\boldsymbol{E}_\omega$ is equal to the electrostatic field strength $\boldsymbol{E}_0$, i.e., $\boldsymbol{E}_\omega = \boldsymbol{E}_0$. From this, it can be determined that the rotating electric field strength $\boldsymbol{E}_\omega$ and the electrostatic field strength $\boldsymbol{E}_0$ are electric field strengths of the same nature. The difference between the two is that the rotating electric field strength $\boldsymbol{E}_\omega$ is moving, while the electrostatic field strength $\boldsymbol{E}_0$ is stationary.

Because the electric field of the high energy accelerator can make the speed $\boldsymbol{v}$ of the moving electrons increasing, so the moving electric charge $\boldsymbol{q}$ can not only produce the magnetic field $\boldsymbol{B}$, but also has a certain electric field (rotating electric field strength $\boldsymbol{E}_\omega$) at the same time, and it is this rotating electric field strength $\boldsymbol{E}_\omega$ that makes the electrons in the high energy accelerator's electric field constantly be accelerated.

When the velocity $\boldsymbol{v}$ of the moving electron approaches the speed of light $\boldsymbol{c}$. The electric field of the high energy accelerator is unable to bring the electron's velocity $\boldsymbol{v}$ to the speed of light because the value of the rotating electric field strength $\boldsymbol{E}_\omega$ is close to zero.

If the angle $\boldsymbol{\theta} = \boldsymbol{0}$, then the radius $\boldsymbol{r}$ coincides with the direction line of motion of the electric charge $\boldsymbol{q}$. Since the function $\sin 0^0 = 0$, the rotating electric field strength $\boldsymbol{E}_\omega = \boldsymbol{0}$.

If the angle $\boldsymbol{\theta} = 90^0$, then the radius $\boldsymbol{r}$ is perpendicular to the direction of motion of the electric charge $\boldsymbol{q}$. Since the function $\sin 90^0 = 1$, the rotating electric field strength $\boldsymbol{E}_\omega$ has a maximum value at this time, i.e:

$$\boldsymbol{E}_\omega = -\frac{1}{4\pi\varepsilon_0}\frac{q}{r^2}\sqrt{1 - \frac{v^2}{c^2}}\,, \qquad r > 1(l_m) \qquad (19.2.11)$$

If the electric charge velocity $\boldsymbol{v} = \mathbf{0}$, then the above equation becomes the electrostatic field strength equation (19.2.8), i.e. $\boldsymbol{E}_\omega = \boldsymbol{E}_0$.

19.2.3 Curl $\nabla\times\varphi_\omega$ of rotating electric potential strength φ_ω equals rotating electric field strength E_ω

When electric charge $\boldsymbol{q}$ is moving with velocity $\boldsymbol{v}$ in the cosmic vacuum frame of reference, according to equation (18.3.7), the rotational electric potential $\boldsymbol{\varphi}_\omega$ generated by the moving electric charge $\boldsymbol{q}$ is:

$$\boldsymbol{\varphi}_\omega = +\frac{1}{4\pi\varepsilon_0}\frac{q}{r}\sqrt{1-\frac{v^2}{c^2}}\sin\theta\,, \qquad r > 1(l_m) \tag{19.2.12}$$

The $\boldsymbol{\theta}$ in equation is the angle between the radius $\boldsymbol{r}$ and the electric charge velocity $\boldsymbol{v}$.

Assume that the distance $\boldsymbol{r}$ from the motion electric charge $\boldsymbol{q}$ to each point on the circular line $\boldsymbol{L_R} = \boldsymbol{2\pi R}$ is equal. Since the rotational electric potential $\boldsymbol{\varphi}_\omega$ is constant on the circular line $\boldsymbol{L_R}$, the integral $\oint \boldsymbol{\varphi}_\omega \boldsymbol{dl}$ of the rotational electric potential B along the circular line $\boldsymbol{L_R}$ is:

$$\oint \varphi_\omega dl = \varphi_\omega L_R = +\frac{1}{4\pi\varepsilon_0}\frac{q}{r}\sqrt{1-\frac{v^2}{c^2}}\sin\theta\cdot 2\pi R \tag{19.2.13}$$

Simplify the above equation to get the relationship equation:

$$\oint \varphi_\omega dl = +\frac{1}{2\varepsilon_0}\frac{qR}{r}\sqrt{1-\frac{v^2}{c^2}}\sin\theta \tag{19.2.14}$$

Based on the above equation and Stokes' theorem, the relationship equation can be obtained:

$$\oint \varphi_\omega dl = \iint \nabla\times\varphi_\omega\, dS = +\frac{1}{2\varepsilon_0}\frac{qR}{r}\sqrt{1-\frac{v^2}{c^2}}\sin\theta \tag{19.2.15}$$

We use the notation $\boldsymbol{E}_\omega$ to denote the Curl Curl $\boldsymbol{\nabla}\times\boldsymbol{\varphi}_\omega$ of the rotational electric potential $\boldsymbol{\varphi}_\omega$, i.e:

$$E_\omega = \nabla\times\varphi_\omega \tag{19.2.16}$$

This book defines Curl $\boldsymbol{E}_\omega = -\boldsymbol{\nabla}\times\boldsymbol{\varphi}_\omega$ as the rotating electric field strength. Substituting the above equation into equation (19.2.15) gives the relation equation:

$$\iint E_\omega dS = \frac{1}{2\varepsilon_0}\frac{qR}{r}\sqrt{1-\frac{v^2}{c^2}}\sin\theta \tag{19.2.17}$$

When the radius $\boldsymbol{R}$ of the circular line is kept constant, if the surface $\boldsymbol{S}$ is Wang's circular crown $\boldsymbol{S_W} = \boldsymbol{2\pi Rr}$, then Wang's circular crown $\boldsymbol{S_W}$ is just a function of the radius $\boldsymbol{r}$. Differentiating Wang's circular crown $\boldsymbol{S_W} = \boldsymbol{2\pi Rr}$ with respect to radius $\boldsymbol{r}$ yields the relation equation:

$$dS_W = 2\pi R dr$$

Substituting the above equation into equation (19.2.17) gives the relationship equation:

$$\iint E_\omega\cdot 2\pi R dr = \frac{1}{2\varepsilon_0}\frac{qR}{r}\sqrt{1-\frac{v^2}{c^2}}\sin\theta \tag{19.2.18}$$

Simplifying the above equation and eliminating the constant $\boldsymbol{R}$ yields the relational equation:

$$\int E_\omega dr = \frac{1}{4\pi\varepsilon_0}\frac{qR}{r}\sqrt{1-\frac{v^2}{c^2}}\sin\theta \tag{19.2.19}$$

Take the derivative of the above equation with respect to the radius $\boldsymbol{r}$ to obtain the relational equation:

$$E_\omega = -\frac{1}{4\pi\varepsilon_0}\frac{qR}{r^2}\sqrt{1-\frac{v^2}{c^2}}\sin\theta$$

The above equation is the rotating electric field strength equation (19.2.10). The above equation shows that the rotating electric field strength $\boldsymbol{E}_\omega$ is proportional to the electric charge $\boldsymbol{q}$ and inversely proportional to the radius $\boldsymbol{r}$ squared. The rotating electric field strength $\boldsymbol{E}_\omega$ becomes smaller as the velocity $\boldsymbol{v}$ increases.

Substituting equation (19.2.10) into equation (19.2.16) gives the relationship equation:

$$\nabla \times \boldsymbol{\varphi}_{\omega} = -\frac{1}{4\pi\varepsilon_0}\frac{q}{r^2}\sqrt{1-\frac{v^2}{c^2}}\sin\theta = E\sqrt{1-\frac{v^2}{c^2}} \tag{19.2.20}$$

The above equation is the Curl equation for the rotational electric potential $\boldsymbol{\varphi}_{\omega}$.

19.2.4 The magnitude of the electrostatic field strength E_0' on the ground changes periodically

Assume that the electric charge q is moving with velocity v in the cosmic vacuum, and assume that θ is the angle between the distance r and the direction of displacement of electric charge q. According to equation (19.2.10), the rotating electric field strength E_{ω} generated by electric charge q is:

$$E_{\omega} = -\frac{1}{4\pi\varepsilon_0}\frac{q}{r^2}\sqrt{1-\frac{v^2}{c^2}}\sin\theta = E\sqrt{1-\frac{v^2}{c^2}}$$

The relationship between the rotating electric field strength E_{ω} and the electrostatic field strength E_0 is given by:

$$E_{\omega} = E_0\sqrt{1-\frac{v^2}{c^2}}\sin\theta \tag{19.2.21}$$

Suppose that the ground (S' inertial system) is moving with velocity u in the cosmic vacuum frame of reference. Since the electric charge q_0', which is at rest on the ground, is moving with velocity u in the cosmic vacuum, the electrostatic field strength E_0' of the ground, according to the above equation, is essentially the rotating electric field strength E_{ω} in the cosmic vacuum frame of reference, i.e:

$$E_0' = E_{\omega} = E_0\sqrt{1-\frac{u^2}{c^2}}\sin\theta \tag{19.2.22}$$

The θ in the formula is the angle between the distance r and the velocity u. The above equation shows that the electrostatic field strength E_0' of the ground becomes smaller as the velocity u increases. The following two inferences can be drawn from the above equation.

Corollary 1. The strength of the electrostatic field E_0' at any point on the ground varies periodically with the rotation of the earth.

Since the sun revolves around the center of the Milky Way Galaxy and there is rotation and revolution of the earth, any point on the earth's equator moves at a different speed u in the cosmic system during the 24 hours of the day.

Assume that the velocity u at point P at the equator is maximum at 12 o'clock during the day and minimum at 12 o'clock at night. According to Equation (19.2.22) , the electrostatic field strength E_{01}' at point P at 12 o'clock in the daytime is less than the electrostatic field strength E_{02}' at 12 o'clock in the nighttime, i.e. $E_{01}' < E_{02}'$. One can verify equation (19.2.22) by a physical experiment with a 12-hour time difference.

The above causes periodic changes in the magnitude of the electrostatic field strength E_0' on the ground. In real life, the spot position of laser irradiation undergoes periodic changes, and the periodic changes in the electrostatic field strength E_0' are similar to the periodic changes in the spot position.

Corollary 2. At the same moment, the strength E_{01}' of the electrostatic field at the north and south poles of the earth, is greater than the strength E_{02}' of the electrostatic field at the earth's equator.

Since different latitudes have different linear velocities u, the strength of the electrostatic field E_0' is not equal at different latitudes. Suppose that the north and south poles of the earth are moving with velocity u_1 in the cosmic vacuum. According to equation (19.2.22), the electrostatic field strength E_{01}' at the north and south poles is:

$$E_{01}' = E_0\sqrt{1-\frac{u_1^2}{c^2}}\sin\theta \tag{19.2.23}$$

Assume that the Earth's equator is moving with velocity u_2 in the cosmic vacuum. According to equation (19.2.22), the electrostatic field strength E_{02}' at the Earth's equator is:

$$E'_{02} = E_0 \sqrt{1 - \frac{u_2^2}{c^2} \sin\theta} \qquad (19.2.24)$$

Since the velocity u_1 at the north and south poles is less than the velocity u_2 at the equator, i.e., $u_1 < u_2$. Therefore the electrostatic field strength E'_{01} at the north and south poles is greater than the electrostatic field strength E'_{02} at the equator, i.e:

$$E'_{01} > E'_{02} \qquad (19.2.25)$$

If the above formula is confirmed experimentally, then it proves that the cosmic vacuum frame of reference is the absolute coordinate system in Newton's theory.

19.2.5 The rate of change $d\varphi_\omega/dt$ of the rotating electric potential is equal to the time-varying rotating electric field strength E_ω

According to the rotational electric potential equation (18.3.7), the rotational electric potential φ_ω generated by the motion electric charge q is:

$$\varphi_\omega = +\frac{1}{4\pi\varepsilon_0}\frac{q}{r}\sqrt{1 - \frac{v^2}{c^2}\sin\theta}, \quad r > 1(l_m) \qquad (19.2.26)$$

When the electric charge velocity $v = 0$, the rotational electric potential φ_ω becomes the static electric potential φ_0, i.e:

$$\varphi_\omega = \varphi_0 = \frac{1}{4\pi\varepsilon_0}\frac{q}{r} \qquad (19.2.27)$$

Taking the derivative of equation (19.2.26) with respect to time t yields the relation equation:

$$\frac{d\varphi_\omega}{dt} = \frac{\partial\varphi_\omega}{\partial q}\frac{dq}{dt} + \frac{\partial\varphi_\omega}{\partial v}\frac{dv}{dt} + \frac{\partial\varphi_\omega}{\partial r}\frac{dr}{dt}$$

$$= \frac{\sin\theta}{4\pi\varepsilon_0}\left(\frac{1}{r}\sqrt{1 - \frac{v^2}{c^2}}\frac{dq}{dt} - \frac{qv}{rc\sqrt{c^2 - v^2}}\frac{dv}{dt} - \frac{q}{r^2}\sqrt{1 - \frac{v^2}{c^2}}\frac{dr}{dt}\right) \qquad (19.2.28)$$

The command acceleration $a = \frac{dv}{dt}$ and velocity $v_r = \frac{dr}{dt}$. The above equation becomes:

$$\frac{d\varphi_\omega}{dt} = \frac{\partial\varphi_\omega}{\partial q}\frac{dq}{dt} + \frac{\partial\varphi_\omega}{\partial v}a + \frac{\partial\varphi_\omega}{\partial r}v_r$$

$$= \frac{\sin\theta}{4\pi\varepsilon_0}\left(\frac{1}{r}\sqrt{1 - \frac{v^2}{c^2}}\frac{dq}{dt} - \frac{qva}{rc^2\sqrt{1 - \frac{v^2}{c^2}}} - \frac{qv_r}{r^2}\sqrt{1 - \frac{v^2}{c^2}}\right) \qquad (19.2.29)$$

The above equation is the formula for the rate of change of rotational electric potential. The a in the formula is the acceleration of the motion electric charge q, and v_r is the rate of change of radius r. Divide the above equation by the speed of light c to get the relation equation:

$$\frac{1}{c}\frac{d\varphi_\omega}{dt} = \frac{1}{c}\frac{\partial\varphi_\omega}{\partial q}\frac{dq}{dt} + \frac{a}{c}\frac{\partial\varphi_\omega}{\partial v} + \frac{v_r}{c}\frac{\partial\varphi_\omega}{\partial r}$$

$$= \frac{1}{c}\frac{\sin\theta}{4\pi\varepsilon_0}\left(\frac{1}{r}\sqrt{1 - \frac{v^2}{c^2}}\frac{dq}{dt} - \frac{qva}{rc^2\sqrt{1 - \frac{v^2}{c^2}}} - \frac{qv_r}{r^2}\sqrt{1 - \frac{v^2}{c^2}}\right) \qquad (19.2.30)$$

When the radius r is a variable, and the electric charge q and the velocity v are constants. Since the rate of electric charge change $\frac{dq}{dt} = 0$ and the acceleration $a = 0$, the above equation becomes:

$$\frac{1}{c}\frac{d\varphi_\omega}{dt} = \frac{v_r}{c}\frac{\partial\varphi_\omega}{\partial r} = -\frac{v_r}{c}\frac{1}{4\pi\varepsilon_0}\frac{q}{r^2}\sqrt{1 - \frac{v^2}{c^2}\sin\theta} \qquad (19.2.31)$$

According to equation (19.2.10), the rotating electric field E_ω is:

$$E_\omega = -\frac{1}{4\pi\varepsilon_0}\frac{q}{r^2}\sqrt{1-\frac{v^2}{c^2}}\sin\theta$$

Substituting the above equation into equation (19.2.31) gives the relationship equation:

$$\frac{1}{c}\frac{d\varphi_\omega}{dt} = \frac{v_r}{c}E_\omega \tag{19.2.32}$$

Since the ratio $\frac{v_r}{c}$ of velocities has no physical unit, $\frac{v_r}{c}E_\omega$ in the formula belongs to the electric field strength. From this it can be determined that the rates of change $\frac{1}{c}\frac{d\varphi_\omega}{dt}$, $\frac{a}{c}\frac{d\varphi_\omega}{dt}$ and $\frac{v_r}{c}\frac{\partial\varphi_\omega}{\partial r}$ are all electric field strengths. The above equation is the electric field strength produced by a change in radius r.

When the velocity v is a variable, and the electric charge q and radius r are constants. Since the rate of electric charge change $\frac{dq}{dt} = 0$ and the velocity $v_r = 0$, equation (19.2.30) becomes:

$$\frac{1}{c}\frac{d\varphi_\omega}{dt} = \frac{a}{c}\frac{\partial\varphi_\omega}{\partial v} = -\frac{1}{4\pi\varepsilon_0 c^2}\frac{qva}{r\sqrt{c^2-v^2}}\sin\theta \tag{19.2.33}$$

When the electric charge velocity v remains constant. Since the acceleration $a = 0$, the above equation becomes:

$$\frac{1}{c}\frac{d\varphi_\omega}{dt} = \frac{a}{c}\frac{\partial\varphi_\omega}{\partial v} = 0$$

The above equation shows that a charge q in uniform motion does not generate time-varying rotating electric field strength, and only a charge q in accelerated motion generates time-varying rotating electric field strength. From this it can be determined that the constant electric current I does not produce the time-varying electromagnetic field E.

19.3 Derivation of the magnetic flux density formula for the motion electric charge *q*

19.3.1 Definition of Magnetic-electric field strength E_v

If the two wires electric current I_1, I_2 are in the same direction, then the magnetic force between the two wires is suction $(F > 0)$. If the two wires electric current I_1, I_2 opposite direction, then the magnetic field force between the two wires is repulsive $(F < 0)$.

According to equation (19.2.4), the electromagnetic field strength E can be decomposed into two component electric field strengths, viz:

$$E = -\sqrt{\left(-\frac{\sin\theta}{4\pi\varepsilon_0}\frac{q_\omega}{r^2}\right)^2 + \left(+\frac{\sin\theta}{4\pi\varepsilon_0}\frac{q_v}{r^2}\right)^2} = -\sqrt{(E_\omega)^2 + (E_v)^2}$$

Since the electric charge q_v is the magnetic electric charge contained in the motion electric charge q, this book defines the second component electric field strength E_v inside the root sign as the Magnetic-electric field strength, i.e:

$$E_v = +\frac{1}{4\pi\varepsilon_0}\frac{q_v}{r^2}\sin\theta, \qquad r > 1(l_m) \tag{19.3.1}$$

The above formula is the Magnetic-electric field strength formula. The above equation shows that positive magnetic electric charge $+q_v$ produces Magnetic-electric field strength $E_v > 0$, and negative magnetic electric charge $-q_v$ produces Magnetic-electric field strength $E_v < 0$.

According to equation (19.2.6), the magnetic electric charge q_v is:

$$q_v = \frac{qv}{c}$$

Substituting the above equation into equation (19.3.1) yields the relationship equation:

$$E_v = +\frac{1}{4\pi\varepsilon_0}\frac{qv}{cr^2}\sin\theta \tag{19.3.2}$$

The above formula is the Magnetic-electric field strength formula. The $\boldsymbol{\theta}$ in the formula is the angle between the electric charge velocity $\boldsymbol{v}$ and the distance $\boldsymbol{r}$. When the velocity of the electric charge $\boldsymbol{v} = \mathbf{0}$, the Magnetic-electric field strength $\boldsymbol{E_v}$ and the magnetic flux density $\boldsymbol{B}$ produced by the moving electric charge $\boldsymbol{q}$ are both equal to zero, i.e:

$$\boldsymbol{E_v} = \mathbf{0}, \qquad \boldsymbol{B} = \mathbf{0}$$

Since Equation (19.3.2) contains two vectors, velocity $\boldsymbol{v}$ and distance $\boldsymbol{r}$, the direction of the Magnetic-electric field strength $\boldsymbol{E_v}$, is perpendicular to the plane defined by both the velocity $\boldsymbol{v}$ and the distance $\boldsymbol{r}$.

When the angle $\boldsymbol{\theta} = \mathbf{90^0}$. Since the distance $\boldsymbol{r}$ is perpendicular to the velocity $\boldsymbol{v}$, the Magnetic-electric field strength $\boldsymbol{E_v}$ is maximum at this time, i.e:

$$\boldsymbol{E_v} = +\frac{\mathbf{1}}{\mathbf{4}\boldsymbol{\pi\varepsilon_0}}\frac{\boldsymbol{qv}}{\boldsymbol{cr^2}} \qquad (\mathbf{19.3.3})$$

The above equation shows that the constant Magnetic-electric field strength $\boldsymbol{E_v}$ is proportional to the electric charge $\boldsymbol{q}$ and the velocity $\boldsymbol{v}$.

When the angle $\boldsymbol{\theta} = \mathbf{0}$. Since the distance $\boldsymbol{r}$ coincides with the direction of velocity $\boldsymbol{v}$, the Magnetic-electric field strength $\boldsymbol{E_v} = \mathbf{0}$.

19.3.2 Magnetic-electric field strength lines and Magnetic-electric field strength circles lines

Magnetic-electric field strength Line.

When the radius $\boldsymbol{r}$ varies continuously from the position $\boldsymbol{r} = 1$ to infinity in one direction, or converges to the electric charge $\boldsymbol{q}$, these innumerable radius $\boldsymbol{r}$ points form a straight line. Since each point on the straight line has a Magnetic-electric field strength $\boldsymbol{E_v}$, this book defines the straight line as a Magnetic-electric field strength line. This Magnetic-electric field strength line is the solid line that intersects the circular line in 19-2.

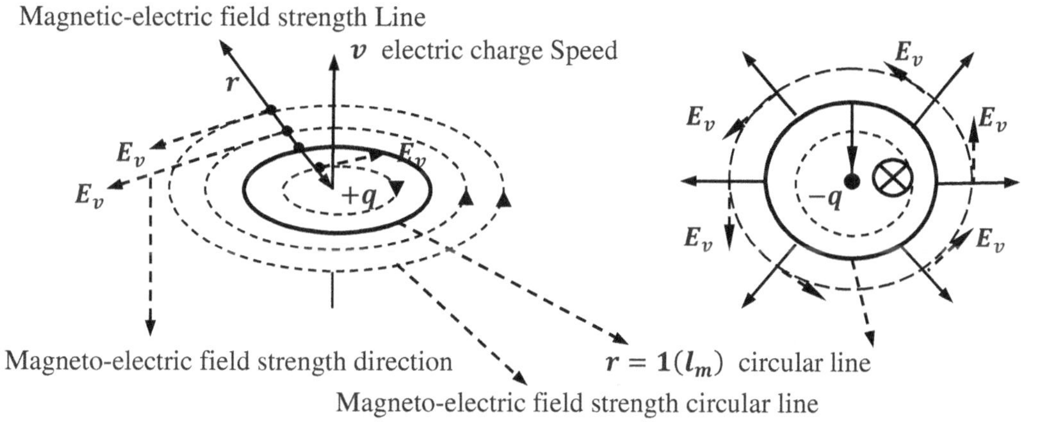

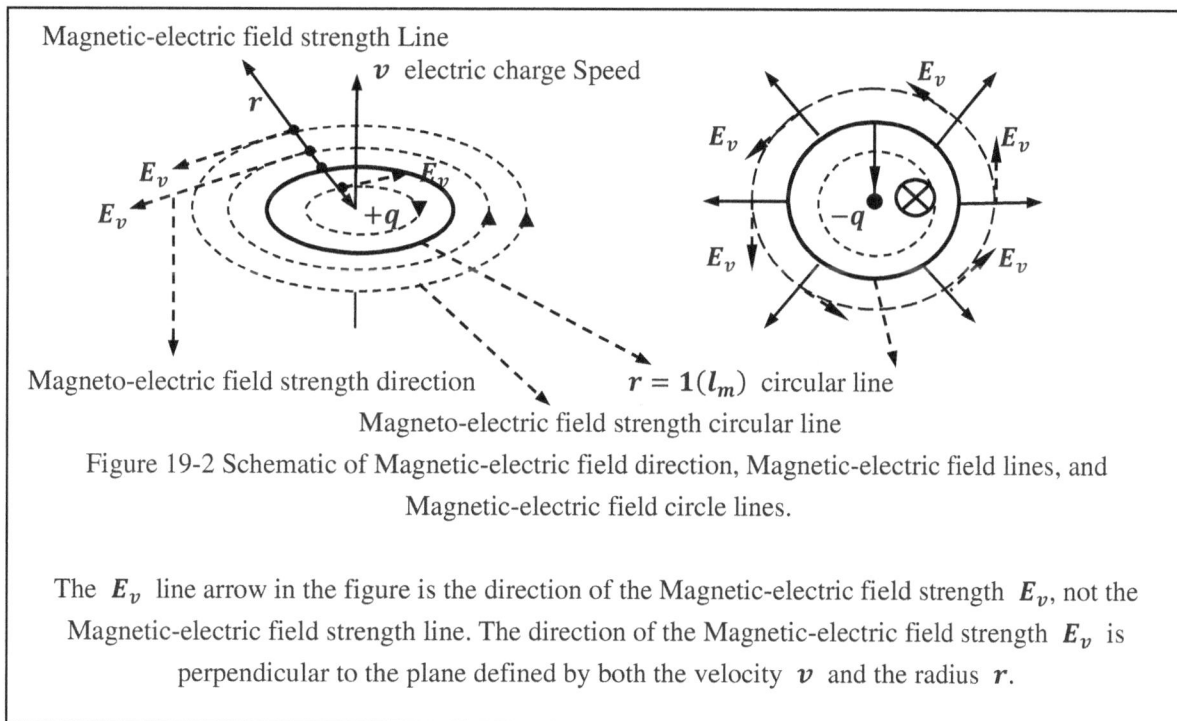

Figure 19-2 Schematic of Magnetic-electric field direction, Magnetic-electric field lines, and Magnetic-electric field circle lines.

The $\boldsymbol{E_v}$ line arrow in the figure is the direction of the Magnetic-electric field strength $\boldsymbol{E_v}$, not the Magnetic-electric field strength line. The direction of the Magnetic-electric field strength $\boldsymbol{E_v}$ is perpendicular to the plane defined by both the velocity $\boldsymbol{v}$ and the radius $\boldsymbol{r}$.

Note that the direction of the Magnetic-electric field strength is not the same as the direction of the Magnetic-electric field strength lines. The direction of the Magnetic-electric field strength is perpendicular to the plane defined by both the radius $\boldsymbol{r}$ and the velocity $\boldsymbol{v}$. Whereas the direction of the Magnetic-electric field strength line is the direction of the vector radius $\boldsymbol{r}$. The direction of the Magnetic-electric field strength $\boldsymbol{E_v}$ is determined by the right-hand rule.

Magnetic-electric field strength lines possess the following properties:

(1) Electric charge $\boldsymbol{q}$ is the source of Magnetic-electric field strength lines.

(2) Magnetic-electric field strength lines are not closed.

(3) Magnetic-electric field lines are continuous non-intersecting curves.

(4) The Magnetic-electric field strength $\boldsymbol{E_v}$ is not equal at each point on the Magnetic-electric field strength line.

(5) The direction of the Magnetic-electric field strength $\boldsymbol{E_v}$ at each point on the Magnetic-electric field strength line is perpendicular to the plane defined by both the velocity $\boldsymbol{v}$ and the radius $\boldsymbol{r}$.

(6) Every line of Magnetic-electric field strength is a vortex line.

Circular line of Magnetic-electric field strength

When the distance $\boldsymbol{r}$ from the electric charge $\boldsymbol{q}$ to each point of the circular line $\boldsymbol{L_R} = \boldsymbol{2\pi R}$ is equal, the Magnetic-electric field strength $\boldsymbol{E_v}$ is equal at each point on the circular line $\boldsymbol{L_R}$. In this book, the circular line with equal Magnetic-electric field strength $\boldsymbol{E_v}$ is defined as the Magnetic-electric field strength circular line. This Magnetic-electric field strength circular line is the circular dashed line in Figure 19-2.

The Magnetic-electric field strength circular line has the following properties:

(1), electric charge $\boldsymbol{q}$ is the source of Magnetic-electric field strength circular lines. The Magnetic-electric field strength circular line is a cluster of circular lines centered on the electric charge $\boldsymbol{q}$.

(2), The Magnetic-electric field strength $\boldsymbol{E_v}$ is equal at every point on the Magnetic-electric field strength circular line.

(3), The Magnetic-electric field strength circular lines are continuous non-intersecting curves.

(4), the direction of the Magnetic-electric field strength at each point on the Magnetic-electric field strength circle is perpendicular to the plane defined by both the velocity $\boldsymbol{v}$ and the radius $\boldsymbol{r}$. As shown in Figure 19-2.

(5), when the radius $\boldsymbol{r} > 1$, the closer the circular line is to the electric charge $\boldsymbol{q}$, the greater the Magnetic-electric field strength $\boldsymbol{E_v}$ at each point on that circular line.

19.3.3 Negative Curl $-\nabla\times\varphi_v$ of Magnetic-electric potential φ_v equals Magnetic-electric field strength E_v

When electric charge $\boldsymbol{q}$ is moving with velocity $\boldsymbol{v}$ in the cosmic vacuum frame of reference. According to equation (18.4.3), the Magnetic-electric potential $\boldsymbol{\varphi_v}$ generated by electric charge $\boldsymbol{q}$ is:

$$\boldsymbol{\varphi_v = +\frac{1}{4\pi\varepsilon_0}\frac{qv\sin\theta}{cr}}, \qquad r > 1(l_m) \tag{19.3.4}$$

Assume that the distance $\boldsymbol{r}$ from the motion electric charge $\boldsymbol{q}$ to each point on the circular line $\boldsymbol{L_R} = \boldsymbol{2\pi R}$ is equal. Since the Magnetic-electric potential $\boldsymbol{\varphi_v}$ is constant on the circular line $\boldsymbol{L_R}$, the integral of the Magnetic-electric potential $\boldsymbol{\varphi_v}$ along the circular line $\boldsymbol{L_R} = \boldsymbol{2\pi R}$ is:

$$\boldsymbol{\oint \varphi_v dl = \varphi_v L_R = +\frac{R}{2\varepsilon_0}\frac{qv\sin\theta}{cr}}, \qquad r > 1(l_m) \tag{19.3.5}$$

Based on the above equation and Stokes' theorem, the relationship equation can be obtained:

$$\boldsymbol{\oint \varphi_v dl = \iint \nabla\times\varphi_v\, dS = +\frac{R}{2\varepsilon_0}\frac{qv\sin\theta}{cr}}, \qquad r > 1(l_m) \tag{19.3.6}$$

The surface $\boldsymbol{S}$ in equation is bounded by the circular line $\boldsymbol{L_R} = \boldsymbol{2\pi R}$. We use the symbol $\boldsymbol{E_v}$ to denote the negative Curl $\boldsymbol{\nabla\times\varphi_v}$ of the Magnetic-electric potential $\boldsymbol{\varphi_v}$, i.e:

$$\boldsymbol{E_v = -\nabla\times\varphi_v} \tag{19.3.7}$$

This book defines Curl $\boldsymbol{E_v = -\nabla\times\varphi_v}$ as the Magnetic-electric field strength. Substituting the above equation into equation (19.3.6) yields the relation equation:

$$\boldsymbol{\iint E_v\, dS = -\frac{R}{2\varepsilon_0}\frac{qv\sin\theta}{cr}}, \qquad r > 1(l_m) \tag{19.3.8}$$

When the radius $\boldsymbol{R}$ of the circular line is kept constant, if the surface $\boldsymbol{S}$ is Wang's circular crown $\boldsymbol{S_W} = \boldsymbol{2\pi Rr}$, then Wang's circular crown $\boldsymbol{S_W}$ is just a function of the radius $\boldsymbol{r}$. Differentiating Wang's circular crown $\boldsymbol{S_W} = \boldsymbol{2\pi Rr}$ with respect to radius $\boldsymbol{r}$ yields the relation equation:

$$dS_W = 2\pi R dr$$

Substituting the above equation into equation (19.3.8) gives the relationship equation:

$$\iint E_v \cdot 2\pi R dr = -\frac{R}{2\varepsilon_0}\frac{qv\sin\theta}{cr}, \qquad r > 1(l_m) \tag{19.3.9}$$

Simplifying the above equation and eliminating the constant R yields the relational equation:

$$\int E_v dr = -\frac{1}{4\pi\varepsilon_0}\frac{qv\sin\theta}{cr}, \qquad r > 1(l_m) \tag{19.3.10}$$

Take the derivative of the above equation with respect to the radius r to obtain the relational equation:

$$E_v = +\frac{1}{4\pi\varepsilon_0}\frac{qv}{cr^2}\sin\theta, \qquad r > 1(l_m) \tag{19.3.11}$$

The above equation is the Magnetic-electric field strength equation (19.3.2). The above equation shows that the Magnetic-electric field strength E_v is proportional to the electric charge q and velocity v and inversely proportional to the distance r squared.

When the electric charge velocity $v = 0$, the static electric charge q_0 does not have the Magnetic-electric field strength E_v because the Magnetic-electric field strength $E_v = 0$.

Substituting equation (19.3.11) into equation (19.3.7) gives the relationship equation:

$$\nabla \times \varphi_v = -\frac{1}{4\pi\varepsilon_0}\frac{qv}{cr^2}\sin\theta, \qquad r > 1(l_m) \tag{19.3.12}$$

The above equation is the Curl $\nabla \times \varphi_v$ equation for the Magnetic-electric potential φ_v.

19.3.4 Derivation of magnetic flux density formulae

Dividing equation (19.3.11) by the speed of light c, gives the relational equation:

$$\frac{1}{c}E_v = +\frac{1}{4\pi\varepsilon_0}\frac{qv}{c^2r^2}\sin\theta, \qquad r > 1(l_m) \tag{19.3.13}$$

The command variable B is equal to $\frac{E_v}{c}$, i.e:

$$B = \frac{1}{c}E_v \tag{19.3.14}$$

Substituting equation (19.3.13) into the above equation gives the relationship equation:

$$B = +\frac{1}{4\pi\varepsilon_0}\frac{qv}{c^2r^2}\sin\theta, \qquad r > 1(l_m) \tag{19.3.15}$$

Since the vacuum dielectric constant ε_0 and the speed of light c are both constants, let the constant μ_0 be:

$$\mu_0 = \frac{1}{\varepsilon_0 c^2} \tag{19.3.16}$$

The constant μ_0 in equation is defined in magnetic field theory as the vacuum permeability.

Theoretically, only the dielectric constant ε_0 exists in a vacuum, not the magnetic permeability μ_0. Since physicists have not derived equation (19.3.15), physicists can only explain magnetic fields in terms of the vacuum permeability μ_0.

Substituting equation (19.3.16) into equation (19.3.15) gives the relationship equation:

$$B = \frac{1}{c}E_v = +\frac{\mu_0}{4\pi}\frac{qv}{r^2}\sin\theta, \qquad r > 1(l_m) \tag{19.3.17}$$

In classical theory, the above equation is the equation for the magnetic flux density produced by a moving electric charge q. The above equation shows that both the Magnetic-electric field strength E_v and the magnetic flux density B are magnetic fields generated by a moving electric charge q.

Taking the derivative of equation (19.3.17) with respect to time t yields the relation equation:

$$\frac{dB}{dt} = \frac{1}{c}\frac{dE_v}{dt} \tag{19.3.18}$$

The above equation shows that. The rate of change $\frac{dB}{dt}$ of magnetic flux density, is equal to the ratio of the rate of change $\frac{dE_v}{dt}$ of the Magnetic-electric field strength to the speed of light c.

19.3.5 Classification of electromagnetic field strengths

In the cosmic vacuum frame of reference, if the velocity $\boldsymbol{v} = \mathbf{0}$ of electric charge $\boldsymbol{q}$, then the electric field produced by the static electric charge $\boldsymbol{q_0}$ is the static electrostatic field $\boldsymbol{E_0}$. Since the velocity $\boldsymbol{v} = \mathbf{0}$ of electric charge $\boldsymbol{q}$ in the cosmic vacuum has uniqueness, the static electrostatic field strength $\boldsymbol{E_0}$ has uniqueness in the new physics. Alternatively, the magnitude of the static electrostatic field strength $\boldsymbol{E_0}$ is independent of the velocity $\boldsymbol{u}$ of the $\boldsymbol{S'}$ inertial system and the state of motion of the observer.

In classical electromagnetic field theory, however, the static electrostatic field strength $\boldsymbol{E'_0}$ is usually the electric field observed when the observer is at rest relative to the electric charge $\boldsymbol{q}$. Since electric charge $\boldsymbol{q}$ has a velocity $\boldsymbol{v'} = \mathbf{0}$ in different $\boldsymbol{S'}$ inertial systems, the static electrostatic field strength $\boldsymbol{E'_0}$ is not unique.

Alternatively, in classical electromagnetic field theory, the magnitude of the static electrostatic field strength $\boldsymbol{E'_0}$ is related only to the state of motion of the observer, and is independent of the velocity $\boldsymbol{u}$ of the $\boldsymbol{S'}$ inertial system in the cosmic vacuum.

Although the electric charge $\boldsymbol{q}$ is at rest in the $\boldsymbol{S'}$ inertial system, the electric charge $\boldsymbol{q}$ always moves with the motion of the $\boldsymbol{S'}$ inertial system. From this point of view, the static electrostatic field strength $\boldsymbol{E_0}$ used in this book is equal to the strength of the electrostatic field in the cosmic vacuum frame of reference and not equal to the strength of the electrostatic field in the $\boldsymbol{S'}$ inertial system $\boldsymbol{E'_0}$.

To summarize, electric charge $\boldsymbol{q}$ can be classified into the following four types.

(1) Static electric charge $\boldsymbol{q_0}$, an electric charge that is stationary in the cosmic vacuum .

(2) Motion electric charge $\boldsymbol{q}$, the electric charge moving in the cosmic vacuum.

(3) Rotating electric charge $\boldsymbol{q_\omega}$, is the electric charge contained in the moving electric charge $\boldsymbol{q}$, which generates the rotating electric field strength $\boldsymbol{E_\omega}$.

(4) Magnetic electric charge $\boldsymbol{q_v}$, is the electric charge contained in the moving electric charge $\boldsymbol{q}$, which generates the Magnetic-electric field strength $\boldsymbol{E_v}$.

According to the definition of electric field strength, electric field strength can be divided into the following five types.

(1) Static electrostatic field strength $\boldsymbol{E_0}$, is the electric field strength generated by staticelectric charge $\boldsymbol{q_0}$.

(2) Electromagnetic field strength $\boldsymbol{E}$, is the electric field strength generated by a moving electric charge $\boldsymbol{q}$.

(3) Rotating electric field strength $\boldsymbol{E_\omega}$, is the electric field strength generated by rotatingelectric charge $\boldsymbol{q_\omega}$.

(4) Magnetic-electric field strength $\boldsymbol{E_v}$, is the electric field generated by magnetic electric charge $\boldsymbol{q_v}$.

(5) Magnetic flux density $\boldsymbol{B}$, is the magnetic field produced by magnetic electric charge $\boldsymbol{q_v}$.

It should be noted that the magnetic flux density $\boldsymbol{B}$ is essentially the Magnetic-electric field strength $\boldsymbol{E_v}$, with a difference in physical units of one speed of light $\boldsymbol{c}$. Or they are two different manifestations of the same thing.

19.4 Magnetic flux density line and magnetic flux density circular line (Magnetic induction line)

19.4.1 Definition of magnetic flux density line

According to magnetic flux density equation (19.3.17), the magnetic flux density $\boldsymbol{B}$ produced by a moving electric charge $\boldsymbol{q}$ is:

$$\boldsymbol{B} = +\frac{\mu_0}{4\pi}\frac{\boldsymbol{q}\boldsymbol{v}}{\boldsymbol{r}^2}\sin\boldsymbol{\theta}\,, \qquad \boldsymbol{r} > 1(\boldsymbol{l_m})$$

When the radius $\boldsymbol{r}$ varies continuously from the position $\boldsymbol{r} = 1$ to infinity in one direction, or converges to the electric charge $\boldsymbol{q}$, these innumerable points of radius $\boldsymbol{r}$ form a straight line. Since each point on the line has magnetic flux density $\boldsymbol{B}$, this book defines the line as a magnetic flux density line. This magnetic flux density line is the solid line that intersects the circular line in Figure 19-3.

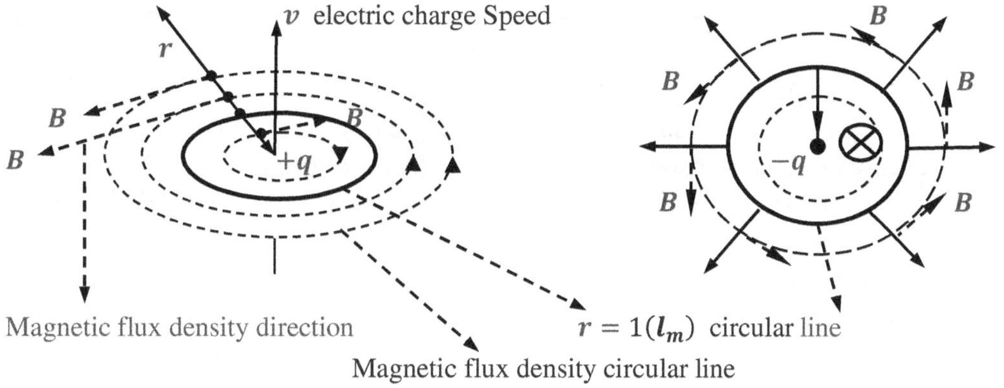

Figure 19-3 Schematic diagram of the magnetic flux density direction, the magnetic flux density line, and the magnetic flux density circular line.

The arrow on line B is the direction of magnetic flux density B, not the magnetic flux density line. The direction of magnetic flux density B is perpendicular to the plane defined by the velocity v and the radius r.

Note that the direction of the magnetic flux density $\boldsymbol{B}$ is perpendicular to the plane defined by both the radius $\boldsymbol{r}$ and the velocity $\boldsymbol{v}$. as shown in Figure 19-4.

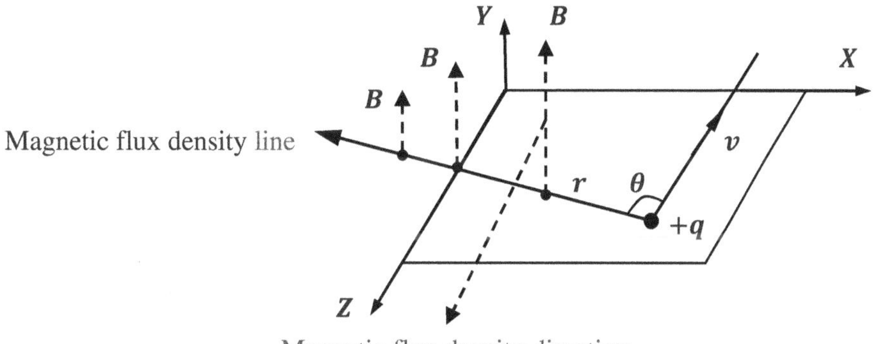

Figure 19-4 Schematic diagram of magnetic flux density line and magnetic flux density direction.

It is important to note that the magnetic flux density line defined in this book is not the magnetic induction line of classical theory. physicists define a magnetic induction line as being perpendicular to the plane defined by the velocity $\boldsymbol{v}$ and the radius $\boldsymbol{r}$. Physicists consider magnetic induction lines to be closed lines with no beginning and no end. the magnetic induction line is the circular line in Figure 19-3. The magnetic flux density line defined in this book is the straight line in Figure 19-3.

19.4.2 Magnetic induction line is essentially a magnetic flux density circular line

When the distance $\boldsymbol{r}$ from the electric charge $\boldsymbol{q}$ to each point of the circular line $\boldsymbol{L_R} = \boldsymbol{2\pi R}$ is equal, the magnetic flux density $\boldsymbol{B}$ is equal at each point on the circular line $\boldsymbol{L_R}$. In this book, a circular line with equal magnetic flux density $\boldsymbol{B}$ is defined as a magnetic flux density circular line. This circular line is the circular dashed line in Figure 19-3.

The magnetic flux density circular line has the following properties:

(1) The moving electric charge q is the source of the magnetic flux density circular line.

(2) The magnetic flux density circle line is a cluster of closed curves centered on the motion electric charge q.

(3) The direction of the tangent at each point on the magnetic flux density circle line is perpendicular to the vector radius r of the point.

(4) The magnetic flux density circle line is characterized by no intersections, no interruptions, and a continuous head and tail. Equal values of magnetic flux density circle lines can form a magnetic flux density sphere.

(5) The magnetic flux density of each point on the same magnetic flux density circle line is equal.

(6) When the radius $r > 1$, the closer the circular line is to the electric charge q, the greater the magnetic flux density B at each point on the circular line.

Since the direction of the tangent line at each point on the magnetic induction line coincides with the direction of the magnetic flux density at that point, the magnetic induction line as defined by physicists is a magnetic flux density circular line as defined in this book.

19.4.3 Integration $\oint Bdl$ of magnetic flux density B along circular line $L_R=2\pi R$

Assume that $L = f(r)$ is a curve in the electromagnetic field, and take the line element dl on the curve $L = f(r)$. according to equation (19.3.17) , the product Bdl of the magnetic flux density B and the line element dl is:

$$Bdl = \frac{\mu_0}{4\pi}\frac{q\sin\theta}{r^2}dl, \qquad r > 1(l_m)$$

Integrating the above equation along the curve $L = f(r)$ yields the relation equation:

$$\int Bdl = \int \frac{\mu_0}{4\pi}\frac{q\sin\theta}{r^2}dl \tag{19.4.1}$$

The above equation is the magnetic flux density line flow equation. The integral $\int Bdl$ for the magnetic flux density B is of two types.

The first type is an integral whose radius r is a variable.

Assume that the radius r is variable on the curve $L_X = f(r)$. Assume that the curve L_X starts at point a and ends at point b. According to equation (19.4.1), the integral of the magnetic flux density B along the curve $L_X = f(r)$ is:

$$\int_{r_a}^{r_b} Bdl = +\frac{\mu_0 qv\sin\theta}{4\pi}\int_{r_a}^{r_b}\frac{1}{r^2}dl = -\frac{\mu_0 qv\sin\theta}{4\pi}\left(\frac{1}{r_b}-\frac{1}{r_a}\right) \tag{19.4.2}$$

The above equation is the magnetic flux density line flow equation for a change in radius r.

The curve L_X is closed if the curve $L_X = f(r)$ goes from the starting point a to the point b and returns from the point b to the starting point a. The integral $\oint Bdl$ of the magnetic flux density B on the closed curve L_X is:

$$\oint Bdl = \int_{r_a}^{r_b} Bdl + \int_{r_b}^{r_a} Bdl = 0 \tag{19.4.3}$$

The above equation shows that the integral $\oint Bdl$ of the magnetic flux density B over the closed curve L_X is always equal to 0, if the radius r varies over the closed curve L_X.

The 2nd type is an integral whose radius r remains constant on the circular line $L_R = 2\pi R$.

If the distance r from the electric charge q to each point on the circular line $L_R = 2\pi R$ is equal, then the magnetic flux density B is constant on the circular line L_R. According to Equation (19.3.17), the integral $\oint Bdl$ of magnetic flux density B along the circular line $L_R = 2\pi R$ is:

$$\oint Bdl = BL_R = +2\pi R\cdot\frac{\mu_0}{4\pi}\frac{qv}{r^2}\sin\theta, \qquad r > 1(l_m) \tag{19.4.4}$$

When the angle $\theta = 90^0$ in Figure 18-1, the above equation becomes since the radius $R = r$:

$$\oint Bdl = +\frac{\mu_0}{2}\frac{qv}{r}, \qquad r > 1(l_m) \tag{19.4.5}$$

The above formula is the maximum formula for the magnetic flux density B circular line integral. At this point, the distance $Z = 0$, and the motion electric charge q is located in the center of the circular plane $P = \pi R^2$. The above equation shows that the maximum magnetic flux density circular line integral $\oint Bdl$ is proportional to the electric charge q and the velocity v, and inversely proportional to the radius r.

19.4.4 Magnetic flux density lines and magnetic induction lines produced by electric current *I*

According to electromagnetic field theory, the direction of the magnetic flux density B_I produced by the electric current I is perpendicular to the plane defined by the distance r and electric charge velocity v. as shown in Figure 19-5.

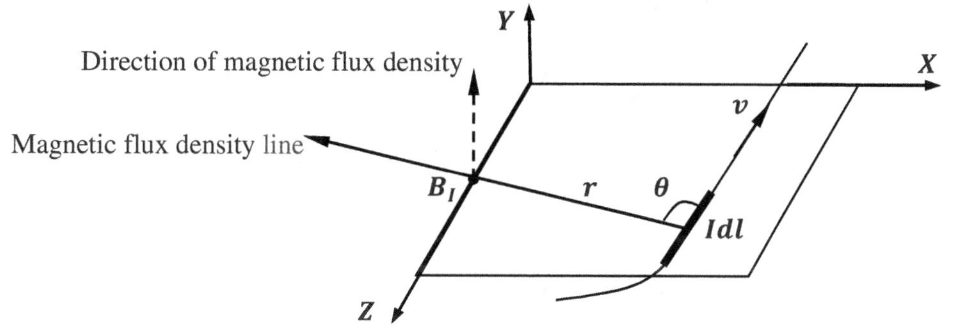

Figure 19-5 Schematic of the direction and magnetic flux density line of the magnetic flux density B_I generated by Current element Idl.

Note that the magnetic flux density line is the solid line with an arrow in Figure 19-5. the magnetic flux density B_I direction line is the dashed line with an arrow in Figure 19-5.

The magnetic induction line generated by the conductor electric current I is a magnetic flux density circular line centered on the conductor, i.e., the circular line in Figure 19-6.

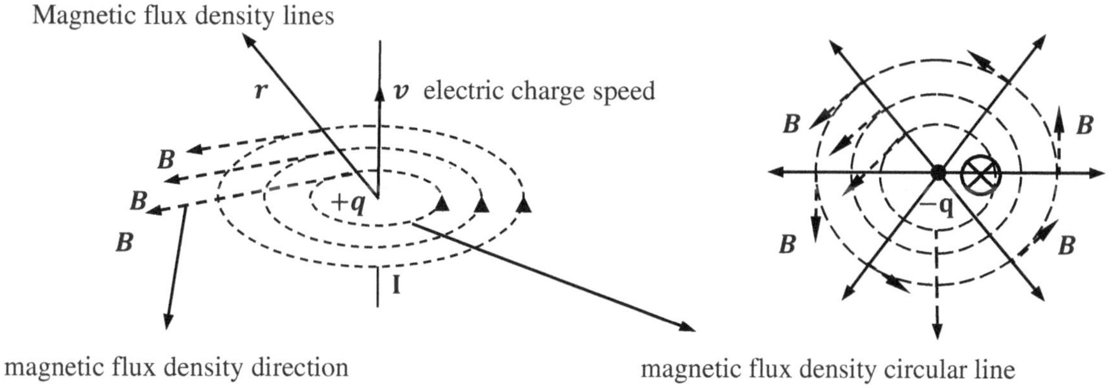

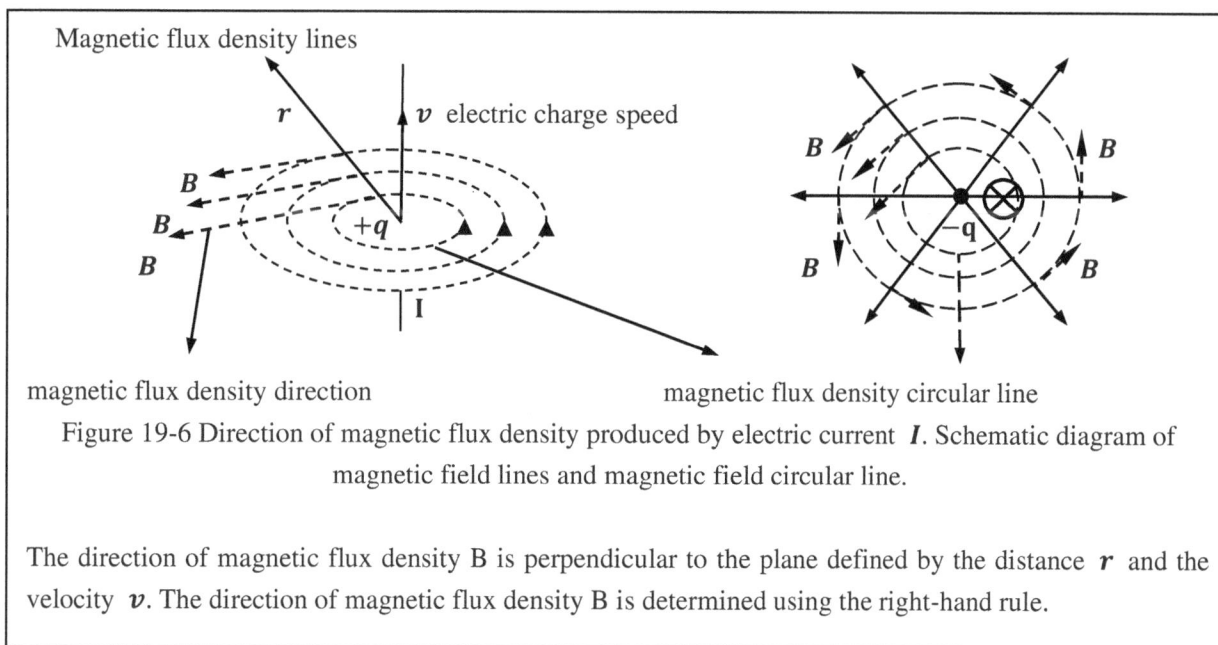

Figure 19-6 Direction of magnetic flux density produced by electric current I. Schematic diagram of magnetic field lines and magnetic field circular line.

The direction of magnetic flux density B is perpendicular to the plane defined by the distance r and the velocity v. The direction of magnetic flux density B is determined using the right-hand rule.

Note that different magnetic induction lines in the figure do not have equal magnetic flux densities, and that the magnetic flux density B_I at each point on each magnetic induction line is equal.

(1) Hold an energized wire in your right hand so that your thumb is pointing in the direction of the electric current I, and the four bent fingers are pointing in the direction of the magnetic flux density B_I.

(2) Hold the energizing solenoid in your right hand so that the four bent fingers point in the direction of electric

current I. Then the thumb is pointing in the direction of the N pole of the magnetic field inside the energizing solenoid.

According to the fact that electric current in the same direction attracts and electric current in the opposite direction repels, it can be determined that the magnetic field force F between two magnetic induction lines with the same direction of rotation is the suction force, and the magnetic field force F between two magnetic induction lines with the opposite direction of rotation is the repulsive force.

19.5 The displacement velocity v of the energy element ε_h is a prerequisite for the production of the magnetic field B

19.5.1 The plane defined by the electric charge velocity v and radius r is the interface dividing the N and S poles of the magnetic field

There are two different manifestations of the magnetic field B, one produced by a magnet and the other by electric current I (i.e., motion electric charge q). electric current I produces the magnetic field shown in Figure 19-7

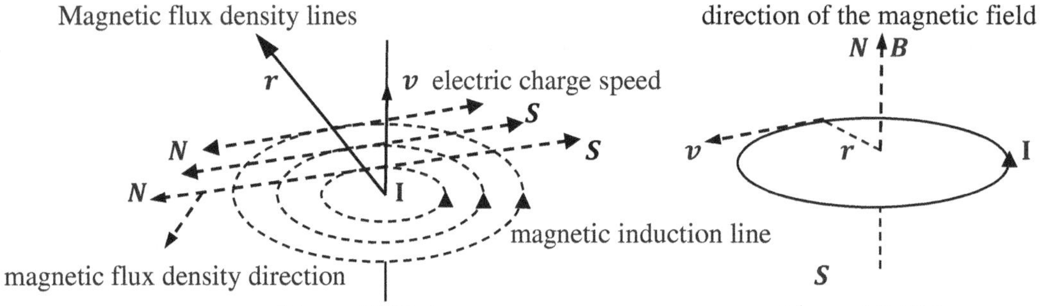

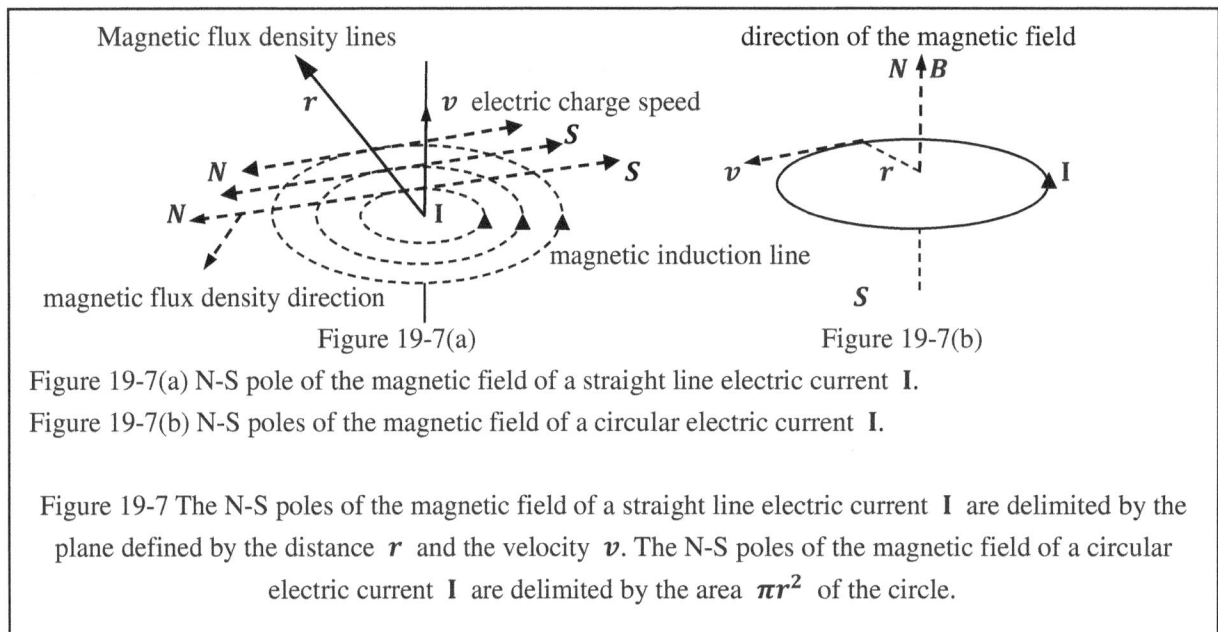

Figure 19-7(a) Figure 19-7(b)

Figure 19-7(a) N-S pole of the magnetic field of a straight line electric current **I**.

Figure 19-7(b) N-S poles of the magnetic field of a circular electric current **I**.

Figure 19-7 The N-S poles of the magnetic field of a straight line electric current **I** are delimited by the plane defined by the distance r and the velocity v. The N-S poles of the magnetic field of a circular electric current **I** are delimited by the area πr^2 of the circle.

According to Figure 19-7, the magnetic induction line generated by electric current I is the magnetic flux density circular line centered on electric current I. Different radii r correspond to different magnetic flux density circular lines.

If the direction of electric current I is changed, then the direction of magnetic field B becomes the opposite direction. The classical theory has 2 rules regarding the direction of the N pole of the magnetic field.

Provision 1: At any point in a magnetic field, the direction of force on the N pole of a small magnetic needle is the direction of the magnetic field at that point.

Rule 2: The direction of the N pole of the magnetic field is the direction of the tangent to the magnetic induction line.

Since the direction of the tangent at each point on the magnetic induction line is perpendicular to the plane defined by the electron velocity v and radius r, the direction of the N-pole of the magnetic field is perpendicular to the plane defined by both the radius r and the velocity v.

From this it can be determined that the plane defined by the velocity v and radius r of the electric charge is the interface that divides the N and S poles of the magnetic field. This is shown in Figure 19-7.

19.5.2 The displacement velocity v of the energy element ε_h is a prerequisite for the production of the magnetic field B

It is important to note that Figure 19-6 and 19-7 show a commonality in that the magnetic field N-pole direction is always perpendicular to the displacement direction of the energy element ε_h. Or rather, the magnetic field N-pole direction is always perpendicular to the displacement velocity v of the energy element ε_h. Since electrons are made up of the energy element ε_h, we can think analytically about the process of the formation of the magnetic field from the perspective of the energy element ε_h.

When an electron moves in the cosmic vacuum with speed v, according to the axiom of invariance of the speed of light for the energy element ε_h (Axiom 4), the spiral speed of light c of the energy element ε_h contained in the electron is:

$$c = \sqrt{v_\omega^2 + v^2} \tag{19.5.1}$$

The above equation shows that the energy element ε_h contains two directions of motion within it, one in the direction of the internal rotation speed v_ω and the other in the direction of the displacement velocity v.

Suppose r is the distance between the energy elements ε_{hA} and ε_{hB}. Assume that v_A is the displacement velocity of the energy element ε_{hA}, v_B is the displacement velocity of the energy element ε_{hB}, β is the angle between the displacement velocities v_A and v_B, and θ is the angle between the distance r and the displacement velocity v_A.

According to the directional force equation (13.5.8), the directional force $f_{\varepsilon h}$ between an energy element ε_{hA} and a neighboring energy element ε_h is:

$$f_{\varepsilon h} = T\frac{\varepsilon_{hA} \cdot \varepsilon_h}{r^2}\sin\theta\cos\beta$$

The T in equation is the directional force constant. If the angle $\beta = 0$ between the displacement velocity v_A and v_B, then the directional force $f_{\varepsilon h}$ between the energy elements ε_{hA} and ε_{hB} moving in the same direction is the suction force $f_{\varepsilon h(+)} > 0$.

If the angle $\beta = 180^0$ between the displacement velocities v_A and v_B, then the directional force $f_{\varepsilon h}$ between the energy elements ε_{hA} and ε_{hB} moving in the opposite direction is the repulsive force $f_{\varepsilon h(-)} < 0$.

If the angle $\beta = 90^0$ between the displacement velocity v_A and v_B, then the directional force $f_{\varepsilon h} = 0$ between the energy element ε_{hA} and ε_{hB} moving in perpendicular direction.

We know that the magnetic force F_B between electric currents in the same direction is a suction force and the magnetic force F_B between electric currents in the opposite direction is a repulsion force. Since electric current is formed by moving electrons, which are formed by the energy element ε_h, the displacement velocity v of the electrons is also the displacement velocity of the energy element ε_h.

If electric current I_A and electric current I_B are oriented in the same direction, then the free electrons e, and the energy element ε_h, in both electric current I_A and I_B, are displaced in the same direction with velocity v. Since the directional force $f_{\varepsilon h}$ between the energy element ε_h in electric current I_A and the energy element ε_h in electric current I_B is the suction force $f_{\varepsilon h(+)}$, the magnetic field force F_B between electric current I_A and I_B is the suction force.

If electric current I_A and electric current I_B are in opposite directions, then the energy elements ε_h in electric current I_A and I_B are moving in opposite directions.

Since the directional force $f_{\varepsilon h}$ between the energy element ε_h in electric current I_A and the energy element ε_h in I_B is repulsive $f_{\varepsilon h(-)}$, the magnetic field force F_B between electric current I_A and I_B is repulsive.

In summary, the displacement velocity v of the energy element ε_h is a prerequisite for the production of the magnetic field B.

19.5.3 The magnetic force F_I on an energized wire L in a horseshoe magnet can be explained using the directional force axiom for the energy element ε_h

According to electromagnetic field theory, the Lorentz magnetic field force F_q can change the direction of the motion electric charge q. The formula for the Lorentz magnetic force F_q is:

$$F_q = qvB\sin\theta \qquad (19.5.2)$$

The B in the formula is the uniform magnetic field, v is the velocity of the motion electric charge q in the uniform magnetic field, and θ is the angle between the velocity v and the magnetic field B.

The direction of the Lorentz forceF_q is determined by the left hand rule. The palm of the left hand is flattened, the thumb is perpendicular to the four fingers, so that the magnetic induction line passes through the palm of the hand (i.e., the palm of the hand is facing the N pole of the magnetic field), and the four fingers point in the direction of the motion of the positive electric charge $+q$, then the direction pointed to by the thumb is the direction of the Lorentz force F_q.

According to the electromagnetic field theory, the magnetic force F_I of the magnetic field B on the energized wire L is the amperometric force. In a uniform magnetic field B, the amperometric force F_I on the energized wire L is:

$$F_I = ILB\sin\theta \qquad (19.5.3)$$

The θ in the formula is the angle between the direction of electric current and the direction of magnetic field.

The direction of the amperometric force F_I is determined by the left hand rule. The palm of the left hand is flattened, the thumb is perpendicular to the four fingers, the magnetic induction line passes through the palm of the hand (i.e., the palm of the hand is facing the N pole of the magnetic field), and the four fingers point in the direction of the electric current I. The direction pointed to by the thumb is the direction of the amperometric force F_I. This is shown in Figure 19-8.

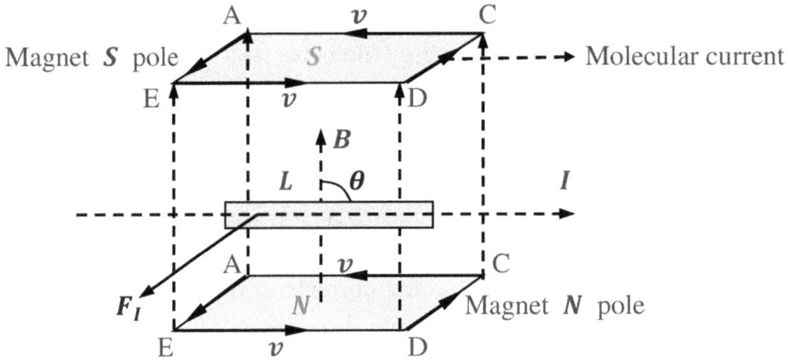

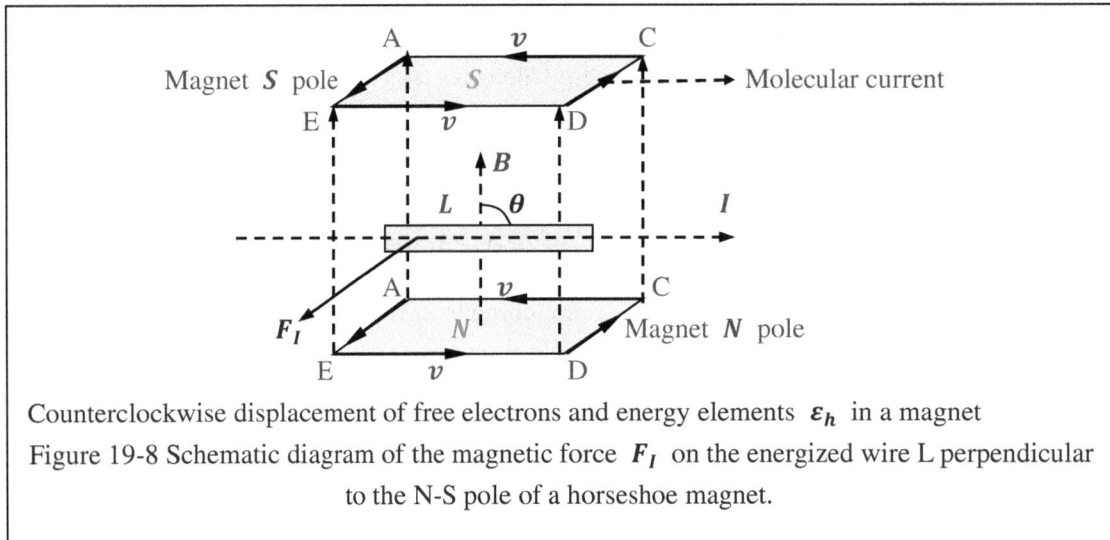

Counterclockwise displacement of free electrons and energy elements ε_h in a magnet

Figure 19-8 Schematic diagram of the magnetic force F_I on the energized wire L perpendicular to the N-S pole of a horseshoe magnet.

The energized wire L is placed in a uniform magnetic field, when the wire L is perpendicular to the direction of the magnetic field B (i.e., the angle $\theta = 90^0$, the amperometric force F_I on the wire L is maximal. When the wire L is oriented in the same direction as the magnetic field B (i.e., angle $\theta = 0$), the amperometric force $F_I = 0$ on the wire L.

We can use the axiom of directional force for the energy element ε_h to explain the force on an energized wire L in a horseshoe magnet. Assume that the molecularelectric current in the magnet are all rotating counterclockwise. Assume that the rectangular ACED electric current is the total electric current I_B formed by countless molecular electric currents in the magnet. Assume that the energized wire L is perpendicular to the direction of the magnetic field (i.e., angle $\theta = 90^0$). This is shown in Figure 19-8.

When the energized wire L is parallel to the AC line segment and the direction of electric current I is to the right. Since the displacement direction of the energy element ε_h in the upper and lower AE line segments and the upper and lower CD line segments are both perpendicular to the displacement direction of the energy element ε_h in the electric current I, the directional force $f_{\varepsilon h}$ of the energy element ε_h in the AE line segment and the CD line segment on the energized conductor L is equal to zero.

Since the direction of displacement of the energy element ε_h in the upper and lower AC line segments is opposite to the direction of displacement of the energy element ε_h in the electric current I, the directional forces $f_{\varepsilon h}$ of the energy element ε_h in the AC line segments on the energized conductor L are all repulsive forces.

If the energized wire L lies in the midline of the magnetic field, then the direction of the combined upper and lower repulsive forces $f_{\varepsilon h(-)}$, is pointing outward from the book. This is shown in Figure 19-8.

Since the displacement direction of the energy element ε_h in the upper and lower ED line segments is the same as that of the energy element ε_h in the electric current I, the directional force $f_{\varepsilon h}$ of the energy element ε_h in the ED line segments on the energized conductor L are both suction forces.

If the energized wire L is located in the midline of the magnetic field, then the direction of the combined upper and lower suction forces $f_{\varepsilon h(+)}$ is pointing out from the book. This is shown in Figure 19-8.

Since the directional force $f_{\varepsilon h}$ on the energized wire L by both the AE and ED line segments is directed vertically outward from the book, the combined force F of the directional forces on the energized wire L is directed vertically outward from the book. Obviously, the direction of this combined force F is the direction of the amperometric force.

Conversely, when the energized conductor L is parallel to the AC line segment and the direction of electric current I is to the left. The direction of the combined force F on the energized wire L is from the outside of the book to the inside of the book.

When the energized wire L is perpendicular to the ACDE plane, i.e., the direction of electric current I is upward or downward (i.e., the angle $\theta = 0$, or $\theta = 180^0$).

Since the direction of displacement of the energy element ε_h in the ACDE line segment is all perpendicular to the direction of displacement of the energy element ε_h in the electric current I, the directional force $f_{\varepsilon h}$ on the wire L is equal to zero.

At this time the magnetic force F on the energized wire L parallel to the direction of the magnetic field is always equal to zero.

Chapter 20 Derivation of the formula $F_B = qvB \cdot \sin\delta$ for the Lorentz magnetic force

Introduction: This chapter derives a formula for the Lorentzian magnetic force generated by the motion electric charge q based on the magnetic electric charge q_v formula.

20.1 Derivation of the formula for the electromagnetic field force F_{EB}

20.1.1 Magnetic-electric field force F_v between moving electric charges q_1 and q_2

Assume that the velocities of the motions electric charge q_1 and q_2 in the cosmic vacuum are v_1 and v_2, respectively.

Suppose r is the distance between the motions electric charge q_1 and q_2.

Suppose θ is the angle between distance r and velocity v_1, and β is the angle between velocity v_2 and velocity v_1.

Assume that the distance r, the velocities v_1, v_2, and the angles θ and β are all in the vertical plane. This is shown in Figure 20-1.

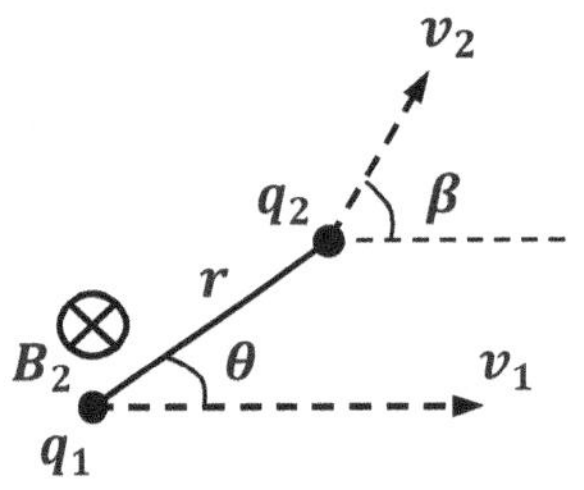

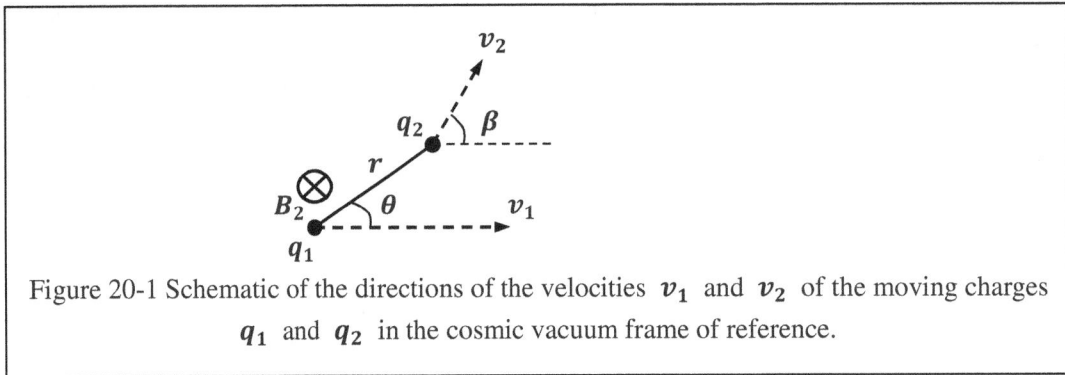

Figure 20-1 Schematic of the directions of the velocities v_1 and v_2 of the moving charges q_1 and q_2 in the cosmic vacuum frame of reference.

According to the magnetic flux density B equation, the magnetic field B_2 generated by the motion electric charge q_2 at the position of electric charge q_1 is:

$$B_2 = \frac{\mu_0}{4\pi}\frac{q_2 v_2}{r^2}\sin\theta \qquad (20.1.1)$$

Since the direction of the magnetic field B_2 is perpendicular to the plane defined by both the velocity v_2 and the radius r, the velocity v_1 is perpendicular to the direction of the magnetic field B_2. This is shown in Figure 20-1.

According to equation (18.3.4), the magnetic electric charge q_{v1} and q_{v2} generated by the motion electric charges q_1 and q_2, respectively, are:

$$q_{v1} = q_1\frac{v_1}{c}, \qquad q_{v2} = q_2\frac{v_2}{c} \qquad (20.1.2)$$

According to equation (19.3.1), the Magnetic-electric fields E_{v1} and E_{v2} generated by the motion electric charges q_1 and q_2, respectively, are:

$$E_{v1} = \frac{1}{4\pi\varepsilon_0}\frac{q_{v1}}{r^2}\sin\theta\,, \quad E_{v2} = \frac{1}{4\pi\varepsilon_0}\frac{q_{v2}}{r^2}\sin\theta \tag{20.1.3}$$

Since the function $\sin\theta$ in equation has no physical units, the physical units of the Magnetic-electric fields E_{v1} and E_{v2} are the same as the physical units of the electric field strength E.

According to the electric field theory, the electric field force F received by the electric charge q in the electric field E is:

$$F = qE \tag{20.1.4}$$

Multiply the magnetic charge q_{v1} by the magnetic-electric field E_{v2}. Assuming that the velocities v_1 and v_2 are moving in the same direction, and since the angle $\beta = 0$, the magnetic-electric field force F_v between the magnetic charges q_{v1} and q_{v2} is:

$$F_v = q_{v1}E_{v2} = \frac{1}{4\pi\varepsilon_0}\frac{q_{v1}\cdot q_{v2}}{r^2}\sin\theta \tag{20.1.5}$$

This book defines the electric field force F_v between the magnetic electric charge q_{v1} and q_{v2} as the Magnetic-electric field force. Substituting equation (20.1.2) into the above equation yields the relational equation:

$$F_v = \frac{1}{4\pi\varepsilon_0 c^2}\frac{q_1v_1\cdot q_2v_2}{r^2}\sin\theta \tag{20.1.6}$$

The command constant μ_0 is:

$$\mu_0 = \frac{1}{\varepsilon_0 c^2} \tag{20.1.7}$$

In classical theory, the constant μ_0 is defined as the vacuum magnetic permeability. Since physicists need to use the vacuum permeability μ_0 in order to derive the electromagnetic wave equation, the vacuum permeability μ_0 is considered to be an objectively existing physical quantity.

However, since this book derives the electromagnetic wave equation without using the vacuum permeability μ_0, it can be determined from equation (20.1.7) that the vacuum permeability μ_0 is a physical quantity subjectively specified by physicists and is not an objectively real physical quantity.

Substituting equation (20.1.7) into equation (20.1.6) gives the relationship equation:

$$F_v = \frac{\mu_0}{4\pi}\frac{q_1v_1\cdot q_2v_2}{r^2}\sin\theta \tag{20.1.8}$$

Substituting equation (20.1.1) into the above equation gives the relationship equation:

$$F_v = q_1v_1B_2 \tag{20.1.9}$$

According to Figure 20-1, the velocity v_1 is perpendicular to the direction of the magnetic field B_2. Assuming that δ is the angle between the velocity v_1 and the magnetic field B_2, the above equation becomes:

$$F_v = q_1v_1B_2\sin\delta \tag{20.1.10}$$

When the velocity v_1 is perpendicular to the direction of the magnetic field B_2, the above equation becomes equation (20.1.9) since the angle $\delta = 90^0$.

According to the magnetic field theory, the Lorentzian magnetic force F_B on the motion electric charge q in a uniform magnetic field B is:

$$F_B = qvB\sin\delta$$

The v in the formula is the velocity of the electric charge q, and δ is the angle between the velocity v and the magnetic field B. Comparing the above equation with equation (20.1.10), it can be determined that the Lorentzian magnetic field force F_B is essentially the Magnetic-electric field force F_v.

Since electric current I is generated by free electron moves, we can think of the electric charge q_1 and q_2 moves as electric current I_1 and I_2.

Assume that β is the angle between velocity v_1 and velocity v_2. Since the direction of electric current I is the direction of motion of electric charge q, the angle β is the angle between electric current I_1 and I_2.

When the electric current I_1 and I_2 in the two wires are in the same direction, the magnetic force $F_B > 0$ (suction force) between the two wires. At this time, since the electric charge q_1 and q_2 are moving in the same direction, the angle $\beta = 0$ between the electric charge velocity v_1 and v_2.

When the direction of electric current I_1 and I_2 is opposite, the magnetic force $F_B < 0$ (repulsive force) between the two wires. At this time, since the electric charge q_1 and q_2 are moving in opposite direction, the angle $\beta = 180^0$ between the electric charge velocity v_1 and v_2.

When the direction of electric current I_1 and I_2 are perpendicular, the magnetic force between the two wires $F_B = 0$. At this time, since the direction of motion of electric charge q_1 and q_2 are perpendicular to each other, the angle $\beta = 90^0$ between the electric charge velocity v_1 and v_2.

According to the above facts, the Magnetic-electric field force F_v between two motion electric charge q_1 and q_2 is:

$$F_v = \frac{1}{4\pi\varepsilon_0 c^2}\frac{q_1 v_1 \cdot q_2 v_2}{r^2}\sin\theta\cos\beta \qquad (20.1.11)$$

The above formula is the Magnetic-electric field force formula for the motion electric charge. In the formula, β is the angle between the velocity v_1 and v_2, and θ is the angle between the velocity v_1 and the distance r.

Since the electric charge q has a positive electric charge $+q$ and a negative electric charge $-q$, and the magnetic force between the same kind of electric charges moving in the same direction is a suction force ($F_B > 0$), equation (20.1.11) is taken to have a positive sign.

Further, assume that the motions electric charge q_1 and q_2 are the same kind of electric charge .

When the angle $\beta = 0$, the electric charges q_1 and q_2 are moving in the same direction. At this time, the magnetic force $F_B > 0$ (suction force) between electric charge q_1 and q_2. For example, the electromagnetic force between electric current in the same direction is suction.

When the angle $\beta = 180^0$, the electric charge q_1 and q_2 are moving in opposite direction. At this time, the magnetic force $F_B < 0$ (repulsive force) between electric charge q_1 and q_2. For example the electromagnetic force between electric current in opposite direction is repulsive force.

When the angle $\beta = 90^0$, the motion direction of electric charge q_1 and q_2 are perpendicular to each other, and the magnetic field force $F_B = 0$ between electric charge q_1 and q_2 at this time.

Substituting the vacuum permeability equation (20.1.7) into equation (20.1.11), the Magnetic-electric field force F_v becomes:

$$F_v = F_B = \frac{\mu_0}{4\pi}\frac{q_1 v_1 \cdot q_2 v_2}{r^2}\sin\theta\cos\beta \qquad (20.1.12)$$

The above equation is the magnetic field force equation for a moving electric charge. The above equation shows that the magnetic field force F_B between moving electric charges is equal to the Magnetic-electric field force F_v.

20.1.2 Rotating electric field force F_ω between moving electric charge q_1 and q_2

According to equation (18.3.3), the rotational electric charges $q_{\omega 1}$ and $q_{\omega 2}$ generated by the motion electric charges q_1 and q_2, respectively, are:

$$q_{\omega 1} = q_1\sqrt{1-\frac{v_1^2}{c^2}}, \qquad q_{\omega 2} = q_2\sqrt{1-\frac{v_2^2}{c^2}} \qquad (20.1.13)$$

According to equation (19.2.9), the rotating electric fields $E_{\omega 1}$ and $E_{\omega 2}$ generated by the motionelectric charges q_1 and q_2, respectively, are:

$$E_{\omega 1} = -\frac{1}{4\pi\varepsilon_0}\frac{q_{\omega 1}}{r^2}\sin\theta, \qquad E_{\omega 2} = -\frac{1}{4\pi\varepsilon_0}\frac{q_{\omega 2}}{r^2}\sin\theta \qquad (20.1.14)$$

Multiply the rotating electric field $E_{\omega 2}$ by the rotating electric charge $q_{\omega 1}$. If the effect of the angle β on the electric field force is not taken into account, then the electric field force F_ω between the rotating electric charge $q_{\omega 1}$ and $q_{\omega 2}$ is:

$$F_\omega = q_{\omega 1}E_{\omega 2} = -\frac{1}{4\pi\varepsilon_0}\frac{q_{\omega 1}q_{\omega 2}}{r^2}\sin\theta \qquad (20.1.15)$$

This book defines the electric field force F_ω between the rotating electric charge q_ω as the rotating electric field force.

According to the axiom of directional force of energy element ε_h (Axiom 5), when energy elements ε_{h1} and

ε_{h2} move perpendicular to each other, the directional force $f_{\varepsilon h}=0$ between energy elements ε_{h1} and ε_{h2}.

When electric charge q_1 and q_2 move perpendicular to each other. Since the direction of motion of the energy element ε_{h1} in the electric charge q_1 is perpendicular to the direction of motion of the energy element ε_{h2} in the electric charge q_2, the directional force $f_{\varepsilon h}=0$ between the motion electric charge q_1 and q_2.

At this time, since the angle $\beta=90^0$ between the velocities v_1 and v_2, the Magnetic-electric field force $F_v=0$, and the rotational electric field force $F_\omega=0$. From this, we can determine the rotational electric field force F_ω between the motion electric charge q_1 and q_2 is:

$$F_\omega=-\frac{1}{4\pi\varepsilon_0}\frac{q_{\omega1}q_{\omega2}}{r^2}\sin\theta\cos\beta\,,\qquad r>1(l_m)\tag{20.1.16}$$

Substituting equation (20.1.13) into the above equation gives the relationship equation:

$$F_\omega=-\frac{1}{4\pi\varepsilon_0}\sqrt{1-\frac{v_1^2}{c^2}}\sqrt{1-\frac{v_2^2}{c^2}}\cdot\frac{q_1q_2}{r^2}\sin\theta\cos\beta\,,\qquad r>1(l_m)\tag{20.1.17}$$

The above equation is the rotating electric field force equation. Since electric charge q has positive electric charge $+q$ and negative electric charge $-q$, the above equation takes a negative sign. The above equation shows that the magnitude of the rotating electric field force F_ω is proportional to the electric charge q and inversely proportional to the square of the radius r. The rotating electric field force F_ω becomes smaller as the velocity v increases.

Note that when the same kind of electric charge moves in the same direction, the rotational electric field force F_ω between the same kind of electric charge is a suction force ($F_\omega>0$). On the contrary, the rotational electric field force F_ω between electric charges of the same type is repulsive ($F_\omega<0$).

20.1.3 Absence of electric field force between magnetic electric charge q_v and rotating electric charge q_ω

According to equation (20.1.13) and equation (20.1.2), the rotational electric charge $q_{\omega1}$ and the magnetic electric charge q_{v1} generated by the moving electric charge q_1 are, respectively:

$$q_{\omega1}=q_1\sqrt{1-\frac{v_1^2}{c^2}}\,,\qquad q_{v1}=q_1\frac{v_1}{c}$$

When the electric charge velocity $v_1=0$, the rotational electric charge $q_{\omega1}$ is equal to the static electric charge q_{01} (i.e., $q_{\omega1}=q_1=q_{01}$), and the magnetic electric charge $q_{v1}=0$.

At this point the electric field $E_{\omega1}$ generated by electric charge q_1 is the electrostatic field E_{01} and the magnetic field $B_1=0$ generated by electric charge q_1.

From this it can be determined that the rotating electric charge q_ω produces the kinematic electric field E_ω and the magnetic electric charge q_v produces the magnetic field B.

According to equation (19.3.17), the magnetic field B_1 is related to the Magnetic-electric field E_{v1} by the equation:

$$cB_2=E_{v2}\tag{20.1.18}$$

Multiply the above equation by static electric charge q_{01}. to get the relationship equation:

$$F_{\omega v}=q_{01}cB_2=q_{01}E_{v2}\tag{20.1.19}$$

The electric field force $F_{\omega v}=q_{01}cB_2$ is analogous to the Lorentz force $F_B=qvB$. It should be noted that q_{01} in equation $F_{\omega v}=q_{01}cB_2$ is the static electric charge and q in equation $F_B=qvB$ is the moving electric charge.

Since the magnetic field B_2 exerts no magnetic force on the static electric charge q_{01}, the electric field force $F_{\omega v}$ is equal to 0, i.e:

$$F_{\omega v}=q_{01}cB_2=q_{01}E_{v2}=0\tag{20.1.20}$$

Substituting the Magnetic-electric field E_{v2} from equation (20.1.3), into the above equation, gives the relational equation:

$$F_{\omega v}=\frac{1}{4\pi\varepsilon_0}\frac{q_{01}q_{v2}}{r^2}\sin\theta=0 \qquad (20.1.21)$$

Since the rotating electric charge $q_{\omega 1}$ possesses the properties of a static electric charge (i.e., it can generate an electric field E), the Magnetic-electric field force $F_{\omega v}$ on the rotating electric charge $q_{\omega 1}$ in the Magnetic-electric field E_{v2} is equal to 0, i.e:

$$F_{\omega v}=q_{\omega 1}E_{v2}=\frac{1}{4\pi\varepsilon_0}\frac{q_{\omega 1}q_{v2}}{r^2}\sin\theta=0 \qquad (20.1.22)$$

The above equation shows that there is no electric field force between the rotational electric charge q_ω and the magnetic electric charge q_v.

Since equation (20.1.22) contains both the Magnetic-electric field $E_{v2}=\frac{1}{4\pi\varepsilon_0}\frac{q_{v2}}{r^2}\sin\theta$ and the rotating electric field $E_{\omega 1}=\frac{1}{4\pi\varepsilon_0}\frac{q_{\omega 1}}{r^2}\sin\theta$, the rotating electric field force $F_{\omega v}$ that the magnetic electric charge q_{v2} is subjected to in the rotating electric field $E_{\omega 1}$ is also equal to $\mathbf{0}$, i.e:

$$F_{\omega v}=q_{\omega 1}E_{v2}=q_{v2}E_{\omega 1}=0 \qquad (20.1.23)$$

Since the rotating electric charge q_ω and the magnetic electric charge q_v are contained within the moving electric charge q, the distance $r=0$ between the rotating electric charge q_ω and the magnetic electric charge q_v.

If there is an electric field force F between the magnetic electric charge q_v and the rotating electric charge q_ω, then there should be an infinite electric field force within the electron.This situation is clearly untrue.

20.1.4 Equation for the electromagnetic field force of a moving electric charge *q*

When electric charge q_1 is moving with velocity v_1 in the cosmic vacuum frame of reference. According to equation (19.2.4), the electromagnetic field strength E_1 produced by the moving electric charge q_1 is:

$$E_1=-\sqrt{(E_{\omega 1})^2+(E_{v1})^2} \qquad (20.1.24)$$

The $E_{\omega 1}$ in equation is the rotational electric field strength generated by the motion electric charge q_1, and E_{v1} is the Magnetic-electric field strength generated by the motion electric charge q_1.

Suppose F_{EB} is the electromagnetic field force on the motionelectric charge q_2 in the electromagnetic field E_1. Since the electromagnetic field E_1 contains the rotating electric field $E_{\omega 1}$ and the Magnetic-electric field E_{v1}, the electromagnetic field force F_{EB} contains the rotating electric field force $F_\omega=q_\omega E_\omega$ and the Magnetic-electric field force $F_v=q_vE_v$.

According to the electric field force equation $F=qE$, there are two ways to calculate the electromagnetic field force F_{EB}.

One is based on the rotational electric charge $q_{\omega 2}$ and the magnetic electric charge q_{v2}, i.e:

$$F_{EB}=q_{\omega 2}E_1+q_{v2}E_1 \qquad (20.1.25)$$

The other is calculated based on the motionelectric charge q_2, viz:

$$F_{EB}=q_2E_1=E_1\sqrt{q_{\omega 2}^2+q_{v2}^2} \qquad (20.1.26)$$

It should be noted that the electromagnetic field force F_{EB} calculated from the above equation is wrong. We can prove this conclusion by the following analysis....

If the directions of motion of electric charges q_1 and q_2 are the same, then the angle $\beta=0$ between the directions of motion of electric charges q_1 and q_2. According to equation (20.1.17), the rotational electric field force F_ω between the moving electric charges q_1 and q_2 is:

$$F_\omega=-\frac{1}{4\pi\varepsilon_0}\sqrt{1-\frac{v_1^2}{c^2}}\sqrt{1-\frac{v_2^2}{c^2}}\cdot\frac{q_1q_2}{r^2}\sin\theta\cos\beta\,,\quad r>1(l_m) \qquad (20.1.27)$$

Since the rotating electric field force F_ω contains only one vector (i.e., radius r), the rotating electric field force $F_\omega<0$ (repulsive force) between electric charges $+q_1$ and $+q_2$. The direction of the repulsive force F_ω is from $+q_1$ to $+q_2$. as shown in Figure 20-2.

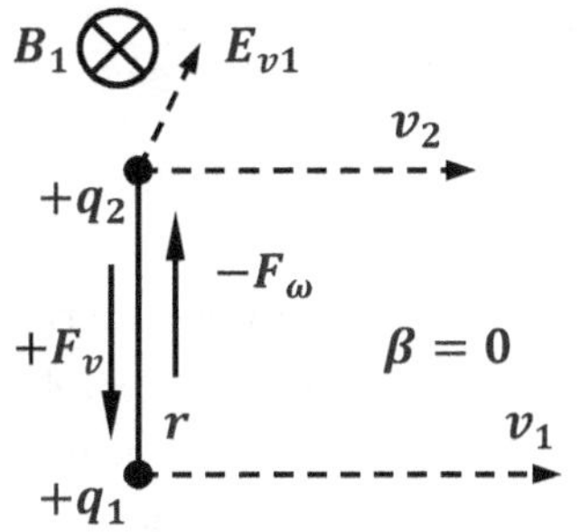

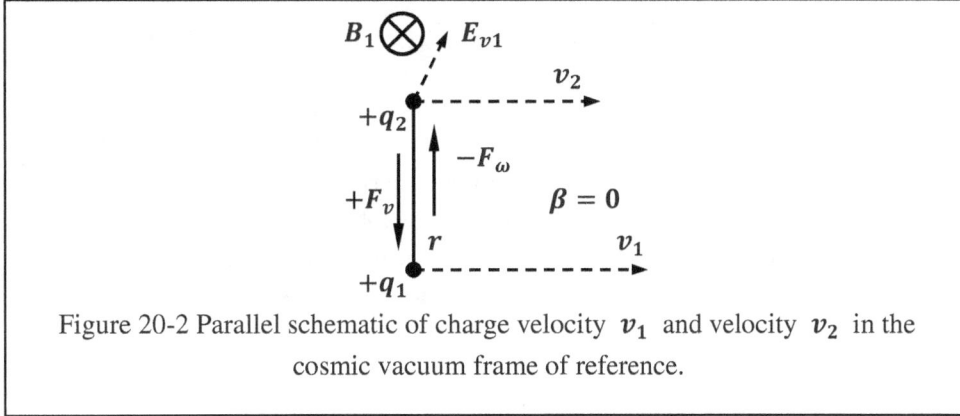

Figure 20-2 Parallel schematic of charge velocity v_1 and velocity v_2 in the cosmic vacuum frame of reference.

According to equation (20.1.11), the Magnetic-electric field force F_v between the motionelectric charge q_1 and q_2 is:

$$F_v = \frac{1}{4\pi\varepsilon_0 c^2}\frac{q_1 v_1 \cdot q_2 v_2}{r^2}\sin\theta\cos\beta$$

Assume that the velocity v_1 is parallel to the velocity v_2. Since the angle $\beta = 0$, the above equation becomes:

$$F_v = \frac{1}{4\pi\varepsilon_0 c^2}\frac{q_1 v_1 \cdot q_2 v_2}{r^2}\sin\theta \qquad (20.1.28)$$

According to equation (19.3.2), the Magnetic-electric field strength E_{v1} generated by the motion electric charge q_1 is:

$$E_{v1} = +\frac{1}{4\pi\varepsilon_0}\frac{q_1 v_1}{cr^2}\sin\theta \qquad (20.1.29)$$

Since the Magnetic-electric field strength E_{v1} contains two vectors, the velocity v_1 and the radius r, the direction of the Magnetic-electric field strength E_{v1} is perpendicular to the plane defined by the velocity v_1 and radius r. Substituting the above equation into equation (20.1.28), the Magnetic-electric field force F_v between the moving electric charges q_1 and q_2 is:

$$F_v = q_2 v_2 \frac{E_{v1}}{c} \qquad (20.1.30)$$

Since the Magnetic-electric field force F_v contains two vectors, the velocity v_2 and the Magnetic-electric field strength E_{v1}, the Magnetic-electric field force $F_v > 0$ (suction force) between electric charges $+q_1$ and $+q_2$. The direction of the suction force F_v is from $+q_2$ to $+q_1$. as shown in Figure 20-2.

Since both rotational electric field force F_ω and Magnetic-electric field force F_v are located on the line between electric charges $+q_1$ and $+q_2$, and the rotational electric field force F_ω between the same charges is repulsive (i.e., $F_\omega < 0$), the electromagnetic field force F_{EB} between the motion electric charge $+q_1$ and $+q_2$ is:

$$F_{EB} = F_v - F_\omega \qquad (20.1.31)$$

From this it can be determined that the electromagnetic field force $F_{EB} = q_2 E_1$ of equation (20.1.26) is incorrect.

According to equation (20.1.11) the Magnetic-electric field force F_v is:

$$F_v = \frac{1}{4\pi\varepsilon_0 c^2}\frac{q_1 v_1 \cdot q_2 v_2}{r^2}\sin\theta\cos\beta$$

According to equation (20.1.17) the rotating electric field force F_ω is:

$$F_\omega = -\frac{1}{4\pi\varepsilon_0}\sqrt{1-\frac{v_1^2}{c^2}}\sqrt{1-\frac{v_2^2}{c^2}}\cdot\frac{q_1 q_2}{r^2}\sin\theta\cos\beta$$

Substituting the Magnetic-electric field force F_v and the rotational electric field force F_ω into equation (20.1.31) gives the relation equation:

$$F_{EB}=\frac{1}{4\pi\varepsilon_0}\frac{q_1q_2\sin\theta\cos\beta}{r^2}\left(\frac{v_1\cdot v_2}{c^2}-\sqrt{1-\frac{v_1^2}{c^2}}\sqrt{1-\frac{v_2^2}{c^2}}\right) \qquad (20.1.32)$$

The above formula is the electromagnetic field force formula for the motion electric charge. In the formula, θ is the angle between the velocity v_1 and the distance r, and β is the angle between the velocities v_1 and v_2. Since the electric current I in the wire is generated by the motion electric charge q, the above formula can be used to explain the force between the energized wires from a microscopic point of view.

When the electric current I_1 and I_2 in the two wires are in the same direction, the angle $\beta=0$ between the directions of motion of electric charge q_1 and q_2. According to equation (20.1.32), the electromagnetic field force F_{EB} between the moving electric charge q_1 and q_2 is:

$$F_{EB}=\frac{1}{4\pi\varepsilon_0}\frac{q_1q_2\sin\theta}{r^2}\left(\frac{v_1\cdot v_2}{c^2}-\sqrt{1-\frac{v_1^2}{c^2}}\sqrt{1-\frac{v_2^2}{c^2}}\right) \qquad (20.1.33)$$

Since the velocities v_1 and v_2 of the moving electric charge are large, $\left(\frac{v_1\cdot v_2}{c^2}-\sqrt{1-\frac{v_1^2}{c^2}}\sqrt{1-\frac{v_2^2}{c^2}}\right)>0$ in equation. At this point, the electromagnetic field force $F_I>0$ (suction force) between the electric current I_1 and I_2 in the same direction, since the electromagnetic field force $F_{EB}>0$ between the electric charge q_1 and q_2.

When the electric current I_1 and I_2 in the two wires are in opposite directions, the angle $\beta=180^0$ between the directions of motion of electric charge q_1 and q_2. According to equation (20.1.32), the electromagnetic field force F_{EB} between the motion electric charge q_1 and q_2 is:

$$F_{EB}=-\frac{1}{4\pi\varepsilon_0}\frac{q_1q_2\sin\theta}{r^2}\left(\frac{v_1\cdot v_2}{c^2}-\sqrt{1-\frac{v_1^2}{c^2}}\sqrt{1-\frac{v_2^2}{c^2}}\right) \qquad (20.1.34)$$

Since the electromagnetic field force $F_{EB}<0$ between the opposite direction motion electric charge q_1 and q_2, the electromagnetic field force F_I between the two opposite direction electric current I_1 and I_2 is repulsive.

20.2 The spin velocity v_ω of a moving electron is a prerequisite for the generation of the rotational electric field force F_ω

20.2.1 Rotating electric charge q_ω produces a rotating electric field force F_ω is the electric field force of motion

According to equation (20.1.17), the rotating electric field force F_ω between two moving electric charges q_1 and q_2 is:

$$F_\omega=-\frac{1}{4\pi\varepsilon_0}\sqrt{1-\frac{v_1^2}{c^2}}\sqrt{1-\frac{v_2^2}{c^2}}\cdot\frac{q_1q_2}{r^2}\sin\theta\cos\beta \qquad (20.2.1)$$

In the formula, θ is the angle between the velocity v_1 and the distance r, β is the angle between the velocity v_1 and v_2. If the angle $\theta=90^0$ and angle $\beta=0$ (or $\beta=180^0$), then the rotating electric field force F_ω reaches the maximum value, i.e:

$$F_\omega=\mp\frac{1}{4\pi\varepsilon_0}\sqrt{1-\frac{v_1^2}{c^2}}\sqrt{1-\frac{v_2^2}{c^2}}\cdot\frac{q_1q_2}{r^2} \qquad (20.2.2)$$

When the velocity $v_1=v_2=0$, the above equation becomes the electrostatic field force equation. From this, it can be determined that the rotating electric field force F_ω generated by the rotating electric charge q_ω is the electric field force of motion.

If the angle $\theta=0$, or the angle $\beta=90^0$, then the rotating electric field force F_ω is equal to 0, i.e:

$$F_\omega = 0 \tag{20.2.3}$$

Assume that the moving electric charges q_1 and q_2 are the same kind of electric charge. When the angle $\beta = 0$, the two electric charges are moving in the same direction. At this time, the rotational electric field force F_ω is repulsive because the rotational electric field force $F_\omega < 0$ between electric charges q_1 and q_2. This is shown in Figure 20-3a.

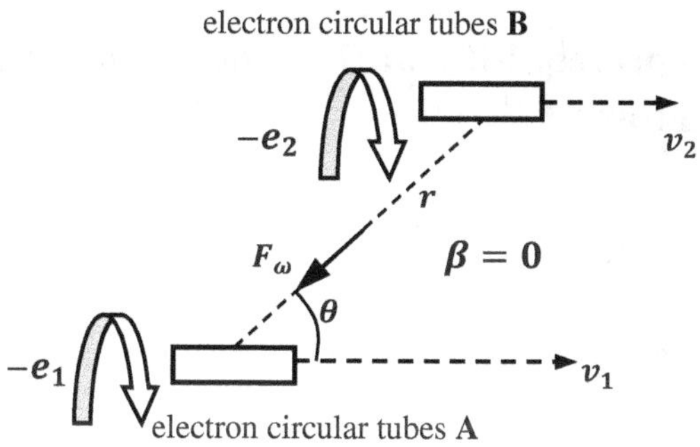

Figure 20-3a Schematic diagram of same direction displacement of electron circular tubes A and B.

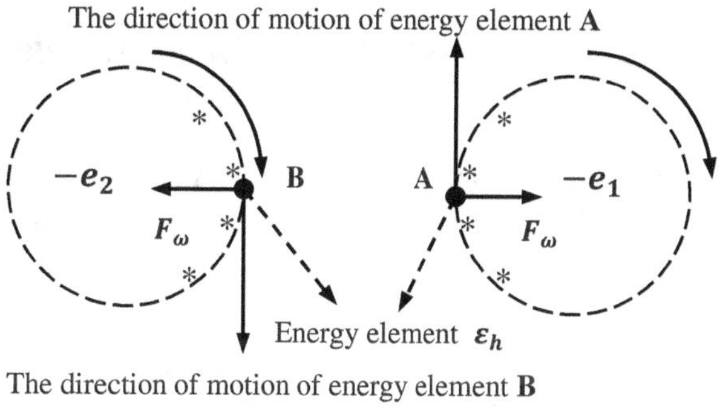

Figure 20-3b Schematic diagram of electron circular tubes A and B spinning in the same direction.

Although the rotational electric field force F_ω between electric charges q_1 and q_2 is a repulsive force, the electromagnetic force F_{EB} between the two energized wires behaves as a suction force because the Magnetic-electric field force F_v between the electric charges q_1 and q_2 moving in the same direction is greater than the rotational electric field force F_ω, i.e. $F_v > F_\omega$.

If the angle $\beta = 180^0$, then electric charge q_1 and q_2 are moving in opposite direction. At this time, the rotational electric field force F_ω is suction force because the rotational electric field force $F_\omega > 0$ between the moving electric charge q_1 and q_2. This is shown in Figure 20-4a.

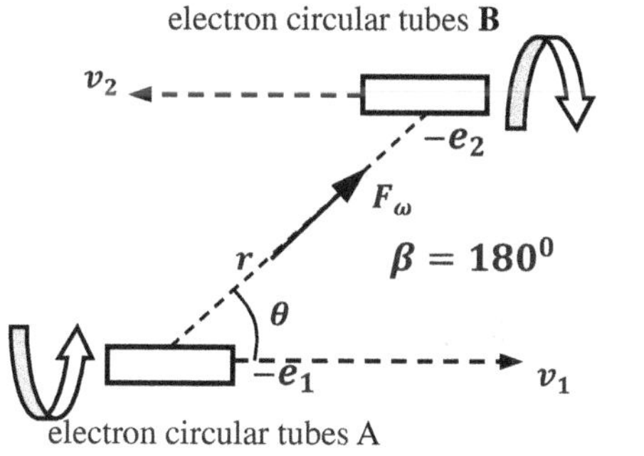

Figure 20-4a Schematic diagram of displacement of electron circular tubes A and B in opposite directions.

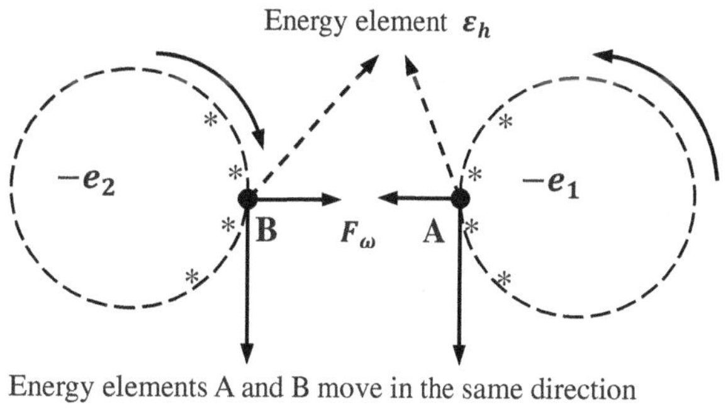

Figure 20-4b Schematic diagram of electron circular tubes A and B rotating in opposite directions.

Although the rotational electric field force F_ω between the electric charges q_1 and q_2 moving in opposite directions is a suction force, the electromagnetic force F_{EB} between the two wires behaves as a repulsive force because the Magnetic-electric field force F_v between the electric charges q_1 and q_2 is greater than the rotational electric field force F_ω, i.e., $F_v > F_\omega$.

If the angle $\beta = 90^0$, then electric charge q_1 and q_2 are moving perpendicular to each other. At this time, since the rotational electric field force $F_\omega = 0$ and Magnetic-electric field force $F_v = 0$ between electric charge q_1 and q_2, the electromagnetic force $F_{EB} = 0$ between the two wires.

When the moving electric charges q_1 and q_2 are positive and negative electric charges. If the angle $\beta = 0$, then the rotational electric field force $F_\omega > 0$ (suction force) between positive and negative electric charge q_1 and q_2. If angle $\beta = 180^0$ then rotational electric field force $F_\omega < 0$ (repulsive force) between positive and negative electric charge q_1 and q_2.

20.2.2 The spin velocity v_ω of a moving electron is a prerequisite for the generation of the rotational electric field force F_ω

Using the directional force axiom (Axiom 5) for the energy element ε_h, we can explain the reason for the rotational electric field force F_ω generated by the motion electric charges q_1 and q_2.

Assume that the electron circular tube e_1 and the electron circular tube e_2 are of the same type of electron. If the direction of displacement of the electron circular tubes e_1 and e_2 is the same, then the direction of rotation of the electron circular tubes e_1 and e_2 is the same.

Assuming that the electron circular tubes e_1 and e_2 both rotate clockwise, the energy elements ε_h on the neighboring faces of the electron circular tubes e_1 and e_2 rotate in opposite directions, i.e., energy element A rotates downward and energy element B rotates upward. This is shown in Figure 20-3b.

Since the angle $\beta = 180^0$ between the direction of motion of energy elements A and B, the directional force $f_{\varepsilon h} < 0$ (repulsive force) between energy element A and energy element B according to the axiom of directional force of energy element ε_h (Axiom 5).

Since the directional force $f_{\varepsilon h}$ of the energy element ε_h on the semicircular plane of A and the energy element ε_h on the semicircular plane of B are repulsive, the rotational electric field force F_ω between the electrons e_1 and e_2 moving in the same direction is repulsive, i.e:

$$F_\omega = -\frac{1}{4\pi\varepsilon_0}\sqrt{1-\frac{v_1^2}{c^2}}\sqrt{1-\frac{v_2^2}{c^2}}\cdot\frac{q_1 q_2}{r^2}\sin\theta \qquad (20.2.4)$$

If the electrons e_1 and e_2 are moving in opposite directions, then the electron circular tube e_1 is rotating clockwise and the electron circular tube e_2 is rotating counterclockwise. At this time, the energy element ε_h on the neighboring faces of the electron circular tubes e_1 and e_2 is rotating in the same direction, i.e., the energy elements A and B are both rotating downward. This is shown in Figure 20-4b.

Since the angle $\beta = 0$ between the directions of motion of energy elements A and B. Therefore, according to the axiom of directional force of energy element ε_h (Axiom 5). At this point the directional force $f_{\varepsilon h} > 0$ (suction force) between energy elements A and B.

Since the directional force $f_{\varepsilon h}$ between the energy element ε_h on the A semicircular plane and the energy element ε_h on the B semicircular plane is a suction force, the rotational electric field force F_ω between the electrons e_1 and e_2 moving in the opposite direction is a suction force, i.e:

$$F_\omega = +\frac{1}{4\pi\varepsilon_0}\sqrt{1-\frac{v_1^2}{c^2}}\sqrt{1-\frac{v_2^2}{c^2}}\cdot\frac{q_1 q_2}{r^2}\sin\theta \qquad (20.2.5)$$

20.2.3 The magnitude of the electrostatic field force F_0' on the ground changes periodically

Assume that electric charge q_1 and q_2 are moving in the same direction with velocity v in the cosmic vacuum, and assume that θ is the angle between the distance r and the direction of displacement of electric charge q_1. When the electric charge q_1 and q_2 move with equal velocity v and in the same direction of motion, according to equation (20.2.4), the rotating electric field force F_ω between the electric charge q_1 and q_2 is:

$$F_\omega = +\frac{1}{4\pi\varepsilon_0}\frac{q_1 q_2}{r^2}\left(1-\frac{v^2}{c^2}\right)\sin\theta \qquad (20.2.6)$$

The rotating electric field force F_ω is related to the electrostatic field force F_0 by the equation:

$$F_\omega = F_0\left(1-\frac{v^2}{c^2}\right)\sin\theta \tag{20.2.7}$$

Suppose that the ground (S' inertial system) is moving with velocity u in the cosmic vacuum frame of reference. Since the electric charge q'_0, which is at rest on the ground, is moving with velocity u in the cosmic vacuum, the electrostatic field force F'_0 on the ground, according to equation (20.2.7), is essentially the rotating electric field force F_ω in the cosmic vacuum frame of reference, viz:

$$F'_0 = F_\omega = F_0\left(1-\frac{u^2}{c^2}\right)\sin\theta \tag{20.2.8}$$

The θ in the formula is the angle between the distance r and the velocity u. The above equation shows that the electrostatic field force F'_0 on the ground becomes smaller as the velocity u increases. Based on the above equation, the following two inferences can be made.

Corollary 1. The electrostatic field force F'_0 at any point on the ground varies periodically with the rotation of the earth.

Since the sun revolves around the center of the Milky Way Galaxy and there is rotation and revolution of the earth, any point on the earth's equator moves at a different speed u in the cosmic system during the 24 hours of the day.

Assume that the velocity u at point P at the equator is maximum at 12 o'clock during the day and minimum at 12 o'clock at night. According to Equation (20.2.8) , the electrostatic field force F'_{01} at point P at 12 o'clock in the daytime is less than the electrostatic field force F'_{02} at 12 o'clock in the nighttime, i.e., $F'_{01} < F'_{02}$. One can verify equation (20.2.8) by a physical experiment with a 12-hour time difference.

The above causes periodic changes in the magnitude of the electrostatic field force F'_0 on the ground. In real life, the spot position of laser irradiation undergoes periodic changes, and the periodic change in the magnitude of the electrostatic field force F'_0 is analogous to the periodic change in the spot position.

Corollary 2. At the same moment, the electrostatic field force F'_{01} at the north and south poles of the earth, has a greater value than the electrostatic field force F'_{02} at the earth's equator.

Since different latitudes have different linear velocities u, the electrostatic field forces F'_0 are not equal at different latitudes. Suppose that the north and south poles of the earth are moving with velocity u_1 in the cosmic vacuum. According to equation (20.2.8), the electrostatic field force F'_{01}at the north and south poles is:

$$F'_{01} = F_0\left(1-\frac{u_1^2}{c^2}\right)\sin\theta \tag{20.2.9}$$

Suppose that the earth's equator is moving with velocity u_2 in the cosmic vacuum. According to equation (20.2.8), the electrostatic field force F'_{02} at the Earth's equator is:

$$F'_{02} = F_0\left(1-\frac{u_2^2}{c^2}\right)\sin\theta \tag{20.2.10}$$

Since the velocity u_1 at the North and South Poles is less than the velocity u_2 at the Equator, i.e., $u_1 < u_2$, the electrostatic field force F'_{01} at the North and South Poles is greater than that at the Equator, i.e., F'_{02}, i.e.,:

$$F'_{01} > F'_{02} \tag{20.2.11}$$

If the above formula is confirmed experimentally, then it proves that the cosmic vacuum frame of reference is the absolute coordinate system in Newton's theory.

20.3 Derivation of the Lorentz force equation

20.3.1 Motion electric charge q_1 and q_2 produce a magnetic field force F_B equal to the Magnetic-electric field force F_v

According to equation (20.1.11), the Magnetic-electric field force F_v between the motionelectric charge q_1 and q_2 is:

$$F_v = \frac{1}{4\pi\varepsilon_0 c^2}\frac{q_1 v_1 \cdot q_2 v_2}{r^2}\sin\theta\cos\beta \quad (20.3.1)$$

In the formula, θ is the angle between distance r and velocity v_1, β is the angle between velocity v_1 and v_2. If the angle $\theta = 90^0$ and angle $\beta = 0$ (or $\beta = 180^0$), then the Magnetic-electric field force F_v reaches its maximum value, i.e:

$$F_v = \pm\frac{1}{4\pi\varepsilon_0 c^2}\frac{q_1 v_1 \cdot q_2 v_2}{r^2} \quad (20.3.2)$$

When the displacement velocity $v_1 = 0$, or $v_2 = 0$, the Magnetic-electric field force $F_v = 0$.

If the angle $\theta = 0$, or the angle $\beta = 90^0$, then the Magnetic-electric field force F_v is equal to 0, i.e:

$$F_v = 0 \quad (20.3.3)$$

Assume that the moving electric charges q_1 and q_2 are the same kind of electric charge. When the angle $\beta = 0$, the two electric charges are moving in the same direction. The Magnetic-electric field force F_v is a suction force because the Magnetic-electric field force $F_v > 0$ between electric charges q_1 and q_2. This is shown in Figure 20-5.

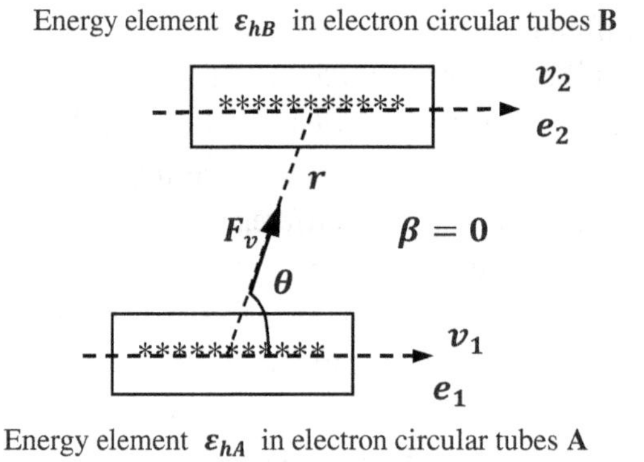

Figure 20-5 Schematic of the displacement of the energy element ε_h in the same direction in electron circular tubes A and B.

Since the Magnetic-electric field force F_v is greater than the rotational electric field force F_ω, i.e., $F_v > F_\omega$, the electromagnetic force F_{EB} between the two energized wires behaves as a suction force.

If the angle $\beta = 180^0$, then electric charge q_1 and q_2 are moving in opposite direction. At this time, the Magnetic-electric field force F_v is repulsive because the Magnetic-electric field force $F_v < 0$ between the moving electric charge q_1 and q_2. This is shown in Figure 20-6.

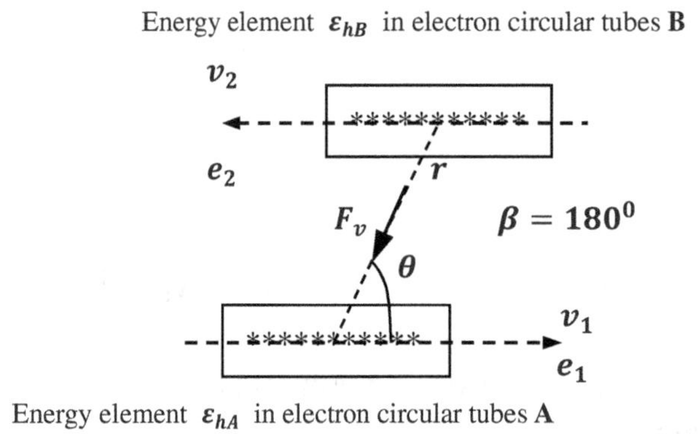

Figure 20-6 Schematic diagram of the reverse motion of the energy element ε_h in electron circular tube A and the energy element ε_h in electron circular tube B.

Although the rotational electric field force F_ω between the electric charge q_1 and q_2 moving in opposite directions is a suction force. However, since the Magnetic-electric field force F_v is larger than the rotating electric field force F_ω, i.e., $F_v > F_\omega$, the electromagnetic force F_{EB} between the two wires behaves as a repulsive force.

If the angle $\beta = 90^0$, then the electric charges q_1 and q_2 are moving perpendicular to each other. According to the axiom of directional force of energy element ε_h (Axiom 5), the rotational electric field force $F_\omega = 0$ and the Magnetic-electric field force $F_v = 0$ between electric charge q_1 and q_2, which makes the electromagnetic force $F_{EB} = 0$ between the two wires.

According to equation (20.1.7), the vacuum permeability μ_0 is:

$$\mu_0 = \frac{1}{\varepsilon_0 c^2}$$

Substituting the above equation into equation (20.3.1) yields the relationship equation:

$$F_v = F_B = \frac{\mu_0}{4\pi} \frac{q_1 v_1 \cdot q_2 v_2}{r^2} \sin\theta \cos\beta \qquad (20.3.4)$$

The above equation is the formula for the magnetic field force F_B generated by the motion electric charges q_1 and q_2. The above equation shows that the magnetic field force F_B generated by the motion electric charge q_1 and q_2 is equal to the Magnetic-electric field force F_v.

Since electric current I is formed by moving electrons, the Magnetic-electric field force F_v between electric current I_1 and I_2 is a suction force if electric current I_1 and I_2 are in the same direction. This is shown in Figure 20-7a.

If electric currents I_1 and I_2 are circular line electric currents in the same direction, then the Magnetic-electric field force F_v between circular line electric currents I_1 and I_2 is a suction force. This is shown in Figure 20-7b.

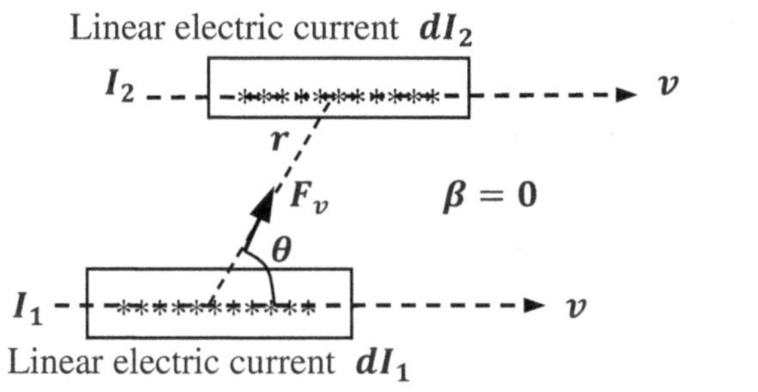

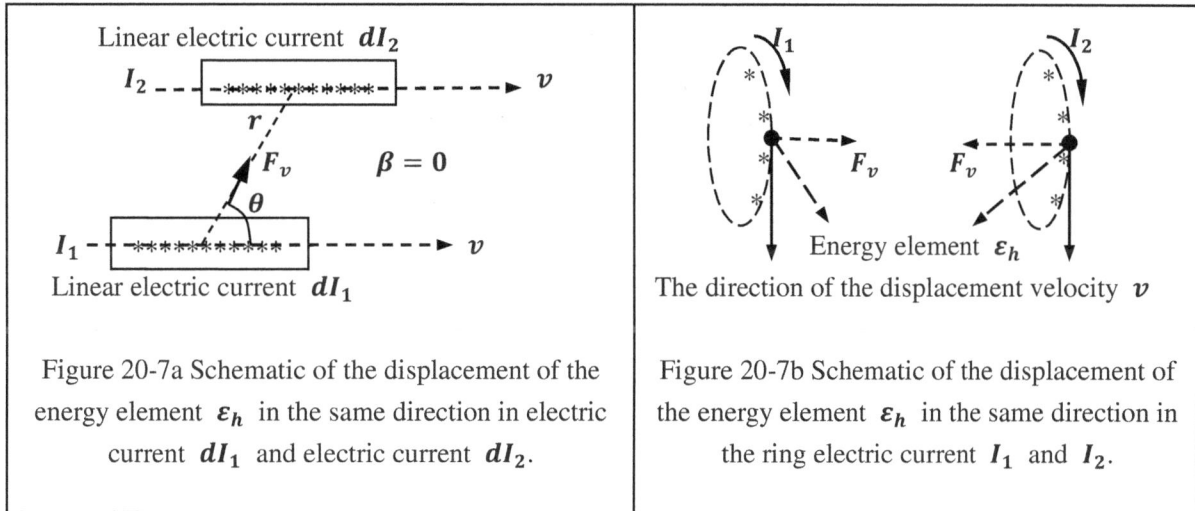

Figure 20-7a Schematic of the displacement of the energy element ε_h in the same direction in electric current dI_1 and electric current dI_2.

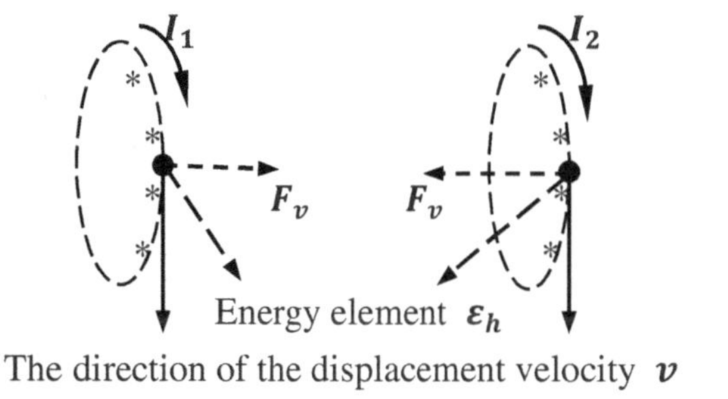

Figure 20-7b Schematic of the displacement of the energy element ε_h in the same direction in the ring electric current I_1 and I_2.

The Magnetic-electric field force F_v between electric current I_1 and I_2 is repulsive if electric current I_1 and I_2 are in opposite directions. This is shown in Figure 20-8a.

If electric currents I_1 and I_2 are circular line electric currents in opposite directions, then the Magnetic-electric field force F_v between circular line electric currents I_1 and I_2 is a repulsive force. This is shown in Figure 20-8b.

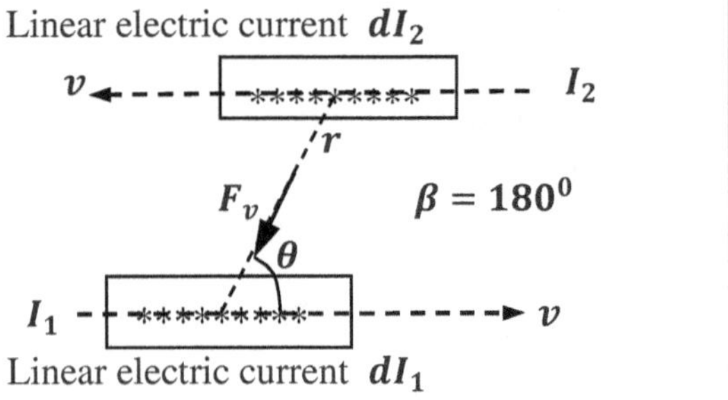

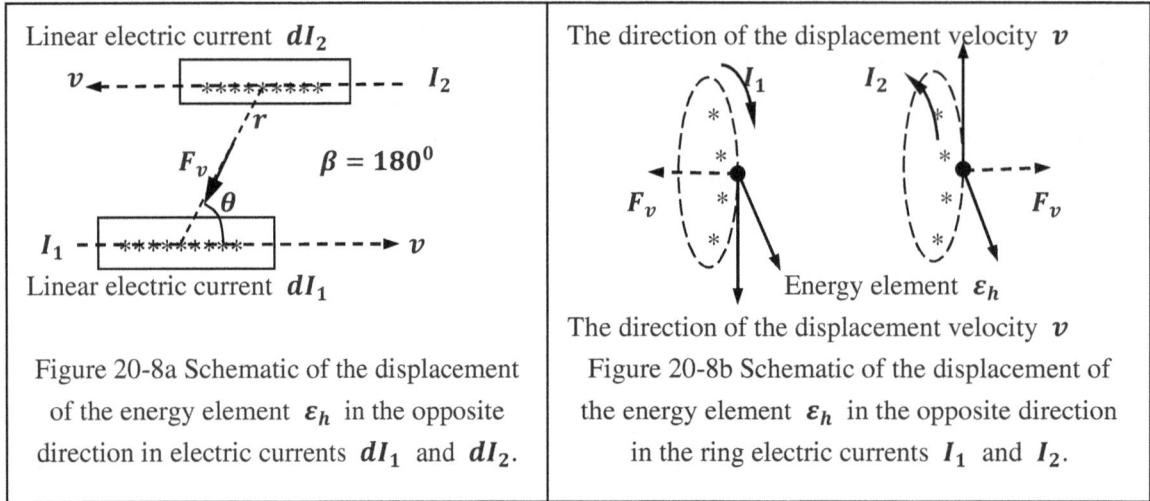

Figure 20-8a Schematic of the displacement of the energy element ε_h in the opposite direction in electric currents dI_1 and dI_2.

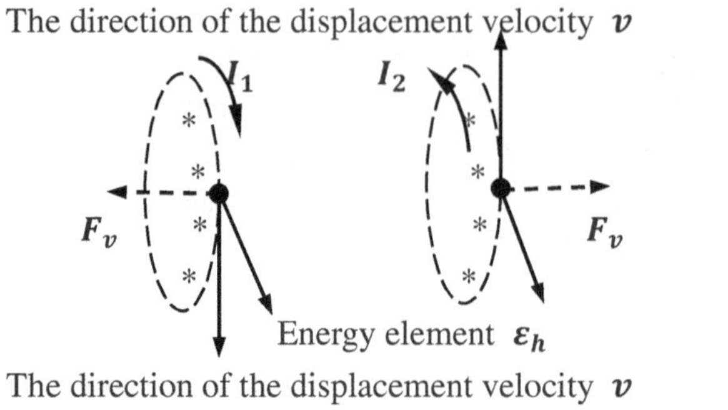

Figure 20-8b Schematic of the displacement of the energy element ε_h in the opposite direction in the ring electric currents I_1 and I_2.

For the same direction of the spiral tube electric current I_1 and I_2, since the energy element ε_h inside the spiral electric current I_1 and I_2 is moving in the same direction, according to the axiom of the directional force of the energy element ε_h, at this time, the magnetic electric force F_v between the spiral electric current I_1 and I_2 is the suction force. As shown in Figure 20-9a.

It should be noted that there is also a rotational electric field force F_ω between electric currents I_1 and I_2. Since the free electrons e_1 and e_2 rotate in the same direction, the rotational electric field force F_ω is a repulsive force. It is this repulsive force F_ω that makes the two magnets N-pole and S-pole repulsive between their edges.

For the opposite direction spiral tube electric current I_1 and I_2, since the displacement direction of the energy element ε_h in the electric current I_1 and I_2 is opposite, according to the axiom of the directional force of the energy element ε_h, the magnetic field force F_v between the spiral current I_1 and I_2 is a repulsive force. As shown in Figure 20-9b.

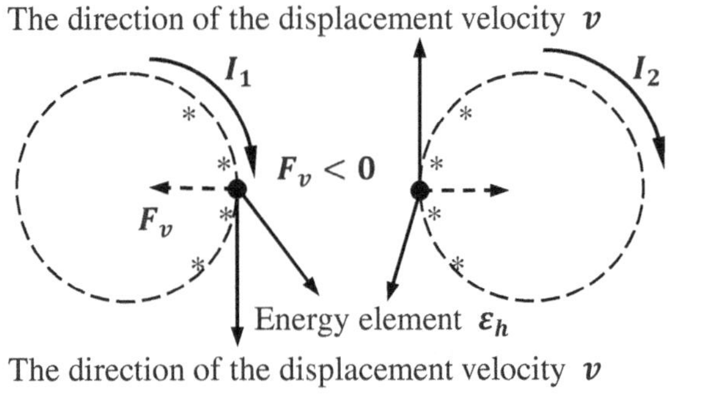

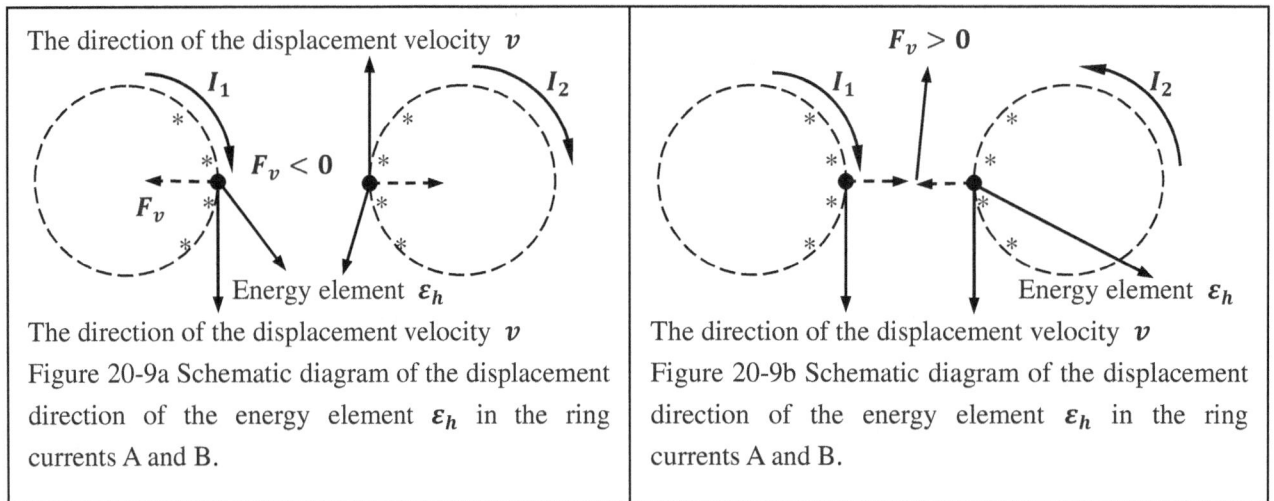

Figure 20-9a Schematic diagram of the displacement direction of the energy element ε_h in the ring currents A and B.

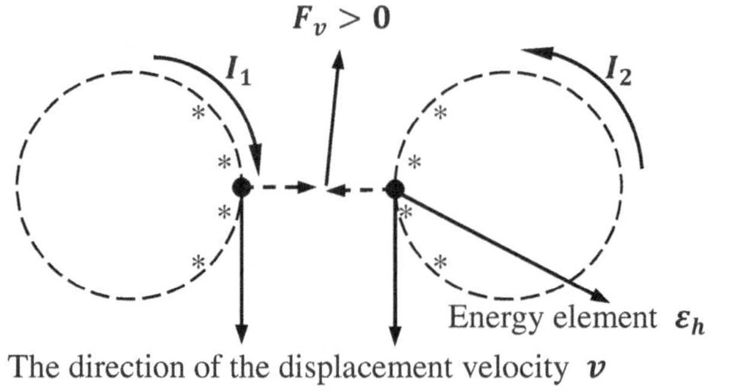

Figure 20-9b Schematic diagram of the displacement direction of the energy element ε_h in the ring currents A and B.

20.3.2 Derivation of the Lorentz force formula $F_B=qvB\cdot sin\delta$

Electrons emitted from the cathode of an electron tube will have a certain speed of motion under voltage. The path of motion of the electrons can be shown on an oscilloscope. Experiments have shown that in the absence of an external magnetic field, the electrons move in a straight line.

If an electron tube is placed between the N-S poles of a hoof magnet, the path of the electrons shown on the oscilloscope will be bent. This indicates that the moving electric charge is subjected to the force of a magnetic field, a force often called the Lorentz force, which was first proposed by the Dutch physicist H.A. Lorentz.

The Lorentz force has the following three properties.

(1) The direction of the Lorentz force is always perpendicular to the direction of motion of electric charge $\boldsymbol{q}$.

(2) The Lorentz force does not do work on electric charge $\boldsymbol{q}$.

(3) The Lorentz force does not change the rate or kinetic energy of the moving electric charge $\boldsymbol{q}$. It can only change the direction of motion of the electric charge to deflect it.

Magnetic field forces include the Lorentz force $\boldsymbol{F_B}$ and the amperometric force $\boldsymbol{F_I}$.

The force $\boldsymbol{F_B}$ of the magnetic field $\boldsymbol{B}$ on the motion electric charge $\boldsymbol{q}$ is defined as the Lorentz force, and the Lorentz force $\boldsymbol{F_B}$ is perpendicular to the plane defined by both the magnetic field $\boldsymbol{B}$ and the velocity $\boldsymbol{v}$ of the electric charge.

The force $\boldsymbol{F_I}$ of the magnetic field $\boldsymbol{B}$ on the electric current $\boldsymbol{I}$ is defined as the amperometric force. The amperometric force $\boldsymbol{F_I}$ is perpendicular to both the direction of the magnetic field $\boldsymbol{B}$ and the direction of the electric current $\boldsymbol{I}$. The direction of the amperometric force $\boldsymbol{F_I}$ can be determined using the left-hand rule.

The left-hand rule:

The left hand is spread out flat with the thumb perpendicular to the four fingers. Place the left hand in a magnetic field so that the line of magnetic induction passes perpendicularly through the palm of the hand (the palm faces the N pole of the magnetic field), and the four fingers point in the direction of the electric current I. The direction of the thumb is the direction of the force on the conductor.

The Lorentz force formula can be derived using the Magnetic-electric field force formula. According to the magnetic flux density formula, the magnetic flux density $\boldsymbol{B_1}$ produced by the moving electric charge $\boldsymbol{q_1}$ is:

$$\boldsymbol{B_1 = \frac{\mu_0}{4\pi}\frac{q_1 v_1}{r^2}\sin\theta} \qquad (20.3.5)$$

The $\boldsymbol{\theta}$ in equation is the angle between the distance $\boldsymbol{r}$ and the velocity $\boldsymbol{v_1}$. The direction of the magnetic flux density $\boldsymbol{B_1}$ is shown in Figure 20-10.

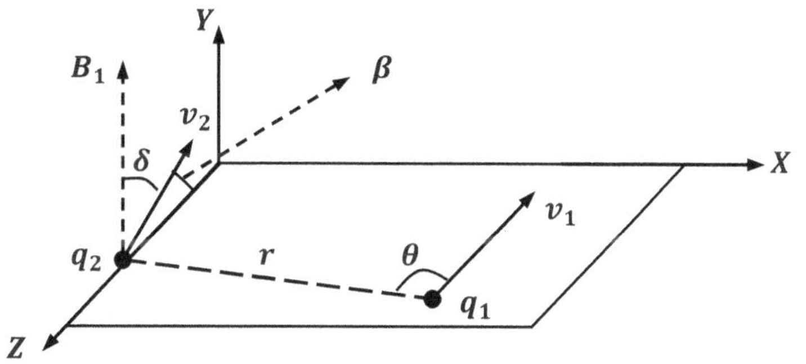

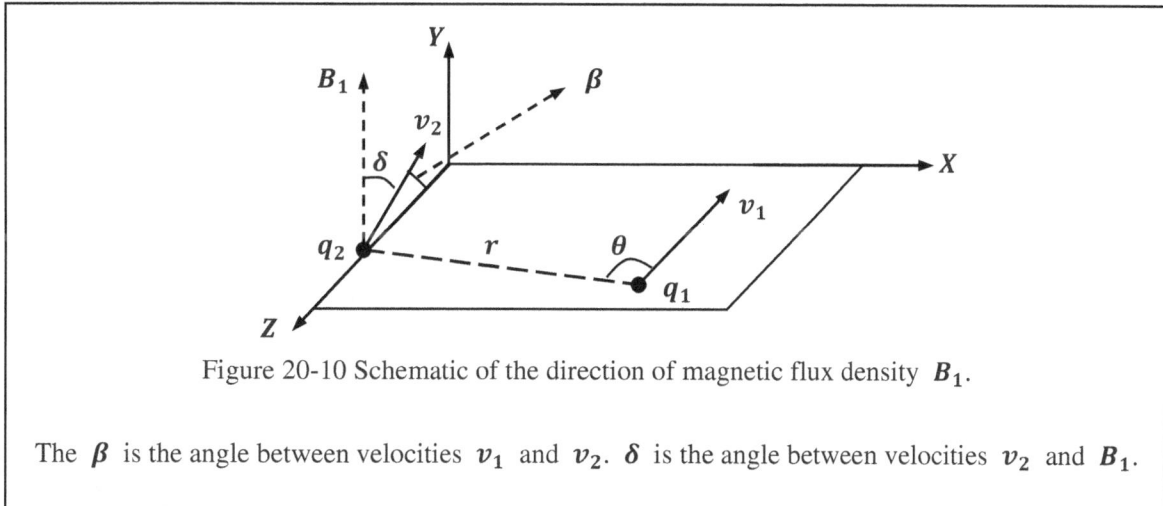

Figure 20-10 Schematic of the direction of magnetic flux density $\boldsymbol{B_1}$.

The $\boldsymbol{\beta}$ is the angle between velocities $\boldsymbol{v_1}$ and $\boldsymbol{v_2}$. $\boldsymbol{\delta}$ is the angle between velocities $\boldsymbol{v_2}$ and $\boldsymbol{B_1}$.

Substituting equation (20.3.5) into equation (20.3.4) yields the relationship equation:

$$\boldsymbol{F_v = q_2 v_2 B_1 \cos\beta} \qquad (20.3.6)$$

The $\boldsymbol{\beta}$ in the formula is the angle between the velocity $\boldsymbol{v_2}$ and the velocity $\boldsymbol{v_1}$. If the velocity $\boldsymbol{v_2}$ is in the same direction as the velocity $\boldsymbol{v_1}$, then the angle $\boldsymbol{\beta = 0}$. At this point, the angle $\boldsymbol{\delta = 90^0}$ between the velocity $\boldsymbol{v_2}$ and the magnetic flux density $\boldsymbol{B_1}$.

If the velocity $\boldsymbol{v_2}$ is perpendicular to the velocity $\boldsymbol{v_1}$, then the angle $\boldsymbol{\beta = 90^0}$ between the velocity $\boldsymbol{v_2}$ and the velocity $\boldsymbol{v_1}$. At this time, the angle $\boldsymbol{\delta = 0}$ between the velocity $\boldsymbol{v_2}$ and the magnetic flux density $\boldsymbol{B_1}$. From this, we can determine the angle $\boldsymbol{\beta}$ between the electric charge velocity $\boldsymbol{v_2}$ and the magnetic flux density $\boldsymbol{B_1}$ can be expressed as:

$$\boldsymbol{\beta = 90^0 - \delta} \qquad (20.3.7)$$

Substituting the above equation into equation (20.3.6) gives the relationship equation:

$$F_v = q_2 v_2 B_1 \sin\delta \tag{20.3.8}$$

If we make the electric charge $q_2 = q$, make the velocity $v_2 = v$, make the magnetic field $B_1 = B$, and make the magnetic field force $F_v = F_B$, then the above equation becomes:

$$F_B = qvB\sin\delta = F_v \tag{20.3.9}$$

The above equation is the classical Lorentz force equation. In the formula δ is the angle between the velocity v of the electric charge q and the magnetic flux density B. The above formula shows that the Lorentz force F_B and the Magnetic-electric field force F_v are both the same force, i.e.:

$$F_v = qvB\sin\delta = \frac{\mu_0}{4\pi}\frac{q_1 v_1 q_2 v_2}{r^2}\sin\theta\cos\beta \tag{20.3.10}$$

To summarize, the electromagnetic field forces generated by the moving electric charge q in the cosmic vacuum are classified into the following four types.

(1) The electrostatic field force F_0 between the static electric charge q_1 and q_2, viz:

$$F_0 = -\frac{1}{4\pi\varepsilon_0}\frac{q_1 q_2}{r^2} \tag{20.3.11}$$

(2) The rotating electric field force F_ω between the rotating electric charge $q_{\omega 1}$ and $q_{\omega 2}$, viz:

$$F_\omega = -\frac{1}{4\pi\varepsilon_0}\frac{q_{\omega 1} q_{\omega 2}}{r^2}\sin\theta\cos\beta \tag{20.3.12}$$

(3) The Magnetic-electric field force F_v between The magnetic electric charge q_{v1} and q_{v2}, viz:

$$F_v = \frac{1}{4\pi\varepsilon_0}\frac{q_1 v_1 q_2 v_2}{c^2 r^2}\sin\theta\cos\beta \tag{20.3.13}$$

(4) The magnetic field force F_B between The magnetic electric charge q_{v1} and q_{v2}, viz:

$$F_B = \frac{\mu_0}{4\pi}\frac{q_1 v_1 q_2 v_2}{r^2}\sin\theta\cos\beta \tag{20.3.14}$$

The θ in the equation is the angle between the velocity v_1 and the distance r, and β is the angle between the velocities v_1 and v_2.

20.3.3 Reasons why electrons in atoms do not radiate electromagnetic waves

According to classical physics, electrons rotating around the nucleus of an atom are constantly in accelerated motion under the force of the Coulomb electric field and radiate electromagnetic waves outward. Since the electrons are constantly losing energy, the electrons must fall to the nucleus in a helical orbit. This suggests that the atom (and all matter) would quickly collapse into a very compact state.

But the fact is that the atom is a stable system, and Bohr's model of the atom cannot explain why electrons revolving around the nucleus do not radiate electromagnetic waves. This book utilizes new electromagnetic theory that can explain why the Bohr atomic model does not radiate electromagnetic waves.

Assume that M and m are the masses of the nucleus and the electron, respectively, and r is the distance between the masses M and m. Assume that the nucleus M is stationary and v is the speed of rotation of the electron m around the nucleus. According to the new gravitational force formula (9.4.3), the gravitational force F_M between the nucleus M and the electron m is:

$$F_M = K\frac{Mm}{r^3} \tag{20.3.15}$$

According to kinematics, the centripetal force F_0 for the electron m to rotate around the nucleus M is:

$$F_0 = m\frac{v^2}{r} \tag{20.3.16}$$

According to the rotating electric field force equation (20.1.17), the maximum rotating electric field force F_ω between the moving electric charges q_1 and q_2 is:

$$F_\omega = -\frac{1}{4\pi\varepsilon_0}\frac{q_1 q_2}{r^2}\sqrt{1-\frac{v_1^2}{c^2}}\sqrt{1-\frac{v_2^2}{c^2}}$$

Suppose Q is the electric charge of the nucleus, and q is the electric charge of the electron . Since the velocity of the nucleus M is equal to 0, the Magnetic-electric field force $F_v = 0$ between the static electric charge Q and the moving electric charge q. According to the above equation, the rotational electric field force F_E between the nucleus and the electrons of the hydrogen atom is:

$$F_E = \frac{1}{4\pi\varepsilon_0}\frac{Qq}{r^2}\sqrt{1-\frac{v^2}{c^2}} \tag{20.3.17}$$

Since atoms are stable systems, the centripetal force F_O between the nucleus and the electrons should be equal to the sum of both the gravitational force F_M and the rotational electric field force F_E, viz:

$$F_O = F_M + F_E$$

Based on the above equation, the relationship equation can be obtained:

$$m\frac{v^2}{r} = K\frac{Mm}{r^3} + \frac{1}{4\pi\varepsilon_0}\frac{Qq}{r^2}\sqrt{1-\frac{v^2}{c^2}} \tag{20.3.18}$$

According to the tangential angular momentum conservation equation (5.2.9), the tangential angular momentum H_v of the electron m around the nucleus M is:

$$H_v = rmv = T \tag{20.3.19}$$

The T in the formula is a constant. Multiplying equation (20.3.18) by mr^3 gives the relational equation:

$$m^2r^2v^2 - KMm^2 = \frac{Qqm}{4\pi\varepsilon_0}r\sqrt{1-\frac{v^2}{c^2}} \tag{20.3.20}$$

Substituting equation (20.3.19) into the above equation yields the relationship equation:

$$cT^2 - cKMm^2 = \frac{Qq}{4\pi\varepsilon_0}\sqrt{m^2r^2c^2 - T^2} \tag{20.3.21}$$

Since $cT^2 - cKMm^2$ on the left side of the equal sign is a constant, the radius r on the right side of the equal sign must remain constant or the above equation will not hold. If radius r is a variable, then the electric field force should be inversely proportional to radius r cubed, not inversely proportional to radius r squared.

Since the radius r of the electron around the nucleus of a hydrogen atom remains constant, the moving electron does not radiate electromagnetic waves. In the vacuum of the universe, since the nucleus M moves with the motion of the earth, this gives the energy of the atom a fine structure.

Chapter 21 Biot-Savart Law and Ampere's law and the derivation of the Hall effect equation

Introduction: According to the analytical discussion in Chapter 19, the moving electric charge q generates the Magnetic-electric field E_v (i.e., magnetic flux density B) and the rotating electric field E_ω. Since the electric current is composed of the moving electric charge q, the current element Idl is not only capable of generating the Magnetic-electric field E_{vI}, but also capable of generating the rotating electric field $E_{\omega I}$.

Physicists only analyze and discuss the magnetic flux density B_I and the magnetic field force F_{BI} generated by the current element Idl. They do not analyze the rotating electric field $E_{\omega I}$ and the rotating electric field force $F_{\omega I}$ generated by current element Idl.

21.1 Electromagnetic field strength E_{I0} generated by electric current I

21.1.1 Rotational electric charge $q_{\omega I}$ and magnetic electric charge q_{vI} contained in current element Idl

Assume that n is the density of free electrons per unit volume of the wire and S is the cross section of the wire. When there is electric current I passing through the wire. Since the volume of current element Idl is $dV = Sdl$, the number of free electrons dN contained in current element Idl is:

$$dN = nSdl \tag{21.1.1}$$

The free electric charge q contained in current element Idl is:

$$q = edN = enSdl \tag{21.1.2}$$

According to equation (18.3.3), the rotational electric charge q_ω contained in the motion electric charge q is:

$$q_\omega = q\sqrt{1-\frac{v^2}{c^2}}$$

Substituting equation (21.1.2) into the above equation gives the relationship equation:

$$q_{\omega I} = \sqrt{1-\frac{v^2}{c^2}}\,enSdl \tag{21.1.3}$$

The $q_{\omega I}$ in equation is the rotational electric charge contained in the current element Idl.

According to equation (18.3.4), the magnetic electric charge q_v contained in the moving electric charge q is:

$$q_v = q\frac{v}{c}$$

Substituting equation (21.1.2) into the above equation gives the relationship equation:

$$q_{vI} = \frac{v}{c}enSdl = \frac{Idl}{c} \tag{21.1.4}$$

The q_{vI} in equation is the magnetic electric charge contained in the current element Idl.

21.1.2 Rotational electric charge $q_{\omega S}$ and magnetic electric charge q_{vS} contained in the conductor cross-section S

According to equation (21.1.1), the number of free electrons dN contained in current element Idl is:

$$dN = nSdl$$

The number of free electrons $\frac{dN}{dl}$ contained in the cross section S of the conductor is:

$$\frac{dN}{dl} = nS \tag{21.1.5}$$

The electric charge q_S contained in the conductor cross section S is:

$$q_S = enS \tag{21.1.6}$$

The rotational electric charge $q_{\omega S}$ contained in the conductor cross section S is:

$$q_{\omega S} = enS\sqrt{1-\frac{v^2}{c^2}} \tag{21.1.7}$$

The magnetic electric charge q_{vS} contained in the conductor cross section S is:

$$q_{vS} = enS\frac{v}{c} = \frac{I}{c} \tag{21.1.8}$$

21.1.3 Electromagnetic field strength E_{I0} generated by electric current I

According to the electromagnetic field strength formula (19.1.6), a moving positive electric charge $+q$ produces an electromagnetic field strength E of:

$$E = -\frac{\sin\theta\, q}{4\pi\varepsilon_0 r^2}$$

According to the above equation, the electromagnetic field strength E_e produced by a moving electron e at the cosmic vacuum point P is:

$$E_e = -\frac{1}{4\pi\varepsilon_0}\frac{e}{r^2}\sin\theta \tag{21.1.9}$$

When there is electric current I passing through the wire, the number of free electrons contained in the cross-section S of the wire is nS according to equation (21.1.5).

Assume that r is the average distance from the nS free electrons in the cross-section of the wire to point P. According to equation (21.1.9), the electromagnetic field strength E_{I0} generated by nS free electrons in the cross-section of the wire at point P is:

$$E_{I0} = nSE_e - \frac{1}{4\pi\varepsilon_0}\frac{enS}{r^2}\sin\theta \tag{21.1.10}$$

According to equation (15.2.16), the electric charge e of a moving electron can be decomposed into rotational electric charge e_ω and magnetic electric charge e_v, i.e:

$$e = \sqrt{e_\omega^2 + e_v^2}$$

Substituting the above equation into equation (21.1.10) gives the relationship equation:

$$E_{I0} = -\frac{1}{4\pi\varepsilon_0}\frac{nS\sqrt{e_\omega^2 + e_v^2}}{r^2}\sin\theta \tag{21.1.11}$$

According to equation (15.2.6) and equation (15.2.15), the rotational electric charge e_ω and the magnetic electric charge e_v contained in the moving electric charge e are, respectively:

$$e_\omega = e\sqrt{1-\frac{v^2}{c^2}}, \quad e_v = e\frac{v}{c} \tag{21.1.12}$$

Substituting the above equation into equation (21.1.11) gives the relationship equation:

$$E_{I0} = -\sqrt{\left(-\frac{\sin\theta\, enS}{4\pi\varepsilon_0\, r^2}\sqrt{1-\frac{v^2}{c^2}}\right)^2 + \left(\frac{\sin\theta\, evnS}{4\pi\varepsilon_0\, r^2 c}\right)^2} \tag{21.1.13}$$

Substituting electric current $I = evnS$ into the above equation gives the relationship equation:

$$E_{I0} = -\sqrt{\left(-\frac{\sin\theta\, I}{4\pi\varepsilon_0\, r^2 v}\sqrt{1-\frac{v^2}{c^2}}\right)^2 + \left(\frac{\sin\theta\, I}{4\pi\varepsilon_0\, r^2 c}\right)^2} \tag{21.1.14}$$

The above equation is the electric current electromagnetic field equation. The above equation shows that the electromagnetic field strength $\boldsymbol{E_{I0}}$ generated by electric current $\boldsymbol{I}$ can be decomposed into two component electromagnetic field strengths.

The first component electromagnetic field strength is the rotating electric field strength $\boldsymbol{E_{I0\omega}}$, viz:

$$\boldsymbol{E_{I0\omega} = -\frac{1}{4\pi\varepsilon_0}\frac{I}{r^2 v}\sqrt{1-\frac{v^2}{c^2}}\sin\theta} \qquad (21.1.15)$$

The above formula is the electric current rotating electric field strength formula. Note that the above equation $\boldsymbol{E_{I0\omega}}$ is the strength of the rotating electric field generated by electric current $\boldsymbol{I}$ at the cosmic vacuum point P.

The second component electromagnetic field strength is the Magnetic-electric field strength $\boldsymbol{E_{I0v}}$, ie:

$$\boldsymbol{E_{I0v} = \frac{1}{4\pi\varepsilon_0}\frac{I}{r^2 c}\sin\theta} \qquad (21.1.16)$$

The above formula is the electric current Magnetic-electric field strength formula. The above formula shows that the electric current Magnetic-electric field strength $\boldsymbol{E_{I0v}}$ is proportional to electric current $\boldsymbol{I}$. Note that the above formula $\boldsymbol{E_{I0v}}$ is the Magnetic-electric field strength generated by electric current $\boldsymbol{I}$ at point P of the cosmic vacuum.

21.2 Derivation of the Biot-Savart Law

21.2.1 Electromagnetic field E_I generated by current element *Idl*

According to equation (21.1.9), the electromagnetic field strength $\boldsymbol{E_e}$ generated by a moving electron $\boldsymbol{e}$ at the cosmic vacuum point P is:

$$\boldsymbol{E_e = -\frac{1}{4\pi\varepsilon_0}\frac{e}{r^2}\sin\theta}$$

According to equation (21.1.2), the electric charge $\boldsymbol{q}$ contained in current element $\boldsymbol{Idl}$ is:

$$\boldsymbol{q = edN = enSdl}$$

Suppose $\boldsymbol{r}$ is the distance from point O at the center of current element $\boldsymbol{Idl}$ to point P. That is, the distance $\boldsymbol{r}$ is the average distance from $\boldsymbol{dN = nSdl}$ free electrons in current element $\boldsymbol{Idl}$ to point P. When the line element $\boldsymbol{dl}$ is at rest in the cosmic vacuum, the free electric charge $\boldsymbol{q}$ in the current element $\boldsymbol{Idl}$ produces an electromagnetic field strength $\boldsymbol{E_I}$ at the point P as:

$$\boldsymbol{E_I = E_e \cdot dN = -\frac{1}{4\pi\varepsilon_0}\frac{enSdl}{r^2}\sin\theta} \qquad (21.2.1)$$

The above equation is the current element electromagnetic field equation. According to equation (15.2.16), the electric charge $\boldsymbol{e}$ of a moving electron can be decomposed into rotational electric charge $\boldsymbol{e_\omega}$ and magnetic electric charge $\boldsymbol{e_v}$, i.e:

$$\boldsymbol{e = \sqrt{e_\omega^2 + e_v^2}}$$

Substituting the above equation into equation (21.2.1) gives the relationship equation

$$\boldsymbol{E_I = -\sqrt{\left(-\frac{\sin\theta}{4\pi\varepsilon_0}\frac{e_\omega nSdl}{r^2}\right)^2 + \left(\frac{\sin\theta}{4\pi\varepsilon_0}\frac{e_v nSdl}{r^2}\right)^2}} \qquad (21.2.2)$$

The above equation shows that the electromagnetic field strength $\boldsymbol{E_I}$ generated by current element $\boldsymbol{Idl}$, can be decomposed into two component electromagnetic field strengths.

21.2.2 Rotating electric field strength $E_{I\omega}$ generated by current element *Idl*

Since $\boldsymbol{e_\omega}$ is the rotational electric charge of the electron, this book defines the first component inside the root of equation (21.2.2) as the rotational electric field strength $\boldsymbol{E_{I\omega}}$ of the current element, viz:

$$\boldsymbol{E_{I\omega} = -\frac{1}{4\pi\varepsilon_0}\frac{e_\omega nSdl}{r^2}\sin\theta} \qquad (21.2.3)$$

According to equation (15.2.6), the rotational electric charge $\boldsymbol{e_\omega}$ of the electron is:

$$e_\omega = e\sqrt{1-\frac{v^2}{c^2}}$$

Substituting the above equation into equation (21.2.3) gives the relationship equation:

$$E_{I\omega} = -\frac{1}{4\pi\varepsilon_0}\frac{enSdl}{r^2}\sqrt{1-\frac{v^2}{c^2}}\sin\theta \qquad (21.2.4)$$

Substituting electric current $I = evnS$ into the above equation gives the relationship equation:

$$E_{I\omega} = -\frac{1}{4\pi\varepsilon_0}\frac{Idl}{r^2 v}\sqrt{1-\frac{v^2}{c^2}}\sin\theta \qquad (21.2.5)$$

The above equation is the current element rotating electric field equation. The above equation shows that the current element rotating electric field $E_{I\omega}$ becomes smaller as the free electron velocity v increases.

21.2.3 Derivation of the Biot-Savart Law

Since e_v is the magnetic electric charge of the electron, this book defines the second component within the root of equation (21.2.2) as the Magnetic-electric field strength E_{Iv} of the current element, viz:

$$E_{Iv} = \frac{1}{4\pi\varepsilon_0}\frac{e_v nSdl}{r^2}\sin\theta \qquad (21.2.6)$$

According to equation (15.2.15), the magnetic electric charge e_v of the electron is:

$$e_v = e\frac{v}{c}$$

Substituting the above equation into equation (21.2.6) gives the relationship equation:

$$E_{Iv} = \frac{1}{4\pi\varepsilon_0}\frac{evnSdl}{r^2 c}\sin\theta \qquad (21.2.7)$$

Substituting electric current $I = evnS$ into the above equation gives the relationship equation:

$$E_{Iv} = \frac{1}{4\pi\varepsilon_0}\frac{Idl}{r^2 c}\sin\theta \qquad (21.2.8)$$

The above formula is the Magnetic-electric field strength formula for the current element.

It should be noted that since the current element Idl contains the displacement velocity v of the electron, according to the vector calculation law, the direction of the Magnetic-electric field strength E_{Iv} is perpendicular to the plane determined by both the distance r and the velocity v, as shown in Figure 21-1.

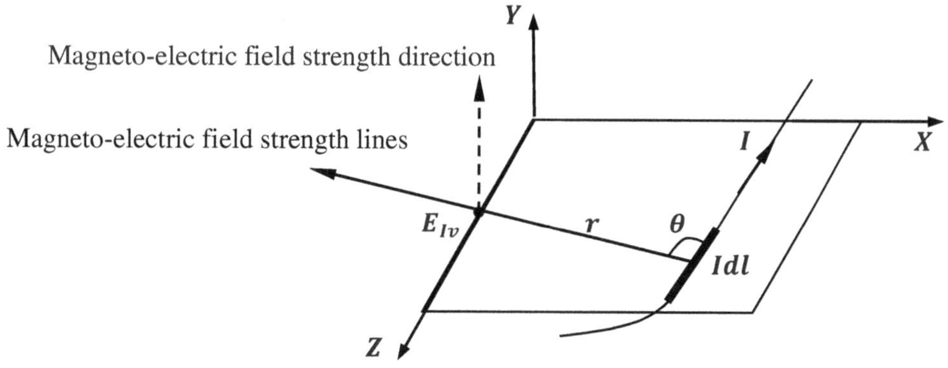

Figure 21-1 Schematic diagram of magneto-electric field strength lines and magneto-electric field strength direction generated by current element Idl.

Divide equation (21.2.8) by the speed of light c to obtain the relational equation:

$$\frac{E_{Iv}}{c} = \frac{1}{4\pi\varepsilon_0 c^2}\frac{Idl}{r^2}\sin\theta \qquad (21.2.9)$$

According to equation (19.3.16), the vacuum permeability μ_0 is:

$$\mu_0 = \frac{1}{\varepsilon_0 c^2} \tag{21.2.10}$$

Substituting the above equation into equation (21.2.9) yields the relationship equation

$$dB = \frac{E_{Iv}}{c} = \frac{\mu_0}{4\pi}\frac{Idl}{r^2}\sin\theta \tag{21.2.11}$$

The above equation is the classical Biot-Savart Law. dB in the equation is the magnetic flux density generated by the current element Idl. The direction of dB is the same as the direction of the current element's Magnetic-electric field E_{Iv}.

21.3 Derivation of Ampere's law

21.3.1 Rotational electric field force $F_{q\omega}$ between current element $I_1\,dl_1$ and $I_2\,dl_2$

When the line elements dl_1 and dl_2 are stationary in the cosmic vacuum. Assume that the displacement velocities of electric charges q_1 and q_2 in the cosmic vacuum frame of reference are v_1 and v_2, respectively. Assume that β is the angle between the direction of electric current I_2 and the direction of electric current I_1, and r is the distance between current element $I_1 dl_1$ and the center point of $I_2 dl_2$. Assume that θ is the angle between the distance r and the direction of electric current I_1. as shown in Figure 21-2.

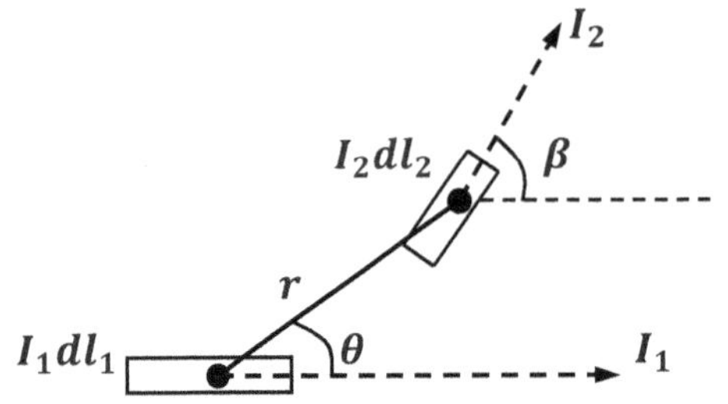

Figure 21-2 Schematic diagram of the movement direction of current elements I_1dl_1 and I_2dl_2.

It should be noted that the direction of the magnetic field B_2 generated by the current element I_1dl_1 at the center point O of the line element dl_2 is oriented perpendicular to the plane defined by both the distance r and the velocity v_1.

Assume that the free electron densities contained in the unit volume of the wires l_1 and l_2 are n_1 and n_2, respectively, and assume that the cross sections of the current element I_1dl_1 and I_2dl_2 are S_1 and S_2, respectively.

According to equation (21.1.2), the free electric charges q_1 and q_2 contained in current element I_1dl_1 and I_2dl_2 are, respectively:

$$q_1 = e_1 n_1 S_1 dl_1\,, \qquad q_2 = e_2 n_2 S_2 dl_2 \tag{21.3.1}$$

Since the rotational electric charge q_ω and the magnetic electric charge q_v are contained in the motion electric charge , the rotational electric charges $q_{\omega 1}$ and $q_{\omega 2}$ contained in the current elements I_1dl_1 and I_2dl_2 are respectively:

$$q_{\omega 1} = e_1\sqrt{1-\frac{v_1^2}{c^2}}\,n_1 S_1 dl_1\,, \qquad q_{\omega 2} = e_2\sqrt{1-\frac{v_2^2}{c^2}}\,n_2 S_2 dl_2 \tag{21.3.2}$$

The magnetic electric charge q_{v1} and q_{v2} contained in the current elements I_1dl_1 and I_2dl_2, respectively:

$$q_{v1} = e_1\frac{v_1}{c}n_1 S_1 dl_1\,, \qquad q_{v2} = e_2\frac{v_2}{c}n_2 S_2 dl_2 \tag{21.3.3}$$

According to the rotating electric field force equation (20.1.17), the rotating electric field force F_ω between the moving electric charges q_1 and q_2 is:

$$F_\omega = -\frac{1}{4\pi\varepsilon_0}\sqrt{1-\frac{v_1^2}{c^2}}\sqrt{1-\frac{v_2^2}{c^2}}\cdot\frac{q_1 q_2}{r^2}\sin\theta\cos\beta\,, \qquad r>1$$

According to the above equation, the rotational electric field force $F_{e\omega}$ between the moving electrons e_1 and e_2 is:

$$F_{e\omega} = -\frac{1}{4\pi\varepsilon_0}\sqrt{1-\frac{v_1^2}{c^2}}\sqrt{1-\frac{v_2^2}{c^2}}\cdot\frac{e_1 e_2}{r^2}\sin\theta\cos\beta\,, \qquad r>1 \tag{21.3.4}$$

According to equation (21.1.1), the number dN_1 and dN_2 of free electrons contained in current elements $I_1 dl_1$ and $I_2 dl_2$ are:

$$dN_1 = n_1 S_1 dl_1\,, \qquad dN_2 = n_2 S_2 dl_2 \tag{21.3.5}$$

Multiplying equation (21.3.4) by the quantities dN_1 and dN_2 gives the relational equation:

$$F_{I\omega} = F_{e\omega}\cdot dN_1\cdot dN_2$$

$$= -\frac{1}{4\pi\varepsilon_0}\sqrt{1-\frac{v_1^2}{c^2}}\sqrt{1-\frac{v_2^2}{c^2}}\frac{e_1 n_1 S_1 dl_1\cdot e_2 n_2 S_2 dl_2}{r^2}\sin\theta\cos\beta \tag{21.3.6}$$

According to the electric current equation $I = evnS$, the electric current I_1 and I_2 are:

$$I_1 = e_1 v_1 n_1 S_1\,, \qquad I_2 == e_2 v_2 n_2 S_2 \tag{21.3.7}$$

Substituting the above equation into equation (21.3.6) gives the relationship equation:

$$F_{I\omega} = -\frac{1}{4\pi\varepsilon_0}\frac{I_1 dl_1\cdot I_2 dl_2}{r^2 v_1 v_2}\sqrt{1-\frac{v_1^2}{c^2}}\sqrt{1-\frac{v_2^2}{c^2}}\sin\theta\cos\beta \tag{21.3.8}$$

The above formula is the current element rotating electric field force formula. When electric current I_1 and I_2 are in the same direction. Since the angle $\beta = 0$, the above equation becomes:

$$F_{I\omega} = -\frac{1}{4\pi\varepsilon_0}\frac{I_1 dl_1\cdot I_2 dl_2}{r^2 v_1 v_2}\sqrt{1-\frac{v_1^2}{c^2}}\sqrt{1-\frac{v_2^2}{c^2}}\sin\theta \tag{21.3.9}$$

The above formula is the formula for the rotational electric field force of the same direction current element. According to the above formula, if the angle $\theta = 0$ between the distance r and the direction of electric current I_1, i.e., current element $I_1 dl_1$ and $I_2 dl_2$ are in a straight line, then the rotating electric field force $F_{I\omega} = 0$ between current element $I_1 dl_1$ and $I_2 dl_2$.

If the angle $\theta = 90^0$ between the distance r and the direction of electric current I_1, i.e., the distance r is perpendicular to the direction of electric current I_1, then the maximum rotational electric field force $F_{I\omega}$ between current element $I_1 dl_1$ and $I_2 dl_2$ is:

$$F_{I\omega} = -\frac{1}{4\pi\varepsilon_0}\frac{I_1 dl_1\cdot I_2 dl_2}{r^2 v_1 v_2}\sqrt{1-\frac{v_1^2}{c^2}}\sqrt{1-\frac{v_2^2}{c^2}} \tag{21.3.10}$$

According to the above equation, the current element rotating electric field force $F_{I\omega}$ becomes smaller with the increase of free electron displacement velocity v_1 and v_2.

Since the rotating electric charge q_ω and the static electric charge q_0 are electric charges of the same nature, the current element rotating electric field force $F_{I\omega}$ is the same as the static electric field force F_0, which is also the same-sex attraction and opposite-sex repulsion.

If current element $I_1 dl_1$ is equal to $I_2 dl_2$ then the above equation becomes:

$$F_{I\omega} = -\frac{1}{4\pi\varepsilon_0}\frac{(Idl)^2}{r^2 v^2}\sqrt{1-\frac{v^2}{c^2}} \tag{21.3.11}$$

Since the direction of the electric current is also the displacement direction of the energy element ε_h, according to the axiom of the directional force of the energy element ε_h (Axiom 5), the rotational electric field force $F_{I\omega}$ between the electric currents in the same direction is a repulsive force, and the rotational electric field force $F_{I\omega}$ between the electric currents in the opposite direction is a suction force.

21.3.2 Magnetic-electric field force F_{Iv} between current element $I_1\, dl_1$ and $I_2\, dl_2$

According to the Magnetic-electric field force equation (20.1.11), the Magnetic-electric field force $\boldsymbol{F_v}$ between the moving electric charges $\boldsymbol{q_1}$ and $\boldsymbol{q_2}$ is:

$$F_v = \frac{1}{4\pi\varepsilon_0 c^2}\frac{q_1 v_1 \cdot q_2 v_2}{r^2}\sin\theta\cos\beta$$

According to the above equation, the Magnetic-electric field force $\boldsymbol{F_{ev}}$ between the moving electrons $\boldsymbol{e_1}$ and $\boldsymbol{e_2}$ is:

$$F_{ev} = +\frac{1}{4\pi\varepsilon_0 c^2}\frac{e_1 v_1 \cdot e_2 v_2}{r^2}\sin\theta\cos\beta \qquad (21.3.12)$$

According to equation (21.3.5), the number of free electrons $\boldsymbol{dN_1}$ and $\boldsymbol{dN_2}$ contained in current element $\boldsymbol{I_1 dl_1}$ and $\boldsymbol{I_2 dl_2}$ are:

$$dN_1 = n_1 S_1 dl_1\,, \qquad dN_2 = n_2 S_2 dl_2$$

Multiplying equation (21.3.12) by the quantities $\boldsymbol{dN_1}$ and $\boldsymbol{dN_2}$ gives the relational equation:

$$\begin{aligned} F_{Iv} &= F_{ev}\cdot dN_1 \cdot dN_2 \\ &= +\frac{1}{4\pi\varepsilon_0 c^2}\frac{v_1 e_1 n_1 S_1 dl_1 \cdot v_2 e_2 n_2 S_2 dl_2}{r^2}\sin\theta\cos\beta \end{aligned} \qquad (21.3.13)$$

Substituting the electric current equation (21.3.7) into the above equation gives the relationship equation:

$$F_{Iv} = \frac{1}{4\pi\varepsilon_0 c^2}\frac{I_1 dl_1 \cdot I_2 l_2}{r^2}\sin\theta\cos\beta \qquad (21.3.14)$$

The above formula is the current element Magnetic-electric field force formula. When electric current $\boldsymbol{I_1}$ and $\boldsymbol{I_2}$ are in the same direction. Since the angle $\boldsymbol{\beta = 0}$, the above equation becomes:

$$F_{Iv} = \frac{1}{4\pi\varepsilon_0 c^2}\frac{I_1 dl_1 \cdot I_2 dl_2}{r^2}\sin\theta \qquad (21.3.15)$$

The above formula is the same direction current element Magnetic-electric field force formula. According to the above formula, the Magnetic-electric field force $\boldsymbol{F_{Iv} = 0}$ between current element $\boldsymbol{I_1 dl_1}$ and $\boldsymbol{I_2 dl_2}$ when the angle $\boldsymbol{\theta = 0}$ between the distance $\boldsymbol{r}$ and the direction of electric current $\boldsymbol{I_1}$. At this time, current element $\boldsymbol{I_1 dl_1}$ and $\boldsymbol{I_2 dl_2}$ are on a straight line.

The Magnetic-electric field force $\boldsymbol{F_{Iv}}$ between current element $\boldsymbol{I_1 dl_1}$ and $\boldsymbol{I_2 dl_2}$ has a maximum value when the angle $\boldsymbol{\theta = 90^0}$ between distance $\boldsymbol{r}$ and electric current $\boldsymbol{I_1}$ direction, i.e., when the distance $\boldsymbol{r}$ is perpendicular to electric current $\boldsymbol{I_1}$ direction, i.e:

$$F_{Iv} = \frac{1}{4\pi\varepsilon_0 c^2}\frac{I_1 dl_1 \cdot I_2 dl_2}{r^2} \qquad (21.3.16)$$

Since the direction of electric current $\boldsymbol{I}$ is also the displacement direction of energy element $\boldsymbol{\varepsilon_h}$, according to the axiom of directional force of energy element $\boldsymbol{\varepsilon_h}$ (Axiom 5), the Magnetic-electric field force $\boldsymbol{F_{Iv}}$ between current element $\boldsymbol{I_1 dl_1}$ and $\boldsymbol{I_2 dl_2}$ in the same direction is a suction force, and the Magnetic-electric field force $\boldsymbol{F_{Iv}}$ between current element $\boldsymbol{I_1 dl_1}$ and $\boldsymbol{I_2 dl_2}$ in the opposite direction is a repulsion force.

21.3.3 Derivation of Ampere's law

The amperometric force $\boldsymbol{F_{IB}}$ is a macroscopic manifestation of the Lorentz force. The amperometric force $\boldsymbol{F_{IB}}$ is perpendicular to the plane determined by the magnetic field B and the electric current $\boldsymbol{I}$. The direction of the amperometric force $\boldsymbol{F_{IB}}$ is determined by the left-hand rule.

According to equation (19.3.16), the vacuum permeability $\boldsymbol{\mu_0}$ is:

$$\mu_0 = \frac{1}{\varepsilon_0 c^2}$$

Substituting the above equation into equation (21.3.14) gives the relationship equation:

$$F_{Iv} = \frac{\mu_0}{4\pi}\frac{I_1 dl_1 \cdot I_2 dl_2}{r^2}\sin\theta\cos\beta \qquad (21.3.17)$$

The above equation is the current element Magnetic-electric field force equation expressed in terms of vacuum permeability $\boldsymbol{\mu_0}$. When the electric current $\boldsymbol{I_1}$ and $\boldsymbol{I_2}$ are in the same direction. Since the angle $\boldsymbol{\beta = 0}$, the above equation becomes:

$$F_{IB} = F_{Iv} = \frac{\mu_0}{4\pi}\frac{I_1 dl_1 \cdot I_2 dl_2}{r^2}\sin\theta \qquad (21.3.18)$$

The above equation is the classical Ampere's law.The above equation shows that the magnetic field force F_{IB} is essentially the Magnetic-electric field force F_{Iv} between current element $I_1 dl_1$ and $I_2 dl_2$, i.e., $F_{IB} = F_{Iv}$.

The magnetic force $F_{IB} = 0$ when the angle $\theta = 0$ between the distance r and the direction of the electric current I_1, i.e., when the current element $I_1 dl_1$ is in a straight line with $I_2 dl_2$.

When the angle $\theta = 90^0$ between the distance r and the direction of electric current I_1, the magnetic force F_{IB} between current element $I_1 dl_1$ and $I_2 dl_2$ has a maximum value, i.e:

$$F_{IB} = F_{Iv} = +\frac{\mu_0}{4\pi}\frac{I_1 dl_1 \cdot I_2 dl_2}{r^2} \qquad (21.3.19)$$

The above equation shows that the magnetic field force $F_{IB} = F_{Iv}$ is proportional to the electric current I.

21.3.4 Derivation of the amperometric force equation *F=BIL sinδ*

According to equation (21.2.8), the Magnetic-electric field strength E_{Iv} generated by current element Idl is:

$$E_{Iv} = \frac{1}{4\pi\varepsilon_0}\frac{Idl}{r^2 c}\sin\theta$$

Note that the direction of the Magnetic-electric field strength E_{Iv} is perpendicular to the plane defined by both the distance r and the current element I. According to the above equation, the Magnetic-electric field strength E_{Iv1} generated by current element $I_1 dl_1$ at the center of line element dl_2 is:

$$E_{Iv1} = \frac{1}{4\pi\varepsilon_0}\frac{I_1 dl_1}{r^2 c}\sin\theta \qquad (21.3.20)$$

In the formula θ is the angle between the distance and the direction of electric current I_1. According to the current element Magnetic-electric field force formula (21.3.14), the Magnetic-electric field force F_{Iv} between the two current elements $I_1 dl_1$ and $I_2 dl_2$ is:

$$F_{Iv} = \frac{1}{4\pi\varepsilon_0}\frac{I_1 dl_1 \cdot I_2 dl_2}{r^2 c^2}\sin\theta\cos\beta \qquad (21.3.21)$$

The β in the formula is the angle between the direction of electric current I_2 and the direction of electric current I_1. As shown in Figure 21-3.

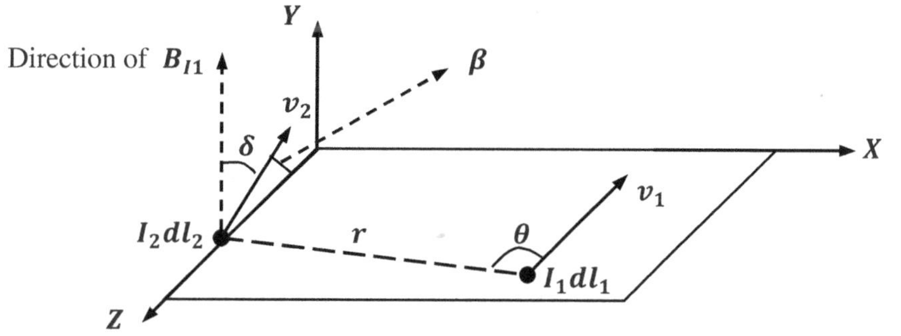

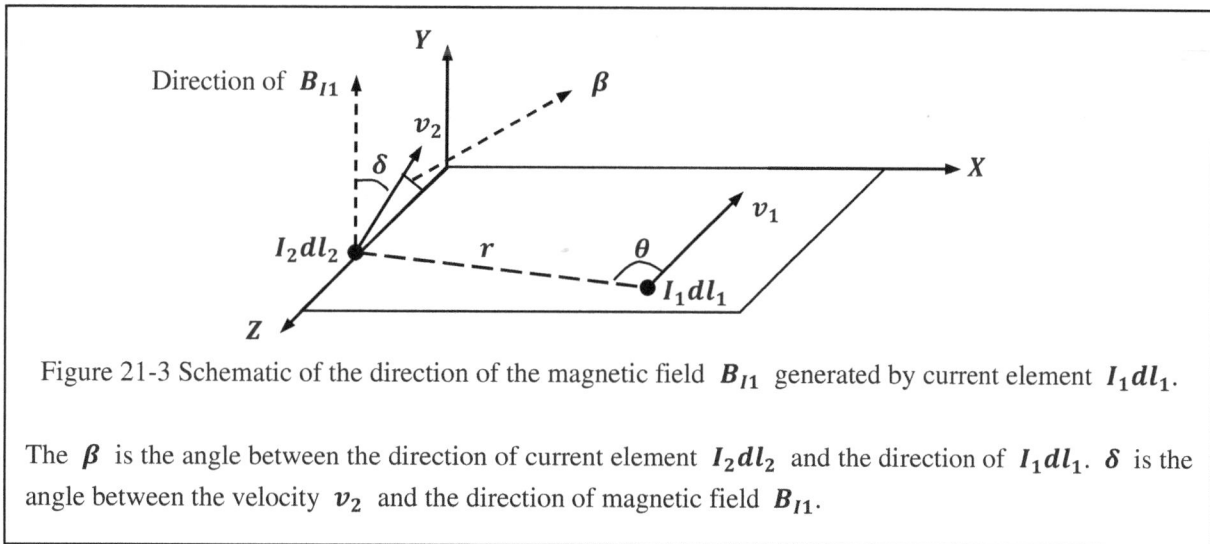

Figure 21-3 Schematic of the direction of the magnetic field B_{I1} generated by current element $I_1 dl_1$.

The β is the angle between the direction of current element $I_2 dl_2$ and the direction of $I_1 dl_1$. δ is the angle between the velocity v_2 and the direction of magnetic field B_{I1}.

Substituting equation (21.3.20) into equation (21.3.21) gives the relationship equation:

$$F_{Iv} = \frac{E_{Iv1}}{c} I_2 dl_2 \cos\beta \qquad (21.3.22)$$

According to equation (19.3.14), the relationship between the magnetic flux density B and the Magnetic-electric field strength $\boldsymbol{E_v}$ is:

$$\boldsymbol{B} = \frac{\mathbf{1}}{\boldsymbol{c}}\boldsymbol{E_v}$$

According to the above equation, the Magnetic-electric field strength $\boldsymbol{E_{Iv1}}$ generated by electric current $\boldsymbol{I_1}$ can be equivalently transformed into magnetic flux density $\boldsymbol{B_{I1}}$, viz:

$$\boldsymbol{B_{I1}} = \frac{\mathbf{1}}{\boldsymbol{c}}\boldsymbol{E_{Iv1}} \tag{21.3.23}$$

The $\boldsymbol{B_{I1}}$ in equation is the magnetic flux density generated at the center point of the current element $\boldsymbol{I_1 dl_1}$ at the line element $\boldsymbol{dl_2}$. Substituting the above equation into equation (21.3.22) gives the relational equation:

$$\boldsymbol{F_{Iv}} = \boldsymbol{B_{I1} I_2 dl_2 \cos\beta} \tag{21.3.24}$$

The $\boldsymbol{F_{Iv}}$ in the formula is the Magnetic-electric field force between current element $\boldsymbol{I_1 dl_1}$ and $\boldsymbol{I_2 dl_2}$. If we make current element $\boldsymbol{I_2 dl_2} = \boldsymbol{Idl}$, and make magnetic field $\boldsymbol{B_{I1}} = \boldsymbol{B}$, then the above equation becomes:

$$\boldsymbol{F_{Iv}} = \boldsymbol{BIdl\cos\beta} \tag{21.3.25}$$

The direction of the magnetic flux density $\boldsymbol{B}$ in equation is perpendicular to the plane defined by both the electric current $\boldsymbol{I_1}$ and the distance $\boldsymbol{r}$. The direction of the magnetic flux density B in equation The $\boldsymbol{\beta}$ in equation is the angle between the direction of electric current $\boldsymbol{I_2}$ and the direction of electric current $\boldsymbol{I_1}$. This is shown in Figure 21-3.

According to Figure 21-3, if electric current $\boldsymbol{I_2}$ is in the same direction as $\boldsymbol{I_1}$, then the angle $\boldsymbol{\beta} = \mathbf{0}$ between $\boldsymbol{I_2}$ and $\boldsymbol{I_1}$. The angle $\boldsymbol{\delta} = \mathbf{90^0}$ between electric current $\boldsymbol{I_2}$ and B.

If electric current $\boldsymbol{I_2}$ is perpendicular to $\boldsymbol{I_1}$, then the angle between $\boldsymbol{I_2}$ and $\boldsymbol{I_1}$ is $\boldsymbol{\beta} = \mathbf{90^0}$. At this point, the angle between electric current $\boldsymbol{I_2}$ and $\boldsymbol{B}$ is $\boldsymbol{\delta} = \mathbf{0}$. Thus the angle $\boldsymbol{\beta}$ between electric current $\boldsymbol{I_2}$ and $\boldsymbol{B}$ can be expressed as:

$$\boldsymbol{\beta} = \mathbf{90^0} - \boldsymbol{\delta} \tag{20.3.26}$$

Substituting the above equation into equation (21.3.25) gives the relationship equation:

$$\boldsymbol{F_{Iv}} = \boldsymbol{BIdl\sin\delta} \tag{21.3.27}$$

The $\boldsymbol{F_{Iv}}$ in equation is the magnetic force on current element $\boldsymbol{Idl}$ in magnetic field $\boldsymbol{B}$. Assume that $\boldsymbol{B}$ is a uniform magnetic field. Suppose $\boldsymbol{L}$ is a straight wire, $\boldsymbol{dl}$ is a line element on the straight wire $\boldsymbol{L}$. According to the above equation the magnetic field force $\boldsymbol{F}$ on the current carrying wire $\boldsymbol{L}$ in the uniform magnetic field $\boldsymbol{B}$ is:

$$\boldsymbol{F} = \int_0^L \boldsymbol{F_{Iv}}\,\boldsymbol{dl} = \boldsymbol{BIL\sin\delta} \tag{21.3.28}$$

The above formula is the classical amperometric force formula. In the formula, $\boldsymbol{B}$B is the uniform magnetic flux density, $\boldsymbol{L}$ is the length of the current-carrying wire in the magnetic field $\boldsymbol{B}$, and $\boldsymbol{\delta}$ is the angle between the wire $\boldsymbol{L}$ and the magnetic flux density $\boldsymbol{B}$.

The amperometric force $\boldsymbol{F}$ is perpendicular to the plane defined by both magnetic flux density $\boldsymbol{B}$ and electric current $\boldsymbol{I}$. The direction of the amperometric force $\boldsymbol{F}$ can be determined by the left-hand rule.

It should be noted that the amperometric force $\boldsymbol{F}$ in equation (21.3.28) is the magnetic force of the cosmic vacuum frame of reference, and since the Earth's velocity $\boldsymbol{u}$ of motion in the cosmic vacuum is very small, the amperometric force $\boldsymbol{F'}$ of the terrestrial frame of reference ($\boldsymbol{S'}$ inertial system) is approximately equal to the amperometric force $\boldsymbol{F}$ of the cosmic vacuum frame of reference.

21.3.5 Electrostatic field force F_{0I} on energized wire L in a uniform electrostatic field E_0

Assume that $\boldsymbol{E_0}$ is a stationary uniformly strong electrostatic field, and that the electric field $\boldsymbol{E_0}$ is similar to the electric field of a flat capacitor. Assume that the wire $\boldsymbol{L}$ is stationary in the electrostatic field $\boldsymbol{E_0}$, assume that $\boldsymbol{\delta}$ is the angle between the direction of the electric current $\boldsymbol{I}$ and the direction of the electrostatic field $\boldsymbol{E_0}$, and assume that $\boldsymbol{v}$ is the velocity of the free electrons in the electric current $\boldsymbol{I}$.

When the direction of electric current $\boldsymbol{I}$ is parallel to the direction of the electrostatic field $\boldsymbol{E_0}$, i.e., the angle $\boldsymbol{\delta} = \mathbf{0}$, the electrostatic field force $\boldsymbol{F}$ applied to the energized wire $\boldsymbol{L}$ is equal to zero.

If the voltage produced by the electrostatic field $\boldsymbol{E_0}$ at the ends of the wire $\boldsymbol{L}$ is greater than the voltage $\boldsymbol{U}$ at which the electric current $\boldsymbol{I}$ is produced, then the electric current $\boldsymbol{I}$ in the energized wire will increase.

If the voltage produced by the electrostatic field $\boldsymbol{E_0}$ at the ends of the wire $\boldsymbol{L}$ is less than the voltage $\boldsymbol{U}$ that produces electric current I, then the electric current I in the energized wire will remain constant.

When the angle $\delta = 90^0$ between the direction of electric current I and the direction of electrostatic field E_0. Since the electrostatic field force on the free electron e is perpendicular to the wire L, the electric field force F_{0I} on the wire L has a maximum value.

The rotating electric charge q_ω contained in current element Idl is subjected to the electrostatic field force dF_{0I} in the electrostatic field E_0 as:

$$dF_{0I} = q_\omega E_0 \sin\delta = e_\omega nSdl \cdot E_0 \sin\delta \qquad (21.3.29)$$

According to equation (15.2.6), the rotational electric charge e_ω contained in the kinematic electric charge e is:

$$e_\omega = e\sqrt{1 - \frac{v^2}{c^2}}$$

Substituting the above equation into equation (21.3.29) gives the relationship equation:

$$dF_{0I} = enSdl \cdot E_0 \sqrt{1 - \frac{v^2}{c^2}} \sin\delta \qquad (21.3.30)$$

When dl is the line element of the straight wire L, the electrostatic field force F_{0I} on the energized wire L in the electrostatic field E_0 is:

$$F_{0I} = \int dF_{0I} = enSL \cdot E_0 \sqrt{1 - \frac{v^2}{c^2}} \sin\delta \qquad (21.3.31)$$

The above formula is the electrostatic field force formula for energized wire. When the energized wire L is perpendicular to the direction of electrostatic field E_0, i.e., the angle $\delta = 90^0$, the maximum electrostatic field force F_{0I} on the energized wire L is:

$$F_{0I} = enSL \cdot E_0 \sqrt{1 - \frac{v^2}{c^2}} \qquad (21.3.32)$$

The direction of the electrostatic field force F_{0I} is in the direction of the uniform electrostatic field E_0. The above equation shows that the electrostatic field force F_{0I} on the energized wire L becomes smaller as the electron velocity v increases.

Since the electrostatic field E_0 has force only on the rotating electric charge e_ω and no force on the magnetic electric charge e_v, the force of the electrostatic field E_0 on the magnetic electric charge q_v in the energized wire L is always equal to zero.

21.4 Derivation of the Hall effect equation

21.4.1 Lorentz force F_{Be} on a moving electron -e in a uniform magnetic field B

The Hall effect was discovered by the American physicist Hall in 1879. Suppose d is the width of a flat metal plate, b is the length, and h is the height. Assume that a uniform magnetic field B is perpendicular to the db plane of the metal plate.

Assume that I is the electric current contained in the metal plate cross section $S = hd$. If the electric current I passing through the cross-section $S = hd$ of a metal plate is perpendicular to the magnetic field B, then there exists a steady voltage V_H (Hall voltage) between the left plane bh and the right plane bh of the metal plate. This phenomenon is known as the Hall effect. It is shown in Figure 21-4.

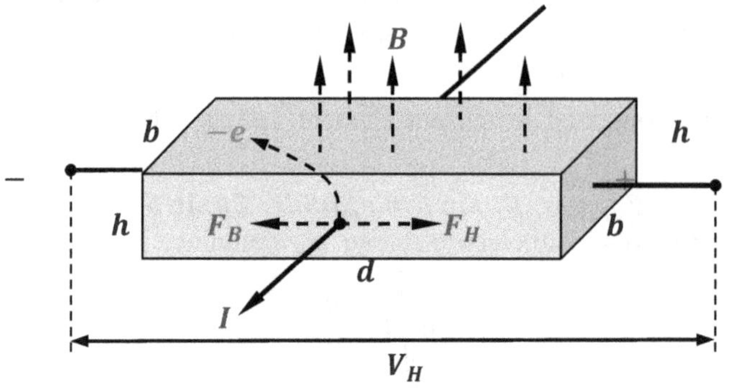

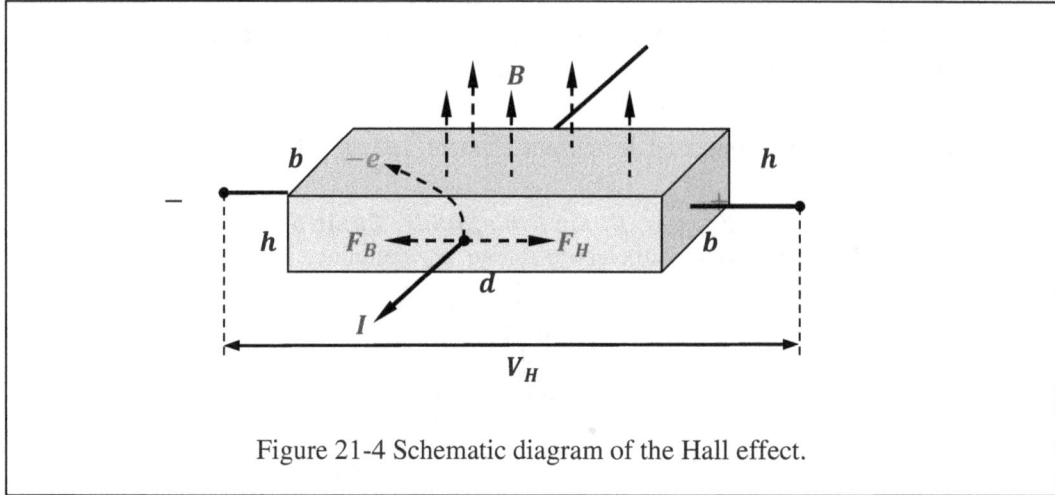

Figure 21-4 Schematic diagram of the Hall effect.

Suppose $\boldsymbol{v}$ is the displacement velocity of electron $-\boldsymbol{e}$ in the cosmic vacuum frame of reference. According to the Lorentz force formula (20.3.9), the Lorentz force $\boldsymbol{F_{Be}}$ on a moving electron $-\boldsymbol{e}$ in a uniform magnetic field $\boldsymbol{B}$ is:

$$\boldsymbol{F_{Be} = -evB \sin \delta} \qquad (21.4.1)$$

The $\boldsymbol{\delta}$ in the formula is the angle between the electron velocity $\boldsymbol{v}$ and the magnetic field $\boldsymbol{B}$. Since the electric current $\boldsymbol{I}$ through the cross section $\boldsymbol{S = hd}$ is perpendicular to the direction of the uniform magnetic field $\boldsymbol{B}$, the angle $\boldsymbol{\delta = 90^0}$. The above equation becomes:

$$\boldsymbol{F_{Be} = -evB} \qquad (21.4.2)$$

Under the action of the Lorentz force $\boldsymbol{F_{Be}}$, the electron $-\boldsymbol{e}$ displaced with velocity $\boldsymbol{v}$ moves toward the plane $\boldsymbol{bh}$ to the left of the metal plate. as shown in Figure 21-4.

When the displacement electron $-\boldsymbol{e}$ accumulates to a certain number in the left plane $\boldsymbol{bh}$. There is a potential difference $\boldsymbol{V_H}$ (Hall voltage) between the left plane $\boldsymbol{bh}$ and the right plane $\boldsymbol{bh}$.

21.4.2 Lorentz force F_B contained in energized cross section *S=hd*

Assume that $\boldsymbol{n}$ is the density of free electrons per unit volume of the conductor plate. The number of free electrons $\boldsymbol{N}$ contained in the energized cross section $\boldsymbol{S = hd}$ is:

$$\boldsymbol{N = nS = nhd} \qquad (21.4.3)$$

The free electric charge $\boldsymbol{q}$ contained in the energized cross section $\boldsymbol{S = hd}$ is:

$$\boldsymbol{q = eN = enhd} \qquad (21.4.4)$$

Assume that the energized conductor plate and the uniform magnetic field $\boldsymbol{B}$ are stationary in the cosmic vacuum. Assume that $\boldsymbol{v}$ is the displacement velocity of the free electron $\boldsymbol{e}$ in the cosmic vacuum, and $\boldsymbol{\delta}$ is the angle between the displacement direction of electron $\boldsymbol{e}$ and the direction of the magnetic field $\boldsymbol{B}$.

According to the Lorentz magnetic force formula (20.3.9), the magnetic field force $\boldsymbol{F_B}$ on the moving electric charge $\boldsymbol{q}$ in the uniform magnetic field $\boldsymbol{B}$ is:

$$\boldsymbol{F_B = qvB \sin \delta} \qquad (21.4.5)$$

The direction of Lorentz force $\boldsymbol{F_B}$ is determined by the left hand rule. When the angle $\boldsymbol{\delta = 90^0}$ between the direction of electron $\boldsymbol{e}$ displacement and the direction of uniform magnetic field $\boldsymbol{B}$, the above formula becomes:

$$\boldsymbol{F_B = qvB} \qquad (21.4.6)$$

Substituting equation (21.4.4) into the above equation gives the relationship equation:

$$\boldsymbol{F_B = enhd \cdot vB} \qquad (21.4.7)$$

Substituting electric current $\boldsymbol{I = envS = envhd}$ into the above equation gives the relationship equation:

$$\boldsymbol{F_B = IB} \qquad (21.4.8)$$

The $\boldsymbol{F_B}$ in equation is the Lorentz force contained in the energized cross section $\boldsymbol{S = hd}$ and $\boldsymbol{I}$ is the electric current contained in the energized cross section $\boldsymbol{S = hd}$.

21.4.3 Derivation of a new Hall effect equation

Since the Lorentzian magnetic force $\boldsymbol{F_B}$ causes the free electrons $-\boldsymbol{e}$ in the conductor plate to move to the left plane $\boldsymbol{bh}$, the left plane $\boldsymbol{bh}$ carries a negative electric charge, and the right plane $\boldsymbol{bh}$ carries a positive electric charge, resulting in an voltage $\boldsymbol{V_H}$ (i.e. Hall voltage) between left plane $\boldsymbol{bh}$ and right plane $\boldsymbol{bh}$.

Since the number of free electrons $\boldsymbol{N}$ in the energized cross-section $\boldsymbol{S = hd}$ always remains the same, the Lorentzian magnetic force $\boldsymbol{F_B}$ only causes a change in the distribution of the electric charge within the cross-section $\boldsymbol{S}$, and does not change the number of electric charges within the cross-section $\boldsymbol{S}$.

According to equation (15.2.16), the electric charge $\boldsymbol{e}$ of a moving electron can be decomposed into rotational electric charge $\boldsymbol{e_\omega}$ and magnetic electric charge $\boldsymbol{e_v}$, i.e:

$$\boldsymbol{e = \sqrt{e_\omega^2 + e_v^2}} \qquad (21.4.9)$$

If the moving charge $\boldsymbol{q}$ contains $\boldsymbol{N}$ electrons $\boldsymbol{e}$, then the moving charge $\boldsymbol{q}$ is:

$$\boldsymbol{q = eN = \sqrt{(e_\omega N)^2 + (e_v N)^2}} \qquad (21.4.10)$$

The rotational electric charge $\boldsymbol{q_\omega}$ contained in the motion electric charge $\boldsymbol{q}$ is:

$$\boldsymbol{q_\omega = e_\omega N} \qquad (21.4.11)$$

Since the rotating electric charge $\boldsymbol{q_\omega}$ is the electric charge that generates the electric field, the Hall electric field $\boldsymbol{E_H}$ is generated by the rotating electric charge $\boldsymbol{q_\omega}$.

The magnetic electric charge $\boldsymbol{q_v}$ contained in the motion electric charge $\boldsymbol{q}$ is:

$$\boldsymbol{q_v = e_v N} \qquad (21.4.12)$$

Although the magnetic electric charge $\boldsymbol{q_v}$ does not produce the electric field $\boldsymbol{E_H}$, the magnetic electric charge $\boldsymbol{q_v}$ causes the free electron $-\boldsymbol{e}$ to move in the magnetic field $\boldsymbol{B}$ to the left plane $\boldsymbol{bh}$, thus producing the Hall voltage $\boldsymbol{V_H}$.

The Hall voltage $\boldsymbol{V_H}$ produces a uniform electric field $\boldsymbol{E_H}$ (Hall's electric field) in the conductor plate , and the direction of the Hall's electric field $\boldsymbol{E_H}$ points from the right plane $\boldsymbol{bh}$ to the left plane $\boldsymbol{bh}$. Since the distance between the left plane $\boldsymbol{bh}$ and the right plane $\boldsymbol{bh}$ is $\boldsymbol{d}$, the Hall's electric field $\boldsymbol{E_H}$ is:

$$\boldsymbol{E_H = \frac{V_H}{d}} \qquad (21.4.13)$$

The Hall electric field $\boldsymbol{E_H}$ is similar to the electric field of a planar capacitor. The direction of the Hall electric field $\boldsymbol{E_H}$ is perpendicular to the plane defined by both electric current $\boldsymbol{I}$ and the magnetic field $\boldsymbol{B}$.

Since the Hall electric field $\boldsymbol{E_H}$ has electric field force $\boldsymbol{F_H}$ only for rotational electric charge $\boldsymbol{q_\omega}$ and not for magnetic electric charge $\boldsymbol{q_v}$, the electric field force $\boldsymbol{F_H}$ of the Hall electric field $\boldsymbol{E_H}$ on the free electrons in the cross-section $\boldsymbol{S = hd}$ is:

$$\boldsymbol{F_H = q_\omega E_H = \frac{q_\omega V_H}{d}} \qquad (21.4.14)$$

According to equation (18.3.3), the rotational electric charge $\boldsymbol{q_\omega}$ contained in the motion electric charge $\boldsymbol{q}$ is:

$$\boldsymbol{q_\omega = q\sqrt{1 - \frac{v^2}{c^2}}}$$

Substituting the above equation into equation (21.4.14) gives the relationship equation:

$$\boldsymbol{F_H = \sqrt{1 - \frac{v^2}{c^2}}\frac{qV_H}{d}} \qquad (21.4.15)$$

The Hall electric field force $\boldsymbol{F_H}$ is the electric field force contained in the cross section $\boldsymbol{S = hd}$, and the direction of the Hall electric field force $\boldsymbol{F_H}$ is opposite to the direction of the Lorentz force $\boldsymbol{F_B}$. This is shown in Figure 21-4.

When the free electrons $-\boldsymbol{e}$ gather to a certain number in the left plane $\boldsymbol{bh}$. Since negative electrons $-\boldsymbol{e}$ are repulsive to each other, the negative electrons $-\boldsymbol{e}$ gathered on the left plane $\boldsymbol{bh}$ will prevent other negative electrons $-\boldsymbol{e}$ from approaching the left plane $\boldsymbol{bh}$.

Since the Lorentz force $\boldsymbol{F_B}$ makes the negative electron $-\boldsymbol{e}$ in the cross-section $\boldsymbol{S = hd}$ move to the left, and the Hall electric field force $\boldsymbol{F_H}$ prevents the negative electron $-\boldsymbol{e}$ from moving to the left, the Hall electric field force $\boldsymbol{F_H}$ is equal to the Lorentz force $\boldsymbol{F_B}$ when the negative electron $-\boldsymbol{e}$ no longer moves to the left, i. e:

$$F_H = F_B \tag{21.4.16}$$

Substituting equation (21.4.8) and equation (21.4.15) into the above equation yields the relationship equation:

$$\sqrt{1-\frac{v^2}{c^2}}\frac{qV_H}{d} = IB \tag{21.4.17}$$

Substituting the free electric charge $q = enhd$ contained in the cross section $S = hd$ into the above equation gives the relational equation:

$$V_H = \frac{1}{ne\sqrt{1-\frac{v^2}{c^2}}}\frac{IB}{h} \tag{21.4.18}$$

The above equation is the new Hall effect equation. The above formula shows that the Hall voltage V_H becomes larger as the electron displacement velocity v increases. If the electron displacement velocity v is much less than the speed of light, then the above equation becomes:

$$V_H = \frac{1}{ne}\frac{IB}{h} \tag{21.4.19}$$

The above equation is the classical Hall effect equation.

Substituting the electric current $I = envhd$ in the cross section $S = hd$ into equation (21.4.18) gives the relational equation:

$$V_H = \frac{1}{ne\sqrt{1-\frac{v^2}{c^2}}}\frac{envhd \cdot B}{h}$$

Simplifying the above equation gives the relationship equation:

$$V_H = \frac{vBd}{\sqrt{1-\frac{v^2}{c^2}}} \tag{21.4.20}$$

The above equation shows that the Hall voltage V_H is proportional to the magnetic susceptibility B and the width d of the metal plate, and the Hall voltage V_H becomes larger with the increase of the electron displacement velocity v. The Hall voltage V_H is independent of the length b and height h of the plate.

Although the Hall voltage V_H in the above equation is independent of the free electron density n, the Hall voltage V_H is related to the free electron density n since the number of free electrons $N = nS$ contained in the cross-section $S = hd$ is proportional to the density n.

21.4.4 New Hall coefficient equation

According to equation (21.4.18), the new Hall coefficient R_H is:

$$R_H = \frac{1}{ne\sqrt{1-\frac{v^2}{c^2}}} \tag{21.4.21}$$

The above equation shows that the Hall coefficient R_H becomes larger as the electron displacement velocity v increases.

When the room temperature increases, the moving electrons absorb energy, which increases the electron displacement velocity v. When the room temperature decreases the electrons radiate energy, which makes the electron displacement velocity v smaller. Below we analyze and determine the relationship between the electron displacement velocity v and the energy element ε_h.

According to equation (15.1.8), the electron electric charge e is related to the electron mass m_e by the equation:

$$\frac{e}{\eta} = m_e$$

The constant η in equation is the charge-mass coefficient (ratio of electric charge e to mass m_e).

Assume that v_0 is the electron displacement velocity at room temperature. Multiplying the above equation by the square of the velocity v_0 gives the relational equation:

$$\frac{e}{\eta}v_0^2 = m_e v_0^2 \tag{21.4.22}$$

When the room temperature increases, the moving electron absorbs energy, which increases the room temperature velocity v_0 of the electron, and when the room temperature decreases the electron radiates energy, which makes the room temperature velocity v_0 of the electron smaller,.

If the moving electron absorbs or radiates l energy elements ε_h, i.e., absorbed or radiated energy $\varepsilon = l\varepsilon_h$, then the displacement velocity of the electron changes from the room temperature velocity v_0 to the velocity v, i.e:

$$\frac{e}{\eta}v^2 = m_e v_0^2 \pm l\varepsilon_h \tag{21.4.23}$$

Simplifying the above equation gives the relationship equation:

$$v^2 = \frac{\eta m_e v_0^2 \pm \eta l \varepsilon_h}{e} = v_0^2 \pm l\frac{\varepsilon_h}{m_e} \tag{21.4.24}$$

In the formula l is the amount of energy element ε_h absorbed by the electron. The above equation can be varied as:

$$v = \sqrt{v_0^2 \pm l\frac{\varepsilon_h}{m_e}} \tag{21.4.25}$$

The above equation is the quantum formula for the electron velocity. Substituting the above equation into equation (21.4.21), the Hall coefficient R_H is:

$$R_H = \frac{1}{ne\sqrt{1 - \frac{v_0^2 \pm l\frac{\varepsilon_h}{m_e}}{c^2}}} \tag{21.4.26}$$

The above equation is a quantum formula for the Hall coefficient. Assume that the temperature coefficient g is:

$$g = n\sqrt{1 - \frac{v_0^2 \pm l\frac{\varepsilon_h}{m_e}}{c^2}} \tag{21.4.27}$$

Substituting the above equation into equation (21.4.26) gives the relationship equation:

$$R_H = \frac{1}{ge} \tag{21.4.28}$$

Since the temperature coefficient g in equation contains the energy element ε_h, the variation of the Hall coefficient R_H is quantum in nature.

If the more energy $\varepsilon = l\varepsilon_h$ is absorbed by the electron, then the closer the electron velocity v is to the speed of light c. According to equation (21.4.27) it can be determined that the temperature coefficient g_0 for low temperatures is greater than that for high temperatures g i.e:

$$g_0 > g$$

According to the above equation and equation (21.4.28) it can be determined that the Hall coefficient R_{H0} at low temperatures is less than the Hall coefficient R_H at high temperatures , i.e:

$$R_{H0} = \frac{1}{g_0 e} < R_H = \frac{1}{ge} \tag{21.4.29}$$

Since the Hall coefficient R_H belongs to the resistance R in the electric current loop, the resistance of the conductor plate at low temperatures R_{H0} is less than that at high temperatures R_H. This derivation has been confirmed in low temperature superconductivity experiments.

Following the Hall coefficient formula in classical mechanics, equation (21.4.26) can also be written in the following form:

$$R_H = \frac{1}{\frac{ne^2}{e}\sqrt{1 - \frac{v_0^2 \pm l\frac{\varepsilon_h}{m_e}}{c^2}}} = \frac{1}{ie^2} \tag{21.4.30}$$

According to the above equation coefficient i is:

$$i = \frac{n}{e}\sqrt{1 - \frac{v_0^2 \pm l\frac{\varepsilon_h}{m_e}}{c^2}} \qquad (21.4.31)$$

Since the formula contains the energy element ε_h, the variation of the coefficient i is equally quantum. Physicists have confirmed experimentally that the coefficient i can be an integer, or a fraction.

Chapter 22 Theoretical proof that Gauss's law $\oiint BdS=0$ for magnetic fields is wrong

Introduction: This chapter analyzes and discusses the errors made by physicists in deriving Gauss's law $\oiint \boldsymbol{BdS} = \boldsymbol{0}$ for magnetic fields.

22.1 Electric flux Φ_E through Wang's circular crown $S_W=2\pi Rr$

22.1.1 Derivation of Gauss's law for electrostatic fields

Suppose O is the center of the circular line $L_R = 2\pi R$, and Z is the perpendicular distance from the mass m to the center O. Suppose Ω is the solid angle possessed by the circular plane $= \pi R^2$, and θ is the angle between the distance r and the perpendicular distance Z. Assume that the surface S is bordered by the circular line $L_R = 2\pi R$. This is shown in Figure 22-1.

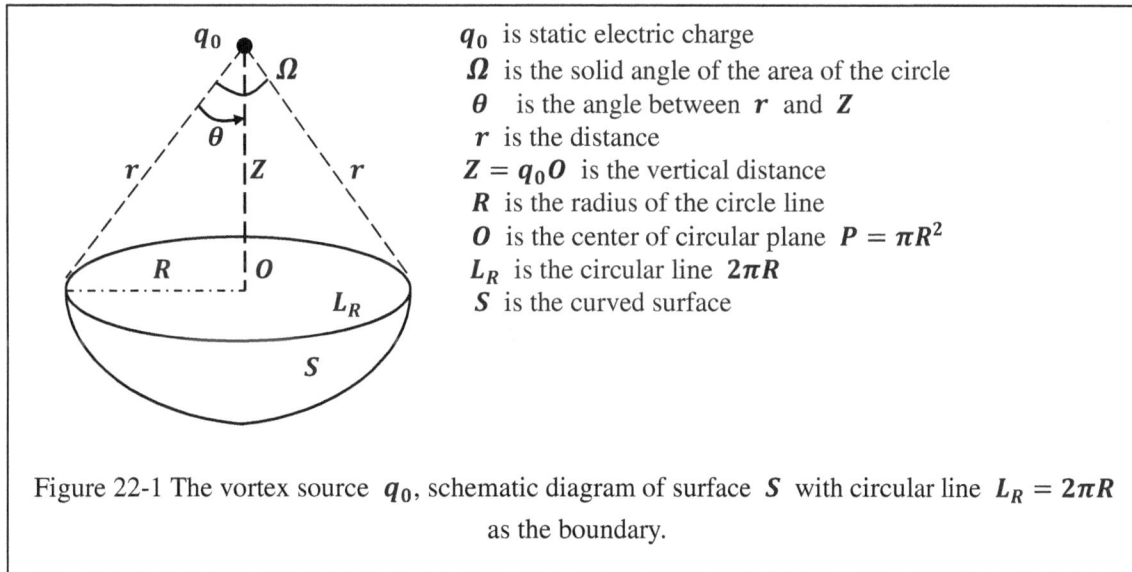

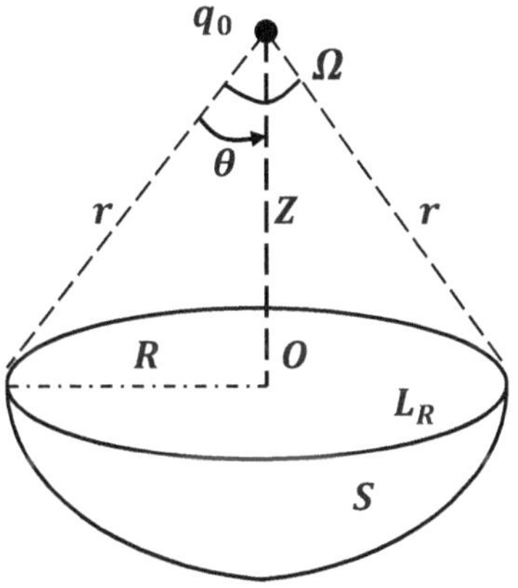

q_0 is static electric charge
Ω is the solid angle of the area of the circle
θ is the angle between r and Z
r is the distance
$Z = q_0 O$ is the vertical distance
R is the radius of the circle line
O is the center of circular plane $P = \pi R^2$
L_R is the circular line $2\pi R$
S is the curved surface

Figure 22-1 The vortex source q_0, schematic diagram of surface S with circular line $L_R = 2\pi R$ as the boundary.

According to equation (17.4.2), the constant D_q is:

$$D_q = \frac{1}{4\pi\varepsilon_0} \tag{22.1.1}$$

Substituting the above equation into equation (17.3.7), the static electric potential φ_0 is:

$$\begin{cases} \varphi_0 = +\dfrac{1}{4\pi\varepsilon_0}\dfrac{q_0}{r}, & r > 1 \\ \varphi_0 = 0, & r = 1 \\ \varphi_0 = -\dfrac{1}{4\pi\varepsilon_0}\dfrac{q_0}{r}, & 0 < r < 1 \end{cases} \tag{22.1.2}$$

The above equation is the new static electric potential equation.

Assume that the distance r from the electric charge q to each point on the circular line $L_R = 2\pi R$ is equal. Since the static electric potential φ_0 remains constant on the circular line L_R, the integral $\oint \varphi_0 dl$ of the static electric potential φ_0 along the circular line L_R is:

$$\oint \varphi_0 dl = \varphi_0 L_R = 2\pi R \cdot \frac{1}{4\pi\varepsilon_0}\frac{q_0}{r} \tag{22.1.3}$$

According to equation (17.3.3), the integral $\oint \varphi_0 dl$ of the static electric potential φ_0 along the circular line $L_R = 2\pi R$ is:

$$\oint \varphi_0 dl = \iint E_0\, dS = 2\pi R \cdot D_q \frac{q_0}{r}$$

Substituting equation (22.1.1) into the above equation gives the relationship equation:

$$\oint \varphi_0 dl = \iint E_0\, dS = 2\pi R \cdot \frac{1}{4\pi\varepsilon_0}\frac{q_0}{r} \tag{22.1.4}$$

The surface S in equation is bordered by the circular line $L_R = 2\pi R$. According to equation (17.4.4), the electrostatic field strength E_0 generated by the static electric charge q_0 is:

$$E_0 = -\frac{1}{4\pi\varepsilon_0}\frac{q_0}{r^2} \tag{22.1.5}$$

When the surface S is bordered by a circular line $L_R = 2\pi R$. According to the definition of flux and equation (22.1.4), the electrostatic flux Φ_{E0} through the surface S is:

$$\Phi_{E0} = \oint \varphi_0 dl = \iint E_0\, dS = 2\pi R \cdot \frac{1}{4\pi\varepsilon_0}\frac{q_0}{r} \tag{22.1.6}$$

Note that the electrostatic potential φ_0 is equal at every point on the circular line $L_R = 2\pi R$. The direction of the electric flux Φ_{E0} of a positive charge is from the positive charge $+q_0$ to infinity. The direction of the electric flux Φ_{E0} of a negative charge is from infinity pointing toward the negative charge $-q_0$.

when the radius R of the circular line is kept constant. According to equation (22.1.6), the electrostatic flux Φ_{E0} through the surface S is:

$$\Phi_{E0} = \frac{R}{2\varepsilon_0}\frac{q_0}{r} \tag{22.1.7}$$

The surface $S = 2\pi r^2$ when the radius of the circular line $R = r$. Substituting $R = r$ into the above equation, the electrostatic flux Φ_{E0} through the surface $S = 2\pi r^2$ is:

$$\Phi_{E0} = \frac{q_0}{2\varepsilon_0} \tag{22.1.8}$$

when the surface S is a closed sphere $S_O = 4\pi r^2$. Since the closed sphere $S_O = 4\pi r^2$ is twice as large as the half sphere $S = 2\pi r^2$, the electrostatic flux Φ_O through the closed sphere $S_O = 4\pi r^2$ is:

$$\Phi_O = \oiint E_0 dS_O = 2\Phi_{E0} = \frac{q_0}{\varepsilon_0} \tag{22.1.9}$$

The above equation is the classical Gauss's law for electrostatic fields. The above equation shows that the electrostatic flux Φ_O through the closed sphere $S_O = 4\pi r^2$ is independent of the magnitude of the radius r of the closed sphere. Note that the static electric charge q_0 is inside the sphere $S_O = 4\pi r^2$. If the static electric charge q_0 lies outside the sphere $S_O = 4\pi r^2$, then the electrostatic flux $\Phi_O = 0$.

22.1.2 Electric flux Φ_E produced by motion electric charge q

When electric charge q is moving with velocity v in the cosmic vacuum frame of reference. Assume that θ is the angle between the distance r and the direction of motion of electric charge q. According to equation (19.1.6), the electromagnetic field strength E produced by the moving electric charge q is:

$$E = -\frac{1}{4\pi\varepsilon_0}\frac{q}{r^2}\sin\theta \tag{22.1.10}$$

Assume that the angle change interval $\Delta\theta = \theta_2 - \theta_1$, where θ_1 is the starting point of the angle change and θ_2 is the end point of the angle change. Taking any angular element $d\theta$ on the pinch angle change interval $\Delta\theta$ and differentiating the electric field strength E with respect to the pinch angle θ, the relation equation can be obtained:

$$dE = \frac{\partial E}{\partial \theta} d\theta = -\frac{1}{4\pi\varepsilon_0}\frac{q\cos\theta}{r^2} d\theta \tag{22.1.11}$$

Suppose $\boldsymbol{S}$ is a surface in an electric field, and take any surface element $\boldsymbol{dS}$ on the surface $\boldsymbol{S}$. Suppose $\boldsymbol{\delta}$ is the angle between the normal line $\boldsymbol{n_0}$ of the surface element $\boldsymbol{dS}$ and the electric field $\boldsymbol{E}$. By the definition of flux, the electric flux $\boldsymbol{d\Phi_E}$ through the surface element $\boldsymbol{dS}$ is:

$$d\Phi_E = dE\cos\delta\, dS \tag{22.1.12}$$

When the angle $\boldsymbol{\delta = 0}$ between the normal line $\boldsymbol{n_0}$ of the surface element $\boldsymbol{dS}$ and the direction of the electric field, the electric flux $\boldsymbol{d\Phi_E}$ through the surface element $\boldsymbol{dS}$ is:

$$d\Phi_E = dEdS \tag{22.1.13}$$

Substituting equation (22.1.11) into the above equation gives the relationship equation:

$$d\Phi_E = -\frac{1}{4\pi\varepsilon_0}\frac{q\cos\theta}{r^2} d\theta dS \tag{22.1.14}$$

The electric flux $\boldsymbol{\Phi_E}$ through the surface $\boldsymbol{S}$ is:

$$\Phi_E = \iint d\Phi_E = -\frac{1}{4\pi\varepsilon_0}\iint \frac{q}{r^2}\left(\int_{\theta_1}^{\theta_2}\cos\theta\, d\theta\right) dS \tag{22.1.15}$$

The above formula is the electric flux formula for the motion electric charge.

22.1.3 Electric flux Φ_E through Wang's circular crown $S_W=2\pi Rr$

Suppose O is the center of the circular line $\boldsymbol{L_R = 2\pi R}$, and $\boldsymbol{Z}$ is the perpendicular distance from the mass $\boldsymbol{m}$ to the center O. Suppose $\boldsymbol{\Omega}$ is the solid angle possessed by the circular plane $\boldsymbol{P = \pi R^2}$, and $\boldsymbol{\theta}$ is the angle between the distance $\boldsymbol{r}$ and the perpendicular distance $\boldsymbol{Z}$. Assume that the surface $\boldsymbol{S}$ is bordered by the circular line $\boldsymbol{L_R = 2\pi R}$. Suppose that the electric charge $\boldsymbol{q}$ is moving with velocity $\boldsymbol{v}$ towards the center O of the circular line $\boldsymbol{L_R = 2\pi R}$. This is shown in Figure 22-2.

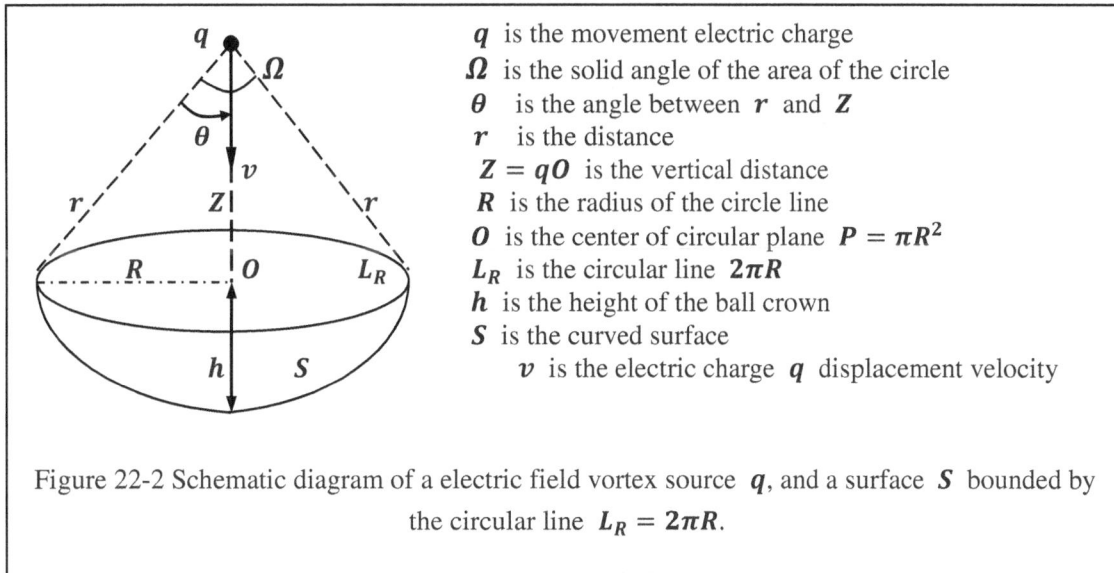

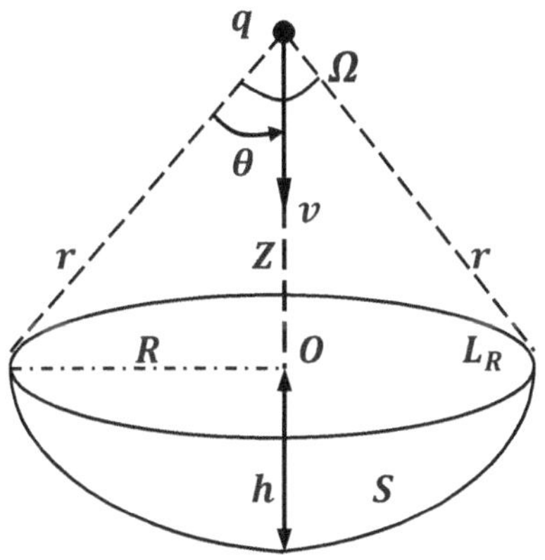

$\boldsymbol{q}$ is the movement electric charge
$\boldsymbol{\Omega}$ is the solid angle of the area of the circle
$\boldsymbol{\theta}$ is the angle between $\boldsymbol{r}$ and $\boldsymbol{Z}$
$\boldsymbol{r}$ is the distance
$\boldsymbol{Z = qO}$ is the vertical distance
$\boldsymbol{R}$ is the radius of the circle line
$\boldsymbol{O}$ is the center of circular plane $\boldsymbol{P = \pi R^2}$
$\boldsymbol{L_R}$ is the circular line $\boldsymbol{2\pi R}$
$\boldsymbol{h}$ is the height of the ball crown
$\boldsymbol{S}$ is the curved surface
$\boldsymbol{v}$ is the electric charge $\boldsymbol{q}$ displacement velocity

Figure 22-2 Schematic diagram of a electric field vortex source $\boldsymbol{q}$, and a surface $\boldsymbol{S}$ bounded by the circular line $\boldsymbol{L_R = 2\pi R}$.

Theoretically, there are an infinite number of surfaces $\boldsymbol{S}$ bordered by the circular line $\boldsymbol{L_R = 2\pi R}$. The spherical crown $\boldsymbol{S_\Omega = 2\pi hr}$ is one of them, as is Wang's circular crown $\boldsymbol{S_W = 2\pi Rr}$.

If the surface $\boldsymbol{S}$ in (22.1.15) uses Wang's circular crown $\boldsymbol{S_W = 2\pi Rr}$, then when the radius $\boldsymbol{r}$ is kept constant. Since both the electric field strength $\boldsymbol{E}$ and Wang's circular crown $\boldsymbol{S_W}$ are constants, the integral of the electric field strength $\boldsymbol{E}$ over Wang's circular crown $\boldsymbol{S_W}$ is:

$$\Phi_E = \iint EdS = -\frac{1}{4\pi\varepsilon_0}\frac{q\cdot 2\pi Rr}{r^2}\int_{\theta_1}^{\theta_2}\cos\theta\, d\theta = -\frac{Rq}{2\varepsilon_0 r}\int_{\theta_1}^{\theta_2}\cos\theta\, d\theta \tag{22.1.16}$$

Integrating the above equation gives the relationship equation:

$$\boldsymbol{\Phi}_E = \frac{1}{2\varepsilon_0}\frac{Rq(\sin\theta_2 - \sin\theta_1)}{r} \qquad (22.1.17)$$

Assume that the distance Z line is the starting line of the angle change interval $\Delta\boldsymbol{\theta} = \boldsymbol{\theta}_2 - \boldsymbol{\theta}_1$, i.e., the angle starting point $\boldsymbol{\theta}_1 = \mathbf{0}$ and the angle ending point $\boldsymbol{\theta}_2 = \boldsymbol{\theta}$, so the above equation becomes:

$$\boldsymbol{\Phi}_E = \frac{R}{2\varepsilon_0}\frac{q\sin\theta}{r} \qquad (22.1.18)$$

The $\boldsymbol{\Phi}_E$ in equation is the formula for the electric flux through Wang's circular crown $\boldsymbol{S}_W = \mathbf{2}\boldsymbol{\pi}\boldsymbol{R}\boldsymbol{r}$. Theoretically, the electric flux $\boldsymbol{\Phi}_E$ is obtained using Stokes' theorem, i.e:

$$\oint \varphi dl = \iint \nabla \times \varphi\, dS = \iint EdS = \boldsymbol{\Phi}_E \qquad (22.1.19)$$

The above equation shows that the kinematic electric flux $\boldsymbol{\Phi}_E$ is equal to the loop integral $\oint \varphi dl$ of the kinematic electric potential $\boldsymbol{\varphi}$.

22.1.4 Equation for electric field flux expressed using distance Z

Since the distance Z is perpendicular to the plane of the circular line $\boldsymbol{P} = \boldsymbol{\pi}\boldsymbol{R}^2$, the distance $\boldsymbol{r}$ from the motion electric charge $\boldsymbol{q}$ to any point on the circular line $\boldsymbol{L}_R = \mathbf{2}\boldsymbol{\pi}\boldsymbol{R}$ is:

$$r = \sqrt{R^2 + Z^2}$$

The function $\sin\boldsymbol{\theta}$ is:

$$\sin\theta = \frac{R}{r} = \frac{R}{\sqrt{R^2 + Z^2}} \qquad (22.1.20)$$

Substituting the above equation into equation (22.1.18), there is the relationship equation:

$$\boldsymbol{\Phi}_E = \iint EdS = \frac{1}{2\varepsilon_0}\frac{qR^2}{r^2} = \frac{1}{2\varepsilon_0}\frac{qR^2}{R^2 + Z^2} \qquad (22.1.21)$$

The above equation shows that the electric flux $\boldsymbol{\Phi}_E$ through Wang's circular crown $\boldsymbol{S}_W = \mathbf{2}\boldsymbol{\pi}\boldsymbol{R}\boldsymbol{r}$ is proportional to the radius $\boldsymbol{R}$ squared and inversely proportional to the distance $\boldsymbol{r}$ squared.

When Wang's circular crown $\boldsymbol{S}$ is half a sphere $\boldsymbol{S} = \mathbf{2}\boldsymbol{\pi}\boldsymbol{r}^2$. Since the angle $\boldsymbol{\theta} = \mathbf{90}^0$, the distance $\boldsymbol{Z} = \mathbf{0}$, and the radius $\boldsymbol{R} = \boldsymbol{r}$, the electric flux $\boldsymbol{\Phi}_E$ through the half sphere $\boldsymbol{S} = \mathbf{2}\boldsymbol{\pi}\boldsymbol{r}^2$ is:

$$\boldsymbol{\Phi}_E = \iint EdS = \frac{q}{2\varepsilon_0} \qquad (22.1.2)$$

Since half a sphere $\boldsymbol{S} = \mathbf{2}\boldsymbol{\pi}\boldsymbol{r}^2$ corresponds to the circular line plane $\boldsymbol{P} = \boldsymbol{\pi}\boldsymbol{r}^2$, the above equation is also the electric flux through the largest circular line plane $\boldsymbol{P} = \boldsymbol{\pi}\boldsymbol{r}^2$. Since the closed sphere $\boldsymbol{S}_O = \mathbf{4}\boldsymbol{\pi}\boldsymbol{r}^2$ is two times the half sphere $\boldsymbol{S} = \mathbf{2}\boldsymbol{\pi}\boldsymbol{r}^2$, the electric flux $\boldsymbol{\Phi}_{OE}$ through the closed sphere $\boldsymbol{S}_O = \mathbf{4}\boldsymbol{\pi}\boldsymbol{r}^2$ is:

$$\boldsymbol{\Phi}_{OE} = \oiint EdS_O = 2\boldsymbol{\Phi}_E = \frac{q}{\varepsilon_0} \qquad (22.1.23)$$

The above equation is the electric field flux formula for the motion electric charge $\boldsymbol{q}$. The above equation shows that the electric flux $\boldsymbol{\Phi}_{OE}$ through the closed sphere $\boldsymbol{S}_O = \mathbf{4}\boldsymbol{\pi}\boldsymbol{r}^2$ is independent of the distance $\boldsymbol{r}$. The above equation is valid if the moving electric charge $\boldsymbol{q}$ is inside the closed sphere $\boldsymbol{S}_O = \mathbf{4}\boldsymbol{\pi}\boldsymbol{r}^2$. At this time the electric flux $\boldsymbol{\Phi}_E \neq \mathbf{0}$ through the closed sphere $\boldsymbol{S}_O = \mathbf{4}\boldsymbol{\pi}\boldsymbol{r}^2$.

When the motion electric charge $\boldsymbol{q}$ is located outside the closed sphere $\boldsymbol{S}_O = \mathbf{4}\boldsymbol{\pi}\boldsymbol{r}^2$. Since the electric field lines penetrating into and out of the closed sphere $\boldsymbol{S}_O = \mathbf{4}\boldsymbol{\pi}\boldsymbol{r}^2$ are equal, the electric flux $\boldsymbol{\Phi}_{OE}$ is:

$$\boldsymbol{\Phi}_{OE} = \oiint EdS_O S = 0 \qquad (22.1.24)$$

Note that the condition for the above equation to hold is that the motion electric charge $\boldsymbol{q}$, lies outside the closed sphere $\boldsymbol{S}_O = \mathbf{4}\boldsymbol{\pi}\boldsymbol{r}^2$.

22.2 Rotational electric flux $\Phi_{E\omega}$ and magnetic electric flux Φ_{Ev}

22.2.1 The electric flux Φ_E can be decomposed into two component electric fluxes

When the electric charge q is moving with velocity v in the cosmic vacuum frame of reference. According to equation (18.3.2), the moving electric charge q can be decomposed into the rotational electric charge q_ω and the magnetic electric charge q_v, i.e:

$$q = \sqrt{q_\omega^2 + q_v^2}$$

Substituting the above equation into equation (22.1.18) gives the relationship equation:

$$\Phi_E = \frac{R}{2\varepsilon_0}\frac{\sqrt{q_\omega^2 + q_v^2}\sin\theta}{r} \tag{22.2.1}$$

Based on the above equation, the relationship equation can be obtained:

$$\Phi_E = \sqrt{\left(\frac{q_\omega R\sin\theta}{2\varepsilon_0 r}\right)^2 + \left(\frac{q_v R\sin\theta}{2\varepsilon_0 r}\right)^2} \tag{22.2.2}$$

The above equation shows that the electric flux Φ_E can be decomposed into two component electric fluxes.

22.2.2 Derivation of the rotating electric flux equation

According to equation (19.2.9) the rotating electric field strength E_ω is:

$$E_\omega = -\frac{1}{4\pi\varepsilon_0}\frac{q_\omega}{r^2}\sin\theta$$

The above equation shows that if the distance r is perpendicular to the direction of electron displacement, i.e., the angle $\theta = 90^0$, then the rotating electric field strength E_ω is maximized. If the distance r is located on the line of electron motion, i.e., the angle $\theta = 0$, then the rotating electric field strength $E_\omega = 0$.

Assume that the pinch angle change interval $\Delta\theta = \theta_2 - \theta_1$, where θ_1 is the starting point of the pinch angle change and θ_2 is the end point of the pinch angle change. Taking any angular element $d\theta$ on the pinch angle change interval $\Delta\theta$ and differentiating the rotating electric field strength E_ω with respect to the pinch angle θ, the relation equation can be obtained:

$$dE_\omega = \frac{\partial E_\omega}{\partial\theta}d\theta = -\frac{1}{4\pi\varepsilon_0}\frac{q_\omega\cos\theta}{r^2}d\theta \tag{22.2.3}$$

According to the definition of flux, the rotational electric flux $d\Phi_{E\omega}$ through the surface element dS is:

$$d\Phi_{E\omega} = dE_\omega\cos\delta\, dS \tag{22.2.4}$$

When the angle $\delta = 0$ between the normal line n_0 of the surface element dS and the rotating electric field intensity line, the rotating electric flux $d\Phi_{E\omega}$ through the surface element dS is:

$$d\Phi_{E\omega} = dE_\omega dS \tag{22.2.5}$$

Substituting equation (22.2.3) into the above equation gives the relationship equation:

$$d\Phi_{E\omega} = -\frac{1}{4\pi\varepsilon_0}\frac{q_\omega\cos\theta}{r^2}d\theta dS \tag{22.2.6}$$

Integrating the above equation, the rotating electric flux $\Phi_{E\omega}$ through the surface S is:

$$\Phi_{E\omega} = \iint E_\omega dS = -\frac{1}{4\pi\varepsilon_0}\iint\frac{q_\omega}{r^2}\left(\int_{\theta_1}^{\theta_2}\cos\theta\, d\theta\right)dS \tag{22.2.7}$$

If the surface S adopts Wang's circular crown $S_W = 2\pi Rr$, then when the radius R and the distance r are kept constant, since the rotating electric field strength E_ω and Wang's circular crown S_W are constants, the rotating electric flux $\Phi_{E\omega}$ through Wang 's circular crown $S_W = 2\pi Rr$ the rotating electric flux $\Phi_{E\omega}$ is:

$$\Phi_{E\omega} = -\frac{1}{4\pi\varepsilon_0}\frac{q_\omega\cdot 2\pi Rr}{r^2}\int_{\theta_1}^{\theta_2}\cos\theta\, d\theta \tag{22.2.8}$$

Integrating the above equation gives the relationship equation:

$$\boldsymbol{\Phi}_{E\omega} = \frac{1}{2\varepsilon_0} \frac{Rq_\omega(\sin\theta_2 - \sin\theta_1)}{r} \qquad (22.2.9)$$

Assume that the distance Z line is the starting line of the angle change interval $\Delta\boldsymbol{\theta} = \boldsymbol{\theta}_2 - \boldsymbol{\theta}_1$, i.e., the angle starting point $\boldsymbol{\theta}_1 = \mathbf{0}$ and the angle ending point $\boldsymbol{\theta}_2 = \boldsymbol{\theta}$, so the above equation becomes:

$$\boldsymbol{\Phi}_{E\omega} = \iint E_\omega dS = \frac{1}{2\varepsilon_0} \frac{q_\omega R \sin\theta}{r} \qquad (22.2.10)$$

According to equation (18.3.3), the rotational electric charge $\boldsymbol{q}_\omega$ contained in the moving electric charge $\boldsymbol{q}$ is:

$$q_\omega = q\frac{v_\omega}{c} = q\sqrt{1 - \frac{v^2}{c^2}}$$

Substituting the above equation into equation (22.2.10) gives the relationship equation:

$$\boldsymbol{\Phi}_{E\omega} = \iint E_\omega dS = \frac{R}{2\varepsilon_0} \frac{q\sin\theta}{r} \sqrt{1 - \frac{v^2}{c^2}} = \sqrt{1 - \frac{v^2}{c^2}} \boldsymbol{\Phi}_E \qquad (22.2.11)$$

The above equation is the rotational electric flux equation. The above equation shows that the rotational electric flux $\boldsymbol{\Phi}_{E\omega}$ through Wang's circular crown $\boldsymbol{S}_W = \mathbf{2\pi Rr}$ becomes smaller as the electric charge velocity $\boldsymbol{v}$ increases. When the electric charge velocity $\boldsymbol{v} = \mathbf{0}$, the rotating electric flux $\boldsymbol{\Phi}_{E\omega}$ changes to the electrostatic field flux $\boldsymbol{\Phi}_{E0}$, i.e:

$$\boldsymbol{\Phi}_{E\omega} = \boldsymbol{\Phi}_{E0}$$

22.2.3 Gauss's law for rotating electric fields

When the angle $\boldsymbol{\theta} = \mathbf{90^0}$. Since the radius $\boldsymbol{R} = \boldsymbol{r}$, Wang's circular crown $\boldsymbol{S}_W = \mathbf{2\pi Rr}$ is half a sphere $\boldsymbol{S} = \mathbf{2\pi r^2}$, so the rotating electric flux $\boldsymbol{\Phi}_{E\omega}$ through half a sphere $\boldsymbol{S} = \mathbf{2\pi r^2}$ is:

$$\boldsymbol{\Phi}_{E\omega} = \iint E_\omega dS = \frac{q}{2\varepsilon_0} \sqrt{1 - \frac{v^2}{c^2}} \qquad (22.2.12)$$

Since the closed sphere $\boldsymbol{S}_O = \mathbf{4\pi r^2}$ is twice as large as half a sphere $\boldsymbol{S} = \mathbf{2\pi r^2}$, the rotating electric flux $\boldsymbol{\Phi}_{OE\omega}$ through the closed sphere $\boldsymbol{S}_O = \mathbf{4\pi r^2}$ is:

$$\boldsymbol{\Phi}_{OE} = \oiint E_\omega dS_O = 2\boldsymbol{\Phi}_{E\omega} = \frac{q}{\varepsilon_0} \sqrt{1 - \frac{v^2}{c^2}} \qquad (22.2.13)$$

The above equation is Gauss's law for rotating electric field. The above equation shows that the magnetic electric flux $\oiint \boldsymbol{E}_v \boldsymbol{dS}_O$ through the closed sphere $\boldsymbol{S}_O$ is independent of the magnitude of the radius $\boldsymbol{r}$. The condition for the above equation to hold is that the motion electric charge $\boldsymbol{q}$ is located inside the closed sphere $\boldsymbol{S}_O = \mathbf{4\pi r^2}$.

When the electric charge velocity $\boldsymbol{v} = \mathbf{0}$, the rotational electric flux $\boldsymbol{\Phi}_{OE\omega}$ through the closed sphere $\boldsymbol{S}_O = \mathbf{4\pi r^2}$ changes to the electrostatic field flux $\boldsymbol{\Phi}_O$ at the closed sphere, i.e:

$$\boldsymbol{\Phi}_{OE\omega} = \boldsymbol{\Phi}_O = \frac{q_0}{\varepsilon_0}$$

The above equation shows that the rotating electric field $\boldsymbol{E}_\omega$ is an active field and the rotating electric field lines are not closed curves.

When the motion electric charge $\boldsymbol{q}$ is located outside the closed sphere $\boldsymbol{S}_O = \mathbf{4\pi r^2}$. Since the rotating electric field lines penetrating into and out of the closed sphere $\boldsymbol{S}_O = \mathbf{4\pi r^2}$ are equal, the rotating electric flux $\boldsymbol{\Phi}_{OE\omega}$ is:

$$\boldsymbol{\Phi}_{OE\omega} = \oiint E_\omega dS_O = 0 \qquad (22.2.14)$$

The rotational electric flux $\boldsymbol{\Phi}_{OE\omega}$ is equal to 0 for the same reason that Gauss's law for magnetic fields is equal to 0.

22.2.4 Derivation of magnetic electric flux equations

According to equation (19.3.2) the Magnetic-electric field strength $\boldsymbol{E}_v$ generated by the moving electric charge $\boldsymbol{q}$

is

$$E_v = +\frac{1}{4\pi\varepsilon_0}\frac{qv}{cr^2}\sin\theta \tag{22.2.15}$$

The above equation shows that if the distance r is perpendicular to the direction of electron displacement, i.e., the angle $\theta = 90^0$, then the Magnetic-electric field strength E_v reaches its maximum. If the distance r is located on the line of electron motion, i.e., the angle $\theta = 0$, then the Magnetic-electric field strength $E_v = 0$.

Assume that the angle change interval $\Delta\theta = \theta_2 - \theta_1$, where θ_1 is the starting point of the angle change and θ_2 is the end point of the angle change. Taking any angular element $d\theta$ on the pinch angle change interval $\Delta\theta$ and differentiating the Magnetic-electric field strength E_v with respect to the pinch angle θ, the relation equation can be obtained:

$$dE_v = \frac{\partial E_v}{\partial\theta}d\theta = \frac{1}{4\pi\varepsilon_0}\frac{qv\cos\theta}{cr^2}d\theta \tag{22.2.16}$$

According to the definition of flux, the magnetic electric flux $d\Phi_{Ev}$ through the surface element dS is:

$$d\Phi_{Ev} = dE_v\cos\delta\, dS \tag{22.2.17}$$

When the angle $\delta = 0$ between the normal line n_0 of the surface element dS and the rotating electric field intensity line, the magnetic electric flux $d\Phi_{Ev}$ through the surface element dS is:

$$d\Phi_{Ev} = dE_v dS \tag{22.2.18}$$

Substituting equation (22.2.16) into the above equation gives the relationship equation:

$$d\Phi_{Ev} = \frac{1}{4\pi\varepsilon_0}\frac{qv\cos\theta}{cr^2}d\theta dS \tag{22.2.19}$$

Integrating the above equation over the surface S and the angle change interval $\Delta\theta = \theta_2 - \theta_1$, the magnetic electric flux Φ_{Ev} through the surface S is then:

$$\Phi_{Ev} = \iint E_v dS = \frac{1}{4\pi\varepsilon_0 c}\iint \frac{qv}{r^2}\left(\int_{\theta_1}^{\theta_2}\cos\theta\, d\theta\right)dS \tag{22.2.20}$$

If the surface S adopts Wang's circular crown $S_W = 2\pi Rr$, then when the radius R and the distance r are kept constant, since the Magnetic-electric field strength E_v and Wang's circular crown S_W are constants, the magnetic electric flux Φ_{Ev} through Wang 's circular crown $S_W = 2\pi Rr$ is:

$$\Phi_{Ev} = \frac{1}{4\pi\varepsilon_0 c}\frac{qv\cdot 2\pi Rr}{r^2}\left(\int_{\theta_1}^{\theta_2}\cos\theta\, d\theta\right) \tag{22.2.21}$$

Integrating the above equation gives the relationship equation:

$$\Phi_{Ev} = -\frac{R}{2\varepsilon_0 c}\frac{qv(\sin\theta_2 - \sin\theta_1)}{r} \tag{22.2.22}$$

Assume that the distance Z line is the starting line of the angle change interval $\Delta\theta = \theta_2 - \theta_1$, i.e., the angle starting point $\theta_1 = 0$ and the angle ending point $\theta_2 = \theta$, so the above equation becomes:

$$\Phi_{Ev} = \iint E_v dS = -\frac{R}{2\varepsilon_0 c}\frac{qv\sin\theta}{r} \tag{22.2.23}$$

The above equation is the magnetic electric flux equation. The above equation shows that the magnetic electric flux Φ_{Ev} through Wang's circular crown $S_W = 2\pi Rr$ is proportional to the electric charge velocity v. When the electric charge velocity $v = 0$, the static electric charge q_0 does not have magnetic electric flux since the magnetic electric flux $\Phi_{Ev} = 0$.

According to equation (22.1.20), the function $\sin\theta$ is:

$$\sin\theta = \frac{R}{r} = \frac{R}{\sqrt{R^2 + Z^2}}$$

Substituting the above equation into equation (22.2.23), there is the relationship equation:

$$\Phi_{Ev} = \iint E_v dS = -\frac{1}{2\varepsilon_0}\frac{qvR^2}{cr^2} = -\frac{1}{2\varepsilon_0 c}\frac{qvR^2}{R^2 + Z^2} \tag{22.2.24}$$

22.2.5 Gauss's law for Magnetic-electric fields

When the surface $\boldsymbol{S}$ is half sphere $\boldsymbol{S}=\boldsymbol{2\pi r^2}$. Since the angle $\boldsymbol{\theta}=\boldsymbol{90^0}$ and the radius $\boldsymbol{R}=\boldsymbol{r}$, the magnetic flux $\boldsymbol{\Phi_{Ev}}$ through half a sphere $\boldsymbol{S}=\boldsymbol{2\pi r^2}$ according to equation (22.2.23) is:

$$\boldsymbol{\Phi_{Ev}}=-\frac{\boldsymbol{qv}}{\boldsymbol{2\varepsilon_0 c}} \tag{22.2.25}$$

When the surface $\boldsymbol{S}$ is a closed sphere $\boldsymbol{S_O}=\boldsymbol{4\pi r^2}$. Since the closed sphere $\boldsymbol{S_O}$ is twice the size of half a sphere, the magnetic flux $\oiint \boldsymbol{E_v dS_O}$ through the closed sphere $\boldsymbol{S_O}=\boldsymbol{4\pi r^2}$ is:

$$\oiint \boldsymbol{E_v dS_O}=2\iint \boldsymbol{E_v dS}=-\frac{\boldsymbol{qv}}{\boldsymbol{\varepsilon_0 c}} \tag{22.2.26}$$

The above equation is Gauss's law for Magnetic-electric field. The above equation shows that the magnetic electric flux $\oiint \boldsymbol{E_v dS_O}$ through the closed sphere $\boldsymbol{S_O}$ is independent of the magnitude of the radius $\boldsymbol{r}$. The condition for the above equation to hold is that the moving electric charge $\boldsymbol{q}$ lies inside the closed sphere $\boldsymbol{S_O}$. Since the magnetic electric flux $\oiint \boldsymbol{E_v dS_O}\neq \boldsymbol{0}$ through the closed sphere $\boldsymbol{S_O}$, the Magnetic-electric field $\boldsymbol{E_v}$ is an active field.

If the electric charge $\boldsymbol{q}=\boldsymbol{0}$ contained in the closed sphere $\boldsymbol{S_O}=\boldsymbol{4\pi r^2}$, or the motion electric charge $\boldsymbol{q}$ is located outside the closed sphere $\boldsymbol{S_O}=\boldsymbol{4\pi r^2}$, then the magnetic electric flux $\oiint \boldsymbol{E_v dS_O}$ is:

$$\oiint \boldsymbol{E_v dS_O}=\boldsymbol{0} \tag{22.2.27}$$

The magnetic electric flux $\oiint \boldsymbol{E_v dS_O}$ equals 0 for the same reason that Gauss's law for magnetic fields equals 0.

22.3 Gauss's law $\oiint BdS=0$ for magnetic fields is an erroneous theoretical proof

22.3.1 Gauss's law for the electric field *E* in Maxwell's system of equations is Gauss's law for the rotating electric field E_ω

According to the electric flux equation (22.1.23), the electric flux $\boldsymbol{\Phi_{OE}}$ through through the closed sphere $\boldsymbol{S_O}=\boldsymbol{4\pi r^2}$ is:

$$\boldsymbol{\Phi_{OE}}=\oiint \boldsymbol{EdS_O}=\frac{\boldsymbol{q}}{\boldsymbol{\varepsilon_0}} \tag{22.3.1}$$

when the electric charge $\boldsymbol{q}$ is moving with velocity $\boldsymbol{v}$ in the cosmic vacuum frame of reference. According to equation (18.3.2), the moving electric charge $\boldsymbol{q}$ can be decomposed into the rotational electric charge $\boldsymbol{q_\omega}$ and the magnetic electric charge $\boldsymbol{q_v}$, i.e:

$$\boldsymbol{q}=\sqrt{\boldsymbol{q_\omega^2}+\boldsymbol{q_v^2}}$$

Substituting the above equation into equation (22.3.1) gives the relationship equation:

$$\oiint \boldsymbol{EdS_O}=\sqrt{\left(\frac{q_\omega}{\varepsilon_0}\right)^2+\left(\frac{q_v}{\varepsilon_0}\right)^2} \tag{22.3.2}$$

The above equation shows that the electric flux $\oiint \boldsymbol{EdS_O}$ through the closed sphere $\boldsymbol{S_O}=\boldsymbol{4\pi r^2}$ can be decomposed into a rotational electric flux $\boldsymbol{\Phi_{OE\omega}}$ and a magnetic electric flux $\boldsymbol{\Phi_{OEv}}$, viz:

$$\oiint \boldsymbol{EdS_O}=\sqrt{\left(\frac{q_\omega}{\varepsilon_0}\right)^2+\left(\frac{q_v}{\varepsilon_0}\right)^2}=\sqrt{\boldsymbol{\Phi_{OE\omega}^2}+\boldsymbol{\Phi_{OEv}^2}} \tag{22.3.3}$$

In the 19th century, the British physicist Maxwell established a set of equations describing the relationships of electromagnetic fields. The set of Maxwell's equations in integral form is:

$$\begin{cases} \oiint E dS = \dfrac{1}{\varepsilon_0} q & (1) \\ \oiint B dS = 0 & (2) \\ \oint E dl = -\iint \dfrac{\partial B}{\partial t} dS & (3) \\ \oint B dl = \mu_0 I + \mu_0 \varepsilon_0 \iint \dfrac{\partial E}{\partial t} dS & (4) \end{cases}$$

(22.3.4)

The above system of equations describes the law of change of the electromagnetic field in a given volume or area.

According to equation (22.2.13), the rotating electric flux $\boldsymbol{\Phi_{OE\omega}}$ through a closed sphere $\boldsymbol{S_O = 4\pi r^2}$ is:

$$\boldsymbol{\Phi_{OE\omega}} = \oiint \boldsymbol{E_\omega dS} = \frac{q}{\varepsilon_0}\sqrt{1 - \frac{v^2}{c^2}}$$

Since equation (1) in Maxwell's system of equations is an equation that describes the change in the electric field $\boldsymbol{E}$, not the change in the magnetic field $\boldsymbol{B}$, equation (1) should be modified to:

$$\oiint \boldsymbol{E_\omega dS} = \frac{q}{\varepsilon_0}\sqrt{1 - \frac{v^2}{c^2}} \tag{22.3.5}$$

When the electric charge $\boldsymbol{q = 0}$ is contained within the closed sphere $\boldsymbol{S_O = 4\pi r^2}$, the rotational electric flux $\boldsymbol{\Phi_{OE\omega}}$ is:

$$\oiint \boldsymbol{E_\omega dS_O} = 0 \tag{22.3.6}$$

When the electric charge velocity $\boldsymbol{v = 0}$, equation (22.3.5) becomes Gauss's theorem for the electrostatic field, i.e:

$$\oiint \boldsymbol{E_0 dS} = \frac{q_0}{\varepsilon_0} \tag{22.3.7}$$

The system of Maxwell's equations in differential form is:

$$\begin{cases} \nabla \cdot E = \dfrac{\rho}{\varepsilon} & (5) \\ \nabla \cdot B = 0 & (6) \\ \nabla \times E = -\dfrac{\partial B}{\partial t} & (7) \\ \nabla \times B = \mu J + \mu\varepsilon \dfrac{E}{\partial t} & (8) \end{cases}$$

(22.3.8)

The above system of equations describes the pattern of change of the electromagnetic field at each point. $\boldsymbol{\rho}$ is the density of free charge $\boldsymbol{q}$ per unit volume of the medium, $\boldsymbol{\varepsilon}$ is the dielectric constant of the medium, and $\boldsymbol{\mu}$ is the magnetic permeability of the medium.

Since $\boldsymbol{\rho}$ is related to the free charge $\boldsymbol{q}$, the density $\boldsymbol{\rho}$ of the free charge can be decomposed into the rotational charge density $\boldsymbol{\rho_\omega}$ and the magnetic charge density $\boldsymbol{\rho_v}$, viz:

$$\rho = \sqrt{\left(\rho\sqrt{1 - \frac{v^2}{c^2}}\right)^2 + \left(\rho\frac{v}{c}\right)^2} = \sqrt{(\rho_\omega)^2 + (\rho_v)^2} \tag{22.3.9}$$

Since equation (5) in Maxwell's system of equations is a divergence formula that describes the variation of the electric field $\boldsymbol{E}$, not a divergence formula that describes the variation of the magnetic field $\boldsymbol{B}$, equation (5) should be modified to::

$$\nabla \cdot \boldsymbol{E}_\omega = \frac{\rho_\omega}{\varepsilon} = \frac{\rho}{\varepsilon}\sqrt{1 - \frac{v^2}{c^2}} \qquad (22.3.10)$$

If the bulk density $\boldsymbol{\rho} = \mathbf{0}$ of the free charge at a point in space, then the divergence of the rotating electric field $\boldsymbol{E}_\omega$ at that point is:

$$\nabla \cdot \boldsymbol{E}_\omega = 0 \qquad (22.3.11)$$

When the electric charge velocity $\boldsymbol{v} = \mathbf{0}$, equation (22.3.10) becomes the divergence equation for the electrostatic field, i.e:

$$\nabla \cdot \boldsymbol{E}_0 = \frac{\rho}{\varepsilon} \qquad (22.3.12)$$

The above equation is the divergence equation for the electrostatic field.

22.3.2 Derivation of the new magnetic flux formula

Suppose O is the center point of the plane of the circular line $\boldsymbol{L_R} = \boldsymbol{2\pi R}$. Assume that the distance $\boldsymbol{r}$ from the electric charge $\boldsymbol{q}$ to each point of the circular line $\boldsymbol{L_R} = \boldsymbol{2\pi R}$ is equal, and that the electric charge $\boldsymbol{q}$ moves toward the circular line $\boldsymbol{L_R} = \boldsymbol{2\pi R}$ with velocity $\boldsymbol{v}$. Assume that $\boldsymbol{\theta}$ is the angle between the distance $\boldsymbol{r}$ and the velocity $\boldsymbol{v}$, and that the surface $\boldsymbol{S}$ has the circular line $\boldsymbol{L_R} = \boldsymbol{2\pi R}$ as its boundary. As shown earlier in Figure 22-2.

According to equation (19.3.17), the magnetic flux density $\boldsymbol{B}$ produced by the moving electric charge $\boldsymbol{q}$ at each point on the circular line $\boldsymbol{L_R} = \boldsymbol{2\pi R}$ is:

$$B = +\frac{\mu_0}{4\pi}\frac{qv}{r^2}\sin\theta$$

The direction of the magnetic flux density $\boldsymbol{B}$ is perpendicular to the plane defined by both the radius $\boldsymbol{r}$ and the velocity $\boldsymbol{v}$.

Assume that the angle change interval $\Delta\boldsymbol{\theta} = \boldsymbol{\theta}_2 - \boldsymbol{\theta}_1$, where $\boldsymbol{\theta}_1$ is the starting point of the angle change and $\boldsymbol{\theta}_2$ is the end point of the angle change. Taking the angular element $\boldsymbol{d\theta}$ on the angle change interval $\Delta\boldsymbol{\theta}$ and differentiating the magnetic flux density $\boldsymbol{B}$ with respect to the angle $\boldsymbol{\theta}$, the relation equation can be obtained:

$$dB = \frac{\partial B}{\partial \theta}d\theta = \frac{\mu_0}{4\pi}\frac{qv\cos\theta}{r^2}d\theta \qquad (22.3.13)$$

Suppose $\boldsymbol{S}$ is a surface in a magnetic field and take the surface element $\boldsymbol{dS}$ on the surface $\boldsymbol{S}$. Suppose $\boldsymbol{\delta}$ is the angle between the normal line $\boldsymbol{n}_0$ of the surface element $\boldsymbol{dS}$ and the magnetic flux density line. According to the definition of magnetic flux, the magnetic flux $\boldsymbol{d\Phi_B}$ through the surface element $\boldsymbol{dS}$ is:

$$d\Phi_B = dB\cos\delta\, dS \qquad (22.3.14)$$

When the angle $\boldsymbol{\delta} = \mathbf{0}$ between the normal line $\boldsymbol{n}_0$ of the surface element $\boldsymbol{dS}$ and the magnetic flux density line, the magnetic flux $\boldsymbol{d\Phi_B}$ through the surface element $\boldsymbol{dS}$ is

$$d\Phi_B = dBdS$$

Substituting equation (22.3.14) into the above equation gives the relationship equation:

$$d\Phi_B = \frac{\mu_0}{4\pi}\frac{qv\cos\theta}{r^2}d\theta dS \qquad (22.3.15)$$

Integrating the above equation, the magnetic flux $\boldsymbol{\Phi_B}$ through the surface $\boldsymbol{S}$ is:

$$\Phi_B = \iint BdS = \frac{\mu_0 qv}{4\pi}\iint \frac{1}{r^2}\left(\int_{\theta_1}^{\theta_2}\cos\theta\, d\theta\right)dS \qquad (22.3.16)$$

Note that it is the magnetic flux density line, not the magnetic induction line, that passes through the surface element $\boldsymbol{dS}$. The surface element $\boldsymbol{dS}$ in the above equation is always perpendicular to the magnetic flux density line, i.e., the angle $\boldsymbol{\delta} = \mathbf{0}$.

If the surface $\boldsymbol{S}$ uses Wang's circular crown $\boldsymbol{S_W} = \boldsymbol{2\pi Rr}$, then when the radius $\boldsymbol{R}$ and the distance $\boldsymbol{r}$ are kept constant. Since the magnetic flux density $\boldsymbol{B}$ and Wang's circular crown $\boldsymbol{S_W}$ are constants, the magnetic flux $\boldsymbol{\Phi_B}$ through Wang's circular crown $\boldsymbol{S_W} = \boldsymbol{2\pi Rr}$ is:

$$\Phi_B = \frac{\mu_0 qv}{4\pi}\cdot\frac{2\pi Rr}{r^2}\int_{\theta_1}^{\theta_2}\cos\theta\, d\theta \qquad (22.3.17)$$

Integrating the above equation gives the relationship equation:

$$\Phi_B = -\frac{\mu_0}{2}\cdot\frac{Rqv(\sin\theta_2 - \sin\theta_1)}{r} \tag{22.3.18}$$

Assume that the distance Z line is the starting line of the angle change interval $\Delta\theta = \theta_2 - \theta_1$, i.e., the angle starting point $\theta_1 = 0$ and the angle ending point $\theta_2 = \theta$, so the above equation becomes:

$$\Phi_B = \iint BdS = -\frac{\mu_0 R}{2}\cdot\frac{qv}{r}\sin\theta \tag{22.3.19}$$

The above equation is the new magnetic flux equation. The above equation shows that the magnetic flux Φ_B through Wang's circular crown $S_W = 2\pi Rr$ is proportional to the electric charge q and the velocity v.

22.3.3 Gauss's law *A* = 0 for magnetic fields is an erroneous theoretical proof

According to Maxwell's equations (22.3.4), Gauss's law for magnetic fields is:

$$\oiint E_0 dS = 0 \tag{22.3.20}$$

Physicists believe that the above equation reflects the nature of the magnetic field. The above equation shows that the magnetic field B is a passive field and that magnetic monopoles do not exist. Note that the above formula is not derived using mathematical methods, but is obtained using the fact that the magnetic field lines penetrating the closed sphere $S_O = 4\pi r^2$ are equal to the magnetic field lines penetrating it.

Speaking further. Since magnetic monopoles do not exist in nature, physicists assume that magnetic induction lines are all closed lines with no head or tail. Any magnetic induction line that enters the closed surface S_O must come out of the closed surface S_O. Otherwise the magnetic induction line would not be closed. Since the magnetic flux into the closed surface S_O is negative and the magnetic flux out is positive, the integral $\oiint BdS = 0$ of the magnetic field B over the closed surface S_O .

Since the magnetic field B is generated by the moving electric charge q, Gauss's law for magnetic fields is wrong when the closed surface S_O contains the moving electric charge q. We can prove this conclusion by the following derivation.

When the electric charge q moves with velocity v toward the circular line $L_R = 2\pi R$. According to equation (18.4.3), the Magnetic-electric potential φ_v generated by the moving electric charge q is:

$$\varphi_v = \frac{1}{4\pi\varepsilon_0 c}\frac{qv}{r}\sin\theta \tag{22.3.21}$$

when the distance r from the moving electric charge q to each point on the circular line $L_R = 2\pi R$ is equal. Since the Magnetic-electric potential φ_v is constant on the circular line L_R, the integral $\oint \varphi_v dl$ of the Magnetic-electric potential φ_v along the circular line $L_R = 2\pi R$ is:

$$\oint \varphi_v dl = \varphi_v L_R = \frac{R}{2\varepsilon_0 c}\frac{qv}{r}\sin\theta \tag{22.3.22}$$

Assume that the surface S is bordered by a closed curve L. According to Stokes' theorem, the relation equation can be obtained:

$$\oint \varphi_v dl = \iint \nabla\times\varphi_v dS \tag{22.3.23}$$

According to equation (19.3.7), the Magnetic-electric field strength E_v is:

$$E_v = -\nabla\times\varphi_v \tag{22.3.24}$$

Substituting the above equation into equation (22.3.23) gives the relationship equation:

$$\oint \varphi_v dl = -\iint E_v dS \tag{22.3.25}$$

When the closed curve L is the circular line $L_R = 2\pi R$, the relationship equation can be obtained by substituting equation (22.3.22) into the above equation:

$$\iint E_v dS = -\frac{R}{2\varepsilon_0 c}\frac{qv}{r}\sin\theta$$

Dividing the above equation by the speed of light c, gives the relationship equation:

$$\iint \frac{1}{c} E_v dS = -\frac{R}{2\varepsilon_0 c^2} \frac{qv}{r} \sin\theta \tag{22.3.26}$$

According to equation (19.3.16), the vacuum permeability μ_0 is:

$$\mu_0 = \frac{1}{\varepsilon_0 c^2} \tag{22.3.27}$$

Substituting equation (22.3.26) in the above equation gives the relationship equation:

$$\iint \frac{1}{c} E_v dS = -\frac{\mu_0 R}{2} \frac{qv}{r} \sin\theta \tag{22.3.28}$$

According to equation (19.3.17), the magnetic flux density B is:

$$B = \frac{1}{c} E_v \tag{22.3.29}$$

Substituting the above equation into equation (22.3.28) gives the relationship equation:

$$\iint BdS = -\frac{\mu_0 R}{2} \frac{qv}{r} \sin\theta \tag{22.3.30}$$

When the radius $R = r$, the surface S is half a sphere $S = 2\pi r^2$. Since the angle $\theta = 90^0$ between the distance r and the velocity v, the above equation becomes:

$$\iint BdS = -\frac{\mu_0 qv}{2} \tag{22.3.31}$$

When the surface S is a closed sphere $S_O = 4\pi r^2$. Since the closed sphere $S_O = 4\pi r^2$ is twice as large as half a sphere $S = 2\pi r^2$, the magnetic flux $\oiint BdS$ through the closed sphere $S_O = 4\pi r^2$ is:

$$\oiint BdS_O = 2\iint BdS = -\mu_0 qv \tag{22.3.32}$$

The above equation is the new Gauss law for magnetic fields. Equation (2) in Maxwell's system of equations should be modified to:

$$\oiint BdS_O = -\mu_0 qv \tag{22.3.33}$$

The above equation is the new Gauss's law for magnetic fields. When the electric charge $q = 0$, or the electric charge velocity $v = 0$, contained in the closed sphere $S_O = 4\pi r^2$, then the above equation becomes:

$$\oiint BdS_O = 0 \tag{22.3.34}$$

The above equation is Gauss's law for the magnetic field in Maxwell's system of equations.

To summarize, Gauss's law for magnetic field holds only if the closed surface S_O contains no motion electric charge q. When the closed surface S_O contains motion electric charge $q \neq 0$. Since the magnetic flux $\oiint BdS \neq 0$, the Gaussian law for the magnetic field $\oiint BdS = 0$ in Maxwell's system of equations is wrong.

According to equation (22.3.9), equation (6) in Maxwell's system of equations should be modified to:

$$\nabla \cdot B = \frac{\rho_v}{\varepsilon} = \frac{\rho v}{\varepsilon c} \tag{22.3.35}$$

The above equation is the divergence equation for the magnetic field. Equation (6) in Maxwell's system of equations is incorrect because the divergence $\nabla \cdot B \neq 0$ for the magnetic field B.

When the free charge body density $\rho = 0$ at a point in space, or the electric charge velocity $v = 0$, then the divergence of the magnetic field B is:

$$\nabla \cdot B = 0 \tag{22.3.36}$$

Chapter 23 Derivation of Faraday's Law of Electromagnetic Induction

Introduction: This chapter derives Faraday's law of electromagnetic induction from the magneticelectric potential formula and Stokes' theorem.

23.1 Derivation of the static field induction equation

In order to discuss accurately and avoid misunderstanding, this book uses different symbols to represent different electrostatic fields. The meanings of the symbols are specified as follows:

(1) Symbol $\boldsymbol{E}_{0t}(\boldsymbol{q}_0, \boldsymbol{r})$ is the electrostatic field corresponding to time $\boldsymbol{t}$ that does not contain the rate of change $\left(\frac{dq_0}{dt}, \frac{dr}{dt}\right)$.

(2) Symbol $\boldsymbol{E}_{0t}\left(\frac{dq_0}{dt}, \frac{dr}{dt}\right)$ is the electrostatic field corresponding to time $\boldsymbol{t}$ that contains the rate of change $\left(\frac{dq_0}{dt}, \frac{dr}{dt}\right)$.

23.1.1 Electromotive force ε equals the integral of the electrostatic field E_{0t} along the curve L

In electrodynamics, the electromotive force $\boldsymbol{\varepsilon}$ is defined as the ratio of the non-static electric work $\boldsymbol{W}$ to the moveselectric charge $\boldsymbol{q}$, viz:

$$\varepsilon = \frac{W}{q} \tag{23.1.1}$$

Non-static forces are forces other than static forces that can act on the electric charge moves. Inside the power supply, non-static forces do work on the electric charge when they move the positive electric charge $+\boldsymbol{q}$ from the negative plate to the positive plate. The process of non-electrostatic force doing work is the process of converting the energy of the non-electrostatic field into the energy of the electric field.

The direction of the electromotive force $\boldsymbol{\varepsilon}$ is specified as pointing from the negative terminal of the power supply through the interior of the power supply to the positive terminal of the power supply, i.e., in the opposite direction to the voltage $\boldsymbol{U}$ at the ends of the power supply. electromotive force $\boldsymbol{\varepsilon}$ is capable of generating voltage $\boldsymbol{U}$ at both ends of the power supply and generating electric current in a closed circuit.

According to equation (17.4.4), the electrostatic field strength $\boldsymbol{E}_{0t}(\boldsymbol{q}_0, \boldsymbol{r})$ generated by the static electric charge $\boldsymbol{q}_0$ is:

$$E_{0t}(q_0, r) = -\frac{1}{4\pi\varepsilon_0}\frac{q_0}{r^2} \tag{23.1.2}$$

According to the Coulomb electric field force formula, the electrostatic field force $\boldsymbol{F}_0$ is:

$$F_0 = qE_{0t} = -\frac{1}{4\pi\varepsilon_0}\frac{q_0 q}{r^2} \tag{23.1.3}$$

When the static electric charge $\boldsymbol{q}_0$ is held constant, suppose that the electric charge $\boldsymbol{q}$ moves from point to point $\boldsymbol{b}$. Since the distance $\boldsymbol{r}$ in equation changes, the work $\boldsymbol{W}_0$ done by the electrostatic field $\boldsymbol{E}_{0t}$ along the curve $\boldsymbol{L}$ is:

$$W_0 = \int_{r_a}^{r_b} qE_{0t}dl = +\frac{q_0 q}{4\pi\varepsilon_0}\left(\frac{1}{r_b} - \frac{1}{r_a}\right) \tag{23.1.4}$$

Substituting the above equation into equation (23.1.1), the static electromotive force ε_{LE0} generated by the electrostatic field E_{0t} is:

$$\varepsilon_{LE0} = \frac{W_0}{q} = \int_{r_a}^{r_b} E_{0t}dl = +\frac{q_0}{4\pi\varepsilon_0}\left(\frac{1}{r_b} - \frac{1}{r_a}\right) \tag{23.1.5}$$

Although the static electromotive force ε_{LE0} meets the definition of the electromotive force, physicists exclude the static electromotive force ε_{LE0} from the electromotive force in order to derive the electromagnetic wave equation.

According to the electric field theory, when the static electric charge q_0 remains constant. If the distance $r_a \neq r_b$, then the time-varying electrostatic field E_{0t} is variable at points a and b. If the distance $r_a = r_b$, then there is no voltage U at points a and b. At this point electromotive force $\varepsilon_{LE0} = 0$. From this, it can be determined that the electromotive fforce ε is essentially the line integral of the varying electric field, the

23.1.2 Linear electromotive force ε_{LE0} generated by electrostatic field E_{0t}

Theoretically, when the static electric charge q_0 and the distance r vary with time. Different time t_i corresponds to different electrostatic fields $E_{0i}(q_{0i}, r_i)$. The electrostatic field E_{0t} corresponding to time t can be expressed as:

$$E_{0t} = E_{0t}(q_0, r) = -\frac{1}{4\pi\varepsilon_0}\frac{q_0}{r^2} \tag{23.1.6}$$

It should be noted that since the static electric charge q_0 and the distance r vary with time, there is also a rate of change $\left(\frac{dq_0}{dt}, \frac{dr}{dt}\right)$ at the moment t. The above equation is a electrostatic field equation without the rate of change $\left(\frac{dq_0}{dt}, \frac{dr}{dt}\right)$.

Taking the derivative of equation (23.1.6) with respect to time t yields the relation equation:

$$\frac{\partial E_{0t}}{\partial t} = \frac{\partial E_{0t}}{\partial q_0}\frac{dq_0}{dt} + \frac{\partial E_{0t}}{\partial r}\frac{dr}{dt} \tag{23.1.7}$$

Although the above equation contains the rate of change $\left(\frac{dq_0}{dt}, \frac{dr}{dt}\right)$, the above equation is not a electrostatic field equation because the electrostatic field $E_{0t} \neq \frac{\partial E_{0t}}{\partial t}$.

Assume that points a and b are points on the curve L. Integrating the electrostatic field $E_{0t}(q_0, r)$ along the curve L yields the relation equation:

$$\varepsilon_{LE0} = \int_{r_a}^{r_b} E_{0t}(q_0, r)dl \tag{23.1.8}$$

This book defines the line integral ε_{LE0} of the electrostatic field $E_{0t}(q_0, r)$ as the linear electromotive force of the electrostatic field . The above equation is the linear electromotive force formula for the electrostatic field.

When the static electric charge q_0 remains constant and the distance r varies. According to equation (23.1.6) the relation equation. is obtained:

$$\varepsilon_{LE0} = \int_{r_a}^{r_b} E_{0t}(q_0, r)dl = \frac{q_0}{4\pi\varepsilon_0}\left(\frac{1}{r_b} - \frac{1}{r_a}\right) \tag{23.1.9}$$

If the distance $r_a \neq r_b$, then the linear electromotive force $\varepsilon_{LE0} \neq 0$ of the electrostatic field. If the distance $r_a = r_b$, then the linear electromotive force $\varepsilon_{LE0} = 0$ of the electrostatic field.

Suppose that the radius r is changeable on the curve $L = f(r)$. When the curve L reaches the point b from the starting point a and then returns to the point a from the point b. Since the curve L is closed, the integral of the electrostatic field $E_{0t}(q_0, r)$ along the closed curve $L = f(r)$ is:

$$\oint E_{0t}(q_0, r)dl = \int_{r_a}^{r_b} E_{0t}dl + \int_{r_b}^{r_a} E_{0t}dl = 0 \tag{23.1.10}$$

Note that the above equation is the loop theorem for the electrostatic field without the rate of change $\left(\frac{dq_0}{dt}, \frac{dr}{dt}\right)$.

Physicists define the electrostatic field $\boldsymbol{E}_{0t}$ as a conservative field based on the above equation.

23.1.3 Derivation of the electrostatic field E_{0t}

When the staticelectric charge $\boldsymbol{q}_0$ and the distance $\boldsymbol{r}$ vary with time. Since there exists a rate of change $\left(\frac{dq_0}{dt}, \frac{dr}{dt}\right)$ at moment $\boldsymbol{t}$, the electrostatic field $\boldsymbol{E}_{0t}$ at moment $\boldsymbol{t}$ can also be expressed in terms of the rate of change $\left(\frac{dq_0}{dt}, \frac{dr}{dt}\right)$, viz:

$$E_{0t} = E_{0t}\left(\frac{dq_0}{dt}, \frac{dr}{dt}\right) \qquad (23.1.11)$$

We can use the static electric potential $\boldsymbol{\varphi}_0$ to derive the above equation.

According to equation (22.1.2), the static electric potential $\boldsymbol{\varphi}_0$ resulting from the static electric charge $\boldsymbol{q}_0$ is:

$$\varphi_0 = +\frac{1}{4\pi\varepsilon_0}\frac{q_0}{r} \qquad (23.1.12)$$

When the static electric charge $\boldsymbol{q}_0$ and the distance $\boldsymbol{r}$ vary with time. Taking the derivative of the above equation with respect to time $\boldsymbol{t}$ gives the relationship equation:

$$\frac{\partial \varphi_0}{\partial t} = \frac{\partial \varphi_0}{\partial q}\frac{dq}{dt} + \frac{\partial \varphi_0}{\partial r}\frac{dr}{dt} = \frac{1}{4\pi\varepsilon_0}\left(\frac{1}{r}\frac{dq_0}{dt} - \frac{q_0}{r^2}\frac{dr}{dt}\right) \qquad (23.1.13)$$

The command speed $\boldsymbol{v}_r = \frac{dr}{dt}$, the above equation becomes:

$$\frac{\partial \varphi_0}{\partial t} = \frac{1}{4\pi\varepsilon_0}\left(\frac{1}{r}\frac{dq_0}{dt} - \frac{q_0}{r^2}v_r\right) \qquad (23.1.14)$$

When the static electric charge $\boldsymbol{q}_0$ remains constant and the distance $\boldsymbol{r}$ changes. Since the rate of change of electric charge $\frac{dq_0}{dt} = \boldsymbol{0}$, the above equation becomes:

$$\frac{\partial \varphi_0}{\partial t} = -\frac{1}{4\pi\varepsilon_0}\frac{q_0}{r^2}\cdot v_r \qquad (23.1.15)$$

Dividing the above equation by the velocity $\boldsymbol{v}_r$ gives the relationship equation:

$$E'_{0t} = \frac{1}{v_r}\frac{\partial \varphi_0}{\partial t} = -\frac{1}{4\pi\varepsilon_0}\frac{q_0}{r^2}$$

Comparing the above equation with equation (23.1.2), it can be determined that the rate of change $\frac{\partial \varphi_0}{\partial t}$ of the static electric potential $\boldsymbol{\varphi}_0$, as a ratio to the velocity, belongs to the electrostatic field $\boldsymbol{E}_{0t}$. From this, dividing equation (23.1.14) by the velocity $\boldsymbol{v}_r$ yields the relational equation:

$$E'_{0t} = \frac{1}{v_r}\frac{\partial \varphi_0}{\partial t} = \frac{1}{v_r}\frac{1}{4\pi\varepsilon_0}\left(\frac{1}{r}\frac{dq_0}{dt} - \frac{q_0}{r^2}v_r\right) \qquad (23.1.16)$$

Theoretically, there is a serious error in the above equation. When the distance $\boldsymbol{r}$ is kept constant and the static electric charge $\boldsymbol{q}_0$ changes. Since the rate of change $\frac{\partial \varphi_0}{\partial t} \neq \boldsymbol{0}$ of static electric potential $\boldsymbol{\varphi}_0$ and the velocity $\boldsymbol{v}_r = \boldsymbol{0}$, the above formula becomes:

$$E'_{0t} = \frac{1}{v_r}\frac{\partial \varphi_0}{\partial t} = \infty \qquad (23.1.17)$$

It is clear that when the distance $\boldsymbol{r}$ is kept constant and the staticelectric charge $\boldsymbol{q}_0$ varies. The electrostatic field $\boldsymbol{E}_{0t}$ will not be equal to infinity. From this it can be determined that equation (23.1.16) is wrong.

Since the ratio of the rate of change $\frac{\partial \varphi_0}{\partial t}$ of the static electric potential $\boldsymbol{\varphi}_0$ to the velocity belongs to the electrostatic field $\boldsymbol{E}_{0t}$, this velocity should be independent of the rate of change of the distance $\boldsymbol{v}_r$. Otherwise the result that the electrostatic field $\boldsymbol{E}_{0t}$ is equal to infinity would occur.

Since the speed of light $\boldsymbol{c}$ is constant in vacuum, the ratio of the rate of change $\frac{\partial \varphi_0}{\partial t}$ to the speed of light $\boldsymbol{c}$ belongs to the electrostatic field $\boldsymbol{E}_{0t}$. Dividing equation (23.1.14) by the speed of light $\boldsymbol{c}$ gives the relational equation:

$$E_{0t} = \frac{1}{c}\frac{\partial \varphi_0}{\partial t} = \frac{1}{c}\frac{\partial \varphi_0}{\partial q}\frac{dq}{dt} + \frac{1}{c}\frac{\partial \varphi_0}{\partial r}\frac{dr}{dt} = \frac{1}{4\pi\varepsilon_0 c}\left(\frac{1}{r}\frac{dq_0}{dt} - \frac{q_0}{r^2}v_r\right) \qquad (23.1.18)$$

When the static electric charge q_0 remains constant and the distance r is variable. Since the rate of change $\frac{dq_0}{dt} = 0$ of electric charge, the above equation becomes:

$$E_{0t} = \frac{1}{c}\frac{\partial \varphi_0}{\partial t} = -\frac{1}{4\pi\varepsilon_0}\frac{q_0 v_r}{r^2 c} \quad (23.1.19)$$

According to equation (18.3.4), the magnetic electric charge q_v contained in the moving electric charge q is:

$$q_v = q\frac{v}{c}$$

Compare the above equation with $\frac{q_0 v_r}{c}$ in equation (23.1.19). Since $\frac{q_0 v_r}{c}$ is analogous to the magnetic electric charge q_v, it can be determined that the ratio C belongs to the electrostatic field strength.

Since the electrostatic field E_{0t} generated by the static electric charge q_0 at time t has a unique nature, the rate of change $\frac{1}{c}\frac{\partial \varphi_0}{\partial t}$ is equal to the electrostatic field $E_{0t}\left(\frac{dq_0}{dt}, \frac{dr}{dt}\right)$, i.e.:

$$E_{0t}\left(\frac{dq_0}{dt}, \frac{dr}{dt}\right) = \frac{1}{c}\frac{\partial \varphi_0}{\partial t} = \frac{1}{4\pi\varepsilon_0 c}\left(\frac{1}{r}\frac{dq_0}{dt} - \frac{q_0}{r^2}v_r\right) \quad (23.1.20)$$

Assume that the staticelectric charge q_0 is at an equal distance r to each point on the circular line $L_R = 2\pi R$. At moment t, since the static electric field $E_{0t}\left(\frac{dq_0}{dt}, \frac{dr}{dt}\right)$ is constant on the circular line L_R, the integral of the static electric field $E_{0t}\left(\frac{dq_0}{dt}, \frac{dr}{dt}\right)$ along the circular line $L_R = 2\pi R$ is:

$$\oint E_{0t}\left(\frac{dq_0}{dt}, \frac{dr}{dt}\right) dl = E_{0t} \cdot L_R = \frac{R}{2\varepsilon_0 c}\left(\frac{1}{r}\frac{dq_0}{dt} - \frac{q_0}{r^2}v_r\right) \quad (23.1.21)$$

In summary, there are two equivalent formulas for expressing the electrostatic field E_{0t} when the electric charge q_0 and the distance r vary with time.

One is equation $E_{0t}(q_0, r)$ and the other is equation $E_{0t}\left(\frac{dq_0}{dt}, \frac{dr}{dt}\right)$. Both equation $E_{0t}(q_0, r)$ and equation $E_{0t}\left(\frac{dq_0}{dt}, \frac{dr}{dt}\right)$ are expressions for the rotating electric field E_{0t} since the electrostatic field E_{0t} generated by the motion electric charge q_0 at moment t possesses uniqueness.

23.1.4 The process by which physicists derive the electromagnetic wave equation is unscientific and unreasonable

When the distance r varies on the closed curve and the electrostatic charge q_0 remains constant. According to equation (23.1.10), the integral of the electrostatic field $E_{0t}(q_0, r)$ along the closed curve L is:

$$\oint E_{0t}(q_0, r) dl = 0 \quad (23.1.22)$$

If the electrostatic field E_{0t} remains constant on the closed curve L, then the loop integral of the electrostatic field E_{0t} is:

$$\oint E_{0t}(q_0, r) dl = E_{0t}\oint dl \quad (23.1.23)$$

In other words, assume that each point on the circular line $L_R = 2\pi R$ is equidistant r from the electrostatic charge q_0. Since the static electric field E_0 is constant on the circular line L_R, the integral of the static electric field E_0 along the circular line $L_R = 2\pi R$ is:

$$\oint E_{0t}(q_0, r) dl = E_{0t} L_R = -\frac{R}{2\varepsilon_0}\frac{q_0}{r^2} \quad (23.1.24)$$

Since the electrostatic field $E_{0t}(q_0, r)$ in equation is constant on the circular line $L_R = 2\pi R$, the loop integral $\oint E_{0t}(q_0, r) dl$ is not the Electromotive Force ε. Compare the above equation with equation (23.1.21). Although both formulas are loop integrals of the electrostatic field E_{0t}, the electrostatic field strength E_{0t} in the two formulas is different. The electrostatic field $E_{0t}(q_0, r)$ in the above equation does not include the rate of change $\left(\frac{dq_0}{dt}, \frac{dr}{dt}\right)$, while the electrostatic field $E_{0t}\left(\frac{dq_0}{dt}, \frac{dr}{dt}\right)$ in equation (23.1.21) includes the rate of change $\left(\frac{dq_0}{dt}, \frac{dr}{dt}\right)$.

Since the time-varying magnetic field B generates electric current and the electric field E can make the electric charge $\boldsymbol{q}$ moves, the British physicist Maxwell proposed that the time-varying magnetic field $\boldsymbol{B}$ excites a new time-varying electromagnetic field $\boldsymbol{E}_t$ in the space around it. the time-varying electromagnetic field $\boldsymbol{E}_t$ generates an induced Electromotive Force $\boldsymbol{\varepsilon}$, i.e. :

Time-varying magnetic field $\boldsymbol{B}$ → excitation of time-varying electromagnetic field $\boldsymbol{E}_t$ → action on free electric charge $\boldsymbol{q}$ → excitation of induced electromotive force $\boldsymbol{\varepsilon}$

Physicists believe that the time-varying magnetic field $\boldsymbol{B}$ excites the time-varying electromagnetic field $\boldsymbol{E}_t$, which in turn excites the time-varying magnetic field $\boldsymbol{B}$. The two are continually excited by each other to travel into the distance. In Maxwell's system of equations, Faraday's law of electromagnetic induction is:

$$\oint \boldsymbol{E}_t d\boldsymbol{l} = -\iint \frac{\partial \boldsymbol{B}}{\partial t} d\boldsymbol{S} \tag{23.1.25}$$

It should be noted that the time-varying electromagnetic field $\boldsymbol{E}_t$ contains not only $(\boldsymbol{q},\boldsymbol{r},\boldsymbol{v})$, but also the rate of change $\left(\frac{dq}{dt},\frac{dr}{dt},\frac{dv}{dt}\right)$. Since physicists do not distinguish between the expressions $\boldsymbol{E}_t(\boldsymbol{q},\boldsymbol{r},\boldsymbol{v})$ and $\boldsymbol{E}_t\left(\frac{dq}{dt},\frac{dr}{dt},\frac{dv}{dt}\right)$ for the time-varying electromagnetic field $\boldsymbol{E}_t$, the time-varying electromagnetic field $\boldsymbol{E}_t$ in equation can only be the expression $\boldsymbol{E}_t(\boldsymbol{q},\boldsymbol{r},\boldsymbol{v})$. It is this reason that makes the process by which physicists derive the equations for electromagnetic waves unscientific and irrational.

To derive the electromagnetic wave equation, physicists use mathematical tricks to represent the time-varying electromagnetic field $\boldsymbol{E}_t$ as:

$$\boldsymbol{E}_t = -\nabla\varphi - \frac{\partial \boldsymbol{A}}{\partial t} \tag{23.1.26}$$

The $\nabla\varphi$ in the equation is the gradient of the scalar potential φ and $\frac{\partial A}{\partial t}$ is the rate of change of the vector potential $\boldsymbol{A}$. The electromagnetic wave equation is derived from the above equation.

In a time-varying electromagnetic field $\boldsymbol{E}_t$, assume that the radius $\boldsymbol{r}$ is variable on the curve $\boldsymbol{L} = \boldsymbol{f}(\boldsymbol{r})$. When the curve $\boldsymbol{L}$ reaches the point $\boldsymbol{b}$ from the starting point $\boldsymbol{a}$ and then returns to the point $\boldsymbol{a}$ from the point $\boldsymbol{b}$. Since the curve $\boldsymbol{L}$ is a closed curve, according to equation (23.1.26). The integral $\oint \boldsymbol{E}_t d\boldsymbol{l}$ of the time-varying electromagnetic field $\boldsymbol{E}_t$ along the closed curve $\boldsymbol{L} = \boldsymbol{f}(\boldsymbol{r})$ is:

$$\oint \boldsymbol{E}_t d\boldsymbol{l} = \int_{r_a}^{r_b} \boldsymbol{E}_t d\boldsymbol{l} + \int_{r_b}^{r_a} \boldsymbol{E}_t d\boldsymbol{l} = 0 \tag{23.1.27}$$

The above equation shows that the time-varying electromagnetic field $\boldsymbol{E}_t$ should also be a conservative field with a field source. Based on the above equation, the relation equation can be obtained:

$$\oint \boldsymbol{E}_t d\boldsymbol{l} \neq -\iint \frac{\partial \boldsymbol{B}}{\partial t} d\boldsymbol{S} \tag{23.1.28}$$

It is clear that the above equation contradicts equation (23.1.25).

Theoretically, since both the time-varying electromagnetic field $\boldsymbol{E}_t$ and the electrostatic field $\boldsymbol{E}_{0t}$ cause the electric charge $\boldsymbol{q}$ moves to produce electric current, the time-varying electromagnetic field $\boldsymbol{E}_t$ should also be an electric field with a field source. Alternatively, the time-varying electromagnetic field $\boldsymbol{E}_t$ is generated by the moving electric charge $\boldsymbol{q}$. Otherwise it is impossible to explain who produces the time-varying electromagnetic field $\boldsymbol{E}_t$.

Physicists believe that the time-varying electromagnetic field $\boldsymbol{E}_t$ is generated not only by the motionelectric charge $\boldsymbol{q}$, but also by the time-varying magnetic field $\boldsymbol{B}$. This view of the physicists is contradicted by their definition of the time-varying electromagnetic field $\boldsymbol{E}_t$ as a passive field.

To summarize, in order to avoid the loop integral $\oint \boldsymbol{E}_t d\boldsymbol{l} = \boldsymbol{0}$ of the time-varying electromagnetic field $\boldsymbol{E}_t$, physicists define the electrostatic field $\boldsymbol{E}_{0t}$ as a conservative field with a field source and the time-varying electromagnetic field $\boldsymbol{E}_t$ as a non-conservative field without a field source.

It should be noted that the time-varying electromagnetic field $\boldsymbol{E}_t$ defined in this book does not distinguish between conservative and non-conservative fields. Alternatively, both the electric field $\boldsymbol{E}_t$ and the magnetic field $\boldsymbol{B}$ as defined in this book are fields with field sources.

23.1.5 Flux electromotive force ε_{SE0} for electrostatic field E_{0t}

Assume that the static electric charge q_0 and the distance r remain constant. Assume that each point on the circular line $L_R = 2\pi R$ arrives at the static electric charge q_0 at an equal distance r. Assume that the surface S is bounded by the closed circular line L_R. According to the electrostatic field flux equation (22.1.6) , the electrostatic field flux Φ_{E0t} through the surface S is:

$$\Phi_{E0t} = \iint E_{0t}\, dS = \frac{1}{2\varepsilon_0}\frac{q_0 R}{r} \tag{23.1.29}$$

Taking the derivative of the above equation with respect to time t gives the relationship equation:

$$\frac{\partial \Phi_{E0t}}{\partial t} = \iint \frac{\partial E_{0t}(q_0, r)}{\partial t} dS = \frac{\partial \Phi_{E0t}}{\partial q_0}\frac{dq_0}{dt} + \frac{\partial \Phi_{E0t}}{\partial r}\frac{dr}{dt} = \frac{R}{2\varepsilon_0}\left(\frac{1}{r}\frac{dq_0}{dt} - \frac{q_0}{r^2}v_r\right) \tag{23.1.30}$$

The above equation is the equation for the rate of change of flux for a electrostatic field E_{0t}. The velocity $v_r = \frac{dr}{dt}$ in the formula is the rate of change of distance r. When the distance r changes and the static electric charge q_0 remains constant. Since the rate of change of static electric charge $\frac{dq_0}{dt} = 0$, the above equation becomes:

$$\frac{\partial \Phi_{E0t}}{\partial t} = -\frac{R}{2\varepsilon_0}\frac{q_0}{r^2}v_r \tag{23.1.31}$$

Dividing the above equation by the velocity v_r gives the relationship equation:

$$\frac{1}{v_r}\frac{\partial \Phi_{E0t}}{\partial t} = -\frac{R}{2\varepsilon_0}\frac{q_0}{r^2} \tag{23.1.32}$$

According to equation (23.1.9), the linear electromotive force ε_{LE0} (voltage U) generated by the static electric charge q_0 is:

$$\varepsilon_{LE0} = +\frac{q_0}{4\pi\varepsilon_0}\left(\frac{1}{r_b} - \frac{1}{r_a}\right) = U(\text{voltage})$$

Compare the above equation with equation (23.1.32). Since distance r and radius R are both distances, both $\frac{R}{r^2}$ and $\left(\frac{1}{r_b} - \frac{1}{r_a}\right)$ have the same physical units. From this it can be determined that the ratio of flux rate of change $\frac{d\Phi_{E0t}}{dt}$ to velocity belongs to the electromotive force ε. Dividing Equation (23.1.30) by the velocity v_r gives the relationship equation:

$$\frac{1}{v_r}\frac{\partial \Phi_{E0t}}{\partial t} = \frac{1}{v_r}\frac{R}{2\varepsilon_0}\left(\frac{1}{r}\frac{dq_0}{dt} - \frac{q_0}{r^2}v_r\right) \tag{23.1.33}$$

Theoretically, there is a serious error in the above equation. When the distance r is kept constant and the static electric charge q_0 changes. Since the rate of flux change $\frac{\partial \Phi_{E0t}}{\partial t} \neq 0$, velocity $v_r = 0$, the above formula becomes:

$$\varepsilon = \frac{1}{v_r}\frac{\partial \Phi_{E0t}}{\partial t} = \infty \tag{23.1.34}$$

However, the electromotive force ε is not equal to infinity when the distance r is held constant and the static electric charge q_0 varies. From this it can be determined that equation (23.1.33) is wrong.

Since dividing the flux rate of change $\frac{\partial \Phi_{E0t}}{\partial t}$ by the velocity v_r would lead to an infinity result for $\frac{1}{v_r}\frac{\partial \Phi_{E0t}}{\partial t}$, the flux rate of change $\frac{\partial \Phi_{E0t}}{\partial t}$ cannot be divided by the velocity v_r.

Since the speed of light c is constant in vacuum, the ratio of therate of flux change $\frac{\partial \Phi_{E0t}}{\partial t}$ to the speed of light c belongs to the electromotive force ε. Dividing equation (23.1.30) by the speed of light c gives the relational equation:

$$\varepsilon_{SE0} = \frac{1}{c}\frac{\partial \Phi_{E0t}}{\partial t} = \frac{1}{c}\iint \frac{\partial E_{0t}(q_0, r)}{dt} dS = \frac{R}{2\varepsilon_0 c}\left(\frac{1}{r}\frac{dq_0}{dt} - \frac{q_0}{r^2}v_r\right) \tag{23.1.35}$$

This book defines ε_{SE0} as the flux electromotive force of an electrostatic field.The above formula is the flux electromotive force formula of an electrostatic field.

When the distance $\boldsymbol{r}$ changes and the static electric charge $\boldsymbol{q_0}$ remains constant. Since the rate of change $\frac{dq_0}{dt} = \mathbf{0}$ for static electric charge $\boldsymbol{q_0}$, equation (23.1.35) becomes:

$$\varepsilon_{SE0} = \frac{1}{c}\frac{\partial \Phi_{E0t}}{\partial t} = -\frac{R}{2\varepsilon_0}\frac{q_0}{r^2}\cdot\frac{v_r}{c} \tag{23.1.36}$$

The above equation shows that the direction of the flux electromotive force ε_{SE0} is opposite to the direction of the velocity $\boldsymbol{v_r}$. According to equation (18.3.4), the magnetic electric charge $\boldsymbol{q_v}$ contained in the moving electric charge $\boldsymbol{q}$ is:

$$q_v = q\frac{v}{c}$$

Compare the above equation with $\frac{q_0 v_r}{c}$ in equation (23.1.36), where $\frac{q_0 v_r}{c}$ is analogous to the magnetic electric charge $\boldsymbol{q_v}$.

According to equation (23.1.5), the linear electromotive force ε_{LE0} (voltage $\boldsymbol{U}$) generated by the static electric charge $\boldsymbol{q_0}$ is:

$$\varepsilon_{LE0} = \frac{q_0}{4\pi\varepsilon_0}\left(\frac{1}{r_b} - \frac{1}{r_a}\right) = U$$

Compare the above equation with equation (23.1.36). Since distance $\boldsymbol{r}$ and radius $\boldsymbol{R}$ are both distances, $\frac{R}{r^2}$ and $\left(\frac{1}{r_b} - \frac{1}{r_a}\right)$ have the same physical units. From this it can be determined that the flux electromotive force $\varepsilon_{SE0} = \frac{1}{c}\iint \frac{\partial E_{0t}}{\partial t}$ of the electrostatic field belongs to the electromotive force $\boldsymbol{\varepsilon}$ (voltage $\boldsymbol{U}$).

23.1.6 Effect of distance Z variation on static flux electromotive force ε_{SE0}

Suppose O is the center of the circular line $L_R = 2\pi R$, **Z** is the perpendicular distance from the static electric charge $\boldsymbol{q_0}$ to the center O, and $\boldsymbol{\theta}$ is the angle between distance $\boldsymbol{r}$ and distance Z. Assume that $\boldsymbol{\Omega}$ is the steradian angle possessed by the circular plane $P = \pi R^2$. As shown earlier in Figure 22-1.

According to Figure 22-1, the distance $\boldsymbol{r}$ is:

$$r = \sqrt{R^2 + Z^2} \tag{23.1.37}$$

Taking the derivative of the above equation with respect to the distance Z yields the relationship equation:

$$\frac{\partial r}{\partial Z} = \frac{Z}{\sqrt{R^2 + Z^2}} = \frac{Z}{r} \tag{23.1.38}$$

when the staticelectric charge $\boldsymbol{q_0}$ and the distance $\boldsymbol{r}$ vary. According to equation (23.1.29), the electrostatic field flux $\boldsymbol{\Phi_{E0t}}$ is:

$$\Phi_{E0t} = \frac{1}{2\varepsilon_0}\frac{q_0 R}{r} \tag{23.1.39}$$

Taking the derivative of the above equation with respect to the distance $\boldsymbol{r}$ gives the relationship equation:

$$\frac{\partial \Phi_{E0t}}{\partial r} = -\frac{1}{2\varepsilon_0}\frac{q_0 R}{r^2} \tag{23.1.40}$$

When the static electric charge $\boldsymbol{q_0}$ and the distance Z are variables, taking the derivative of equation (23.1.39) with respect to time $\boldsymbol{t}$ yields the relation equation:

$$\frac{\partial \Phi_{E0t}}{\partial r} = \frac{\partial \Phi_{E0t}}{\partial q_0}\frac{dq_0}{dt} + \frac{\partial \Phi_{E0t}}{\partial r}\frac{\partial r}{\partial Z}\frac{dZ}{dt} = \frac{R}{2\varepsilon_0}\left(\frac{1}{r}\frac{dq_0}{dt} - \frac{q_0}{r^2}\cdot\frac{Z}{r}\cdot\frac{dZ}{dt}\right) \tag{23.1.41}$$

The above equation is the electrostatic flux rate of change equation. The command velocity $v_Z = \frac{dZ}{dt}$. Divide the above equation by the speed of light $\boldsymbol{c}$ to get the relational equation:

$$\varepsilon_{SE0} = \frac{1}{c}\frac{\partial \Phi_{E0t}}{\partial r} = \frac{R}{2\varepsilon_0 c}\left(\frac{1}{r}\frac{dq_0}{dt} - \frac{q_0 Z}{r^3}v_Z\right) \tag{23.1.42}$$

When the distance Z varies and the static electric charge $\boldsymbol{q_0}$ remains constant. Since the rate of change of static electric charge $\frac{dq_0}{dt} = \mathbf{0}$, the above equation becomes:

$$\varepsilon_{SE0} = \frac{1}{c}\frac{\partial \Phi_{E0t}}{\partial r} = -\frac{q_0 v_Z}{2\varepsilon_0 c}\frac{RZ}{r^3} \tag{23.1.43}$$

The ε_{SE0} in equation is the electromotive force (voltage U) possessed on the circular line $L_R = 2\pi R$. The above equation shows that the direction of the flux electromotive force ε_{SE0} is opposite to the direction of the velocity v_Z.

23.1.7 Derivation of the electrostatic field induction equation

Assume that each point on the circular line $L_R = 2\pi R$ has an equal distance r to the static electric charge q_0. Since the electrostatic field strength E_{0t} is constant on the circular line L_R, the integral $\oint E_{0t} dl$ of the electrostatic field strength E_{0t} along the circular line $L_R = 2\pi R$ is:

$$\oint E_{0t}(q_0, r) dl = E_{0t} L_R = -\frac{1}{2\varepsilon_0} \cdot \frac{q_0 R}{r^2} \tag{23.1.44}$$

Assume that the surface S is bordered by the closed circular line L_R. According to equation (23.1.35), the static flux electromotive force ε_{SE0} is:

$$\varepsilon_{SE0} = \frac{1}{c}\iint \frac{\partial E_{0t}(q_0, r)}{\partial t} dS = \frac{R}{2\varepsilon_0 c}\left(\frac{1}{r}\frac{dq_0}{dt} - \frac{q_0}{r^2} v_r\right) \tag{23.1.45}$$

Comparing the above equation with equation (23.1.44), the integral $\oint E_{0t}(q_0, r) dl$ of the electrostatic field E_{0t} on the circular line $L_R = 2\pi R$, is not equal to the static flux electromotive force ε_{SE0}, viz:

$$\oint E_{0t}(q_0, r) dl \neq \frac{1}{c}\iint \frac{\partial E_{0t}(q_0, r)}{\partial t} dS$$

Since the left side of the inequality sign does not contain the rate of change $\left(\frac{dq_0}{dt}, \frac{dr}{dt}\right)$ and the right side of the inequality sign contains the rate of change $\left(\frac{dq_0}{dt}, \frac{dr}{dt}\right)$. So the above equations are necessarily unequal.

Assume that the electric charge q is an equal distance r to each point on the circular line $L_R = 2\pi R$. According to equation (23.1.21), the integral of the electrostatic field $E_{0t}\left(\frac{dq_0}{dt}, \frac{dr}{dt}\right)$ along the circular line $L_R = 2\pi R$ is:

$$\oint E_{0t}\left(\frac{dq_0}{dt}, \frac{dr}{dt}\right) dl = \frac{R}{2\varepsilon_0 c}\left(\frac{1}{r}\frac{dq_0}{dt} - \frac{q_0}{r^2} v_r\right)$$

According to equation (23.1.35), the ratio $\frac{1}{c}\iint \frac{\partial E_{0t}(q_0, r)}{\partial t} dS$ is:

$$\frac{1}{c}\iint \frac{\partial E_{0t}(q_0, r)}{\partial t} dS = \frac{R}{2\varepsilon_0 c}\left(\frac{1}{r}\frac{dq_0}{dt} - \frac{q_0}{r^2} v_r\right)$$

The relationship equation is obtained from the above two formulas:

$$\oint E_{0t}\left(\frac{dq_0}{dt}, \frac{dr}{dt}\right) dl = \frac{1}{c}\iint \frac{\partial E_{0t}(q_0, r)}{\partial t} dS \tag{23.1.46}$$

We can derive the above equation from the static electric potential φ_0 and Stokes' theorem. According to equation (23.1.12), the static electric potential φ_0 resulting from the static electric charge q_0 is:

$$\varphi_0 = \frac{1}{4\pi\varepsilon_0}\frac{q_0}{r}$$

Assume that the static electric potential φ_0 integrates along the closed curve L. Assume that the surface S has the closed curve L as its boundary. According to Stokes' theorem, the relation equation can be obtained:

$$\oint \varphi_0 dl = \iint \nabla \times \varphi_0 \, dS \tag{23.1.47}$$

According to equation (17.3.2), the electrostatic field strength E_{0t} is equal to the Curl $\nabla \times \varphi_0$ of the electrostatic field electric potential φ_0, i.e:

$$E_{0t} = \nabla \times \varphi_0 \tag{23.1.48}$$

Bringing the above equation into equation (23.1.47) gives the relationship equation:

$$\Phi_{E0t} = \oint \varphi_0 dl = \iint E_{0t}(q_0, r) \, dS \tag{23.1.49}$$

Taking the derivative of the above equation with respect to time t gives the relationship equation:

$$\frac{\partial \boldsymbol{\Phi}_{E0t}}{\partial t} = \oint \frac{\partial \varphi_0}{\partial t} dl = \iint \frac{\partial E_{0t}(q_0, r)}{\partial t} dS \tag{23.1.50}$$

Dividing the above equation by the speed of light c, gives the relationship equation:

$$\varepsilon_{SE0} = \frac{1}{c}\frac{\partial \boldsymbol{\Phi}_{E0t}}{\partial t} = \oint \frac{1}{c}\frac{\partial \varphi_0}{\partial t} dl = \iint \frac{1}{c}\frac{\partial E_{0t}(q_0, r)}{\partial t} dS \tag{23.1.51}$$

The above equation shows that the static flux electromotive force ε_{SE0} is equal to the integral of the ratio $\frac{1}{c}\frac{d\varphi_0}{dt}$ along the closed curve L. According to the time-varying electrostatic field equation (23.1.20), the time-varying electrostatic field E_{0t} is:

$$E_{0t}\left(\frac{dq_0}{dt}, \frac{dr}{dt}\right) = \frac{1}{c}\frac{\partial \varphi_0}{\partial t}$$

Bringing the above equation into equation (23.1.51) gives the relationship equation:

$$\varepsilon_{SE0} = \oint E_{0t}\left(\frac{dq_0}{dt}, \frac{dr}{dt}\right) dl = \iint \frac{1}{c}\frac{\partial E_{0t}(q_0, r)}{\partial t} dS \tag{23.1.52}$$

The above equation is the induction equation for an electrostatic field. This equation is similar to the Faraday formula for electromagnetic induction.

23.2 Derivation of the induction formula for time-varying electromagnetic fields

Symbol Description:

In order to discuss accurately and avoid misunderstanding, this book uses different symbols to represent different electromagnetic fields. The meaning of the symbols is as follows:

(1) Symbol $E_t(q, r, v)$ is the time-varying electromagnetic field corresponding to time t that does not contain the rate of change $\left(\frac{dq}{dt}, \frac{dr}{dt}, \frac{dv}{dt}\right)$.

(2) Symbol $E_t\left(\frac{dq}{dt}, \frac{dr}{dt}, \frac{dv}{dt}\right)$ is the time-varying electromagnetic field corresponding to time t that contains the rate of change $\left(\frac{dq}{dt}, \frac{dr}{dt}, \frac{dv}{dt}\right)$.

23.2.1 Linear electromotive force ε_{LEt} for a time-varying electromagnetic field $E_t\,(q,r,v)$

when electric charge q is moving with velocity v in the cosmic vacuum. According to equation (18.3.2), the moving electric charge q can be decomposed into the rotational electric charge q_ω and the magnetic electric charge q_v, i.e:

$$q = \sqrt{q_\omega^2 + q_v^2} \tag{23.2.1}$$

According to equation (18.3.3) and equation (18.3.4), the rotational electric charge q_ω and the magnetic electric charge q_v contained in the motion electric charge q are, respectively:

$$q_\omega = q\sqrt{1 - \frac{v^2}{c^2}}, \qquad q_v = q\frac{v}{c} \tag{23.2.2}$$

Since the motion electric charge q generates a rotating electric field E_ω and a Magnetic-electric field E_v, this book defines the vortex field E generated by the motion electric charge q as an electromagnetic field. Note that the electromagnetic field E contains both the rotating electric field E_ω and the Magnetic-electric field E_v.

According to equation (19.1.7), the electromagnetic field strength E produced by the motion electric charge q is:

$$E = -\frac{1}{4\pi\varepsilon_0}\frac{q}{r^2}\sin\theta \tag{23.2.3}$$

The θ in the formula is the angle between the distance r and the velocity v. Substituting equation (23.2.1) into the above equation gives the relationship equation:

$$E = -\frac{1}{4\pi\varepsilon_0}\frac{\sqrt{q_\omega^2 + q_v^2}}{r^2}\sin\theta = -\sqrt{E_\omega^2 + E_v^2} \tag{23.2.4}$$

The above equation shows that the electromagnetic field E generated by the motion electric charge q can be decomposed into a rotating electric field E_ω and a Magnetic-electric field E_v.

Theoretically, when the electric charge q, velocity v and distance r vary with time. Different time t_i corresponds to different electromagnetic field $E_i(q_i, r_i, v_i)$. The time-varying electromagnetic field $E_t(q, r, v)$ at a certain time t can be expressed as follows:

$$E_t = E_t(q, r, v) = -\frac{1}{4\pi\varepsilon_0}\frac{q}{r^2}\sin\theta \tag{23.2.5}$$

Since the electric charge q, the distance r and the velocity v vary with time, there is also a rate of change $\left(\frac{dq}{dt}, \frac{dr}{dt}, \frac{dv}{dt}\right)$ at the moment t. The above equation is the time-varying electromagnetic field equation without the rate of change $\left(\frac{dq}{dt}, \frac{dr}{dt}, \frac{dv}{dt}\right)$.

Since the electrostatic field E_{0t} can generate the static linear electromotive force ε_{LE0}, the time-varying electromagnetic field $E_t(q, r, v)$ must also be capable of generating the linear electromotive force ε_{LE}, viz:

$$\varepsilon_{LE} = \int_{r_a}^{r_b} E_t(q, r, v)dl \tag{23.2.6}$$

This book defines the line integral ε_{LE} of a time-varying electromagnetic field $E_t(q, r, v)$ as the linear electromotive force of the time-varying electromagnetic field . The above equation is the equation of linear electromotive force of time-varying electromagnetic field.

Assume that points a and b are points on the curve $L = f(r)$. When the electric charge q and velocity v are kept constant and the distance r can be varied on the curve L. The time-varying electromagnetic field $E_t(q, r, v)$ produces a linear electromotive force ε_{LE} on the curve $L = f(r)$ as:

$$\varepsilon_{LE} = \int_{r_a}^{r_b} E_t(q, r, v)dl = \frac{q}{4\pi\varepsilon_0}\left(\frac{1}{r_b} - \frac{1}{r_a}\right)\sin\theta \tag{23.2.7}$$

When the curve $L = f(r)$ reaches point b from point a and then returns to point a from point b. Since the curve L is closed, the integral of the time-varying electromagnetic field $E_t(q, r, v)$ along the closed curve $L = f(r)$ is:

$$\oint E_t(q, r, v)dl = \int_{r_a}^{r_b} E_t dl + \int_{r_b}^{r_a} E_t dl = 0 \tag{23.2.8}$$

The above equation is the loop theorem for time-varying electromagnetic fields. The above equation is similar to the loop theorem for electrostatic fields.

23.2.2 Derivation of the time-varying electromagnetic field formula E_t

Theoretically, when the electric charge q, speed v and distance r change with time, at time t there is not only (q, r, v), but also the rate of change $\left(\frac{dq}{dt}, \frac{dr}{dt}, \frac{dv}{dt}\right)$. The time-varying electromagnetic field E_t can also be expressed as:

$$E_t = E_t\left(\frac{dq}{dt}, \frac{dr}{dt}, \frac{dv}{dt}\right)$$

Taking the derivative of equation (23.2.5) with respect to time t yields the relation equation:

$$\frac{\partial E_t}{\partial t} = \frac{\partial E_t}{\partial q}\frac{dq}{dt} + \frac{\partial E_t}{\partial r}\frac{dr}{dt} + \frac{\partial E_t}{\partial v}\frac{dv}{dt} \tag{23.2.9}$$

Although the above equation contains the rate of change $\left(\frac{dq}{dt}, \frac{dr}{dt}, \frac{dv}{dt}\right)$, the above equation is not a time-varying electromagnetic field equation because the time-varying electromagnetic field $E_t \neq \frac{\partial E_t}{\partial t}$.

We can use the electromagnetic potential φ to derive the time-varying electromagnetic field equation

$E_t\left(\frac{dq}{dt},\frac{dr}{dt},\frac{dv}{dt}\right)$.

When the electric charge q is moving with velocity v in the cosmic vacuum. according to equation (18.2.8), the electromagnetic potential φ generated by the moving electric charge q is:

$$\varphi=\frac{1}{4\pi\varepsilon_0}\frac{q}{r}\sin\theta \tag{23.2.10}$$

When the electric charge q, distance r and speed v varies with time, the above formula to the time t derivative, you can get the relationship between the formula:

$$\frac{\partial\varphi}{\partial t}=\frac{\partial\varphi}{\partial q}\frac{dq}{dt}+\frac{\partial\varphi}{\partial r}\frac{dr}{dt}+\frac{\partial\varphi}{\partial v}\frac{dv}{dt}=\frac{\sin\theta}{4\pi\varepsilon_0}\left(\frac{1}{r}\frac{dq}{dt}-\frac{q}{r^2}\frac{dr}{dt}\right)+\frac{\partial\varphi}{\partial v}\frac{dv}{dt} \tag{23.2.11}$$

The command velocity $v_r=\frac{dr}{dt}$ and acceleration $a=\frac{dv}{dt}$. The above equation becomes:

$$\frac{\partial\varphi}{\partial t}=\frac{\sin\theta}{4\pi\varepsilon_0}\left(\frac{1}{r}\frac{dq}{dt}-\frac{q}{r^2}v_r\right)+\frac{\partial\varphi}{\partial v}a \tag{23.2.12}$$

When the electric charge q and velocity v remain constant and the distance r changes. Since rate of change of electric charge $\frac{dq}{dt}=0$ and acceleration $a=0$, the above equation becomes:

$$\frac{\partial\varphi}{\partial t}=-\frac{1}{4\pi\varepsilon_0}\frac{q}{r^2}\sin\theta\cdot v_r \tag{23.2.13}$$

Dividing the above equation by the velocity v_r gives the relationship equation:

$$\frac{1}{v_r}\frac{\partial\varphi}{\partial t}=-\frac{1}{4\pi\varepsilon_0}\frac{q}{r^2}\sin\theta=E_t$$

The above equation shows that the ratio of the rate of change $\frac{\partial\varphi}{\partial t}$ of the electromagnetic potential, to the velocity belongs to the electromagnetic field strength, E_t. Dividing equation (23.2.12) by the velocity v_r, yields the relation equation:

$$E_t=\frac{1}{v_r}\frac{\partial\varphi}{\partial t}=\frac{1}{v_r}\frac{\sin\theta}{4\pi\varepsilon_0}\left(\frac{1}{r}\frac{dq}{dt}-\frac{q}{r^2}v_r\right)+\frac{1}{v_r}\frac{\partial\varphi}{\partial v}a \tag{23.2.14}$$

Theoretically, there is a serious error in the above equation. When the distance r is kept constant and the electric charge q and the velocity v change. Since the rate of change $\frac{\partial\varphi}{\partial t}\neq 0$ of electromagnetic potential and velocity $v_r=0$, the above equation becomes:

$$E_t=\frac{1}{v_r}\frac{\partial\varphi}{\partial t}=\infty \tag{23.2.15}$$

However, the electromagnetic field E_t is not equal to infinity when the distance r is held constant and the electric charge q and velocity v are varied. From this it can be determined that equation (23.2.14) is wrong.

Since dividing the rate of change $\frac{\partial\varphi}{\partial t}$ of the electromagnetic potential by the velocity v_r would lead to an infinity result for $\frac{1}{v_r}\frac{\partial\varphi}{\partial t}$, the rate of change $\frac{\partial\varphi}{\partial t}$ cannot be divided by the velocity v_r.

Since the speed of light c is constant in the cosmic vacuum frame of reference, the ratio of the rate of change $\frac{\partial\varphi}{\partial t}$ to the speed of light c is equal to the time-varying electromagnetic field E_t. Dividing equation (23.2.12) by the speed of light c yields the relational equation:

$$E_t=\frac{1}{c}\frac{\partial\varphi}{\partial t}=\frac{1}{c}\frac{\partial\varphi}{\partial q}\frac{dq}{dt}+\frac{1}{c}\frac{\partial\varphi}{\partial r}\frac{dr}{dt}+\frac{1}{c}\frac{\partial\varphi}{\partial v}\frac{dv}{dt}=\frac{\sin\theta}{4\pi\varepsilon_0 c}\left(\frac{1}{r}\frac{dq}{dt}-\frac{q}{r^2}v_r\right)+\frac{1}{c}\frac{d\varphi}{dv}a \tag{23.2.16}$$

When the electric charge q and velocity v are kept constant and the distance r is variable. Since the rate of change $\frac{dq}{dt}=0$ of electric charge q and acceleration $a=0$, the above equation becomes:

$$E_t=\frac{1}{c}\frac{\partial\varphi}{\partial t}=-\frac{v_r}{c}\cdot\frac{1}{4\pi\varepsilon_0}\frac{q}{r^2}\sin\theta \tag{23.2.17}$$

According to equation (18.3.4), the magnetic electric charge q_v contained in the moving electric charge q is:

$$q_v=q\frac{v}{c}$$

Comparing the above equation with $\frac{qv_r}{c}$ in equation (23.2.17), $\frac{qv_r}{c}$ is similar to the magnetic electric charge $\boldsymbol{q_v}$. Since the ratio of velocities $\frac{v_r}{c}$ has no physical units, $\frac{1}{c}\frac{d\varphi}{dt}$ in equation (23.2.17) belongs to the time-varying electromagnetic field $\boldsymbol{E_t}$. From this it can be determined that $\frac{1}{c}\frac{\partial\varphi}{\partial t}$, $\frac{1}{c}\frac{\partial\varphi}{\partial q}\frac{dq}{dt}$, $\frac{1}{c}\frac{\partial\varphi}{\partial r}\frac{dr}{dt}$, and $\frac{1}{c}\frac{\partial\varphi}{\partial v}\frac{dv}{dt}$ in equation (23.2.16) all belong to the electric field strength.

Since the time-varying electromagnetic field $\boldsymbol{E_t}\left(\frac{dq}{dt},\frac{dr}{dt},\frac{dv}{dt}\right)$ generated by electric charge $\boldsymbol{q}$ at time $\boldsymbol{t}$ possesses uniqueness, the ratio $\frac{1}{c}\frac{d\varphi}{dt}$ is equal to the time-varying electromagnetic field $\boldsymbol{E_t}$, viz:

$$\boldsymbol{E_t\left(\frac{dq}{dt},\frac{dr}{dt},\frac{dv}{dt}\right)=\frac{1}{c}\frac{\partial\varphi}{\partial t}=\frac{\sin\theta}{4\pi\varepsilon_0 c}\left(\frac{1}{r}\frac{dq}{dt}-\frac{q}{r^2}v_r\right)+\frac{1}{c}\frac{d\varphi}{dv}a} \tag{23.2.18}$$

Assume that the distance $\boldsymbol{r}$ from the electric charge $\boldsymbol{q}$ to each point on the circular line $\boldsymbol{L_R=2\pi R}$ is equal. At moment $\boldsymbol{t}$, since the time-varying electromagnetic field $\boldsymbol{E_t}\left(\frac{dq}{dt},\frac{dr}{dt},\frac{dv}{dt}\right)$ is constant on the circular line $\boldsymbol{L_R}$, the integral of the time-varying electromagnetic field $\boldsymbol{E_t}\left(\frac{dq}{dt},\frac{dr}{dt},\frac{dv}{dt}\right)$ along the circular line $\boldsymbol{L_R=2\pi R}$ is:

$$\boldsymbol{\oint E_t\left(\frac{dq}{dt},\frac{dr}{dt},\frac{dv}{dt}\right)dl=E_t\cdot L_R=\frac{R\sin\theta}{2\varepsilon_0 c}\left(\frac{1}{r}\frac{dq}{dt}-\frac{q}{r^2}v_r\right)+\frac{2\pi R}{c}\frac{d\varphi}{dv}a} \tag{23.2.19}$$

In summary, when the electric charge $\boldsymbol{q}$, the distance $\boldsymbol{r}$ and the velocity $\boldsymbol{v}$ vary with time. There are two equivalent formulas for expressing the time-varying electromagnetic field $\boldsymbol{E_t}$.

One is the formula $\boldsymbol{E_t(q,r,v)}$, and there are two loop integral formulas for the time-varying electromagnetic field $\boldsymbol{E_t(q,r,v)}$, viz:

$$\boldsymbol{\oint E_t(q,r,v)dl=0}$$

The distance $\boldsymbol{r}$ in the above equation is variable on the closed curve $\boldsymbol{L}$.

$$\boldsymbol{\oint E_t(q,r,v)dl=E_t\cdot L_R=-\frac{1}{2\varepsilon_0}\cdot\frac{qR}{r^2}\sin\theta} \tag{23.2.20}$$

The distance $\boldsymbol{r}$ in the above equation remains constant on the circular line $\boldsymbol{L_R=2\pi R}$.

The other is equation $\boldsymbol{E_t}\left(\frac{dq}{dt},\frac{dr}{dt},\frac{dv}{dt}\right)$. The loop integral of the time-varying electromagnetic field $\boldsymbol{E_t}\left(\frac{dq}{dt},\frac{dr}{dt},\frac{dv}{dt}\right)$ is:

$$\boldsymbol{\oint E_t\left(\frac{dq}{dt},\frac{dr}{dt},\frac{dv}{dt}\right)dl=\frac{R\sin\theta}{2\varepsilon_0 c}\left(\frac{1}{r}\frac{dq}{dt}-\frac{1}{r^2}v_r\right)+\frac{2\pi R}{c}\frac{\partial\varphi}{\partial v}a} \tag{23.2.21}$$

Since the time-varying electromagnetic field $\boldsymbol{E_t}$ produced by the moving electric charge $\boldsymbol{q}$ at the moment $\boldsymbol{t}$ is unique, Equations $\boldsymbol{E_t(q,r,v)}$ and $\boldsymbol{E_t}\left(\frac{dq}{dt},\frac{dr}{dt},\frac{dv}{dt}\right)$ are both expressions for the time-varying electromagnetic field $\boldsymbol{E_t}$.

23.2.3 Flux electromotive force ε_{SE} for time-varying electromagnetic field E_t

Assume that the distance $\boldsymbol{r}$ from electric charge $\boldsymbol{q}$ to each point on the circular line $\boldsymbol{L_R=2\pi R}$ is equal. Assume that the surface $\boldsymbol{S}$ is bordered by the circular line $\boldsymbol{L_R=2\pi R}$. At time $\boldsymbol{t}$, since the time-varying electromagnetic field $\boldsymbol{E_t(q,r,v)}$ is constant on the circular line $\boldsymbol{L_R}$, according to equation (22.1.18) , the electromagnetic field flux $\boldsymbol{\Phi_{Et}}$ through the surface $\boldsymbol{S}$ is:

$$\boldsymbol{\Phi_{Et}=\iint E_t(q,r,v)\,dS=\frac{R}{2\varepsilon_0}\frac{q}{r}\sin\theta} \tag{23.2.22}$$

Taking the derivative of the above equation with respect to time $\boldsymbol{t}$ gives the relationship equation:

$$\boldsymbol{\frac{\partial\Phi_{Et}}{\partial t}=\iint\frac{\partial E_t(q,r,v)}{\partial t}dS=\frac{\partial\Phi_{Et}}{\partial q}\frac{dq}{dt}+\frac{\partial\Phi_{Et}}{\partial r}\frac{dr}{dt}+\frac{\partial\Phi_{Et}}{\partial v}\frac{dv}{dt}}$$
$$\boldsymbol{=\frac{R}{2\varepsilon_0}\left(\frac{1}{r}\frac{dq}{dt}-\frac{q}{r^2}\frac{dr}{dt}\right)\sin\theta+\frac{\partial\Phi_{Et}}{\partial v}\frac{dv}{dt}} \tag{23.2.23}$$

The above equation is the equation for the flux rate of change of the time-varying electromagnetic field $\boldsymbol{E_t(q,r,v)}$. Commanding the velocity $\boldsymbol{v_r=\frac{dr}{dt}}$ and the acceleration $\boldsymbol{a=\frac{dv}{dt}}$, the relation equation can be obtained:

$$\frac{\partial \Phi_{Et}}{\partial t} = \iint \frac{\partial E_t(q,r,v)}{\partial t} dS = \frac{R}{2\varepsilon}\left(\frac{1}{r}\frac{dq}{dt} - \frac{q}{r^2}v_r\right)\sin\theta + \frac{\partial \Phi_{Et}}{\partial v}a \qquad (23.2.24)$$

When the electric charge q and velocity v remain constant and the distance r changes. Since rate of change of electric charge $\frac{dq}{dt} = 0$ and acceleration $a = 0$, the above equation becomes:

$$\frac{\partial \Phi_{Et}}{\partial t} = -\frac{R}{2\varepsilon_0}\frac{q}{r^2}v_r \sin\theta \qquad (23.2.25)$$

Dividing the above equation by the velocity v_r gives the relationship equation:

$$\varepsilon'_{LE} = \frac{1}{v_r}\frac{\partial \Phi_{Et}}{\partial t} = -\frac{R}{2\varepsilon_0}\frac{q}{r^2}\sin\theta \qquad (23.2.26)$$

According to equation (23.2.7), the linear electromotive force ε_{LE} (voltage U) generated by the electric charge q is:

$$\varepsilon_{LE} = +\frac{q}{4\pi\varepsilon_0}\left(\frac{1}{r_b} - \frac{1}{r_a}\right) = U(\text{voltage}) \qquad (23.2.27)$$

Compare the above equation with equation (23.2.26). Since distance r and radius R are both distances, $\frac{R}{r^2}$ and $\left(\frac{1}{r_b} - \frac{1}{r_a}\right)$ have the same physical units. From this it can be determined that the ratio of the rate of change of flux $\frac{\partial \Phi_{Et}}{\partial t}$ to the velocity belongs to the electromotive force ε_{LE}. Dividing equation (23.2.24) by the velocity v_r gives the relationship equation that:

$$\varepsilon'_{LE} = \frac{1}{v_r}\frac{\partial \Phi_{Et}}{\partial t} = \frac{1}{v_r}\frac{R}{2\varepsilon_0}\left(\frac{1}{r}\frac{dq}{dt} - \frac{q}{r^2}v_r\right)\sin\theta + \frac{1}{v_r}\frac{\partial \Phi_{Et}}{\partial v}a \qquad (23.2.28)$$

Theoretically, there is a serious error in the above equation. When the distance $r$remains constant and the electric charge q changes. Since the rate of change of electric charge $\frac{\partial \Phi_{Et}}{\partial t} \neq 0$ and the velocity $v_r = 0$, the above formula becomes:

$$\varepsilon'_{LE} = \frac{1}{v_r}\frac{\partial \Phi_{Et}}{\partial t} = \infty \qquad (23.2.29)$$

However, when the distance r is held constant and the electric charge q and velocity v vary, the electromotive force ε_{LE} is not equal to infinity. From this it can be determined that equation (23.2.28) is wrong.

Since the rate of flux change $\frac{\partial \Phi_{Et}}{\partial t}$ divided by the velocity v_r would result in an infinity result for $\frac{1}{v_r}\frac{\partial \Phi_{Et}}{\partial t}$, the rate of flux change $\frac{\partial \Phi_{Et}}{\partial t}$ cannot be divided by the velocity v_r.

Since the speed of light c is constant in the cosmic vacuum frame of reference, the ratio of the flux rate of change $\frac{\partial \Phi_{Et}}{\partial t}$ to the speed of light c belongs to the electromotive force ε_{LE}. Dividing equation (23.2.24) by the speed of light c yields the relational equation:

$$\varepsilon_{SE} = \frac{1}{c}\frac{\partial \Phi_{Et}}{\partial t} = \frac{1}{c}\iint \frac{\partial E_t(q,r,v)}{\partial t} dS = \frac{R}{2\varepsilon_0 c}\left(\frac{1}{r}\frac{dq}{dt} - \frac{q}{r^2}v_r\right)\sin\theta + \frac{1}{c}\frac{\partial \Phi_{Et}}{\partial v}a \qquad (23.2.30)$$

This book defines ε_{SE} as the flux electromotive force of the time-varying electromagnetic field E_t. The above equation is the flux electromotive force equation of the time-varying electromagnetic field.

When the electric charge q and velocity v remain constant and the distance r changes. Since the rate of change of electric charge $\frac{dq}{dt} = 0$ and the acceleration $a = 0$, equation (23.2.30) becomes:

$$\varepsilon_{SE} = \frac{1}{c}\iint \frac{\partial E_t(q,r,v)}{\partial t} dS = -\frac{v_r}{c}\cdot\frac{1}{2\varepsilon_0}\cdot\frac{qR}{r^2}\sin\theta \qquad (23.2.31)$$

The above equation shows that the direction of the flux electromotive force ε_{SE} is opposite to the direction of the velocity v_Z. The $\frac{qv_r}{c}$ in equation is analogous to the magnetic electric charge q_v.

According to equation (23.1.5), the linear electromotive force ε_{LE0} (voltage U) generated by the static electric charge q_0 is:

$$\varepsilon_{LE0} = +\frac{q_0}{4\pi\varepsilon_0}\left(\frac{1}{r_b} - \frac{1}{r_a}\right) = U(\text{voltage})$$

Compare equation (23.2.31) with the above equation. Since distance r and radius R are both distances, both $\frac{R}{r^2}$ and $\left(\frac{1}{r_b} - \frac{1}{r_a}\right)$ have the same physical units. Since $\frac{v_r}{c}\sin\theta$ has no physical units, the flux electromotive force $\varepsilon_{SE} = \frac{1}{c}\iint \frac{\partial E_t(q,r,v)}{\partial t} dS$ is the electromotive force ε (voltage U).

23.2.4 Derivation of the induction formula for time-varying electromagnetic fields

Assume that the distance r from the electric charge q to each point on the circular line $L_R = 2\pi R$ is equal. At time t, since the time-varying electromagnetic field $E_t(q,r,v)$ is constant on the circular line L_R, the integral of the time-varying electromagnetic field $E_t(q,r,v)$ along the circular line $L_R = 2\pi R$ is:

$$\oint E_t(q,r,v)dl = E_t L_R = -\frac{1}{2\varepsilon_0}\frac{qR}{r^2}\sin\theta \tag{23.2.32}$$

Assume that the surface S is bordered by the closed circular line L_R. According to equation (23.2.30), the static flux electromotive force ε_{SE0} is:

$$\varepsilon_{SE} = \frac{1}{c}\iint \frac{\partial E_t(q,r,v)}{\partial t} dS = \frac{R}{2\varepsilon_0 c}\left(\frac{1}{r}\frac{dq}{dt} - \frac{q}{r^2}v_r\right)\sin\theta + \frac{1}{c}\frac{\partial \Phi_{Et}}{\partial v}a$$

Comparing the above equation with equation (23.2.32), the integral $\oint E_t(q,r,v)dl$ of the time-varying electromagnetic field $E_t(q,r,v)$ over the circular line $L_R = 2\pi R$ is not equal to the flux electromotive force ε_{SE} of the time-varying electromagnetic field $E_t(q,r,v)$, viz:

$$\oint E_t(q,r,v)dl \neq \frac{1}{c}\iint \frac{\partial E_t(q,r,v)}{\partial t} dS \tag{23.2.33}$$

Since the left side of the inequality sign does not contain the rate of change $\left(\frac{dq}{dt}, \frac{dr}{dt}, \frac{dv}{dt}\right)$ and the right side of the inequality sign contains the rate of change $\left(\frac{dq}{dt}, \frac{dr}{dt}, \frac{dv}{dt}\right)$. So the above equations are necessarily unequal.

Assume that the electric charge q is equidistant r to each point on the circular line $L_R = 2\pi R$. According to equation (23.2.21), the integral of the time-varying electromagnetic field $E_t\left(\frac{dq}{dt}, \frac{dr}{dt}, \frac{dv}{dt}\right)$ along the circular line $L_R = 2\pi R$ is:

$$\oint E_t\left(\frac{dq}{dt}, \frac{dr}{dt}, \frac{dv}{dt}\right) dl = \frac{R\sin\theta}{2\varepsilon_0 c}\left(\frac{1}{r}\frac{dq}{dt} - \frac{q}{r^2}v_r\right) + \frac{2\pi R}{c}\frac{d\varphi}{dv}a$$

According to equation (23.2.30), the ratio $\frac{1}{c}\iint \frac{dE_{\omega t}(q,r,v)}{dt} dS$ is:

$$\frac{1}{c}\iint \frac{\partial E_t(q,r,v)}{\partial t} dS = \frac{R\sin\theta}{2\varepsilon_0 c}\left(\frac{1}{r}\frac{dq}{dt} - \frac{q}{r^2}v_r\right) + \frac{1}{c}\frac{d\Phi_{Et}}{dv}a$$

The relationship equation is obtained from the above two formulas:

$$\oint E_t\left(\frac{dq}{dt}, \frac{dr}{dt}, \frac{dv}{dt}\right) dl = \frac{1}{c}\iint \frac{\partial E_t(q,r,v)}{\partial t} dS \tag{23.2.34}$$

The above equation can be derived from the electromagnetic potential φ and Stokes' theorem. According to equation (23.2.10), the electromagnetic potential φ generated by electric charge q is:

$$\varphi = \frac{1}{4\pi\varepsilon_0}\frac{q}{r}\sin\theta \tag{23.2.35}$$

Assume that the electromagnetic potential φ integrates along the closed curve L. Assume that the surface S has the closed curve L as its boundary. According to Stokes' theorem, the relation equation can be obtained:

$$\oint \varphi dl = \iint \nabla \times \varphi\, dS \tag{23.2.36}$$

According to equation (19.1.2), the electromagnetic field strength E is equal to the Curl $\nabla \times \varphi$ of the electromagnetic potential φ, viz:

$$E_t = \nabla \times \varphi \tag{23.2.37}$$

Substituting the above equation into equation (23.2.36) gives the relationship equation:

$$\boldsymbol{\Phi}_{Et} = \oint \varphi dl = \iint E_t dS \tag{23.2.38}$$

Taking the derivative of the above equation with respect to time t gives the relationship equation:

$$\frac{\partial \boldsymbol{\Phi}_{Et}}{\partial t} = \oint \frac{\partial \varphi}{\partial t} dl = \iint \frac{\partial E_t(q,r,v)}{\partial t} dS \tag{23.2.39}$$

Dividing the above equation by the speed of light c, gives the relationship equation:

$$\varepsilon_{SE} = \frac{1}{c}\frac{\partial \boldsymbol{\Phi}_{Et}}{\partial t} = \oint \frac{1}{c}\frac{\partial \varphi}{\partial t} dl = \iint \frac{1}{c}\frac{\partial E_t(q,r,v)}{\partial t} dS \tag{23.2.40}$$

The above equation shows that the flux electromotive force ε_{SE} of the electromagnetic field is equal to the loop integral of the rate of change $\frac{1}{c}\frac{d\varphi}{dt}$. According to the time-varying electromagnetic field equation (23.2.18), the time-varying electromagnetic field E_t is:

$$E_t\left(\frac{dq}{dt},\frac{dr}{dt},\frac{dv}{dt}\right) = \frac{1}{c}\frac{\partial \varphi}{\partial t} \tag{23.2.41}$$

Substituting the above equation into equation (23.2.40) yields the relationship equation:

$$\varepsilon_{SE} = \oint E_t\left(\frac{dq}{dt},\frac{dr}{dt},\frac{dv}{dt}\right) dl = \iint \frac{1}{c}\frac{\partial E_t(q,r,v)}{\partial t} dS \tag{23.2.42}$$

The above equation is the induction equation for a time-varying electromagnetic field. This equation is similar to Faraday's formula for electromagnetic induction. Note that equation $E_t\left(\frac{dq}{dt},\frac{dr}{dt},\frac{dv}{dt}\right)$ and equation $E_t(q,r,v)$ are different expressions for the time-varying electromagnetic field.

23.2.5 Effect of distance Z variation on flux electromotive force ε_{SE}

Suppose O is the center of the circular line $L_R = 2\pi R$, and Z is the perpendicular distance from electric charge q to the center O. In the cosmic vacuum frame of reference, the electric charge q moves downward along the distance Z with velocity v. Assume that θ is the angle between the distance r and the velocity v. Assume that Ω is the steradian angle possessed by the circular plane $P = \pi R^2$. As shown earlier in Figure 22-2.

According to Figure 22-2, the distance r is:

$$r = \sqrt{R^2 + Z^2} \tag{23.2.43}$$

Taking the derivative of the above equation with respect to the distance Z yields the relationship equation:

$$\frac{\partial r}{\partial Z} = \frac{Z}{\sqrt{R^2 + Z^2}} = \frac{Z}{r} \tag{23.2.44}$$

According to Figure 22-2, the function $\sin\theta$ is:

$$\sin\theta = \frac{R}{r} \tag{23.2.45}$$

Substituting the above equation into equation (23.2.22) yields the relationship equation:

$$\boldsymbol{\Phi}_{Et} = \frac{R^2}{2\varepsilon_0}\frac{q}{r^2} \tag{23.2.46}$$

Taking the derivative of the above equation with respect to the distance r gives the relationship equation:

$$\frac{\partial \boldsymbol{\Phi}_{Et}}{\partial r} = -\frac{R^2}{\varepsilon_0}\frac{q}{r^3} \tag{23.2.47}$$

Taking the derivative of equation (23.2.46) with respect to electric charge q yields the relation equation:

$$\frac{\partial \boldsymbol{\Phi}_{Et}}{\partial q} = \frac{R^2}{2\varepsilon_0}\frac{1}{r^2} \tag{23.2.48}$$

when the electric charge q, velocity v and distance Z vary. Taking the derivative of equation (23.2.46) with respect to time t gives the relation equation:

$$\frac{\partial \boldsymbol{\Phi}_{Et}}{\partial t} = \frac{\partial \boldsymbol{\Phi}_{Et}}{\partial q}\frac{dq}{dt} + \frac{\partial \boldsymbol{\Phi}_{Et}}{\partial r}\frac{\partial r}{\partial Z}\frac{dZ}{dt} + \frac{\partial \boldsymbol{\Phi}_{Et}}{\partial v}\frac{dv}{dt} = \frac{R^2}{2\varepsilon_0}\frac{1}{r^2}\frac{dq}{dt} - \frac{R^2}{\varepsilon_0}\frac{q}{r^3}\cdot\frac{Z}{r}\cdot\frac{dZ}{dt} + \frac{d\boldsymbol{\Phi}_E}{dv}\frac{dv}{dt} \tag{23.2.49}$$

The above equation is the equation for the rate of change of flux of a time-varying electromagnetic field. Dividing the formula by the speed of light c gives the relational formula:

$$\varepsilon_{SE} = \frac{1}{c}\frac{\partial \Phi_{Et}}{\partial t} = \frac{R^2}{2\varepsilon_0 c}\left(\frac{1}{r^2}\frac{dq}{dt} - 2\frac{qZ}{r^4}\frac{dZ}{dt}\right) + \frac{1}{c}\frac{\partial \Phi_{Et}}{\partial v}\frac{dv}{dt}$$

The command velocity $v_Z = \frac{dZ}{dt}$ and acceleration $a = \frac{dv}{dt}$. From this the above equation becomes:

$$\varepsilon_{SE} = \frac{1}{c}\frac{\partial \Phi_{Et}}{\partial t} = \frac{R^2}{2\varepsilon_0 c}\left(\frac{1}{r^2}\frac{dq}{dt} - 2\frac{qZ}{r^4}v_Z\right) + \frac{1}{c}\frac{\partial \Phi_{Et}}{\partial v}a \qquad (23.2.50)$$

The above equation is the flux electromotive force equation for a time-varying electromagnetic field. When the electric charge q and velocity v are kept constant and the distance Z changes. Since the rate of change $\frac{dq}{dt} = 0$ of electric charge and acceleration $a = 0$, the above equation becomes:

$$\varepsilon_{SE} = -\frac{qv_Z}{\varepsilon_0 c}\cdot\frac{R^2 Z}{r^4} \qquad (23.2.51)$$

The $\frac{qv_Z}{c}$ in equation is analogous to magnetic electric charge $q_v = \frac{qv}{c}$. According to equation (23.1.5), the linear electromotive force ε_{LE0} (voltage U) generated by the static electric charge q_0 is:

$$\varepsilon_{LE0} = \frac{q_0}{4\pi\varepsilon_0}\left(\frac{1}{r_b} - \frac{1}{r_a}\right) = U(\text{voltage})$$

Compare equation (23.2.51) with the above equation. Since the ratio $\frac{v_Z}{c}$ of velocities has no physical units, $\frac{qv_Z}{\varepsilon_0 c}$ and $\frac{q_0}{4\pi\varepsilon_0}$ have the same physical units. Since the distance r and the radius R and the distance Z are both distances, $\frac{R^2 Z}{r^4}$ has the same physical units as $\left(\frac{1}{r_b} - \frac{1}{r_a}\right)$. From this it can be determined that the flux electromotive force ε_{SE} of the time-varying electromagnetic field belongs to the electromotive force ε (voltage U).

23.3 Derivation of the rotating electric field induction formula

23.3.1 Linear rotation electromotive force $\varepsilon_{LE\omega}$

When electric charge q is moving in the cosmic vacuum frame of reference with velocity v According to equation (19.2.10), the rotating electric field strength E_ω generated by the moving electric charge q is:

$$E_\omega(q, r, v) = -\frac{1}{4\pi\varepsilon_0}\frac{q}{r^2}\sqrt{1 - \frac{v^2}{c^2}}\sin\theta \qquad (23.3.1)$$

The θ in the formula is the angle between the distance r and the velocity v. The direction of the rotating electric field E_ω is the same as the direction of the distance r.

When the electric charge q, velocity v and distance r vary with time. Since different times t_i correspond to different rotating electric field strengths $E_{\omega i}(q_i, r_i, v_i)$, the time-varying rotating electric field $E_{\omega t}$ corresponding to time t can be expressed as:

$$E_{\omega t} = E_{\omega t}(q, r, v) = -\frac{1}{4\pi\varepsilon_0}\frac{q}{r^2}\sqrt{1 - \frac{v^2}{c^2}}\sin\theta \qquad (23.3.2)$$

Theoretically, at time t there exists not only a time-varying rotating electric field $E_{\omega i}(q, r, v)$, but also a time-varying rotating electromagnetic field $E_{\omega t}$ containing a rate of change $\left(\frac{dq}{dt}, \frac{dr}{dt}, \frac{dv}{dt}\right)$, i.e:

$$E_{\omega t} = E_{\omega t}\left(\frac{dq}{dt}, \frac{dr}{dt}, \frac{dv}{dt}\right)$$

Taking the derivative of equation (23.3.2) with respect to time t yields the relation equation:

$$\frac{\partial E_{\omega t}}{\partial t} = \frac{\partial E_{\omega t}}{\partial q}\frac{dq}{dt} + \frac{\partial E_{\omega t}}{\partial r}\frac{dr}{dt} + \frac{\partial E_{\omega t}}{\partial v}\frac{dv}{dt} \qquad (23.3.3)$$

The above formula is not a time-varying rotating electric field formula because the rotating electric field $\boldsymbol{E}_{\omega t} \neq \frac{\partial E_{\omega t}}{\partial t}$, although it includes the rate of change $\left(\frac{dq}{dt}, \frac{dr}{dt}, \frac{dv}{dt}\right)$.

Assume that the radius $\boldsymbol{r}$ is variable on the curve $\boldsymbol{L} = \boldsymbol{f}(\boldsymbol{r})$. Assume that points $\boldsymbol{a}$ and $\boldsymbol{b}$ are points on the curve $\boldsymbol{L}$. The integral $\boldsymbol{\varepsilon}_{LE\omega}$ of the time-varying rotating electric field $\boldsymbol{E}_{\omega t}(\boldsymbol{q}, \boldsymbol{r}, \boldsymbol{v})$ along the curve $\boldsymbol{L}$ is:

$$\boldsymbol{\varepsilon}_{LE\omega} = \int_{r_a}^{r_b} \boldsymbol{E}_{\omega t}(\boldsymbol{q}, \boldsymbol{r}, \boldsymbol{v}) d\boldsymbol{l} \tag{23.3.4}$$

This book defines the line integral $\boldsymbol{\varepsilon}_{LE\omega}$ as the linear rotational electromotive force . The above equation is the linear rotational electromotive force equation. When the electric charge $\boldsymbol{q}$ and velocity $\boldsymbol{v}$ are kept constant and the distance $\boldsymbol{r}$ is varied. According to equation (23.3.2) the relation equation can be obtained:

$$\boldsymbol{\varepsilon}_{LE\omega} = \int_{r_a}^{r_b} \boldsymbol{E}_{\omega t}(\boldsymbol{q}, \boldsymbol{r}, \boldsymbol{v}) d\boldsymbol{l} = \frac{\boldsymbol{q}}{4\pi\varepsilon_0} \sqrt{1 - \frac{v^2}{c^2}} \left(\frac{1}{r_b} - \frac{1}{r_a}\right) \sin\theta \tag{23.3.5}$$

If distance $\boldsymbol{r}_a \neq \boldsymbol{r}_b$, then linear rotational electromotive force $\boldsymbol{\varepsilon}_{LE\omega} \neq \boldsymbol{0}$. If distance $\boldsymbol{r}_a = \boldsymbol{r}_b$, then linear rotational electromotive force $\boldsymbol{\varepsilon}_{LE\omega} = \boldsymbol{0}$.

When the electric charge velocity $\boldsymbol{v} = \boldsymbol{0}$. Since the angle $\boldsymbol{\theta} = \boldsymbol{90^0}$, the rotational electromotive force $\boldsymbol{\varepsilon}_{LE\omega}$ is equal to the line integral of the electrostatic field $\boldsymbol{\varepsilon}_{LE0}$, i.e:

$$\boldsymbol{\varepsilon}_{LE\omega} = \boldsymbol{\varepsilon}_{LE0} = \int_{r_a}^{r_b} \boldsymbol{E}_{0t} d\boldsymbol{l} = \frac{q_0}{4\pi\varepsilon_0}\left(\frac{1}{r_b} - \frac{1}{r_a}\right) \tag{23.3.6}$$

Suppose that the radius $\boldsymbol{r}$ is changeable on the curve $\boldsymbol{L} = \boldsymbol{f}(\boldsymbol{r})$. When the curve $\boldsymbol{L}$ reaches point $\boldsymbol{b}$ from point $\boldsymbol{a}$ and returns to point $\boldsymbol{a}$ from point $\boldsymbol{b}$ Since the curve $\boldsymbol{L}$ is closed, the integral of the time-varying rotating electric field $\boldsymbol{E}_{\omega t}(\boldsymbol{q}, \boldsymbol{r}, \boldsymbol{v})$ along the closed curve $\boldsymbol{L}$ is:

$$\oint \boldsymbol{E}_{\omega t}(\boldsymbol{q}, \boldsymbol{r}, \boldsymbol{v}) d\boldsymbol{l} = \int_{r_a}^{r_b} \boldsymbol{E}_{\omega t} d\boldsymbol{l} + \int_{r_b}^{r_a} \boldsymbol{E}_{\omega t} d\boldsymbol{l} = \boldsymbol{0} \tag{23.3.7}$$

The above equation is the loop integral equation for a time-varying rotating electric field, which is analogous to the loop integral equation for a time-varying electrostatic field.

Assume that the electric charge $\boldsymbol{q}$ is at an equal distance $\boldsymbol{r}$ to each point on the circular line $\boldsymbol{L}_R = \boldsymbol{2\pi R}$. At time $\boldsymbol{t}$, since the time-varying rotating electric field $\boldsymbol{E}_{\omega t}(\boldsymbol{q}, \boldsymbol{r}, \boldsymbol{v})$ is constant on the circular line $\boldsymbol{L}_R$, the integral of the time-varying rotating electric field $\boldsymbol{E}_{\omega t}(\boldsymbol{q}, \boldsymbol{r}, \boldsymbol{v})$ along the circular line $\boldsymbol{L}_R = \boldsymbol{2\pi R}$ is:

$$\oint \boldsymbol{E}_{\omega t}(\boldsymbol{q}, \boldsymbol{r}, \boldsymbol{v}) d\boldsymbol{l} = \boldsymbol{E}_{\omega t} \cdot \boldsymbol{L}_R = -\frac{R}{2\varepsilon_0} \cdot \frac{q}{r^2} \sqrt{1 - \frac{v^2}{c^2}} \sin\theta \tag{23.3.8}$$

Since the time-varying rotating electric field $\boldsymbol{E}_{\omega t}(\boldsymbol{q}, \boldsymbol{r}, \boldsymbol{v})$ is constant on the circular line $\boldsymbol{L}_R = \boldsymbol{2\pi R}$, the loop integral $\oint \boldsymbol{E}_{\omega t}(\boldsymbol{q}, \boldsymbol{r}, \boldsymbol{v}) d\boldsymbol{l}$ is not the rotating electromotive force $\boldsymbol{\varepsilon}_{LE\omega}$.

23.3.2 Derivation of the time-varying rotating electric field formula $E_{\omega t}$

Theoretically, there are two different formulas for expressing the time-varying rotating electric field $\boldsymbol{E}_{\omega t}$.

The first is a time-varying rotating electric field formula that does not include the rate of change $\left(\frac{dq}{dt}, \frac{dr}{dt}, \frac{dv}{dt}\right)$, i.e:

$$\boldsymbol{E}_{\omega t} = \boldsymbol{E}_{\omega t}(\boldsymbol{q}, \boldsymbol{r}, \boldsymbol{v})$$

The second one is a time-varying rotating electric field formula containing the rate of change $\left(\frac{dq}{dt}, \frac{dr}{dt}, \frac{dv}{dt}\right)$, i.e:

$$\boldsymbol{E}_{\omega t} = \boldsymbol{E}_{\omega t}\left(\frac{dq}{dt}, \frac{dr}{dt}, \frac{dv}{dt}\right)$$

We can derive the above equation from the rotational electric potential $\boldsymbol{\varphi}_\omega$.

When the electric charge $\boldsymbol{q}$ is moving with velocity $\boldsymbol{v}$ in the cosmic vacuum. According to the rotational electric potential equation (18.3.7), the rotational electric potential $\boldsymbol{\varphi}_\omega$ generated by the moving electric charge $\boldsymbol{q}$ is:

$$\boldsymbol{\varphi}_\omega = \frac{1}{4\pi\varepsilon_0} \frac{q}{r} \sqrt{1 - \frac{v^2}{c^2}} \sin\theta \tag{23.3.9}$$

The $\boldsymbol{\theta}$ in the formula is the angle between the distance $\boldsymbol{r}$ and the velocity $\boldsymbol{v}$. Taking the derivative of the above equation with respect to time $\boldsymbol{t}$ gives the relationship equation:

$$\frac{\partial\varphi_\omega}{\partial t}=\frac{\partial\varphi_\omega}{\partial q}\frac{dq}{dt}+\frac{\partial\varphi_\omega}{\partial r}\frac{dr}{dt}+\frac{\partial\varphi_\omega}{\partial v}\frac{dv}{dt}$$

Command the velocity $v_r=\frac{dr}{dt}$ and the acceleration $a=\frac{dv}{dt}$. Since dividing the rate of change $\frac{\partial\varphi_\omega}{\partial t}$ by the velocity v_r would result in infinity for $\frac{1}{v_r}\frac{\partial\varphi_\omega}{\partial t}$, the rate of change $\frac{\partial\varphi_\omega}{\partial t}$ cannot be divided by the velocity v_r. Divide the above equation by the speed c of light, to obtain the relational equation:

$$\frac{1}{c}\frac{\partial\varphi_\omega}{\partial t}=\frac{1}{c}\frac{\partial\varphi_\omega}{\partial q}\frac{dq}{dt}+\frac{1}{c}\frac{\partial\varphi_\omega}{\partial r}\frac{dr}{dt}+\frac{1}{c}\frac{\partial\varphi_\omega}{\partial v}\frac{dv}{dt}=\frac{\sin\theta}{4\pi\varepsilon_0 c}\sqrt{1-\frac{v^2}{c^2}}\left(\frac{1}{r}\frac{dq}{dt}-\frac{q}{r^2}v_r\right)+\frac{1}{c}\frac{\partial\varphi_\omega}{\partial v}a \quad (23.3.10)$$

When the electric charge q and velocity v are kept constant and the distance r is variable. Since the rate of change of electric charge $\frac{dq}{dt}=0$ and the acceleration $a=0$, the above equation becomes:

$$\frac{1}{c}\frac{\partial\varphi_\omega}{\partial t}=-\frac{v_r}{c}\cdot\frac{1}{4\pi\varepsilon_0}\frac{q}{r^2}\sqrt{1-\frac{v^2}{c^2}}\sin\theta=\frac{v_r}{c}E_\omega \quad (23.3.11)$$

Since the velocity ratio $\frac{v_r}{c}$ has no physical unit, $\frac{1}{c}\frac{\partial\varphi_\omega}{\partial t}$ belongs to the rotating electric field strength E_ω. From this it can be determined that $\frac{1}{c}\frac{\partial\varphi_\omega}{\partial t}$, $\frac{1}{c}\frac{\partial\varphi_\omega}{\partial q}\frac{dq}{dt}$, $\frac{1}{c}\frac{\partial\varphi_\omega}{\partial r}\frac{dr}{dt}$, and $\frac{1}{c}\frac{\partial\varphi_\omega}{\partial v}\frac{dv}{dt}$ in equation (23.3.10) belong to the time-varying rotating electric field.

Since the time-varying rotating electric field $E_{\omega t}$ generated by the moving electric charge q at time t possesses uniqueness, the time-varying rotating electric field $E_{\omega t}$ is equal to the ratio $\frac{1}{c}\frac{\partial\varphi_\omega}{\partial t}$, viz:

$$E_{\omega t}\left(\frac{dq}{dt},\frac{dr}{dt},\frac{dv}{dt}\right)=\frac{1}{c}\frac{\partial\varphi_\omega}{\partial t}=\frac{\sin\theta}{4\pi\varepsilon_0 c}\sqrt{1-\frac{v^2}{c^2}}\left(\frac{1}{r}\frac{dq}{dt}-\frac{q}{r^2}v_r\right)+\frac{1}{c}\frac{\partial\varphi_\omega}{\partial v}a \quad (23.3.12)$$

Assume that the electric charge q is at an equal distance r to each point on the circular line $L_R=2\pi R$. At time t, since the time-varying rotating electric field $E_{\omega t}\left(\frac{dq}{dt},\frac{dr}{dt},\frac{dv}{dt}\right)$ is constant on the circular line L_R, the integral of the time-varying rotating electric field $E_{\omega t}\left(\frac{dq}{dt},\frac{dr}{dt},\frac{dv}{dt}\right)$ along the circular line $L_R=2\pi R$ is:

$$\oint E_{\omega t}\left(\frac{dq}{dt},\frac{dr}{dt},\frac{dv}{dt}\right)dl=E_{\omega t}\cdot L_R=\frac{R\sin\theta}{2\varepsilon_0 c}\sqrt{1-\frac{v^2}{c^2}}\left(\frac{1}{r}\frac{dq}{dt}-\frac{q}{r^2}v_r\right)+\frac{2\pi R}{c}\frac{d\varphi_\omega}{dv}a$$

When the electric charge q and velocity v are kept constant and the distance r is variable. Since the rate of change of electric charge $\frac{dq}{dt}=0$ and the acceleration $a=0$, the above equation becomes:

$$\oint E_{\omega t}\left(\frac{dq}{dt},\frac{dr}{dt},\frac{dv}{dt}\right)dl=-\frac{v_r}{c}\cdot\frac{1}{2\varepsilon_0}\cdot\frac{qR}{r^2}\sqrt{1-\frac{v^2}{c^2}}\sin\theta \quad (23.3.13)$$

To summarize: when the electric charge q, the distance r and the velocity v vary with time. The time-varying rotating electric field $E_{\omega t}$ has two equivalent expressions.

The first one is equation $E_{\omega t}(q,r,v)$. Note that there are two loop integral formulas for the time-varying rotating electric field $E_{\omega t}(q,r,v)$, viz:

$$\oint E_{\omega t}(q,r,v)dl=0 \quad (23.3.14)$$

The distance r in the above equation is variable on the closed curve L.

$$\oint E_{\omega t}(q,r,v)dl=E_{\omega t}\cdot L_R=-\frac{1}{2\varepsilon_0}\cdot\frac{qR}{r^2}\sqrt{1-\frac{v^2}{c^2}}\sin\theta \quad (23.3.15)$$

The distance r in the above equation remains constant on the circular line $L_R=2\pi R$.

The second is equation $\boldsymbol{E}_{\omega t}\left(\frac{dq}{dt},\frac{dr}{dt},\frac{dv}{dt}\right)$. The loop integral of the time-varying rotating electric field $\boldsymbol{E}_{\omega t}\left(\frac{dq}{dt},\frac{dr}{dt},\frac{dv}{dt}\right)$ is:

$$\oint \boldsymbol{E}_{\omega t}\left(\frac{dq}{dt},\frac{dr}{dt},\frac{dv}{dt}\right)dl=\frac{R\sin\theta}{2\varepsilon_0 c}\sqrt{1-\frac{v^2}{c^2}}\left(\frac{1}{r}\frac{dq}{dt}-\frac{q}{r^2}v_r\right)+\frac{2\pi R}{c}\frac{d\varphi_\omega}{dv}a \qquad (23.3.16)$$

Since the time-varying rotating electric field $\boldsymbol{E}_{\omega t}$ generated by the moving electric charge $\boldsymbol{q}$ at time $\boldsymbol{t}$ possesses uniqueness, both equation $\boldsymbol{E}_{\omega t}(\boldsymbol{q},\boldsymbol{r},\boldsymbol{v})$ and equation $\boldsymbol{E}_{\omega t}\left(\frac{dq}{dt},\frac{dr}{dt},\frac{dv}{dt}\right)$ are expressions for the time-varying rotating electric field $\boldsymbol{E}_{\omega t}$.

23.3.3 Flux electromotive force $\varepsilon_{SE\omega}$ for rotating electric field $E_{\omega t}(q,r,v)$

Assume that the electric charge $\boldsymbol{q}$, velocity $\boldsymbol{v}$ and distance $\boldsymbol{r}$ vary with time. Assume that the distance $\boldsymbol{r}$ from the moving electric charge $\boldsymbol{q}$ to each point on the circular line $\boldsymbol{L_R}=\boldsymbol{2\pi R}$ is equal, and assume that the surface $\boldsymbol{S}$ is bounded by the circular line $\boldsymbol{L_R}=\boldsymbol{2\pi R}$.

According to equation (22.2.11), the rotating electric field flux $\boldsymbol{\Phi}_{E\omega t}$ across the surface $\boldsymbol{S}$ at time $\boldsymbol{t}$ is:

$$\boldsymbol{\Phi}_{E\omega t}=\iint \boldsymbol{E}_{\omega t}(q,r,v)\,dS=\frac{R}{2\varepsilon_0}\frac{q}{r}\sqrt{1-\frac{v^2}{c^2}}\sin\theta$$

Take the derivative of the above equation with respect to time $\boldsymbol{t}$ to obtain the relationship equation:

$$\frac{\partial\boldsymbol{\Phi}_{E\omega t}}{\partial t}=\iint\frac{\partial \boldsymbol{E}_{\omega t}}{\partial t}dS=\frac{\partial\boldsymbol{\Phi}_{E\omega t}}{\partial q}\frac{dq}{t}+\frac{\partial\boldsymbol{\Phi}_{E\omega t}}{\partial r}\frac{dr}{dt}+\frac{\partial\boldsymbol{\Phi}_{E\omega t}}{\partial v}\frac{dv}{dt}$$

$$=\frac{R\sin\theta}{2\varepsilon_0}\sqrt{1-\frac{v^2}{c^2}}\left(\frac{1}{r}\frac{dq}{dt}-\frac{q}{r^2}\frac{dr}{dt}\right)+\frac{d\boldsymbol{\Phi}_{E\omega t}}{dv}\frac{dv}{dt} \qquad (23.3.17)$$

The command velocity $v_r=\frac{dr}{dt}$ and acceleration $a=\frac{dv}{dt}$. Since the rate of change $\frac{\partial\Phi_{E\omega t}}{\partial t}$ divided by the velocity v_r would result in the ratio $\frac{1}{v_r}\frac{\partial\Phi_{E\omega t}}{\partial t}$ being equal to infinity, the rate of change $\frac{\partial\Phi_{E\omega t}}{\partial t}$ can not be divided by the velocity v_r. Dividing the above equation by the speed of light $\boldsymbol{c}$ yields the relationship equation:

$$\frac{1}{c}\frac{\partial\boldsymbol{\Phi}_{E\omega t}}{\partial t}=\frac{1}{c}\iint\frac{\partial \boldsymbol{E}_{\omega t}(q,r,v)}{\partial t}dS=\frac{R\sin\theta}{2\varepsilon_0 c}\sqrt{1-\frac{v^2}{c^2}}\left(\frac{1}{r}\frac{dq}{dt}-\frac{q}{r^2}v_r\right)+\frac{1}{c}\frac{d\boldsymbol{\Phi}_{E\omega t}}{dv}a \qquad (23.3.18)$$

When the distance $\boldsymbol{r}$ changes and the electric charge $\boldsymbol{q}$ and velocity $\boldsymbol{v}$ remain constant. Since the rate of change of electric charge $\frac{dq}{dt}=0$ and the acceleration $a=0$, the above equation becomes:

$$\frac{1}{c}\iint\frac{\partial \boldsymbol{E}_{\omega t}(q,r,v)}{\partial t}dS=-\frac{v_r}{c}\cdot\frac{R}{2\varepsilon_0}\cdot\frac{q}{r^2}\sqrt{1-\frac{v^2}{c^2}}\sin\theta \qquad (23.3.19)$$

According to equation (23.1.5), the linear electromotive force ε_{LE0} (voltage U) generated by the static electric charge q_0 is:

$$\varepsilon_{LE0}=\int_{r_a}^{r_b}E_{0t}dl=\frac{q_0}{4\pi\varepsilon_0}\left(\frac{1}{r_b}-\frac{1}{r_a}\right)=U(\text{voltage})$$

Compare equation (23.3.19) with the above equation. Since the velocity ratios $\frac{v_r}{c}$ and $\sqrt{1-\frac{v^2}{c^2}}\sin\theta$ have no physical units, $\frac{v_r}{c}\frac{q}{2\varepsilon_0}\sqrt{1-\frac{v^2}{c^2}}\sin\theta$ and $\frac{q_0}{4\pi\varepsilon_0}$ have the same physical units. Since distance $\boldsymbol{r}$ and radius $\boldsymbol{R}$ are both distances, $\frac{R}{r^2}$ and $\left(\frac{1}{r_b}-\frac{1}{r_a}\right)$ have the same physical units. From this it can be determined that the ratio $\frac{1}{c}\iint\frac{\partial E_{\omega t}(q,r,v)}{\partial t}dS$ belongs to the electromotive force ε (voltage U).

This book uses the symbol $\varepsilon_{SE\omega}$ to denote the flux electromotive force of the time-varying rotating electric field $E_{\omega t}(q,r,v)$, viz:

$$\varepsilon_{SE\omega} = \frac{1}{c}\iint \frac{\partial E_{\omega t}(q,r,v)}{\partial t} dS \qquad (23.3.20)$$

The above equation is the flux electromotive force equation for a rotating electric field.

23.3.4 Derivation of the rotating electric field induction formula

Assume that the electric charge q is at an equal distance r to each point on the circular line $L_R = 2\pi R$. Since the rotating electric field $E_{\omega t}$ is constant on the circular line L_R, the integral of the rotating electric field $E_{\omega t}$ along the circular line $L_R = 2\pi R$ is:

$$\oint E_{\omega t}(q,r,v)dl = E_{\omega t}L_R = -\frac{1}{2\varepsilon_0}\frac{qR}{r^2}\sqrt{1-\frac{v^2}{c^2}}\sin\theta \qquad (23.3.21)$$

Since the rotating electric field $E_{\omega t}$ is constant on the circular line $L_R = 2\pi R$, the rotating electric field strength $E_{\omega t}(q,r,v)$ does not have a potential difference on the circular line $L_R = 2\pi R$. From this it can be determined that the loop integral $\oint E_{\omega t}(q,r,v)dl$ is not equal to the flux electromotive force $\varepsilon_{SE\omega}$, i.e:

$$\oint E_{\omega t}(q,r,v)dl \neq \frac{1}{c}\iint \frac{\partial E_{\omega t}(q,r,v)}{\partial t} dS \qquad (23.3.22)$$

Since the left side of the inequality sign does not contain the rate of change $\left(\frac{dq}{dt},\frac{dr}{dt},\frac{dv}{dt}\right)$ and the right side of the inequality sign contains the rate of change $\left(\frac{dq}{dt},\frac{dr}{dt},\frac{dv}{dt}\right)$. So the above equations are necessarily unequal.

Assume that the electric charge q is at an equal distance r to each point on the circular line $L_R = 2\pi R$. According to equation (23.3.16), the integral of the rotating electric field $E_{\omega t}\left(\frac{dq}{dt},\frac{dr}{dt},\frac{dv}{dt}\right)$ along the circular line $L_R = 2\pi R$ is:

$$\oint E_{\omega t}\left(\frac{dq}{dt},\frac{dr}{dt},\frac{dv}{dt}\right)dl = \frac{R\sin\theta}{2\varepsilon_0 c}\sqrt{1-\frac{v^2}{c^2}}\left(\frac{1}{r}\frac{dq}{dt}-\frac{q}{r^2}v_r\right)+\frac{2\pi R}{c}\frac{d\varphi_\omega}{dv}a$$

According to equation (23.3.18) , the ratio $\frac{1}{c}\iint \frac{\partial E_{\omega t}(q,r,v)}{\partial t} dS$ is:

$$\frac{1}{c}\iint \frac{\partial E_{\omega t}(q,r,v)}{\partial t} dS = \frac{R\sin\theta}{2\varepsilon_0 c}\sqrt{1-\frac{v^2}{c^2}}\left(\frac{1}{r}\frac{dq}{dt}-\frac{q}{r^2}v_r\right)+\frac{1}{c}\frac{\partial \Phi_{E\omega t}}{\partial v}a$$

The relationship equation can be obtained from the above two formulas:

$$\oint E_{\omega t}\left(\frac{dq}{dt},\frac{dr}{dt},\frac{dv}{dt}\right)dl = -\frac{1}{c}\iint \frac{dE_{\omega t}(q,r,v)}{dt} dS \qquad (23.3.23)$$

We can also derive the above equation from the rotational electric potential φ_ω and Stokes' theorem. According to equation (19.2.12), the rotational electric potential φ_ω generated by the motion electric charge q is:

$$\varphi_\omega = \frac{1}{4\pi\varepsilon_0}\frac{q}{r}\sqrt{1-\frac{v^2}{c^2}}\sin\theta$$

Assume that the rotational electric potential φ_ω integrates along the closed curve L. Suppose that the surface S is bordered by the closed loop L. By Stokes' theorem, the relation equation:

$$\oint \varphi_\omega dl = \iint \nabla\times\varphi_\omega\, dS \qquad (23.3.24)$$

According to equation (19.2.16), the rotating electric field strength $E_{\omega t}$ is equal to the Curl $\nabla\times\varphi_\omega$ of the rotating electric potential φ_ω, viz:

$$E_{\omega t} = \nabla\times\varphi_\omega \qquad (23.3.25)$$

Bringing the above equation into equation (23.3.24) gives the relationship equation:

$$\Phi_{E\omega t} = \oint \varphi_\omega dl = \iint E_{\omega t} dS \tag{23.3.26}$$

Taking the derivative of the above equation with respect to time t gives the relationship equation:

$$\frac{\partial \Phi_{E\omega t}}{\partial t} = \oint \frac{\partial \varphi_\omega}{\partial t} dl = \iint \frac{\partial E_{\omega t}}{\partial t} dS \tag{23.3.27}$$

Divide the above equation by the speed of light c, to get the relationship equation:

$$\frac{1}{c}\frac{\partial \Phi_{E\omega t}}{\partial t} = \oint \frac{1}{c}\frac{\partial \varphi_\omega}{\partial t} dl = \iint \frac{1}{c}\frac{\partial E_{\omega t}}{\partial t} dS \tag{23.3.28}$$

According to equation (23.3.12), the time-varying rotating electric field $E_{\omega t}$ is:

$$E_{\omega t}\left(\frac{dq}{dt}, \frac{dr}{dt}, \frac{dv}{dt}\right) = \frac{1}{c}\frac{\partial \varphi_\omega}{\partial t} \tag{23.3.29}$$

Bringing the above equation into equation (23.3.28) gives the relationship equation:

$$\oint E_{\omega t}\left(\frac{dq}{dt}, \frac{dr}{dt}, \frac{dv}{dt}\right) dl = \iint \frac{1}{c}\frac{\partial E_{\omega t}(q,r,v)}{\partial t} dS \tag{23.3.30}$$

The above formula is similar to Faraday's formula for electromagnetic induction.

23.4 Derivation of the Magnetic-electric field induction formula

23.4.1 Linear electromotive force for Magnetic-electric fields ε_{LEv}

When electric charge q is moving with velocity v in the cosmic vacuum frame of reference. According to Equation (19.3.2), the Magnetic-electric field strength E_v generated by the moving electric charge q is:

$$E_v(q,r,v) = +\frac{1}{4\pi\varepsilon_0 c}\frac{qv}{r^2}\sin\theta \tag{23.4.1}$$

The θ in the formula is the angle between the distance r and the velocity v. The direction of the Magnetic-electric field E_v is perpendicular to the plane defined by the velocity v and the distance r.

When the electric charge q, velocity v and distance r vary with time. Since different times t_i correspond to different Magnetic-electric field strengths $E_{vt}(q_i, r_i, v_i)$, the time-varying Magnetic-electric field E_{vt} corresponding to time t can be expressed as:

$$E_{vt} = E_{vt}(q,r,v) = \frac{1}{4\pi\varepsilon_0 c}\frac{qv}{r^2}\sin\theta \tag{23.4.2}$$

Theoretically, there exists at time t not only a time-varying Magnetic-electric field $E_{vt}(q,r,v)$, but also a time-varying Magnetic-electric field E_{vt} containing a rate of change $\left(\frac{dq}{dt}, \frac{dr}{dt}, \frac{dv}{dt}\right)$, i.e:

$$E_{vt} = E_{vt}\left(\frac{dq}{dt}, \frac{dr}{dt}, \frac{dv}{dt}\right)$$

Taking the derivative of equation (23.4.2) with respect to time t yields the relation equation:

$$\frac{\partial E_{vt}}{\partial t} = \frac{\partial E_{vt}}{\partial q}\frac{dq}{dt} + \frac{\partial E_{vt}}{\partial r}\frac{dr}{dt} + \frac{\partial E_{vt}}{\partial v}\frac{dv}{dt} \tag{23.4.3}$$

Although the above equation contains the rate of change $\left(\frac{dq}{dt}, \frac{dr}{dt}, \frac{dv}{dt}\right)$, the above equation is not a time-varying Magnetic-electric field equation since the time-varying Magnetic-electric field $E_{vt} \neq \frac{\partial E_{vt}}{\partial t}$.

Suppose that the distance r is variable on the curve $L = f(r)$. Assume that points a and b are points on the curve L. The integral ε_{LEv} of the time-varying Magnetic-electric field $E_{vt}(q,r,v)$ along the curve L is:

$$\varepsilon_{LEv} = \int_{r_a}^{r_b} E_{vt}(q,r,v) dl \tag{23.4.4}$$

This book defines ε_{LEv} as the linear magnetic electromotive force. The above equation is the linear magnetic electromotive force equation. When the electric charge q and velocity v are kept constant and the distance r is varied. According to equation (23.4.2) the relation equation can be obtained:

$$\varepsilon_{LEv} = \int_{r_a}^{r_b} E_{vt}(q,r,v)dl = -\frac{1}{4\pi\varepsilon_0}\frac{qv}{c}\left(\frac{1}{r_b} - \frac{1}{r_a}\right)\sin\theta \qquad (23.4.5)$$

If distance $r_a \neq r_b$, then linear magnetic electromotive force $\varepsilon_{LEv} \neq 0$. If distance $r_a = r_b$, then linear magnetic electromotive force $\varepsilon_{LEv} = 0$.

When the curve L reaches point b from point a and returns to point a from point b. Since the curve L is closed, the integral of the time-varying Magnetic-electric field $E_{vt}(q,r,v)$ along the closed curve L is:

$$\oint E_{vt}(q,r,v)dl = \int_{r_a}^{r_b} E_{vt}dl + \int_{r_b}^{r_a} E_{vt}dl = 0 \qquad (23.4.6)$$

The above equation is the loop integral equation for the time-varying Magnetic-electric field $E_{vt}(q,r,v)$, which is analogous to the loop integral equation for an electrostatic field. The distance r in the above equation is variable on the closed curve L.

Assume that the electric charge q is at an equal distance r to each point on the circular line $L_R = 2\pi R$. At time t, since the time-varying Magnetic-electric field $E_{vt}(q,r,v)$ is constant on the circular line L_R, the integral of the time-varying Magnetic-electric field $E_{vt}(q,r,v)$ along the circular line $L_R = 2\pi R$ is:

$$\oint E_{vt}(q,r,v)dl = E_{vt}\cdot L_R = \frac{R}{2\varepsilon_0}\cdot\frac{qv}{cr^2}\sin\theta \qquad (23.4.7)$$

Since the time-varying Magnetic-electric field $E_{vt}(q,r,v)$ is constant on the circular line $L_R = 2\pi R$, the loop integral $\oint E_{vt}(q,r,v)dl$ is not the magnetic electromotive force ε_{LEv}.

23.4.2 Derivation of the time-varying Magnetic-electric field E_{vt}

Theoretically, there are two different ways to express the time-varying Magnetic-electric field E_{vt}.

The first one is the time-varying Magnetic-electric field $E_{vt}(q,r,v)$ without the rate of change $\left(\frac{dq}{dt},\frac{dr}{dt},\frac{dv}{dt}\right)$

The second one is the time-varying Magnetic-electric field E_{vt} containing the rate of change $\left(\frac{dq}{dt},\frac{dr}{dt},\frac{dv}{dt}\right)$, i.e:

$$E_{vt} = E_{vt}\left(\frac{dq}{dt},\frac{dr}{dt},\frac{dv}{dt}\right)$$

We can derive the above equation from the Magnetic-electric potential φ_v.

When electric charge q is moving with velocity v in the cosmic vacuum frame of reference. According to equation (18.4.3), the Magnetic-electric potential φ_v generated by electric charge q is:

$$\varphi_v = \frac{1}{4\pi\varepsilon_0 c}\frac{qv}{r}\sin\theta \qquad (23.4.8)$$

The θ in the formula is the angle between the distance r and the velocity v. Taking the derivative of the above equation with respect to time t gives the relationship equation:

$$\frac{\partial\varphi_v}{\partial t} = \frac{\partial\varphi_v}{\partial q}\frac{dq}{dt} + \frac{\partial\varphi_v}{\partial r}\frac{dr}{dt} + \frac{\partial\varphi_v}{\partial v}\frac{dv}{dt} = \frac{\sin\theta}{4\pi\varepsilon_0 c}\left(\frac{v}{r}\frac{dq}{dt} - \frac{qv}{r^2}\frac{dr}{dt} + \frac{q}{r}\frac{dv}{dt}\right) \qquad (23.4.9)$$

Command the velocity $v_r = \frac{dr}{dt}$ and the acceleration $a = \frac{dv}{dt}$. Since dividing the rate of change $\frac{\partial\varphi_v}{\partial t}$ by the velocity v_r would result in the ratio $\frac{1}{v_r}\frac{\partial\varphi_v}{\partial t}$ being equal to infinity, the rate of change $\frac{\partial\varphi_v}{\partial t}$ cannot be divided by the velocity v_r. Dividing the above equation by the speed of light c yields the relational equation:

$$\frac{1}{c}\frac{\partial\varphi_v}{\partial t} = \frac{1}{c}\frac{\partial\varphi_v}{\partial q}\frac{dq}{dt} + \frac{1}{c}\frac{\partial\varphi_v}{\partial r}\frac{dr}{dt} + \frac{1}{c}\frac{\partial\varphi_v}{\partial v}\frac{dv}{dt} = \frac{\sin\theta}{4\pi\varepsilon_0 c^2}\left(\frac{v}{r}\frac{dq}{dt} - \frac{qv}{r^2}v_r + \frac{q}{r}a\right) \qquad (23.4.10)$$

When the electric charge q and velocity v are kept constant and the distance r is variable. Since the rate of change of electric charge $\frac{dq}{dt} = 0$ and the acceleration $a = 0$, the above equation becomes:

$$\frac{1}{c}\frac{\partial\varphi_v}{\partial t} = -\frac{v_r}{c}\cdot\frac{1}{4\pi\varepsilon_0}\frac{qv}{cr^2}\sin\theta = -\frac{v_r}{c}E_v \qquad (23.4.11)$$

Since the velocity ratio $\frac{v_r}{c}$ has no physical unit, the ratio $\frac{1}{c}\frac{\partial\varphi_v}{\partial t}$ belongs to the Magnetic-electric field strength E_v

From this, it can be determined that the ratios $\frac{1}{c}\frac{\partial\varphi_v}{\partial t}$, $\frac{1}{c}\frac{\partial\varphi_v}{\partial q}\frac{dq}{dt}$, $\frac{1}{c}\frac{\partial\varphi_v}{\partial r}\frac{dr}{dt}$, and $\frac{1}{c}\frac{\partial\varphi_v}{\partial v}\frac{dv}{dt}$ in equation (23.4.10) all belong to time-varying Magnetic-electric fields. /dt are all time-varying Magnetic-electric fields.

Since the time-varying Magnetic-electric field $\boldsymbol{E}_{vt}$ generated by the motion electric charge $\boldsymbol{q}$ at time $\boldsymbol{t}$ possesses uniqueness, the time-varying Magnetic-electric field $\boldsymbol{E}_{vt}$ is equal to the ratio $\frac{1}{c}\frac{d\varphi_v}{dt}$, i.e:

$$\boldsymbol{E}_{vt}\left(\frac{dq}{dt},\frac{dr}{dt},\frac{dv}{dt}\right)=\frac{1}{c}\frac{\partial\varphi_v}{\partial t}=\frac{\mathrm{si}\,\theta}{4\pi\varepsilon_0c^2}\left(\frac{v}{r}\frac{dq}{dt}-\frac{qv}{r^2}v_r+\frac{q}{r}a\right) \tag{23.4.12}$$

Assume that the electric charge $\boldsymbol{q}$ is equidistant $\boldsymbol{r}$ to each point on the circular line $\boldsymbol{L_R}=\boldsymbol{2\pi R}$. At time $\boldsymbol{t}$, since the time-varying Magnetic-electric field $\boldsymbol{E}_{vt}\left(\frac{dq}{dt},\frac{dr}{dt},\frac{dv}{dt}\right)$ is constant on the circular line $\boldsymbol{L_R}$, the integral of the time-varying Magnetic-electric field $\boldsymbol{E}_{vt}\left(\frac{dq}{dt},\frac{dr}{dt},\frac{dv}{dt}\right)$ along the circular line $\boldsymbol{L_R}=\boldsymbol{2\pi R}$ is:

$$\oint \boldsymbol{E}_{vt}\left(\frac{dq}{dt},\frac{dr}{dt},\frac{dv}{dt}\right)dl=\boldsymbol{E}_{vt}L_R=\frac{R\sin\theta}{2\varepsilon_0c^2}\left(\frac{v}{r}\frac{dq}{dt}-\frac{qv}{r^2}v_r+\frac{q}{r}a\right)$$

When the electric charge $\boldsymbol{q}$ and velocity $\boldsymbol{v}$ are kept constant and the distance $\boldsymbol{r}$ is variable. Since the rate of change of electric charge $\frac{dq}{dt}=\boldsymbol{0}$ and the acceleration $\boldsymbol{a}=\boldsymbol{0}$, the above equation becomes:

$$\oint \boldsymbol{E}_{vt}\left(\frac{dq}{dt},\frac{dr}{dt},\frac{dv}{dt}\right)dl=-\frac{v_r}{c}\cdot\frac{R}{2\varepsilon_0}\cdot\frac{qv}{cr^2}\sin\theta \tag{23.4.13}$$

To summarize: when the electric charge $\boldsymbol{q}$, distance $\boldsymbol{r}$ and velocity $\boldsymbol{v}$ vary with time. The time-varying Magnetic-electric field $\boldsymbol{E}_{vt}$ has two equivalent expressions.

The first is the equation $\boldsymbol{E}_{vt}(\boldsymbol{q},\boldsymbol{r},\boldsymbol{v})$. Note that there are two loop integral formulas for the time-varying Magnetic-electric field $\boldsymbol{E}_{vt}(\boldsymbol{q},\boldsymbol{r},\boldsymbol{v})$, viz:

$$\oint \boldsymbol{E}_{vt}(q,r,v)dl=0 \tag{23.4.14}$$

The distance $\boldsymbol{r}$ in the above equation is variable on the closed curve $\boldsymbol{L}$.

$$\oint \boldsymbol{E}_{vt}(q,r,v)dl=\boldsymbol{E}_{vt}\cdot L_R=\frac{R}{2\varepsilon_0}\cdot\frac{qv}{cr^2}\sin\theta \tag{23.4.15}$$

The distance $\boldsymbol{r}$ in the above equation remains constant on the circular line $\boldsymbol{L_R}=\boldsymbol{2\pi R}$.

The second is equation $\boldsymbol{E}_{vt}\left(\frac{dq}{dt},\frac{dr}{dt},\frac{dv}{dt}\right)$. The loop integral of the time-varying Magnetic-electric field $\boldsymbol{E}_{vt}\left(\frac{dq}{dt},\frac{dr}{dt},\frac{dv}{dt}\right)$ is:

$$\oint \boldsymbol{E}_{vt}\left(\frac{dq}{dt},\frac{dr}{dt},\frac{dv}{dt}\right)dl=\frac{R\sin\theta}{2\varepsilon_0c^2}\left(\frac{v}{r}\frac{dq}{dt}-\frac{qv}{r^2}v_r+\frac{q}{r}a\right) \tag{23.4.16}$$

Since the time-varying Magnetic-electric field $\boldsymbol{E}_{vt}$ generated by the moving electric charge $\boldsymbol{q}$ at time $\boldsymbol{t}$ possesses uniqueness, the formula $\boldsymbol{E}_{vt}(\boldsymbol{q},\boldsymbol{r},\boldsymbol{v})$ and the formula $\boldsymbol{E}_{vt}\left(\frac{dq}{dt},\frac{dr}{dt},\frac{dv}{dt}\right)$ are both expressions for the time-varying Magnetic-electric field $\boldsymbol{E}_{vt}$.

23.4.3 Flux electromotive force of Magnetic-electric field ε_{SE_v}

Assume that the electric charge $\boldsymbol{q}$, velocity $\boldsymbol{v}$ and distance $\boldsymbol{r}$ vary with time. Assume that the distance $\boldsymbol{r}$ from the moving electric charge $\boldsymbol{q}$ to each point on the circular line $\boldsymbol{L_R}=\boldsymbol{2\pi R}$ is equal, and assume that the surface $\boldsymbol{S}$ is bounded by the circular line $\boldsymbol{L_R}=\boldsymbol{2\pi R}$.

According to equation (22.2.23), the Magnetic-electric field flux $\boldsymbol{\Phi}_{Evt}$ across the surface $\boldsymbol{S}$ at time $\boldsymbol{t}$ is:

$$\boldsymbol{\Phi}_{Evt}=\iint \boldsymbol{E}_{vt}(q,r,v)\,dS=-\frac{R}{2\varepsilon_0c}\frac{qv}{r}\sin\theta \tag{23.4.17}$$

Take the derivative of the above equation with respect to time $\boldsymbol{t}$ to get the relationship equation:

$$\begin{aligned}\frac{\partial\boldsymbol{\Phi}_{Evt}}{\partial t}&=\iint\frac{\partial \boldsymbol{E}_{vt}(q,r,v)}{\partial t}dS=\frac{\partial\boldsymbol{\Phi}_{Evt}}{\partial q}\frac{dq}{t}+\frac{\partial\boldsymbol{\Phi}_{Evt}}{\partial dr}\frac{dr}{dt}+\frac{\partial\boldsymbol{\Phi}_{Evt}}{\partial v}\frac{dv}{dt}\\&=-\frac{R\sin\theta}{2\varepsilon_0c}\left(\frac{v}{r}\frac{dq}{dt}-\frac{qv}{r^2}\frac{dr}{dt}+\frac{q}{r}\frac{dv}{dt}\right)\end{aligned} \tag{23.4.18}$$

The command velocity $\boldsymbol{v_r} = \frac{dr}{dt}$ and acceleration $\boldsymbol{a} = \frac{dv}{dt}$. Since the rate of change $\frac{\partial \Phi_{Evt}}{\partial t}$ divided by the velocity $\boldsymbol{v_r}$ would result in the ratio $\frac{1}{v_r}\frac{\partial \Phi_{Evt}}{\partial t}$ being equal to infinity, the rate of change $\frac{\partial \Phi_{Evt}}{\partial t}$ can not be divided by the velocity $\boldsymbol{v_r}$. Dividing the above equation by the speed of light $\boldsymbol{c}$ yields the relationship equation:

$$\frac{1}{c}\frac{\partial \Phi_{Evt}}{\partial t} = \frac{1}{c}\iint \frac{\partial E_{vt}(q,r,v)}{\partial t} dS = -\frac{R\sin\theta}{2\varepsilon_0 c^2}\left(\frac{v}{r}\frac{dq}{dt} - \frac{qv}{r^2}v_r + \frac{q}{r}a\right) \quad (23.4.19)$$

When the electric charge $\boldsymbol{q}$ and velocity $\boldsymbol{v}$ remain constant and the distance $\boldsymbol{r}$ changes. Since the rate of change of electric charge $\frac{dq}{dt} = \mathbf{0}$ and the acceleration $\boldsymbol{a} = \mathbf{0}$, the above equation becomes:

$$\frac{1}{c}\iint \frac{\partial E_{vt}(q,r,v)}{\partial t} dS = \frac{v_r}{c}\cdot\frac{R}{2\varepsilon_0}\cdot\frac{qv}{cr^2}\sin\theta \quad (23.4.20)$$

According to equation (23.1.5), the linear electromotive force $\boldsymbol{\varepsilon_{LE0}}$ (voltage U) generated by the static electric charge $\boldsymbol{q_0}$ is:

$$\varepsilon_{LE0} = \frac{q_0}{4\pi\varepsilon_0}\left(\frac{1}{r_b} - \frac{1}{r_a}\right) = U(\text{voltage})$$

Compare equation (23.4.20) with the above equation. Since the ratios $\frac{v_r}{c}$ and $\frac{v}{c}\sin\theta$ of velocity have no physical units, both $\frac{v_r}{c}\frac{q}{2\varepsilon_0}\frac{v}{c}\sin\theta$ and $\frac{q_0}{4\pi\varepsilon_0}$ have the same physical units. Since distance $\boldsymbol{r}$ and radius $\boldsymbol{R}$ are both distances, $\frac{R}{r^2}$ and $\left(\frac{1}{r_b} - \frac{1}{r_a}\right)$ have the same physical units. From this it can be determined that the ratio $\frac{1}{c}\iint \frac{\partial E_{vt}(q,r,v)}{\partial t} dS$ belongs to the electromotive force $\boldsymbol{\varepsilon}$ (voltage $\boldsymbol{U}$).

The symbol $\boldsymbol{\varepsilon_{SEv}}$ is used in this book to denote the flux electromotive force of the time-varying Magnetic-electric field $\boldsymbol{E_{vt}(q,r,v)}$, viz:

$$\varepsilon_{SEv} = \frac{1}{c}\iint \frac{\partial E_{vt}(q,r,v)}{\partial t} dS \quad (23.4.21)$$

The above equation is the flux electromotive force equation for a Magnetic-electric field.

23.4.4 Derivation of the Magnetic-electric field induction formula

Assume that the distance $\boldsymbol{r}$ from the electric charge $\boldsymbol{q}$ to each point on the circular line $\boldsymbol{L_R} = \boldsymbol{2\pi R}$ is equal. Since the Magnetic-electric field $\boldsymbol{E_{vt}(q,r,v)}$ is constant on the circular line $\boldsymbol{L_R}$, the integral of the Magnetic-electric field $\boldsymbol{E_{vt}(q,r,v)}$ along the circular line $\boldsymbol{L_R} = \boldsymbol{2\pi R}$ is:

$$\oint E_{vt}(q,r,v)dl = E_{vt}L_R = \frac{R}{2\varepsilon_0}\cdot\frac{qv}{cr^2}\sin\theta \quad (23.4.22)$$

Comparing the above equation with equation (23.4.20), it can be determined that the integral of the Magnetic-electric field strength $\boldsymbol{E_{vt}(q,r,v)}$ over the circular line $\boldsymbol{L_R} = \boldsymbol{2\pi R}$ is not equal to the flux electromotive force $\boldsymbol{\varepsilon_{SEv}}$, viz:

$$\oint E_{vt}(q,r,v)dl \neq \frac{1}{c}\iint \frac{\partial E_{vt}(q,r,v)}{\partial t} dS$$

Since the left side of the inequality sign does not contain the rate of change $\left(\frac{dq}{dt},\frac{dr}{dt},\frac{dv}{dt}\right)$ and the right side of the inequality sign contains the rate of change $\left(\frac{dq}{dt},\frac{dr}{dt},\frac{dv}{dt}\right)$. So the above equations are necessarily unequal.

Assume that the electric charge $\boldsymbol{q}$ is at an equal distance $\boldsymbol{r}$ to each point on the circular line $\boldsymbol{L_R} = \boldsymbol{2\pi R}$. According to equation (23.4.16), the integral of the Magnetic-electric field $\boldsymbol{E_{vt}}\left(\frac{dq}{dt},\frac{dr}{dt},\frac{dv}{dt}\right)$ along the circular line $\boldsymbol{L_R} = \boldsymbol{2\pi R}$ is:

$$\oint E_{vt}\left(\frac{dq}{dt},\frac{dr}{dt},\frac{dv}{dt}\right)dl = \frac{R\sin\theta}{2\varepsilon_0 c^2}\left(\frac{v}{r}\frac{dq}{dt} - \frac{qv}{r^2}v_r + \frac{q}{r}a\right)$$

According to equation (23.4.19), the ratio $\frac{1}{c}\iint \frac{\partial E_{vt}(q,r,v)}{\partial t} dS$ is:

$$\frac{1}{c}\iint \frac{\partial E_{vt}(q,r,v)}{\partial t} dS = -\frac{R\sin\theta}{2\varepsilon_0 c^2}\left(\frac{v}{r}\frac{dq}{dt} - \frac{qv}{r^2}v_r + \frac{q}{r}a\right)$$

The relationship equation can be obtained from the above two formulas:

$$\oint E_{vt}\left(\frac{dq}{dt},\frac{dr}{dt},\frac{dv}{dt}\right) dl = -\frac{1}{c}\iint \frac{\partial E_{vt}(q,r,v)}{\partial t} dS \tag{23.4.23}$$

We can also derive the above equation from the Magnetic-electric potential φ_v and Stokes' theorem. According to equation (23.4.8), the Magnetic-electric potential φ_v generated by the electric charge q is:

$$\varphi_v = \frac{1}{4\pi\varepsilon_0}\frac{qv}{cr}\sin\theta$$

Assume that the Magnetic-electric potential φ_v integrates along the closed curve L. Assume that the surface S is bordered by the closed loop L. According to Stokes' theorem, the relation equation can be obtained:

$$\oint \varphi_v dl = \iint \nabla\times\varphi_v\, dS \tag{23.4.24}$$

According to equation (19.3.7) , the Magnetic-electric field strength E_{vt} is equal to the negative Curl $\nabla\times\varphi_v$ of the Magnetic-electric potential φ_v, viz:

$$E_{vt}(q,r,v) = -\nabla\times\varphi_v \tag{23.4.25}$$

Substituting the above equation into equation (23.4.24) gives the relationship equation:

$$\Phi_{Evt} = \oint \varphi_v dl = -\iint E_{vt}(q,r,v)\, dS \tag{23.4.26}$$

Take the derivative of the above equation with respect to time t to obtain the relationship equation:

$$\frac{\partial\Phi_{Evt}}{\partial t} = \oint \frac{\partial\varphi_v}{\partial t} dl = -\iint \frac{\partial E_{vt}(q,r,v)}{\partial t} dS \tag{23.4.27}$$

Divide the above equation by the speed of light c, to get the relational equation:

$$\frac{1}{c}\frac{\partial\Phi_{Evt}}{\partial t} = \oint \frac{1}{c}\frac{\partial\varphi_v}{\partial t} dl = -\iint \frac{1}{c}\frac{\partial E_{vt}(q,r,v)}{\partial t} dS \tag{23.4.28}$$

According to equation (23.4.12), the time-varying Magnetic-electric field $E_{vt}\left(\frac{dq}{dt},\frac{dr}{dt},\frac{dv}{dt}\right)$ is:

$$E_{vt}\left(\frac{dq}{dt},\frac{dr}{dt},\frac{dv}{dt}\right) = \frac{1}{c}\frac{\partial\varphi_v}{\partial t} \tag{23.4.29}$$

Substituting the above equation into equation (23.4.28) gives the relationship equation:

$$\oint E_{vt}\left(\frac{dq}{dt},\frac{dr}{dt},\frac{dv}{dt}\right) dl = -\frac{1}{c}\iint \frac{\partial E_{vt}(q,r,v)}{\partial t} dS \tag{23.4.30}$$

The above formula is the Magnetic-electric field induction formula. The formula is similar to Faraday's formula for electromagnetic induction.

23.4.5 Effect of change in distance Z on magnetic flux electromotive force

According to Figure 22-2, the function $\sin\theta$ is:

$$\sin\theta = \frac{R}{r} \tag{23.4.31}$$

Substituting the above equation into equation (23.4.17) yields the relationship equation:

$$\Phi_{Evt} = \iint E_{vt}(q,r,v)\, dS = -\frac{R^2}{2\varepsilon_0 c}\frac{qv}{r^2} \tag{23.4.32}$$

When electric charge q, velocity v and distance r are variables. Taking the derivative of the above equation with respect to time t gives the relationship equation:

$$\begin{aligned}\frac{\partial\Phi_{Evt}}{\partial t} = \iint \frac{\partial E_{vt}}{\partial t} dS &= \frac{\partial\Phi_{Evt}}{\partial q}\frac{dq}{dt} + \frac{\partial\Phi_{Evt}}{\partial r}\frac{dr}{dt} + \frac{\partial\Phi_{Evt}}{\partial v}\frac{dv}{dt} \\ &= -\frac{R^2}{2\varepsilon_0 c}\left(\frac{v}{r^2}\frac{dq}{dt} - 2\frac{qv}{r^3}\cdot\frac{dr}{dt} + \frac{q}{r^2}\frac{dv}{dt}\right)\end{aligned} \tag{23.4.33}$$

Dividing the above equation by the speed of light c, yields the relational equation that:

$$\varepsilon_{SEv}=\frac{1}{c}\frac{\partial \Phi_{Evt}}{\partial t}=\frac{1}{c}\iint\frac{\partial E_{vt}}{\partial t}dS=-\frac{R^2}{2\varepsilon_0 c^2}\left(\frac{v}{r^2}\frac{dq}{dt}-2\frac{qv}{r^3}\cdot\frac{dr}{dt}+\frac{q}{r^2}\frac{dv}{dt}\right) \quad (23.4.34)$$

The above equation is the magnetic flux electromotive force equation.

By taking the derivative of equation (23.4.32) with respect to the distance r, we can obtain the relationship equation:

$$\frac{\partial \Phi_{Evt}}{\partial r}=\frac{R^2}{\varepsilon_0 c}\cdot\frac{qv}{r^3} \quad (23.4.35)$$

According to Figure 22-2, the distance r is:

$$r=\sqrt{R^2+Z^2} \quad (23.4.36)$$

Taking the derivative of the above equation with respect to the distance Z yields the relationship equation:

$$\frac{\partial r}{\partial Z}=\frac{Z}{\sqrt{R^2+Z^2}}=\frac{Z}{r} \quad (23.4.37)$$

Substituting equation (23.4.36) into equation (23.4.32) gives the relationship equation:

$$\Phi_{Evt}=\iint E_{vt}\,dS=-\frac{R^2}{2\varepsilon_0 c}\cdot\frac{qv}{R^2+Z^2} \quad (23.4.38)$$

When electric charge q velocity v and distance Z are variables. Taking the derivative of the above equation with respect to time t gives the relationship equation:

$$\begin{aligned}\frac{\partial \Phi_{Evt}}{\partial t}=\iint\frac{\partial E_{vt}}{\partial t}dS&=\frac{\partial \Phi_{Evt}}{\partial q}\frac{dq}{dt}+\frac{\partial \Phi_{Evt}}{\partial v}\frac{dv}{dt}+\frac{\partial \Phi_{Evt}}{\partial r}\frac{\partial r}{\partial Z}\frac{dZ}{dt}\\&=-\frac{R^2}{2\varepsilon_0 c}\left(\frac{v}{r^2}\frac{dq}{dt}-2\frac{qv}{r^3}\cdot\frac{Z}{r}\cdot\frac{dZ}{dt}+\frac{q}{r^2}\frac{dv}{dt}\right)\end{aligned} \quad (23.4.39)$$

Dividing the above equation by the speed of light c, gives the relationship equation:

$$\varepsilon_{SEv}=\frac{1}{c}\iint\frac{\partial E_{vt}}{\partial t}dS=-\frac{R^2}{2\varepsilon_0 c^2}\left(\frac{v}{r^2}\frac{dq}{dt}-2\frac{qvZ}{r^4}\frac{dZ}{dt}+\frac{q}{r^2}\frac{dv}{dt}\right) \quad (23.4.40)$$

The above equation is the magnetic flux electromotive force equation including the rate of change $\left(\frac{dq}{dt},\frac{dZ}{dt},\frac{dv}{dt}\right)$.

When the electric charge q and velocity v are held constant and the distance Z is varied, the command velocity $v_Z=\frac{dZ}{dt}$. Since the rate of change of electric charge $\frac{dq}{dt}=0$ and the acceleration $a=0$, equation (23.4.40) becomes:

$$\varepsilon_{SEv}=\frac{q}{\varepsilon_0}\cdot\frac{vv_Z}{c^2}\cdot\frac{R^2 Z}{r^4} \quad (23.4.41)$$

The ε_{SEv} in equation is the magnetic flux electromotive force , and the velocity $v_Z=\frac{dZ}{dt}$ is the rate of change of the distance Z .

According to equation (23.1.5), the linear electromotive force ε_{LE0} (voltage U) generated by the static electric charge q_0 is:

$$\varepsilon_{LE0}=\frac{q_0}{4\pi\varepsilon_0}\left(\frac{1}{r_b}-\frac{1}{r_a}\right)=U(\text{voltage})$$

Compare equation (23.4.41) with the above equation. Since the ratio of velocities $\frac{vv_Z}{c^2}$ has no physical units, $\frac{q}{\varepsilon_0}\cdot\frac{vv_Z}{c^2}$ has the same physical units as $\frac{q_0}{4\pi\varepsilon_0}$. Since distance r and radius R and distance Z are both distances, $\frac{R^2Z}{r^4}$ has the same physical units as $\left(\frac{1}{r_b}-\frac{1}{r_a}\right)$. From this it can be determined that the magnetic flux electromotive force ε_{SEv} belongs to the electromotive force ε (voltage U) .

Equation (23.4.41) shows that the greater the velocity v_Z of the circular wire $L_R=2\pi R$ approaching or moving away from the electric charge q, the greater the magnetic flux electromotive force ε_{SEv} across the circular wire $L_R=2\pi R$, and hence the greater the induced voltage U and the induced electric current I generated in the conductor circular wire $L_R=2\pi R$.

23.5 Derivation of Faraday's law of electromagnetic induction $\varepsilon_B = n\iint \partial B/\partial t\, dS$

23.5.1 Derivation of Faraday's law of electromagnetic induction $\oint E_{vt} \cdot dl = -\iint \partial B/\partial t\, dS$

Physicists believe that a time-varying magnetic field $\boldsymbol{B}$ produces a time-varying electromagnetic field $\boldsymbol{E_t}$. The equation between the time-varying magnetic field $\boldsymbol{B}$ and the time-varying electromagnetic field $\boldsymbol{E_t}$ in Maxwell's system of equations is:

$$\oint \boldsymbol{E_t dl} = -\iint \frac{\boldsymbol{\partial B}}{\boldsymbol{\partial t}} \boldsymbol{dS} \tag{23.5.1}$$

Since the time-varying electromagnetic field $\boldsymbol{E_t}$ contains the rotating electric field $\boldsymbol{E_{\omega t}}$ and the Magnetic-electric field $\boldsymbol{E_{vt}}$, while the time-varying magnetic field $\boldsymbol{B}$ contains only the Magnetic-electric field $\boldsymbol{E_{vt}}$, the above formula is wrong. The above formula should be changed to:

$$\oint \boldsymbol{E_{vt}} \left(\frac{\boldsymbol{dq}}{\boldsymbol{dt}}, \frac{\boldsymbol{dr}}{\boldsymbol{dt}}, \frac{\boldsymbol{dv}}{\boldsymbol{dt}}\right) \boldsymbol{dl} = -\iint \frac{\boldsymbol{\partial B(q,r,v)}}{\boldsymbol{\partial t}} \boldsymbol{dS} \tag{23.5.2}$$

The above formula is the magnetic induction formula. We can derive the above equation using the magnetic flux formula.

Assume that the surface $\boldsymbol{S}$ is bordered by the circular line $\boldsymbol{L_R = 2\pi R}$. Assume that the distance $\boldsymbol{r}$ from the electric charge $\boldsymbol{q}$ to each point on the circular line $\boldsymbol{L_R = 2\pi R}$ is equal. According to equation (22.3.19) , the magnetic flux $\boldsymbol{\Phi_B}$ through the surface $\boldsymbol{S}$ is:

$$\boldsymbol{\Phi_B} = \iint \boldsymbol{BdS} = -\frac{\boldsymbol{\mu_0 R}}{\boldsymbol{2}} \cdot \frac{\boldsymbol{qv}}{\boldsymbol{r}} \sin \boldsymbol{\theta} \tag{23.5.3}$$

Taking the derivative of the above equation with respect to time $\boldsymbol{t}$ gives the relationship equation:

$$\iint \frac{\boldsymbol{\partial B}}{\boldsymbol{\partial t}} \boldsymbol{dS} = \frac{\boldsymbol{\partial \Phi_B}}{\boldsymbol{\partial q}} \frac{\boldsymbol{dq}}{\boldsymbol{dt}} + \frac{\boldsymbol{\partial \Phi_B}}{\boldsymbol{\partial r}} \frac{\boldsymbol{dr}}{\boldsymbol{dt}} + \frac{\boldsymbol{\partial \Phi_B}}{\boldsymbol{\partial v}} \frac{\boldsymbol{dv}}{\boldsymbol{dt}} = -\frac{\boldsymbol{\mu_0 R} \sin \boldsymbol{\theta}}{\boldsymbol{2}} \left(\frac{\boldsymbol{v}}{\boldsymbol{r}} \frac{\boldsymbol{dq}}{\boldsymbol{dt}} - \frac{\boldsymbol{qv}}{\boldsymbol{r^2}} v_r + \frac{q}{r} a\right) \tag{23.5.4}$$

The above equation is the equation for the rate of change of magnetic flux. According to equation (19.3.17), the magnetic flux density $\boldsymbol{B}$ is:

$$\boldsymbol{B} = \frac{\boldsymbol{1}}{\boldsymbol{c}} \boldsymbol{E_{vt}}$$

Substituting the above equation into equation (23.5.4) gives the relationship equation:

$$\frac{\boldsymbol{1}}{\boldsymbol{c}} \iint \frac{\boldsymbol{\partial E_{vt}(q,r,v)}}{\boldsymbol{\partial t}} \boldsymbol{dS} = \iint \frac{\boldsymbol{\partial B}}{\boldsymbol{\partial t}} \boldsymbol{dS} = -\frac{\boldsymbol{\mu_0 R} \sin \boldsymbol{\theta}}{\boldsymbol{2}} \left(\frac{\boldsymbol{v}}{\boldsymbol{r}} \frac{\boldsymbol{dq}}{\boldsymbol{dt}} - \frac{\boldsymbol{qv}}{\boldsymbol{r^2}} v_r + \frac{q}{r} a\right) \tag{23.5.5}$$

Substituting equation (23.4.30) into the above equation gives the relationship equation:

$$\oint \boldsymbol{E_{vt}} \left(\frac{\boldsymbol{dq}}{\boldsymbol{dt}}, \frac{\boldsymbol{dr}}{\boldsymbol{dt}}, \frac{\boldsymbol{dv}}{\boldsymbol{dt}}\right) \boldsymbol{dl} = -\iint \frac{\boldsymbol{\partial B(q,r,v)}}{\boldsymbol{\partial t}} \boldsymbol{dS} \tag{23.5.6}$$

The above equation is the magnetic field induction equation. $\boldsymbol{E_{vt}}$ in the formula is the time-varying Magnetic-electric field and $\boldsymbol{B(q,r,v)}$ is the time-varying magnetic field.

23.5.2 Dynamic electromotive force and induced electromotive force

The British physicist Faraday determined through electromagnetic experiments that the induced electromotive force $\boldsymbol{\varepsilon_B}$ is proportional to the rate of change of magnetic flux $\frac{d\Phi_B}{dt}$ through a closed circuit.

The induced electromotive force $\boldsymbol{\varepsilon_B}$ is distinguished by physicists between the induced electromotive force and the dynamic electromotive force.

Dynamic electromotive force is the electromotive force $\boldsymbol{\varepsilon_B}$ generated by the relative motion of the constant magnetic field $\boldsymbol{B}$ and the conductor loop $\boldsymbol{L}$.

Alternatively, the dynamic electromotive force $\boldsymbol{\varepsilon_B}$ is the electromotive force for which the electric charge $\boldsymbol{q}$ and velocity $\boldsymbol{v}$ remain constant and the distance $\boldsymbol{r}$ varies .

According to Figure 22-2, the function $\sin \boldsymbol{\theta}$ is:

$$\sin\theta = \frac{R}{r}$$

Substituting the above equation into equation (23.5.3) yields the relationship equation:

$$\boldsymbol{\Phi_B} = \iint \boldsymbol{B dS} = -\frac{\mu_0 R^2}{2}\cdot\frac{qv}{r^2} \quad (23.5.7)$$

Taking the derivative of the above equation with respect to time t gives the relationship equation:

$$\varepsilon_B = \frac{\partial \Phi_B}{\partial t} = \frac{\partial \Phi_B}{\partial q}\frac{dq}{dt} + \frac{\partial \Phi_B}{\partial r}\frac{dr}{dt} + \frac{\partial \Phi_B}{\partial v}\frac{dv}{dt} = -\frac{\mu_0 R^2}{2}\left(\frac{v}{r^2}\frac{dq}{dt} - 2\frac{qv}{r^3}v_r + \frac{q}{r^2}a\right) \quad (23.5.8)$$

When the electric charge q and the velocity v remain constant and the distance r changes. Since the rate of change of charge $\frac{dq}{dt} = 0$ and the acceleration $a = 0$, the above equation becomes:

$$\varepsilon_B = \frac{\partial \Phi_B}{\partial t} = \mu_0 R^2 \cdot \frac{qv}{r^3}v_r \quad (23.5.9)$$

The above equation is the dynamic electromotive force equation.

When the N-pole of the magnet is moving away from the metal coil. Since the velocity $v_r > 0$, the dynamic electromotive force $\varepsilon_B > 0$. The induced electric current generated by the positive electromotive force $+\varepsilon_B$ prevents the magnet N-pole from moving away from the metal coil.

When the N-pole of the magnet is close to the metal coil. Since the velocity $v_r < 0$, the dynamic electromotive force $\varepsilon_B < 0$. The induced electric current generated by the negative electromotive force $-\varepsilon_B$ prevents the magnet N-pole from approaching the metal coil.

When electric charge q, velocity v and distance r remain constant and radius R changes. Taking the derivative of equation (23.5.7) with respect to time t gives the relation equation:

$$\varepsilon_B = \frac{\partial \Phi_B}{\partial t} = \frac{\partial \Phi_B}{\partial R}\frac{dR}{dt} = -\mu_0 R \cdot \frac{qv}{r^2}v_R \quad (23.5.10)$$

The ε_B in equation is the dynamic electromotive force resulting from the variation of the circular line $L_R = 2\pi R$ in a constant magnetic field B .

When the radius R becomes small, the dynamic electromotive force $\varepsilon_B > 0$. Since the velocity $v_R < 0$, the dynamic electromotive force $\varepsilon_B > 0$. The induced electric current generated by the positive electromotive force $+\varepsilon_B$ prevents the metal circle line $L_R = 2\pi R$ from becoming smaller.

When the radius R becomes large. Since the velocity $v_R > 0$, the dynamic electromotive force $\varepsilon_B < F_v < 0$. The induced electric current generated by the negative electromotive force $-\varepsilon_B$ prevents the metal circle line $L_R = 2\pi R$ from becoming larger.

If the conductor loop L is stationary with respect to the magnetic field B, then the electromotive force ε_B generated by the time-varying magnetic field B is an induced electromotive force. or, in other words, the induced electromotive force ε_B is the electromotive force generated by the change of electric charge q and velocity v.

When the electric charge q and velocity v change and the distance r remains constant. Since velocity $v_r = 0$, equation (23.5.8) becomes:

$$\varepsilon_B = \frac{\partial \Phi_B}{\partial t} = -\frac{\mu_0 R^2}{2}\left(\frac{v}{r^2}\frac{dq}{dt} + \frac{q}{r^2}a\right) \quad (23.5.11)$$

The above equation is the induced electromotive force equation.

23.5.3 Derivation of Faraday's law of electromagnetic induction $\varepsilon_B = -n\iint \partial B/\partial t\, dS$

According to equation (23.5.7), the magnetic flux Φ_B is:

$$\Phi_B = \iint B\, dS = -\frac{\mu_0 R^2}{2}\cdot\frac{qv}{r^2} \quad (23.5.12)$$

Taking the derivative of the above equation with respect to the distance r gives the relationship equation:

$$\frac{\partial \Phi_B}{\partial r} = \mu_0 R^2 \cdot \frac{qv}{r^3} \quad (23.5.13)$$

Substituting equation (23.4.36), i.e., $\boldsymbol{r} = \sqrt{\boldsymbol{R}^2 + \boldsymbol{Z}^2}$, into equation (23.5.12) yields the relationship equation:

$$\boldsymbol{\Phi_B} = \iint \boldsymbol{B\,dS} = -\frac{\mu_0 R^2}{2} \cdot \frac{qv}{R^2 + Z^2} \tag{23.5.14}$$

Taking the derivative of the above equation with respect to time $\boldsymbol{t}$ gives the relationship equation:

$$\varepsilon_B = \frac{\partial \Phi_B}{\partial t} = \iint \frac{\partial B}{\partial t} dS = \frac{\partial \Phi_B}{\partial q}\frac{dq}{dt} + \frac{\partial \Phi_B}{\partial v}\frac{dv}{dt} + \frac{\partial \Phi_B}{\partial r}\frac{\partial r}{\partial Z}\frac{dZ}{dt}$$

$$= -\frac{\mu_0 R^2}{2}\left(\frac{v}{r^2}\frac{dq}{dt} - 2\frac{qvZ}{r^4}\frac{dZ}{dt} + \frac{q}{r^2}\frac{dv}{dt}\right) \tag{23.5.15}$$

The surface $\boldsymbol{S}$ in equation is bordered by the circle line $\boldsymbol{L_R} = \boldsymbol{2\pi R}$. Command acceleration $\boldsymbol{a} = \frac{dv}{dt}$, velocity $\boldsymbol{v_Z} = \frac{dZ}{dt}$. The above equation becomes:

$$\varepsilon_B = \iint \frac{\partial B}{\partial t} dS = -\frac{\mu_0 R^2}{2}\left(\frac{v}{r^2}\frac{dq}{dt} - 2\frac{qvZ}{r^4}v_Z + \frac{q}{r^2}a\right) \tag{23.5.16}$$

The above formula is the rate of change of magnetic flux formula. Note that $\boldsymbol{\varepsilon_B} = \iint \frac{\partial B}{\partial t} \boldsymbol{dS}$ in the formula is the rate of change of magnetic flux through a closed circular line $\boldsymbol{L_R} = \boldsymbol{2\pi R}$.

If the time-varying magnetic field $\boldsymbol{B}$ passes through $\boldsymbol{n}$ circular lines $\boldsymbol{L_R} = \boldsymbol{2\pi R}$, (e.g., a magnet passes through an $\boldsymbol{n}$-turn coil), then the induced electromotive force $\boldsymbol{\varepsilon_B}$ produced by the time-varying magnetic field in $\boldsymbol{n}$ closed circular lines $\boldsymbol{L_R}$ is:

$$\varepsilon_B = n\iint \frac{\partial B}{\partial t} dS = -n\frac{\mu_0 R^2}{2}\left(\frac{v}{r^2}\frac{dq}{dt} - 2\frac{qvZ}{r^4}v_Z + \frac{q}{r^2}a\right) \tag{23.5.17}$$

The above equation is Faraday's law of electromagnetic induction. The $\boldsymbol{\varepsilon_B}$ in the formula is the rate of change of magnetic flux of the time-varying magnetic field $\boldsymbol{B}$ across $\boldsymbol{n}$ closed circular lines $\boldsymbol{L_R} = \boldsymbol{2\pi R}$.

When the electric charge $\boldsymbol{q}$ and velocity $\boldsymbol{v}$ remain constant and the distance $\boldsymbol{Z}$ changes. Since the rate of change of electric charge $\frac{dq}{dt} = \boldsymbol{0}$ and the acceleration $\boldsymbol{a} = \boldsymbol{0}$, equation (23.5.17) becomes:

$$\varepsilon_B = n \cdot \mu_0 R^2 \cdot \frac{qvv_Z \cdot Z}{r^4} \tag{23.5.18}$$

When the distance Z becomes small. Since the approach velocity $\boldsymbol{v_Z} < \boldsymbol{0}$, the electromotive force $\boldsymbol{\varepsilon_B} < \boldsymbol{0}$. The induced electric current generated by the negative electromotive force $-\boldsymbol{\varepsilon_B}$ prevents the magnet's N-pole from approaching the metal coil closer.

When the distance Z becomes large. Since the away velocity $\boldsymbol{v_Z} > 0$, the electromotive force $\boldsymbol{\varepsilon_B} > \boldsymbol{0}$. The induced electric current generated by the positive electromotive force $+\boldsymbol{\varepsilon_B}$ prevents the magnet N-pole from moving away from the metal coil.

When the electric charge $\boldsymbol{q}$, velocity $\boldsymbol{v}$ and distance Z are kept constant, since the rate of change of electric charge $\frac{dq}{dt} = \boldsymbol{0}$, the acceleration $\boldsymbol{a} = \boldsymbol{0}$ and the velocity $\boldsymbol{v_Z} = \boldsymbol{0}$, equation (23.5.17) becomes:

$$\varepsilon_B = n\iint \frac{\partial B}{\partial t} dS = 0 \tag{23.5.19}$$

Chapter 24 Derivation of the electromagnetic wave equation

Introduction: This chapter derives the electromagnetic wave equations based on the loop integral $\oint \boldsymbol{E_t dl}$ of the time-varying electromagnetic field $\boldsymbol{E_t}$ and Stokes' theorem.

24.1 Time-varying electromagnetic field E_t described by vector potential A and scalar potential φ

24.1.1 Physicist-defined equations for time-varying electromagnetic fields

In early electromagnetic theory, the electric field $\boldsymbol{E}$ and the magnetic field $\boldsymbol{B}$ were independent of each other. Since a moving electric charge $\boldsymbol{q}$ excites both the electric field $\boldsymbol{E}$ and the magnetic field $\boldsymbol{B}$, physicists used the time-varying electromagnetic field $\boldsymbol{E_t}$ to describe the electromagnetic field excited by a moving electric charge $\boldsymbol{q}$.

To derive the electromagnetic wave equation, physicists define a vector potential $\boldsymbol{A}$ and a scalar potential $\boldsymbol{\varphi}$ to describe the time-varying electromagnetic field $\boldsymbol{E_t}$.

According to Maxwell's system of equations, the Faraday's law of electromagnetic induction in integral form is:

$$\oint \boldsymbol{E_t dl} = -\iint \frac{\partial \boldsymbol{B}}{\partial t} \boldsymbol{dS} \qquad (24.1.1)$$

The $\boldsymbol{E_t}$ in the formula is the time-varying electromagnetic field, $\boldsymbol{B}$ is the time-varying magnetic field, and the surface $\boldsymbol{S}$ is bordered by the closed curve $\boldsymbol{L}$. According to Stokes' theorem, the above equation changes to:

$$\oint \boldsymbol{E_t dl} = \iint \nabla \times \boldsymbol{E_t dS} = -\iint \frac{\partial \boldsymbol{B}}{\partial t} \boldsymbol{dS} \qquad (24.1.2)$$

To derive the electromagnetic wave formula, physicists use mathematical tricks to define the time-varying electromagnetic field $\boldsymbol{E_t}$ as:

$$\boldsymbol{E_t} = -\nabla\varphi - \frac{\partial \boldsymbol{A}}{\partial t} \qquad (24.1.3)$$

The $\boldsymbol{\varphi}$ in equation is the scalar potential, and $\nabla\boldsymbol{\varphi}$ is the gradient of the scalar potential $\boldsymbol{\varphi}$. The $\boldsymbol{A}$ in the formula is the vector potential, $\frac{\partial \boldsymbol{A}}{\partial t}$ is the rate of change of the vector potential. The process by which physicists derive the above equation is as follows.

According to the divergence $\nabla \cdot \boldsymbol{B} = \boldsymbol{0}$ of the magnetic field $\boldsymbol{B}$, a vector potential $\boldsymbol{A}$ can be introduced such that the magnetic induction $\boldsymbol{B}$ is equal to the Curl $\nabla \times \boldsymbol{A}$ of the vector potential $\boldsymbol{A}$, i.e:

$$\boldsymbol{B} = \nabla \times \boldsymbol{A} \qquad (24.1.4)$$

Integrating the vector potential $\boldsymbol{A}$ along the closed curve $\boldsymbol{L}$ gives the relational equation according to Stokes' theorem and the above equation:

$$\oint \boldsymbol{Adl} = \iint \nabla \times \boldsymbol{AdS} = \iint \boldsymbol{BdS} \qquad (24.1.5)$$

The surface $\boldsymbol{S}$ in the formula has the closed curve $\boldsymbol{L}$ as its boundary. The physical meaning of the above equation is that at any time $\boldsymbol{t}$, the integral of the vector potential $\boldsymbol{A}$ along the closed curve $\boldsymbol{L}$ is equal to the magnetic flux through the surface $\boldsymbol{S}$.

Taking the derivative of equation (24.1.4) with respect to time $\boldsymbol{t}$ yields the relation equation:

$$\frac{\partial \boldsymbol{B}}{\partial t} = \nabla \times \frac{\partial \boldsymbol{A}}{\partial t} \qquad (24.1.6)$$

Substituting the above equation into equation (24.1.2) gives the relationship equation:

$$\nabla \times \boldsymbol{E}_t = -\nabla \times \frac{\partial \boldsymbol{A}}{\partial t} \qquad (24.1.7)$$

According to the rules of vector arithmetic, the above equation can be changed to:

$$\nabla \times \boldsymbol{E}_t + \nabla \times \frac{\partial \boldsymbol{A}}{\partial t} = \nabla \times \left(\boldsymbol{E}_t + \frac{\partial \boldsymbol{A}}{\partial t}\right) = \boldsymbol{0} \qquad (24.1.8)$$

The above equation shows that $\left(\boldsymbol{E}_t + \frac{\partial \boldsymbol{A}}{\partial t}\right)$ is a field without Curl.

24.1.2 Physicist-defined scalar potential φ

Physicists believe that the time-varying electromagnetic field $\boldsymbol{E}_t$ is different from the electrostatic field $\boldsymbol{E}_0$. The time-varying electromagnetic field $\boldsymbol{E}_t$ is excited by the motionelectric charge $\boldsymbol{q}$ on the one hand and by the time-varying magnetic field $\boldsymbol{B}$ on the other. And the electric field excited by the time-varying magnetic field $\boldsymbol{B}$ is the electric field with Curl.

Since Curl $\nabla \times \boldsymbol{E}_t \neq \boldsymbol{0}$, the time-varying electromagnetic field $\boldsymbol{E}_t$ is an electric field with Curl. Physicists believe that the time-varying electromagnetic field $\boldsymbol{E}_t$ cannot be described by a single scalar potential $\boldsymbol{\varphi}$. The time-varying electromagnetic field $\boldsymbol{E}_t$ necessarily contains a vector potential $\boldsymbol{A}$ in the case of a change in the electric field $\boldsymbol{E}$ and magnetic field $\boldsymbol{B}$.

Since Curl $\nabla \times \left(\boldsymbol{E}_t + \frac{\partial \boldsymbol{A}}{\partial t}\right) = \boldsymbol{0}$, a scalar potential $\boldsymbol{\varphi}$ can be introduced to describe the time-varying electromagnetic field $\boldsymbol{E}_t$. Since Curl $\nabla \times (\nabla \boldsymbol{\varphi}) = \boldsymbol{0}$ for the gradient $\nabla \boldsymbol{\varphi}$, the relational equation can be obtained according to equation (24.1.8):

$$\nabla \times \left(\boldsymbol{E}_t + \frac{\partial \boldsymbol{A}}{\partial t}\right) = -\nabla \times (\nabla \boldsymbol{\varphi}) = \boldsymbol{0} \qquad (24.1.9)$$

Based on the above formula, the relationship formula can be obtained:

$$\boldsymbol{E}_t = -\nabla \boldsymbol{\varphi} - \frac{\partial \boldsymbol{A}}{\partial t}$$

The above equation is equation (24.1.3). It is important to note that the above equation was specified by physicists using the technique of mathematical transformations and not derived from the time-varying electromagnetic field equation $\boldsymbol{E}_t(\boldsymbol{q}, \boldsymbol{r}, \boldsymbol{v})$.

Physicists consider the scalar potential $\boldsymbol{\varphi}$ to be different from the static electric potential $\boldsymbol{\varphi}_0$ in the following two ways.

Difference 1, the scalar potential $\boldsymbol{\varphi}$ does not have an electric potential energy implication, while the static electric potential $\boldsymbol{\varphi}_0$ has an electric potential energy implication.

Further, when the electric charge $\boldsymbol{q}$ and the velocity $\boldsymbol{v}$ change, the work $\boldsymbol{W}$ done by the electric field force $\boldsymbol{F}$ is related to the route $\boldsymbol{L}$. At this point the scalar potential $\boldsymbol{\varphi}$ loses the meaning of electric potential energy, i.e., the scalar potential $\boldsymbol{\varphi}$ is not equal to the work $\boldsymbol{W}$ done by the electric field force $\boldsymbol{F}$ in moving the electric charge $\boldsymbol{q}_1$ from a point in space to infinity.

Distinction 2, when the vector potential $\boldsymbol{A}$ is independent of time, i.e., $\frac{\partial \boldsymbol{A}}{\partial t} = \boldsymbol{0}$. Since the time-varying electromagnetic field $\boldsymbol{E}_t = -\nabla \boldsymbol{\varphi}$ is the electrostatic field $\boldsymbol{E}_0$, the scalar potential $\boldsymbol{\varphi}$ is the electrostatic potential $\boldsymbol{\varphi}_0$.

Theoretically, since the time-varying electromagnetic field $\boldsymbol{E}_t$ is generated by the moving electric charge $\boldsymbol{q}$, and the electrostatic field $\boldsymbol{E}_0$ is generated by the static electric charge $\boldsymbol{q}_0$, the electrostatic field $\boldsymbol{E}_0$ is the time-varying electromagnetic field $\boldsymbol{E}_t$ with the electric charge velocity $\boldsymbol{v} = \boldsymbol{0}$. However, physicists artificially make the following distinction between the time-varying electromagnetic field $\boldsymbol{E}_t$ and the electrostatic field $\boldsymbol{E}_0$ for the purpose of deducing the electromagnetic wave equation.

Difference 1: The time-varying electromagnetic field $\boldsymbol{E}_t$ is a non-conservative field, while the electrostatic field $\boldsymbol{E}_0$ is a conservative field.

Difference 2, the work done $\boldsymbol{W}$ by the time-varying electromagnetic field force $\boldsymbol{F}$ is related to the path $\boldsymbol{L}$ over which the work is done, while the work done $\boldsymbol{W_0}$ by the electrostatic field force $\boldsymbol{F_0}$ is independent of the path $\boldsymbol{L}$ over which the work is done.

Difference 3, the time-varying electromagnetic field $\boldsymbol{E_t}$ cannot use electric potential $\boldsymbol{\varphi}$ and electric potential energy $\boldsymbol{W}$ to describe the electromagnetic field, while the electrostatic field $\boldsymbol{E_0}$ can use electric potential $\boldsymbol{\varphi}$ and electric potential energy $\boldsymbol{W}$ to describe the electric field.

Difference 4, a time-varying electromagnetic field line is a closed curve with no beginning and no end, while an electrostatic field line is a straight line that starts at a positive electric charge and ends at a negative electric charge.

Classical electrodynamics makes the four strange distinctions above in order to theoretically derive the electromagnetic wave equation.

24.1.3 Gauge transformation theory for vector potentials A and scalar potentials φ

Although the magnetic field $\boldsymbol{B}$ and the time-varying electromagnetic field $\boldsymbol{E_t}$, as well as the vector potential $\boldsymbol{A}$ and the scalar potential $\boldsymbol{\varphi}$ are two equivalent ways of describing the electromagnetic field, the time-varying electromagnetic field $\boldsymbol{E_t}$ described by the vector potential $\boldsymbol{A}$ and the scalar potential $\boldsymbol{\varphi}$ is not unique, i.e., a given $\boldsymbol{E_t}$ and $\boldsymbol{B}$ do not correspond to a unique $\boldsymbol{A}$ and $\boldsymbol{\varphi}$.

Assume that $\boldsymbol{\Psi}$ is an arbitrary scalar function, i.e., $\boldsymbol{\Psi} = (\boldsymbol{x}, \boldsymbol{t})$. Although the vector potential $\boldsymbol{A}$ plus the gradient $\nabla\boldsymbol{\Psi}$ does not affect the magnetic field $\boldsymbol{B}$, the gradient $\nabla\boldsymbol{\Psi}$ affects the time-varying electromagnetic field $\boldsymbol{E_t}$. By transforming the scalar potential $\boldsymbol{\varphi}$ accordingly, the time-varying electromagnetic field $\boldsymbol{E_t}$ can still be kept constant. The vector potential $\boldsymbol{A}$ and the scalar potential $\boldsymbol{\varphi}$ can be transformed as follows:

$$\begin{cases} \boldsymbol{A} \to \boldsymbol{A}' = \boldsymbol{A} + \nabla\boldsymbol{\Psi} \\ \boldsymbol{\varphi} \to \boldsymbol{\varphi}' = \boldsymbol{\varphi} - \dfrac{\partial \boldsymbol{\Psi}}{\partial t} \end{cases} \tag{24.1.10}$$

It is easy to show that the function $(\boldsymbol{B}, \boldsymbol{E_t})$ can be described by the function $(\boldsymbol{A}', \boldsymbol{\varphi}')$, viz:

$$\begin{cases} \nabla \times \boldsymbol{A}' = \nabla \times (\boldsymbol{A} + \nabla\boldsymbol{\Psi}) = \nabla \times \boldsymbol{A} + \nabla \times (\nabla\boldsymbol{\Psi}) = \nabla \times \boldsymbol{A} = \boldsymbol{B} \\ -\nabla\boldsymbol{\varphi}' - \dfrac{\partial \boldsymbol{A}'}{\partial t} = -\nabla\left(\boldsymbol{\varphi} - \dfrac{\partial \boldsymbol{\Psi}}{\partial t}\right) - \dfrac{\partial}{\partial t}(\boldsymbol{A} + \nabla\boldsymbol{\Psi}) \\ \qquad = -\nabla\boldsymbol{\varphi} + \dfrac{\partial}{\partial t}(\nabla\boldsymbol{\Psi}) - \dfrac{\partial \boldsymbol{A}}{\partial t} - \dfrac{\partial}{\partial t}(\nabla\boldsymbol{\Psi}) \\ \qquad = -\nabla\boldsymbol{\varphi} - \dfrac{\partial \boldsymbol{A}}{\partial t} = \boldsymbol{E_t} \end{cases} \tag{24.1.11}$$

Since both the new electromagnetic potential function $(\boldsymbol{A}', \boldsymbol{\varphi}')$ and the electromagnetic potential function $(\boldsymbol{A}, \boldsymbol{\varphi})$ can equivalently describe the electromagnetic field $(\boldsymbol{E_t}, \boldsymbol{B})$, physicists refer to equation (24.1.10) as the gauge transformation of the electromagnetic potential.

In classical theory, since the measurable physical quantity of the electromagnetic field is $(\boldsymbol{E_t}, \boldsymbol{B})$), and different electromagnetic potential functions $(\boldsymbol{A}', \boldsymbol{\varphi}')$ correspond to the same $(\boldsymbol{E_t}, \boldsymbol{B})$, if the electromagnetic field $(\boldsymbol{E_t}, \boldsymbol{B})$ is described by the electromagnetic potential, the pattern of change of the electromagnetic field $(\boldsymbol{E_t}, \boldsymbol{B})$ should be independent of the choice of the electromagnetic potential $(\boldsymbol{A}', \boldsymbol{\varphi}')$. This conclusion is called gauge invariance by physicists.

In other words, gauge invariance means that all physical quantities and physical laws should remain unchanged when the vector potential $\boldsymbol{A}$ and the scalar potential $\boldsymbol{\varphi}$ do gauge transformation.

In classical electrodynamics, the introduction of a vector potential $\boldsymbol{A}$ and a scalar potential $\boldsymbol{\varphi}$ is one way to describe the electromagnetic field $(\boldsymbol{E_t}, \boldsymbol{B})$, and the gauge invariance of the electromagnetic potential $(\boldsymbol{A}, \boldsymbol{\varphi})$ is a constraint imposed on this approach.

In modern physics, electromagnetic potential gauge transformation is introduced by the basic principles of quantum mechanics, electromagnetic potential gauge invariance is an important physical principle.

Mathematically, according to Helmholtz's theorem, a vector potential $\boldsymbol{A}$ is uniquely determined only if its divergence $\nabla \cdot \boldsymbol{A}$ and Curl $\nabla \times \boldsymbol{A}$ are jointly given.

Since the known $\boldsymbol{B} = \nabla \times \boldsymbol{A}$ specifies the Curl of the vector potential $\boldsymbol{A}$, the divergence $\nabla \cdot \boldsymbol{A}$ of the vector potential $\boldsymbol{A}$ must be specified again. This specification does not change $\boldsymbol{E_t}$ and $\boldsymbol{B}$.

In principle the value of divergence $\nabla \cdot \boldsymbol{A}$ can be given arbitrarily, but for simplicity of computation it is common to specify the Coulomb and Lorentz norms.

1 Coulomb specification for:

$$\nabla \cdot \boldsymbol{A} = 0 \tag{24.1.12}$$

The Coulomb specification states that the vector potential $\boldsymbol{A}$ is a spinning passive field (transverse field). The Coulomb specification is characterized as follows:

(1) the longitudinal field of the time-varying electromagnetic field $\boldsymbol{E_t}$ is completely described by the scalar potential $\boldsymbol{\varphi}$ (i.e., the gradient $-\nabla\boldsymbol{\varphi}$ possesses spinlessness)

(2) The transverse field of the time-varying electromagnetic field $\boldsymbol{E_t}$ is described by the vector potential $\boldsymbol{A}$ (i.e., the rate of change $\frac{\partial A}{\partial t}$ has no field origin).

(3) The gradient $-\nabla\boldsymbol{\varphi}$ in the time-varying electric field equation $\boldsymbol{E_t} = -\nabla\boldsymbol{\varphi} - \frac{\partial A}{\partial t}$ corresponds to the Coulomb field $\boldsymbol{E}$, and the rate of change $-\frac{\partial A}{\partial t}$ corresponds to the magnetic field $\boldsymbol{B}$.

2 The Lorentz specification is:

$$\nabla \cdot \boldsymbol{A} + \frac{1}{c^2}\frac{\partial \varphi}{\partial t} = 0 \tag{24.1.13}$$

The vector potential $\boldsymbol{A}$ in the Lorentz specification is a spinning active field (i.e., the vector potential $\boldsymbol{A}$ contains both transverse and longitudinal fields). The Lorentzian specification is characterized by the fact that the fundamental equations for the vector potential $\boldsymbol{A}$ and the scalar potential $\boldsymbol{\varphi}$ are reduced to particularly simple symmetric forms.

Since the Maxwell system of equations lacks the divergence $\nabla \cdot \boldsymbol{A}$ with respect to the vector potential $\boldsymbol{A}$, it is not possible to solve for the vector potential $\boldsymbol{A}$ and the scalar potential $\boldsymbol{\varphi}$ through the Maxwell system of equations, and both Lorentzspecification and Coulombspecification essentially give the divergence $\nabla \cdot \boldsymbol{A}$ with respect to the vector potential $\boldsymbol{A}$.

24.2 Faraday's law of electromagnetic induction *∮ E_t·dl=-∬ ∂B/∂t dS* is false

24.2.1 The time-varying electromagnetic field E_t contains the electric field *E* and the magnetic field *B*

According to equation (23.3.1), the rotating electric field strength $\boldsymbol{E_\omega}$ generated by the moving electric charge $\boldsymbol{q}$ is:

$$\boldsymbol{E_\omega} = -\frac{1}{4\pi\varepsilon_0}\frac{q}{r^2}\sqrt{1-\frac{v^2}{c^2}\sin\theta} \tag{24.2.1}$$

When the electric charge velocity $\boldsymbol{v} = \boldsymbol{0}$, the above equation becomes the electrostatic field strength equation, i.e:

$$\boldsymbol{E_\omega} = \boldsymbol{E_0} = -\frac{1}{4\pi\varepsilon_0}\frac{q}{r^2} \tag{24.2.2}$$

From this it can be determined that the rotating electric field $\boldsymbol{E_\omega}$ possesses the electric field $\boldsymbol{E}$ property.

According to equation (23.4.1), the Magnetic-electric field strength $\boldsymbol{E_v}$ produced by the moving electric charge $\boldsymbol{q}$ is:

$$\boldsymbol{E_v} = \frac{1}{4\pi\varepsilon_0}\frac{qv}{cr^2}\sin\theta \tag{24.2.3}$$

Dividing the formula by the speed of light $\boldsymbol{c}$, gives the relational formula:

$$\frac{1}{c}\boldsymbol{E_v} = \frac{1}{4\pi\varepsilon_0 c^2}\frac{qv}{r^2}\sin\theta \tag{24.2.4}$$

According to equation (19.3.16), the vacuum permeability μ_0 is:

$$\mu_0 = \frac{1}{\varepsilon_0 c^2} \tag{24.2.5}$$

Substituting the above equation into equation (24.2.4) gives the relationship equation:

$$B = \frac{1}{c}E_v = \frac{1}{4\pi\varepsilon_0 c^2}\frac{qv}{r^2}\sin\theta = \frac{\mu_0}{4\pi}\frac{qv}{r^2}\sin\theta \tag{24.2.6}$$

The above equation is the magnetic flux density equation. From this it can be determined that the Magnetic-electric field E_v divided by the speed of light c equals the magnetic flux density B.

According to equation (19.2.4), the electromagnetic field strength E_t generated by the moving electric charge q can be decomposed into the rotational electric field strength E_ω and the Magnetic-electric field strength E_v, i.e:

$$E_t = -\sqrt{(E_\omega)^2 + (E_v)^2} \tag{24.2.7}$$

According to equation (24.1.3), the time-varying electromagnetic field E_t generated by the moving electric charge q is:

$$E_t = -\nabla\varphi - \frac{\partial A}{\partial t} \tag{24.2.8}$$

Comparing the above equation with equation (24.2.7), it can be determined that both the time-varying electromagnetic field $E_t = -\nabla\varphi - \frac{\partial A}{\partial t}$ and the electromagnetic field $E_t = -\sqrt{(E_\omega)^2 + (E_v)^2}$ contain the electric field E and the magnetic field B.

24.2.2 Derivation of the vector potential $A = -\varphi_v / c$ equation

When electric charge q is moving with velocity v in the cosmic vacuum frame of reference. According to equation (18.4.3), the Magnetic-electric potential φ_v generated by electric charge q is:

$$\varphi_v = \frac{1}{4\pi\varepsilon_0}\frac{qv\sin\theta}{cr} \tag{24.2.9}$$

Assume that the distance r from the electric charge q to each point of the circular line $L_R = 2\pi R$ is equal. At time t, since the Magnetic-electric potential φ_v is constant on the circular line L_R, the integral of the Magnetic-electric potential φ_v along the circular line $L_R = 2\pi R$ is:

$$\oint \varphi_v dl = \varphi_v L_R = \frac{1}{4\pi\varepsilon_0}\frac{qv\sin\theta}{cr}\cdot 2\pi R \tag{24.2.10}$$

According to Stokes' theorem, the relation equation can be obtained:

$$\oint \varphi_v dl = \iint \nabla\times\varphi_v\, dS = \frac{1}{4\pi\varepsilon_0}\frac{qv\sin\theta}{cr}\cdot 2\pi R \tag{24.2.11}$$

The surface S in equation is bordered by the circular line $L_R = 2\pi R$. According to equation (19.3.7), the Magnetic-electric field strength E_v is equal to the Curl $\nabla\times\varphi_v$ of the Magnetic-electric potential φ_v, i.e:

$$E_v = -\nabla\times\varphi_v \tag{24.2.12}$$

Substituting the above equation into equation (24.2.11) gives the relationship equation:

$$\oint \varphi_v dl = -\iint E_v\, dS = \frac{1}{4\pi\varepsilon_0}\frac{qv\sin\theta}{cr}\cdot 2\pi R \tag{24.2.13}$$

If the surface S in equation uses Wang's circular crown $S_W = 2\pi Rr$, then Wang's circular crown S_W is just a function of the distance r. Differentiating Wang's circular crown S_W with respect to the distance r gives the relation equation:

$$dS_W = 2\pi R dr$$

Substituting the above equation into equation (24.2.13) gives the relationship equation:

$$\iint E_v\cdot 2\pi R dr = -\frac{1}{4\pi\varepsilon_0}\frac{qv\sin\theta}{cr}\cdot 2\pi R \tag{24.2.14}$$

Simplifying the above equation and eliminating the constant $2\pi R$ yields the relational equation:

$$\int E_v dr = -\frac{1}{4\pi\varepsilon_0}\frac{qv\sin\theta}{cr} \tag{24.2.15}$$

Take the derivative of the above equation with respect to the distance r to obtain the relationship equation:

$$E_v = -\nabla \times \varphi_v = \frac{1}{4\pi\varepsilon_0}\frac{qv}{cr^2}\sin\theta \tag{24.2.16}$$

The above formula is the Magnetic-electric field strength formula. Divide the above formula by the speed of light c to get the relational formula:

$$\frac{1}{c}E_v = -\nabla \times \frac{1}{c}\varphi_v = \frac{1}{4\pi\varepsilon_0 c^2}\frac{qv}{r^2}\sin\theta \tag{24.2.17}$$

Substituting equation (24.2.6) into the above equation gives the relationship equation:

$$B = -\nabla \times \frac{1}{c}\varphi_v = \frac{\mu_0}{4\pi}\frac{qv}{r^2}\sin\theta \tag{24.2.18}$$

The above equation shows that the magnetic flux density B is equal to the ratio of the negative Curl $\nabla \times \varphi_v$ of the Magnetic-electric potential φ_v to the speed of light c.

According to equation (24.1.4), the magnetic flux density B is equal to the Curl $\nabla \times A$ of the vector potential A, viz:

$$B = \nabla \times A$$

Compare the above equation with equation (24.2.18) to get the relationship equation:

$$A = -\frac{1}{c}\varphi_v = -\frac{1}{4\pi\varepsilon_0 c^2}\frac{qv}{r}\sin\theta \tag{24.2.19}$$

The above equation shows that the vector potential A, as defined by physicists, is equal to the ratio of the negative Magnetic-electric potential φ_v to the speed of light c.

When the electric charge q and the velocity v remain constant and the distance r varies. Taking the derivative of equation (24.2.19) with respect to time t gives the relation equation:

$$\frac{\partial A}{\partial t} = \frac{\partial A}{\partial r}\frac{dr}{dt} = \frac{v_r}{4\pi\varepsilon_0 c}\cdot\frac{qv}{cr^2}\sin\theta = \frac{v_r}{c}E_v \tag{24.2.20}$$

The velocity $v_r = \frac{dr}{dt}$ in the equation. Since the ratio $\frac{v_r}{c}$ of velocities has no physical units, $\frac{v_r}{c}E_v$ belongs to the Magnetic-electric field strength. From this it can be determined that the rate of change $\frac{\partial A}{\partial t}$ belongs to the Magnetic-electric field strength E_v.

24.2.3 Theoretical proof of the nonexistence of the time-varying electromagnetic field $E_t = -\nabla\varphi - \partial A/\partial t$

According to equation (24.1.4), the relationship between the time-varying magnetic field B and the vector potential A is:

$$B = \nabla \times A$$

According to Stokes' theorem, the relation equation can be obtained:

$$\oint A dl = \iint \nabla \times A dS = \iint B dS \tag{24.2.21}$$

The surface S in the formula is bordered by the closed curve L. Taking the derivative of the above equation with respect to time t gives the relational equation:

$$\oint \frac{\partial A}{\partial t} dl = \iint \frac{\partial B}{\partial t} dS \tag{24.2.22}$$

According to the Faraday formula for electromagnetic induction from Maxwell's set of equations, the integral $\oint E_t dl$ of the time-varying electromagnetic field E_t along the closed curve L is equal to the negative rate of change of magnetic flux, i.e:

$$\oint E_t dl = -\iint \frac{\partial B}{\partial t} dS \tag{24.2.23}$$

Substituting equation (24.2.22) into the above equation gives the relationship equation:

$$\oint E_t dl = -\oint \frac{\partial A}{\partial t} dl \tag{24.2.24}$$

Based on the above equation, the relationship equation can be obtained:

$$E_t = -\frac{\partial A}{\partial t} \tag{24.2.25}$$

Obviously, the above equation contradicts the equation $E_t = -\nabla\varphi - \frac{\partial A}{\partial t}$ for the time-varying electromagnetic field as defined by physicists. Further, substituting the time-varying electromagnetic field $E_t = -\nabla\varphi - \frac{\partial A}{\partial t}$ into the above formula yields the relational formula:

$$-\nabla\varphi - \frac{\partial A}{\partial t} = -\frac{\partial A}{\partial t} \tag{24.2.26}$$

Based on the above equation, the relationship equation can be obtained:

$$\nabla\varphi = 0 \tag{24.2.27}$$

Since the gradient $\nabla C = 0$ for a constant C, the scalar potential φ as defined by physicists is not a variable but a constant.

Conversely, substituting the time-varying electromagnetic field $E_t = -\nabla\varphi - \frac{\partial A}{\partial t}$ into equation (24.2.23) gives the relation equation:

$$\oint \nabla\varphi dl + \oint \frac{\partial A}{\partial t} dl = \iint \frac{\partial B}{\partial t} dS \tag{24.2.28}$$

If the scalar potential gradient $\nabla\varphi \neq 0$, then according to the above equation, the relation equation can be obtained:

$$\oint \frac{\partial A}{\partial t} dl \neq \iint \frac{\partial B}{\partial t} dS \tag{24.2.29}$$

It is clear that the above formula and formula (24.2.22) are contradictory to each other. Since equation (24.2.22) is derived from equation $B = \nabla \times A$, equation $B = \nabla \times A$ is wrong.

The above analytical discussion shows that the time-varying electromagnetic field $E_t = -\nabla\varphi - \frac{\partial A}{\partial t}$ as defined by physicists suffers from the following two contradictory problems.

Problem 1, if the scalar potential gradient $\nabla\varphi = 0$, then physicists use mathematical tricks to obtain the time-varying electromagnetic field formula $E_t = -\nabla\varphi - \frac{\partial A}{\partial t}$ does not exist at all:

Problem 2, when the scalar potential gradient $\nabla\varphi \neq 0$. Since the equation (24.1.4) derived by physicists, i.e., equation $B = \nabla \times A$ is wrong, the Faraday formula for electromagnetic induction in Maxwell's system of equations does not hold, i.e:

$$\oint E_t dl \neq -\iint \frac{\partial B}{\partial t} dS \tag{24.2.30}$$

It should be noted that according to equation (24.2.6), the magnetic flux density B is related to the Magnetic-electric field E_v:

$$B = \frac{1}{c} E_v$$

Since the Magnetic-electric field E_v does not contain the rotating electric field E_ω, the rate of change of magnetic flux $\iint \frac{\partial B}{\partial t} dS$ does not contain the electric field E. Since the time-varying electromagnetic field E_t, as defined by physicists, contains the electric field E, equation (24.2.23) does not hold, viz:

$$\oint E_t dl \neq -\iint \frac{\partial B}{\partial t} dS \tag{24.2.31}$$

24.2.4 Modifications to Faraday's formula for electromagnetic induction $\oint E_t dl = -\iint \partial B/\partial t\, dS$

When electric charge q is moving with velocity v in the cosmic vacuum frame of reference. According to equation (18.4.3), the Magnetic-electric potential $\boldsymbol{\varphi_v}$ generated by electric charge $\boldsymbol{q}$ is:

$$\boldsymbol{\varphi_v = \frac{1}{4\pi\varepsilon_0 c}\frac{qv}{r}\sin\theta}$$

According to Stokes' theorem, the relation equation can be obtained:

$$\oint \boldsymbol{\varphi_v dl} = \iint \boldsymbol{\nabla \times \varphi_v\, dS} \tag{24.2.32}$$

According to equation (19.3.7), the negative Curl $-\boldsymbol{\nabla \times \varphi_v}$ of the Magnetic-electric potential $\boldsymbol{\varphi_v}$ is equal to the Magnetic-electric field strength $\boldsymbol{E_v}$, viz:

$$\boldsymbol{E_v = -\nabla \times \varphi_v}$$

Substituting the above equation into equation (24.2.32) gives the relationship equation:

$$\oint \boldsymbol{\varphi_v dl} = -\iint \boldsymbol{E_v(q,r,v)\, dS} \tag{24.2.33}$$

The surface $\boldsymbol{S}$ in equation has the closed curve $\boldsymbol{L}$ as its boundary. Taking the derivative of the above equation with respect to time $\boldsymbol{t}$ gives the relational equation:

$$\oint \boldsymbol{\frac{\partial \varphi_v}{\partial t} dl} = -\iint \boldsymbol{\frac{\partial E_v(q,r,v)}{\partial t} dS} \tag{24.2.34}$$

Dividing the above equation by the speed of light $\boldsymbol{c}$, gives the relationship equation:

$$\oint \boldsymbol{\frac{1}{c}\frac{\partial \varphi_v}{\partial t} dl} = -\iint \boldsymbol{\frac{1}{c}\frac{\partial E_v(q,r,v)}{\partial t} dS} \tag{24.2.35}$$

According to equation (24.2.6), the magnetic flux density $\boldsymbol{B}$ is equal to the time-varying Magnetic-electric field $\boldsymbol{E_{vt}}$ divided by the speed of light $\boldsymbol{c}$, i.e:

$$\boldsymbol{B = \frac{1}{c}E_{vt}} \tag{24.2.36}$$

Substituting the above equation into equation (24.2.35) yields the relationship equation:

$$\oint \boldsymbol{\frac{1}{c}\frac{\partial \varphi_v}{\partial t} dl} = -\iint \boldsymbol{\frac{\partial B(q,r,v)}{\partial t} dS} \tag{24.2.37}$$

According to equation (23.4.12), the time-varying Magnetic-electric field $\boldsymbol{E_{vt}}$ is:

$$\boldsymbol{E_{vt}\left(\frac{dq}{dt},\frac{dr}{dt},\frac{dv}{dt}\right) = \frac{1}{c}\frac{\partial \varphi_v}{\partial t}}$$

Substituting the above equation into equation (24.2.37) yields the relationship equation:

$$\oint \boldsymbol{E_{vt} dl} = \oint \boldsymbol{\frac{1}{c}\frac{\partial \varphi_v}{\partial t} dl} = -\iint \boldsymbol{\frac{\partial B(q,r,v)}{\partial t} dS} \tag{24.2.38}$$

The above equation is the new Faraday's law of electromagnetic induction. Note that since the time-varying Magnetic-electric field $\boldsymbol{E_{vt}}$ does not contain a rotating electric charge $\boldsymbol{q_\omega}$, the rate of change of magnetic flux $\iint \boldsymbol{\frac{\partial B}{\partial t} dS}$ is not equal to the loop integral of the time-varying electromagnetic field $\boldsymbol{E_t}$, i.e:

$$\oint \boldsymbol{E_t dl} = \oint \boldsymbol{\frac{1}{c}\frac{\partial \varphi}{\partial t} dl} \neq -\iint \boldsymbol{\frac{\partial B}{\partial t} dS} \tag{24.2.39}$$

The $\boldsymbol{E_t}$ in the equation is the time-varying electromagnetic field $\boldsymbol{\frac{1}{c}\frac{d\varphi}{dt}}$. The above equation shows that physicists' interpretation of Faraday's law of electromagnetic induction is wrong.

According to Maxwell's system of equations, the Faraday's law of electromagnetic induction in integral form is:

$$\oint \boldsymbol{E_t dl} = -\oint \boldsymbol{\left(\nabla\varphi + \frac{\partial A}{\partial t}\right) dl} = -\iint \boldsymbol{\frac{\partial B(q,r,v)}{\partial t} dS}$$

The time-varying electromagnetic field $\boldsymbol{E_t = -\nabla\varphi - \frac{\partial A}{\partial t}}$ in equation is assumed by physicists. Comparing the above equation with equation (24.2.38), it can be determined that the time-varying electromagnetic field $\boldsymbol{E_t}$ should be the time-varying Magnetic-electric field $\boldsymbol{E_{vt} = \frac{1}{c}\frac{\partial \varphi_v}{\partial t}}$. From this, it can be determined that the time-varying electromagnetic field $\boldsymbol{E_t = -\nabla\varphi - \frac{\partial A}{\partial t}}$ assumed by physicists is wrong.

24.3 Derivation of the electromagnetic wave equation

24.3.1 Divergence $\nabla \cdot E_t = 0$ of the time-varying electromagnetic field E_t in the cosmic vacuum

According to equation (23.2.22), the electromagnetic field flux $\boldsymbol{\Phi}_{Et}$ through the surface $\boldsymbol{S}$ is:

$$\boldsymbol{\Phi}_{Et} = \iint \boldsymbol{E}_t(q, r, v)\, d\boldsymbol{S} = \frac{R}{2\varepsilon_0} \frac{q}{r} \sin\theta \tag{24.3.1}$$

Assume that the closed sphere $\boldsymbol{S} = 4\pi r^2$ has the point P as the center of the sphere, and assume that the interior of the closed sphere $\boldsymbol{S}$ does not contain an electric charge $\boldsymbol{q}$. When the closed sphere $\boldsymbol{S}$ shrinks toward the point P. Since the number of electromagnetic field lines entering and leaving the closed sphere $\boldsymbol{S}$ is equal, the electromagnetic field flux $\boldsymbol{\Phi}_E$ through the closed sphere $\boldsymbol{S} = 4\pi r^2$ is equal to 0, i.e:

$$\boldsymbol{\Phi}_{Et} = \oiint \boldsymbol{E}_t d\boldsymbol{S} = 0 \tag{24.3.2}$$

Note that the above equation holds provided that the electric charge $\boldsymbol{q}$ lies outside the closed sphere $\boldsymbol{S} = 4\pi r^2$.

By the definition of divergence, the divergence $\nabla \cdot \boldsymbol{E}_t$ of the electromagnetic field strength $\boldsymbol{E}_t$ at point P in space is:

$$\nabla \cdot \boldsymbol{E}_t = \lim_{\Delta V \to 0} \frac{\oiint \boldsymbol{E}_t d\boldsymbol{S}}{\Delta V} \tag{24.3.3}$$

Substituting equation (24.3.2) into the above equation gives the relationship equation:

$$\nabla \cdot \boldsymbol{E}_t = 0 \tag{24.3.4}$$

The above equation shows that the divergence $\nabla \cdot \boldsymbol{E}_t$ of the electromagnetic field at any point in space is equal to zero.

24.3.2 Wave equation for the electromagnetic field E_t

According to equation (23.2.42), the loop integral $\oint \boldsymbol{E}_t d\boldsymbol{l}$ of the time-varying electromagnetic field $\boldsymbol{E}_t$ is:

$$\oint \boldsymbol{E}_t d\boldsymbol{l} = \iint \frac{1}{c} \frac{\partial \boldsymbol{E}_t}{\partial t} d\boldsymbol{S} \tag{24.3.5}$$

According to Stokes' theorem, the relation equation can be obtained:

$$\oint \boldsymbol{E}_t d\boldsymbol{l} = \iint \nabla \times \boldsymbol{E}_t d\boldsymbol{S} = -\iint \frac{1}{c} \frac{\partial \boldsymbol{E}_t}{\partial t} d\boldsymbol{S} \tag{24.3.6}$$

Based on the above equation, the relationship equation can be obtained:

$$\nabla \times \boldsymbol{E}_t = -\frac{1}{c} \frac{\partial \boldsymbol{E}_t}{\partial t} \tag{24.3.7}$$

Taking Curl from the above equation yields the relationship equation:

$$\nabla \times (\nabla \times \boldsymbol{E}_t) = -\nabla \times \left(\frac{1}{c} \frac{\partial \boldsymbol{E}_t}{\partial t}\right) \tag{24.3.8}$$

According to the curl theory, the curl $\nabla \times (\nabla \times \boldsymbol{E}_t)$ on the left side of the equal sign in the above formula is:

$$\nabla \times (\nabla \times \boldsymbol{E}_t) = \nabla(\nabla \cdot \boldsymbol{E}_t) - \nabla^2 \boldsymbol{E}_t \tag{24.3.9}$$

The $\nabla(\nabla \cdot \boldsymbol{E}_t)$ in equation is the gradient of divergence $\nabla \cdot \boldsymbol{E}_t$ at point P in space . According to equation (24.3.4), divergence $\nabla \cdot \boldsymbol{E}_t$ at any point in space is equal to 0, i.e:

$$\nabla \cdot \boldsymbol{E}_t = 0$$

Since the gradient $\nabla \boldsymbol{C} = 0$ for constant C, the gradient of divergence $\nabla \cdot \boldsymbol{E}_t$ is equal to 0, i.e:

$$\nabla(\nabla \cdot \boldsymbol{E}_t) = 0 \tag{24.3.10}$$

Substituting the above formula into the formula (24.3.9), the relationship formula can be obtained:

$$\nabla \times (\nabla \times \boldsymbol{E}_t) = -\nabla^2 \boldsymbol{E}_t \tag{24.3.11}$$

In addition, the curl $\nabla \times \left(\frac{1}{c} \frac{\partial \boldsymbol{E}_t}{\partial t}\right)$ on the right side of the equal sign in the formula (24.3.8) is:

$$\nabla \times \left(\frac{1}{c} \frac{\partial \boldsymbol{E}_t}{\partial t}\right) = \frac{1}{c} \frac{\partial}{\partial t} \nabla \times \boldsymbol{E}_t \tag{24.3.12}$$

Substituting equation (24.3.7) into the above equation gives the relationship equation:

$$\nabla \times \left(\frac{1}{c}\frac{\partial \boldsymbol{E}_t}{\partial t}\right) = -\frac{1}{c^2}\frac{\partial^2 \boldsymbol{E}_t}{\partial t^2} \qquad (24.3.13)$$

Substituting the above formula and formula (24.3.11) into the formula (24.3.8), the relationship formula can be obtained:

$$\nabla^2 \boldsymbol{E}_t - \frac{1}{c^2}\frac{\partial^2 \boldsymbol{E}_t}{\partial t^2} = 0 \qquad (24.3.14)$$

The above equation is the wave equation for the electromagnetic field $\boldsymbol{E}_t$. The above equation shows that the propagation speed of an electromagnetic wave in the cosmic vacuum frame of reference is equal to the speed of light $\boldsymbol{c}$.

Chapter 25 displacement current $I_D=\varepsilon_0\iint \partial E_0/\partial t\, dS$ does not exist

Introduction: This chapter analyzes and discusses the errors made by physicists in deriving the displacement current I_D.

25.1 Derivation of Ampere circuital theorem $\oint Bdl=\mu_0 I$

25.1.1 Circular line integral of magnetic flux density B

Suppose R is the radius of the circular line $L_R = 2\pi R$ and O is the center point of the plane of the circular line $L_R = 2\pi R$. Suppose that electric charge q is moving with velocity v toward the circular line $L_R = 2\pi R$ in the cosmic vacuum frame of reference. Assume that the distance r is the distance from electric charge q to the circular line $L_R = 2\pi R$, and Z is the distance from electric charge q to the point O. Assume that θ is the angle between the distance r and the distance Z. as shown in Figure 25-1.

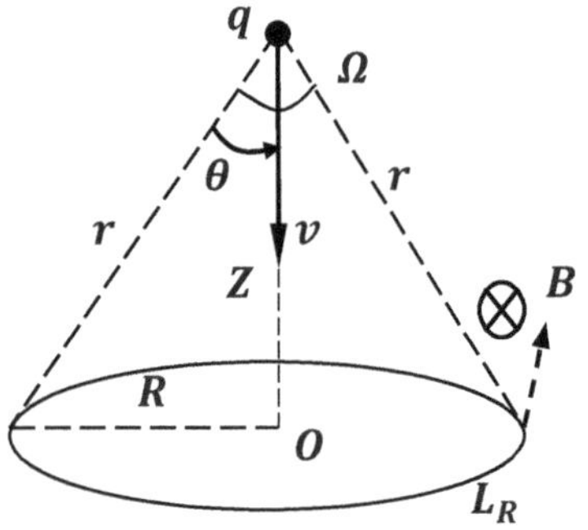

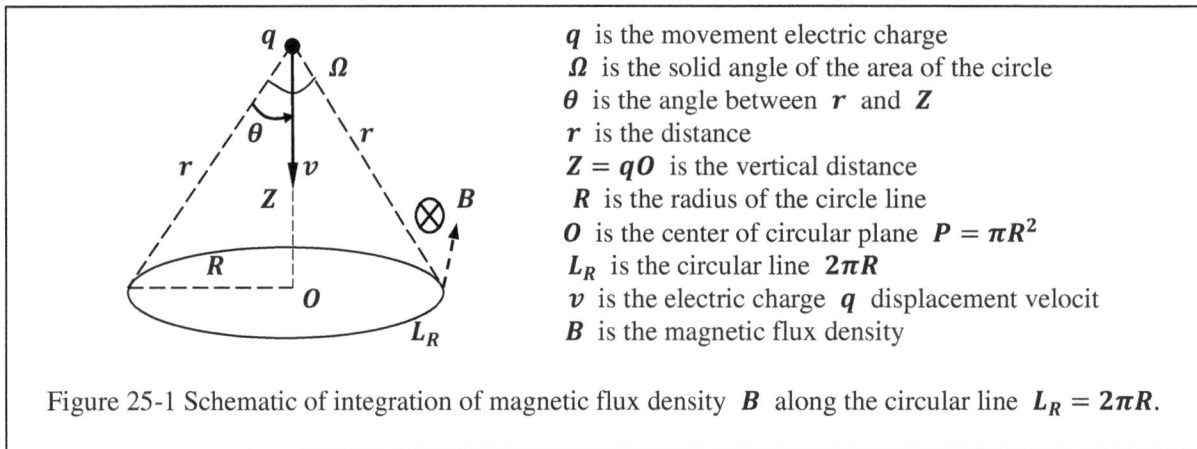

q is the movement electric charge
Ω is the solid angle of the area of the circle
θ is the angle between r and Z
r is the distance
$Z = qO$ is the vertical distance
R is the radius of the circle line
O is the center of circular plane $P = \pi R^2$
L_R is the circular line $2\pi R$
v is the electric charge q displacement velocit
B is the magnetic flux density

Figure 25-1 Schematic of integration of magnetic flux density B along the circular line $L_R = 2\pi R$.

According to equation (19.3.17), the magnetic flux density B at each point on the circular line $L_R = 2\pi R$ is:

$$B = \frac{\mu_0}{4\pi}\frac{qv}{r^2}\sin\theta \qquad (25.1.1)$$

Note that the magnetic flux density B is a function of the angle θ. If the angle $\theta = 90^0$, then the above equation becomes:

$$B = \frac{\mu_0}{4\pi}\frac{qv}{r^2} \qquad (25.1.2)$$

Since the magnetic flux density B remains constant on the circular line $L_R = 2\pi R$. Therefore the integral of magnetic flux density B along the circular line $L_R = 2\pi R$ is:

$$\oint Bdl = BL_R = \frac{\mu_0 R}{2}\cdot\frac{qv}{r^2}\sin\theta \qquad (25.1.3)$$

The above equation is the integration formula of the magnetic field B. When the radius of the circle line $R = r$. Since the electric charge q is located in the center of the circular plane $P = \pi R^2$, the angle $\theta = 90^0$, so the above formula becomes:

$$\oint Bdl = \frac{\mu_0}{2}\frac{qv}{r} \qquad (25.1.4)$$

Note that the electric charge q lies on a point, i.e., the electric charge q is not the charge contained in the energized wire L.

25.1.2 Magnetic flux density *dB* generated by current element *Idl* at point P

Assume S is the cross-section of the wire and n is the free electron density per unit volume of the wire. Suppose dl is the length of current element Idl. Since the volume of current element Idl is $dV = Sdl$, the number dN of free electrons contained in current element Idl is:

$$dN = ndV = nSdl \tag{25.1.5}$$

Since dl is the length of the wire, not the moves distance dl' of the free electrons, the free electric charge q contained in the length of the wire dl is:

$$q = edN = enSdl \tag{25.1.6}$$

Assume that the free electrons are moving at the same time with velocity v. Multiplying the above equation by the displacement velocity v of the electrons gives the relational equation:

$$Idl = qv = venSdl \tag{25.1.7}$$

According to equation (16.2.3), the magnetic electric charge momentum P_{ev} of the electron is:

$$P_{ev} = ev$$

Substituting the above equation into equation (25.1.7) gives the relationship equation:

$$Idl = P_{ev} \cdot nSdl \tag{25.1.8}$$

Since $nSdl$ in equation is the number of free electrons contained in current element Idl, current element Idl belongs to the electric charge momentum.

Theoretically, the electric charge q possessed by current element Idl can be viewed as concentrated at the center of current element Idl, as shown in Figure 25-2.

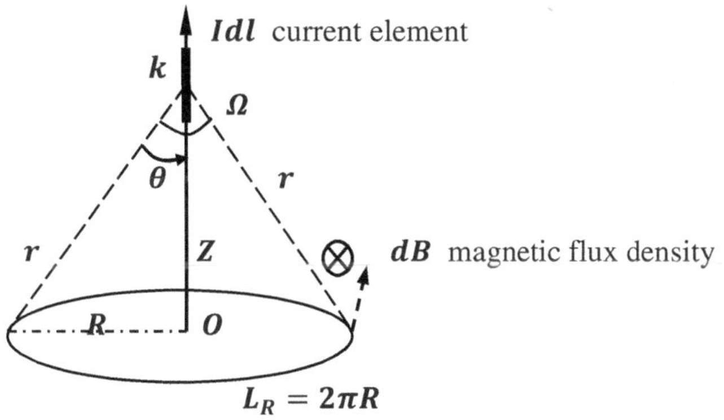

Figure 25-2 Schematic diagram of the integration of magnetic flux density dB along the circular line $L_R = 2\pi R$.

Substituting Equation (25.1.6) into Equation (25.1.1) gives the relationship equation:

$$dB = \frac{\mu_0}{4\pi}\frac{venSdl}{r^2}\sin\theta \tag{25.1.9}$$

Substituting the electric current equation $I = venS$ into the above equation gives the relationship equation:

$$dB = \frac{\mu_0}{4\pi}\frac{Idl}{r^2}\sin\theta \tag{25.1.10}$$

The above equation is the classical Biot-Savart Law. dB in the equation is the magnetic flux density generated by the current element Idl at the point **P**. Note that the magnetic flux density dB is a function of the angle θ.

If the angle $\theta = 90^0$, then the above equation becomes:

$$dB = \frac{\mu_0}{4\pi}\frac{Idl}{r^2} \tag{25.1.11}$$

The current element Idl in equation is located at the center O of the circular line plane $P = \pi r^2$, i.e., at a distance $Z = 0$. Comparing the above equation with equation (25.1.2), it can be determined that the magnetic flux density $dB = B$ and the current element $Idl = qv$.

25.1.3 The magnetic flux density *B* generated by the energized wire D_{AC} at point P and the integral $\oint Bdl$ of the magnetic field *B* along the circular line $L_R = 2\pi R$

Suppose A is the starting point of energized wire D_{AC} and C is the end point of wire D_{AC}. The length of the wire $D_{AC} = AC$. Suppose r is the distance from the current element Idl to point P Assume that θ is the angle between the distance r and the direction of electric current I. Assume that $R = PO$ is the perpendicular distance from point P to wire D_{AC}. As shown in Figure 25-3.

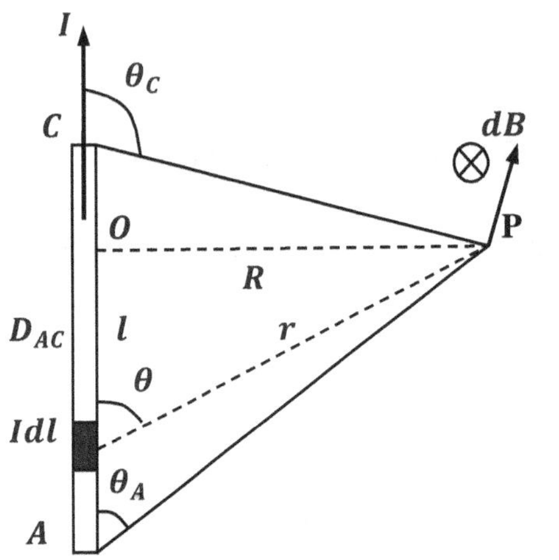

Figure 25-3 Schematic diagram of the magnetic flux density dB generated by current element Idl at point P.

Integrate the magnetic flux density dB in equation (25.1.10) along the wire D_{AC} to obtain the relation equation:

$$B = \int_A^C dB = \frac{\mu_0 I}{4\pi}\int_A^C \frac{\sin\theta}{r^2}dl \tag{25.1.12}$$

The B in equation is the magnetic flux density generated by the energized wire D_{AC} at point P.

Assume that the length from point O to the center of current element Idl is l. According to Figure 25-3 the length l is:

$$l = R\,\mathrm{ctg}\,\theta \tag{25.1.13}$$

Differentiating the above equation with respect to the angle θ gives the relation equation:

$$dl = -\frac{R}{\sin^2\theta}d\theta \tag{25.1.14}$$

According to Figure 25-3 the distance r is:

$$r = \frac{R}{\sin\theta} \tag{25.1.15}$$

Substituting the differential dl and the distance r into equation (25.1.12) gives the relation equation:

$$B = -\frac{\mu_0 I}{4\pi}\int_A^C \frac{\sin\theta}{\left(\frac{R}{\sin\theta}\right)^2}\frac{Rd\theta}{\sin^2\theta}$$

Simplifying the above equation yields the relationship equation:

$$B = -\frac{\mu_0}{4\pi}\frac{I}{R}\int_{\theta_A}^{\theta_C}\sin\theta\, d\theta \tag{25.1.16}$$

Solving the above equation yields the relationship equation:

$$B = \frac{\mu_0}{4\pi}\frac{I}{R}(\cos\theta_A - \cos\theta_C) \tag{25.1.17}$$

The B in equation is the magnetic flux density produced by the energized wire D_{AC} at point P . Note that the magnetic flux density B is a function of the angle θ_A and the angle θ_C.

Assume that the perpendicular distance R from each point on the circular line $L_R = 2\pi R$ to the wire D_{AC} is equal. Since the magnetic flux density B is constant on the circular line L_R, the integral $\oint Bdl$ of the magnetic flux density B along the circular line $L_R = 2\pi R$ is:

$$\oint Bdl = BL_R = \frac{\mu_0 I}{2}(\cos\theta_A - \cos\theta_C) \tag{25.1.18}$$

The above equation is the integral formula for the magnetic field loop of the energized wire D_{AC}. Note that the above equation possesses the following three properties.

The first property is that the loop integral $\oint Bdl$ is not related to any surface S.

If the surface S is bordered by a circular line $L_R = 2\pi R$, then the loop integral $\oint Bdl$ remains the same no matter what shape the surface S chooses. Yet when physicists derive the displacement electric current I_D, physicists assume that the loop integral $\oint Bdl$ is related to the shape of the surface S. This view is erroneous.

The 2nd property is that the loop integral $\oint Bdl$ is a function of the pinch angle θ_A and the pinch angle θ_C.

When the circular line $L_R = 2\pi R$ is held constant and the pinch angle θ_A and pinch angle θ_C are varied. Since the magnetic field B is a function of the pinch angle θ_A and the pinch angle θ_C, the wire D_{AC} has many different loop integrals $\oint Bdl$.

The third property is that the loop integral $\oint Bdl$ applies to any electric current I .

Since the magnetic field B is a function of the electric current I, the loop integral $\oint Bdl$ applies not only to constant electric current but also to varying electric current.

Alternatively, when electric current I is constant electric current, the integral $\oint Bdl$ describes the loop integral of the constant magnetic field B. When electric current I is time-varying electric current, the integral $\oint Bdl$ describes the loop integral of the time-varying magnetic field B at a given instant.

25.1.4 Derivation of the Ampere circuital theorem $\oint Bdl=\mu_0 I$

When the energized wire D_{AC} is a straight wire of infinite length, since the angle $\theta_A = 0$ and the angle $\theta_C = 180^0$, equation (25.1.18) becomes:

$$\oint Bdl = \mu_0 I \tag{25.1.19}$$

The above equation is the Ampere circuital theorem. note that the Ampere circuital theorem has the following three properties.

The first property is that the electric current I in the Ampere circuital theorem is a straight electric current of infinite length .

Since the electric current I in the Ampere circuital theorem is a straight electric current of infinite length, the Ampere circuital theorem cannot be applied to loop integrals of short energized conductors. For example, the Ampere circuital theorem cannot be used to analyze the circuit of a parallel plate capacitor. Yet physicists use an infinitely long electric current as a short electric current when deriving displacement currents.

The 2nd property is that the Ampere circuital theorem $\oint Bdl = \mu_0 I$ is a function of the electric current I and has nothing to do with the surface S through which the electric current I passes.

If the surface S is bordered by a circular line $L_R = 2\pi R$, then Ampere circuital theorem $\oint Bdl = \mu_0 I$ remains constant regardless of any shape chosen for the surface S. Yet when physicists derive the displacement electric current I_D, physicists assume that Ampere circuital theorem $\oint Bdl = \mu_0 I$ is related to the shape of surface S. This view is erroneous.

The third property is that the Ampere circuital theorem applies to time-varying electric current I .

Since the Ampere circuital theorem is a function of the electric current I, the Ampere circuital theorem applies not only to constant electric current but also to varying electric current.

However physicists believe that the Ampere circuital theorem applies only to constant electric current and not to varying electric current. Since this view denies that the Ampere circuital theorem is a function of electric current I, this view of physicists is wrong.

25.2 Errors in Maxwell-Ampere's law

25.2.1 The derivation of Maxwell-Ampere's law

In the 19th century, the British physicist Maxwell established a set of equations describing the relationship between electromagnetic fields. Maxwell's system of equations consists of four equations. The system of Maxwell's equations in integral form is:

$$\begin{cases} \oiint \boldsymbol{E}d\boldsymbol{S} = \dfrac{q}{\varepsilon_0} & (1) \\ \oiint \boldsymbol{B}d\boldsymbol{S} = 0 & (2) \\ \oint \boldsymbol{E}d\boldsymbol{l} = -\iint \dfrac{\partial \boldsymbol{B}}{\partial t}\boldsymbol{S} & (3) \\ \oint \boldsymbol{B}d\boldsymbol{l} = \mu_0 \boldsymbol{I} + \mu_0\varepsilon_0 \iint \dfrac{\partial \boldsymbol{E}}{\partial t} d\boldsymbol{S} & (4) \end{cases} \quad (25.2.1)$$

The Maxwell-Ampere law containing the displacement current $I_D = \varepsilon_0 \iint \frac{\partial E}{\partial t} dS$ is:

$$\oint \boldsymbol{B}d\boldsymbol{l} = \mu_0 \boldsymbol{I} + \mu_0\varepsilon_0 \iint \frac{\partial \boldsymbol{E}}{\partial t} d\boldsymbol{S} \quad (25.2.2)$$

The above equation shows that the magnetic field $\boldsymbol{B}$ is generated by two types of electric current. One is the magnetic field produced by the conducting electric current $\boldsymbol{I}$ and the other is the magnetic field produced by the displacement current I_D. Physicists derive the above equation as follows.

Suppose that there is a parallel-plate capacitor on a closed circuit. Suppose that S_1 is a surface with the circular line $L_R = 2\pi R$ as its boundary line. Assume that the line AC is a wire passing through the circular line L_R. $\boldsymbol{E}$ is the electric field strength between the two parallel plates of the capacitor. This is shown in Figure 25-4.

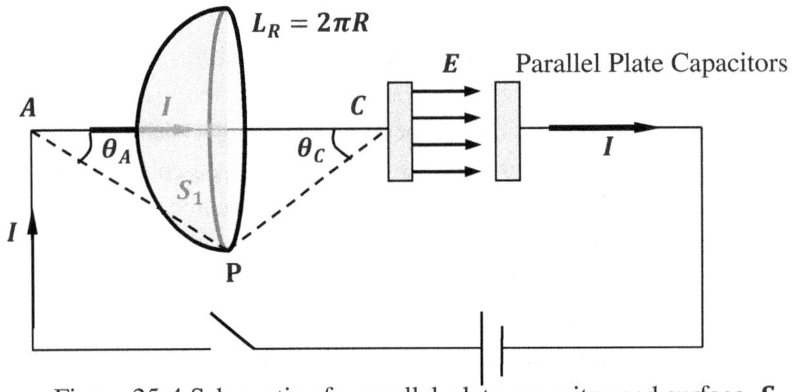

Figure 25-4 Schematic of a parallel-plate capacitor and surface S_1.

When the switch is closed, the free electrons are driven by the power supply to move to the capacitor plates, and since the capacitor can hold a certain amount of electric charge, the movement of free electrons in the circuit is manifested by the conduction of electric current $\boldsymbol{I} = \boldsymbol{enSv}$.

According to the electromagnetic induction experiment, electric current $\boldsymbol{I}$ produces a magnetic field $\boldsymbol{B}$. When the magnetic field $\boldsymbol{B}$ integrates along the circular line $\boldsymbol{L_R} = \boldsymbol{2\pi R}$. Since the electric current $\boldsymbol{I} \neq \boldsymbol{0}$ across the surface $\boldsymbol{S_1}$, the loop integral of the magnetic field $\boldsymbol{B}$ is, according to the amperometric loop theorem:

$$\oint \boldsymbol{Bdl} = \boldsymbol{\mu_0 I} \qquad (25.2.3)$$

(It should be noted that the electric current $\boldsymbol{I}$ in equation is the electric current in the wire, not the electric current across the surface $\boldsymbol{S_1}$.)

Since the Ampere Loop Theorem is only relevant to closed loops and has no bearing on the shape and size of the surface $\boldsymbol{S}$, there are many surfaces $\boldsymbol{S}$ with the circular line $\boldsymbol{L_R} = \boldsymbol{2\pi R}$ as a border.

Suppose that $\boldsymbol{S_2}$ is a surface bordered by the circular line $\boldsymbol{L_R} = \boldsymbol{2\pi R}$. This is shown in Figure 25-5.

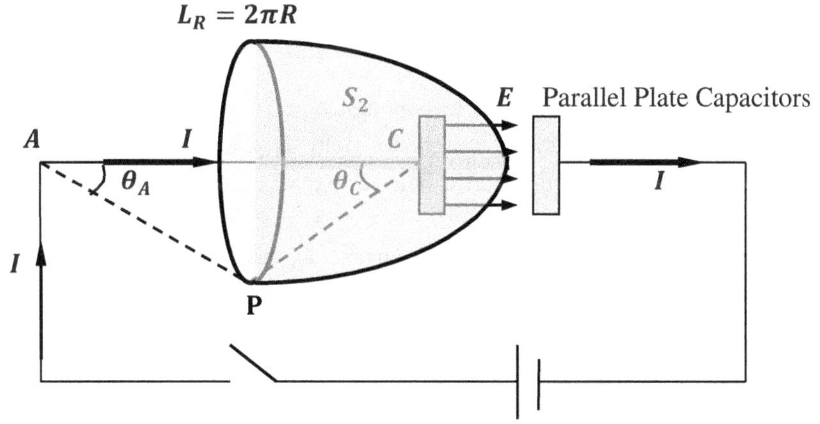

Figure 25-5 Schematic of a parallel-plate capacitor and surface $\boldsymbol{S_2}$.

Since there is no electric current in the two pole plates of the capacitor , the electric current $\boldsymbol{I} = \boldsymbol{0}$ across the surface $\boldsymbol{S_2}$. According to the Ampere loop theorem: the loop integral of the magnetic field $\boldsymbol{B}$ is:

$$\oint \boldsymbol{Bdl} = \boldsymbol{0} \qquad (25.2.4)$$

(Point to be made. Since physicists consider the electric current $\boldsymbol{I}$ in the wire as the electric current across the surface $\boldsymbol{S}$, the above equation is wrong.)

Although surfaces $\boldsymbol{S_1}$ and $\boldsymbol{S_2}$ have the same boundary line $\boldsymbol{L_R} = \boldsymbol{2\pi R}$, the loop integrals of surfaces $\boldsymbol{S_1}$ and $\boldsymbol{S_2}$ are not equal. Alternatively, since electric current $\boldsymbol{I}$ is disconnected between the two pole plates of the capacitor, Ampere's Loop Theorem $\oint \boldsymbol{Bd} = \boldsymbol{\mu_0 I}$ fails inside the capacitor. From this physicists consider the Ampere Loop Theorem to be imperfect.

Since there are only varying positive and negative electric charges on the two parallel plates of a capacitor, there is a varying electric field $\boldsymbol{E}$ between the two parallel plates. In order to keep the electric current uninterrupted between the two parallel plates of the capacitor, physicists believe that the varying electric field $\boldsymbol{E}$ produces the magnetic field $\boldsymbol{B}$ as well.

Or, physicists believe that the time-varying electric field $\boldsymbol{E}$ between two parallel polar plates is equivalent to electric current . The hypothesis that the time-varying electric field $\boldsymbol{E}$ produces the time-varying magnetic field $\boldsymbol{B}$ was first proposed by the British physicist Maxwell.

When the capacitor is charging. Since the electric field $\boldsymbol{E}$ between the two parallel pole plates changes with time, the electric flux $\iint \boldsymbol{EdS}$ between the two pole plates also changes. The electric flux $\iint \boldsymbol{EdS}$ between the two parallel pole plates is:

$$\iint EdS = \frac{q}{\varepsilon_0} \tag{25.2.5}$$

The q in the formula is the charge on the pole plate. Taking the derivative of the above equation with respect to time t gives the relational equation:

$$\varepsilon_0 \iint \frac{\partial E}{\partial t} dS = \frac{dq}{dt} \tag{25.2.6}$$

Since physicists define the rate of change of electric charge, $\frac{dq}{dt}$, as electric current , physicists define $\varepsilon_0 \iint \frac{dE}{dt} dS$ as the displacement current I_D, i.e:

$$I_D = \varepsilon_0 \iint \frac{\partial E}{\partial t} dS = \frac{dq}{dt} \tag{25.2.7}$$

Physicists have argued that although the conductionelectric current I is disconnected inside the two parallel polar plates. However, since the displacement current I_D allows the electric current in the circuit to remain continuous and uninterrupted, the Ampere Loop Theorem should be modified to:

$$\oint Bdl = \mu_0 I + \mu_0 \varepsilon_0 \iint \frac{\partial E}{\partial t} dS = \mu_0 I + \mu_0 I_D$$

The above equation is Maxwell-Ampere's law. In electromagnetism, a time-varying electric field E excites a time-varying magnetic field B, and a time-varying magnetic field B excites a vortex electric field E. The time-varying magnetic field B and the time-varying electric field E excite each other to propagate an electromagnetic wave far away in space.

Physicists believe that the only thing that displacement current I_D and conduction electric current I have in common is that they can both excite a magnetic field B in space, and the differences between the two are as follows:

(1) Displacement current I_D is essentially a changing electric field E, whereas conduction electric current I is the directed motion of a free electric charge.

(2) Conducted electric current I generates Joule heat when it passes through a conductor, whereas displacement current I_D does not generate Joule heat, and displacement current has no chemical effect.

(3) Displacement current I_D can exist in vacuum, conductors, and dielectrics, whereas conduction electric current I can exist only in conductors.

(4) The magnetic effect of displacement current I_D obeys the amperometric loop theorem.

(5) Conducted electric current I can be measured directly in magnitude, while displacement current I_D cannot be measured directly in magnitude.

Although conduction electric current I and displacement current I_D are both electric current , there is a huge difference in their properties, a difference that is theoretically implausible. Since displacement current I_D cannot be directly measured in magnitude, physicists cannot experimentally prove that displacement current I_D exists objectively.

We analyze and discuss the errors in displacement current $I_D = \varepsilon_0 \iint \frac{dE}{dt} dS$ below.

25.2.2 Maxwell-Ampere's law is contradictory inside the two parallel pole plates of a capacitor

Assume that surface S_1 is outside the capacitor, and assume that the circular line $L_R = 2\pi R$ and surface S_2 are between the two parallel pole plates of the capacitor. This is shown in Figure 25-6.

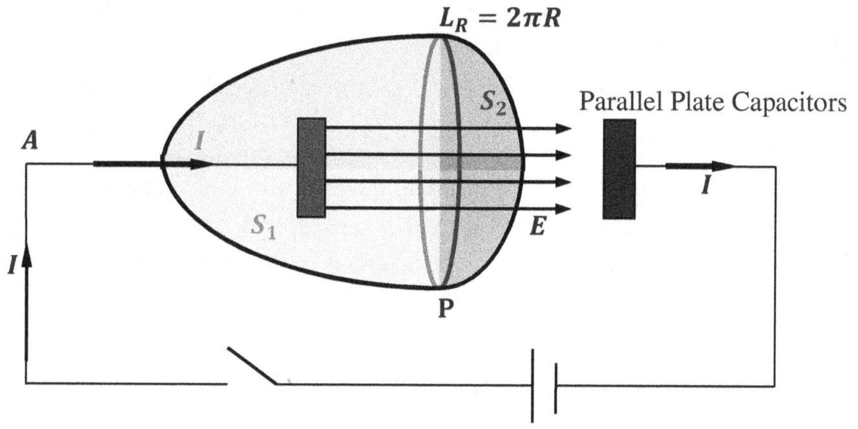

Figure 25-6 Schematic of the position of circular line $L_R = 2\pi R$ in the middle of the parallel pole plates of a capacitor.

Since the conduction electric current $I \neq 0$ and displacement current $I_D \neq 0$ through surface S_1, according to the Maxwell-Ampere loop theorem. The magnetic field loop integral of surface S_1 is:

$$\oint Bdl = \mu_0 I + \mu_0 \varepsilon_0 \iint \frac{\partial E}{\partial t} dS \tag{25.2.8}$$

Since the conductionelectric current $I = 0$ and displacement current $I_D \neq 0$ between the two parallel pole plates of the capacitor, according to the Maxwell-Ampere loop theorem. The magnetic field loop integral of the surface S_2 is:

$$\oint Bdl = \mu_0 \varepsilon_0 \iint \frac{\partial E}{\partial t} dS \tag{25.2.9}$$

Compare the above equation with equation (25.2.8). Although surfaces S_1 and S_2 have the same boundary line $L_R = 2\pi R$, the integrals of the magnetic field B along the boundary line $L_R = 2\pi R$ are not equal. Since the Maxwell-Ampere law is contradictory between two parallel polar plates, the Maxwell-Ampere loop theorem is wrong.

25.2.3 Errors in the displacement current I_D derivation process

The 1st error is that physicists consider the rate of change of charge $\frac{dq}{dt}$ as electric current I.

Assume that S is the cross-section of the wire and n is the density of free electrons contained per unit volume of the wire. Assume that v is the displacement velocity of the free electrons. According to equation (16.4.14), electric current I is:

$$I = q_S v = enSv \tag{25.2.10}$$

According to equation (16.4.12), the free electric charge q_S in the cross-section S of the conductor possesses a magnetic electric charge momentum P_{Sv} of:

$$P_{Sv} = q_S v = enSv$$

Comparing the above equation with equation (25.2.10) gives the relationship equation:

$$I = P_{Sv} = enSv \tag{25.2.11}$$

The above equation shows that electric current I is essentially the magnetic electric charge momentum P_{Sv} contained in the cross-section S of the wire.

According to equation (25.2.7), the displacement current I_D is:

$$I_D = \varepsilon_0 \iint \frac{\partial E}{\partial t} dS = \frac{dq}{dt}$$

Since displacement current I_D is a scalar (i.e., the charge velocity v is not included in the rate of change of electric charge $\frac{dq}{dt}$) and electric current $\boldsymbol{I} = \boldsymbol{enSv}$ is a vector, physicists who define the rate of change of electric charge $\frac{dq}{dt}$ as electric current $\boldsymbol{I}$ is wrong.

Since the rate of change of electric charge $\frac{dq}{dt}$ is not equal to current, physicists are wrong to define $\boldsymbol{\varepsilon_0} \iint \frac{\partial E}{\partial t} \boldsymbol{dS} = \frac{dq}{dt}$ as displacement current $\boldsymbol{I_D}$.

The 2nd mistake is that physicists secretly switch physical quantities in their derivations, i.e., they transform the infinitely long electric current into the electric current of the $\boldsymbol{AC}$ line segment

According to Figure 25-6, the conductor of the $\boldsymbol{AC}$ line segment is a wire passing through the circular line $\boldsymbol{L_R} = \boldsymbol{2\pi R}$. According to equation (25.1.17), the magnetic field $\boldsymbol{B}$ produced by the wire of the $\boldsymbol{AC}$ line segment at any point on the circular line $\boldsymbol{L_R} = \boldsymbol{2\pi R}$ is:

$$\boldsymbol{B} = \frac{\boldsymbol{\mu_0}}{\boldsymbol{4\pi}}\frac{\boldsymbol{I}}{\boldsymbol{R}}(\cos\boldsymbol{\theta}_{\mathrm{A}} - \cos\boldsymbol{\theta_C})$$

When the $\boldsymbol{AC}$ line segment of the wire for the infinite length of straight wire, due to the angle $\boldsymbol{\theta}_{\mathrm{A}} = \boldsymbol{0}$, angle $\boldsymbol{\theta}_{\mathrm{C}} = \boldsymbol{180^0}$, so the above formula becomes:

$$\oint \boldsymbol{Bdl} = \boldsymbol{\mu_0 I}$$

The above equation is the Ampere circuital theorem for infinitely long electric current.

Since physicists use the Ampere circuital theorem to derive the displacement electric current $\boldsymbol{I_D}$, physicists implicitly switched the electric current $\boldsymbol{I}$ of the $\boldsymbol{AC}$ line segment to the infinitely long electric current $\boldsymbol{I}$ during the derivation.

The 3rd error is that physicists consider the loop integral $\oint \boldsymbol{Bdl}$ of the magnetic field $\boldsymbol{B}$ as a function of the surface $\boldsymbol{S}$.

When the magnetic field $\boldsymbol{B}$ generated by the AC line segment integrates along the circular line $\boldsymbol{L_R} = \boldsymbol{2\pi R}$. Since the magnetic field $\boldsymbol{B}$ remains constant on the circular line $\boldsymbol{L_R}$, the loop integral of the magnetic field $\boldsymbol{B}$ is:

$$\oint \boldsymbol{Bdl} = \boldsymbol{BL_R} = \frac{\boldsymbol{\mu_0 I}}{\boldsymbol{2}}(\cos\boldsymbol{\theta}_{\mathrm{A}} - \cos\boldsymbol{\theta_C}) \qquad (25.2.12)$$

The above equation shows that the loop integral $\oint \boldsymbol{Bdl}$ is a function of the electric current $\boldsymbol{I}$, the pinch angle $\boldsymbol{\theta}_{\mathrm{A}}$, and the pinch angle $\boldsymbol{\theta_C}$, and not a function of the surface $\boldsymbol{S}$.

Since electric current $\boldsymbol{I}$, pinch angle $\boldsymbol{\theta}_{\mathrm{A}}$ and pinch angle $\boldsymbol{\theta_C}$ are independent of surface $\boldsymbol{S}$, the loop integral $\oint \boldsymbol{Bdl}$ of magnetic field $\boldsymbol{B}$ is independent of surface $\boldsymbol{S}$.

In other words, if the surface $\boldsymbol{S}$ is bordered by a circular line $\boldsymbol{L_R} = \boldsymbol{2\pi R}$, then the loop integral $\oint \boldsymbol{Bdl}$ remains constant regardless of any shape chosen for the surface $\boldsymbol{S}$. Yet physicists, in deriving the displacement electric current $\boldsymbol{I_D}$, assume that the loop integral $\oint \boldsymbol{Bdl}$ is related to the shape of the surface $\boldsymbol{S}$. This view is erroneous.

The fourth error is that the amplitude $\boldsymbol{A_B}$ of the magnetic field wave and the amplitude $\boldsymbol{A_E}$ of the electric field wave are always equal.

According to Equation (16.4.17), the total electric current $\boldsymbol{I_S}$ contained in the cross-section $\boldsymbol{S}$ of the conductor is:

$$\boldsymbol{I_S} = \boldsymbol{P_S} = \boldsymbol{c} \cdot \boldsymbol{q_S} = \boldsymbol{c} \cdot \boldsymbol{enS}$$

Since the number $\boldsymbol{enS}$ of free electrons contained in the cross-section $\boldsymbol{S}$ of the wire remains constant, the total electric current $\boldsymbol{I_S}$ in the wire is a constant.

According to equation (16.4.18), the total electric current $\boldsymbol{I_S}$ contained in the cross-section $\boldsymbol{S}$ of the conductor is:

$$\boldsymbol{I_S} = \sqrt{\boldsymbol{I_\omega^2} + \boldsymbol{I^2}}$$

Both the conducting electric current $\boldsymbol{I}$ and the spin electric current $\boldsymbol{I_\omega}$ are inside the wire. The conduction electric current $\boldsymbol{I}$ produces the magnetic field $\boldsymbol{B}$, while the spin electric current $\boldsymbol{I_\omega}$ produces the electric field $\boldsymbol{E}$.

Electromagnetic waves contain mutually perpendicular magnetic field waves and electric field waves. Magnetic field waves are generated by conduction electric current $\boldsymbol{I}$ and electric field waves are generated by spin electric current $\boldsymbol{I_\omega}$.

The amplitude A_B of the magnetic field wave and the amplitude A_E of the electric field wave vary with the velocity v of the free electron. The relationship between the electromagnetic wave amplitude A, the magnetic field wave amplitude A_B, and the electric field wave amplitude A_E is:

$$A = \sqrt{A_B^2 + A_E^2} \tag{25.2.13}$$

Since the total electric current $I_S = c \cdot enS$ in a wire is a constant, the amplitude A of an electromagnetic wave is also a constant. The amplitude A_B of the magnetic field wave and the amplitude A_E of the electric field wave are perpendicular to each other in the wavelength $\lambda/2$, and the direction of change of the amplitude A_B is opposite to the direction of change of the amplitude A_E.

According to Maxwell-Ampere's law, the total electric current I_S is:

$$I_S = I_D + I \tag{25.2.14}$$

Since physicists believe that the time-varying magnetic field B produces the time-varying electric field E, and the time-varying electric field E in turn produces the time-varying magnetic field B, the amplitude A_B of the magnetic field wave, and the amplitude A_E of the electric field wave, in Maxwell-Ampere's law are always equal, viz:

$$A_B = A_E \tag{25.2.15}$$

We know that electric current I (directionally moving free electrons) produces both a magnetic field B and an electric field E.

If electric current I varies at low voltage, then the electric field E generated by the variation of electric current I is greater than the magnetic field B, i.e., $E > B$. At this time, the amplitude of the electric field wave A_E is greater than the amplitude of the magnetic field wave A_B, i.e., $A_E > A_B$.

If electric current I varies in a high voltage state, then the electric field E generated by the variation of electric current I is smaller than the magnetic field B, i.e., $E < B$. At this time, the amplitude of the electric field wave A_E is smaller than the amplitude of the magnetic field wave A_B, i.e., $A_E < A_B$.

From this, it can be determined that Maxwell-Ampere's law of amplitude formula $A_B = A_E$ does not conform to the law of change of electromagnetic wave amplitude, while formula $A = \sqrt{A_B^2 + A_E^2}$ conforms to the law of change of electromagnetic wave amplitude.

25.2.4 Using the rate of change $\iint \partial E_0/\partial t\, dS$ of electric flux can prove that the equation displacement current $I_D=\varepsilon_0\iint \partial E_0/\partial t\, dS$ is wrong

According to equation (22.1.2), the static electric potential φ_0 generated by the static electric charge q_0 on the capacitor pole plate is:

$$\varphi_0 = \frac{1}{4\pi\varepsilon_0}\frac{q_0}{r} \tag{25.2.16}$$

Assume that the distance r from the center of the capacitor pole plate to each point of the circular line $L_R = 2\pi R$ is equal. This is shown in Figure 25-7.

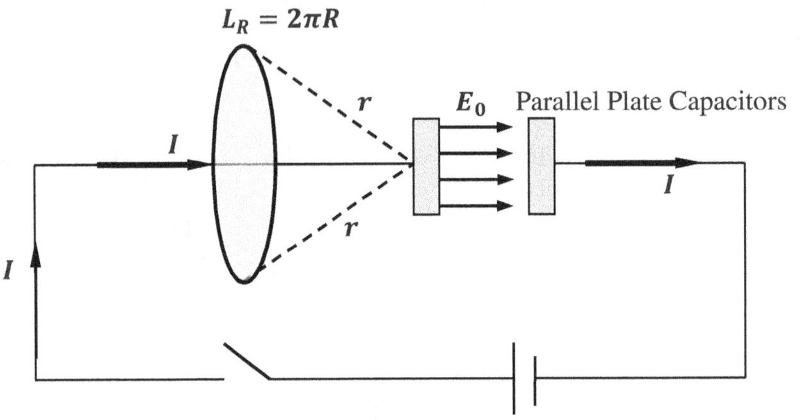

Figure 25-7 Schematic of the integration of the time-varying electrostatic field E_0 of a capacitor along the circular line $L_R = 2\pi R$.

Since the static electric potential φ_0 is constant on the circular line L_R, the static electric potential φ_0 integrates along the circular line $L_R = 2\pi R$ as:

$$\oint \varphi_0 dl = \varphi_0 L_R = 2\pi R \cdot \frac{1}{4\pi\varepsilon_0}\frac{q_0}{r} \tag{25.2.17}$$

According to Stokes' theorem, the relation equation can be obtained:

$$\oint \varphi_0 dl = \iint \nabla \times \varphi_0 dS = \frac{R}{2\varepsilon_0}\frac{q_0}{r} \tag{25.2.18}$$

According to equation (17.3.2), the Curl $\nabla \times \varphi_0$ of the electrostatic potential is equal to the electrostatic field strength E_0, i.e:

$$E_0 = \nabla \times \varphi_0 \tag{25.2.19}$$

Substituting the above equation into equation (25.2.18) gives the relationship equation:

$$\oint \varphi_0 dl = \iint E_0 dS = \frac{R}{2\varepsilon_0}\frac{q_0}{r} \tag{25.2.20}$$

Taking the derivative of the above equation with respect to time t gives the relationship equation:

$$\iint \frac{\partial E_0}{\partial t} dS = \frac{R}{2\varepsilon_0}\left(\frac{1}{r}\frac{dq_0}{dt} - \frac{q_0}{r^2}\frac{dr}{dt}\right) \tag{25.2.21}$$

When the static electric charge q_0 on the pole plates of the capacitor varies and the distance r from the pole plates to the circular line $L_R = 2\pi R$ remains constant. Since the rate of change $\frac{dr}{dt} = 0$ of the distance r, the rate of change of the capacitor's electric flux is:

$$\iint \frac{\partial E_0}{\partial t} dS = \frac{R}{2\varepsilon_0 r}\frac{dq_0}{dt} \tag{25.2.22}$$

Since the static electric charge q_0 changes, the integral $\iint \frac{\partial E_0}{\partial t} dS$ is the equation for the rate of change of electric flux for a time-varying electric field. The above equation can be varied as:

$$\frac{2r}{R}\varepsilon_0 \iint \frac{\partial E_0}{\partial t} dS = \frac{dq_0}{dt} \tag{25.2.23}$$

According to equation (25.2.7), the displacement current I_D between the two parallel pole plates of the capacitor is:

$$I_D = \varepsilon_0 \iint \frac{\partial E_0}{\partial t} dS = \frac{dq_0}{dt}$$

Comparing the above equation with equation (25.2.23) gives the relationship equation:

$$I_D \neq \frac{2r}{R}\varepsilon_0 \iint \frac{\partial E_0}{\partial t} dS \tag{25.2.24}$$

The above formula shows that the displacement current $I_D = \varepsilon_0 \iint \frac{\partial E_0}{\partial t} dS$ formula is wrong.

25.2.5 Using the time-varying magnetic field *B* it can be shown that the displacement current $I_D = \varepsilon_0 \iint \partial E_0/\partial t \, dS$ formula is wrong

According to equation (19.3.17), the time-varying magnetic flux density B is:

$$B = \frac{1}{c} E_{vt}$$

Substituting equation (23.4.12) into the above equation yields the relationship equation:

$$B = \frac{1}{c} E_{vt} = \frac{\sin\theta}{4\pi\varepsilon_0 c^3}\left(\frac{v}{r}\frac{dq}{dt} - \frac{qv}{r^2} v_r + \frac{q}{r} a\right) \tag{25.2.25}$$

Assume that the distance r from the electric charge q to each point of the circular line $L_R = 2\pi R$ is equal. At time t, since the time-varying magnetic field B is constant on the circular line L_R, the integral of the time-varying magnetic field B along the circular line $L_R = 2\pi R$ is:

$$\oint B dl = BL_R = 2\pi R \frac{\sin\theta}{4\pi\varepsilon_0 c^3}\left(\frac{v}{r}\frac{dq}{dt} - \frac{qv}{r^2} v_r + \frac{q}{r} a\right) \tag{25.2.26}$$

When the distance r and velocity v remain constant and the electric charge q changes. Since the rate of change $v_r = 0$ of distance r, acceleration $a = 0$, the above equation becomes:

$$\oint B dl = \frac{R}{2\varepsilon_0 c^3} \cdot \frac{v}{r}\frac{dq}{dt} \sin\theta \tag{25.2.27}$$

According to equation (25.2.7), the displacement current I_D of the capacitor is:

$$I_D = \varepsilon_0 \iint \frac{\partial E}{\partial t} dS = \frac{dq}{dt}$$

Multiply the above equation by the vacuum permeability μ_0. According to Maxwell-Ampere's law, the loop integral of the magnetic field B is:

$$\oint B dl = \mu_0 \varepsilon_0 \iint \frac{\partial E}{\partial t} dS = \mu_0 I_D = \frac{1}{\varepsilon_0 c^2}\frac{dq}{dt} \tag{25.2.28}$$

Comparing the above equation with equation (25.2.27) gives the relationship equation:

$$\frac{R}{2\varepsilon_0 c^3} \cdot \frac{v}{r}\frac{dq}{dt} \sin\theta \neq \frac{1}{\varepsilon_0 c^2}\frac{dq}{dt}$$

Based on the above equation, the relationship equation can be obtained:

$$\oint B dl \neq \mu_0 I_D = \frac{1}{\varepsilon_0 c^2}\frac{dq}{dt}$$

The above equation shows that the displacement current equation $\oint B dl = \mu_0 I_D$ is wrong.

Chapter 26 A New Interpretation of the Michelson-Morley Experiment

Introduction: This chapter presents a new analysis of the Michelson-Morley experiment from the point of view of the cosmic vacuum frame of reference, based on the Galilean velocity transformation formula $\boldsymbol{v}' = \boldsymbol{v} \pm \boldsymbol{u}$. The authors find that physicists' interpretations of the results of the Michelson experiment were wrong.

26.1 Physicists' interpretation of the results of the Michelson-Morley experiment is wrong

26.1.1 In the cosmic vacuum frame of reference, the beam splitting point O_A, the penetration point O_{A1} and the reflection point O_{A2} are three different points in space.

During air shows, airplanes usually spray multiple colors of smoke. Although the aircraft spraying the colored smoke moves forward at high speed, the starting point of the colored smoke in the air does not move forward at the same time as the aircraft.

Similarly, suppose that point A in the vacuum of the universe is the starting point of light beam A (similar to the starting point of colored smoke). If we consider the light source S of the interferometer as an airplane, then at a certain point in time, the starting point A of the light beam A output from the light source S will not follow the light source S in synchronized forward moves.

Note that the starting point A belongs to a stationary space point in the cosmic vacuum frame of reference, i.e., the starting point A in the cosmic vacuum will not move forward in synchronization with the interferometer moves.

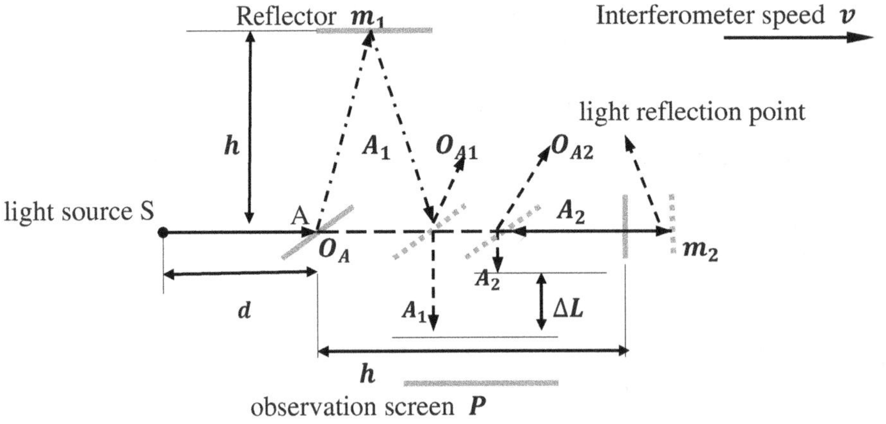

Figure 26-1 The time t_{A1} for vertical light beam A_1 to reach point O_A is not equal to the time t_{A2} for horizontal light beam A_2 to reach point O_{A2}.

Assume that the interferometer is moves to the right in the cosmic vacuum with velocity $\boldsymbol{v}$. Assume that $\boldsymbol{d}$ is the distance from the light source S to the point O of the beamsplitter, and that $\boldsymbol{h}$ is the distance from the point O of the beamsplitter to the two planar mirrors, point m_1 and point m_2.

Assume that light beam A is split into a longitudinal light beam A_1 and a transverse light beam A_2 at point O_A of the beamsplitter O. This is shown in Figure 26-1.

Note that the beam splitting point O_A, the penetration point O_{A1}, the reflection point O_{A2}, the reflection point m_{A1}, and the reflection point m_{A2} all belong to stationary spatial points in the cosmic vacuum frame of reference, and are not spatial points in the terrestrial frame of reference that can move.

26.1.2 Time t_{A1} required for beam splitter O to move from point O_A to point O_{A1}

Since the interferometer is moving in the cosmic vacuum frame of reference, according to the Galilean velocity transformation formula, the optical path length L_d of light beam A from the starting point A to the beam splitting point O_A is.

$$L_d = \frac{cd}{c - v} \tag{26.1.1}$$

In the cosmic vacuum frame of reference, the longitudinal time t_{A1} required for the longitudinal light beam A_1 to travel upward from the point O_A to the point m_{A1} and then return from the point m_{A1} to the point O_{A1} is.

$$t_{A1} = \frac{2h}{\sqrt{c^2 - v^2}} \tag{26.1.2}$$

Note that the longitudinal time t_{A1} belongs to Newton's absolute time, not to the time of special relativity. The longitudinal time t_{A1} has two meanings.

In the first meaning, time t_{A1} is the time required for the longitudinal light beam A_1 to reach point m_{A1} from point O_A upward and return from point m_{A1} to point O_{A1}. At this point the longitudinal light beam A_1 propagates from point O_{A1} toward the display P. This is shown in Figure 26-1.

In the second meaning, the time t_{A1} is the time required for beam splitter O to moves from point O_A to point O_{A1}. The distance $l_{O_AO_{A1}}$ that beam splitter O moves from point O_A to point O_{A1} is.

$$l_{O_AO_{A1}} = vt_{A1} = \frac{2hv}{\sqrt{c^2 - v^2}} \tag{26.1.3}$$

We can use experimental data to prove the fact that beam splitter O moves from point O_A to point O_{A1}.

The wavelength of sodium light in the experiment is $\lambda = 5.9 \times 10^{-7}m$; the distance from the plane mirrors m_1 and m_2 to the point O_A is $h = 5.5m$, and the velocity of the Earth's motion relative to the cosmic background radiation (CMB) is $v_1 \approx 20.9 \times 10^{-4}c$.

The number n of wavelengths contained in the moves distance $l_{O_AO_{A1}}$ of beam splitter O is.

$$n = \frac{l_{O_AO_{A1}}}{\lambda} = \frac{2hv}{\lambda\sqrt{c^2 - v^2}} = \frac{2 \times 5.5 \times 20.9 \times 10^{-4}c}{5.9 \times 10^{-7}c} = 3.89 \times 10^4$$

From the above equation, the spatial distance $l_{O_AO_{A1}}$ contains **38,900** wavelengths of sodium light.

In addition, the Sun moves around the galactic center with a velocity $v = 8 \times 10^{-4}c$. The number of wavelengths n contained in the moves distance $l_{O_AO_{A1}}$ of the beam splitter O is.

$$n = \frac{l_{O_AO_{A1}}}{\lambda} = \frac{2hv}{\lambda\sqrt{c^2 - v^2}} = \frac{2 \times 5.5 \times 8 \times 10^{-4}c}{5.9 \times 10^{-7}c} = 1.49 \times 10^4$$

From the above equation, the spatial distance $l_{O_AO_{A1}}$ contains **14900** wavelengths of sodium light.

Based on the above two calculations, it can be determined that the beam splitting point O_A and the penetration point O_{A1} are not the same cosmic vacuum point in the cosmic vacuum reference system.

The longitudinal length $L_{O_AO_{A1}}$ of the longitudinal light beam A_1 from point O_A up to point m_{A1} and back to point O_{A1} from point m_{A1} is.

$$L_{O_AO_{A1}} = ct_{A1} = \frac{2ch}{\sqrt{c^2 - v^2}} \tag{26.1.4}$$

Note that the longitudinal length $L_{O_AO_{A1}}$ starts at the beam splitting point O_A.

In the cosmic vacuum frame of reference, the longitudinal light beam A_1 leaves the starting point A and passes through the path $A \to O_A \to m_{A1} \to O_{A1}$ with the optical path length $L_{AO_AO_{A1}}$ as.

$$L_{AO_AO_{A1}} = L_d + L_{O_AO_{A1}} = \frac{cd}{c-v} + \frac{2ch}{\sqrt{c^2-v^2}} \qquad (26.1.5)$$

Note that the length $L_{AO_AO_{A1}}$ of the longitudinal optical path starts at point A. The end point is the transmission point O_{A1}.

From the above equation, if the length of the longitudinal light path A_1 is equal to $L_{AO_AO_{A1}}$, then the longitudinal light light beam A_1 reaches the transmitting point O_{A1} and propagates from the point O_{A1} to the direction P of the display. As shown in Figure 26-1.

26.1.3 Time t_{A2} required for beam splitter O to move from point O_A to point O_{A2}

In the cosmic vacuum frame of reference, the transverse time t_{A2} required for the transverse light beam A_2 to reach the point m_{A2} from the point O_A to the right and to return from the point m_{A2} to the point O_{A2} is.

$$t_{A2} = \frac{h}{c-v} + \frac{h}{c+v} = \frac{2ch}{c^2-v^2} \qquad (26.1.6)$$

Note that time t_{A2} belongs to Newton's absolute time, not to the time of special relativity. Transverse time t_{A2} is not equal to longitudinal time t_{A1}.

The transverse time t_{A2} has two meanings.

In the first meaning, time t_{A2} is the time it takes for the transverse light beam A_2 to reach point m_{A2} from point O_A to the right and return from point m_{A2} to point O_{A2}. At this point the transverse light beam A_2 propagates from point O_{A2} toward the display P. This is shown in Figure 26-1.

In the second meaning, the time t_{A2} is the time required for beam splitter O to moves from point O_A to point O_{A2}. The distance $l_{O_AO_{A2}}$ for beam splitter O to moves from point O_A to point O_{A2} is.

$$l_{O_AO_{A2}} = vt_{A2} = \frac{2chv}{c^2-v^2} \qquad (26.1.7)$$

Since moves distance $l_{O_AO_{A1}} < l_{O_AO_{A2}}$, the transmission point O_{A1} is not the same spatial point as the reflection point O_{A2}. This is shown in Figure 26-1.

The transverse length $L_{O_AO_{A2}}$ of the transverse beam A_2 from the point O_A to the right to the point m_{A2} and from m_{A2} back to the point O_{A2} is.

$$L_{O_AO_{A2}} = ct_{A2} = \frac{2c^2h}{c^2-v^2} \qquad (26.1.8)$$

Note that the length $L_{O_AO_{A2}}$ of the transverse optical path starts at point O_A. The end point is the reflection point O_{A2}.

In the cosmic vacuum frame of reference, the transverse light beam A_2 leaves the point A and passes through the path $A \to O_A \to m_{A2} \to O_{A2}$ with the transverse length $L_{AO_AO_{A2}}$ is.

$$L_{AO_AO_{A2}} = L_d + L_{O_AO_{A2}} = \frac{cd}{c-v} + \frac{2c^2h}{c^2-v^2} \qquad (26.1.9)$$

Note that the transverse length $L_{AO_AO_{A2}}$ starts at point A and ends at the reflection point O_{A2}. From the above equation, if the length of transverse optical path A_2 is equal to $L_{AO_AO_{A2}}$, then the transverse light beam A_2 reaches the point O_{A2} and propagates from the point O_{A2} to the direction P of the display. As shown in figure 26-1.

It should be noted that since the transverse time t_{A2} is greater than the longitudinal time t_{A1}, the moves distance $l_{O_AO_{A2}} > l_{O_AO_{A1}}$ of the beam splitter O is greater. It is thus determined that the reflection point O_{A2} is not the same point in space as the transmission point O_{A1}. The distance $l_{O_{A1}O_{A2}}$ between the transmission point O_{A1} and the reflection point O_{A2} is.

$$l_{O_{A1}O_{A2}} = v(t_{A2} - t_{A1}) = v\left(\frac{2ch}{c^2-v^2} - \frac{2h}{\sqrt{c^2-v^2}}\right) \approx \frac{2hv^3}{c^3} \qquad (26.1.10)$$

From the above equation, it can be seen that the longitudinal light beam A_1 and the transverse light beam A_2 do not propagate from the light transmission point O_{A1} to the display P at the same time. That is, the longitudinal light beam A_1 propagates downward first, and the transverse light beam A_2 propagates downward later, and the time difference Δt between the propagation of light beam A_1 and light beam A_2 toward the display P is.

$$\Delta t = t_{A2} - t_{A1} = \frac{2ch}{c^2 - v^2} - \frac{2h}{\sqrt{c^2 - v^2}} \approx \frac{2hv^2}{c^3} \tag{26.1.11}$$

Note that the time difference Δt belongs to Newton's absolute time, not to the time of special relativity.

26.1.4 Optical path length difference ΔL between longitudinal light beam A_1 and transverse light beam A_2

If beam splitter O goes from point O_A moves to point O_{A1}, then longitudinal light beam A_1 travels upward from point O_A to point m_{A1}, returns from point m_{A1} to point O_{A1}, and propagates from point O_{A1} toward screen P. This is shown in Figure 26-1.

If beam splitter O moves from point O_A to point O_{A1}, then transverse light beam A_2 has not yet reached point O_{A2}. If beam splitter O continues to moves from point O_{A1} for a tiny time Δt, then transverse light beam A_2 reaches point O_{A2} and propagates from point O_{A2} toward observation screen P, as shown in Figure 26-1.

Since both the longitudinal optical path length $L_{AO_AO_{A1}} = L_d + L_{O_AO_{A1}}$ and the transverse optical path length $L_{AO_AO_{A2}} = L_d + L_{O_AO_{A2}}$ start at point A, the difference ΔL between the transverse optical path length $L_{AO_AO_{A2}}$ and the longitudinal optical path length $L_{AO_AO_{A1}}$ in the cosmic vacuum frame of reference is.

$$\Delta L = L_{AO_AO_{A2}} - L_{AO_AO_{A1}} = \frac{2hc^2}{c^2 - v^2} - \frac{2hc}{\sqrt{c^2 - v^2}} \approx \frac{2hv^2}{c^2} \tag{26.1.12}$$

Since the wavelength of sodium light in the experiment $\lambda = 5.9 \times 10^{-7}m$, the distance $h = 5.5m$ from the plane reflectors m_1 and m_2 to the point O_A, and the velocity $v_1 \approx 20.9 \times 10^{-4}c$ of the Earth relative to the cosmic background radiation (CMB), the number of wavelengths n contained in the optical path length difference ΔL is.

$$n = \frac{\Delta L}{\lambda} = \frac{2hv^2}{c^2\lambda} = \frac{2 \times 5.5 \times (20.9 \times 10^{-4}c)^2}{5.9 \times 10^{-7}c^2} = 81.4 \tag{26.1.13}$$

From the above equation, if the longitudinal light beam A_1 propagates downward from point O_{A1} by a distance $\Delta L = 81.4\lambda$, then the transverse light beam A_2 arrives at point O_{A2} and propagates from point O_{A2} toward the display P.

Since the distance $l_{O_{A1}O_{A2}}$ between points O_{A1} and O_{A2} is small, it can be assumed that the longitudinal light beam A_1 and the transverse light beam A_2 propagate along the same straight line toward the display P. At this point the optical path length difference $\Delta L = 81.4\lambda$ between light beam A_1 and light beam A_2.

26.1.5 Conditions for optical path length difference ΔL=0 between longitudinal light beam A_1 and transverse light beam A_2

From Figure 26-1, if the longitudinal light beam A_1 is perpendicular to the velocity v, and the transverse light beam A_2 is parallel to the velocity v, then according to equation (26.1.12), the length of the transverse optical path $L_{AO_AO_{A2}}$ is larger than that of the longitudinal optical path $L_{AO_AO_{A1}}$, that is.

$$L_{AO_AO_{A2}} > L_{AO_AO_{A1}} \tag{26.1.14}$$

From equation (26.1.13), the optical path length difference $\Delta L = 81.4\lambda$ between the transverse light beam A_2 and the longitudinal light beam A_1.

Assume that θ is the angle between the light beam A_2 and the velocity v. If the interferometer starts to rotate from the position where the angle $\theta = 0$, then the length of the longitudinal optical path $L_{AO_AO_{A1}}$ starts to increase and the length of the transverse optical path $L_{AO_AO_{A2}}$ starts to decrease.

If the angle $\theta = 90^0$, then the transverse light beam A_1 and the longitudinal light beam A_2 have exchanged their states with respect to the direction of velocity v. At this time, since light beam A_2 is perpendicular to the velocity v and light beam A_1 is parallel to the velocity v, the length of the longitudinal path $L_{AO_AO_{A2}}$ is smaller than the length of the transverse path $L_{AO_AO_{A1}}$, i.e..

$$L_{AO_AO_{A2}} < L_{AO_AO_{A1}} \qquad \mathbf{(26.1.15)}$$

From equation (26.1.13), the optical path length difference between the longitudinal light beam A_2 and the transverse light beam A_1 at this point in time $\Delta L = -81.4\lambda$.

According to equations (26.1.14) and (26.1.15), it can be determined that if the optical path length $L_{AO_AO_{A2}}$ of the transverse light beam A_2, changes from $L_{AO_AO_{A2}} > L_{AO_AO_{A1}}$ to $L_{AO_AO_{A2}} < L_{AO_AO_{A1}}$, then there should exist an angle θ, such that the optical path length $L_{AO_AO_{A2}} = L_{AO_AO_{A1}}$.

If the angle $\theta = 45^0$, then the projected velocities v_1 and v_2 of the velocity v on the two optical paths h are equal, i.e., $v_1 = v_2$. At this time, since the longitudinal length $L_{AO_AO_{A1}}$ is equal to the transverse length $L_{AO_AO_{A2}}$, i.e., $L_{AO_AO_{A1}} = L_{AO_AO_{A2}}$, the longitudinal light beam A_1 intersects the transverse light beam A_2 at the point O_{A1}, i.e., the longitudinal light beam A_1 and the transverse light beam A_2 arrive at the point O_{A1} at the same time and propagate to the display P simultaneously from the point O_{A1}. As shown in Figure 26-2.

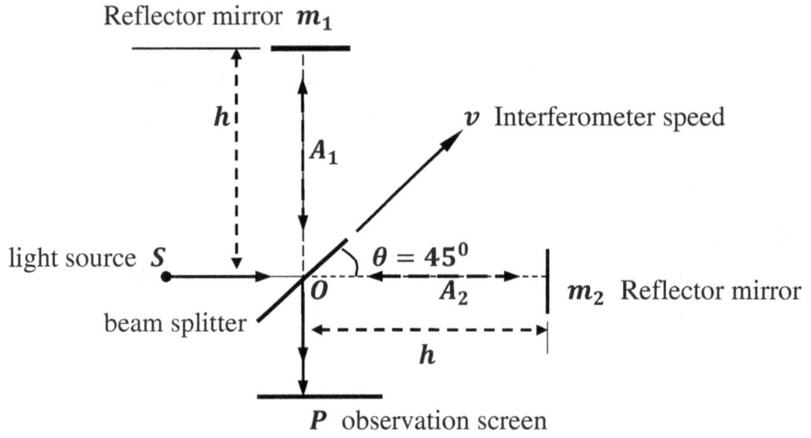

Figure 26-2 Optical path diagram of interferometer rotation angle $\theta = 45^0$

If the angle $\theta = 45^0$, then the optical path length difference $\Delta L = 0$ between the longitudinal light beam A_1 and the transverse light beam A_2.

At this time, since the longitudinal light beam A_1 and the transverse light beam A_2 arrive at the beam splitter O point at the same time, the interference fringes on the display P are produced by both light beam A_1 and light beam A_2.

If the angle $\theta < 45^0$, then the optical path length difference $\Delta L < 0$ between the longitudinal light beam A_1 and the transverse light beam A_2.

At this time, the longitudinal light beam A_1 arrives at the beam splitter O point first, and the transverse light beam A_2 arrives at the beam splitter O point later, so the interference fringes on the display P are not formed by the light beam A_1 and the light beam A_2.

If the angle $\theta > 45^0$, then the optical path length difference $\Delta L > 0$ between the longitudinal light beam A_1 and the transverse light beam A_2.

At this time, since the transverse light beam A_2 arrives at the beam splitter O point first and the longitudinal light beam A_1 arrives at the beam splitter O point later, the interference fringes on the display P are not formed by the light beam A_1 and the light beam A_2.

26.2 If angle θ<45^0, then longitudinal light beam B_1 interferes with transverse light beam A_2 instead of longitudinal light beam A_1

26.2.1 Longitudinal beams A_1, B_1 and C_1 do not propagate at the same point toward display P.

In the cosmic vacuum, there are many points in space between the light source S and the beam splitter O. Since light is particulate in nature and light particles are emitted one by one, the light particles have a certain distance between them.

There are many light beams with different starting points between the light source S and the beam splitter O. Suppose A, B, and C are three different points in space.

If a light source S emits light beams A, B, and C at three spatial points A, B, and C respectively, then light beams A, B, and C are light beams with different starting points, i.e., light beam A leaves the light source S at point A, light beam B leaves the light source S at point B, and light beam C leaves the light source S at point C. Notice that light beams A, B, and C are superimposed to propagate in the direction of the beamsplitter O. This is shown in Figure 26-3.

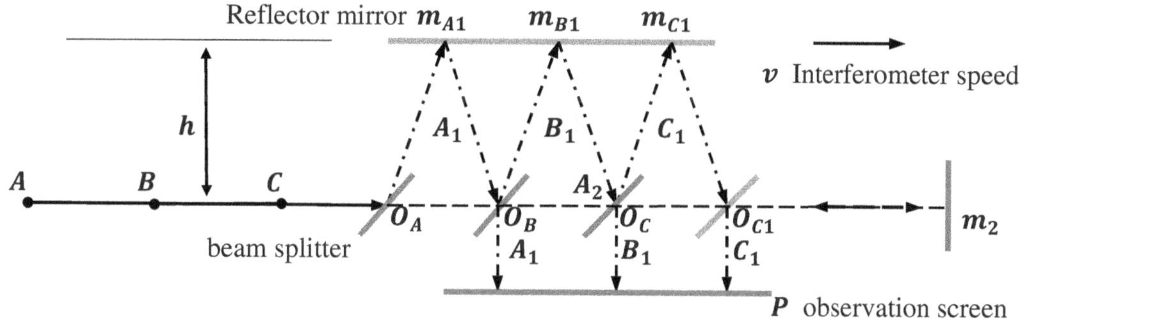

Figure 26-3 Light beams A_1, B_1, C_1 propagate towards the viewing screen **P** at different locations.

Since the beam splitter O follows the interferometer moves, the time and spatial points at which light beams A, B, and C reach the beam splitter O are different. Obviously, light beam A reaches beam splitter O_A first, light beam B reaches beam splitter O_B second, and light beam C reaches beam splitter O_C third.

Suppose light beam A is split into longitudinal light beam A_1 and transverse light beam A_2 at point O_A. Longitudinal light beam A_1 travels upward to point m_{A1}, then returns from point m_{A1} to point O_{A1}, and then propagates from point O_{A1} toward display P. (For clarity in Figure 26-3, point O_B is used in place of point O_{A1} in the Figure).

Transverse light beam A_2 reaches point m_{A2} to the right, returns from point m_{A2} to point O_{A2}, and then propagates from point O_{A2} toward display P.

Assume that light beam B is split into longitudinal light beam B_1 and transverse light beam B_2 at point O_B. Longitudinal light beam B_1 travels upward to point m_{B1}, then returns from point m_{B1} to point O_{B1}, and then propagates from point O_{B1} toward display P. (For clarity in Figure 26-3, point O_C is used in place of point O_{B1} in the diagram).

Transverse light beam B_2 reaches point m_{B2} to the right, returns from point m_{B2} to point O_{B2}, and then propagates from point O_{B2} toward display P.

Since light beam A leaves the light source S earlier than light beam B leaves the light source S, light beam B is after light beam A. If light beam A arrives at point O_A, then light beam B cannot split into longitudinal light beam B_1 and transverse light beam B_2 at point O_A since light beam B has not yet arrived at point O_A.

Suppose that light beam C is split into longitudinal light beam C_1 and transverse light beam C_2 at point O_C. Longitudinal light beam C_1 reaches upward to point m_{C1}, then returns from point m_{C1} to point O_{C1}, and then propagates from point O_{C1} to display P.

The transverse light beam C_2 reaches point m_{C2} to the right, then returns from point m_{C2} to point O_{C2}, and then propagates from point O_{C2} toward the display P.

Since light beam B leaves the light source S at a time earlier than light beam C leaves the light source S, light beam C is behind light beam B. When the light beam B reaches the point O_B, the light beam C cannot be split into the longitudinal light beam C_1 and the transverse light beam C_2 at the point O_B since the light beam C has not yet reached the point O_B.

The beam splitting points O_A, O_B, and O_C are three spatial points with different positions because the starting points of the A, B, and C beams are different. This is shown in Figure 26-3.

26.2.2 Light beams A, B, and C are equal in longitudinal length and equal in transverse length

As shown in Figure 26-1, light beam A is split into longitudinal light beam A_1 and transverse light beam A_2 at point O_A. Longitudinal light beam A_1 propagates from point O_{A1} toward display P, and transverse light beam A_2 propagates from point O_{A2} toward display P.

From equation (26.1.2), the time t_{A1} required for the longitudinal light beam A_1 to reach the point O_A from the starting point A, then up to the point m_{A1}, and then return from the point m_{A1} to the point O_{A1} is.

$$t_{A1} = \frac{2h}{\sqrt{c^2 - v^2}} \tag{26.2.1}$$

From equation (26.1.5), the longitudinal length $L_{AO_AO_{A1}}$ of light beam A_1 from the starting point A to the point O_{A1} is.

$$L_{AO_AO_{A1}} = \frac{cd}{c - v} + \frac{2ch}{\sqrt{c^2 - v^2}} \tag{26.2.2}$$

From equation (26.1.6), the time t_{A2} required for the transverse light beam A_2 to reach the point m_{A2} to the right and to return from the point m_{A2} to the point O_{A2} is.

$$t_{A2} = \frac{2ch}{c^2 - v^2} \tag{26.2.3}$$

From equation (26.1.9), the transverse length $L_{AO_AO_{A2}}$ of light beam A_2 from the starting point A to the point O_{A2} is.

$$L_{AO_AO_{A2}} = \frac{cd}{c - v} + \frac{2c^2h}{c^2 - v^2} \tag{26.2.4}$$

From equations (26.2.1) and (26.2.3), the difference Δt between time t_{A1} and time t_{A2} is.

$$\Delta t = t_{A2} - t_{A1} = \frac{2ch}{c^2 - v^2} - \frac{2h}{\sqrt{c^2 - v^2}} \approx \frac{2v^2h}{c^3} \tag{26.2.5}$$

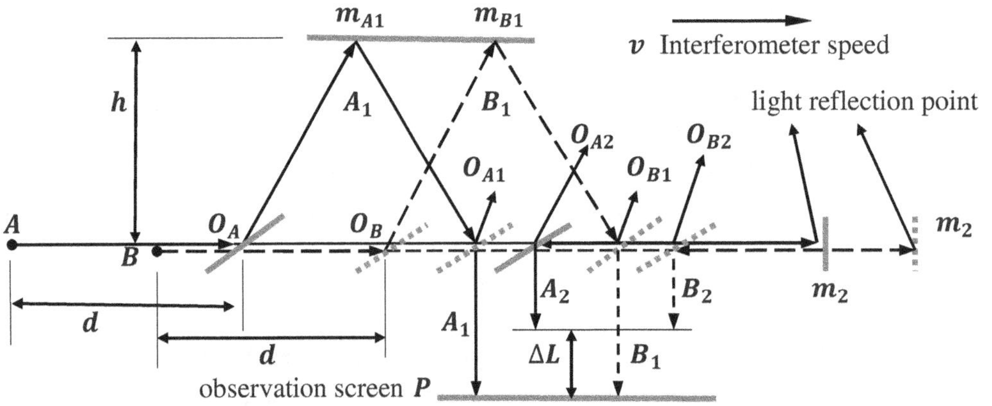

Figure 26-4 The optical paths of vertical light beam A_1 and B_1 are equal, and the optical paths of horizontal light beam A_2 and B_2 are equal

When the longitudinal light beam A_1 reaches the point O_{A1} from the point O_A, the transverse light beam A_2 has not yet reached the point O_{A2} because the time t_{A1} is less than the time t_{A2}.

After light beam A_1 has moves for Δt time from point O_{A1} toward display P, light beam A_2 arrives at point O_{A2} and propagates from O_{A2} toward display P. This is shown in Figure 26-4.

Although light beam A and light beam B overlap to propagate toward the beam splitter O, the starting points of the two are different. Since light beam A leaves the light source S earlier than light beam B leaves the light source S, light beam B cannot be split into longitudinal light beam B_1 and transverse light beam B_2 at point O_A.

Suppose that light beam B splits into longitudinal light beam B_1 and transverse light beam B_2 at point O_B. Longitudinal light beam B_1 travels upward to point m_{B1}, returns from point m_{B1} to point O_{B1}, and then propagates from point O_{B1} toward display P. As shown in Figure 26-4.

Transverse light beam B_2 reaches point m_{B2} to the right, then returns from point m_{B2} to point O_{B2}, and then propagates from point O_{B2} toward display P. As shown in Figure 26-4.

After light beam A and light beam B leave the light source S. Since the interferometer is in motion, the optical path length L_d of light beam A and light beam B from the starting point A and starting point B to the beam splitter O is.

$$L_d = \frac{cd}{c-v} = l_{AO_A} = l_{BO_B} \tag{26.2.6}$$

Where d is the distance from the light source S to the beam splitter O. Since the beam splitting points O_A and O_B have equal distances h to the reflector sub m_1, the moves distance l_{AB} of the light source S is equal to the moves distance of the beam splitter O, i.e..

$$l_{AB} = l_{O_AO_B} = l_{O_{A1}O_{B1}} = l_{O_{A2}O_{B2}} \tag{26.2.7}$$

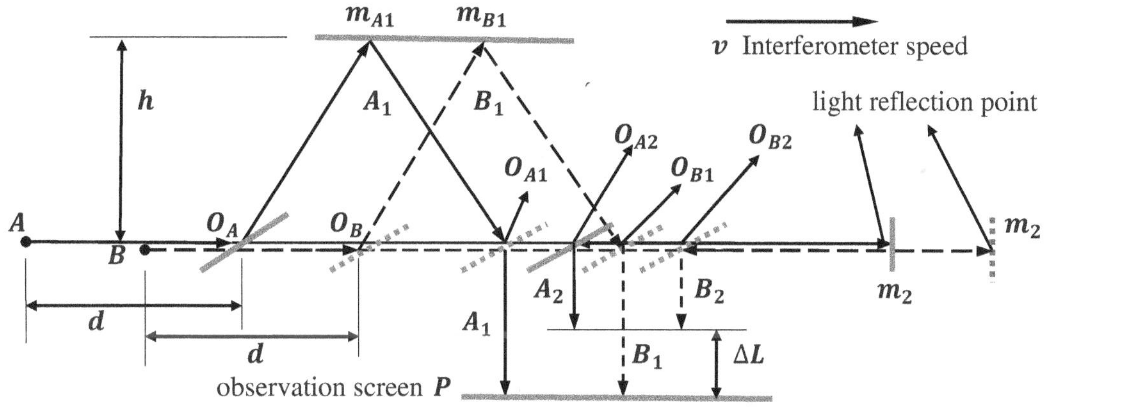

Figure 26-5 When the light output point **B** gradually approaches the light output point A, the reflection point O_{B1} gradually approaches the reflection point O_{A2}.

From Figure 26-5, the longitudinal length $L_{BO_BO_{B1}}$ of the longitudinal light beam B_1 from the starting point B to the point O_B, then upward from the point O_B to the point m_{B1}, and then back to the point O_{B1} from the point m_{B1} is.

$$L_{BO_BO_{B1}} = \frac{cd}{c-v} + \frac{2ch}{\sqrt{c^2-v^2}} \tag{26.2.8}$$

Comparing the longitudinal length $L_{BO_BO_{B1}}$ of the above equation with the longitudinal length $L_{AO_AO_{A1}}$ of equation (26.2.2), it is possible to determine that the lengths of the longitudinal optical paths of light beam A and light beam B are equal, that is.

$$L_{AO_AO_{A1}} = L_{BO_BO_{B1}} \tag{26.2.9}$$

From Figure. 26-5, the transverse light beam B_2 reaches the point O_B from the point B, then reaches the point

m_{B2} to the right, and then returns to the point O_{B2} from the point m_{B2} with the transverse length $L_{BO_BO_{B2}}$ as.

$$L_{BO_BO_{B2}} = \frac{cd}{c-v} + \frac{2c^2h}{c^2-v^2} \tag{26.2.10}$$

Compare the above equation transverse length $L_{BO_BO_{B2}}$ with equation (26.2.4) transverse length $L_{AO_AO_{A2}}$. It can be determined that the lengths of the transverse optical paths of light beam A and light beam B are equal, that is.

$$L_{AO_AO_{A2=}} = L_{BO_BO_{B2}} \tag{26.2.11}$$

26.2.3 Moves distance and corresponding optical path length for beam splitter O

When the position of the starting point A of light beam A remains unchanged, the positions of the beam splitting point O_A, the transreflecting point O_{A1} and the reflecting point O_{A2} also remain unchanged. If the starting point B of light beam B is gradually approaching point A, then the beam splitter point O_B will gradually approach the beam splitter point O_A, the transreflector point O_{B1} will gradually approach the transreflector point O_{A1}, and the reflector point O_{B2} will gradually approach the reflector point O_{A2}, as shown in Figure 26-5.

According to Figure 26-5, the distance $l_{O_{A1}O_{A2}} < l_{O_{A1}O_{B1}}$. The distance $l_{O_{A2}O_{B1}}$ between the reflecting point O_{A2} and the transreflecting point O_{B1} will gradually decrease as point B approaches point A.

When point B is moves to the left to the position of point B_0, assume that the reflection point O_{A2} coincides with the transmission point O_{B1}. This is shown in Figure 26-6.

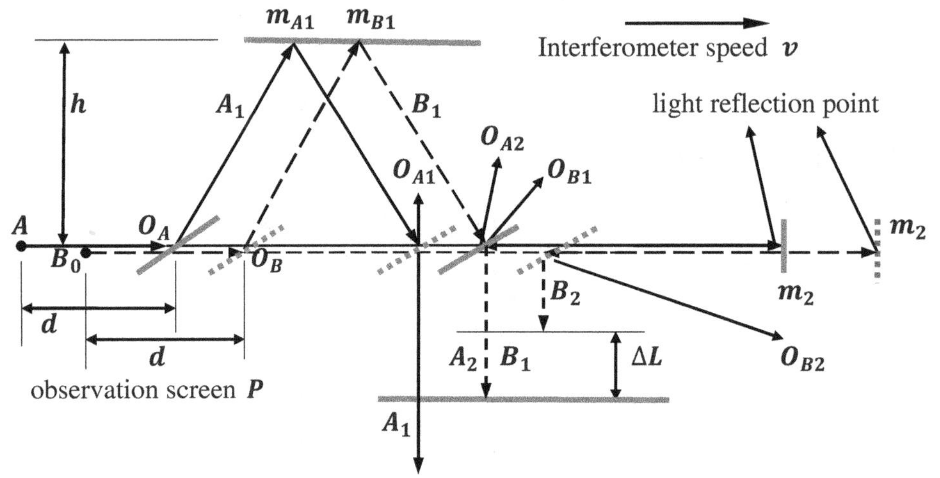

Figure 26-6 Beam B_1 replaces beam A_1 in optical interference with beam A_2.

At this point, the distance $l_{O_{A2}O_{B1}}$ between the reflection point O_{A2} and the transmission point O_{B1} is.

$$l_{O_{A2}O_{B1}} = 0 \tag{26.2.12}$$

When point B is at the B_0 point position, the distance l_{AB} between beam output points, $l_{O_AO_B}$ between beam splitting points, $l_{O_{A1}O_{B1}}$ between transmission points, and $l_{O_{A2}O_{B2}}$ between reflection points are all equal, i.e..

$$l_{AB} = l_{O_AO_B} = l_{O_{A1}O_{A2}} = l_{O_{B1}O_{B2}} = l_{O_{A1}O_{B1}} = l_{O_{A2}O_{B2}} \tag{26.2.13}$$

At this time, since the reflection point O_{A2} and the transreflection point O_{B1} are the same point, the transverse light beam A_2 and the longitudinal light beam B_1 arrive at the reflection point O_{A2} (i.e., the transreflection point O_{B1}) at the same time and propagate toward the display P.

From Figure 26-6, light beam A_1 propagates toward display P at the transreflective point O_{A1}, and the longitudinal length $L_{O_AO_{A1}}$ of light beam A_1 is.

$$L_{O_AO_{A1}} = \frac{2ch}{\sqrt{c^2-v^2}} \tag{26.2.14}$$

The transverse light beam A_2 propagates toward the display P at the reflection point O_{A2} and the transverse length $L_{O_AO_{A2}}$ of light beam A_2 is.

$$L_{O_AO_{A2}} = \frac{2c^2h}{c^2 - v^2} \tag{26.2.15}$$

The optical path length difference ΔL corresponding to the distance $l_{O_{A1}O_{A2}}$ between the transmitted point O_{A1} and the reflected point O_{A2} is.

$$\Delta L = L_{O_AO_{A2}} - L_{O_AO_{A1}} = \frac{2c^2h}{c^2 - v^2} - \frac{2ch}{\sqrt{c^2 - v^2}} \tag{26.2.16}$$

According to equation (26.2.13). The optical path length difference ΔL is also the optical path length difference corresponding to the distance $l_{AB} = l_{O_{A1}O_{A2}} = l_{O_{A1}O_{B1}}$.

26.2.4 In the Michelson-Morley experiment, light beam A_1 is replaced by light beam B_1, i.e., light beam B_1 produces interference fringes with light beam A_2

From Figure 26-6, the optical path length $L_{AO_AO_{A2}}$ of the transverse light beam A_2 from point A to the reflection point O_{A2} is.

$$L_{AO_AO_{A2}} = \frac{cd}{c - v} + \frac{2c^2h}{c^2 - v^2} \tag{26.2.17}$$

When light beam B leaves the light source S at point B_0, light beam A has already passed through point B_0. If light beam A_2 takes point B_0 as its new starting point, the optical path length $L_{BO_AO_{A2}}$ of light beam A_2 from point B_0 to the reflection point O_{A2} is.

$$L_{BO_AO_{A2}} = L_{AO_AO_{A2}} - \Delta L \tag{26.2.18}$$

Substituting equations (26.2.17) and (26.2.16) into the above equation yields the relationship.

$$L_{BO_AO_{A2}} = L_{AO_AO_{A2}} - \Delta L = \frac{cd}{c - v} + \frac{2ch}{\sqrt{c^2 - v^2}} \tag{26.2.19}$$

From Figure 26-6, the longitudinal length $L_{BO_BO_{B1}}$ of light beam B_1 from point B_0 to transmission point O_{B1} (i.e., reflection point O_{A2}) is.

$$L_{BO_BO_{B1}} = \frac{cd}{c - v} + \frac{2ch}{\sqrt{c^2 - v^2}} \tag{26.2.20}$$

Comparing the above equation with equation (26.2.19), if the starting point of light beam B is point B_0, then the optical path length difference $\Delta L_{A_2B_1}$ between transverse light beam A_2 and longitudinal light beam B_1 is.

$$\Delta L_{A_2B_1} = L_{BO_AO_{A2}} - L_{BO_BO_{B1}} = 0 \tag{26.2.21}$$

The above equation shows that the transverse light beam A_2 and the longitudinal light beam B_1 reach the transmission point O_{B1} (i.e., the reflection point O_{A2}) at the same time. From this, it can be determined that in the Michelson-Morley experiment, the longitudinal light beam A_1 is replaced by the longitudinal light beam B_1, i.e., the longitudinal light beam B_1 produces interference fringes with the transverse light beam A_2.

It should be noted that when the distance l_{AB} between point B and point A continues to decrease, i.e., $l_{AB} < l_{AB_0}$, the reflection point O_{A2} and the transmission point O_{B1} do not coincide because the distance $l_{O_{A1}O_{A2}}$ begins to be greater than $l_{O_{A1}O_{B1}}$, i.e., $l_{O_{A1}O_{A2}} > l_{O_{A1}O_{B1}}$. At this point the longitudinal light beam B_1 and the transverse light beam A_2 are not interfering light.

If point C is exactly at the position of point B_0, the longitudinal light beam A_1 is replaced by the longitudinal light beam C_1, i.e., the longitudinal light beam C_1 produces an interference fringe with the transverse light beam A_2.

Since physicists believe that the longitudinal light beam A_1 and the transverse light beam A_2 should propagate to the display P at the same point O, Lorentz uses the method of length reduction of the moving object, so that the light beam A_1 and the light beam A_2 both propagate to the point O at the same time, and then produce interference fringes on the display.

Special relativity, on the other hand, uses the method of constant speed of light to make both light beam A_1 and light beam A_2 propagate to point O at the same time and then produce interference fringes on the display.

26.3 A new interpretation of the Michelson-Morley experiment

26.3.1 Optical path length L_{AB} between start point A and start point B_0

From equation (26.2.2), the optical path length $L_{AO_AO_{A1}}$ of the longitudinal light beam A_1 from the starting point A to O_{A1} is.

$$L_{AO_AO_{A1}} = \frac{cd}{c-v} + \frac{2ch}{\sqrt{c^2-v^2}}$$

From equation (26.2.4), the optical path length $L_{AO_AO_{A2}}$ of the transverse light beam A_2 from the starting point A to O_{A2} is.

$$L_{AO_AO_{A2}} = \frac{cd}{c-v} + \frac{2c^2h}{c^2-v^2}$$

The optical path length difference ΔL between the transverse light beam A_2 of light beam A and the longitudinal light beam A_1 is.

$$\Delta L = L_{AO_AO_{A2}} - L_{AO_AO_{A1}} = \frac{2c^2h}{c^2-v^2} - \frac{2ch}{\sqrt{c^2-v^2}} \qquad (26.3.1)$$

From equation (26.2.8), the optical path length $L_{BO_BO_{B1}}$ of the longitudinal light beam B_1 from the starting point B_0 to O_{B1} is.

$$L_{BO_BO_{B1}} = \frac{cd}{c-v} + \frac{2ch}{\sqrt{c^2-v^2}}$$

When the reflection point O_{A2} coincides with the transmission point O_{B1}. Since the optical path length $L_{AO_AO_{A1}}$ is equal to $L_{BO_BO_{B1}}$, the distance L_{AB} between the start point A and the start point B_0 is equal to the optical path length difference ΔL, i.e..

$$L_{AB} = \Delta L = \frac{2c^2h}{c^2-v^2} - \frac{2ch}{\sqrt{c^2-v^2}} \qquad (26.3.2)$$

When the reflection point O_{A2} coincides with the transmission point O_{B1}, the optical path length difference ΔL between light beam A and light beam B can be calculated using the experimental data.

In the experiment, the wavelength of sodium light $\lambda = 5.9 \times 10^{-7}m$, and the distance $h = 5.5m$ from the reflectors m_1 and m_2 to the point O. The velocity of the earth relative to the cosmic background radiation (CMB) $v_1 \approx 20.9 \times 10^{-4}c$. The number of wavelengths n included in the optical path length difference ΔL is.

$$n = \frac{\Delta L}{\lambda} \approx \frac{2hv^2}{c^2\lambda} = \frac{2 \times 5.5 \times (20.9 \times 10^{-4}c)^2}{5.9 \times 10^{-7}c^2} = 81.4 \qquad (26.3.3)$$

From the above equation, the optical path length $L_{AB} = 81.4\lambda$ of the between point A and point B_0.

26.3.2 Transverse light beam A_2 can use point B_0 as starting point

In the cosmic vacuum frame of reference, according to equation (26.2.4), the optical path length $L_{AO_AO_{A2}}$ of the transverse light beam A_2 starting at point A is.

$$L_{AO_AO_{A2}} = \frac{cd}{c-v} + \frac{2c^2h}{c^2-v^2}$$

Subtracting equation (26.3.2) from the above equation yields the relationship.

$$L_{BO_AO_{A2}} = L_{AO_AO_{A2}} - L_{AB} = \frac{cd}{c-v} + \frac{2ch}{\sqrt{c^2-v^2}} \qquad (26.3.4)$$

The optical path length $L_{BO_AO_{A2}}$ is the optical path length of transverse light beam A_2 starting from point B_0. From equation (26.2.8), the optical path length $L_{BO_BO_{B1}}$ of longitudinal light beam B_1 starting from point B_0 is.

$$L_{BO_BO_{B1}} = \frac{cd}{c-v} + \frac{2ch}{\sqrt{c^2-v^2}}$$

Comparing the above equation with equation (26.3.4), the optical path length $L_{BO_AO_{A2}}$ of the transverse light beam A_2 is equal to the optical path length $L_{BO_BO_{B1}}$ of the longitudinal light beam B_1 when the transverse light beam A_2 starts at the point B_0, i.e.

$$L_{BO_BO_{B1}} = L_{BO_AO_{A2}} = L_{AO_AO_{A2}} - L_{AB} = \frac{cd}{c-v} + \frac{2ch}{\sqrt{c^2 - v^2}} \tag{26.3.5}$$

26.3.3 Conditions for the coincidence of the transmission point O_{A1} of light beam A with the reflection point O_{A2}

As shown in Figure 26-1, the longitudinal light beam A_1 propagates toward the display P at point O_{A1}, and the transverse light beam A_2 propagates toward the display P at point O_{A2}.

From equation (26.3.1), the length difference ΔL between the length of transverse optical path $L_{AO_AO_{A2}}$ and the length of longitudinal optical path $L_{AO_AO_{A1}}$ is.

$$\Delta L = L_{AO_AO_{A2}} - L_{AO_AO_{A1}} = \frac{2hc^2}{c^2 - v^2} - \frac{2hc}{\sqrt{c^2 - v^2}}$$

Since the length difference $\Delta L > 0$, the length of transverse optical path $L_{AO_AO_{A2}}$ is larger than the length of longitudinal optical path $L_{AO_AO_{A1}}$, ie.

$$L_{AO_AO_{A2}} > L_{AO_AO_{A1}} \tag{26.3.6}$$

Since the transverse optical path length $L_{AO_AO_{A2}}$ and the longitudinal optical path length $L_{AO_AO_{A1}}$ have the same starting point, and the difference of their optical path lengths $\Delta L > 0$, the transmission point O_{A1} and the reflection point O_{A2} are not the same point.

When the interferometer gradually increases from the angle $\theta = 0$. The length of longitudinal optical path $L_{AO_AO_{A1}}$ becomes longer and the length of transverse optical path $L_{AO_AO_{A2}}$ becomes shorter.

When the angle $\theta = 90^0$. Longitudinal light beam A_1 and transverse light beam A_2 exchange state, that is, light beam A_2 is perpendicular to the velocity v, light beam A_1 is parallel to the velocity v. At this time, the length of the optical path $L_{AO_AO_{A2}}$ is smaller than the length of the optical path $L_{AO_AO_{A1}}$, i.e:

$$L_{AO_AO_{A2}} < L_{AO_AO_{A1}} \tag{26.3.7}$$

According to the above equation and equation (26.3.6), it can be determined that when the optical path length changes from $L_{AO_AO_{A2}} > L_{AO_AO_{A1}}$ to $L_{AO_AO_{A2}} < L_{AO_AO_{A1}}$. There should be a rotation angle θ so that the optical path length $L_{AO_AO_{A2}} = L_{AO_AO_{A1}}$.

When the rotation angle of the interferometer $\theta = 45^0$, the length of the optical path $L_{AO_AO_{A2}} = L_{AO_AO_{A1}}$, because the velocity v is projected on the longitudinal and transverse optical paths with the velocity v_1 and v_2 are equal (i.e. $v_1 = v_2$).

At this time, since the transmission point O_{A1} and the reflection point O_{A2} coincide, the longitudinal light beam A_1 and the transverse light beam A_2 intersect at the beam splitter O point.

When the interferometer rotation angle $\theta \neq 45^0$, the optical path length difference $\Delta L = L_{AO_AO_{A2}} - L_{AO_AO_{A1}} \neq 0$ between the longitudinal light beam A_1 and the transverse light beam A_2. At this time, the longitudinal beam A_1 and the transverse beam A_2 will not arrive at the beam splitter O point at the same time, nor will they propagate from the beam splitter O point to the display P at the same time.

When the rotation angle of the interferometer is $\theta < 45^0$. The longitudinal light beam A_1 arrives at the beam splitter O point before the transverse light beam A_2, and propagates from the beam splitter O point to the display P.

When the rotation angle of the interferometer $\theta > 45^0$. The transverse light beam A_2 arrives at the beam splitter O point before the longitudinal light beam A_1 and propagates from the beam splitter O point toward the display P.

26.3.4 Conditions for coincidence of the reflection point O_{A2} of light beam A with the transmission point O_{B1} of light beam B

As shown in Figure 26-4, the transverse light beam A_2 of light beam A propagates toward the display P at point

$\boldsymbol{O_{A2}}$, and the longitudinal light beam $\boldsymbol{B_1}$ of light beam B propagates toward the display P at point $\boldsymbol{O_{B1}}$.

From equation (26.3.1), the length difference $\boldsymbol{\Delta L}$ between the transverse optical path length $\boldsymbol{L_{AO_AO_{A2}}}$ and the longitudinal optical path length $\boldsymbol{L_{AO_AO_{A1}}}$ of light beam A is.

$$\Delta L = L_{AO_AO_{A2}} - L_{AO_AO_{A1}} = \frac{2c^2h}{c^2 - v^2} - \frac{2ch}{\sqrt{c^2 - v^2}}$$

Since the optical path length $\boldsymbol{L_{BO_BO_{B1}}} = \boldsymbol{L_{AO_AO_{A1}}}$, the length difference $\boldsymbol{\Delta L}$ between the optical path length $\boldsymbol{L_{AO_AO_{A2}}}$ and the optical path length $\boldsymbol{L_{BO_BO_{B1}}}$ is when the transmitting point $\boldsymbol{O_{B1}}$ of the light light beam B, and the reflecting point $\boldsymbol{O_{A2}}$ of the light light beam A, are coincident is:

$$\Delta L = L_{AO_AO_{A2}} - L_{BO_BO_{B1}} = \frac{2c^2h}{c^2 - v^2} - \frac{2ch}{\sqrt{c^2 - v^2}} \qquad (26.3.8)$$

The optical path length difference $\boldsymbol{\Delta L}$ in equation is equal to the optical path length $\boldsymbol{L_{AB}}$ between the starting point A of light beam A and the starting point $\boldsymbol{B_0}$ of light beam B, i.e., $\boldsymbol{\Delta L = L_{AB}}$. From equation (26.3.3), the optical path lengths $\boldsymbol{L_{AB}}$ at points A and $\boldsymbol{B_0}$ are equal to the wavelengths of $\boldsymbol{81.4}$ sodium light, i.e., $\boldsymbol{L_{AB} \approx 81.4\lambda}$.

26.3.5 A new interpretation of the Michelson-Morley experiment

When the angle of rotation of the interferometer $\boldsymbol{\theta = 0}$. From Figure 26-6, it can be seen that the longitudinal light beam $\boldsymbol{A_1}$ and the transverse light beam $\boldsymbol{A_2}$ do not propagate toward the display P at the same point. It can thus be determined that no interference fringes are produced between the longitudinal light beam $\boldsymbol{A_1}$ and the transverse light beam $\boldsymbol{A_2}$.

However, physicists believed that the longitudinal light beam $\boldsymbol{A_1}$ and the transverse light beam $\boldsymbol{A_2}$ should intersect at the point O of the beamsplitter and produce interference fringes in the display P. This view of the physicists led to the discovery of the length contraction effect.

Since the time at which light beam B leaves the light source S lags behind the time at which light beam A leaves the light source S, the longitudinal light beam $\boldsymbol{B_1}$ has a chance to intersect the transverse light beam $\boldsymbol{A_2}$ at a certain point.

If the transmission point $\boldsymbol{O_{B1}}$ of light beam B coincides with the reflection point $\boldsymbol{O_{A2}}$ of light beam A, then, as can be seen from Figure 26-6, the transverse light beam $\boldsymbol{A_2}$ intersects the longitudinal light beam $\boldsymbol{B_1}$ at the point $\boldsymbol{O_{A2}}$, and then they both simultaneously propagate from the point $\boldsymbol{O_{A2}}$ toward the display P and produce interference fringes on the display P.

If the transverse light beam $\boldsymbol{A_2}$ uses the point $\boldsymbol{B_0}$ as the new starting point, then according to equation (26.3.5), the optical path length $\boldsymbol{L_{BO_AO_{A2}}}$ of the transverse light beam $\boldsymbol{A_2}$ is equal to that $\boldsymbol{L_{BO_BO_{B1}}}$ of the longitudinal light beam $\boldsymbol{B_1}$, i.e

$$L_{BO_AO_{A2}} = L_{AO_AO_{A2}} - L_{AB} = L_{BO_BO_{B2}} = \frac{cd}{c - v} + \frac{2c^2h}{c^2 - v^2}$$

The above equation shows that the transverse light beam $\boldsymbol{A_2}$ and the longitudinal light beam $\boldsymbol{B_1}$ propagate toward the display P at the same time at the point $\boldsymbol{O_{A2}}$ (i.e., the point $\boldsymbol{O_{B1}}$).

Since the transverse light beam $\boldsymbol{A_2}$ and the longitudinal light beam $\boldsymbol{B_1}$ are propagating toward the display P at the same point at the same time, the transverse light beam $\boldsymbol{A_2}$ and the longitudinal light beam $\boldsymbol{B_1}$ will produce interference fringes on the display P.

Since the transverse light beam $\boldsymbol{A_2}$ and the longitudinal light beam $\boldsymbol{A_1}$ are not propagating toward the display P at the same point, the transverse light beam $\boldsymbol{A_2}$ and the longitudinal light beam $\boldsymbol{A_1}$ will not produce interference fringes on the display P.

When the interferometer starts to rotate from angle $\boldsymbol{\theta = 0}$. The optical path length $\boldsymbol{L_{AO_AO_{A2}}}$ of the transverse light beam $\boldsymbol{A_2}$ keeps getting shorter, and the optical path length $\boldsymbol{L_{BO_BO_{B1}}}$ of the longitudinal light beam $\boldsymbol{B_1}$ keeps getting longer.

When the rotation angle is $\boldsymbol{0 < \theta < 45^0}$, the optical path length $\boldsymbol{L_{AB}}$ between the start point A and the start point $\boldsymbol{B_0}$ decreases as the angle $\boldsymbol{\theta}$ increases.

At this time, since the start point B_0 is constantly close to the start point A, the optical path length $L_{BO_AO_{A2}}$ of the transverse optical path A_2 is always equal to the optical path length $L_{BO_BO_{B1}}$ of the longitudinal optical path B_1. That is to say, the optical path length difference between the two is always equal to 0, i.e.

$$\left(L_{AO_AO_{A2}} - L_{AB}\right) - L_{BO_BO_{B1}} = 0 \qquad (26.3.9)$$

When the rotation angle of the interferometer $\theta = 45^0$, the length of the optical path $L_{AO_AO_{A2}} = L_{AO_AO_{A1}}$, because the velocity v is projected on the longitudinal and transverse optical paths with the velocity v_1 and v_2 are equal (i.e. $v_1 = v_2$).

At this time, since the transmission point O_{A1} and the reflection point O_{A2} coincide, the longitudinal light beam A_1 and the transverse light beam A_2 intersect at the beam splitter O point.

Obviously, the longitudinal light beam B_1 of light beam B at this point is also the longitudinal light beam A_1 of light beam A. That is to say, at this point the start point A and the start point B_0 coincide. The length of the optical path between the two is $L_{AB} = 0$.

When the rotation angle of the interferometer is $45^0 < \theta \le 90^0$. The start point A and the start point B_0 are not coincident. The length of the optical path between them is $L_{AB} < 0$.

The transverse light beam B_2 of light beam B and the longitudinal light beam A_1 of light beam A at this point propagate simultaneously from point O_{B2} (i.e., point O_{A1}) toward the display P. It can thus be determined that the optical range difference between the transverse light beam B_2 and the longitudinal light beam A_1 is still equal to 0, i.e:

$$L_{BO_BO_{B2}} - \left(L_{AO_AO_{A1}} + L_{AB}\right) = 0 \qquad (26.3.10)$$

In summary, when the interferometer rotation angle is $0 < \theta \le 45^0$. Since the transverse optical path length $L_{BO_AO_{A2}}$ of light beam A is always equal to the longitudinal optical path length $L_{BO_BO_{B1}}$ of light beam B, the difference ΔL between the transverse light beam A_2 and the longitudinal light beam B_1 is always equal to 0. At this time, the transverse light beam A_2 and the longitudinal light beam B_1 will produce the interference fringes on display P.

When the interferometer rotation angle is $45^0 < \theta \le 90^0$. Since the transverse optical path length $L_{BO_BO_{B2}}$ of light beam B is always equal to the longitudinal optical path length $L_{AO_AO_{A1}}$ of light beam A, the difference ΔL between the transverse light beam B_2 and the longitudinal light beam A_1 is always equal to 0. At this time, the transverse light beam B_2 and the longitudinal light beam A_1 will produce the interference fringes on display P.

It is the two reasons above that cause the results of the Michelson-Murray experiment to always equal zero.

Reference

[1] (China) Jianhua Wang, NEW THEORY OF PLANETARY MOTION AND NEW FORMULA OF UNIVERSAL GRAVITATION, (ISBN 978-1631815836) . Publisher American Academic Press, in 2022.

[2] R.P.Feynman. Feynman Lectures on Physics: Shanghai Science and Technology Press, China, 2013

[3] П.Д. Landau / E.M. Li Fuschitz. "Mechanics": China Higher Education Press, 2007China Science and Technology Press, 2012

[4] (US) M. Yang. "Optics and Lasers": China Science Press, 2007

[5] 〔Su〕 B.M. Yaursky.A.A.Jetlaff. "Physics Handbook": China Science Press, 1986

www.ingramcontent.com/pod-product-compliance
Lightning Source LLC
Chambersburg PA
CBHW080135220326
41598CB00032B/5074